Karl-Heinz Spatschek

High Temperature Plasmas

Related Titles

Diver, D.

Plasma Formulary for Physics, Astronomy, and Technology

2012

ISBN: 978-3-527-40905-1

Turcu, I. C. E., Dance, J. B.

X-Rays From Laser Plasmas

Generation and Applications

1998

ISBN: 978-0-471-98397-2

Stacey, W. M.

Fusion

An Introduction to the Physics and Technology of Magnetic Confinement Fusion

2010

ISBN: 978-3-527-40967-9

Shore, S. N.

Astrophysical Hydrodynamics

An Introduction

2007

ISBN: 978-3-527-40669-2

Smirnov, B. M.

Plasma Processes and Plasma Kinetics

2007

ISBN: 978-3-527-40681-4

Smirnov, B. M.

Physics of Ionized Gases

2001

ISBN: 978-0-471-17594-0

Karl-Heinz Spatschek

High Temperature Plasmas

Theory and Mathematical Tools
for Laser and Fusion Plasmas

WILEY-VCH

WILEY-VCH Verlag GmbH & Co. KGaA

The Author

Prof. K.-H. Spatschek
Institut für Theoretische Physik I
Universität Düsseldorf
khs@tp1.uni-duesseldorf.de

Cover picture
Temperature distribution pattern on the divertor target plates of TEXTOR during DED operation, measured by an infrared camera.
Photograph by Marcin Jakubowski

Library of Congress Card No.: applied for

British Library Cataloguing-in-Publication Data:
A catalogue record for this book is available from the British Library.

Bibliographic information published by the Deutsche Nationalbibliothek
The Deutsche Nationalbibliothek lists this publication in the Deutsche Nationalbibliografie; detailed bibliographic data are available on the Internet at http://dnb.d-nb.de.

Typesetting le-tex publishing services GmbH, Leipzig
Cover Design Grafik-Design Schulz, Fußgönheim
Printing and Binding Fabulous Printers Pte Ltd, Singapore

Printed in Singapore
Printed on acid-free paper

ISBN Print 978-3-527-41041-5

ISBN ePDF 978-3-527-63813-0
ISBN ePub 978-3-527-63812-3
ISBN Mobi 978-3-527-63814-7
ISBN oBook 978-3-527-63811-6

Contents

Preface

High-temperature plasma physics is a rapidly growing field of physics. In the areas of magnetic confinement, laser-plasma interaction, and plasma astrophysics, the scientific progress is enormous. During the last decades, the record value of fusion power reached in magnetic confinement experiments has significantly increased (even faster than the memory capacity of semiconductor chips). Large tokamaks in USA, Europe, and Japan, came close to the energy break even point while operating without tritium. The international test reactor ITER, being built in Cadarache (France), should demonstrate within the next 15 years reactor capability of ignition, producing net power. A new area or laser-plasma interaction started approx. 40 years ago when laser fields became strong enough to directly ionize matter. Meanwhile, since the invention of the Chirped Pulse Amplification (CPA) technique (approx. 20 years ago), short pulse laser-matter interaction lead to an entirely new area of research, namely what is called relativistic optics. Intensities of more than 10^{19} W/cm^2 are now available in terawatt tabletop laser systems to study matter under extreme conditions. In the future, in the ultra-relativistic regime with socalled ultra-high intensities, short-pulse lasers may permit the test of nonlinear quantum-electrodynamics (QED). The development of modern ionospheric, magnetospheric and solar physics relies to a large extent on our understanding of plasma physics. Main topics of astrophysics, such as, star formation, emission of electromagnetic radiation from astronomical objects, theory of cosmic rays, and cosmological models, benefit to a large extent from plasma physics.

The present book concentrates on high-temperature plasma physics, with applications to fusion and laser plasmas and perspectives for plasma astrophysics. It discusses the fundamentals and provides mathematical tools for handling problems of high-temperature plasma theory. The transferability of ideas between the different and important areas is one of the peculiarities characterizing high-temperature plasma physics.Theoretical plasma physics is being taught in physics departments as graduate and advanced undergraduate courses following the theory courses on mechanics, electrodynamics, quantum theory, and statistical physics. The present book is laid out for the basic plasma physics contribution in a theoretical physics curriculum. Of course, it should be supplemented by other, more detailed presentations, such as, “Strongly coupled plasmas”, “Non-neutral plasmas”, “Low temperature plasmas", "Plasma astrophysics", "Fusion plasmas", "Laser produced plasmas", and so on.

There are two persons who indirectly influenced the choice of topics (fusion and laser plasmas) of this book and whom I want to thank (although they are not responsible for the presentation at all). The personal contacts with Radu Balescu (a statistical physicists who tremendously advanced the field of transport in magnetically confined plasmas; much to our regret he passed away June 1st, 2006) and Akira Hasegawa (who is one of the leading contributors to nonlinear plasma physics, and the "father of the optical soliton") certainly influenced my scientific understanding and choice of research topics. I also want to thank the many students, collaborators, and colleagues with whom I had the privilege to work. Just to mention a few of them, lately I collaborated with Sadrilla Abdullaev, Wolfgang Laedke, Götz Lehmann, and Andreas Wingen; part of their work has been included into the present monograph. Unfortunately, I cannot refer to all the individuals who have contributed during the last years. The same is true for the huge amount of literature being available in the wide range of material being reviewed here. I only tried to reference to those papers and books which probably could lead readers to other related work. I apologize to those who were not adequately mentioned.

SI units are used throughout the text (unless otherwise so specified), following general recommendations of international societies and publishers. Theoreticians and plasma physicists very often prefer the Gaussian system. Therefore, a simple translation table is presented in Appendix A. In the majority of cases, temperatures are given in energy equivalents eV (as mostly done in high-temperature plasma physics). However, at some places, when the main emphasis is on the statistical methods, the Boltzmann constant is maintained and temperature is measured in Kelvin, as usually being done in statistical physics. I hope this ambiguity will satisfy both communities, and not lead to confusion. Finally, the transliteration of names, for example with Cyrillic letters, has not been standardized (see, for example, Lyapunov, Liapunov, or Ljapunov) to retain the custom in literature.

The technical help of my collaborators for many years, Elvira Gröters and Eckhard Zügge, is gratefully acknowledged. I would like to express my deepest gratitude to Wiley-VCH, in particular Christoph von Friedeburg, who initiated the project, and Ulrike Werner, who took on the role afterwards and became a judicious advisor. During the production process, the team of le-tex publishing services GmbH was extremely helpful in solving technical problems. The book never would have been completed in time without the unfailing encouragement and patience of my wife Tuta to whom I am most thankful.

Düsseldorf, August 2011

Karl-Heinz Spatschek

1 Introduction

A plasma is a many body system consisting of a huge number of interacting (charged and neutral) particles. More precisely, one defines a plasma as a quasineutral gas of charged and neutral particles which exhibits collective behavior. In many cases, one considers quasineutral gases consisting of charged particles, namely electrons and positively charged ions of one species. Plasmas may contain neutral atoms. In this case, the plasma is called partially or incompletely ionized. Otherwise the plasma is said to be completely or fully ionized. The term plasma is not limited to the most common electron–ion case. One may have electron–positron plasmas, quark-gluon plasmas, and so on. Semiconductors contain plasma consisting of electrons and holes.

Most of the matter in the universe occurs in the plasma state. It is often said that 99% of the matter in the universe is in the plasma state. However, this estimate may not be very accurate. Remember the discussion on dark matter. Nevertheless, the plasma state is certainly dominating the universe.

Modern plasma physics emerged in 1950s, when the idea of thermonuclear reactor was introduced. Fortunately, modern plasma physics completely dissociated from weapons development. The progression of modern plasma physics can be retraced from many monographs, for example, [1–15].

In simple terms, plasmas can be characterized by two parameters, namely, charged particle density n and temperature T. The density varies over roughly 28 orders of magnitude, for example, from 10^6 to 10^{34} m^{-3}. The kinetic energy $k_B T$, where k_B is the Boltzmann constant, can vary over approximately seven orders, for example, from 0.1 to 10^6 eV.

The notation "plasma" was introduced by Langmuir, Tonks, and their collaborators in the 1920s when they studied processes in electronic lamps filled with ionized gases, that is, low-pressure discharges. The word "plasma" seems to be a misnomer [16]. The Greek $\pi\lambda\acute{\alpha}\sigma\mu\alpha$ means something modeled or fabricated. However, a plasma does not tend to generally conform to external influences. On the contrary, because of collective behavior, it often behaves as if it had a mind of its own [16].

Plasmas appear in space and astrophysics [17–19], laser–matter interaction [20, 21], technology [22], fusion [14, 23–26], and so on. Technical plasmas, magnetic fusion plasmas, and laser-generated plasmas represent the main applications of

High Temperature Plasmas, Theory and Mathematical Tools for Laser and Fusion Plasmas, First Edition.
Karl-Heinz Spatschek.

plasma physics on earth. Space plasmas, as they appear, for example, in the magnetosphere of the earth are very important for our life on earth. A continuous stream of charged particles, mainly electrons and protons, called the solar wind impinges on the earth's magnetosphere which shields us from this radiation. Typical solar wind parameters are $n = 5 \times 10^6\,\mathrm{m}^{-3}$, $k_B T_i = 10\,\mathrm{eV}$, and $k_B T_e = 50\,\mathrm{eV}$. However, also temperatures of the order $k_B T = 1\,\mathrm{keV}$ may appear. The drift velocity is approximately 300 km/s.

We could present numerous other examples of important plasmas, for example, stellar interiors and atmospheres which are hot enough to be in the plasma state, stars in galaxies, which are not charged but behave like particles in a plasma, free electrons and holes in semiconductors which also constitute a plasma. However, before discussing some of these objects in more detail, let us concentrate first on basic theoretical tools for their description.

1.1 Quasineutrality and Debye Shielding

Negative charge fluctuations $\delta\rho = -e\delta n$ (e is the elementary charge) generate electrostatic potential fluctuations $\delta\phi$ (we use SI units; see Appendix A)

$$\nabla^2 \delta\phi = \frac{1}{\varepsilon_0} e \delta n \,. \tag{1.1}$$

A rough estimate gives

$$\nabla^2 \delta\phi \sim \frac{\delta\phi}{l^2} \,, \tag{1.2}$$

where l is the characteristic fluctuation scale. Thus,

$$\delta\phi \approx \frac{1}{\varepsilon_0} e \delta n l^2 \,. \tag{1.3}$$

On the other hand, the characteristic potential energy $-e\delta\phi$ cannot be larger than the mean kinetic energy of particles which we roughly approximate by $k_B T$ (we measure the temperature in kelvin, k_B is the Boltzmann constant, and we ignore numerical factors). Thus,

$$\frac{\delta n}{n} \lesssim \frac{\varepsilon_0 k_B T}{n e^2 l^2} \,. \tag{1.4}$$

We recognize that a typical length appears, namely, the Debye length (more details will be given within the next chapters)

$$\lambda_D = \sqrt{\frac{\varepsilon_0 k_B T}{n_e e^2}} \,, \tag{1.5}$$

such that

$$\frac{\delta n}{n} \lesssim \frac{\lambda_D^2}{l^2} . \tag{1.6}$$

A plasma is quasineutral on distances much larger than the Debye radius. If the plasma size is comparable with λ_D, then it is not a "real" plasma, but rather just a heap of charged particles.

The Debye length is the shielding length in a plasma (at the moment, we will not discuss which species, electrons or ions are dominating in the shielding process). Let us again start from the Poisson equation

$$\nabla^2 \phi = \frac{1}{\varepsilon_0} e (n_e - n_i) . \tag{1.7}$$

Assuming Boltzmann distributed electrons and ions (let us assume with the same temperature T which we measure in eV, that is, $k_B T \to T$), we have

$$n_e = n_{e0} e^{e\phi/T} \approx n_{e0} \left(1 + e\frac{\phi}{T}\right) , \quad n_i = n_{e0} e^{-e\phi/T} \approx n_{e0} \left(1 - e\frac{\phi}{T}\right) . \tag{1.8}$$

Anticipating spherical symmetry, we obtain

$$\nabla^2 \phi = \frac{d^2\phi}{dr^2} + \frac{2}{r}\frac{d\phi}{dr} = \frac{2e^2 n_{e0}}{\varepsilon_0 T} \phi . \tag{1.9}$$

This is a homogeneous linear differential equation. The amplitude parameter is free. It is easy to check that for $r \neq 0$, a solution is

$$\Phi = \frac{e}{r} e^{-\sqrt{2} r/\lambda_D} . \tag{1.10}$$

Thus, the potential in a plasma is exponentially shielded with the Debye length as the shielding distance.

However, the calculation is not yet complete. Thus far, the central charge q, which creates the shielding cloud, is missing. This is reflected in the fact that the amplitude is still free. It is obvious that instead of Eqs. (1.7) and (1.9), we should solve the following inhomogeneous linearized Poisson–Boltzmann equation, that is,

$$\begin{aligned} \nabla^2 \phi &= -\frac{1}{\varepsilon_0} q \delta(\boldsymbol{r}) - \frac{1}{\varepsilon_0} \sum_{s=e,i} q_s n_s \\ &\approx \kappa^2 \phi - \frac{1}{\varepsilon_0} q \delta(\boldsymbol{r}) , \end{aligned} \tag{1.11}$$

where the index s denotes the species, electrons and ions, the test charge q is sitting at $\boldsymbol{r} = 0$, and

$$\kappa^2 = \sum_s \frac{n_s q_s^2}{\varepsilon_0 T} \mathrel{\hat{=}} \frac{2}{\lambda_D^2} . \tag{1.12}$$

Its solution is the screened Debye potential of the test charge q, that is,

$$\phi = \frac{1}{4\pi\varepsilon_0}\frac{q}{r}e^{-\kappa r}, \tag{1.13}$$

as expected. The easiest direct way for finding the solution is by Fourier transformation (see Appendix B)

$$\phi_k = \int d^3 r e^{-i\mathbf{k}\cdot\mathbf{r}}\phi(\mathbf{r}) \tag{1.14}$$

which immediately leads to

$$\phi_k = \frac{1}{\varepsilon_0}\frac{q}{k^2+\kappa^2}. \tag{1.15}$$

Back-transformation gives

$$\begin{aligned}\phi(\mathbf{r}) &= \frac{1}{(2\pi)^3}\int d^3 k e^{i\mathbf{k}\cdot\mathbf{r}}\phi_k = \frac{1}{4\pi\varepsilon_0}\frac{1}{\pi}\int_{-1}^{1} dx \int_0^{\infty} dk k^2 \frac{q}{k^2+\kappa^2}e^{ikrx} \\ &= \frac{1}{4\pi\varepsilon_0}\frac{2}{\pi}\int_0^{\infty} dk k \frac{q}{k^2+\kappa^2}\frac{\sin(kr)}{r} = \frac{1}{4\pi\varepsilon_0}\frac{q}{r}e^{-\kappa r},\end{aligned} \tag{1.16}$$

since

$$\int_0^{\infty}\frac{x^{2m+1}\sin(ax)}{(x^2+z)^{n+1}}dx = \frac{(-1)^{n+m}}{n!}\frac{\pi}{2}\frac{d^n}{dz^n}\left(z^m e^{-a\sqrt{z}}\right), \tag{1.17}$$

which we applied for $n = m = 0$, $a = r$, $x = k$, and $z = \kappa^2$.

In conclusion, let us calculate the induced space charge. We decompose the potential

$$\phi(r) = \frac{1}{4\pi\varepsilon_0}\frac{q}{r}e^{-\kappa r} = \phi^{Cb} + \phi^{ind}, \tag{1.18}$$

where

$$\phi^{ind} = \frac{1}{4\pi\varepsilon_0}\frac{q}{r}(e^{-\kappa r}-1). \tag{1.19}$$

The induced space charge ρ^{ind}, which is responsible for the shielding (compared to the "naked" Coulomb potential ϕ^{Cb}) follows from

$$\nabla^2\phi^{ind} \equiv \frac{1}{r^2}\frac{d}{dr}\left(r^2\frac{d\phi^{ind}}{dr}\right) = -\frac{1}{\varepsilon_0}\rho^{ind}. \tag{1.20}$$

A straightforward calculation leads to

$$\rho^{ind} = -\frac{q\kappa^2}{4\pi r}e^{-\kappa r}. \tag{1.21}$$

Integrating the induced space charge density, which expresses the polarization, we obtain the expected result

$$\int d^3 r \rho^{ind} = -q \,. \qquad (1.22)$$

The sign of the shielding charge distribution is opposite to the sign of the test charge q. Shielding causes a redistribution of charges compared to the ideal plasma situation (in the latter, the interaction potential is neglected).

The picture of the Debye shielding is valid only if there are enough particles in the charge cloud. We can compute the number N_D of particles in a Debye sphere,

$$N_D = n \frac{4}{3} \pi \lambda_D^3 \,; \qquad (1.23)$$

effective shielding over a Debye length requires

$$N_D \gg 1 \,. \qquad (1.24)$$

A plasma with characteristic dimension L can be considered quasineutral, provided

$$\lambda_D \ll L \,. \qquad (1.25)$$

It is easy to show that the largest spherical volume of a plasma that spontaneously could become depleted of electrons has the radius of a few Debye lengths. Let us consider a sphere of uniformly distributed ions with density $n_i(\mathbf{r}) = \text{const}$. Starting from the center of the sphere, the radial coordinate r is being introduced. The charge $Q = 4\pi n_i e r^3/3$ is enclosed up to r, and the electric field only has a radial component

$$E_r = \frac{1}{4\pi\varepsilon_0} \frac{Q}{r^2} = \frac{n_i e r}{3\varepsilon_0} \,. \qquad (1.26)$$

From here, we find the electrostatic field energy W in the sphere of radius r as

$$W = \frac{\varepsilon_0}{2} \int_0^r E_r^2 4\pi r^2 dr = \frac{2\pi r^5 n_i^2 e^2}{45\varepsilon_0} \,. \qquad (1.27)$$

Let us now discuss the scenario that the ion-filled sphere has been created by evacuating electrons. Before leaving the sphere, the electrons had the kinetic energy

$$E_{kin} = \frac{3}{2} n_e T_e \times \frac{4}{3} \pi r^3 \,. \qquad (1.28)$$

The electrostatic energy W did not exist when (neutralizing) electrons were initially in the sphere to balance the ion charge. In other words, W must be equivalent to the work done by electrons on leaving the sphere. E_{kin} was available to the electrons. By equating

$$W = E_{kin}, \qquad (1.29)$$

we find the maximum radius or the largest spherical volume that could spontaneously become depleted of electrons. A short calculation leads to

$$r_{max}^2 = 45\varepsilon_0 \frac{T_e}{n_e e^2} \tag{1.30}$$

or

$$r_{max} \approx 7\lambda_{De} \,. \tag{1.31}$$

1.2 Degree of Ionization

A plasma may be partially or full ionized, respectively. The degree of ionization depends on several parameters. Let us first estimate the degree of ionization on the basis of thermal ionization in a system consisting of hydrogen (H) atoms, electrons (*e*), and protons (*p*) (ions).

1.2.1 The Saha Equation

The Saha equation gives a relationship between the free particles (for example, electrons and protons) and those bound in atoms (H). To derive the Saha equation, we assume thermodynamic equilibrium and *collisional ionization*. We choose a consistent set of energy levels. Let us choose $E = 0$ when the electron is free (unbound) and has velocity zero, and $E = E_n < 0$ when the electron is in a bound state of the hydrogen atom ($Z = 1$). Using the simple Bohr energy formula for $n = 1$, we ignore the higher n levels,

$$E_n = \frac{Z}{n^2} \times (-13.6\,\text{eV}) \,. \tag{1.32}$$

The first excited state is already close to the bound-free-limit when compared to the ground state. When sufficient energy is available to excite an electron from the ground state $n = 1$ to $n = 2$, only a little bit more (one third) is needed to directly ionize.

For independent particles, we first calculate the *single particle* partition functions for electrons, protons, and hydrogen atoms which have the form[1)] [27]

$$Z = \sum_n e^{-E(n)/k_B T} \,. \tag{1.33}$$

Here, the sum is over all states (bound and free) with energies $E(n)$. The sums in the partition functions are actually integrals for the free particles since the particles have a continuous momentum distribution. The degeneracy of states (or statistical

1) We keep the Boltzmann constant k_B as it is done in most books on statistical physics.

weights) g_i with $g_e = g_p = 2$ and $g_H = 4$ (for hydrogen) has to be taken into account. Therefore, for free electrons and free protons (ions), we have

$$Z_j = \frac{1}{h^3} \int g_j e^{[-p^2/(2m_j)]/(k_B T_j)} d^3 r d^3 p \quad \text{for} \quad j = e, p \; . \tag{1.34}$$

Because of isotropy, $d^3 p = 4\pi p^2 dp$, the integrals can be performed, that is,

$$Z_e = \frac{2V}{h^3} \left(2\pi m_e k_B T_e\right)^{3/2} , \tag{1.35}$$

$$Z_p = \frac{2V}{h^3} \left(2\pi m_p k_B T_p\right)^{3/2} . \tag{1.36}$$

Here, m_e and m_p are the electron and proton masses, respectively. For the temperature, we have in equilibrium $T = T_e = T_p = T_H$. A similar calculation leads for freely moving hydrogen atoms consisting of a bound electron–proton pair (in the ground state) to

$$Z_H = \frac{4V}{h^3} \left(2\pi m_H k_B T_H\right)^{3/2} e^{E_i/k_B T_H} \tag{1.37}$$

with $E_0 = -13.6\,\text{eV} \equiv -E_i$, where E_i is the ionization energy.

Let Z be the total partition function of the system (do not mix with Z for the atomic number) consisting of $N_e \equiv N_p$ free electrons (and protons) out of $N = N_H + N_p$ atoms in a given ensemble. For indistinguishable particles, we have

$$Z\left(V, T, N_e, N_p, N_H\right) = \frac{Z_e^{N_e}}{N_e!} \frac{Z_p^{N_p}}{N_p!} \frac{Z_H^{N_H}}{N_H!} . \tag{1.38}$$

From here, we get the free energy

$$F = -k_B T \ln Z \; . \tag{1.39}$$

The actual particle densities which are realized in nature are the ones that give a minimum of the free energy.

To find the most probable state, we should differentiate Eq. (1.39). For large numbers N, we use the Stirling formula

$$\ln N! \approx N \ln N - N \; , \tag{1.40}$$

leading to

$$\begin{aligned} -\frac{F}{k_B T} \approx\; & N_e \ln Z_e + N_p \ln Z_p + N_H \ln Z_H - N_e \ln N_e \\ & + N_e - N_p \ln N_p + N_p - N_H \ln N_H + N_H \; . \end{aligned} \tag{1.41}$$

Making use of $N_H = N - N_p$ and $N_p = N_e$, we find from the vanishing of the derivative

$$\frac{\partial F}{\partial N_e} \sim \ln Z_e + \ln Z_p - \ln Z_H - \ln N_e - \ln N_e + \ln(N - N_e) = 0 \; . \tag{1.42}$$

This results into

$$\frac{Z_e Z_p}{Z_H} = \frac{N_e^2}{N - N_e} . \tag{1.43}$$

With the expressions for the partition functions and $m_H \approx m_p$, we arrive at

$$\frac{V}{h^3} (2\pi m_e k_B T)^{3/2} e^{-E_i/k_B T} = \frac{N_e^2}{N - N_e} . \tag{1.44}$$

Introducing particle densities $N_e/V = n_e$, $N_p/V = n_p$, and $N_H/V = n_H \equiv n_n$, the Saha (equilibrium) equation can be written in the form

$$\begin{aligned} \frac{n_i n_e}{n_n} &= \frac{2g_1}{g_0} \left(\frac{m_e}{2\pi}\right)^{3/2} \hbar^{-3} (k_B T_e)^{3/2} e^{-E_i/k_B T_e} \equiv K(T_e) \\ &\approx 2.4 \times 10^{15} \, (T_e \, [\mathrm{K}])^{3/2} \, e^{-E_i/(T_e \, [\mathrm{eV}])} \quad [\mathrm{cm}^{-3}] \\ &\approx 3 \times 10^{21} \, (T_e \, [\mathrm{eV}])^{3/2} \, e^{-E_i/(T_e \, [\mathrm{eV}])} \quad \left[\mathrm{cm}^{-3}\right] . \end{aligned} \tag{1.45}$$

Once again, n_e is the electron density, n_n is the neutral particle (hydrogen) density, n_i is the density of single ionized particles (protons), g_ν is the statistical weight ($g_1 = 2$ for electrons and protons, $g_0 = 4$ for hydrogen), $k_B = 1.3807 \times 10^{-16}$ erg/K is the Boltzmann constant, T_e is the electron temperature, m_e is the electron mass, and $E_i = 13.6$ eV is the ionization energy. $1\,\mathrm{eV} \mathrel{\hat{=}} 1.6022 \times 10^{-12}$ erg, $1\,\mathrm{eV} \mathrel{\hat{=}} 1.1605 \times 10^4$ K.

The Saha formula for hydrogen may be written in the form

$$\frac{\alpha_i \alpha_e}{\alpha_n} = \frac{2g_1}{n g_0} \frac{1}{\lambda_e^3} e^{-E_i/k_B T_e} \tag{1.46}$$

where the ionization degrees

$$\alpha_e = \frac{N_e}{N} , \quad \alpha_i = \frac{N_i}{N} , \quad \alpha_n = \frac{N_H}{N} , \tag{1.47}$$

and the thermal de Broglie wavelength

$$\lambda_e = \sqrt{\frac{h^2}{2\pi m_e k_B T}} \tag{1.48}$$

have been introduced. Note that

$$N = N_H + N_i \equiv N_H + N_e , \quad N_e \equiv N_i , \quad n = \frac{N}{V} . \tag{1.49}$$

Thus, the Saha equation determines the degree of ionization in terms of $T \equiv T_e$ and n. When, besides the temperature T, the pressure p is given, we should use an equation of state, that is,

$$pV = \sum_\mu N_\mu k_B T_\mu = (1 + \alpha_e) N k_B T . \tag{1.50}$$

We can then replace

$$n = \frac{p}{(1 + \alpha_e) k_B T} , \tag{1.51}$$

leading to

$$\frac{\alpha_i \alpha_e}{\alpha_n (1 + \alpha_e)} = \frac{2g_1}{g_0} \frac{k_B T}{p \lambda_e^3} e^{-E_i / k_B T_e} . \tag{1.52}$$

Noting

$$\alpha_n \equiv \alpha_H = 1 - \alpha_e , \quad \alpha_i \equiv \alpha_e \tag{1.53}$$

for hydrogen, we may write

$$\frac{\alpha_e^2}{(1 - \alpha_e)(1 + \alpha_e)} = \frac{k_B T}{p \lambda_e^3} e^{-E_i / k_B T} \equiv \frac{1}{p \mathcal{K}_p(T)} . \tag{1.54}$$

The solution of this quadratic equation is

$$\alpha_e = \frac{1}{\sqrt{1 + p \mathcal{K}_p(T)}} , \quad \mathcal{K}_p(T) = \frac{\lambda_e^3}{k_B T} e^{E_i / k_B T} . \tag{1.55}$$

This formula allows one to determine the degree of ionization α_e for a given gas pressure p as a function of temperature T. The numerical evaluation is straightforward. Note

$$\begin{aligned} m_e &\approx 9.109 \times 10^{-31}\,\mathrm{kg} , & h &\approx 6.626 \times 10^{-34}\,\mathrm{J\,s} \\ 1\,\mathrm{eV} &\approx 1.602 \times 10^{-19}\,\mathrm{J} , & 1\,\mathrm{bar} &= 10^5\,\mathrm{N/m^2} \\ E_i &\approx 13.6057\,\mathrm{eV} , & k_B &\approx 1.3807 \times 10^{-16}\,\mathrm{erg/deg(K)} . \end{aligned}$$

In a similar way, one can derive the degree of ionization for a prescribed total density $n = N/V$ by

$$\alpha_e = \frac{1}{2n\mathcal{K}_n(T)} \left[\sqrt{1 + 4n\mathcal{K}_n(T)} - 1 \right], \quad \mathcal{K}_n(T) = \frac{\mathcal{K}_p(T)}{k_B T} . \tag{1.56}$$

The dependence of n_i / n_n on temperature for a fixed pressure p is shown in Figure 1.1. Another form of the Saha equation starts from the partition functions Z_i and Z_{i+1} for the atom in its initial and final stages of ionization. The ratio of the number of atoms in stage $i + 1$ to the number of atoms in stage i is

$$\frac{N_{i+1}}{N_i} = \frac{2Z_{i+1}}{n_e Z_i} \left(\frac{2\pi m_e k_B T}{h^2} \right)^{3/2} e^{-\chi_i / k_B T} , \tag{1.57}$$

where χ_i is the ionization energy needed to remove an electron from ionization stage i to $i + 1$. Because a free electron is produced in the ionization process, the number density of free electrons appears on the right-hand side. As n_e increases,

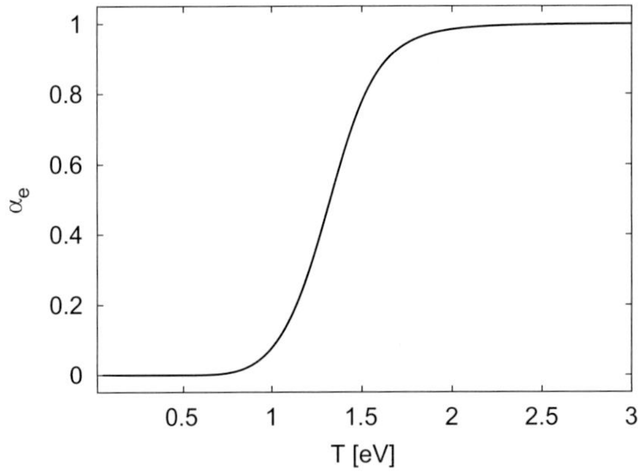

Figure 1.1 Evaluation of the Saha equation (1.55) for hydrogen ($E_i = 13.6\,\text{eV}$) with pressure $p = 1\,\text{bar}$.

the number of atoms in the higher stage of ionization decreases since there are more electrons with which the ion may recombine. Sometimes, the abbreviated form

$$\frac{Z_i}{N_i} = \frac{Z_{i+1} Z_e}{N_{i+1} N_e} \tag{1.58}$$

is said to be the most useful form. Convince yourself that all forms presented here are equivalent.

The Saha equation can also be seen as a restatement of the equilibrium condition for the chemical potentials (note the definition of the chemical potential in terms of a derivative of the free energy and compare with the previous derivation of the most probable state),

$$\mu_i = \mu_{i+1} + \mu_e \,. \tag{1.59}$$

This equation states that the potential of an atom of ionization state i to ionize is the same as the sum of the potentials of an electron and an atom of ionization state $i+1$. The potentials are equal and therefore the system is in equilibrium and no net change of ionization will occur.

The Saha equation named after the Indian astrophysicist Meghnad Saha who first derived it in 1920 [28].

1.2.2 Thomson Cross Section and Rate Equation

The cross section for the collisional ionization process can be easily estimated. The following estimate treats the collision process on the atomic scale (Bohr radius) in a (in principle not valid) classical model. Nevertheless, the results are close to the exact quantum-mechanical calculation.

When an electron approaches an H-atom on atomic dimension, we consider the scattering from the nucleus. Let us assume large velocities such that the scattering angle is small. The electron is accelerated in the Coulomb field of a proton. If ρ is the transverse distance being approximated by the *constant* scattering parameter ρ, the strength of the force is approximately proportional to $e^2/(\rho^2 + v_e^2 t^2)$, where v_e is the characteristic velocity. The closest approach is at $t = 0$. The perpendicular component requires an additional factor, which is one for the closest approach and decays approximately as $\rho/(\rho^2 + v_e^2 t^2)^{1/2}$ for larger t. Thus, the transverse (to the initial velocity) velocity change is

$$\frac{dv_\perp}{dt} \approx \frac{1}{4\pi\varepsilon_0} \frac{e^2}{m_e} \frac{\rho}{\left(\rho^2 + v_e^2 t^2\right)^{3/2}} . \tag{1.60}$$

Figure 1.2 shows some definitions of angles during the scattering, where θ is called the scattering angle, and α has been introduced to calculate the perpendicular component of the Coulomb force. The velocity of the electron is $v \mathrel{\hat{=}} v_e$. Integrating, we get the perpendicular velocity

$$v_\perp \approx \frac{1}{4\pi\varepsilon_0} \frac{e^2}{m_e} \int_{-\infty}^{\infty} \frac{\rho}{\left(\rho^2 + v_e^2 t^2\right)^{3/2}} dt = \frac{1}{4\pi\varepsilon_0} \frac{2e^2}{m_e v_e \rho} . \tag{1.61}$$

The energy difference

$$\Delta E_\perp \equiv \frac{m_e v_\perp^2}{2} \approx \frac{1}{(4\pi\varepsilon_0)^2} \frac{e^4}{E\rho^2} \equiv \varepsilon , \quad E = \frac{m_e v_e^2}{2} , \tag{1.62}$$

is available for collisional ionization. Rewriting the last relation as

$$\rho^2 = \frac{1}{(4\pi\varepsilon_0)^2} \frac{e^4}{E\varepsilon} , \tag{1.63}$$

we obtain the relation between collision parameter ρ, initial energy E, and energy change ε. The differential cross section for an energy change $d\varepsilon$ for given initial energy E is

$$d\sigma = |2\pi\rho\, d\rho| = \frac{1}{(4\pi\varepsilon_0)^2} \frac{\pi e^4}{E\varepsilon^2} d\varepsilon . \tag{1.64}$$

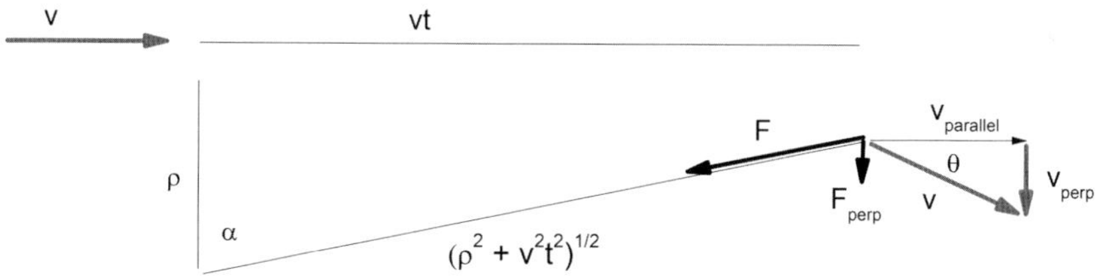

Figure 1.2 The velocity v of an electron approaching a proton at collision parameter ρ at two instances of time. On the left the velocity at $t = -\infty$ is shown. At a later time t, scattering occurred with scattering angle θ. F designates the electrostatic force between the electron and the proton at time t.

The hydrogen atom will be ionized if $\varepsilon > E_i$, where E_i is the ionization energy. Integrating from E_i to E, we obtain the Thomson cross section

$$\sigma_T = \frac{1}{(4\pi\varepsilon_0)^2}\frac{\pi e^4}{E^2 E_i}(E - E_i) . \tag{1.65}$$

The maximum cross section occurs for $E = 2E_i$. It is

$$\sigma_{T,max} = \frac{1}{(4\pi\varepsilon_0)^2}\frac{\pi e^4}{4E_i^2} \approx \pi a_B^2 \approx 10^{-16}\,\text{cm}^2 , \tag{1.66}$$

where a_B is the Bohr radius. The latter is given by

$$a_B = \frac{4\pi\varepsilon_0\hbar^2}{m_e e^2} \approx 0.529 \times 10^{-10}\,\text{m} , \tag{1.67}$$

and the ionization energy is

$$E_i = \frac{m_e e^4}{8\varepsilon_0^2 h^2} . \tag{1.68}$$

At large energies, the cross section decays proportional to $1/E$. More exact quantum calculations lead to a decay proportional to $\ln(E)/E$ for $E \gg E_i$ which better agrees with experiments.

Multiplying the cross section with the velocity and the density of scatterers, we obtain the probability of collisional ionization per time. Totally, the change is also proportional to the density of incident particles. For *collisional ionization* and (triple) recombination, the electron particle balance is then

$$\frac{dn_e}{dt} = n_e n_H \langle\sigma_T v_e\rangle - \beta n_e^2 n_p , \tag{1.69}$$

where we abbreviate $\langle\sigma_T v_e\rangle \equiv \alpha$ for the mean product of cross section with velocity. The recombination coefficient β can be found in the stationary state from

$$\beta = \frac{n_H\langle\sigma_T v_e\rangle}{n_e n_p} \equiv \frac{\alpha}{K(T)} , \tag{1.70}$$

where in the last reformulation, we used the Saha equation; $K(T)$ is on the right-hand side of the Saha formula.

1.2.3
The Corona Formula

When investigating all possible mechanisms for ionization and recombination in a plasma, a modified ionization formula is often used for the intermediate range of electron temperature and density which is of interest for laboratory experiments and astrophysics. That formula portrays, under certain situations, the physical phenomena better than previous calculations. In some hydrogen discharges, it may yield the tenfold value for the neutral plasma component, as would be expected from the Saha equation.

Let us give some arguments why, besides collisional ionization, other processes are sometimes important. Thus far, we considered collisional ionization, for example, of hydrogen. The process is

$$\mathrm{H} + \mathrm{e}^- \to \mathrm{H}^+ + \mathrm{e}^- + \mathrm{e}^- \tag{1.71}$$

The inverse process is (triple) recombination

$$\mathrm{H}^+ + \mathrm{e}^- + \mathrm{e}^- \to \mathrm{H} + \mathrm{e}^- \tag{1.72}$$

The Saha equation covers these processes in thermodynamic equilibrium.

Another ionization and recombination channel is *photoionization*

$$\mathrm{H} + \hbar\omega \to \mathrm{H}^+ + \mathrm{e}^- \tag{1.73}$$

and photorecombination

$$\mathrm{H}^+ + \mathrm{e}^- \to \mathrm{H} + \hbar\omega \tag{1.74}$$

We can combine all these processes in the rate equation

$$\frac{dn_e}{dt} = \alpha n_e n_H - \beta n_e^2 n_p + \mu n_H - \gamma n_e n_p \,, \tag{1.75}$$

with coefficients α (collisional ionization from the ground state), β (collisional three-body recombination), μ (photoionization), and γ (electron–ion radiative recombination), respectively. At thermal equilibrium, assuming a detailed balance, we should have

$$\mu = \gamma \frac{n_e n_p}{n_H} \equiv \gamma K(T) \,. \tag{1.76}$$

Under laboratory conditions, the radiation can often leave the plasma. That is why the photon density can be much lower than in equilibrium. For this reason, one can often neglect the photoionization. When the plasma is rarefied, one can also neglect the triple recombination. The corresponding condition is

$$\frac{\mu}{\alpha} \ll n_e \ll \frac{\gamma}{\beta} \,. \tag{1.77}$$

In this limit, the equilibrium electron concentration is given by the balance between photorecombination and collisional ionization, and we arrive at the Elwert formula

$$\frac{n_p}{n_H} = \frac{\alpha}{\gamma} \,. \tag{1.78}$$

This relation is also called the corona equilibrium formula. In many astrophysical situations, this balance is appropriate. The corona model is usually assumed to be applicable if [29]

$$10^{12} t_I^{-1} < n_e\left[\mathrm{cm}^{-3}\right] < 10^{16}\, T_e\,[\mathrm{eV}]^{7/2} \,, \tag{1.79}$$

where t_I is the normalized ionization time.

For the γ-coefficient, we may use the formula [30]

$$\gamma \approx 2.7 \times 10^{-13} T_e^{-1/2} \left[\frac{\text{cm}^3}{\text{s}}\right] \quad \text{when} \quad 1 < T_e\,[\text{eV}] < 15\,. \tag{1.80}$$

Collisional three-body recombination is often approximated by [29]

$$\beta \approx 8.75 \times 10^{-27} (T_e\,[\text{eV}])^{-4.5} \left[\frac{\text{cm}^6}{\text{s}}\right]. \tag{1.81}$$

The interested reader is referred to [31] for more details.

1.3
Characteristic Parameters

Temperature and density are two characteristic parameters of plasmas. For the temperature, the attributes hot and cold are, of course, relative. Similar is the situation for the density when we talk about high and low densities. In this section, we want to establish a better understanding for the relevant orders of magnitude.

In magnetic confinement, hot means high temperatures fulfill the Lawson criterion. Then, we are in the temperature range

$$10\,\text{keV} \le k_B T \le 20\,\text{keV}\,, \tag{1.82}$$

and the Lawson criterion [32] reads[2)]

$$n k_B T \tau_E \ge 3 \times 10^{21}\,\text{m}^{-3}\,\text{keV}\,\text{s}\,. \tag{1.83}$$

Here, τ_E is the energy confinement time.

The first question we want to deal with is whether a *relativistic* treatment is required. As a rough estimate for the need of a relativistic treatment, we postulate

$$\frac{v_{the}^2}{c^2} \ge 0.01\,, \tag{1.84}$$

with the electron thermal velocity

$$v_{the} = \left(\frac{k_B T_e}{m_e}\right)^{1/2}. \tag{1.85}$$

It is straightforward to find the temperature range for the necessity of a relativistic treatment in the form

$$k_B T_e \ge 0.01 m_e c^2 \approx 0.005\,\text{MeV} = 5\,\text{keV} \gtrapprox 50\,000\,000\,\text{K}\,. \tag{1.86}$$

2) John D. Lawson, an engineer, was noted for his 1955 paper, published in 1957 [32].

Here, we made use of the approximate values

$$m_e c^2 \approx 0.5\,\text{MeV}\,, \quad 1\,\text{eV} \mathrel{\hat{\approx}} 10\,000\,\text{K}\,. \tag{1.87}$$

More exact values can be obtained when using more precise constants; see, for example, [29].

Quantum effects become important when the degeneracy parameter $n_e \lambda_{dB}^3$ becomes larger than one. In the *nonrelativistic* case, we use the thermal de Broglie wavelength

$$\lambda_{dB} \equiv \lambda_e = \frac{h}{\sqrt{2\pi m_e k_B T_e}}\,. \tag{1.88}$$

Then,

$$n_e \gg \left(\frac{m_e k_B T_e}{\hbar^2}\right)^{3/2} \tag{1.89}$$

follows for the need of a quantum-mechanical description.

In the *relativistic* case, we use the energy expression

$$E^2 = c^2 p^2 + m_e^2 c^4 \tag{1.90}$$

together with $E \sim k_B T_e$ as well as $\lambda_{dB} \sim h/p$. In the *ultrarelativistic* case, from $p \sim k_B T_e/c$,

$$n_e \gg \left(\frac{k_B T_e}{\hbar c}\right)^3 \tag{1.91}$$

follows. For the general case Eq. (1.90), the generalization of the estimate Eq. (1.91) is straightforward to obtain. In Figure 1.3, we have plotted the borderline between the classical and the quantum-mechanical behaviors. The areas of applicability of other model zones are also shown.

The *ideal* gas approximation proves to be good if the nonideality parameter $|E_{pot}|/|E_{kin}|$ is small. Since

$$E_{pot} \sim \frac{1}{4\pi\varepsilon_0}\frac{e^2}{\lambda_n} \sim n_e^{1/3} \tag{1.92}$$

and

$$E_{kin} = \frac{3}{2} k_B T_e \quad \text{[classical]}\,, \tag{1.93}$$

we can express, with the electron Debye length $\lambda_{De} = (\varepsilon_0 k_B T_e / n_e e^2)^{1/2}$, the *classical ideality condition* as

$$n_e \lambda_{De}^3 \gg 1\,. \tag{1.94}$$

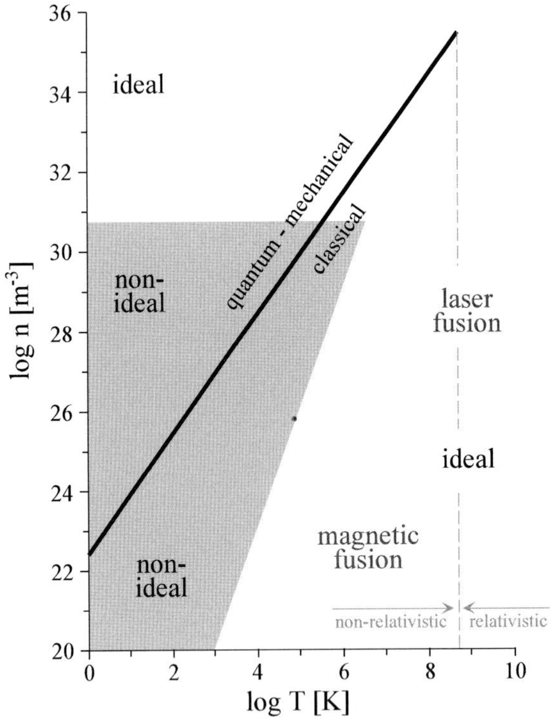

Figure 1.3 Overview over different model approximations in physical parameter space.

When a quantum-mechanical description is appropriate, the estimate is different. Then, we approximate

$$E_{kin} \approx \frac{p_F^2}{2m_e} \sim \frac{\hbar^2 n_e^{2/3}}{m_e}, \tag{1.95}$$

with the Fermi momentum p_F to find the *quantum-mechanical ideality condition*

$$\lambda_B \equiv \frac{e^2 m_e}{4\pi\varepsilon_0 n_e^{1/3} \hbar^2} \ll 1. \tag{1.96}$$

The expression λ_B on the left-hand side is called the Brueckner parameter. Note the different conditions for ideality (in terms of the density) in the classical and quantum-mechanical regimes, respectively. In the classical case, we have

$$\frac{E_{Coulomb}}{k_B T_e} \sim n_e^{1/3}, \tag{1.97}$$

whereas in the quantum case,

$$\frac{E_{Coulomb}}{\frac{p_F^2}{2m_e}} \sim n_e^{-1/3} \tag{1.98}$$

follows.

A global overview over different model approximations in physical parameter space is shown in Figure 1.3.

We conclude this subsection by the quite surprising remark that "low" temperatures in quantum-mechanically degenerate systems, that is,

$$k_B T_e \ll \varepsilon_F \equiv \sqrt{m_e^2 c^4 + c^2 p_F^2}\,, \tag{1.99}$$

may be extremely "high", in the sense that relativistic effects are important. The latter appear for

$$p_F > m_e c \tag{1.100}$$

with

$$p_F = \left(\frac{3n_e}{8\pi}\right)^{1/3} h\,. \tag{1.101}$$

Thus, for

$$n_e \gg \left(\frac{m_e c}{h}\right)^3, \tag{1.102}$$

we have to perform relativistic calculations even if the temperatures are "low."

In the following, we discuss two important applications in more detail.

1.3.1 Typical Parameters of (Magnetic) Fusion Plasmas

When developing a theory, we have to specify the parameter regime to which it should apply. In the following, we show typical parameters[3)] for a magnetic fusion plasma,

$$T \approx 10\,\mathrm{keV}\,, \quad B \approx 4\,\mathrm{T} \hat{=} 4 \times 10^4\,\mathrm{G}\,, \quad n_e \approx n_i := n \approx 10^{14}\,\mathrm{cm}^{-3}\,. \tag{1.103}$$

For these parameters, the Debye length is then of the order

$$\lambda_D \approx 7.43 \times 10^2\, T^{1/2} n^{-1/2}\,[\mathrm{cm}] \approx 7 \times 10^{-3}\,\mathrm{cm}\,. \tag{1.104}$$

Thermal de Broglie wavelength, mean particle distance and thermal velocity are

$$\lambda_{dB} \approx 2.76 \times 10^{-8}\, T^{-1/2}\,[\mathrm{cm}] \approx 3 \times 10^{-10}\,\mathrm{cm}\,, \tag{1.105}$$

$$\lambda_n \approx n^{-1/3}\,[\mathrm{cm}] \approx 10^{-5}\,\mathrm{cm}\,, \tag{1.106}$$

$$v_{the} \approx 4.19 \times 10^7\, T^{1/2} \left[\frac{\mathrm{cm}}{\mathrm{s}}\right] \approx 4 \times 10^9\,\frac{\mathrm{cm}}{\mathrm{s}}\,. \tag{1.107}$$

3) Only within the next few formulas, should B be inserted in gauss units, as it is often done in fusion applications [29]. Thus, when not stated otherwise, in this subsection, T is in electron volt and B in gauss.

Larmor radii follow via the gyrofrequency $\Omega = qB/m$ (in SI units) divided by the thermal velocity,

$$\rho_i \approx 1.02 \times 10^2 T_i^{1/2} B^{-1} \,[\mathrm{cm}] \approx 0.2\,\mathrm{cm}\,, \tag{1.108}$$

$$\rho_e \approx 2.38 T_e^{1/2} B^{-1} \,[\mathrm{cm}] \approx 0.5 \times 10^{-2}\,\mathrm{cm}\,. \tag{1.109}$$

A plasma is magnetized for $\delta \equiv \rho/L \ll 1$, that is,

$$\delta_i \approx \sqrt{\frac{m_i}{m_e}}\,\delta_e \ll 1\,. \tag{1.110}$$

The total collision cross section $\sigma_{tot} \equiv \pi R^2$ (see previous and following discussions) is often estimated by making use of

$$\frac{1}{4\pi\varepsilon_0}\frac{e^2}{R} \approx T\,. \tag{1.111}$$

Then, the mean free path of a particle

$$\lambda_{mfp} \sim v_{th}\nu^{-1} \sim \frac{1}{\sigma_{tot} n} \quad \text{[estimate]} \tag{1.112}$$

turns out to vary as

$$\lambda_{mfp} \sim \frac{T^2}{n}\,. \tag{1.113}$$

The more exact formula [29] leads, for example for electrons, to

$$\lambda_{mfp} \approx 10^6\,\mathrm{cm}\,. \tag{1.114}$$

From the mean free paths, estimates of the collision frequencies also follow. Here, we present the kinetic collision frequencies $\nu \sim \lambda_{mfp}/v_{th}$ as

$$\nu_e \approx 2.91 \times 10^{-6} n_e \ln\Lambda\, T_e^{-3/2} \left[\mathrm{s}^{-1}\right] \approx 3 \times 10^3\,\mathrm{s}^{-1}\,, \tag{1.115}$$

$$\nu_i \approx 4.78 \times 10^{-8} n_i \ln\Lambda\, T_i^{-3/2} \left[\mathrm{s}^{-1}\right] \approx 5 \times 10^1\,\mathrm{s}^{-1}\,, \tag{1.116}$$

where $\ln\Lambda$ is the Coulomb logarithm (see later chapters). The gyrofrequencies have the orders of magnitude

$$|\Omega_e| \approx 1.76 \times 10^7 B \left[\frac{\mathrm{rad}}{\mathrm{s}}\right] \approx 8 \times 10^{11}\,\frac{\mathrm{rad}}{\mathrm{s}}\,, \tag{1.117}$$

$$\Omega_i \approx 9.58 \times 10^3 \frac{\mathrm{rad}}{\mathrm{s}} \sim 4 \times 10^8\,\frac{\mathrm{rad}}{\mathrm{s}}\,. \tag{1.118}$$

In a plasma, a collective electron plasma frequency also appears,

$$\omega_{pe} = \sqrt{\frac{ne^2}{\varepsilon_0 m_e}} \sim 5.64 \times 10^4 n_e^{1/2} \left[\frac{\mathrm{rad}}{\mathrm{s}}\right] \approx 5 \times 10^{11}\,\frac{\mathrm{rad}}{\mathrm{s}}\,. \tag{1.119}$$

Similarly, an ion plasma frequency exists, that is,

$$\omega_{pi} \approx 1.32 \times 10^3 n_i^{1/2} \left[\frac{\mathrm{rad}}{\mathrm{s}}\right] \approx 10^{10}\,\frac{\mathrm{rad}}{\mathrm{s}}\,. \tag{1.120}$$

1.3.2
Parameters of the Sun

In our opinion, the most important burning star is the sun which provides us with the necessary energy for life. Thermonuclear fusion in gravitational confinement is its energy source. Geological observations do not show any indications for significant changes in the behavior of the sun during the last million years. Thus, we may assume that the sun is in hydrostatic equilibrium. The gravitational energy can be estimated by (we assume an approximately constant mass density $\rho = \text{const}$)

$$E_{Gr} = -\int_0^R \frac{G m(r)\rho(r)}{r} 4\pi r^2 dr$$
$$= -\frac{(4\pi)^2}{3} G\rho^2 \int_0^R r^4 dr = -\frac{3}{5}\frac{GM^2}{R} \sim \mathcal{O}\left(-\frac{GM^2}{R}\right) . \qquad (1.121)$$

Here, $m(r)$ is the mass of the sphere up to radius r; it is the "enclosed mass." In the following, we shall use

$$E_{Gr} \approx -\frac{GM_\odot^2}{R_\odot} . \qquad (1.122)$$

The exact numerical factors are not needed for a rough estimate. In Eq. (1.122), $R_\odot$ is the radius and $M_\odot$ is the mass of the sun. Astrophysical observations provide us with the values $M_\odot = 1.99 \times 10^{30}$ kg and $R_\odot = 6.96 \times 10^8$ m. The age of the sun is estimated as $t_\odot \approx 4.55$ Gyr. Measurements of the luminosity $L_\odot$ deliver the basis for estimates of the surface temperature $T_\odot$. The approximate value for the luminosity is $L_\odot = 3.86 \times 10^{26}$ W. Making use of the luminosity formula $L_\odot = 4\pi R_\odot^2 \sigma T_0^4$ with the Stefan–Boltzmann constant σ, we find for the surface temperature $T_O \approx T_\odot \equiv 5780 \approx 6000$ K. This value is quite different from the temperature in the interior of the sun. The latter should be much larger in order to enable the nuclear fusion process.

The physical parameters of the interior of the sun follow from solar models [18, 19]. Solar models are meanwhile quite advanced. Typical values for the central temperature, central mass density, and pressure at the center, respectively, are

$$T_c = 15.6 \times 10^6 \text{ K} , \qquad (1.123)$$

$$\rho_c = 1.48 \times 10^5 \text{ kg m}^{-3} , \qquad (1.124)$$

$$P_c = 2.29 \times 10^{16} \text{ Pa} . \qquad (1.125)$$

The large central temperature becomes plausible from the following estimate based on the stationarity condition

$$\frac{dP}{dr} = -\rho \frac{GM_r}{r^2} , \qquad (1.126)$$

where $M_r \equiv m(r)$ is the "enclosed mass," G the gravitational constant, and P is the pressure. This is a minimal model; many effects are neglected.

Multiplying Eq. (1.126) by $4\pi r^3$ and integrating over r from zero to $R \approx R_\odot$ leads to

$$\int_0^R dr 4\pi r^3 \frac{dP}{dr} = -\int \frac{GM_r}{r} dm \equiv E_{Gr} , \tag{1.127}$$

with $dm = 4\pi r^2 \rho dr$. E_{Gr} is the potential (gravitational) energy of the whole mass distribution and ρ is the mass density.

Next, we integrate on the left-hand side by parts. For (nearly) vanishing pressure at the surface $R \approx R_\odot$, we obtain

$$-3\langle P\rangle V = E_{Gr} , \tag{1.128}$$

with the averaged pressure

$$\langle P\rangle = \frac{\int P(r) dV}{\int dV} . \tag{1.129}$$

The averaged pressure is the mean pressure value averaged over the whole volume. With the new notation, Eq. (1.128) is known as the virial theorem, usually written in the form

$$\langle P\rangle = -\frac{1}{3}\frac{E_{Gr}}{V} . \tag{1.130}$$

The contents of the virial theorem is the maximum information we can get without specifying an equation of state.

Making use of the virial theorem, we find for the sun $\langle P\rangle \approx 10^{14}$ Pa. Note that we obtain the mean pressure averaged over the whole volume. The pressure in the center is much larger. When we are interested in more details, we need the equation of state. For rough values, a classical calculation is still possible. In addition, although deviations from the ideal gas law will become significant, for simplicity, we still use the ideal gas approximation

$$\langle P\rangle = nkT = \frac{\langle\rho\rangle}{\bar{m}} kT . \tag{1.131}$$

This leads to the mean temperature for the hydrostatic equilibrium $\langle T\rangle \approx 6 \times 10^6$ K. That temperature is much larger than the surface temperature (by a factor of approx. 10^3), and reasonably close to the temperature at center $T_c \approx 15.6 \times 10^6$ K. In any case, the low surface temperature is very important. Larger surface temperatures would lead to a X-ray bombardment on the earth, making life in its present form impossible.

1.4
Individual and Collective Effects

In this section, we discuss the differences between individual and collective effects in more detail. We have already mentioned that binary collisions are typical individual effects. Their strength is measured in terms of a collision frequency ν. The latter should be compared with the plasma frequency ω_p, a typical collective phenomenon. Let us now first work out the physical meaning of the electron plasma frequency ω_{pe}.

1.4.1
The Plasma Frequency

A "Gedankenexperiment" starts from a sphere of radius r uniformly filled with electrons and protons such that the whole system is globally neutral. If, on the other hand, only one species, for example, electrons, would be present, the sphere would have a charge q_e resulting in a radial field at the surface,

$$q_e = -\frac{4\pi}{3} r^3 e n_e \,, \quad |E| = \frac{1}{4\pi\varepsilon_0} \frac{q_e}{r^2} \,. \tag{1.132}$$

The field strength could be enormously large, depending on the size of the sphere and the electron density. However, as stated already, we start our "Gedankenexperiment" with homogeneously distributed electrons and protons such that outside the sphere no electric field exists. Next, let us expand the electron sphere from radius r to radius $r + x$ with $x \ll r$, creating an electron shell of thickness x. Since the total number N of electrons and protons, respectively, is constant, the electron density is decreased to

$$n_e = n_{e0} + \delta n_e \approx \frac{N}{\frac{4\pi}{3} r^3 \left(1 + 3\frac{x}{r}\right)} \approx n_{e0} - 3 n_{e0} \frac{x}{r} \,. \tag{1.133}$$

Within the sphere of radius r, we now have a positive excess charge

$$\Delta q \approx 4\pi n_{e0} e r^2 x \,, \tag{1.134}$$

creating, at r, the radial electric field component

$$E \approx \frac{1}{4\pi\varepsilon_0} 4\pi e n_{e0} x \,. \tag{1.135}$$

A shell electron feels the radial force

$$K \equiv m_e \frac{d^2 x}{dt^2} = -eE = -\frac{1}{\varepsilon_0} e^2 n_{e0} x \,, \tag{1.136}$$

which pulls it back into the original sphere. An overshooting will occur, resulting in oscillations with the electron plasma frequency

$$\omega_{pe} = \sqrt{\frac{e^2 n_e}{\varepsilon_0 m_e}} \,. \tag{1.137}$$

In this scenario, we kept the ions fixed, having in mind their large mass compared to the electron mass and the high-frequency of the electron oscillations.

The Debye length is connected to the plasma frequency via

$$\lambda_D = \frac{v_{th}}{\omega_p} \,. \tag{1.138}$$

1.4.2
General Remark on Individual Collisions

Next, we will discuss the individual collision frequencies in more detail. Binary collisions correspond to the classical two-body problem. At this point, we should mention that the two-body problem with Coulomb interaction can be reduced to an effective one-particle problem if we go to the center of mass (COM) system. Then, the reduced mass appears. If one mass is much larger than the other (or if we fix the position of the scatterer, which effectively means that we assume an infinitely large mass of the scatterer), the difference between COM and laboratory system disappears. The next considerations can be interpreted as the calculations for a fixed scatterer. The two-body problem with a *fixed scattering center* is depicted in Figure 1.4.

A charge $q = +e$ scatters a beam of electrons with charge $-e$. The incoming particle current density is $j = n_e v_e$. Per time unit, we have $2\pi\rho d\rho j$ particles being scattered when passing through a ring area $2\pi\rho d\rho$. Each particle is scattered at a specific angle θ. Asymptotically, the change of *longitudinal* (i.e., in the direction of initial propagation) momentum of a single particle is

$$\Delta p_0 = -m_e v_e (1 - \cos\theta) . \tag{1.139}$$

We can define the total force on the *beam* (in the presence of a fixed scatterer), whose magnitude is

$$F_e = -\int_0^\infty m v_e (1 - \cos\theta) j 2\pi\rho d\rho \mathrel{\hat{=}} - m_e v_e j \sigma . \tag{1.140}$$

Here, we assumed fixed initial velocities of magnitude v_e. The expression

$$\sigma = \int_0^\infty (1 - \cos\theta) 2\pi\rho d\rho \tag{1.141}$$

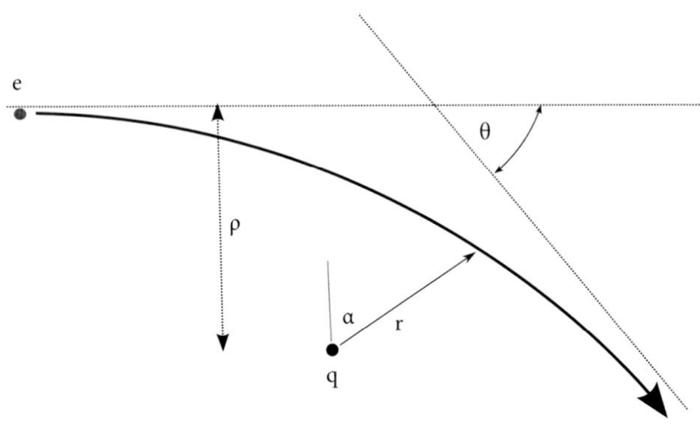

Figure 1.4 Sketch of a two-particle collision.

is called the transport cross section (for momentum transfer). To evaluate the latter, we need the functional dependence $\theta = \theta(\rho)$. That can be taken from standard textbooks of classical mechanics [33],

$$\tan\frac{\theta}{2} = \frac{1}{4\pi\varepsilon_0}\frac{e^2}{m_e\rho v_e^2} \,. \tag{1.142}$$

As a side note, this relation can be made easily plausible for small scattering angles $\theta \ll 1$. Then, the straight-path approximation may be assumed in evaluating integrals. For example, we get for the transverse momentum change

$$\begin{aligned} \Delta p_{0\perp} \approx m_e v_e \theta &= \int_{-\infty}^{\infty} F_\perp \, dt \approx \frac{1}{4\pi\varepsilon_0}\int_{-\infty}^{\infty} \frac{e^2\rho}{\left(\rho^2 + v_e^2 t^2\right)^{3/2}} dt \\ &= \frac{1}{4\pi\varepsilon_0}\frac{2e^2}{\rho v_e} \,, \end{aligned} \tag{1.143}$$

where we have approximated $\sin\theta \approx \theta$ and $\cos\alpha \approx \rho/\sqrt{\rho^2 + v_e^2 t^2}$. The angle θ is the scattering angle and α is the angle between the total force and the perpendicular component.

Evaluating the cross section Eq. (1.141), we make us of $1 - \cos\theta \approx \theta^2/2$ to obtain

$$\sigma = \frac{1}{(4\pi\varepsilon_0)^2}\frac{4\pi e^4}{m_e^2 v_e^4}\int_0^\infty \frac{1}{\rho} d\rho \,. \tag{1.144}$$

Apparently, the integral is logarithmically divergent. As a lower limit, we define

$$\rho_{min} = \frac{1}{4\pi\varepsilon_0}\frac{e^2}{m_e v_e^2} \,, \tag{1.145}$$

which is the collision parameter for 90° scattering. Note that $\tan(\pi/4) = 1$. For more details, see the chapter on transport theory. For very small collision parameters (strong collisions), a classical treatment will break down. The upper limit $\rho_{max} = \lambda_D$ is postulated as the Debye length due to the screening of the potential of the scatterer. Thus, we evaluate, instead of Eq. (1.144),

$$\sigma = \frac{1}{(4\pi\varepsilon_0)^2}\frac{4\pi e^4}{m_e^2 v_e^4}\int_{\rho_{min}}^{\rho_{max}} \frac{1}{\rho} d\rho \,. \tag{1.146}$$

The expression

$$\ln\Lambda = \ln\frac{\rho_{max}}{\rho_{min}} = \ln 16\pi^2\varepsilon_0^{5/2}\frac{3(k_B T)^{3/2}}{e^3\sqrt{n}} \tag{1.147}$$

is proportional to the logarithm of the number of particles in the Debye sphere and is called the Coulomb logarithm. In magnetic fusion plasmas, it is of the order 10–20. In Eq. (1.147), we have used

$$\frac{1}{2}m_e v_e^2 \approx \frac{3}{2}k_B T \,. \tag{1.148}$$

Of course, within an approximate treatment, also slightly different estimates are in use. The result for the cross section is now

$$\sigma = \frac{1}{(4\pi\varepsilon_0)^2}\frac{4\pi e^4}{m_e^2 v_e^4}\ln\Lambda\ . \tag{1.149}$$

We find the order of magnitude

$$\sigma \sim \frac{10^{-12}}{(E\,[\text{eV}])^2}\ \text{cm}^2\ , \tag{1.150}$$

where $E = 1/2 m_e v_e^2$ is the kinetic energy.

1.4.3
Collision Frequencies for Momentum and Energy Transfer

The transport cross section allows us to find the *momentum* transfer rate between particles. Let us consider an electron beam in a cold plasma. Then, we still assume the scatterers as fixed. Let n_i be the ion density. When we have n_i scatterers per unit volume, the mean force acting on a single particle of the beam will be

$$\boldsymbol{F}_e = -m_e \boldsymbol{v}_e j \sigma n_i \frac{1}{n_e} = -m_e v_e \sigma n_i \boldsymbol{v}_e\ . \tag{1.151}$$

Due to this force, the electron mean velocity is decelerated,

$$\frac{d\boldsymbol{v}_e}{dt} = \frac{\boldsymbol{F}_e}{m_e} = -n_i \sigma v_e \boldsymbol{v}_e \equiv -\nu_{ei}\boldsymbol{v}_e\ . \tag{1.152}$$

Here, $\boldsymbol{v}_e$ is the relative velocity between electron and ion. For a fixed scatterer, the total energy of an electron does not change. The longitudinal velocity component of the scattered electron changes according to Eq. (1.139), or approximately as

$$\Delta v_e = -v_e(1 - \cos\theta) \approx -v_e\frac{\theta^2}{2}\ . \tag{1.153}$$

Thus, the angular spread grows, that is,

$$\frac{d\theta^2}{dt} = 2\sigma n_i v_e\ . \tag{1.154}$$

The characteristic time

$$\tau_{ei} = \frac{1}{n_i \sigma v_e} \equiv \frac{1}{\nu_{ei}} \tag{1.155}$$

is the inverse of the collision frequency ν_{ei}.

The collision frequency follows from the cross section after multiplying $n_i v_e$, that is, we obtain the approximate result

$$\nu_{ei} = \frac{1}{(4\pi\varepsilon_0)^2}\frac{4\pi e^4 n_i}{m_e^2 v_e^3}\ln\Lambda\ . \tag{1.156}$$

The ratio between the individual collision frequency and the collective plasma frequency is proportional to the inverse of the number of particles in a Debye sphere,

$$\frac{\nu_{ei}}{\omega_{pe}} \sim \frac{1}{n\lambda_{De}^3} \tag{1.157}$$

for $n_i \approx n_e \approx n$.

Note that after the time $T \approx \nu_{ei}^{-1}$ the quantity θ^2 suffers a change of its order of magnitude. We may calculate characteristic values from

$$\nu_{ei} \sim 6 \times 10^{-5} \frac{n_i\,[\mathrm{cm}^{-3}]}{(E\,[\mathrm{eV}])^{3/2}} \,. \tag{1.158}$$

The electron mean free path is

$$\lambda_{mfpe} = \frac{1}{n_i \sigma} \sim 10^{12} \frac{(E\,[\mathrm{eV}])^2}{n_i\,[\mathrm{cm}^{-3}]}\ \mathrm{cm}\,. \tag{1.159}$$

For a Maxwellian, we may use

$$\langle E^2 \rangle = \frac{15}{4}(k_B T)^2\,. \tag{1.160}$$

With that, we determine the temperature dependencies of the characteristic quantities. Measuring lengths in cm, densities in cm^{-3}, and temperatures in eV, we obtain the characteristic values

$$\sigma \sim \frac{3 \times 10^{-13}}{T^2}\ \mathrm{cm}^2\,, \tag{1.161}$$

$$\nu_{ei} \sim 3 \times 10^{-5} \frac{n}{T^{3/2}}\ \mathrm{s}^{-1}\,, \tag{1.162}$$

$$\lambda_{mfpe} \sim 3 \times 10^{12} \frac{T^2}{n}\ \mathrm{cm}\,. \tag{1.163}$$

1.4.3.1 Calculation in the Center of Mass System

When the masses of scatterer (m_2) and scattered particle (m_1) are similar or $m_1 \gg m_2$, we transform to the center of mass system with center of gravity $\boldsymbol{R}$. Using $\boldsymbol{r} = \boldsymbol{r}_2 - \boldsymbol{r}_1$ for the difference of the position vectors, and the reduced mass $m_{12} = m_1 m_2/(m_1 + m_2)$, we have

$$\boldsymbol{r}_1 = \boldsymbol{R} - \frac{m_2}{m_1 + m_2}\boldsymbol{r}\,, \quad \boldsymbol{r}_2 = \boldsymbol{R} + \frac{m_1}{m_1 + m_2}\boldsymbol{r}\,, \tag{1.164}$$

and find

$$m_{12}\ddot{\boldsymbol{r}} = -Z_1 Z_2 e^2 \frac{\boldsymbol{r}}{r^3}\,. \tag{1.165}$$

In the following, we assume that the scatterer has charge $Z_2 e$, while for the scattered particle, the charge is $-Z_1 e$. One obtains an effective two-body problem similar to the treatment for fixed scatterers. The mass has to be replaced by the reduced mass, and the position of the scatterer is replaced by the relative distance. We can immediately translate the previous formulas, for example, Eq. (1.149) can now be written as

$$\sigma = \frac{1}{(4\pi\varepsilon_0)^2} \frac{4\pi Z_1^2 Z_2^2 e^4}{m_{12}^2 v^4} \ln \Lambda \,, \tag{1.166}$$

and the general formula for the collision frequency (to deflect the incident particle by an angle 90° in the presence of n_2 scatterers per unit volume)

$$\nu_{12} = v \sigma n_2 = \frac{1}{(4\pi\varepsilon_0)^2} \frac{4\pi Z_1^2 Z_2^2 e^4 n_2 \ln \Lambda}{m_{12}^2 v^3} \,. \tag{1.167}$$

Note that we calculated the frequency for velocity deflections. Thus, ν_{12} directly gives the frequency for momentum transfer in the center of gravity system. We shall denote the *momentum* scattering frequencies as $\nu_{ee}, \nu_{ii}, \nu_{ei}$, and ν_{ie} for the various possible interactions between species. The reciprocals are denoted by $\tau \sim \nu^{-1}$.

Energy scattering is characterized by the time required for an incident particle to transfer its kinetic energy to the target particle. The energy transfer collision frequencies are denoted by $\nu_{ee}^E, \nu_{ii}^E, \nu_{ei}^E$, and ν_{ie}^E, respectively.

As a typical velocity, we take $v \sim v_{th}$. We reference all collision frequencies to ν_{ee}. Note for the reduced mass $m_{ee} \sim m_e/2 \sim m_e$ and $v \sim T^{1/2}/m_e^{1/2}$ for this case. When we calculate ν_{ei}, we have $m_{ei} \sim m_e$ and $v \sim T^{1/2}/m_e^{1/2}$, that is, (except for a factor two) the same values, and therefore

$$\nu_{ei} \sim \nu_{ee} \tag{1.168}$$

follows.

Next, we consider ν_{ii}. Now, $m_{ii} \sim m_i/2$ and $v \sim T^{1/2}/m_i^{1/2}$, and therefore

$$\nu_{ii} \sim \sqrt{\frac{m_e}{m_i}} \nu_{ee} \tag{1.169}$$

follows. In all cases, the differences between the lab frame and the center of gravity system are tolerable since we ignore factors of order two in the present estimates.

Care is required when calculating ν_{ie}. The transformation into the lab frame is necessary, straightforward, but not immediately obvious. An easier way to estimate ν_{ie} is momentum conservation in lab frame which leads to $m_i \Delta \boldsymbol{v}_i = -m_e \Delta \boldsymbol{v}_e$ where Δ means the change in quantity as a result of the collision. For a head-on collision, we approximately have $\Delta \boldsymbol{v}_e \approx 2\boldsymbol{v}_i$ and therefore $|\Delta \boldsymbol{v}_i|/|\boldsymbol{v}_i| \approx 2m_e/m_i$. Thus, in order to have $|\Delta \boldsymbol{v}_i|/|\boldsymbol{v}_i|$ of order unity, it is necessary to have m_i/m_e collisions. Hence,

$$\nu_{ie} \sim \frac{m_e}{m_i} \nu_{ee} \,. \tag{1.170}$$

Next, we consider *energy* changes. If a moving electron makes a head-on collision with an electron at rest, then the incident electron stops while the originally standing electron flies off with the same momentum and energy that the incident electron had. From here, we conclude

$$\nu_{ee}^{E} \sim \nu_{ee} . \tag{1.171}$$

A similar picture holds for an ion hitting an ion,

$$\nu_{ii}^{E} \sim \nu_{ii} \sim \sqrt{\frac{m_e}{m_i}}\nu_{ee} . \tag{1.172}$$

Finally, we compare energy changes during electron–ion and ion–electron collisions. The electron momentum change is $-2m_e\boldsymbol{v}_e$ during a collision of an electron with an ion. From conservation of momentum, we find $m_i\boldsymbol{v}_i = 2m_e\boldsymbol{v}_e$. The energy transferred to the ion is $1/2m_iv_i^2 = 4(m_e/m_i)m_ev_e^2/2$. An electron has to make m_i/m_e collisions for transferring all its energy to the ions, and hence

$$\nu_{ei}^{E} \sim \frac{m_e}{m_i}\nu_{ee} . \tag{1.173}$$

Similarly, during an ion–electron collision, an electron initially at rest will fly off with twice the incident ion velocity. The electron gains energy $1/2m_ev_e^2 \sim 2(m_e/m_i)m_iv_i^2$. Again, approximately m_i/m_e collisions are necessary for the ion to transfer all its energy to the electrons, that is,

$$\nu_{ie}^{E} \sim \frac{m_e}{m_i}\nu_{ee} . \tag{1.174}$$

These rough estimates have an important consequence, namely, that we may describe – even in nonequilibrium situations – an electron–ion plasma by a two-fluid model. The relaxation to approximate Maxwellians occurs fast in each component (fastest for the electrons), and the exchange between the components occurs on a much slower scale.

1.4.4
Friction Force in Thermal Plasmas

We now consider a thermal *e*–*i*-plasma. The velocity distribution functions are assumed as

$$f_i(\boldsymbol{v}') = \left(\frac{m_i}{2\pi T_i}\right)^{3/2} \exp\left(-\frac{m_i(\boldsymbol{v}' - \boldsymbol{u}_i)^2}{2T_i}\right) \tag{1.175}$$

and

$$f_e(\boldsymbol{v}') = \left(\frac{m_e}{2\pi T_e}\right)^{3/2} \exp\left(-\frac{m_e(\boldsymbol{v}' - \boldsymbol{u}_e)^2}{2T_e}\right) . \tag{1.176}$$

The electric current density in this configuration is

$$\boldsymbol{j} = n_i e\boldsymbol{u}_i - n_e e\boldsymbol{u}_e . \tag{1.177}$$

It is appropriate to transform into the system of the mean velocity $\boldsymbol{u}_e$ of the electrons and to define the relative velocity

$$\boldsymbol{u}_{rel} = \boldsymbol{u}_i - \boldsymbol{u}_e \equiv \boldsymbol{u} \,. \tag{1.178}$$

The distribution functions in this frame will be

$$f_i(\boldsymbol{v}) = \left(\frac{m_i}{2\pi T_i}\right)^{3/2} \exp\left(-\frac{m_i\left(\boldsymbol{v}-\boldsymbol{u}_{rel}\right)^2}{2T_i}\right) \tag{1.179}$$

and

$$f_e(\boldsymbol{v}) = \left(\frac{m_e}{2\pi T_e}\right)^{3/2} \exp\left(-\frac{m_e v^2}{2T_e}\right). \tag{1.180}$$

Since the ion thermal velocity is much smaller than the electron thermal velocity, the ion velocity distribution is much narrower than the electron distribution function, and the ions can be considered approximately as a monoenergetic beam. When impinging on n_e electrons per unit volume, we calculate the net force on the ions making use of the previous expressions and applying the actio $=$ reactio principle. Using Eq. (1.151), for a moment we go to the system where the ions are at rest, such that the electrons have velocity $\boldsymbol{v}_e - \boldsymbol{u}$. Then, $\boldsymbol{F}_e$ can be used in the form Eq. (1.151). Setting $\boldsymbol{F}_i = -\boldsymbol{F}_e$, we find

$$\boldsymbol{F}_i = -m_e\left|\boldsymbol{u}-\boldsymbol{v}_e\right| \sigma n_e(\boldsymbol{u}-\boldsymbol{v}_e) \,, \tag{1.181}$$

with ($Z_1 = 1, Z_2 = Z$)

$$\sigma \approx \frac{1}{(4\pi\varepsilon_0)^2} \frac{4\pi Z^2 e^4}{m_e^2\left|\boldsymbol{u}-\boldsymbol{v}_e\right|^4} \ln\Lambda \,. \tag{1.182}$$

By the way, similar formulas would hold for a fast electron component in a thermal plasma. Let us assume that we have a fraction of scatterers in a certain velocity domain. Then, we replace

$$n_e \to dn_e = n_e f_e(\boldsymbol{v}')d^3v' \,. \tag{1.183}$$

We rewrite the friction force for dn_e scatterers per unit volume. Having in mind the relative speed $\boldsymbol{u} - \boldsymbol{v}'$ of the scattered particles, we obtain

$$d\boldsymbol{F}_i = -\frac{1}{(4\pi\varepsilon_0)^2} \frac{4\pi Z^2 \ln\Lambda\, e^4 n_e}{m_e} \frac{\boldsymbol{u}-\boldsymbol{v}'}{\left|\boldsymbol{u}-\boldsymbol{v}'\right|^3} f_e\left(\boldsymbol{v}'\right) d^3v' \,. \tag{1.184}$$

When averaging over the possible velocities of the scatterers, one has to evaluate the integral

$$\boldsymbol{I} = \int \frac{\boldsymbol{u}-\boldsymbol{v}'}{\left|\boldsymbol{u}-\boldsymbol{v}'\right|^3} f_e(\boldsymbol{v}')d^3v' \,. \tag{1.185}$$

Its form reminds us of the integral over a charge distribution appearing when solving the Poisson equation [34]. If the distribution function f is isotropic, the integral corresponds to an electric field caused by a spherically symmetric charge distribution with "radius vector" $\boldsymbol{u}$. We know that the radial electric field at distance u is only caused by the charge within the sphere of "radius" u, that is,

$$I = \frac{\boldsymbol{u}}{u^3} \int_0^u f_e(v') 4\pi v'^2 dv' \,. \tag{1.186}$$

For readers who are not familiar with the discussion of the Poisson integral, we briefly present an "evaluation for pedestrians." Let us start with

$$\int \frac{\boldsymbol{u} - \boldsymbol{v}'}{|\boldsymbol{u} - \boldsymbol{v}'|^3} f_e(\boldsymbol{v}') d^3 v' = -\nabla_u \int \frac{1}{|\boldsymbol{u} - \boldsymbol{v}'|} f_e(\boldsymbol{v}') d^3 v' \,. \tag{1.187}$$

Next, we introduce spherical coordinates for the integration on the right-hand side, perform the integration over the azimuthal angle, obtaining after a simple reformulation

$$\begin{aligned} &\int \frac{1}{|\boldsymbol{u} - \boldsymbol{v}'|} f_e(\boldsymbol{v}') d^3 v' \\ &= 2\pi \int_0^\infty dv' v'^2 \int_{-1}^{+1} d\left(\cos\vartheta'\right) \frac{d}{\cos\vartheta'} \sqrt{u^2 + v'^2 - 2uv'\cos\vartheta'} \left(-\frac{f_e(\boldsymbol{v}')}{uv'}\right) \\ &= 4\pi \left\{ \frac{1}{u} \int_0^u v'^2 f_e(\boldsymbol{v}') dv' + \int_u^\infty v' f_e(\boldsymbol{v}') dv' \right\} . \end{aligned} \tag{1.188}$$

Applying the operator $-\nabla_u$ to the right-hand side, we obtain Eq. (1.186).

For small relative velocities compared to the thermal electron velocity, $u \ll v_{the}$, we obtain

$$\int_0^u f_e(v') 4\pi v'^2 dv' \approx f_e(0) \frac{4\pi}{3} u^3 \,, \tag{1.189}$$

and thereby for a Maxwellian

$$I \approx \frac{\sqrt{2}}{3\sqrt{\pi}} \left(\frac{m_e}{T_e}\right)^{3/2} \boldsymbol{u} \,. \tag{1.190}$$

On the other hand, for large velocities u,

$$I \sim \frac{\boldsymbol{u}}{u^3} \,. \tag{1.191}$$

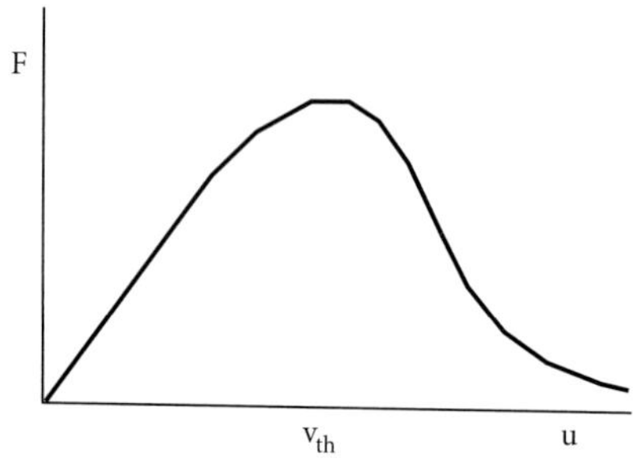

Figure 1.5 Sketch of the friction force as a function of the drift velocity u.

Summarizing for small velocities compared to the thermal electron velocity of the scatterers, the friction force is

$$\boldsymbol{F}_i = -\frac{1}{(4\pi\varepsilon_0)^2}\frac{4\sqrt{2\pi}}{3}\frac{Z^2 \ln\Lambda\, e^4 n_e}{m_e}\left(\frac{m_e}{T_e}\right)^{3/2}\boldsymbol{u}\,, \tag{1.192}$$

that is, it increases in magnitude with velocity. However, when the drift velocity becomes larger than the thermal velocity, we get a decaying friction force, $F \sim u^{-2}$. Remember, u is the mean velocity difference. The friction force reaches some maximum around the electron thermal velocity; see Figure 1.5.

We now apply these results to the problem of current flow through plasma. At equilibrium, the driving force by the electric field must be equilibrated by the friction, that is, for ions

$$\boldsymbol{F}_i + Ze\boldsymbol{E} = 0\,. \tag{1.193}$$

However, when the drift velocity becomes much larger than the electron thermal velocity, collisions cannot stop particle acceleration, and the particles run away. In other words, when the applied electric field becomes large,

$$E > \frac{F_{max}}{Ze}\,, \tag{1.194}$$

we get runaway particles. The critical field is the Dreicer field

$$E_{Dr} \approx \frac{1}{(4\pi\varepsilon_0)^2}\frac{\ln\Lambda\, n_e e^3 Z}{T_e} \sim \ln\Lambda\frac{e}{\lambda_D^2}\,. \tag{1.195}$$

A more general kinetic evaluation gives [15]

$$E_{Dr} \approx \frac{0.43}{(4\pi\varepsilon_0)^2}\frac{2\pi\ln\Lambda\, n_e e^3 Z}{T_e} \approx 5.6\times 10^{-18} n_e Z\frac{\ln\Lambda}{T_e}\left[\frac{\mathrm{V}}{m}\right]. \tag{1.196}$$

In order to elucidate the typical behaviors for $u \ll v_{th}$ and $u \gg v_{th}$, respectively, we solve the momentum equation

$$m_i\frac{d\boldsymbol{u}}{dt} = \boldsymbol{F}_i + Ze\boldsymbol{E} \tag{1.197}$$

in the two regions. In both cases, we anticipate a one-dimensional description. In the small velocity range, the momentum equation is of the form

$$\frac{du}{dt} = -au + b\,, \tag{1.198}$$

(with constants a and b which are easy to determine). Its solution, for $u = u_0 = 0$ at $t = t_0 = 0$, is

$$u(t) = \frac{b}{a}\left(1 - e^{-at}\right) \to \frac{b}{a} \quad \text{for} \quad t \to \infty\,. \tag{1.199}$$

For small velocities (fields), one finds stationary conduction

$$j \sim enu \sim ne\frac{eE}{m\nu_{ei}} \sim \frac{\varepsilon_0 \omega_p^2}{\nu_{ei}} E \sim \sigma E\,. \tag{1.200}$$

The conductivity is inversely proportional to the collision frequency. Since the latter decays with increasing temperature, we find

$$\sigma \sim T_e^{3/2}\,, \tag{1.201}$$

that is, the electric conductivity of hot plasmas becomes very large. The resistivity η is inversely proportional to σ, that is,

$$\eta = \frac{1}{\sigma}\,, \tag{1.202}$$

and thereby it decreases with temperature. The value obtained from kinetics [15] is

$$\eta \approx \frac{1}{(4\pi\varepsilon_0)^2} \frac{8\sqrt{\pi} \ln \Lambda Z e^2 m_e^{1/2}}{3\sqrt{2} T_e^{3/2}}\,. \tag{1.203}$$

For practical applications, we may use

$$\eta \approx 1.03 \times 10^{-4} \frac{Z \ln \Lambda}{T_e^{3/2}}\ \Omega\,\mathrm{m} \tag{1.204}$$

when T_e is measured in eV.

In the large velocity range, the momentum equation is of the form

$$\frac{du}{dt} = -\frac{g}{u^2} + b \tag{1.205}$$

(with constants g and b, which are easy to determine again). Its solution for $u = u_0 > \sqrt{g/b}$ at $t = t_0 = 0$ is

$$u - u_0 + \frac{1}{2}\sqrt{\frac{g}{b}} \ln \left[1 - 2\frac{\sqrt{\frac{g}{b}}}{u + \sqrt{\frac{g}{b}}}\right]_{u_0}^{u} = bt\,. \tag{1.206}$$

It is easy to see that t is a monotonously increasing function of u for $u > \sqrt{g/b}$. In other words,

$$u \to bt \quad \text{for} \quad t \to \infty \,. \tag{1.207}$$

This asymptotic reflects the so called runaway phenomenon.

1.5 Fusion Processes

Nuclear fusion is the source for energy production in burning stars. Fusion powers the sun and stars as hydrogen atoms fuse together to form helium, and matter is converted into energy. At low temperatures, fusion is not possible because the strongly repulsive electrostatic forces between the positively charged nuclei prevent them from getting close enough together for fusion to occur. However, if the conditions are such that the nuclei can overcome the electrostatic forces to the extent that they can come within a very close range of each other. Then, the attractive nuclear force between the nuclei will outweigh the repulsive (electrostatic) force, allowing the nuclei to fuse together.

In the sun, massive gravitational forces create the right conditions for fusion, but on earth they are much harder to achieve. Fusion fuel – different isotopes of hydrogen – must be heated to extreme temperatures of the order of 10^8 °C, and must be kept dense enough and confined for long enough in order to allow the nuclei to fuse. The aim of the controlled fusion research program is to achieve "ignition," which occurs when enough fusion reactions take place for the process to become self-sustaining, with fresh fuel then being added to continue it.

The binding energy per nucleon is the key to the understanding of fusion as an energy source. Light nuclei may fuse to heavier nuclei, releasing energy in the form of kinetic energy and radiation. When plotting the binding energy per nucleon, as depicted in Figure 1.6, we recognize that the curve has a minimum. On the left-hand side of Figure 1.6, the sum of the masses of two fusion components is larger than the mass of the fusion product. The mass defect Δm corresponds to an energy difference $\Delta E \equiv B = \Delta m c^2$. The binding energy per nucleon B/A, where A is the nucleon number (rough atomic weight), has a minimum near $A \approx 56$. The elements close to this atomic number are exceedingly stable.

With current technology, the reaction most readily feasible is between the nuclei of the two heavy forms (isotopes) of hydrogen-deuterium (D) and tritium (T). The deuterium-tritium reaction possesses a favorable cross section. Each D-T fusion event releases approximately 17.6 MeV ($2.8 \times 10{-}12$ J, compared with 200 MeV for a U-235 fission), that is,

$$\text{D} + \text{T} \to {}^4\text{He}\ (3.5\ \text{MeV}) + \text{n}\ (14\ \text{MeV}) \tag{1.208}$$

Deuterium occurs naturally in seawater (30 g/m^3), which makes it very abundant relative to other energy resources. Tritium does not occur naturally and is radioactive, with a halflife of around 12 yr. It can be made in a conventional nuclear reactor,

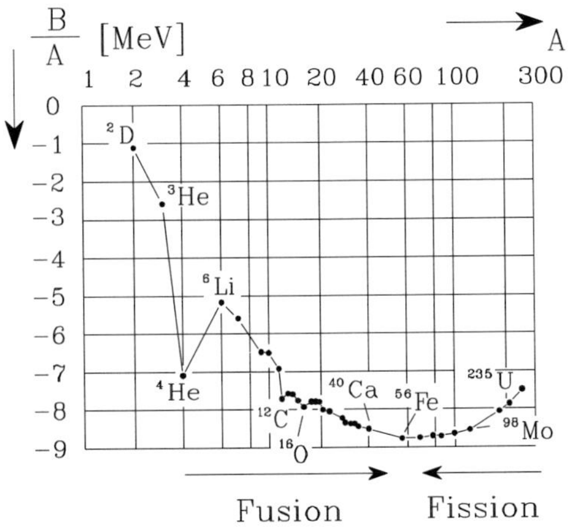

Figure 1.6 Binding energy per nucleon as a function of the number of nucleons *A*.

or in the present context, bred in a fusion system from lithium,

$$^{6}\mathrm{Li} + n \rightarrow {}^{4}\mathrm{He} + \mathrm{T} \tag{1.209}$$

Lithium is found in large quantities (30 ppm) in the Earth's crust and in weaker concentrations in the sea.

Present estimates show that a GW-reactor needs approximately 100 kg D and 400 kg Li per year. In other words, the perspectives of controlled thermonuclear fusion become plausible when estimating that a few liters of seawater and some pieces of rock (order of kg for the lithium extraction) will be sufficient for the energy consumption of a family of four persons during one year. At present, two main experimental approaches are being studied: magnetic confinement and inertial confinement. The first method uses strong magnetic fields to contain the hot plasma. The second involves compressing a small pellet containing fusion fuel to extremely high densities using strong lasers or particle beams.

Although fusion does not generate long-lived radioactive products (and the unburned gases can be treated on site), there appears a short-term radioactive waste problem due to activation of the structural materials. Some component materials will become radioactive during the lifetime of a reactor due to bombardment with high-energy neutrons, and will eventually become radioactive waste. The volume of such waste would be similar to the corresponding volumes from fission reactors. However, the longterm radiotoxicity of the fusion wastes would be considerably lower than that from actinides in used fission fuel. The fast neutron problem triggered other scenarios which are also discussed from time to time in the public [35]. One is the recovery of ^{3}He (consisting of two protons and one neutron) from the surface of the moon. ^{3}He is practically not available on the earth; tritium decays into ^{3}He with a 12 yr halflife. However, cosmic rays, originating from the solar wind, already penetrating for billions of years approximately 1 m into the surface of the moon produced ^{3}He there due to the spallation process. The fusion of D with ^{3}He to ^{4}He is a possible process, without producing significant neutrons. The cross section for the D–^{3}He reaction is, however, smaller by a factor 10 than the cross section for D–T fusion.

1.5.1 Fusion Processes in Burning Stars

Stellar fusion processes were already discussed a long time ago. Among the first authors are Atkinson and Houtermans [36]. Laboratory observations go back to the 1930s. At the end of the 1930s, Weizsäcker and Bethe found the dominating processes responsible for the burning in stars [18].

We have to differentiate between massive and low-mass stars; the borderline is approximately at the order of the mass of our sun. In low-mass stars, the fusion cycle starts with the proton–proton reaction

$$\mathrm{p} + \mathrm{p} \rightarrow \mathrm{D} + \mathrm{e}^+ + \nu \tag{1.210}$$

After that, the deuteron may again react with a proton

$$\mathrm{D} + \mathrm{p} \rightarrow {}^3\mathrm{He} + \gamma\ , \tag{1.211}$$

leading to the already mentioned ^{3}He fusion

$${}^3\mathrm{He} + {}^3\mathrm{He} \rightarrow {}^4\mathrm{He} + 2\mathrm{p} \tag{1.212}$$

For massive stars, that process does not deliver enough energy to sustain the necessary pressure. The compression of massive stars continues, transforming gravitational energy into kinetic energy. Using in the virial theorem $-3\langle P\rangle V \approx E_{gr} < 0$ and $\langle P\rangle V \approx N k_B T$, we obtain for the kinetic energy $E_{kin} \approx 3/2 N k_B T$

$$E_{kin} \approx -\frac{1}{2} E_{gr} > 0\ . \tag{1.213}$$

The latter relation makes, for massive stars, the increase in temperature plausible.

At higher temperatures, a new process becomes possible, the CNO cycle (for carbon-nitrogen-oxygen), or sometimes Bethe–Weizsäcker cycle, with a much higher fusion power than the direct proton–proton reaction just mentioned. The CNO process was proposed by Carl von Weizsäcker and Hans Bethe independently in 1938 and 1939, respectively [18]. Models show that the CNO cycle is the dominant source of energy in stars heavier than about 1.5 times the mass of the sun. The previously mentioned proton–proton chain is more important in stars of the mass of the sun or less. This distinction stems from differences in temperature dependency between the two reactions; the p–p reactions start occurring at lower temperatures, making it dominant in smaller stars. The CNO chain starts occurring at higher temperatures, but its energy output rises much faster with increasing temperatures. That crossover is depicted in Figure 1.7.

In the CNO cycle, four protons fuse, using carbon, nitrogen and oxygen isotopes as a catalyst, to produce one alpha particle, two positrons and two electron

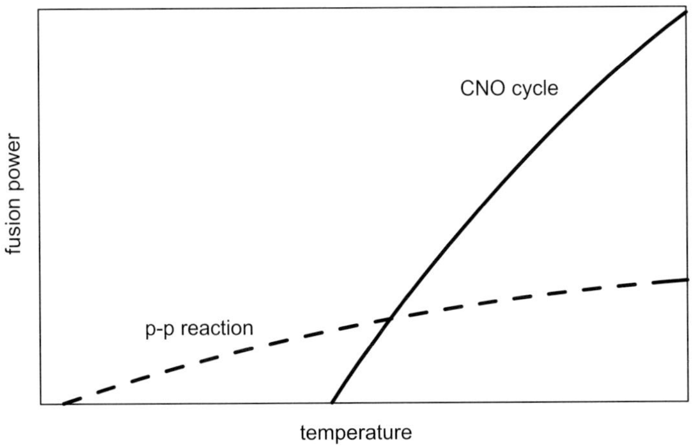

Figure 1.7 Temperature dependence of the fusion power in the p–p reaction and the CNO cycle, respectively.

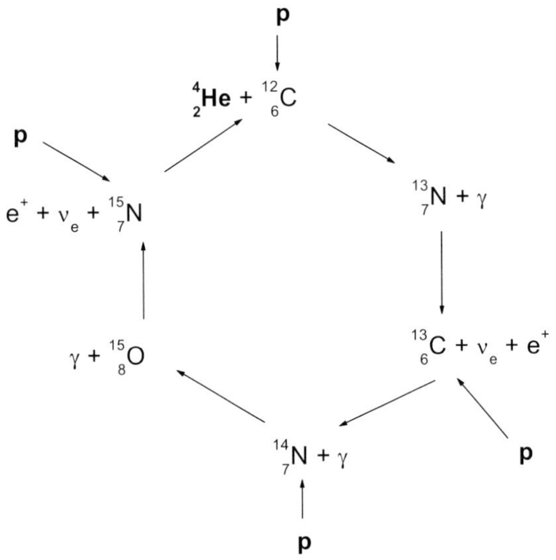

Figure 1.8 Diagram of the CNO cycle. The reaction chain starts at top and should be read clockwise.

neutrinos. The positrons will almost instantly annihilate with electrons, releasing energy in the form of gamma rays. The neutrinos escape from the star carrying away some energy. The carbon, nitrogen, and oxygen isotopes are in effect one nucleus that goes through a number of transformations in an endless loop, as shown in Figure 1.8.

Our sun (mass density $160\,\mathrm{g\,cm^{-3}}$) is confined by gravitation, the primary heating was due to compression. The core temperature is around 15.7×10^{6} K and only 1.7% of ^{4}He nuclei produced in the sun are born in the CNO cycle. The p–p chain reaction is by far dominating. The fusion power density is approximately 0.28 mW cm^{-3}.

During the p–p reaction,

$$p + p \rightarrow D + e^+ + \nu_e \tag{1.214}$$

$$p + D \rightarrow {}^3He + \gamma \tag{1.215}$$

$${}^3He + {}^3He \rightarrow {}^4He + p + p \tag{1.216}$$

the energy gain per produced ^{4}He nucleus is 26 MeV $\hat{=}$ $26 \times 1.6 \times 10^{-13}$ J. The radiation power of the sun is 4×10^{26} W.

It is interesting to note that the p–p reaction is rather slow. A proton needs approximately 5×10^9 yr $= 1.5 \times 10^{17}$ s to fuse with another proton. One needs four protons to effectively create a ^{4}He nucleus with 26 MeV energy gain. For the total power of 4×10^{26} W, one needs

$$\frac{4 \times 4 \times 10^{26}}{26 \times 1.6 \times 10^{-13}} = 4 \times 10^{38}\,\frac{\text{protons}}{\text{s}}\,. \tag{1.217}$$

2×10^{38} neutrinos per second are produced. Part of them will reach the Earth. (The neutrino problem is a separate issue.)

Dividing the power 4×10^{26} W by the mass of the sun, we get the average 0.2 mW/kg. That power density is smaller than the corresponding one of a human body. We may estimate the total number of protons in the sun as 7×10^{56}. One needs $4 \times 10^{38}\,\text{s}^{-1}$ for the total power $10^{38} \times 26 \times 1.6 \times 10^{-13} = 4 \times 10^{26}$ W. $7 \times 10^{56}/4 \times 10^{38} \approx 2 \times 10^{18}$ s $\approx$ 63 Gyr is the upper limit for the lifetime of the sun. Actually, the lifetime may be smaller by a factor of ten since not all protons will fuse.

The sun is essentially a quite stable burning object. It can be considered as a self-regulating thermostat. In case the energy release and thereby the temperature in the interior increases, the total energy $E = E_{kin} + E_{gr}$ increases. Because of Eq. (1.213), we have

$$E \approx \frac{1}{2} E_{gr} \approx -E_{kin}\,. \tag{1.218}$$

Thus, with increasing energy, also the gravitational energy increases and the kinetic energy (temperature) decreases. The sun expands and cools, acting in the opposite direction to the initial disturbance. On the other hand, when the energy production in the interior becomes smaller, the total energy decreases. The sun would contract and increase its temperature; again, a reaction which opposes the original disturbance.

When the fuel for p–p fusion, often called hydrogen burning, is running low in the core, the sun will contract and becomes hotter. Then, the hydrogen burning can also take place in the previously colder outer shells. The next dominating process

Table 1.1 Possible fusion products are shown in dependence of temperature T and mass M of a star.

$T = 10^7$ K $M > 0.08 M_{\odot}$	hydrogen	$\rightarrow$	helium
$T = 10^8$ K $M > 1/2 M_{\odot}$	helium	$\rightarrow$	carbon, oxygen
$T = 5 \times 10^8$ K $M > 8 M_{\odot}$	carbon	$\rightarrow$	oxygen, neon, ...
$T = 10^9$ K	neon	$\rightarrow$	oxygen, magnesium
$T = 2 \times 10^9$ K	oxygen	$\rightarrow$	magnesium, ... sulfur
$T = 3 \times 10^9$ K $M > 11 M_{\odot}$	silicon	$\rightarrow$	iron, ...

in the core will be helium burning. Temperatures of approximately 2×10^8 K will become possible in the center; the maximum mass density will be of the order 10^8 kg m^{-3}. The sun will appear as a red giant. That terminology becomes plausible because of the following typical values for luminosity, surface temperature, and radius, namely, roughly $L = 1000 L_{\odot}$, $T_O = 4000$ K, $R = 70 R_{\odot}$, respectively.

Stars with masses above $0.5 M_{\odot}$ will follow this scenario. Depending on the mass of the star, further steps are possible as depicted in Table 1.1.

As discussed, the highest element in this fusion scenario will be ^{56}Fe. Heavier elements can be generated by neutron capture and subsequent β-decay of the neutron into a proton and an electron.

During the cosmological evolution, the situation occurred in which temperatures were high enough, but the mass densities were too low to enable fusion. Only after the formation of massive stars can fusion become possible.

When all the fuel for fusion runs dry, the stars will further shrink. For stars with intermediate masses like the sun, degenerate electrons may provide the pressure for hindering a complete collapse. The quantum description becomes necessary. Whether it should be nonrelativistic or relativistic is an important question; the answer depends on the mass of the star. In the ultrarelativistic regime, for example, the equilibrium with a pressure delivered by degenerate electrons is fragile. For masses smaller than $1.4 M_{\odot}$ (Chandrasekhar mass), a nonrelativistic description is appropriate, and the final state of our sun will be a compact object with degenerate electrons providing the pressure for equilibrium. Such an object is called a white dwarf; typical parameters are $L = L_{\odot}/100$, $T_O = 16\,000$ K, $R = R_{\odot}/70$, $\rho \approx 10^9$ kg m^{-3}.

For massive stars, the pressure of degenerate electrons will not be sufficient. The star shrinks further, ending, for example, as a neutron star. Degenerate neutrons in the nonrelativistic regime become responsible for the necessary pressure. Mass densities of order 10^{14}–10^{17} kg m^{-3} are typical, temperatures may reach 10^8 K. A typical radius is 17 km.

Similar to the applicability of the model of a white star, not all massive stars will end as a neutron star. An upper limit of the order $M \approx 3 M_{\odot}$ exists. If the masses are larger, the neutrons become ultrarelativistic and the equilibrium is fragile. Such supermassive stars may end as black holes.

When comparing with other energy energy providers, we cite $0.2\,\mathrm{W/cm^3}$ for a coal-fired power plant and $90\,\mathrm{W/cm^3}$ for a light-water reactor. For our sun, the estimate for the center is $3.5 \times 10^{-6}\,\mathrm{W/cm^3}$ (averaged over the whole sun volume we would get $2.7 \times 10^{-7}\,\mathrm{W/cm^3}$).

For details on the fusion window, fusion power rates, cross sections, and so on, we refer to more detailed reports, for example, [14, 18, 19, 37].

1.5.1.1 Ignition Conditions for a Fusion Plasma

We conclude this section by a short summary of the Lawson criterion [32]. The total thermal energy of a plasma consisting of electrons and ions in equal numbers is

$$W = \int dV 3nT \equiv 3\overline{nT}V \,. \tag{1.219}$$

The energy loss rate P_L defines the energy confinement time τ_E via

$$P_L = \frac{W}{\tau_E} \,. \tag{1.220}$$

When in a stationary state the energy loss is balanced by additional heating,

$$P_H = P_L \,, \tag{1.221}$$

we obviously have

$$\tau_E = \frac{W}{P_H} \,. \tag{1.222}$$

However, besides additional (external) heating, we should also take into account alpha particle heating. A fraction of the fusion energy is taken over by the $^4\mathrm{He}(\alpha)$ particles. The α particles are charged, in contrast to the neutrons. The charged particles interact with the plasma, transferring part of the 3.5 MeV ($= E_\alpha$) to electrons and ions. The α particle heating is defined via the heating power

$$P_\alpha = \frac{1}{4}\overline{n^2 \langle \sigma v \rangle} V E_\alpha \,, \tag{1.223}$$

where $\langle \sigma v \rangle$ is the reaction parameter [37]. In a stationary state with α particle heating plus additional heating, the power balance would be

$$P_H + P_\alpha = P_L \tag{1.224}$$

or

$$P_H = \left(\frac{3nT}{\tau_E} - \frac{1}{4} n^2 \langle \sigma v \rangle E_\alpha \right) V \,, \tag{1.225}$$

where, on the right-hand side, for the reason of simplicity, the indication of volume averaging has been omitted. Self-contained burning (or ignition with $P_H = 0$) occurs for

$$n\tau_E \geq \frac{12}{\langle \sigma v \rangle} \frac{T}{E_\alpha} \sim T e^{\mathrm{const}/T^{1/3}} \,. \tag{1.226}$$

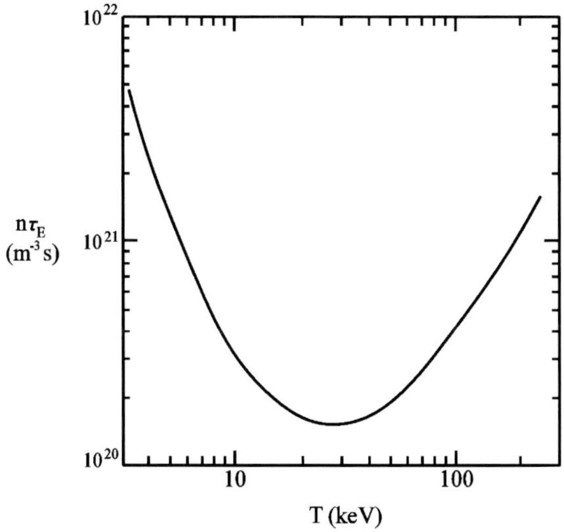

Figure 1.9 $12T/(\langle\sigma v\rangle E_\alpha)$ as a function of T.

The right-hand side is a function of temperature, as shown in Figure 1.9. The minimum of $12T/(\langle\sigma v\rangle E_\alpha)$ occurs at $T \approx 30\,\text{keV}$; taking this value, we find

$$n\tau_E \geq 1.5 \times 10^{20}\,\text{m}^{-3}\,\text{s}\,. \tag{1.227}$$

This relation is known as the Lawson criterion.

Still, τ_E depends on temperature and other variables. In other words, on the transport properties in the plasma. One can take advantage of the fact that in the temperature region

$$10\,\text{keV} \leq T \leq 20\,\text{keV}\,, \tag{1.228}$$

one may approximate

$$\langle\sigma v\rangle \approx 1.1 \times 10^{-24}\,T^2\,\text{m}^3\,\text{s}^{-1} \tag{1.229}$$

when measuring T in keV. Within this approximation, the ignition criterion can be written in the form

$$nT\tau_E \geq 3 \times 10^{21}\,\text{m}^{-3}\,\text{keV}\,\text{s}\,. \tag{1.230}$$

For example, at $n = 10^{20}\,\text{m}^{-3}$, $T = 10\,\text{keV}$, the energy confinement time should be larger than 3 s.

To measure the progress in nuclear fusion research, the Q-factor

$$Q = \frac{5P_\alpha}{P_H} \tag{1.231}$$

has been introduced. Note that the factor $5 = 17.5/3.5 \approx E_{fus}/E_\alpha$ takes care of the whole energy gain and not only the part carried by the α-particles. $Q \to \infty$ means ignition.

2
Single Particle Motion

In this chapter, we consider the motion of particles in given electromagnetic fields. The latter may be spatially inhomogeneous. The time-dependency, if at all, will be assumed as weak when drift motion is considered. The special aspects of (relativistic) motion of a particle in an electromagnetic wave will be considered in a separate section. The chapter is concluded by the motion of a particle under the action of stochastic forces, that is, within the Langevin approach.

2.1
Heuristic Approaches to Guiding Center Motion

When a charged particle with charge q and mass m moves in a constant magnetic field $\boldsymbol{B} = B\hat{z}$, the nonrelativistic equation of motion is

$$m\frac{d\boldsymbol{v}}{dt} = q\boldsymbol{v} \times \boldsymbol{B} . \tag{2.1}$$

The solution for the z-component is trivial, namely, $v_z = \text{const}$. When we introduce the gyrofrequency

$$\Omega = \frac{qB}{m} , \tag{2.2}$$

the other components follow from

$$\frac{d^2 v_x}{dt^2} = -\Omega^2 v_x , \quad \frac{d^2 v_y}{dt^2} = -\Omega^2 v_y . \tag{2.3}$$

Their solutions

$$v_x = a_x \sin(\Omega t + \delta_x) , \quad v_y = a_y \cos\left(\Omega t + \delta_y\right) \tag{2.4}$$

contain the constants of integration a and δ which follow from the initial conditions. When at $t = 0$ we postulate $v_x = 0$ (and thereby $\delta_x = 0$) and

$$v_y(t = 0) = a_y \cos\delta_y \equiv \pm v_\perp , \quad \text{with} \quad v_\perp = \text{const} = \left(v_x^2 + v_y^2\right)^{1/2} , \tag{2.5}$$

High Temperature Plasmas, Theory and Mathematical Tools for Laser and Fusion Plasmas, First Edition.
Karl-Heinz Spatschek.

the solutions become simple. Because of

$$\frac{dv_x}{dt} = \Omega\, v_y, \tag{2.6}$$

we can determine the constants in

$$v_x = a_x \sin \Omega t\,, \quad v_y = (\pm v_\perp) \cos \Omega t - (\pm v_\perp) \tan \delta_y \sin \Omega t \tag{2.7}$$

as $a_x = \pm v_\perp$ and $\delta_y = 0$. This leads to the results

$$v_x = \pm v_\perp \sin \Omega t\,, \quad v_y = \pm v_\perp \cos \Omega t\,, \tag{2.8}$$

which are straightforward to interpret. Particles with positive charge have $\Omega > 0$, and we obtain an anticlockwise helical motion, independent of the sign $\pm v_\perp$. Integration leads to

$$x = x_0 \mp \frac{v_\perp}{\Omega} \cos \Omega t\,, \quad y = y_0 \pm \frac{v_\perp}{\Omega} \sin \Omega t\,. \tag{2.9}$$

The Larmor radius

$$r_L = \frac{v_\perp}{|\Omega|} \tag{2.10}$$

determines the width of the helix.

The circular currents in the x, y-plane are responsible for the diamagnetic behavior of the plasma. The Biot–Savart law

$$\boldsymbol{B}_{Pl}(\boldsymbol{r}) = \frac{\mu_0}{4\pi} \int \frac{\boldsymbol{j}(\boldsymbol{r}') \times (\boldsymbol{r} - \boldsymbol{r}')}{|\boldsymbol{r} - \boldsymbol{r}'|^3} dV'\,, \tag{2.11}$$

where $\boldsymbol{j}$ is the electric current density, allows one to determine the direction of the generated magnetic field $\boldsymbol{B}_{Pl}$. It is oppositely directed to the initial external magnetic field $\boldsymbol{B} = \boldsymbol{B}_{ext}$.

Let us now include an external electric field $\boldsymbol{E}$. Its component $E_\parallel \neq 0$ along $\boldsymbol{B}$ causes an acceleration. In the plane perpendicular to $\boldsymbol{B}$, the motion is governed by

$$\frac{d\boldsymbol{v}_\perp}{dt} = \frac{q}{m} \boldsymbol{v}_\perp \times \boldsymbol{B} + \frac{q}{m} \boldsymbol{E}_\perp\,. \tag{2.12}$$

Splitting the velocity component $\boldsymbol{v}_\perp$ in the form $\boldsymbol{v}_\perp := \boldsymbol{w}_D + \boldsymbol{u}$ with

$$\boldsymbol{w}_D = \frac{\boldsymbol{E}_\perp \times \boldsymbol{B}}{B^2}\,, \tag{2.13}$$

a simple calculation leads to

$$\frac{d\boldsymbol{u}}{dt} = \frac{q}{m} \boldsymbol{u} \times \boldsymbol{B}\,. \tag{2.14}$$

The velocity $\boldsymbol{w}_D$ is called the $E \times B$ velocity. Its direction is independent of the sign of the charge. Since Eq. (2.13) is proportional to $1/B$, one might think that it diverges for $B \to 0$. However, when $\boldsymbol{w}_D \to c$, a relativistic calculation becomes necessary which forbids the divergence.

The splitting into two velocities is the starting point for the "guiding center approximation" [38]. The center of the (fast) gyration is called the "guiding center." It moves in the plane perpendicular to the magnetic field with the $E \times B$ drift velocity (in addition to the motion along the magnetic field). With respect to the guiding center, a rotation with velocity $\boldsymbol{u}$ occurs which generally is fast compared to the drift. In the case of homogeneous and constant straight magnetic and electric fields, the decomposition is exact.

Next, we consider the motion of a particle in an inhomogeneous magnetic field. First, we assume that the magnetic field lines are still straight (parallel to the z-axis). No electric field should be present. The spatial inhomogeneity introduces a new length scale characterized by the inhomogeneity length L,

$$\frac{1}{L} \sim |\nabla \ln B| \,. \tag{2.15}$$

The following considerations will be applicable for

$$\frac{r_L}{L} \ll 1 \,. \tag{2.16}$$

It is the key assumption in the "guiding center approximation." For $\boldsymbol{B} = B(x)\hat{z}$, we may assume the vector potential in the form $\boldsymbol{A} = (0, A_y(x), 0)$. Then, $Q_x \equiv x$ is the only noncyclic coordinate. Starting from the Hamiltonian, one introduces the canonical momentum $\boldsymbol{P} = m\boldsymbol{v} + q\boldsymbol{A}$ such that the Hamiltonian for the particle motion can be written in the form

$$H = \frac{P_x^2}{2m} + V(Q_x) \,, \tag{2.17}$$

with

$$V(Q_x) = \frac{\left[P_y - qA_y(Q_x)\right]^2}{2m} + \frac{P_z^2}{2m} \,. \tag{2.18}$$

We have introduced the constant components P_y and P_z which are conjugate to the cyclic coordinates y and z, respectively. The motion becomes effectively one-dimensional. The solutions depend on $A_y(x)$. Let us assume that the latter has a minimum at $x = 0$. Then, depending on the value of $H = \text{const}$, the motion stays in a finite x-region. Of course, in addition, we have the motion along z and y. Note that $P_y = \text{const}$ does not mean $\dot{y} = \text{const}$ in y-direction. As we shall see, the motion in y-direction is a drift, the so called ∇B-drift.

Let us calculate the averaged forces in the x and y direction. Averaging is over the (by assumption fast) gyro-motion. Starting with the y-component, we obtain from the Lorentz force $\boldsymbol{F} = q\boldsymbol{v} \times \boldsymbol{B}$

$$F_y = -qv_x B_z(x) \,. \tag{2.19}$$

Replacing the exact position x and the velocity by the gyro-motion, after Taylor expansion, we obtain

$$F_y \approx -q v_\perp \sin(\Omega t) \left[B_0 - \frac{v_\perp}{\Omega} \cos(\Omega t) \frac{\partial B}{\partial x} \right], \tag{2.20}$$

where $B_0 = B_z(0)$. Averaging over time leads to $\langle F_y \rangle \approx 0$. Next, the calculation for F_x leads to

$$F_x \approx q v_\perp \cos(\Omega t) \left[B_0 - \frac{v_\perp}{\Omega} \cos(\Omega t) \frac{\partial B}{\partial x} \right], \tag{2.21}$$

and after averaging,

$$\langle F_x \rangle \approx -\frac{q v_\perp^2}{2\Omega} \frac{\partial B}{\partial x} . \tag{2.22}$$

An average force acts on the particle in the plane perpendicular to the magnetic field. Similar to the calculations for a constant electric field leading to the $E \times B$-drift, we now find by replacing $q\boldsymbol{E} \rightarrow \langle F_x \rangle \hat{x}$

$$\boldsymbol{w}_D \approx \frac{1}{q} \frac{\langle \boldsymbol{F} \rangle \times \boldsymbol{B}}{B^2} \approx -\frac{v_\perp^2}{2\Omega B^2} \nabla B \times \boldsymbol{B} , \tag{2.23}$$

which also can be written as

$$\boldsymbol{v}_{\nabla B} = \operatorname{sign}(q) \frac{1}{2} |v_\perp| |r_L| \frac{\boldsymbol{B} \times \nabla B}{B^2} . \tag{2.24}$$

This drift velocity is called the ∇B-drift.

Now, we allow for curved magnetic fields. Assuming that the particles predominantly move along the magnetic field lines, they feel a centrifugal force

$$\boldsymbol{F} = m \boldsymbol{\omega} \times (\boldsymbol{r} \times \boldsymbol{\omega}) , \tag{2.25}$$

leading to

$$\boldsymbol{F} \approx \frac{m v_\parallel^2}{R_c^2} \boldsymbol{R}_c . \tag{2.26}$$

Here, R_c is the local radius of curvature, and $v_\parallel$ is the local velocity along a magnetic field line. Again, similar to the calculation of the $E \times B$–drift (replacing $q\boldsymbol{E}_\perp$ by $\boldsymbol{F}$), we obtain

$$\boldsymbol{v}_R = \frac{m v_\parallel^2}{q B^2} \frac{\boldsymbol{R}_c \times \boldsymbol{B}}{R_c^2} , \tag{2.27}$$

which is called the curvature drift.

In general, the curvature drift appears together with the ∇B-drift. Let us elucidate that on the example of a magnetic field having only a θ-component in cylindrical coordinates, the z-component of $\nabla \times \boldsymbol{B} = 0$ leads to

$$(\nabla \times \boldsymbol{B})_z \equiv \frac{1}{r}\frac{\partial}{\partial r}(r B_\theta) \stackrel{!}{=} 0 \,, \quad \rightarrow \quad B_\theta \sim \frac{1}{r} \,. \tag{2.28}$$

Thus, the strength of the magnetic field varies with the radius, leading to

$$\frac{\nabla B}{B} = -\frac{\boldsymbol{R}_c}{R_c^2} \,. \tag{2.29}$$

A grad B-drift appears in the form

$$\boldsymbol{v}_{\nabla B} = \frac{1}{2}\frac{m}{q} v_\perp^2 \frac{\boldsymbol{R}_c \times \boldsymbol{B}}{R_c^2 B^2} \,. \tag{2.30}$$

Combination with the curvature drift leads to

$$\boldsymbol{v}_R + \boldsymbol{v}_{\nabla B} = \frac{m}{q}\frac{\boldsymbol{R}_c \times \boldsymbol{B}}{R_c^2 B^2}\left[v_\parallel^2 + \frac{1}{2}v_\perp^2\right] . \tag{2.31}$$

Another example, originating from mirror machines, assumes a cylindrically symmetric magnetic field with $B_\theta = 0$. For a specific form of the motion, we may assume $\partial_\theta = 0$. Under this assumption, $\nabla \cdot \boldsymbol{B} = 0$ leads to

$$\frac{1}{r}\frac{\partial}{\partial r}(r B_r) + \frac{\partial B_z}{\partial z} = 0 \,. \tag{2.32}$$

For a magnetic bottle with

$$|B_z| \gg |B_r| \tag{2.33}$$

and only a weak dependence of B_z on r, we find approximately

$$B_r \approx -\frac{1}{2} r \left.\frac{\partial B_z}{\partial z}\right|_{r=0} \neq 0 \,. \tag{2.34}$$

From the r-component of $\boldsymbol{B}$, we get a z-component of the Lorentz force,

$$F_z \approx -q v_\theta B_r \approx \frac{1}{2} q v_\theta r \left.\frac{\partial B_z}{\partial z}\right|_{z=0} \,. \tag{2.35}$$

To estimate its order of magnitude, we approximate $v_\theta \approx \pm v_\perp$ and $r \approx r_L$, finding

$$F_\parallel \approx -\frac{\mu}{B}\left[\boldsymbol{B} \cdot \nabla \boldsymbol{B}\right]_\parallel \,, \quad \mu = \frac{1}{2}\frac{m v_\perp^2}{B} \,. \tag{2.36}$$

The magnetic moment μ appears.

Finally, let us consider inhomogeneous electric fields. We start with spatial inhomogeneity. When assuming

$$\boldsymbol{E} = E_x \hat{x} \equiv E_0 \cos kx \hat{x} \,, \tag{2.37}$$

we may consider it as a Fourier component of a more general form. For $\boldsymbol{B} = B\hat{z}$, the equations of motion follow in the form

$$\frac{dv_x}{dt} = \Omega\, v_y + \frac{q}{m} E_x(x)\,, \quad \frac{dv_y}{dt} = -\Omega\, v_x\,. \tag{2.38}$$

They can be combined as

$$\frac{d^2 v_y}{dt^2} = -\Omega^2 v_y - \Omega^2 \frac{1}{B} E_x(x)\,; \tag{2.39}$$

however, in this formulation, we need the trajectory $x = x(t)$. For strong magnetic fields, we use in lowest order

$$x \approx x_0 - \frac{v_\perp}{\Omega} \cos \Omega\, t\,. \tag{2.40}$$

Still, an exact solution is difficult. Qualitatively, we expect an $E \times B$ motion plus gyration. Averaging over the fast gyration, the averaged velocity component v_x should vanish and thereby

$$\left\langle \frac{d^2 v_x}{dt^2} \right\rangle \stackrel{!}{=} 0 = -\Omega^2 \langle v_x \rangle + \frac{\Omega}{B} \left\langle \frac{dE_x}{dt} \right\rangle \approx -\Omega^2 \langle v_x \rangle + \frac{\Omega}{B} \frac{\partial E_x}{\partial x} \langle v_x \rangle\,, \tag{2.41}$$

$$\left\langle \frac{d^2 v_y}{dt^2} \right\rangle \approx -\Omega^2 \langle v_y \rangle - \Omega^2 \frac{E_0}{B} \left\langle \cos \left[k x_0 - \frac{k v_\perp}{\Omega} \cos \Omega\, t \right] \right\rangle \stackrel{!}{=} 0\,. \tag{2.42}$$

Using the addition theorems and making the assumption of small gyroradii, we find

$$\langle v_y \rangle \approx -\frac{E_0}{B} \cos(k x_0) \left[1 - \frac{1}{4} k^2 r_L^2 \right], \tag{2.43}$$

which is known as the "finite Larmor radius correction" to the $E \times B$-drift, sometimes written as

$$\boldsymbol{v}_D = \left(1 + \frac{1}{4} r_L^2 \nabla^2 \right) \frac{\boldsymbol{E} \times \boldsymbol{B}}{B^2}\,. \tag{2.44}$$

For time-dependent electric fields, we choose

$$\boldsymbol{E} = E_0 e^{i\omega t} \hat{x} + c.c.\,. \tag{2.45}$$

In the x, y-plane, the equations of motion are

$$\frac{d^2 v_x}{dt^2} = -\Omega^2 \left(v_x - v_p \right)\,, \quad \frac{d^2 v_y}{dt^2} = -\Omega^2 \left(v_y - v_E \right)\,. \tag{2.46}$$

Here, we have introduced

$$v_p = \frac{i\omega}{\Omega} \frac{E_x}{B}\,, \quad v_E = -\frac{E_x}{B}\,. \tag{2.47}$$

For $\omega^2 \ll \Omega^2$, the solutions are

$$v_x \approx v_\perp \cos \Omega t + v_p \ , \qquad v_y \approx -v_\perp \sin \Omega t + v_E \ . \tag{2.48}$$

In the y-direction an $E \times B$-drift occurs, whereas in x-direction, we find the polarization drift

$$\boldsymbol{v}_p = \frac{1}{\Omega B} \frac{d\boldsymbol{E}}{dt} \ . \tag{2.49}$$

In summary, in strong magnetic fields, we decompose the motion into a fast gyration $\boldsymbol{v}_\perp$ and a drift motion $\boldsymbol{w}$, that is, $\boldsymbol{v} = \boldsymbol{w} + \boldsymbol{v}_\perp$. For weak inhomogeneities we have a dominating $E \times B$-drift $\boldsymbol{w}_D \equiv \boldsymbol{v}_{E\times B}$ plus gradient and polarization drifts in case of straight magnetic field lines,

$$\boldsymbol{w}_\perp = \boldsymbol{v}_{E\times B} + \frac{\mu}{qB^3} \boldsymbol{B} \times \nabla \frac{B^2}{2} + \frac{m}{qB^2} \boldsymbol{B} \times \dot{\boldsymbol{v}}_{E\times B} \ . \tag{2.50}$$

The dot over $\boldsymbol{v}_{E\times B}$ represents time differentiation. The parallel (to the magnetic field) motion is governed by

$$\frac{dv_\parallel}{dt} = \frac{q}{m} E_\parallel - \frac{\mu}{mB} [(\boldsymbol{B} \cdot \nabla) \boldsymbol{B}]_\parallel \ . \tag{2.51}$$

The solutions are only approximate. When we derive constants of motion within the guiding center approximation, they might not be exact, but are valid for the approximated system. Such constants will be called adiabatic invariants. Let us identify the magnetic moment as an adiabatic invariant. We start from the energy balance

$$\frac{d}{dt} \left[\frac{1}{2} m v^2 \right] = q \boldsymbol{v} \cdot \boldsymbol{E} \tag{2.52}$$

and average over the fast gyration. We note $\langle v^2 \rangle \approx \langle v_\parallel^2 \rangle + \langle v_\perp^2 \rangle$. The term

$$q \langle \boldsymbol{v}_\perp \cdot \boldsymbol{E} \rangle = \frac{q}{T} \int \boldsymbol{E} \cdot \boldsymbol{v}_\perp dt = \frac{q}{T} \oint \boldsymbol{E} \cdot d\boldsymbol{s} \tag{2.53}$$

can be simplified to

$$\frac{q}{T} \int \frac{\partial \boldsymbol{B}}{\partial t} \cdot d\boldsymbol{F} \approx \mu \frac{\partial B}{\partial t} \ . \tag{2.54}$$

Here, the area has been approximated as $F \approx \pi r_L^2$ and the gyration period $T = 2\pi/\Omega$ has been used. The averaged energy balance Eq. (2.52) now reads

$$\begin{aligned} \left\langle \frac{d}{dt} \left(\frac{1}{2} m v_\perp^2 \right) \right\rangle &\approx \mu \frac{\partial B}{\partial t} - m v_\parallel \left\langle \frac{dv_\parallel}{dt} - \frac{q}{m} E_\parallel \right\rangle \\ &\approx \mu \frac{\partial B}{\partial t} + \mu \boldsymbol{v} \cdot \nabla B \equiv \mu \frac{dB}{dt} \ . \end{aligned} \tag{2.55}$$

Rewriting this as

$$\frac{d}{dt}(B\mu) = \mu \frac{dB}{dt}, \tag{2.56}$$

we find

$$\frac{d\mu}{dt} = 0 , \tag{2.57}$$

that is, the invariance of the magnetic moment.

2.2
Systematic Averaging

As we have seen, a magnetic field modifies particle trajectories significantly. Now, we present more systematic procedures to obtain the drift motion. Here, we concentrate on the physically motivated systematic averaging over fast gyro-motion. A different, more mathematically orientated method based on Lie transformations leads to the same results. We shall briefly comment on the latter method (see Appendix D) after presenting the main results in the next subsection.

2.2.1
Systematic Averaging over Fast Gyro-Motion

As we have discussed, the motion of a particle can be decomposed into an averaged drift motion and a fast gyration. It is useful to introduce the vector $\boldsymbol{\rho}$ from the Larmor center to the particle. Its magnitude corresponds to the Larmor radius (also denoted as r_L), see Figure 2.1. Introducing the gyrofrequency $\Omega = qB/m$ with positive or negative sign depending on the charge q of the particle, we write

$$\boldsymbol{\rho} = -\frac{1}{\Omega} \boldsymbol{v} \times \boldsymbol{b} , \tag{2.58}$$

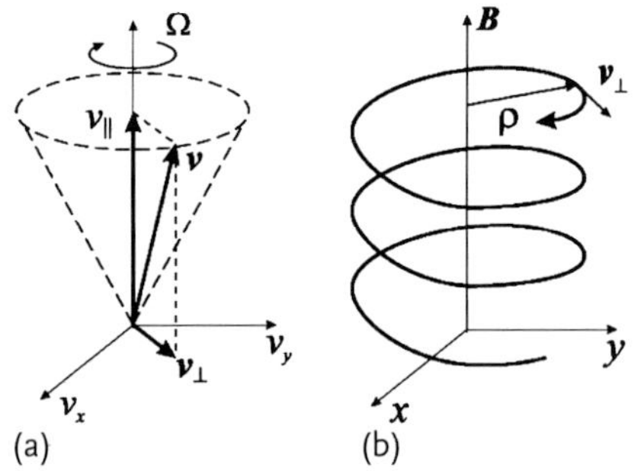

Figure 2.1 Velocity and position of a gyrating particle in a magnetic field. The left part (a) shows the motion in velocity space, whereas the right part (b) depicts the trajectory in configuration space.

where $\boldsymbol{b} = \boldsymbol{B}/B$. We also introduce the radius vector $\boldsymbol{R}$ of the particle guiding center (center of the Larmor helix),

$$\boldsymbol{R} = \boldsymbol{r} - \boldsymbol{\rho} = \boldsymbol{r} + \frac{1}{\Omega}\boldsymbol{v} \times \boldsymbol{b} . \tag{2.59}$$

In this equation, we have to take Ω and $\boldsymbol{b}$ at the particle position. We calculate $d\boldsymbol{R}/dt$:

$$\dot{\boldsymbol{R}} = \dot{\boldsymbol{r}} + \frac{1}{\Omega}\dot{\boldsymbol{v}} \times \boldsymbol{b} + \frac{1}{\Omega}\boldsymbol{v} \times \dot{\boldsymbol{b}} - \frac{\dot{\Omega}}{\Omega^2}\boldsymbol{v} \times \boldsymbol{b} . \tag{2.60}$$

Using Newton's law together with the Lorentz force, we obtain

$$\dot{\boldsymbol{R}} = \boldsymbol{v} + \frac{1}{\Omega}\left(\frac{q}{m}\boldsymbol{E} \times \boldsymbol{b} + \Omega\,(\boldsymbol{v} \times \boldsymbol{b}) \times \boldsymbol{b}\right) + \frac{1}{\Omega}\boldsymbol{v} \times \dot{\boldsymbol{b}} - \frac{\dot{\Omega}}{\Omega^2}\boldsymbol{v} \times \boldsymbol{b} . \tag{2.61}$$

We mention that $(\boldsymbol{v} \times \boldsymbol{b}) \times \boldsymbol{b} = -\boldsymbol{v} + (\boldsymbol{v} \cdot \boldsymbol{b})\boldsymbol{b} = -\boldsymbol{v}_\perp$ and get

$$\dot{\boldsymbol{R}} = v_\parallel \boldsymbol{b} + \frac{q}{\Omega m}\boldsymbol{E} \times \boldsymbol{b} + \frac{1}{\Omega}\boldsymbol{v} \times \dot{\boldsymbol{b}} - \frac{\dot{\Omega}}{\Omega^2}\boldsymbol{v} \times \boldsymbol{b} . \tag{2.62}$$

This equation is exact. It contains terms which are slow and terms which are fast oscillating. To extract the drift, we have to average this equation over the fast Larmor motion. We begin with the first term $\mathrm{I} = v_\parallel \boldsymbol{b}$. It contains no small parameter ε when we define the smallness parameter as the ratio of the Larmor radius r_L over the characteristic length L for the variations of the fields. We Taylor develop all the fields around the guiding center, that is,

$$\boldsymbol{b}(\boldsymbol{r}) = \boldsymbol{b}(\boldsymbol{R} + \boldsymbol{\rho}) \approx \boldsymbol{b}(\boldsymbol{R}) + (\boldsymbol{\rho} \cdot \nabla)\boldsymbol{b}(\boldsymbol{R}) . \tag{2.63}$$

The last term here is of the order ε. However, when we average over the Larmor radius, it vanishes because $\langle \boldsymbol{\rho} \rangle = 0$,

$$\langle \mathrm{I} \rangle = v_\parallel \boldsymbol{b}(\boldsymbol{R}) . \tag{2.64}$$

The second term $\mathrm{II} = q/(m\Omega)\boldsymbol{E} \times \boldsymbol{b}$ is already small because the electric field in a well conducting plasma cannot be too large. Therefore, we can take the fields at the guiding center,

$$\langle \mathrm{II} \rangle = \frac{1}{B^2(\boldsymbol{R})}\boldsymbol{E}(\boldsymbol{R}) \times \boldsymbol{B}(\boldsymbol{R}) . \tag{2.65}$$

We go over to the third term $\mathrm{III} = \Omega^{-1}\boldsymbol{v} \times d\boldsymbol{b}/dt$. Here,

$$\dot{\boldsymbol{b}} = \frac{d\boldsymbol{b}}{dt} = \frac{\partial \boldsymbol{b}}{\partial t} + (\boldsymbol{v} \cdot \nabla)\boldsymbol{b} . \tag{2.66}$$

We assume that the magnetic field is time-independent (static). Then,

$$\mathrm{III} = \frac{1}{\Omega}\boldsymbol{v} \times (\boldsymbol{v} \cdot \nabla)\boldsymbol{b} . \tag{2.67}$$

To average, we use the index notations with summation over repeated indices, introduce the completely antisymmetric tensor of third rank $\varepsilon_{\alpha\beta\gamma}$ in order to find

$$\mathrm{III}_\alpha = \frac{1}{\Omega}\varepsilon_{\alpha\beta\gamma} v_\beta v_\nu \frac{\partial b_\gamma}{\partial x_\nu} . \tag{2.68}$$

This expression contains the small spatial derivative at the guiding center position, and thus we only have to average the velocities,

$$\langle \mathrm{III}_\alpha \rangle = \frac{1}{\Omega}\varepsilon_{\alpha\beta\gamma}\frac{\partial b_\gamma}{\partial x_\nu}\langle v_\beta v_\nu \rangle . \tag{2.69}$$

The averaging here must be invariant with respect to rotations around the direction $\boldsymbol{b}$. Then, the result should be a linear combination of invariant tensors and the components of $\boldsymbol{b}$, namely,

$$\langle v_\beta v_\nu \rangle = A\delta_{\beta\nu} + B b_\beta b_\nu + C\varepsilon_{\beta\nu\mu} b_\mu . \tag{2.70}$$

The constant C must be zero because the average cannot depend on the magnetic field sign. We find factors A and B by making contractions with $\delta_{\beta\nu}$ and $b_\beta b_\nu$, leading to $v^2 = 3A + B$, $v_\parallel^2 = A + B$. Then,

$$\langle v_\beta v_\nu \rangle = \frac{1}{2} v_\perp^2 \left\{ \delta_{\beta\nu} - b_\beta b_\nu \right\} + v_\parallel^2 b_\beta b_\nu . \tag{2.71}$$

Finally,

$$\langle \mathrm{III}_\alpha \rangle = \frac{v_\perp^2}{2\Omega}\varepsilon_{\alpha\beta\gamma}\left\{ \frac{\partial b_\gamma}{\partial x_\beta} - b_\beta b_\nu \frac{\partial b_\gamma}{\partial x_\nu} \right\} + \frac{v_\parallel^2}{\Omega}\varepsilon_{\alpha\beta\gamma} b_\beta b_\nu \frac{\partial b_\gamma}{\partial x_\nu} . \tag{2.72}$$

Returning back to vector notations, we get

$$\langle \mathrm{III} \rangle = \frac{v_\perp^2}{2\Omega}\left\{ \nabla \times \boldsymbol{b} - [\boldsymbol{b}, (\boldsymbol{b}\cdot\nabla)\boldsymbol{b}] \right\} + \frac{v_\parallel^2}{\Omega}[\boldsymbol{b}, (\boldsymbol{b}\cdot\nabla)\boldsymbol{b}] . \tag{2.73}$$

To make the presentation more transparent, we introduced the notation $\boldsymbol{a} \times \boldsymbol{b} \equiv [\boldsymbol{a}, \boldsymbol{b}]$. One can simplify the expression a bit since

$$[\boldsymbol{b}, \nabla \times \boldsymbol{b}] = \underbrace{\nabla \frac{b^2}{2}}_{=0} - (\boldsymbol{b}\cdot\nabla)\,\boldsymbol{b} = -(\boldsymbol{b}\cdot\nabla)\,\boldsymbol{b} , \tag{2.74}$$

to finally obtain

$$\langle \mathrm{III} \rangle = \frac{v_\perp^2}{2\Omega}\boldsymbol{b}\,(\boldsymbol{b}\cdot\nabla\times\boldsymbol{b}) + \frac{v_\parallel^2}{\Omega}[\boldsymbol{b}, (\boldsymbol{b}\cdot\nabla)\,\boldsymbol{b}] . \tag{2.75}$$

Taking the last, fourth term, $\mathrm{IV} = -\dot{\Omega}/\Omega^2 \boldsymbol{v} \times \boldsymbol{b}$, we note

$$\dot{\Omega} = \frac{d\Omega}{dt} = \frac{\partial \Omega}{\partial t} + (\boldsymbol{v}\cdot\nabla)\Omega = (\boldsymbol{v}\cdot\nabla)\Omega . \tag{2.76}$$

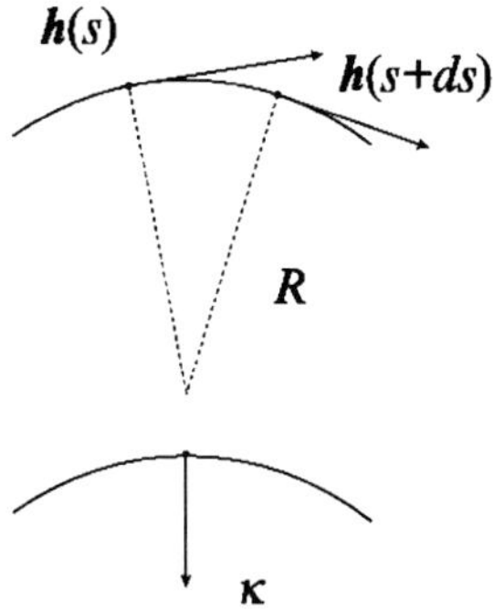

Figure 2.2 Tangent vector and curvature radius of a magnetic field line are shown in the upper part. The lower part of the figure depicts the curvature vector for the same field line.

With similar steps as before, we obtain

$$\langle \text{IV} \rangle = -\left\langle \frac{1}{\Omega^2} v_\nu \frac{\partial \Omega}{\partial x_\nu} \varepsilon_{\alpha\beta\gamma} v_\beta b_\gamma \right\rangle = -\frac{1}{\Omega^2} \frac{\partial \Omega}{\partial x_\nu} \varepsilon_{\alpha\beta\gamma} b_\gamma \langle v_\nu v_\beta \rangle$$
$$= -\frac{1}{\Omega^2} \frac{\partial \Omega}{\partial x_\nu} \varepsilon_{\alpha\beta\gamma} b_\gamma \left\{ \frac{1}{2} v_\perp^2 \delta_{\beta\nu} + \left(v_\parallel^2 - \frac{1}{2} v_\perp^2 \right) b_\beta b_\nu \right\} . \tag{2.77}$$

Because $\varepsilon_{\alpha\beta\gamma} b_\beta b_\gamma = [\boldsymbol{b}, \boldsymbol{b}]_\alpha = 0$, we finally obtain

$$\langle \text{IV} \rangle = -\frac{v_\perp^2}{2\Omega^2} [\nabla \Omega, \boldsymbol{b}] . \tag{2.78}$$

Collecting all the terms, we arrive at the expression

$$\dot{\boldsymbol{R}} = \left(v_\parallel + \frac{v_\perp^2}{2\Omega} \boldsymbol{b} \cdot \nabla \times \boldsymbol{b} \right) \boldsymbol{b} + \frac{1}{B} [\boldsymbol{E}, \boldsymbol{b}] + \frac{v_\parallel^2}{\Omega} [\boldsymbol{b}, (\boldsymbol{b} \cdot \nabla) \boldsymbol{b}] + \frac{v_\perp^2}{2\Omega B} [\boldsymbol{b}, \nabla B] . \tag{2.79}$$

The correction to the parallel velocity $v_\parallel$ in the first term is usually small and can be neglected. The last three terms describe drifts: electric, centrifugal and gradient drifts, respectively. For the centrifugal drift, we mention that

$$(\boldsymbol{b} \cdot \nabla) \boldsymbol{b} = \frac{\partial \boldsymbol{b}}{\partial s} = \boldsymbol{\kappa} \tag{2.80}$$

is the curvature vector for the magnetic field lines, $|\boldsymbol{\kappa}| = 1/R$, where R is the curvature radius; see Figure 2.2. Using the vector $\boldsymbol{\kappa}$, we write the centrifugal drift

$$\boldsymbol{v}_{cd} = \frac{v_\parallel^2}{\Omega} \boldsymbol{b} \times \boldsymbol{\kappa} . \tag{2.81}$$

When $\boldsymbol{n}$ is the normal to the particle trajectory, the centrifugal drift is directed along the binormal $\tilde{\boldsymbol{b}} = \boldsymbol{b} \times \boldsymbol{n}$. This drift depends on the charge sign.

In vacuum, where $\nabla \times \boldsymbol{B} = 0$, the drift velocities of centrifugal and gradient drifts are similar. To prove this, we use the equality $[\boldsymbol{b}, \nabla \times \boldsymbol{b}] = -\boldsymbol{\kappa}$ and

$$\nabla \times \boldsymbol{b} = \nabla \times \frac{\boldsymbol{B}}{B} = \frac{1}{B} \underbrace{\nabla \times \boldsymbol{B}}_{=0} + \left[\nabla \frac{1}{B}, \boldsymbol{B} \right] . \tag{2.82}$$

Furthermore,

$$-\boldsymbol{\kappa} = [\boldsymbol{b}, \nabla \times \boldsymbol{b}] = -\left[\boldsymbol{b}, \left[\frac{\nabla B}{B}, \boldsymbol{b}\right]\right] = -\frac{\nabla B}{B} + \boldsymbol{b}\left(\boldsymbol{b}\cdot\frac{\nabla B}{B}\right), \tag{2.83}$$

and after operating with $\boldsymbol{b}\times$, we have

$$[\boldsymbol{b}, \nabla B] = B\,[\boldsymbol{b}, \boldsymbol{\kappa}]\,. \tag{2.84}$$

Thus, the gradient and the centrifugal drifts can be combined into

$$\boldsymbol{v}_{\text{grad}} + \boldsymbol{v}_{\text{cd}} = \frac{v_\parallel^2 + \frac{1}{2}v_\perp^2}{\Omega}[\boldsymbol{b}, \boldsymbol{\kappa}] = \frac{v_\parallel^2 + \frac{1}{2}v_\perp^2}{\Omega\, R}[\boldsymbol{b}, \boldsymbol{n}]\,. \tag{2.85}$$

Summarizing, the motion of a guiding center is of lowest order

$$\dot{\boldsymbol{R}} = v_\parallel \boldsymbol{b} + \frac{c}{B}[\boldsymbol{E}, \boldsymbol{b}] + \frac{v_\parallel^2}{\Omega}[\boldsymbol{b}, \boldsymbol{\kappa}] + \frac{v_\perp^2}{2\Omega\, B}[\boldsymbol{b}, \nabla B]\,. \tag{2.86}$$

2.2.2 Pseudocanonical Transformations

The drift approximation fits perfectly into Hamiltonian dynamics [39]. This important statement is the basis for a rigorous mathematical justification of the drift equations derived in the previous subsection by the averaging method. Moreover, the Kruskal–Littlejohn theorem [40], based on the fundamental work of Littlejohn [41–44], states that by pseudocanonical transformations, new coordinates can be introduced that are all independent of the gyrophase. In the following, we present some introductory remarks into that fundamental method forming the basis for the dynamics of a charged particle in the drift approximation.

In an Hamiltonian system, the evolution of a function f depending on the phase-space variables q, p can be written in the form $\dot{f} = [f, H]$, where $[\ldots, \ldots]$ is the Lie bracket (in classical mechanics identical with the Poisson bracket). The Lie bracket can be expressed in terms of the so called fundamental Lie brackets

$$\begin{aligned}[a(q,p), b(q,p)] = \sum_{i,j}\Bigg\{&\frac{\partial a}{\partial q_i}\frac{\partial b}{\partial q_j}[q_i, q_j] + \frac{\partial a}{\partial q_i}\frac{\partial b}{\partial p_j}[q_i, p_j] \\ &+ \frac{\partial a}{\partial p_i}\frac{\partial b}{\partial q_j}[p_i, q_j] + \frac{\partial a}{\partial p_i}\frac{\partial b}{\partial p_j}[p_i, p_j]\Bigg\}\,.\end{aligned} \tag{2.87}$$

For

$$[q_i, q_j] = 0\,, \quad [p_i, p_j] = 0\,, \quad [q_i, p_j] = \delta_{ij} \tag{2.88}$$

for all i, j, the Lie bracket is identical with the Poisson bracket

$$[a, b] = \sum_i \left\{\frac{\partial a}{\partial q_i}\frac{\partial b}{\partial p_i} - \frac{\partial a}{\partial p_i}\frac{\partial b}{\partial q_i}\right\}. \tag{2.89}$$

For a canonical transformation to the new variables $Q_i = Q_i(q, p)$, $P_i = P_i(q, p)$, the latter should satisfy

$$[Q_i, Q_j] = [P_i, P_j] = 0\,, \quad [Q_i, P_j] = \delta_{ij}\,. \tag{2.90}$$

The Hamiltonian H transforms into K [33].

In case of a pseudocanonical transformation, the new variables may satisfy

$$[Q_i, Q_j] = F_{ij}(Q, P)\,, \quad [P_i, P_j] = G_{ij}(Q, P)\,, \quad [Q_i, P_j] = H_{ij}(Q, P)\,, \tag{2.91}$$

with certain functions F_{ij}, G_{ij}, and H_{ij}. The equations of motion are

$$\dot{Q}_i = [Q_i, K(Q, P)]\,, \quad \dot{P}_i = [P_i, K(Q, P)]\,, \tag{2.92}$$

for example,

$$\dot{Q}_i = \sum_j \left\{ F_{ij}(Q, P)\frac{\partial K(Q, P)}{\partial Q_j} + H_{ij}(Q, P)\frac{\partial K(Q, P)}{\partial P_j} \right\}. \tag{2.93}$$

The starting point for the Kruskal–Littlejohn theory has to be formulated in the initial coordinates. Instead of the "obvious" coordinates $\boldsymbol{q}$, $\boldsymbol{p}$, we use local coordinates. When we choose a coordinate system being adapted to a magnetic field line, we have the tangent $\hat{b}$, the normal vector $\hat{N}$, and the binormal $\hat{\beta}$, and write

$$\boldsymbol{e}_1 \equiv \hat{N} = \hat{\beta} \times \hat{b}\,, \quad \boldsymbol{e}_2 \equiv \hat{\beta} = \hat{b} \times \hat{N}\,, \quad \boldsymbol{e}_3 \equiv \hat{b} = \hat{N} \times \hat{\beta}\,. \tag{2.94}$$

Next, we consider the path of a particle. Projecting the path into the plane perpendicular to $\boldsymbol{e}_3$, we introduce the tangent vector $\boldsymbol{n}_1(\boldsymbol{q})$ and the orthogonal vector $\boldsymbol{n}_2(\boldsymbol{q})$. It is straightforward to find

$$\boldsymbol{n}_1(\boldsymbol{q}, \varphi) = -\sin\varphi\, \boldsymbol{e}_1(\boldsymbol{q}) - \cos\varphi\, \boldsymbol{e}_2(\boldsymbol{q})\,, \tag{2.95}$$

$$\boldsymbol{n}_2(\boldsymbol{q}, \varphi) = \cos\varphi\, \boldsymbol{e}_1(\boldsymbol{q}) - \sin\varphi\, \boldsymbol{e}_2(\boldsymbol{q})\,. \tag{2.96}$$

Now, the velocity of the particle is

$$\boldsymbol{v} = v_\parallel \boldsymbol{e}_3(\boldsymbol{q}) + v_\perp \boldsymbol{n}_1(\boldsymbol{q}, \varphi)\,. \tag{2.97}$$

Thus, we start with the variables $\boldsymbol{q}$, $v_\parallel$, $v_\perp$, and φ, where φ is the gyrophase, and the "Hamiltonian"

$$H = \frac{m}{2}\left(v_\parallel^2 + v_\perp^2\right) + V(\boldsymbol{q})\,, \quad V(\boldsymbol{q}) = -e\boldsymbol{E} \cdot \boldsymbol{q}\,. \tag{2.98}$$

We use e (including sign) for the charge instead of q in order to avoid confusion with the phase-space coordinate $\boldsymbol{q}$. Again, in the new variables, the equations of motion follow in symplectic form. The explicit expressions are [40]

$$\dot{\boldsymbol{q}} = v_{\|} \boldsymbol{b} + v_{\perp} \boldsymbol{n}_1 , \tag{2.99}$$

$$\dot{v}_{\|} = v_{\perp} \boldsymbol{n}_2 \cdot \boldsymbol{D} + \frac{e}{m} \boldsymbol{b} \cdot \boldsymbol{E} , \tag{2.100}$$

$$\dot{v}_{\perp} = -v_{\|} \boldsymbol{n}_2 \cdot \boldsymbol{D} + \frac{e}{m} \boldsymbol{n}_1 \cdot \boldsymbol{E} , \tag{2.101}$$

$$\dot{\varphi} = \frac{eB}{m} + \boldsymbol{b} \cdot \boldsymbol{D} - \frac{v_{\|}}{v_{\perp}} \boldsymbol{n}_1 \cdot \boldsymbol{D} - \frac{e}{m v_{\perp}} \boldsymbol{n}_2 \cdot \boldsymbol{E} , \tag{2.102}$$

with

$$\boldsymbol{D} := v_{\|} \nabla \times \boldsymbol{b} + v_{\perp} \nabla \times \boldsymbol{n}_1 . \tag{2.103}$$

The transformation from $\boldsymbol{q}, \boldsymbol{p}$ to $\boldsymbol{q}, v_{\|}, v_{\perp}, \varphi$ is pseudocanonical with functional determinant $|J| = v_{\perp}$.

Let us elucidate the form of the equations of motion for constant fields $\boldsymbol{E}$ and $\boldsymbol{B}$. All contributions from $\boldsymbol{D}$ disappear because of $\nabla \times \boldsymbol{b} = 0$ and the fact that $\boldsymbol{n}_1$ is only a function of φ and not of $\boldsymbol{q}$. Therefore, in the case of constant field, the simplifications are

$$\dot{\boldsymbol{q}} = v_{\|} \boldsymbol{b} + v_{\perp} \boldsymbol{n}_1 , \tag{2.104}$$

$$\dot{v}_{\|} = \frac{e}{m} \boldsymbol{b} \cdot \boldsymbol{E} , \tag{2.105}$$

$$\dot{v}_{\perp} = \frac{e}{m} \boldsymbol{n}_1 \cdot \boldsymbol{E} , \tag{2.106}$$

$$\dot{\varphi} = \frac{eB}{m} - \frac{e}{m v_{\perp}} \boldsymbol{n}_2 \cdot \boldsymbol{E} . \tag{2.107}$$

They can be combined to

$$\begin{aligned} \ddot{\boldsymbol{q}} &= \frac{e}{m} (\boldsymbol{b} \cdot \boldsymbol{E}) \boldsymbol{b} + \frac{e}{m} \boldsymbol{n}_1 (\boldsymbol{n}_1 \cdot \boldsymbol{E}) - \boldsymbol{n}_2 v_{\perp} \left[\frac{eB}{m} - \frac{e}{m v_{\perp}} \boldsymbol{n}_2 \cdot \boldsymbol{E} \right] \\ &= \frac{e}{m} \boldsymbol{E} - \boldsymbol{n}_2 v_{\perp} \frac{eB}{m} = \frac{e}{m} [\boldsymbol{E} + \boldsymbol{v} \times \boldsymbol{B}] , \end{aligned} \tag{2.108}$$

which is the familiar equation of motion.

The aim of the guiding center approximation is to decouple the fast (!) gyration from the rest of the dynamics. The central role is played by

$$\dot{\varphi} \approx \frac{eB}{m} \equiv \Omega , \tag{2.109}$$

and the ordering $\omega/\Omega \ll 1$ and $L/r_L \gg 1$, where ω is the characteristic frequency besides gyration, and L is the characteristic length of the field variations. The above ratios define a smallness parameter ε. In the following, the electric field should also not become too large, that is, the $E \times B$ velocity should be much smaller than the actual particle speed.

Let us abbreviate the equations of motion (2.104)–(2.107) as

$$\frac{dX}{dt} = S(X,\varphi)\,, \quad \frac{d\varphi}{dt} = \frac{1}{\varepsilon}\Omega + R(X,\varphi)\,, \tag{2.110}$$

where X summarizes the variables $\boldsymbol{q}$, $v_\parallel$, and $v_\perp$. We have replaced $\Omega \to \Omega/\varepsilon$ to make the following ordering more transparent. The functions R and S depend on all variables.

Instead of X, φ, we want to introduce new variables Y, ϕ such that up to a certain order in ε, the new equations of motion take the form

$$\frac{dY}{dt} = S_0(Y) + \varepsilon S_1(Y) + \ldots\,, \quad \frac{d\phi}{dt} = \frac{1}{\varepsilon}\Omega(Y) + R_0(Y) + \varepsilon R_1(Y) + \ldots\,, \tag{2.111}$$

under the condition that the new functions $S_0, S_1, \ldots, R_0, R_1, \ldots$ do not depend on ϕ.

Let us start with the simple case of constant $\boldsymbol{E}$ and $\boldsymbol{B}$ fields. In that case, the transformations

$$\boldsymbol{Q} = \boldsymbol{q} - \varepsilon \frac{v_\perp}{\Omega} \boldsymbol{n}_2(\varphi) + \mathcal{O}(\varepsilon^2)\,, \tag{2.112}$$

$$U = v_\parallel + \mathcal{O}(\varepsilon^2)\,, \tag{2.113}$$

$$W = v_\perp - \varepsilon \frac{e}{m\Omega} \boldsymbol{E} \cdot \boldsymbol{n}_2(\varphi) + \mathcal{O}(\varepsilon^2)\,, \tag{2.114}$$

$$\phi = \varphi - \varepsilon \frac{e}{m\Omega v_\perp} \boldsymbol{E} \cdot \boldsymbol{n}_1(\varphi) + \mathcal{O}(\varepsilon^2) \tag{2.115}$$

will provide us with the required forms to first order. It is straightforward to calculate the fundamental Lie brackets

$$[Q_i, Q_j] = -\varepsilon \frac{1}{eB} \varepsilon_{ijk} b_k + \mathcal{O}(\varepsilon^2)\,, \tag{2.116}$$

$$[\boldsymbol{Q}, U] = \frac{1}{m} \boldsymbol{b} + \mathcal{O}(\varepsilon^2)\,, \quad [\boldsymbol{Q}, W] = 0 + \mathcal{O}(\varepsilon^2)\,, \tag{2.117}$$

$$[\boldsymbol{Q}, \phi] = 0 + \mathcal{O}(\varepsilon^2)\,, \quad [U, W] = 0 + \mathcal{O}(\varepsilon^2)\,, \tag{2.118}$$

$$[U, \phi] = 0 + \mathcal{O}(\varepsilon^2)\,, \quad [W, \phi] = -\frac{1}{\varepsilon} \frac{\Omega}{mW} + \mathcal{O}(\varepsilon)\,. \tag{2.119}$$

The new Hamiltonian is

$$H = \frac{1}{2} m \left[U^2 + W^2\right] - e\boldsymbol{E} \cdot \boldsymbol{Q} + \mathcal{O}(\varepsilon^2)\,, \tag{2.120}$$

and the equations of motion follow as

$$\dot{Q} = Ub + \varepsilon \frac{E \times B}{B^2} + \mathcal{O}(\varepsilon^2) , \tag{2.121}$$

$$\dot{U} = \frac{e}{m} E \cdot b + \mathcal{O}(\varepsilon^2) , \tag{2.122}$$

$$\dot{W} = 0 + \mathcal{O}(\varepsilon^2) , \tag{2.123}$$

$$\dot{\phi} = \frac{1}{\varepsilon} \Omega + \mathcal{O}(\varepsilon) . \tag{2.124}$$

We have reached the goal of leading order (however only for constant fields):

- The $E \times B$-drift velocity follows in first order in ε.
- Q can be interpreted as the gyrocenter position.
- The first set of equations is decoupled from the gyro-motion.

However, the general case with space- and time-dependent external electric and magnetic fields is the most challenging. It is easy to present the final result (as we shall do below), though it is important that one can come systematically to the transformation equations shown below. They follow the requirement that the guiding center motion decouples from the gyrophase. The original work goes back to Littlejohn [41–44]. The work has been nicely summarized by Balescu [40]. We refer to that book when presenting the final transformations

$$Q = q - \frac{\varepsilon}{\Omega} v_\perp n_2 + \left(\frac{\varepsilon}{\Omega}\right)^2 v_\perp^2 \left\{ \frac{3}{8} b \left[n_2 \cdot (\nabla \times n_1) + n_1 \cdot (\nabla \times n_2) \right] \right. \tag{2.125}$$

$$\left. + \frac{v_\parallel}{v_\perp} (n_2 b + 2 b n_2) \cdot (\nabla \times b) - \frac{1}{4B} (n_2 n_2 - n_1 n_1) \cdot \nabla B \right\} , \tag{2.126}$$

$$U = v_\parallel + \frac{\varepsilon}{\Omega} v_\perp^2 \left\{ \frac{1}{4} \left[n_1 \cdot (\nabla \times n_1) - n_2 \cdot (\nabla \times n_2) + 2 b \cdot (\nabla \times b) \right] \right. \tag{2.127}$$

$$\left. + \frac{v_\parallel}{v_\perp} n_1 \cdot (\nabla \times b) \right\} , \tag{2.128}$$

$$W = v_\perp + \frac{\varepsilon}{\Omega} v_\parallel v_\perp \left\{ -\frac{1}{4} \left[n_1 \cdot (\nabla \times n_1) - n_2 \cdot (\nabla \times n_2) + 2 b \cdot (\nabla \times b) \right] \right. \tag{2.129}$$

$$\left. - \frac{v_\parallel}{v_\perp} n_1 \cdot (\nabla \times b) - \frac{v_\perp}{v_\parallel} \frac{e}{m} n_2 \cdot E \right\} , \tag{2.130}$$

$$\phi = \varphi + \frac{e}{\Omega} \left\{ \frac{v_\parallel^2}{v_\perp} n_2 \cdot (\nabla \times b) - v_\perp b \cdot (\nabla \times n_2) \right. \tag{2.131}$$

$$\left. + \frac{1}{4} v_\parallel \left[n_2 \cdot (\nabla \times n_1) + n_1 \cdot (\nabla \times n_2) \right] + v_\perp \frac{1}{B} n_1 \cdot \nabla B - \frac{e}{m v_\perp} n_1 \cdot E \right\} , \tag{2.132}$$

where $\nabla = \partial/\partial \boldsymbol{q}$, and all fields have to be evaluated at $\boldsymbol{q}$ with phase φ. The new Hamiltonian is

$$H = \frac{m}{2}\left(U^2 + W^2\right) - e\boldsymbol{E}\cdot\boldsymbol{Q} + \mathcal{O}(\varepsilon^2)\,. \tag{2.133}$$

Given the transformation equations, it is straightforward (although tedious) to evaluate the fundamental Lie brackets. In the following, fields are at position $\boldsymbol{Q}$ and $\nabla = \partial/\partial\boldsymbol{Q}$. The result is

$$[Q_i, Q_j] = -\varepsilon(m\Omega)^{-1}\varepsilon_{ijk}b_k\,, \tag{2.134}$$

$$[\boldsymbol{Q}, U] = m^{-1}\boldsymbol{b}^*\,, \tag{2.135}$$

$$[\boldsymbol{Q}, W] = \varepsilon\left(\frac{W}{2mB\Omega}\right)\boldsymbol{b}\times\nabla B\,, \tag{2.136}$$

$$[\boldsymbol{Q}, \phi] = \varepsilon(m\Omega)^{-1}\boldsymbol{b}\times(\nabla\boldsymbol{n}_1\cdot\boldsymbol{n}_2)\,, \tag{2.137}$$

$$[U, W] = -\left(\frac{W}{2mB}\right)\boldsymbol{b}^*\cdot\nabla B\,, \tag{2.138}$$

$$[U, \phi] = -m^{-1}\boldsymbol{b}^*\cdot(\nabla\boldsymbol{n}_1\cdot\boldsymbol{n}_1) + \frac{1}{2m}\boldsymbol{b}\cdot(\nabla\times\boldsymbol{b})\,, \tag{2.139}$$

$$[W, \phi] = -\varepsilon^{-1}\frac{\Omega}{mW} + \varepsilon\frac{W}{2mB\Omega}(\nabla B)\cdot(\nabla\times\boldsymbol{b})\,, \tag{2.140}$$

where

$$\boldsymbol{b}^* = \boldsymbol{b} + \frac{\varepsilon}{\Omega}U\boldsymbol{b}\times(\boldsymbol{b}\cdot\nabla)\boldsymbol{b}\,. \tag{2.141}$$

With these fundamental brackets, Lie theory provides us with the equations of motion

$$\dot{\boldsymbol{Q}} = U\boldsymbol{b} + \frac{\varepsilon}{\Omega}\left\{\frac{W^2}{2B}\boldsymbol{b}\times\nabla B + U^2\boldsymbol{b}\times(\boldsymbol{b}\cdot\nabla)\boldsymbol{b} + \frac{e}{m}\boldsymbol{E}\times\boldsymbol{b}\right\}, \tag{2.142}$$

$$\dot{U} = -\frac{W^2}{2B}\boldsymbol{b}\cdot\nabla B + \frac{e}{m}\boldsymbol{E}\cdot\boldsymbol{b} - \varepsilon\frac{U}{\Omega}[\boldsymbol{b}\times(\boldsymbol{b}\cdot\nabla)\boldsymbol{b}]\cdot\left[\frac{W^2}{2B}\nabla B - \frac{e}{m}\boldsymbol{E}\right], \tag{2.143}$$

$$\dot{W} = \frac{UW}{2B}\boldsymbol{b}\cdot\nabla B + \varepsilon\frac{W}{2B\Omega}\left\{U^2[\boldsymbol{b}\times(\boldsymbol{b}\cdot\nabla)\boldsymbol{b}] - \frac{e}{m}\boldsymbol{b}\times\boldsymbol{E}\right\}\cdot\nabla B\,, \tag{2.144}$$

$$\dot{\phi} = \frac{1}{\varepsilon}\Omega + U\boldsymbol{b}\cdot\boldsymbol{R} - \frac{1}{2}U\boldsymbol{b}\cdot(\nabla\times\boldsymbol{b})\,, \tag{2.145}$$

where $\boldsymbol{R} = \nabla\boldsymbol{e}_1\cdot\boldsymbol{e}_2 = \nabla\boldsymbol{n}_1\cdot\boldsymbol{n}_2$.

A close look at these equations shows that all previous results follow from here. In more detail,

- The separation of the gyro-motion from the "rest" becomes obvious.
- For constant fields, the result reduces to the previous finding with the $E\times B$-drift in first order in ε.

- $\dot{\boldsymbol{Q}}$ represents the leading drift velocities in inhomogeneous fields.
- $\boldsymbol{Q}$ is the guiding-center position.
- U is the ("nontrivial") parallel guiding center velocity.
- The "parallel force component" $m\dot{U}$ contains not only the electric field, but also contributions from ∇B and centrifugal forces.

2.2.3
Magnetic Moment as the First Adiabatic Invariant

The adiabatic invariants can also be calculated systematically. The energy conservation is obtained in the form (we return to charge q instead of e which has been used in the previous subsection)

$$\frac{1}{2}\frac{d}{dt}v^2 = \boldsymbol{v}\cdot\dot{\boldsymbol{v}} = \boldsymbol{v}\cdot\left(\frac{q}{m}\boldsymbol{E} + \frac{q}{m}\boldsymbol{v}\times\boldsymbol{B}\right) = \frac{q}{m}\boldsymbol{v}\cdot\boldsymbol{E}. \tag{2.146}$$

Averaging over fast gyrations, we get to lowest order

$$\frac{1}{2}\frac{d}{dt}\langle v^2\rangle \approx \frac{q}{m}v_{\|}\boldsymbol{b}\cdot\boldsymbol{E}. \tag{2.147}$$

The electric field has to be calculated at the guiding center position. The variation of the parallel velocity component $v_{\|}$ follows from

$$\dot{v}_{\|} = \frac{d}{dt}(\boldsymbol{b}\cdot\boldsymbol{v}) = \boldsymbol{v}\cdot\left[\frac{\partial\boldsymbol{b}}{\partial t} + \boldsymbol{v}\cdot\nabla\boldsymbol{b}\right] + \frac{q}{m}\boldsymbol{E}\cdot\boldsymbol{b}. \tag{2.148}$$

The first term on the right-hand side may be neglected when the magnetic field is not explicitly time-dependent. The second term can be simplified along the following lines. We average over the fast gyration

$$\langle \boldsymbol{v}\,[\boldsymbol{v}\cdot\nabla\boldsymbol{b}]\rangle = \langle v_\alpha v_\beta\rangle\frac{\partial b_\alpha}{\partial x_\beta} \tag{2.149}$$

and apply the Einstein summation convention. Now, introducing the representation

$$\langle v_\alpha v_\beta\rangle = \frac{v_\perp^2}{2}\delta_{\alpha\beta} + \frac{2v_\|^2 - v_\perp^2}{2}b_\alpha b_\beta, \tag{2.150}$$

we can further simplify. This form has been previously derived. Another way to understand the representation is to first show the equivalence by assuming a local coordinate system $\boldsymbol{b} \equiv \hat{\boldsymbol{z}}$. Then, the representation turns out to be true by simply calculating both sides for $\alpha,\beta = x,y,z$. Any other rectangular coordinate system follows by rotation. The tensor representation shown above is invariant under rotations.

Inserting the representation Eq. (2.150) into Eq. (2.149), we get two terms on the right-hand side. The first one reads

$$\frac{v_\perp^2}{2}\frac{\partial b_\alpha}{\partial x_\alpha} = \frac{v_\perp^2}{2}\nabla\cdot\boldsymbol{b}. \tag{2.151}$$

The second one vanishes because of

$$b_\alpha b_\beta \frac{\partial b_\alpha}{\partial x_\beta} = b_\beta \frac{\partial}{\partial x_\beta} \frac{b^2}{2} = 0 \,. \tag{2.152}$$

Finally, we evaluate

$$\nabla \cdot \boldsymbol{b} = \nabla \cdot \frac{\boldsymbol{B}}{B} = -\frac{\boldsymbol{B} \cdot \nabla B}{B^2} = -\frac{1}{B} \boldsymbol{b} \cdot \nabla B \,. \tag{2.153}$$

By combining, we arrive at

$$\dot{v}_\parallel = \frac{q}{m} \boldsymbol{E} \cdot \boldsymbol{b} - \frac{1}{2} \frac{v_\perp^2}{B} \boldsymbol{b} \cdot \nabla B \,. \tag{2.154}$$

When we make use of the energy conservation Eq. (2.146), we can formulate the change of "average perpendicular velocity magnitude" as

$$\frac{d}{dt} \frac{v_\perp^2}{2} \equiv \frac{d}{dt} \left(\frac{v^2}{2} - \frac{v_\parallel^2}{2} \right) = \frac{1}{2} v_\perp^2 v_\parallel \frac{1}{B} \boldsymbol{b} \cdot \nabla B \,. \tag{2.155}$$

The right-hand side needs further consideration. When we calculate the change of the magnetic field strength B during the motion of a guiding center, we get to lowest (averaged) order in stationary magnetic fields

$$\frac{dB}{dt} = \frac{\partial B}{\partial t} + \boldsymbol{v} \cdot \nabla B \approx v_\parallel \boldsymbol{b} \cdot \nabla B \,. \tag{2.156}$$

That means that

$$\frac{d}{dt} v_\perp^2 \approx \frac{v_\perp^2}{B} \frac{dB}{dt} \tag{2.157}$$

or

$$\frac{d}{dt} \left(\frac{v_\perp^2}{B} \right) \approx 0 \,. \tag{2.158}$$

The magnetic moment of a particle is approximately invariant along its trajectory,

$$\mu = \frac{m v_\perp^2}{2B} \approx \text{const} \,. \tag{2.159}$$

When the magnetic field is (explicitly) time-dependent, via the Maxwell equation

$$\nabla \times \boldsymbol{E} = -\frac{\partial \boldsymbol{B}}{\partial t}, \tag{2.160}$$

an electric field is induced. Let us consider the simple case that the magnetic field is uniform. Then, the work of the electric field during one gyration cycle is

$$q \oint \boldsymbol{E} \cdot d\boldsymbol{l} = q \int \nabla \times \boldsymbol{E} \cdot d\boldsymbol{F} \approx q\pi\rho^2 \left(-\frac{\partial \boldsymbol{B}}{\partial t} \right) \cdot \hat{n} \,. \tag{2.161}$$

Note that for gyrating particles of charge q, sign $\{q\hat{n} \cdot \boldsymbol{B}\} = -1$. Therefore, the energy change per unit time is

$$-\frac{|\Omega|}{2\pi} q\pi\rho^2 \frac{\partial \boldsymbol{B}}{\partial t} \cdot \hat{n} = \frac{|q|}{2} \frac{|q|\boldsymbol{B}}{m} \cdot \frac{\partial \boldsymbol{B}}{\partial t} \left(\frac{m v_\perp}{qB}\right)^2 = \mu \frac{\partial B}{\partial t} . \tag{2.162}$$

This work can be used to heat a plasma (adiabatic heating).

2.2.4
On the Second Adiabatic Invariant

Adiabatic invariants are discussed because of the fact that perfect symmetry is almost never attained in reality. This leads to the practical question as to how constants of motion behave when symmetries are "good" but not perfect. A good exercise to study this question is the mathematical pendulum when time symmetry is not exact. We realize the problem of a slowly time-changing resonant frequency $\omega(t)$, that is, the equation of motion is

$$\frac{d^2 x}{dt^2} + \omega^2(t) x = 0 . \tag{2.163}$$

Slowly changing, for the present case, should mean

$$\frac{1}{\omega}\frac{d\omega}{dt} \ll \omega , \tag{2.164}$$

that is, the frequency changes slowly over one period. To solve the equation of motion, we make the ansatz

$$x(t) = \mathrm{Re}\left[A(t) e^{i\int^t \omega(t')dt'}\right], \tag{2.165}$$

which orientates itself on the solution for constant ω. Inserting this ansatz into the pendulum equation leads to

$$i\frac{d\omega}{dt} A + 2i\omega\frac{dA}{dt} + \frac{d^2 A}{dt^2} = 0 . \tag{2.166}$$

The last term on the left-hand side will be neglected for slow variations. Then,

$$-\frac{2}{A}\frac{dA}{dt} \approx \frac{1}{\omega}\frac{d\omega}{dt} ; \tag{2.167}$$

the amplitude A also becomes slowly varying,

$$A \sim \frac{1}{\sqrt{\omega(t)}} , \tag{2.168}$$

and the energy is not an integral of motion. However, the action integral

$$S = \oint v dx \tag{2.169}$$

when the integration is over one period of oscillation, is conserved. As we shall show later, the more general formulation is $\oint P dQ = S = \text{const}$ for the action integral over one period. Let τ be the interval between two times when $dx/dt = 0$ and d^2x/dt^2 has the same sign. We evaluate

$$S = \int_{t_0}^{t_0+\tau} v \frac{dx}{dt} dt = x \left. \frac{dx}{dt} \right|_{t_0}^{t_0+\tau} - \int_{t_0}^{t_0+\tau} x \frac{d^2x}{dt^2} dt = \int_{t_0}^{t_0+\tau} \omega^2 x^2 dt \,. \tag{2.170}$$

Inserting here

$$x(t) = x(t_0)\sqrt{\frac{\omega(t_0)}{\omega(t)}} \cos\left(\int_{t_0}^{t} \omega(t')dt'\right), \tag{2.171}$$

after simple calculations, we obtain

$$S = x^2(t_0)\omega(t_0) \int_0^{2\pi} \cos^2 \xi \, d\xi = \pi x^2(t_0)\omega(t_0) = \text{const} \,, \tag{2.172}$$

when $\xi = \int_{t_0}^{t'} \omega(t'')dt''$ is being used.

Now, we turn over to the generalization for a Hamiltonian H with a slowly changing parameter $\lambda(t)$,

$$H(P, Q; \lambda(t)) = E(t) \,. \tag{2.173}$$

We have

$$\begin{aligned} \frac{dS}{dt} &= \frac{d}{dt}\oint P dQ = \frac{d}{dt} \int_{Q(t)}^{Q(t+\tau)} P\left(E(t), Q, \lambda(t)\right) dQ \\ &= \underbrace{P \left. \frac{dQ}{dt} \right|_{Q(t)}^{Q(t+\tau)}}_{=0} + \int_{Q(t)}^{Q(t+\tau)} \left. \frac{\partial P}{\partial t} \right|_Q dQ \,. \end{aligned} \tag{2.174}$$

Note that

$$\left. \frac{\partial P}{\partial t} \right|_Q = \left. \frac{\partial P}{\partial E} \right|_{Q,\lambda} \frac{dE}{dt} + \left. \frac{\partial P}{\partial \lambda} \right|_{Q,E} \frac{d\lambda}{dt} \tag{2.175}$$

and

$$1 = \frac{\partial H}{\partial P} \left. \frac{\partial P}{\partial E} \right|_{Q,\lambda} , \qquad 0 = \frac{\partial H}{\partial P} \left. \frac{\partial P}{\partial \lambda} \right|_{Q,E} + \frac{\partial H}{\partial \lambda} \,. \tag{2.176}$$

Therefore, we find

$$\frac{dS}{dt} = \oint \left(\frac{\partial H}{\partial P} \right)^{-1} \left[\frac{dE}{dt} - \frac{\partial H}{\partial \lambda} \frac{d\lambda}{dt} \right] dQ \,. \tag{2.177}$$

Making use of Hamilton's equations, we have

$$\frac{dE}{dt} = \frac{\partial H}{\partial P}\frac{dP}{dt} + \frac{\partial H}{\partial Q}\frac{dQ}{dt} + \frac{\partial H}{\partial \lambda}\frac{d\lambda}{dt} = \frac{\partial H}{\partial \lambda}\frac{d\lambda}{dt}, \tag{2.178}$$

and therefore

$$\frac{dS}{dt} = 0\,. \tag{2.179}$$

Next, we apply this to the case of a static magnetic field with inhomogeneity such that the field lines squeeze. The density of field lines is proportional to the strength of magnetic fields. Because the magnetic field lines have zero divergence, they are endless and must bend when they squeeze. For $\partial B_z/\partial z \neq 0$, a radial component B_r is required such that $\boldsymbol{B} = B_z\hat{z} + B_r\hat{r}$. In a frame moving with the guiding center velocity $\boldsymbol{v}_{\perp g}$, the only perpendicular velocity is due to the Larmor rotation, and $\boldsymbol{v}_{\perp g} \perp \boldsymbol{B}$. The parallel velocity is not affected by this change of frame. Introducing s as the distance along the magnetic field, we may write

$$\dot{s} = v_\parallel\,, \quad \dot{\mu} = 0\,, \quad \frac{d}{dt}\frac{mv^2}{2} = m\left(v_\parallel \dot{v}_\parallel + v_\perp \dot{v}_\perp\right). \tag{2.180}$$

From $\dot{\mu} = 0$, we find

$$v_\perp \dot{v}_\perp = \frac{1}{2}v_\perp^2\frac{\dot{B}}{B}\,, \tag{2.181}$$

where

$$\dot{B} = v_\parallel \frac{\partial B}{\partial s}\,. \tag{2.182}$$

Therefore, the energy conservation gives

$$m v_\parallel \dot{v}_\parallel = qE_\parallel v_\parallel - v_\parallel \mu \frac{\partial B}{\partial s}\,. \tag{2.183}$$

Introducing

$$\psi := -\int^s E_\parallel ds, \tag{2.184}$$

we obtain (nonrelativistically)

$$\dot{p}_\parallel = -q\frac{\partial \psi}{\partial s} - \mu\frac{\partial B}{\partial s}\,. \tag{2.185}$$

Here, $p_\parallel = m v_\parallel$ is true in the absence of currents along the magnetic field, since then $A_\parallel = 0$. Thus, in the variables s and $p_\parallel$, the Hamiltonian

$$\mathcal{H} = \mathcal{H}\left(s, p_\parallel\right) = \frac{p_\parallel^2}{2m} + q\psi + \mu B \tag{2.186}$$

leads to the equation of motion in canonical form

$$\dot{s} = \frac{\partial \mathcal{H}}{\partial p_{\parallel}} = v_{\parallel} , \tag{2.187}$$

$$\dot{p}_{\parallel} = -\frac{\partial \mathcal{H}}{\partial s} = -q E_{\parallel} - \mu \frac{\partial B}{\partial s} . \tag{2.188}$$

Applying the concept of adiabatic invariance, the quantity

$$\mathcal{J} = \oint p_{\parallel} ds \tag{2.189}$$

will be an invariant if any time-dependence of a magnetic well is slow compared to the bounce frequency of the trapped particles, and any spatial inhomogeneities of the magnetic field are so gradual that the particle's bounce trajectory changes by only a small amount from one bounce to the next. Note that here we calculate

$$p_{\parallel} = \sqrt{2m(\mathcal{H} - q\psi - \mu B)} . \tag{2.190}$$

We shall use that invariant when discussing the Fermi acceleration process.

2.2.5 On the Third Adiabatic Invariant

The first adiabatic invariant μ is related to the fast gyration of a charged particle, the second one to the bouncing in a magnetic well. Now, let us consider particle motion in a cylindrically symmetric configuration with an axial field. The cylinder is assumed to be symmetric in the poloidal angle θ. The angular momentum is a constant of motion. We assume that A_θ is the only nonzero component of $\mathbf{A}$. The θ-component of the exact equation of motion is in cylindrical coordinates

$$m\left(r\ddot{\theta} + 2\dot{r}\dot{\theta}\right) = -\frac{q}{r}\left(r\dot{A}_\theta + \dot{r}A_\theta + r\dot{r}\frac{\partial A_\theta}{\partial r} + r\dot{z}\frac{\partial A_\theta}{\partial z}\right) . \tag{2.191}$$

The right-hand side can be rewritten in terms of a total derivative such that

$$m\left(r\ddot{\theta} + 2\dot{r}\dot{\theta}\right) = -\frac{q}{r}\frac{d}{dt}(rA_\theta) . \tag{2.192}$$

Multiplying by r and rearranging leads to

$$\frac{d}{dt}\left(mr^2\dot{\theta} + q\,rA_\theta\right) \equiv \frac{d}{dt}[r(mv_\theta + qA_\theta)] \equiv \frac{dP_\theta}{dt} = 0 . \tag{2.193}$$

The axial field is given by

$$B_z = (\nabla \times \mathbf{A}) \cdot \hat{z} = \frac{1}{r}\frac{\partial}{\partial r}(rA_\theta) . \tag{2.194}$$

Upon integration, we obtain an expression for

$$A_\theta(r) = \frac{1}{r}\int_0^r r' B_z(r') dr' . \tag{2.195}$$

When comparing the terms contributing to P_θ, we find

$$\frac{mv_\theta}{|qA_\theta|} \sim \frac{|mv_\perp|}{|qrB_z|} \sim \frac{\rho}{r} \ll 1 . \tag{2.196}$$

Therefore, for dimensions larger than the Larmor radius $\rho \equiv r_L$, we find

$$2\pi r A_\theta = 2\pi \int r' B_z(r') dr' \equiv \int_F \boldsymbol{B} \cdot d\boldsymbol{F} \approx \text{const} . \tag{2.197}$$

The magnetic flux inside a drift shell is adiabatically constant. The surfaces $\psi \equiv rA_\theta = \text{const}$ satisfy

$$\boldsymbol{B} \cdot \nabla\psi \equiv \boldsymbol{B} \cdot \nabla(rA_\theta) = 0 \tag{2.198}$$

and are known as flux surfaces of the magnetic field.

The fact that the magnetic field lines lie within the surfaces $\psi = \text{const}$ follows via the following calculations:

$$\begin{aligned} \boldsymbol{B} \cdot \nabla\psi &= B_r \frac{\partial(rA_\theta)}{\partial r} + B_z \frac{\partial(rA_\theta)}{\partial z} \\ &= -\frac{\partial A_\theta}{\partial z}\frac{\partial(rA_\theta)}{\partial r} + \frac{1}{r}\frac{\partial}{\partial r}(rA_\theta)\frac{\partial(rA_\theta)}{\partial r} = 0 . \end{aligned} \tag{2.199}$$

Thus, the particles are constrained by conservation of angular momentum to move on flux surfaces of the magnetic field, except for excursions of the order of the Larmor radius.

2.2.6 Selected Applications

Electron cyclotron emission An accelerated charged particle radiates. Obviously, the gyration is an accelerated motion. Thus, a charged particle in a strong magnetic field radiates. The radiation frequency is of the order of the gyrofrequency (and the harmonics). Assuming that the radiated intensity obeys a Planck distribution, we estimate that an electron in a magnetic field of 3 T at temperature $k_B T_e \approx 1$ keV radiates with a frequency below the maximum of the Planck distribution. Therefore, we may use the Rayleigh–Jeans law $I \sim \omega^2 T_e$. Since the intensity is proportional to the temperature, for known magnetic field strength, a measurement of I can be used to determine the temperature T_e.

Drift in a simple torus When we bend a cylinder to a torus, one might think one could easily construct a magnetic confinement device for plasma. Let us introduce magnetic field lines being parallel to the original axis of the cylinder. The simplest way to reach the "confining field" is with line currents along the Z-axis; see Figure 2.3. Then, the magnetic field strength falls off the torus axis Z with distance, that is, $\sim 1/R$. As shown in the figure, because of the ∇B-drift, electrons and ions move in opposite directions creating a space charge field parallel to the axis of the torus. Finally, together with the magnetic field, an $E \times B$ motion of the whole plasma to the outside results. Thus, the simple device will not work. An outcome of the dilemma is a helical field structure of the magnetic field lines, as being realized in tokamaks and stellarators.

Magnetic mirror As an example for trapped drift trajectories, we consider particles being trapped in a magnetic bottle consisting of two magnetic mirrors as sketched in Figure 2.4.

A particle starting at the field minimum B_{min} and having $v_{\|0}$ and $v_{\perp 0}$ initially moves towards the magnetic field maximum. Since $\mu = $ const, its perpendicular velocity grows and the parallel velocity decreases. The turnover happens at the point

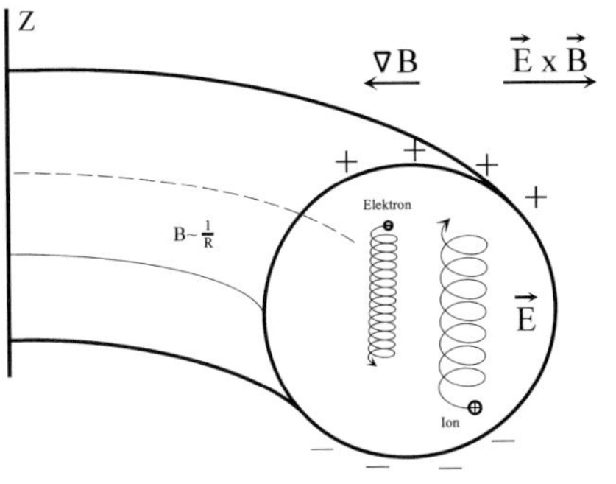

Figure 2.3 Charge separation in a simple torus.

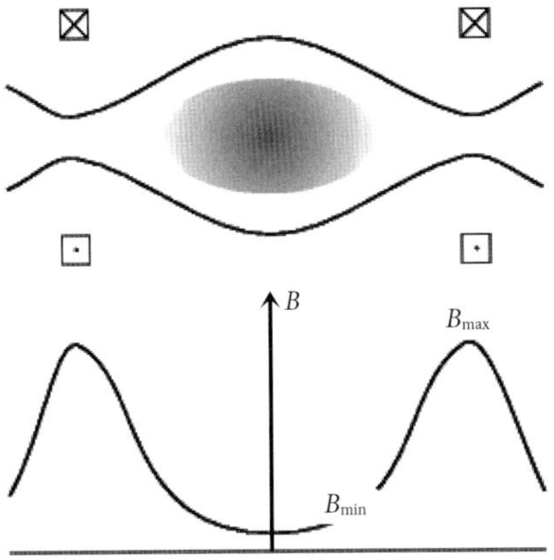

Figure 2.4 The magnetic mirror uses the adiabatic invariance of the magnetic moment for confinement.

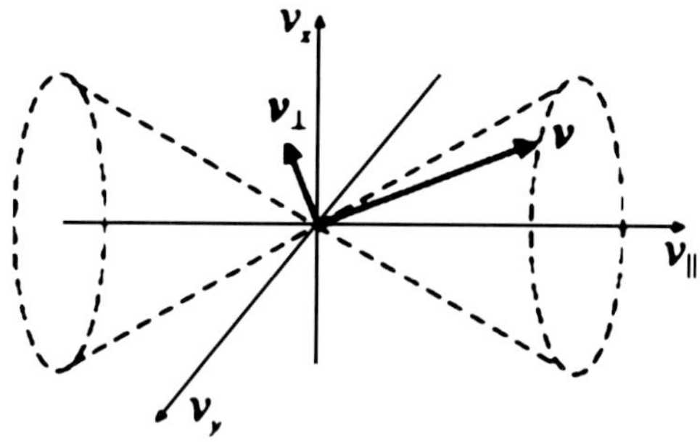

Figure 2.5 Scattering of particles out of a magnetic bottle.

where $v_{\|} = 0$. The condition for the turnover magnetic field strength B is

$$\mu \sim \frac{v_{\perp 0}^2}{B_{min}} = \frac{v_{\perp}^2}{B} \stackrel{!}{=} \frac{v_0^2}{B} \,. \tag{2.200}$$

Consequently, we have

$$B = \frac{v_0^2}{v_{\perp 0}^2} B_{min} \equiv \frac{B_{min}}{\sin^2 \theta} \,. \tag{2.201}$$

The angle θ is called the pitch-angle defined through

$$\sin \theta = \frac{v_{\perp 0}}{v_0} \,. \tag{2.202}$$

We can define a cone angle θ_K via

$$\sin^2 \theta_K = \frac{B_{min}}{B_{max}} \,. \tag{2.203}$$

As shown in Figure 2.5, all particles within the cone

$$\theta < \theta_K \tag{2.204}$$

cannot be confined by the magnetic bottle.

Van Allen belt Our Earth possesses a dipole-like magnetic field. The field lines are strongly deformed by the solar wind. When approaching the poles, the magnetic field strength increases. Particles originating from the solar wind can be scattered into the magnetic structure by collisions. Then, they are trapped in the earth magnetic field, bouncing back an forth. Again by collisions, they may be scattered into the loss cone, and finally cause polar lights (aurora). The curvature of the magnetic field lines causes a curvature drift. Electrons and ions propagate in opposite directions, for example, along the equator because of he curvature drift. For the closed system with rotational symmetry, no space charge field arises (on the average). However, a current is flowing from east to west. This current produces a magnetic field which in turn generates a magnetic field. The latter modifies the original earth magnetic field.

Fermi acceleration mechanism The second adiabatic invariant combined with mirror trapping/detrapping is the basis of the Fermi mechanism for accelerating cosmic ray particles to ultrarelativistic velocities. Consider a particle being initially trapped in a magnetic mirror. Let us assume that $\theta > \theta_K$ is fulfilled for those particles initially. Now, suppose the distance between the magnetic mirrors is slowly reduced. Then, the bounce distance L of the mirror-trapped particles slowly decreases. Because of the (adiabatic) invariance of the second adiabatic invariant, the particle's parallel velocity increases on each bounce. The steady increase in $v_\parallel$ means that the velocity angle decreases. Eventually, $\theta < \theta_K$ and the particle can escape from one end of the mirror with a large parallel velocity. This mechanism provides a slow pumping to very high energy, followed by a sudden and automatic ejection of the energetic particle.

2.3 Motion of a Single Particle (Electron) in an Electromagnetic Wave

A plane light wave propagating in x-direction may be described by a vector potential

$$A(\boldsymbol{r}, t) = \mathrm{Re}\left\{\boldsymbol{A}_0 e^{i\psi}\right\} , \tag{2.205}$$

where $\boldsymbol{r}$ and t are space- and time-coordinate, $\boldsymbol{A}_0 = A_0\hat{e}_y$ for linear polarization (LP) and $\boldsymbol{A}_0 = A_0(\hat{e}_y \pm i\hat{e}_z)$ for circular polarization (CP). The signs + and − depict right- and left-circular polarization, respectively. The phase $\psi = \boldsymbol{k}\cdot\boldsymbol{r} - \omega t$ contains the wavenumber vector $\boldsymbol{k} = k\hat{e}_x$. The dispersion relation in vacuum is $\omega = kc$ with $k = |\boldsymbol{k}| = 2\pi/\lambda$; c is the vacuum velocity of light. Electric and magnetic fields are then obtained as

$$\boldsymbol{E} = \mathrm{Re}\left\{i\omega\boldsymbol{A}_0 e^{i\psi}\right\}, \quad \boldsymbol{B} = \mathrm{Re}\left\{i\boldsymbol{k}\times\boldsymbol{A}_0 e^{i\psi}\right\}, \tag{2.206}$$

and the Poynting vector is $\boldsymbol{S} = \boldsymbol{E}\times\boldsymbol{H}$. The latter determines the intensity (energy per unit area and per unit time) of the light

$$I = |\boldsymbol{S}| = \frac{\omega k}{2\mu_0}A_0^2 \times \begin{cases}[1-\cos(2\psi)], & \text{for LP}\\ 2, & \text{for CP}.\end{cases} \tag{2.207}$$

One should notice that the intensity oscillates with twice the phase for linear polarization because of $\sin^2 x = [1-\cos(2x)]/2$. On the other hand, the intensity is independent of the phase for circular polarization. This makes a significant difference in the interaction with matter. For the averaged (over the oscillations) intensity $I_0 = \langle I\rangle$, we have

$$I_0\lambda^2 = \zeta\frac{\omega k\lambda^2}{2\mu_0}A_0^2 = \zeta\frac{2\pi^2}{\mu_0}cA_0^2 , \tag{2.208}$$

with $\zeta = 1$ for linear and $\zeta = 2$ for circular polarization.

The relativistic threshold intensity is reached when electrons caught by the light wave acquire the velocity of light. For nonrelativistic electrons with $|\boldsymbol{v}| \ll c$, the equation of motion

$$m\frac{d\boldsymbol{v}}{dt} = -e\left(\boldsymbol{E} + \boldsymbol{v} \times \boldsymbol{B}\right) \approx -e\boldsymbol{E} \tag{2.209}$$

has the approximate solution

$$\boldsymbol{v} \approx \mathrm{Re}\left\{\frac{e\boldsymbol{E}}{im\omega}\right\} = \frac{eA_0}{m}\begin{cases}\hat{e}_y \cos\psi & \text{for LP}\\ (\hat{e}_y \cos\psi \mp \hat{e}_z \sin\psi) & \text{for CP ,}\end{cases} \tag{2.210}$$

$$\boldsymbol{r} \approx \mathrm{Re}\left\{\frac{e\boldsymbol{E}}{m\omega^2}\right\} = -\frac{eA_0}{m\omega}\begin{cases}\hat{e}_y \sin\psi & \text{for LP}\\ (\hat{e}_y \sin\psi \mp \hat{e}_z \cos\psi) & \text{for CP .}\end{cases} \tag{2.211}$$

Introducing the dimensionless light amplitude

$$a_0 = \frac{eA_0}{mc} , \tag{2.212}$$

we can rewrite Eq. (2.208) in the form

$$I_0\lambda^2 = \zeta\frac{2\pi^2}{\mu_0}P_0 a_0^2 = \zeta\left[1.37 \times 10^{18}\,\frac{\mathrm{W}}{\mathrm{cm}^2}\,\mu\mathrm{m}^2\right]a_0^2 . \tag{2.213}$$

By definition, the relativistic threshold is reached at $a_0 = 1$, that is, when the quiver velocity approaches c. Of course, the electron trajectories then differ from the simple transverse oscillation derived above. We shall derive them in detail below. Equation (2.213) involves the relativistic power unit

$$P_0 = \frac{4\pi}{\mu_0}\frac{m^2c^3}{e^2} = 8.67\,\mathrm{GW} , \tag{2.214}$$

which may be written as the product of the voltage $mc^2/e = 511\,\mathrm{kV}$ corresponding to the rest energy of the electron and the electric current unit $J_0 = 4\pi/\mu_0(mc)/e = 17\,\mathrm{kA}$, which is related to the Alfvén current $J_A = J_0\beta\gamma$ with $\beta = v/c$ and $\gamma = 1/\sqrt{1-\beta^2}$. Currents larger than J_A cannot be transported in vacuum due to magnetic self-interaction.

Let us briefly comment on the electric current unit J_0 in the SI-system (ampere) to the corresponding cgs-unit. When we construct a unit for the electric current out of the physical constants e (unit for charge Q in terms of the units M for mass, L for length, and T for time), m (unit for mass M), and velocity c, we note in the cgs-system

$$[\mathrm{J}] = \frac{\mathrm{M}^{1/2}\mathrm{L}^{3/2}}{\mathrm{T}^2} \sim m^X c^Y e^Z \sim \mathrm{M}^X\left(\frac{\mathrm{L}}{\mathrm{T}}\right)^Y\left(\frac{\mathrm{M}^{1/2}\mathrm{L}^{3/2}}{\mathrm{T}}\right)^Z , \tag{2.215}$$

leading to

$$X + \frac{1}{2}Z = \frac{1}{2} , \quad Y + \frac{3}{2}Z = \frac{3}{2} , \quad Y + Z = 2 . \tag{2.216}$$

The solution is $X = 1$, $Y = 3$, $Z = -1$, resulting in $J_0 = (mc^3)/e$ which agrees with the previous formula for $\mu_0 = 4\pi/c^2$.

Another remark concerns the relation (2.208). For circular polarization ($\zeta = 2$), we write it in the form

$$I_0 = \frac{1}{\mu_0}\frac{k}{\omega}E^2 = \frac{1}{\mu_0 c}E^2 = \sqrt{\frac{\varepsilon_0}{\mu_0}}E^2 \equiv \frac{E^2}{Z_0}\,, \tag{2.217}$$

with the vacuum impedance

$$Z_0 = \sqrt{\frac{\mu_0}{\varepsilon_0}} = 376.73\ \Omega\,. \tag{2.218}$$

An exact analytic description of electron trajectories is possible for single electrons in a plane light-wave of arbitrary amplitude. For a transparent derivation, it is important to make appropriate use of the symmetries and invariants of the problem. The relativistic Lagrange function of a particle with charge q moving in electromagnetic potentials $\boldsymbol{A}$ and ϕ is given by

$$L(\boldsymbol{r}, \boldsymbol{v}, t) = -mc^2\sqrt{1 - \frac{v^2}{c^2}} + q\boldsymbol{v}\cdot\boldsymbol{A} - q\phi\,. \tag{2.219}$$

For a plane electromagnetic wave, $\phi = 0$ holds. From the Euler–Lagrange equations

$$\frac{d}{dt}\frac{\partial L}{\partial \boldsymbol{v}} - \frac{\partial L}{\partial \boldsymbol{r}} = 0\,, \tag{2.220}$$

we obtain the equation of motion

$$\frac{d\boldsymbol{p}}{dt} = q\,(\boldsymbol{E} + \boldsymbol{v}\times\boldsymbol{B})\,. \tag{2.221}$$

The canonical momentum is $\boldsymbol{p}^{can} \equiv \partial L/\partial \boldsymbol{v} = m\gamma\boldsymbol{v} + q\boldsymbol{A} \equiv \boldsymbol{p} + q\boldsymbol{A}$ with $\gamma = 1/\sqrt{1 - v^2/c^2}$.

For a plane light-wave, two symmetries exist which provide two constants of motion. Planar symmetry implies $\partial L/\partial \boldsymbol{r}_\perp = 0$ and therefore conservation of the canonical momentum in transverse direction,

$$\frac{\partial L}{\partial \boldsymbol{v}_\perp} = \boldsymbol{p}_\perp + q\boldsymbol{A}_\perp = \text{const}\,. \tag{2.222}$$

The second invariant originates from the wave form of $\boldsymbol{A} \equiv \boldsymbol{A}(t - x/c)$. We make use of the relation $dH/dt = -\partial L/\partial t$ for the Hamiltonian function $H(\boldsymbol{x}, \boldsymbol{P}, t) = E(t)$, which expresses the time-dependent energy of the particle. One obtains

$$\frac{dE}{dt} = -\frac{\partial L}{\partial t} = c\frac{\partial L}{\partial x} = c\frac{d}{dt}\frac{\partial L}{\partial v_x} = c\frac{dp_x^{can}}{dt} = c\frac{dp_x}{dt}\,, \tag{2.223}$$

taking into account that $A_x = 0$ for a plane light-wave. From here, we get

$$E - cp_x = C\,, \tag{2.224}$$

where C is a constant. For electrons initially, that is, when no wave is present, being at rest, we have $C = mc^2$. When no potential energy ($\phi = 0$) is present, the kinetic energy is the total energy minus the rest energy, and one finds

$$E_{kin} \equiv E - mc^2 = p_x c \,. \tag{2.225}$$

Note that because of the famous Einstein formula $E = \gamma mc^2$ (m is the rest mass), for the kinetic energy, one also has

$$E_{kin} \equiv E - mc^2 = (\gamma - 1)mc^2 \,. \tag{2.226}$$

The two relations for the kinetic energy become more plausible when one makes use of the canonical equations. Starting from the relativistic Lagrangian, we find the Hamiltonian by first defining the canonical momenta via

$$\frac{\partial L}{\partial \boldsymbol{v}} = \boldsymbol{p}^{can} \equiv \boldsymbol{P} \,. \tag{2.227}$$

The Hamiltonian $H(\boldsymbol{P}, \boldsymbol{q}, t)$ follows through the Legendre transformation

$$H = \sum_k P_k \dot{q}_k - L \,. \tag{2.228}$$

We have

$$\boldsymbol{P}_\perp = \boldsymbol{p}_\perp + q\boldsymbol{A}_\perp \,, \quad \boldsymbol{P}_\parallel = \boldsymbol{p}_\parallel = p_x \hat{x} \,, \quad \boldsymbol{A}_\parallel \equiv A_x \hat{x} = 0 \,, \tag{2.229}$$

$$H = \frac{mc^2}{\sqrt{1 - \frac{v_\perp^2}{c^2} - \frac{v_\parallel^2}{c^2}}} \equiv mc^2 \sqrt{1 + \frac{p_\perp^2}{m^2c^2} + \frac{p_\parallel^2}{m^2c^2}} \,. \tag{2.230}$$

The Hamiltonian H does not depend on the perpendicular coordinates; therefore, $\boldsymbol{P}_\perp = \text{const}$ holds. Since it is constant, we may determine its value from the initial condition, for example, that the electron is at rest before the wave arrives. Then, we set $\boldsymbol{P}_\perp = 0$. Equivalent formulations of the Hamiltonian are

$$H = mc^2 \sqrt{1 + \frac{q^2 A_\perp^2}{m^2c^2} + \frac{P_\parallel^2}{m^2c^2}} \equiv mc^2 \gamma \,. \tag{2.231}$$

The latter formulations show in a very clear way that the relevant variables in the Hamiltonian are $x - ct$ and $P_x \equiv P_\parallel$, that is, $H = H(x - ct, P_x)$. The energy $E \equiv H = mc^2\gamma$ is the total energy. If the electron is initially, that is, before the wave arrives, at rest, the energy is the rest energy mc^2. That means that $H - mc^2$ is the kinetic energy, as used above. Two canonical equations still have to be considered. The first one

$$\frac{dx}{dt} = \frac{\partial H}{\partial P_x} \tag{2.232}$$

leads to the well-known relation $p_x = \gamma m v_x$. The other one,

$$\frac{dP_x}{dt} = -\frac{\partial H}{\partial x}, \tag{2.233}$$

is equivalent to

$$\frac{dp_x}{dt} = -\frac{\partial H}{\partial x} = -\frac{q^2}{2m\gamma}\frac{\partial A^2}{\partial x}. \tag{2.234}$$

Next, we calculate

$$\frac{dH}{dt} = \frac{1}{2\gamma}\frac{q^2}{m}\frac{\partial A^2}{\partial t} + \frac{1}{2\gamma}\frac{q^2}{m}\frac{\partial A^2}{\partial x}\dot{x} + \frac{p_x}{\gamma m}\frac{dp_x}{dt}. \tag{2.235}$$

By inserting Eq. (2.234) on the right-hand side, we recognize

$$\frac{dH}{dt} = \frac{1}{2\gamma}\frac{q^2}{m}\frac{\partial A^2}{\partial t}. \tag{2.236}$$

Since all fields only depend on $\xi = x - ct$, we may replace $\partial/\partial t \mathrel{\hat{=}} -c\partial/\partial x$, and therefore

$$\frac{dH}{dt} = c\frac{dp_x}{dt} \tag{2.237}$$

immediately follows. The latter relation can be integrated,

$$H - cp_x = C\,, \tag{2.238}$$

where the constant $C = mc^2$ for particles initially at rest. In conclusion, for an electron in a plane wave of finite duration, the relativistic equation of motion can be integrated exactly. For an electron initially at rest, the constants of motion give

$$\tilde{\boldsymbol{p}}_\perp \equiv \frac{\boldsymbol{p}_\perp}{mc} = \boldsymbol{a} \equiv \frac{e\boldsymbol{A}_\perp}{mc} = (0, a_y, a_z)\,, \quad \gamma - 1 = \tilde{p}_x\,. \tag{2.239}$$

The γ-factor can be written in the forms

$$\gamma = \frac{1}{\sqrt{1 - \frac{v^2}{c^2}}} = \sqrt{1 + \frac{p^2}{m^2c^2}} \tag{2.240}$$

since $\boldsymbol{p} = \gamma m\boldsymbol{v}$. Starting from

$$\gamma - 1 = \frac{p_x}{mc}, \tag{2.241}$$

we find from its square

$$\gamma = 1 + \frac{p_\perp^2}{2m^2c^2}. \tag{2.242}$$

Thus, we can write

$$\gamma - 1 = \tilde{p}_x = \frac{\tilde{p}_\perp^2}{2} = \frac{a^2}{2} \,. \tag{2.243}$$

An immediate and very important observation at this point is that $E_{kin} \sim \tilde{E}_{kin} \sim \gamma - 1 \sim a^2$ is directly coupled to the light amplitude a and falls back to zero as soon as the electron leaves the light field. The electron cannot gain net energy in a plane light wave. It needs breaking of planar symmetry for net energy gain. This typically occurs in experimental configuration, for example, due to finite beam radius or additional interactions.

From the last equations, with $\boldsymbol{\beta} = \boldsymbol{v}/c$ and $q = -e$, the "equations of motion"

$$\tilde{p}_x = \gamma\beta_x = \frac{\gamma}{c}\frac{dx}{dt} = \frac{a^2}{2} \,, \tag{2.244}$$

$$\tilde{p}_y = \gamma\beta_y = \frac{\gamma}{c}\frac{dy}{dt} = a_y \,, \tag{2.245}$$

$$\tilde{p}_z = \gamma\beta_z = \frac{\gamma}{c}\frac{dz}{dt} = a_z \tag{2.246}$$

follow. Since $\gamma = 1 + a^2/2$, we obtain for $a \gg 1$

$$\beta_x = \frac{\frac{a^2}{2}}{1 + \frac{a^2}{2}} \to 1 \,, \quad \beta_y = \frac{a_y}{1 + \frac{a^2}{2}} \to 0 \,, \quad \beta_z = \frac{a_z}{1 + \frac{a^2}{2}} \to 0 \,, \tag{2.247}$$

and also $\tan\theta = p_\perp/p_x \to 0$. This means that the electron, though oscillating transversely for low field strengths $|\boldsymbol{a}| \ll 1$, moves more and more in the direction of light propagation for relativistic laser intensities with $|a| \gg 1$.

Integration of the equations of motion for a given light pulse $\boldsymbol{a}(t - x/c)$ is straightforward in the variable "Eigenzeit" $\tau = t - x(t)/c$. We have

$$\gamma\frac{d}{dt} = \gamma\frac{d\tau}{dt}\frac{d}{d\tau} = \gamma\left(1 - \frac{1}{c}\frac{dx}{dt}\right)\frac{d}{d\tau} = \left(1 + \frac{a^2}{2} - \frac{a^2}{2}\right)\frac{d}{d\tau} = \frac{d}{d\tau} \,, \tag{2.248}$$

and therefore $d\tau = dt/\gamma$ such that the equations of motion obtain the simple form

$$\frac{dx}{d\tau} = c\frac{a^2}{2} \,, \quad \frac{dy}{d\tau} = c a_y \,, \quad \frac{dz}{d\tau} = c a_z \,. \tag{2.249}$$

For *circular* polarization with

$$\boldsymbol{a}(\boldsymbol{r}, t) = \mathrm{Re}\left\{a_0\left(\hat{e}_y \pm i\hat{e}_z\right)e^{-i\omega\tau}\right\} \,, \tag{2.250}$$

the electron motion is particularly simple. Since $a^2 = a_y^2 + a_z^2 = a_0^2/2$ and $\gamma = 1 + a_0^2/2$, the kinetic energy $E_{kin} = a_0^2/2$ depends on time only through the envelope function $a_0(\tau)$, but not through the rapidly oscillating phase $\psi = -\omega\tau$ of the laser pulse. For *constant* a_0, we obtain $\tau = t/\gamma$, and therefore the trajectory is

$$x(t) = \left(c\frac{a_0^2}{2}\right)\tau = \frac{\frac{a_0^2}{2}}{1+\frac{a_0^2}{2}}ct\,, \tag{2.251}$$

$$y(t) = \frac{ca_0}{\omega}\sin\left(\frac{\omega t}{\gamma}\right)\,, \tag{2.252}$$

$$z(t) = \mp\frac{ca_0}{\omega}\cos\left(\frac{\omega t}{\gamma}\right)\,. \tag{2.253}$$

It describes an electron moving with constant speed on a helix.

For *linear* polarization, the electron motion is more complex. Let us assume a rectangular (over N periods with constant amplitude a_0) pulse, that is,

$$a_y = a_0\cos(\omega\tau) \quad \text{for} \quad 0 < \tau < N\left(2\frac{\pi}{\omega}\right)\,, \tag{2.254}$$

$$a_z \equiv 0\,, \quad a^2 = a_y^2 = a_0^2\cos^2(\omega\tau)\,. \tag{2.255}$$

The electron may be initially at rest and located at $x = y = z = 0$. The trajectories are then obtained in the form

$$x(\tau) = \frac{ca_0^2}{2}\int_0^\tau \cos^2(\omega\tilde{\tau})d\tilde{\tau} = \frac{ca_0^2}{4}\left[\tau + \frac{1}{2\omega}\sin(2\omega\tau)\right]\,, \tag{2.256}$$

$$y(\tau) = ca_0\int_0^\tau \cos(\omega\tilde{\tau})\,d\tilde{\tau} = \frac{ca_0}{\omega}\sin(\omega\tau)\,. \tag{2.257}$$

The parameter τ has to be determined implicitly via $\tau = t - x(t(\tau))/c$ as a function of t. From $\gamma\, d\tau = dt$ and

$$\gamma = 1 + \frac{a^2}{2} = 1 + \frac{a_0^2}{2}\cos^2(\omega\tau)\,, \tag{2.258}$$

we find after direct integration,

$$t = \tau + \frac{a_0^2}{4}\left[\tau + \frac{1}{2\omega}\sin(2\omega\tau)\right]\,. \tag{2.259}$$

Note that t is a *monotonous* function of τ; see Figure 2.6. Using this, apparently the motion consists of an overall drift in x-direction, when taking the nonoscillatory part $t \approx \tau(1 + a_0/4)$,

$$x_d(t) \approx \frac{a_0^2}{a_0^2+4}ct\,, \tag{2.260}$$

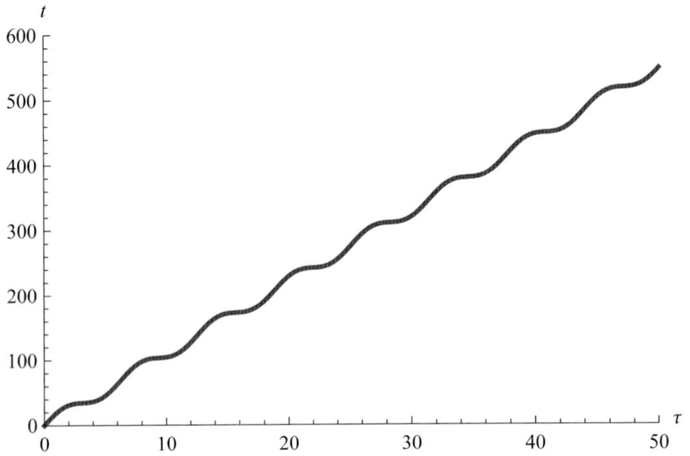

Figure 2.6 Relation (2.259) between time t and τ for some typical parameters.

and a superimposed figure-8 trajectory in the drift frame

$$y(\tau) = \frac{a_0}{k} \sin(\omega \tau) , \tag{2.261}$$

$$x(\tau) - x_d \approx \frac{c a_0^2}{4} \tau - \frac{c a_0^2}{a_0^2 + 4} t + \frac{1}{8\omega} c a_0^2 \sin(2\omega\tau) = \frac{1}{2k} \frac{a_0^2}{4 + a_0^2} \sin(2\omega\tau) , \tag{2.262}$$

with $k = \omega/c$.

In summary, for relativistic laser amplitudes, the motion of an electron is predominantly in laser direction. If an electron is initially at rest, and then overtaken by a finite-length pulse of electromagnetic radiation, after the laser pulse is over, the electron comes to rest again, but is shifted by a distance. Importantly, no net energy has been transferred.

2.4 Lagevin Approach

We now discuss some basic effects of collisions on single particle motion. Much can be learned from the so-called one-dimensional (1d) Langevin equations which model the motion of a heavy particle in a fluid of light particles. The collisions with the molecules of the fluid are caused by a stochastic force $f(t)$. When assuming its average $\langle f(t) \rangle$ to be zero, we explicitly take out the friction force $-m\zeta v$, where ζ is the friction coefficient (corresponding to the collision frequency). Newton's law for the motion of the heavy particle then reads

$$m\dot{v} = -m\zeta v + f(t) . \tag{2.263}$$

Here, we have assumed that no external field is present (therefore, Eq. (2.263) is called a free Langevin equation). The stochastic force is characterized by its average $\langle f(t)\rangle$ and correlation, for example,

$$\left\langle f(t) f\left(t'\right)\right\rangle \equiv \phi\left(t-t'\right) = \lambda\delta\left(t-t'\right) \tag{2.264}$$

for a Delta correlation in the case that the correlation time τ_c is very small (white noise). The solution of Eq. (2.263) is

$$v(t) = v_0 e^{-\zeta t} + \frac{1}{m} e^{-\zeta t} \int_0^t d\tau e^{\zeta\tau} f(\tau)\,. \tag{2.265}$$

It can easily be found by the method of variation of parameters. Since we are not interested in the exact motion but only in the mean behavior, we calculate $\langle[v(t)]^2\rangle$. For $t \gg \zeta^{-1}$, one finds

$$\left\langle v^2\right\rangle \to \frac{\lambda}{2\zeta m^2}\,. \tag{2.266}$$

Note that for $t \gg \zeta^{-1}$, the influence of the initial velocity v_0 has been lost. A single particle should be thermalized after many collisions ($t \to \infty$),

$$\frac{1}{2} m\left\langle v^2\right\rangle \approx \frac{T}{2}\,. \tag{2.267}$$

This immediately leads to the Einstein relation

$$\lambda = 2\zeta m T\,, \tag{2.268}$$

that is, the friction coefficient is proportional to the squared amplitude of the stochastic force. The velocity correlation function also follows straightforwardly,

$$\begin{aligned}\left\langle v(t) v\left(t'\right)\right\rangle &= \frac{\lambda}{2\zeta m^2} e^{-\zeta|t-t'|} + \left(v_0^2 - \frac{\lambda}{2\zeta m^2}\right) e^{-\zeta(t+t')} \\ &\to \frac{\lambda}{2\zeta m^2} e^{-\zeta|t-t'|} \quad \text{for} \quad t, t' \gg \zeta^{-1}\,. \end{aligned} \tag{2.269}$$

It shows that, for long times, we have an exponential decay with the characteristic time ζ^{-1}.

For transport, the mean square displacement $\langle[x(t)]^2\rangle$ is of fundamental interest. Since x follows from v by integration over time, it can be directly related to the velocity correlation function (2.269). In the long-time limit, we obtain

$$\left\langle[x(t)]^2\right\rangle \approx \int_0^t d\tau \int_0^t d\tau' \frac{\lambda}{2\zeta m^2} e^{-\zeta|\tau-\tau'|}\,. \tag{2.270}$$

We may split this integral into two parts, that is,

$$\begin{aligned}
&\int_0^t d\tau \int_0^\tau d\tau' e^{-\zeta(\tau-\tau')} + \int_0^t d\tau \int_\tau^t d\tau' e^{-\zeta(\tau'-\tau)} \\
&= \int_0^t d\tau \frac{1}{\zeta} e^{-\zeta\tau}\left[e^{\zeta\tau}-1\right] + \int_0^t d\tau \frac{1}{(-\zeta)} e^{\zeta\tau}\left[e^{-\zeta t}-e^{-\zeta\tau}\right] \\
&= \frac{t}{\zeta} + \frac{1}{\zeta^2}\left[e^{-\zeta t}-1\right] - \frac{e^{-\zeta t}}{\zeta^2}\left[e^{\zeta t}-1\right] + \frac{t}{\zeta} \\
&\sim \frac{2}{\zeta} t \quad \text{for large } t\,.
\end{aligned} \tag{2.271}$$

Therefore,

$$\left\langle [x(t)]^2 \right\rangle \approx \frac{\lambda}{\zeta^2 m^2} t\,. \tag{2.272}$$

The "standard" definition of the (space-independent) diffusion coefficient D is

$$\langle x^2 \rangle = 2Dt\,, \tag{2.273}$$

and therefore we have

$$D = \frac{\lambda}{2\zeta^2 m^2} = \frac{T}{\zeta m} \equiv \mu T\,, \tag{2.274}$$

where $\mu = 1/\zeta m$ is the mobility. Note that for an (unmagnetized) 1d system, the diffusion coefficient is proportional to the square of the thermal velocity and inversely proportional to the collision frequency.

An important remark concerns Eq. (2.273). Why do we call D the diffusion coefficient? For illustration, let us first have a look at a very simple 1d model. The model consists of a linear chain, and particles may jump at times $\Delta t, 2\Delta t, \ldots$, with a constant step length Δx and equal probabilities 1/2 to the left or to the right. The probability to have r steps in one direction after totally n jumps is

$$P_n(r) = \frac{n!}{r!(n-r)!}\left(\frac{1}{2}\right)^n\,. \tag{2.275}$$

In the following, we introduce the "physical" variables $t = n\Delta t$ and $x = r\Delta x - (n-r)\Delta x$ to calculate the mean square

$$\left\langle x^2 \right\rangle = 4(\Delta x)^2 \sum_{r=0}^{n} \left(r - \frac{n}{2}\right)^2 P_n(r)\,. \tag{2.276}$$

When evaluating the contributions on the right-hand side, we may use with $p = q = 1/2$

$$\sum_{r=0}^{n} P_n(r) = (p+q)^n = 1\,, \tag{2.277}$$

$$\sum_{r=0}^{n} r\,P_n(r) = p\frac{\partial}{\partial p}(p+q)^n = \frac{n}{2}\,, \tag{2.278}$$

$$\sum_{r=0}^{n} r^2\,P_n(r) = p\frac{\partial}{\partial p}p\frac{\partial}{\partial p}(p+q)^n = \frac{n}{2} + \frac{n(n-1)}{4}\,, \tag{2.279}$$

where we have formally written

$$P_n(r) = \frac{n!}{r!(n-r)!}p^r q^{n-r}\,. \tag{2.280}$$

The evaluation leads to

$$\langle x^2 \rangle = \frac{(\Delta x)^2}{\Delta t}t \equiv 2Dt\,. \tag{2.281}$$

Note that $\langle x \rangle = 0$, and therefore the last formula describes the mean square displacement which is linearly growing with time.

The net particle flux $\Gamma = nv$ at a certain point x_0 follows from the balance of the fluxes in positive and negative directions. Here, $n(x)$ describes the (initial) distribution of particles. We count the numbers of jumps in both directions, and from there,

$$\Gamma = \Gamma_+ - \Gamma_- = \frac{1}{2\Delta t}\left[\int\limits_{x_0-\Delta x}^{x_0} n(x)dx - \int\limits_{x_0}^{x_0+\Delta x} n(x)dx\right] \tag{2.282}$$

in the continuum approximation. A Taylor expansion of n up to the first order, that is, in the form

$$n(x) \approx n(x_0) + \frac{dn}{dx_0}x\,, \tag{2.283}$$

will be used in the integrands. This leads to

$$\Gamma \approx -\frac{(\Delta x)^2}{2\Delta t}\frac{dn}{dx}\bigg|_{x_0} \equiv -D\frac{dn}{dx}\,. \tag{2.284}$$

Obviously, the 1d continuity equation now appears in the form

$$\frac{\partial n}{\partial t} = -\frac{\partial \Gamma}{\partial x} = \frac{\partial}{\partial x}\left(D\frac{\partial n}{\partial x}\right)\,. \tag{2.285}$$

More generally, a diffusion flux density can be written as

$$j(x) = -D\nabla n(x) \mathrel{\hat{=}} -D\frac{\partial n}{\partial x} \tag{2.286}$$

in one space dimension. Inserting that into the density continuity equation

$$\frac{\partial n}{\partial t} + \nabla \cdot j = 0 \tag{2.287}$$

leads to the so called diffusion equation

$$\frac{\partial n}{\partial t} - D\frac{\partial^2 n}{\partial x^2} = 0 , \tag{2.288}$$

where a constant diffusion coefficient has been assumed.

Starting at time $t = 0$ with all particles concentrated at $x = 0$, that is, $n(x, t = 0) = N\delta(x)$, the solution of Eq. (2.288) is

$$n(x, t) = \frac{N}{\sqrt{4\pi D t}} e^{-\frac{x^2}{4Dt}} . \tag{2.289}$$

That solution is obtained for $n(x, t = 0) = N\delta(x)$ when we introduce the Fourier transformation

$$\tilde{n}(k, t) = \int_{-\infty}^{\infty} dx\, n(x, t) e^{ikx} . \tag{2.290}$$

The diffusion equation after Fourier transformation is

$$\frac{\partial \tilde{n}}{\partial t} = -Dk^2 \tilde{n} , \tag{2.291}$$

and it has the obvious solution

$$\tilde{n}(k, t) = A e^{-Dk^2 t} . \tag{2.292}$$

The integration constant A can be determined from the initial condition. Since $\tilde{n}(k, t = 0) = N$, we have $A = N$. We can now apply the inverse transformation to obtain

$$n(x, t) = \frac{N}{2\pi} \int_{-\infty}^{\infty} dk e^{-ikx} e^{-Dk^2 t} = \frac{N}{\sqrt{4\pi D t}} e^{-\frac{x^2}{4Dt}} . \tag{2.293}$$

Let us now briefly generalize the above results for the 1d Langevin equation without external fields to a Langevin equation which is more appropriate for expectations in hot and magnetized systems. Then, we have to include the Lorentz force, and the problem essentially becomes 3d. An appropriate generalization, with a magnetic field in z-direction, is

$$\frac{d\boldsymbol{q}(t)}{dt} = \boldsymbol{v}(t) , \tag{2.294}$$

$$\frac{d\boldsymbol{v}_\perp(t)}{dt} = \Omega\, (\boldsymbol{v}_\perp \times \hat{e}_z) - \nu \boldsymbol{v}_\perp + \boldsymbol{a}_\perp(t) , \tag{2.295}$$

$$\frac{dv_z(t)}{dt} = -\nu v_z + a_z(t) . \tag{2.296}$$

We shall come back to this system in the chapter on stochastic transport. Since the present generalization is still a linear (inhomogeneous) system, it can be solved explicitly. By similar arguments as in the simple 1d case, for long times t, a characteristic behavior can be detected when solving for the velocity correlation function and the mean square displacement. The collisional force has an average component $-m\nu\boldsymbol{v}(t)$ and a random, fluctuating component $m\boldsymbol{a}(t)$. Therefore, the collisions are modeled as a force which produces, on average, a friction proportional to the velocity. In the more detailed description, the instantaneous intensity and the direction of the friction force is random, as described by $m\boldsymbol{a}(t)$. The constant ν is identified with the collision frequency. The function $\boldsymbol{a}(t)$ is only defined by its statistical properties. A Gaussian process is completely defined by its first two moments. We assume that $\boldsymbol{a}(t)$ is a stationary, delta-correlated Gaussian process (white noise) such that the first and second moments are $\langle a_r(t)\rangle = 0$ and $\langle a_r(t)a_s(t+\tau)\rangle = A\delta_{rs}\delta(\tau)$ for $r, s = x, y, z$. The constant will be determined later.

Let us assume that the velocity of the test particle at $t = 0$ has the value $\boldsymbol{v}(0) = \boldsymbol{v}_0$ with probability one. We treat the Langevin equation first as an ordinary differential equation for a given realization $\boldsymbol{a}(t)$. Introducing the propagator

$$G(t) = \begin{pmatrix} e^{-\nu t}\cos(\Omega\, t) & e^{-\nu t}\sin(\Omega\, t) & 0 \\ e^{-\nu t}\sin(\Omega\, t) & e^{-\nu t}\cos(\Omega\, t) & 0 \\ 0 & 0 & e^{-\nu t} \end{pmatrix} \tag{2.297}$$

for $t \geq 0$, we can write the solution as

$$\boldsymbol{v} = G(t)\cdot\boldsymbol{v}_0 + \int_0^t d\theta\, G(\theta)\cdot\boldsymbol{a}(t-\theta)\,. \tag{2.298}$$

The first moment follows after straightforward averaging, that is,

$$\langle\boldsymbol{v}(t)\rangle = G(t)\cdot\boldsymbol{v}_0\,. \tag{2.299}$$

The velocity autocorrelation functions for $t_1 > t_2$ are, for example, obtained as

$$\begin{aligned} R_{xx}(t_1, t_2) &\equiv \langle v_x(t_1)v_x(t_2)\rangle \\ &= e^{-\nu(t_1+t_2)}\left[\cos(\Omega\, t_1)\cos(\Omega\, t_2)v_{0x}^2 + \sin(\Omega\, t_1)\sin(\Omega\, t_2)v_{0y}^2\right] \\ &\quad + \frac{A}{2\nu}e^{-\nu(t_1-t_2)}\left[1-e^{-2\nu t_2}\right]\cos\left[\Omega\,(t_1-t_2)\right]. \end{aligned} \tag{2.300}$$

The autocorrelation function R_{yy} is obtained in a similar way. The parallel autocorrelation function R_{zz} follows from R_{xx} by setting $\Omega = 0$. Next, we average over different initial velocities $\boldsymbol{v}_0$. Let us assume that the latter are Maxwell-distributed with the velocity distribution function

$$f(\boldsymbol{v}_0) = \frac{1}{(2\pi)^{3/2}v_{th}^3}\exp\left(-\frac{v_0^2}{2v_{th}^2}\right) \tag{2.301}$$

where $v_{th} = \sqrt{T/m}$. Defining

$$\bar{R}_{xx}(t_1, t_2) = \int d^3 v_0 \, f(v_0) R_{xx}(t_1, t_2) \,, \tag{2.302}$$

we obtain after some algebra

$$\bar{R}_{xx}(t_1, t_2) = \frac{1}{2} \left\{ \frac{A}{\nu} e^{-\nu |t_1 - t_2|} + \left(2 v_{th}^2 - \frac{A}{\nu} \right) e^{-\nu (t_1 + t_2)} \right\} \cos \left[\Omega (t_1 - t_2) \right] . \tag{2.303}$$

In this form, the autocorrelation function is not stationary unless $A = 2 v_{th}^2 \nu$. Introducing $\tau = t_1 - t_2$, the stationary forms are

$$\bar{R}_{xx}(\tau) = v_{th}^2 e^{-\nu |\tau|} \cos(\Omega \tau) \,, \quad \bar{R}_{zz}(\tau) = v_{th}^2 e^{-\nu |\tau|} \,. \tag{2.304}$$

We now determine the instantaneous position

$$x(t) = x(0) + \int_0^t d\theta \, v_x(\theta) \equiv x(0) + \delta x(t) \,, \tag{2.305}$$

and are ready to calculate the mean square displacement

$$\langle \delta x^2(t) \rangle = \int_0^t d t_1 \int_0^t d t_2 \, \bar{R}_{xx}(t_1 - t_2) \,. \tag{2.306}$$

For the inner integral on the right-hand side, we use for fixed t_1 the new variable $\tau = t_1 - t_2$ to obtain via integration by parts

$$\begin{aligned} \langle \delta x^2(t) \rangle &= - \int_0^t d t_1 \int_{t_1}^{t_1 - t} d\tau \, \bar{R}_{xx}(\tau) \\ &= \int_0^t d t_1 (t - t_1) \bar{R}_{xx}(t_1) + \int_0^t d t_1 t_1 \bar{R}_{xx}(t_1 - t) \,. \end{aligned} \tag{2.307}$$

The last term on the right-hand side gives the same result as the first one. That can be easily seen by changing the variable $\bar{t}_1 = t - t_1$ and using the fact that $\bar{R}_{xx}(-\bar{t}_1) = R_{xx}(\bar{t}_1)$. Thus,

$$\langle \delta x^2(t) \rangle = 2 \int_0^t d\tau (t - \tau) \bar{R}_{xx}(\tau) \,. \tag{2.308}$$

We are now in a position to determine the diffusion coefficient

$$D_\perp(t) = \frac{1}{2} \frac{d}{dt} \langle \delta x^2(t) \rangle = \int_0^t d\tau \, \bar{R}_{xx}(\tau) \,. \tag{2.309}$$

We find

$$D_\perp(t) = v_{th}^2 \frac{\nu + \left[\Omega \sin(\Omega t) - \nu \cos(\Omega t)\right] e^{-\nu t}}{\nu^2 + \Omega^2} . \tag{2.310}$$

In the asymptotic limit,

$$D_\perp = \lim_{t \to \infty} D_\perp(t) = v_{th}^2 \frac{\nu}{\nu^2 + \Omega^2} . \tag{2.311}$$

The parallel running diffusion coefficient is

$$D_\parallel(t) = v_{th}^2 \frac{1 - e^{-\nu t}}{\nu} , \tag{2.312}$$

leading to the asymptotic value

$$D_\parallel = \frac{v_{th}^2}{\nu} . \tag{2.313}$$

In summary, we have

$$D_\parallel \approx \frac{T}{\nu m} , \tag{2.314}$$

and

$$D_\perp = \frac{T}{m} \frac{\nu}{\Omega^2 + \nu^2} , \tag{2.315}$$

where $\Omega = qB/m$ is the gyrofrequency. For strong magnetic fields $\Omega \gg \nu$, we approximately have

$$D_\perp \approx \frac{T}{m} \frac{\nu}{\Omega^2} \sim \frac{T\nu}{B^2} . \tag{2.316}$$

Note that now the perpendicular diffusion coefficient is proportional to the collision frequency and inversely proportional to the square of the magnetic field strength. The latter prediction is in favor of good magnetic confinement by strong magnetic fields. Unfortunately, reality is different.

3
Plasma in Thermodynamic Equilibrium

3.1
Basic Approach

The physics of plasmas in thermodynamic equilibrium is a topic in itself. Its full consideration requires more space than available in a single monograph. Nevertheless, the overview over the fundamentals of plasma physics requires, at least, a short introduction into basic principles of plasma equilibrium theory.

A plasma is a many body system of interacting particles. Obviously, the interaction of particles causes the main difficulties compared to an ideal gas. The huge number of particles suggests an application of the methods of statistical physics and a characterization of the system by mean values.

The averaging of a quantity $F(q,p)$ which depends on the phase space variables $\boldsymbol{q}_i$ and $\boldsymbol{p}_i$ of particles $i = 1,\ldots,N$ (abbreviated as q, p) will be denoted by $\langle\ldots\rangle$. The explicit prescription is

$$\langle F\rangle = \int d^{3N}q\, d^{3N}p\, F(q,p)\rho(q,p;t)\ . \tag{3.1}$$

The phase-space density ρ gives the probability of finding a system in a certain phase-space area. Statistical physics constructs an ensemble of systems which differ in the initial conditions, but the dynamics of interacting particles in the various systems is governed by the same Hamiltonian H. Each ensemble member is a realization of the plasma under consideration when only macroscopic boundary conditions are formulated. When the ensemble of classical systems is constructed in that way, the phase-space density of the noninteracting systems obeys a continuity equation

$$\frac{\partial\rho}{\partial t} + \sum_{i=1}^{3N}\left[\frac{\partial(\rho\dot{q}_i)}{\partial q_i} + \frac{\partial(\rho\dot{p}_i)}{\partial p_i}\right] = 0\ . \tag{3.2}$$

Since the density is incompressible, we have

$$\nabla\cdot\boldsymbol{v}_\Gamma \equiv \sum_{i=1}^{3N}\left(\frac{\partial\dot{q}_i}{\partial q_i} + \frac{\partial\dot{p}_i}{\partial p_i}\right) = 0\ . \tag{3.3}$$

High Temperature Plasmas, Theory and Mathematical Tools for Laser and Fusion Plasmas, First Edition.
Karl-Heinz Spatschek.

The last relation follows from the canonical equations. Combining the last two equations, we find

$$\frac{d\rho}{dt} = 0\,, \tag{3.4}$$

that is, the Liouville equation. The latter is often written in the form

$$\frac{\partial \rho}{\partial t} = \{H, \rho\} \tag{3.5}$$

when the Poisson bracket is used. Classical statistical mechanics would "reduce" the problem of determining averaged quantities to "pure" integration once the Liouville equation is solved. Though first, a solution of the Liouville equation for *general* cases is practically impossible, and second the integration over the total phase-space is *in general* totally impracticable. Before reducing the problem to practicable cases, let us briefly comment on quantum plasmas. Then, we have to replace the Liouville equation by the von Neumann equation for the density operator $\hat{\rho}$,

$$i\hbar \frac{\partial \hat{\rho}}{\partial t} = [H, \hat{\rho}]\,. \tag{3.6}$$

The averaging will require one to calculate the trace

$$\langle \hat{A} \rangle = \mathrm{Sp}\left(\hat{A}\hat{\rho}\right). \tag{3.7}$$

The classical limit is obtained by the replacement

$$\mathrm{Sp}\left(\hat{A}\hat{\rho}\right) \longrightarrow \int \dots \int \frac{d^{3N}p\, d^{3N}q}{h^{3N} N!} A\rho\,. \tag{3.8}$$

In thermodynamic equilibrium, the general form of ρ (we will consider classical systems) is restricted to time-independent functions, that is,

$$\frac{\partial \rho}{\partial t} = \dot{\rho} = 0\,. \tag{3.9}$$

Then, the Liouville equation reduces to

$$\{H, \rho\} = 0\,, \tag{3.10}$$

and nontrivial solutions can be found by inspection. When ρ depends on phase-space variables only functionally through H, that is, $\rho(\boldsymbol{q}, \boldsymbol{p}) = \rho(H[\boldsymbol{q}, \boldsymbol{p}])$, or other constants of motion, the time-independent Liouville equation is automatically satisfied.

In statistical physics, different statistical ensembles, for example, the canonical ensemble with

$$\rho = \frac{1}{h^{3N} N! Z} e^{-\beta H}\,, \tag{3.11}$$

or the grand canonical ensemble

$$\rho = \frac{1}{h^{3N} N! Z_g} e^{-\beta H - \alpha N} , \tag{3.12}$$

are in use. A central role is played by the partition function [27]

$$Z = \frac{1}{h^{3N} N!} \int \cdots \int d^{3N} p \, d^{3N} q e^{-\beta H} \tag{3.13}$$

in the canonical case, and

$$Z_g = \sum_{N=0}^{\infty} \frac{1}{h^{3N} N!} \int \cdots \int d^{3N} p \, d^{3N} q e^{-\beta H - \alpha N} \tag{3.14}$$

in the grand canonical case. The canonical partition function Z is related to the free energy F via

$$F = -\theta \ln Z , \tag{3.15}$$

where $\theta = k_B T = 1/\beta$. The free energy is the difference of the internal energy U and the product of temperature and entropy S, that is, $F = U - TS$. Therefore, we have for the differential $dF = dU - TdS - SdT$. The free energy F has the natural variables temperature T, volume V, and particle number N. In more general cases, the volume V is replaced by $\boldsymbol{a}$ and the conjugate variable pressure p by $-\boldsymbol{A}$, such that $pdV \hat{=} -\boldsymbol{A} \cdot d\boldsymbol{a}$. To formulate dF in terms of dT, dV, and dN, we use the laws of thermodynamics, $dS = dQ/T$ and $dQ = dU + pdV - \mu dN$, where μ is the chemical potential. Often, $\alpha = -\beta\mu$ is also used. Putting everything together,

$$dF = \frac{F - U}{T} dT - pdV + \mu dN \tag{3.16}$$

follows. We compare with

$$dF \equiv d(-\theta \ln Z) = -\ln Z d\theta - \theta \frac{1}{Z} dZ , \tag{3.17}$$

where

$$\begin{aligned} dZ &= \frac{\partial Z}{\partial N} dN + \frac{1}{h^{3N} N!} \int \cdots \int d^{3N} p \, d^{3N} q \left[\frac{H}{\theta^2} d\theta - \frac{1}{\theta} \frac{\partial H}{\partial V} dV \right] e^{-\frac{H}{\theta}} \\ &= \frac{\partial Z}{\partial N} dN + \left[\frac{1}{\theta^2} U d\theta + \frac{1}{\theta} p dV \right] . \end{aligned} \tag{3.18}$$

Thus,

$$d(-\theta \ln Z) = \frac{(-\theta \ln Z - U)}{T} dT - pdV + \frac{\partial}{\partial N} (-\theta \ln Z) dN , \tag{3.19}$$

and the comparison with Eq. (3.16) leads to the identification $F = -\theta \ln Z$. In a similar manner, for the grand canonical ensembles, $pV = \theta \ln Z_g$ follows from the partition function Z_g. The grand canonical potential pV has the natural variables T, V, and μ. We stay with the canonical ensemble (in the thermodynamic limit $N \to \infty$, $N/V =$ const, the results do not depend on the choice of the ensemble). Then, the other thermodynamic variables are

$$S = -\left.\frac{\partial F}{\partial T}\right|_{N,V} , \quad p = -\left.\frac{\partial F}{\partial V}\right|_{T,N} , \tag{3.20}$$

$$\mu = \left.\frac{\partial F}{\partial N}\right|_{T,V} , \quad c_V = \left.\frac{\partial (F + TS)}{\partial T}\right|_{N,V} . \tag{3.21}$$

Now, we apply statistical thermodynamics to the simplest model of a plasma consisting of free electrons and protons (capital letters). The corresponding "simple" Hamiltonian is

$$H = \frac{1}{2}\sum_{j=1}^{N}\left[\frac{p_j^2}{m} + \frac{P_j^2}{M}\right] + \frac{1}{2}\sum_{\substack{i,j \\ i \neq j}}^{N}\left[\frac{e^2}{|\boldsymbol{R}_i - \boldsymbol{R}_j|} + \frac{e^2}{|\boldsymbol{r}_i - \boldsymbol{r}_j|}\right] - \sum_{i,j}^{N}\frac{e^2}{|\boldsymbol{R}_i - \boldsymbol{r}_j|} . \tag{3.22}$$

The classical canonical partition function for N protons (mass M) at positions $\boldsymbol{R}_i$ with momenta $\boldsymbol{P}_i$ and N electrons (mass m) at positions $\boldsymbol{r}_i$ with momenta $\boldsymbol{p}_i$ is

$$Z = \frac{1}{h^{6N}(N!)^2}\int \ldots \int d^{3N}p\, d^{3N}P\, d^{3N}r\, d^{3N}R\, e^{-\beta H} , \tag{3.23}$$

where, of course, we have to insert $H = H(\boldsymbol{r}, \boldsymbol{p}, \boldsymbol{R}, \boldsymbol{P})$. The Gibbs factor $\exp(-\beta H)$ factorizes into a part containing positions and a part originating from the momenta. The latter leads to integrals of the form

$$\int_0^\infty x^2 e^{-a^2x^2}\, dx = \frac{\sqrt{\pi}}{4a^3} . \tag{3.24}$$

The part containing the positions is much more complicated unless the interaction of the particles is neglected. The latter approximation is known as the ideal gas approximation. In the ideal gas approximation of a plasma, we obtain

$$Z = \frac{1}{h^{6N}(N!)^2}\left[\frac{2\pi m^{1/2}M^{1/2}}{\beta}\right]^{3N} V^{2N} . \tag{3.25}$$

The free energy

$$F = -k_B T \ln Z \tag{3.26}$$

is then straightforward to calculate. For example, the pressure of a plasma in the ideal gas approximation is

$$p = -\left.\frac{\partial F}{\partial V}\right|_{T,N} = \frac{2Nk_BT}{V} \tag{3.27}$$

since we have $2N$ particles in volume V with temperature T. In a similar way, the mean energy

$$E = -\frac{\partial}{\partial\beta}\ln Z = \frac{3}{2}(2N)k_B T \tag{3.28}$$

follows. The ideal gas approximation will be appropriate when the mean interaction energy is small compared to the kinetic energy, in other words, when the number of particles in the Debye sphere is very large.

The more general case of the partition function for electrons and ions is not exactly solvable. Let us have a short look at what we know for N interacting neutral particles. The partition function is

$$Z = \frac{1}{N!h^{3N}}\int\dots\int d^{3N}p\,d^{3N}q e^{-\frac{H(p,q)}{k_B T}} \equiv J(T)Q(T,V)\,, \tag{3.29}$$

including two-particle interactions, that is, with the Hamiltonian

$$H = \frac{1}{2m}\sum_i p_i^2 + \phi\left(q_1,\dots,q_N\right)\,, \quad \phi = \sum_{i<j}\phi_{ij}\,. \tag{3.30}$$

J is simple to calculate,

$$J(T) = \frac{1}{h^{3N}}\left[\int_{-\infty}^{\infty}\exp\left(-\frac{p^2}{2mk_B T}\right)dp\right]^{3N} = \left[\frac{2\pi m k_B T}{h^2}\right]^{3N/2}, \tag{3.31}$$

though

$$Q(T,V) = \frac{1}{N!}\int\dots\int d^3q_1\dots d^3q_N\exp\left(-\frac{\phi}{k_B T}\right) \tag{3.32}$$

is the problem. The expression $\exp(-\phi/k_B T)$ is called the generalized Boltzmann factor. It determines the two-particle distribution

$$n^{(2)}\left(\boldsymbol{r}_1,\boldsymbol{r}_2\right) := \frac{N(N-1)\int\dots\int d^3r_3\dots d^3r_N\exp\left(-\frac{\phi}{k_B T}\right)}{\int\dots\int d^3r_1\dots d^3r_N\exp\left(-\frac{\phi}{k_B T}\right)}\,, \tag{3.33}$$

which gives the probability of finding two particles at the positions $\boldsymbol{r} = \boldsymbol{r}_1$ and $\boldsymbol{r} = \boldsymbol{r}_2$. The mean energy follows from the two-particle distribution in the form

$$E = -\frac{\partial\ln Z}{\partial\beta} = \frac{3}{2}Nk_B T + \left(\frac{V}{2}\right)\int d^3\left(|\boldsymbol{r}_1-\boldsymbol{r}_2|\right)\phi_{12}n^{(2)}\left(\boldsymbol{r}_1,\boldsymbol{r}_2\right), \tag{3.34}$$

provided the interaction energy ϕ_{12} only depends on the distance of the particles.

For the equation of state, we have to calculate

$$p = k_B T\left.\frac{\partial\ln Q}{\partial V}\right|_{T,N}\,. \tag{3.35}$$

To perform the differentiation with respect to the volume, the latter may be considered as a cube with sides of equal length L, that is, $V = L^3$. Normalizing all space variables by L, we find

$$Q(T,V) = \frac{L^{3N}}{N!} \int_0^1 \ldots \int_0^1 d^3\left(\frac{r_1}{L}\right) \ldots d^3\left(\frac{r_N}{L}\right) \times \exp\left[-\frac{1}{k_B T} \sum_{i<j} \phi_{ij}\left(\frac{|\boldsymbol{r}_i - \boldsymbol{r}_j|\, L}{L}\right)\right] . \tag{3.36}$$

Because of

$$\frac{\partial}{\partial V} = \frac{1}{3V} L \frac{\partial}{\partial L} , \tag{3.37}$$

we may use

$$\begin{aligned} L\frac{\partial \ln Q}{\partial L} &= 3N - \frac{1}{k_B T} \sum_{j<k} \frac{\int \ldots \int d^3\left(\frac{r_1}{L}\right) \ldots d^3\left(\frac{r_N}{L}\right) |\boldsymbol{r}_j - \boldsymbol{r}_k|\, \phi'_{jk} e^{-\beta\phi}}{N! \frac{Q}{L^{3N}}} \\ &= 3N - \frac{N(N-1)}{2k_B T} \frac{\int \ldots \int d^3 r_1 \ldots d^3 r_N \, |\boldsymbol{r}_1 - \boldsymbol{r}_2|\, \phi'_{12} e^{-\beta\phi}}{\int \ldots \int d^3 r_1 \ldots d^3 r_N e^{-\beta\phi}} . \end{aligned} \tag{3.38}$$

Therefore,

$$pV = N k_B T - \frac{1}{6} V \int d^3 (|\boldsymbol{r}_1 - \boldsymbol{r}_2|)\, |\boldsymbol{r}_1 - \boldsymbol{r}_2|\, \phi'_{12} n^{(2)} (\boldsymbol{r}_1, \boldsymbol{r}_2) . \tag{3.39}$$

Again, the two-particle distribution determines the deviation from the ideal gas law.

Thus far, we presented relations for N interacting particles without being able to evaluate the integrals which contain the unknown two-particle distribution. As stated at the beginning, equilibrium statistical thermodynamics is a very complicated topic, especially for interacting neutral particles. It requires effort to determine the integrals which depend on the interaction potentials. The methods developed for neutral systems can be applied, to a large extent, to systems of charged particles, although the latter introduce a lot of additional difficulties. It is not the intent of the present book to go into the details of that important and interesting topic. Instead, in the next subsection, we will only discuss, more or less qualitatively, some specific features of equilibrium plasma physics.

3.2
A Heuristic Derivation of the Modified Equation of State

We have already solved the shielding problem of a stationary test charge. The sign of the shielding charge distribution is opposite to the sign of the test charge q_a.

Shielding causes a redistribution of charges compared to the ideal plasma situation (in the latter, the interaction potential is neglected). The correlation energy follows from the general formula for the electrostatic energy of a charge distribution

$$W = \frac{1}{2}\int d^3 r \rho(\mathbf{r})\phi(\mathbf{r}) \,. \tag{3.40}$$

Now, we evaluate the latter by inserting

$$\rho(\mathbf{r}) = \rho_a^{ind}(r) \,, \quad \phi(\mathbf{r}) = \phi_a^{Cb}(r) \,. \tag{3.41}$$

The use of ϕ_a^{ind} is obvious since we want to calculate the energy change due to the redistribution of charges. In general, $\phi = \phi_a^{Cb} + \phi_a^{ind}$ will appear in the integrand. The second term, however, leads to the self-energy. For the energy shift, we thus have to evaluate the integral

$$\begin{aligned} \Delta E_a &= -\frac{1}{2}\frac{1}{4\pi\varepsilon_0}\frac{q_a^2\kappa^2}{4\pi}\int d^3 r \frac{1}{r^2} e^{-\kappa r} = -\frac{1}{2}\frac{1}{4\pi\varepsilon_0} q_a^2\kappa^2 \int_0^\infty d r e^{-\kappa r} \\ &= -\frac{1}{2}\frac{1}{4\pi\varepsilon_0} q_a^2 \kappa \,. \end{aligned} \tag{3.42}$$

The energy shift of the plasma consisting of N_e electrons and $N_i = N_e = N/2$ protons is

$$U^{int} = -\frac{1}{2}\frac{1}{4\pi\varepsilon_0}\sum_s N_s q_s^2 \kappa = -\frac{N}{2}\frac{1}{4\pi\varepsilon_0} e^2\kappa \sim N\sqrt{\frac{N}{TV}} \,. \tag{3.43}$$

The thermodynamic potential related to the natural variables N, T, and V is the free energy $F = U - TS$. We have

$$U = F + TS = F - T\left(\frac{\partial F}{\partial T}\right)_{N,V} = -T^2\frac{\partial}{\partial T}\left(\frac{F}{T}\right)_{N,V} \,. \tag{3.44}$$

Decomposing any thermodynamic quantity A into

$$A = A^{ideal} + A^{interaction} \equiv A^{id} + A^{int} \,, \tag{3.45}$$

we find

$$\frac{F^{int}}{T} = -\int^T dT \frac{U^{int}}{T^2} \,. \tag{3.46}$$

We have shown above

$$U^{int} = \frac{C}{\sqrt{T}} \tag{3.47}$$

as a function of T. Therefore,

$$\frac{F^{int}}{T} = -C\int^T \frac{dT}{T^{5/2}} = \frac{2}{3}\frac{C}{T^{3/2}} = \frac{2}{3}\frac{U^{int}}{T} \,. \tag{3.48}$$

The result has important consequences. The free energy is modified by the term

$$F^{int} = -\frac{1}{3}\frac{1}{4\pi\varepsilon_0}\sum_s N_s q_s^2 \kappa = -\frac{N}{3}\frac{1}{4\pi\varepsilon_0} e^2 \sqrt{\frac{Ne^2}{\varepsilon_0 k_B T V}} \tag{3.49}$$

for an electron–proton plasma with $N = N_e + N_i$ particles. The pressure correction follows from

$$p^{int} = -\left.\frac{\partial F^{int}}{\partial V}\right|_{T,N} = -\frac{1}{6}\frac{1}{4\pi\varepsilon_0}\sum_s n_s q_s^2 \kappa = -\frac{1}{24\pi}\kappa^3 k_B T \,. \tag{3.50}$$

This is exactly the result one obtains by the diagram technique in the ring approximation.

The present procedure is not only applicable to calculate the pressure correction. Since we have obtained the correction of the free energy (compared to the ideal gas approximation), we also have access to the corrections of the other thermodynamic variables. The chemical potential has to be corrected by

$$\mu_a^{int} = \left.\frac{\partial F^{int}}{\partial N_a}\right|_{T,V} = -\frac{1}{2}\frac{1}{4\pi\varepsilon_0} q_a^2 \kappa \,. \tag{3.51}$$

We may obtain the entropy correction from

$$S^{int} = -\left.\frac{\partial F^{int}}{\partial T}\right|_{V,N} = -\frac{1}{6}\frac{1}{4\pi\varepsilon_0}\sum_a N_a \frac{q_a^2 \kappa}{T} \,. \tag{3.52}$$

When we proceed with the corresponding densities, the following expressions are available, that is,

$$f^{int} \equiv \frac{F^{int}}{V} = -\frac{k_B T}{12\pi}\kappa^3 \,, \tag{3.53}$$

$$u^{int} \equiv \frac{U^{int}}{V} = -\frac{k_B T}{8\pi}\kappa^3 \,, \tag{3.54}$$

$$s^{int} \equiv \frac{S^{int}}{V} = -\frac{k_B}{24\pi}\kappa^3 \,. \tag{3.55}$$

We stop the discussion of thermodynamic variables at this point, again emphasizing that the present introduction only provides a taste of the problems one faces when doing more exact calculations. As a result of the estimates shown here, in the limit of the number of particles in the Debye sphere tending to infinity, the ideal plasma properties become more and more applicable.

3.3 The Holtsmark Distribution for Electric Microfields

We conclude the brief introduction into plasma equilibrium problems by discussing the problem of the probability distribution of a dependent variable. Let us consider the local electric field $\boldsymbol{E}$ (microfield) at position $\boldsymbol{r}$ at time t. It is a fluctuating quantity which consists of contributions from all particles, that is,

$$\boldsymbol{E}(\boldsymbol{r}) \equiv \boldsymbol{E}(\boldsymbol{r};\boldsymbol{r}_1,\ldots) = \sum_{j=1}^{N} \boldsymbol{E}_j\left(\boldsymbol{r},\boldsymbol{r}_j\right). \tag{3.56}$$

Formally, we can write the probability for $\boldsymbol{E}$, which depends on the positions $\boldsymbol{r}_j$ where the particles stay at time t, as the rather complicated integral

$$W(\boldsymbol{E}) = \int\ldots\int \delta\left[\boldsymbol{E} - \boldsymbol{E}(\boldsymbol{r})\right]\rho d^{3N}r d^{3N}p\,, \tag{3.57}$$

which contains the (unknown) Liouville density function. Before introducing some (drastic) simplifications, let us Fourier transform the probability function (which is often done when determining probabilities of dependent functions [45]). The Fourier transform is

$$W(\xi) = \int e^{-i\xi\cdot E} W(E)d^3E = \int\ldots\int e^{-i\xi\cdot E(r)}\rho d^{3N}r d^{3N}p\,. \tag{3.58}$$

For electrostatic Coulomb fields, which we assume here, the integration over the momenta is trivial, and

$$W(\xi) = \int\ldots\int e^{-i\xi\cdot\Sigma E_j} P_N d^{3N}r \tag{3.59}$$

evolves, where P_N is the reduced density in configuration space. Though still, the problem is enormous since P_N is the unknown probability distribution for the N interacting particles. Let us introduce [46]

$$\varepsilon_j = \exp\left(-i\xi\cdot\boldsymbol{E}_j\right) - 1 \tag{3.60}$$

such that

$$W(\xi) = \int\ldots\int \prod_{j=1}^{N}\left(1+\varepsilon_j\right) P_N d^{3N}r\,. \tag{3.61}$$

The still $3N$-dimensional integral necessitates simplifications. The most drastic one is to evaluate the integrals for independent particles for which the probability distribution factorizes,

$$P_N = P_1(\boldsymbol{r}_1)\ldots P_1(\boldsymbol{r}_N)\,. \tag{3.62}$$

Now, let us have a look at the factors $(1+\varepsilon_j)$ forming a product. The latter leads to summands $\varepsilon_1\varepsilon_2\ldots\varepsilon_s$ with s starting at zero and extending up to N. Assuming indistinguishable particles, the combinatorial coefficient

$$\frac{N!}{(N-s)!s!} \tag{3.63}$$

will appear. Each integral can be simplified according to

$$\int \cdots \int \varepsilon_1 \ldots \varepsilon_s P_N d^{3N} r = \left[\int \varepsilon_1 P_1 d^3 r_1 \right]^s . \tag{3.64}$$

Thus, within the independent particle approximation, we obtain

$$W(\xi) = \sum_{s=0}^{N} \frac{N!}{(N-s)!s!} \left[\int \varepsilon_1 P_1 d^3 r_1 \right]^s . \tag{3.65}$$

Consistent with that approximation is

$$P_1 = \frac{1}{V} . \tag{3.66}$$

For large numbers N of particles, we take the limit $N \to \infty$ with

$$\frac{N!}{(N-s)!} \approx N^s \tag{3.67}$$

such that

$$W(\xi) \approx \exp\left\{ n \int \left(e^{-i\xi \cdot E_1} - 1 \right) d^3 r_1 \right\} . \tag{3.68}$$

Here, $n = N/V$ is the particle density. Without loss of generality, we evaluate the field at $\boldsymbol{r} = 0$. Furthermore, the low-frequency component of the microfield distribution will be mainly produced by ions. Assuming a single charged ion at position $\boldsymbol{r}_1$, we have

$$\boldsymbol{E}_1 = -\frac{e^* \boldsymbol{r}_1}{r_1^3} , \quad e^* = \frac{e}{4\pi\varepsilon_0} . \tag{3.69}$$

The integration

$$\int \left(e^{-i\xi \cdot E_1} - 1 \right) d^3 r_1 = -\frac{4}{15} (2\pi e^* \xi)^{3/2} \tag{3.70}$$

is straightforward. The back-transformation leads for the low-frequency part of the microfield distribution to

$$\begin{aligned} W(\boldsymbol{E}) &= \frac{1}{(2\pi)^3} \int e^{i\xi \cdot \boldsymbol{E}} W(\xi) d^3 \xi \\ &= \frac{1}{2\pi^2 E} \int_0^\infty \sin(\xi E) \exp\left\{ -\frac{4n}{15} (2\pi e^* \xi)^{3/2} \right\} \xi \, d\xi . \end{aligned} \tag{3.71}$$

Since we expect an isotropic distribution, we change to

$$W(E) = 4\pi E^2 W(\boldsymbol{E}) \tag{3.72}$$

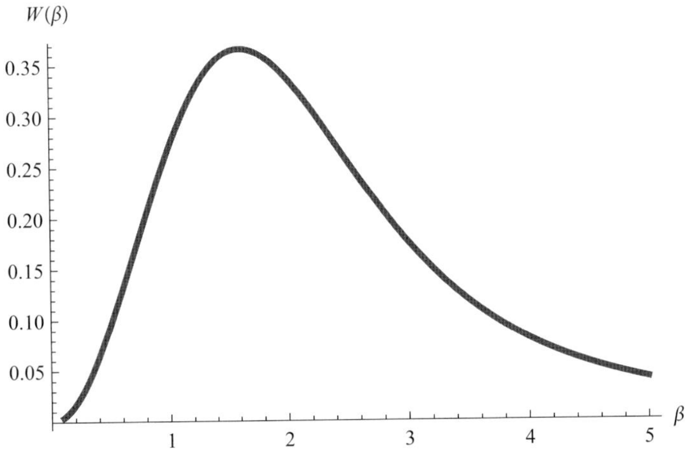

Figure 3.1 Holtsmark distribution of the magnitude of the electric microfield.

for the magnitude of the low-frequency electric microfield. Introducing the mean value

$$E_0 = \frac{e^*}{r_0^2} , \tag{3.73}$$

(with appropriate $r_0 \approx 0.465 r_n$ to simplify the exponent) for normalization $\beta = E/E_0$, one finds

$$W(\beta) = \frac{2}{\pi\beta} \int_0^\infty x \sin x e^{-\left(\frac{x}{\beta}\right)^{3/2}} dx . \tag{3.74}$$

The distribution is shown in Figure 3.1. The asymptotic behaviors are

$$W(\beta) \sim \begin{cases} \frac{4}{3\pi}\beta^2 & \text{for } \beta \to 0 , \\ \frac{3}{2}\beta^{-5/2} & \text{for } \beta \to \infty . \end{cases}$$

Of course, the Holtsmark model [46] is an extremely simple model. It does not take into account the shielding of the fields as well as the interaction of the particles. Ecker and Müller [47] have generalized the Holtsmark calculation for Debye-shielded fields. Interactions were incorporated by Baranger and Moser [48]. The literature on electric microfield distributions and their diagnostic applications is huge; see, for example, [49, 50].

4
Kinetic Description of Nonequilibrium Plasmas

In this chapter, we consider kinetic equations, that is, closed equations for the one-particle distribution function f. We start with some historical remarks, especially regarding the Boltzmann equation which, while in its original form, cannot be used in plasmas where long-range forces dominate. We also review the general Fokker–Planck approach which, in principle, applies to plasmas.

4.1
Historical Remarks on Well-Known Kinetic Equations

4.1.1
The Fokker–Planck Equation

The Landau–Fokker–Planck description is used in many systems. However, it is only applicable under quite restrictive assumptions. For a discussion of the general procedure, we restrict ourselves to a very simple deterministic system. This allows one to elucidate the main strategy without confusing the reader with complicated dynamics. We choose the standard map

$$u_{n+1} = u_n - \frac{K}{2\pi} \sin(2\pi\varphi_n) , \tag{4.1}$$

$$\varphi_{n+1} = \varphi_n + u_{n+1} , \tag{4.2}$$

for the dynamics. In the two-dimensional phase space, u and φ are the action-angle variables. For $K = 0$, the system is unperturbed and integrable. In the presence of (nonintegrable) perturbations ($K \neq 0$), stochastic regions will exist. For large values of K, when more or less complete chaos sets in, large regions of the map will contain "chaotic trajectories," that is, it makes no sense to predict the future evolution of an exact trajectory in that region. However, it is very useful to determine the statistical behavior of trajectories. If the mapping parameter K is large enough, so that no stable islands or cantori exist, a diffusion process occurs which is very much like that found in Brownian motion. For a purely random walk, we expect subsequent values of u_n and φ_n to be independent of one another. Meaning, we

High Temperature Plasmas, Theory and Mathematical Tools for Laser and Fusion Plasmas, First Edition.
Karl-Heinz Spatschek.

can interpret the step length 1 as some collision time. In the completely stochastic region, we expect to find u_n and φ_n equally probable anywhere on the interval $0 \leq u \leq 1$ and $0 \leq \varphi \leq 1$.

Let us now consider the deviation $\Delta u := u_n - u_0$ from the initial value after the "time" $t \hat{=} n \gg 1$. We define a phase average,

$$\langle f \rangle := \int_0^1 d\varphi f \,, \tag{4.3}$$

and assume $\langle \Delta u \rangle \approx 0$. Next, we calculate the variance

$$\langle (\Delta u)^2 \rangle = \left(\frac{K}{2\pi}\right)^2 \sum_{n,n'=0}^{t} \langle \sin(2\pi\varphi_n) \sin(2\pi\varphi_{n'}) \rangle \approx \frac{1}{2}\left(\frac{K}{2\pi}\right)^2 t \,. \tag{4.4}$$

Note that we can interpret $K/2\pi$ as the maximum change in u during one step, that is, $K/2\pi \hat{=} \Delta u_{1max}$. We have just predicted that in such a system, the diffuse scaling $\langle (\Delta u)^2 \rangle \sim t$ appears.

Next, we want to answer the question as to what sense the (reduced) distribution function $P(u, n)$ for u at "time" n can be determined for the stochastic dynamics (if we are only interested in the distribution of the action variable u). Clearly, the subsequent consideration will only be valid in a globally stochastic region (in which islands do not exist or occupy only a negligible phase space volume). In such a region, it is possible to express the evolution of $P(u, n)$ in terms of a Markov process in u (which means that the distribution at "time" $n + \Delta n$ only depends on the previous value at "time" n) with the transition probability W, that is,

$$P(u, n + \Delta n) = \int P(u - \Delta u, n)\, W(u - \Delta u, n, \Delta u, \Delta n) d(\Delta u) \,. \tag{4.5}$$

The transition probability for a jump Δu during the "time" Δn is normalized according to

$$\int W(u, n, \Delta u, \Delta n) d(\Delta u) = 1 \,. \tag{4.6}$$

In order to further simplify Eq. (4.5), we should specify the region we are interested in. To work in the kinetic regime, we should have $\Delta n \gg 1$. Furthermore, we have seen that a diffusive process (in action variable) occurs, and therefore when expanding Eq. (4.5), we anticipate the ordering $\Delta n \sim (\Delta u)^2$ for small changes, that is,

$$|\Delta u| \ll \left| \frac{P}{\frac{dP}{du}} \right| \,. \tag{4.7}$$

This, in combination with the diffusive scaling Eq. (4.7), implies

$$\Delta n \ll \left| \frac{P}{\Delta u_{1max} \frac{dP}{du}} \right|^2 , \tag{4.8}$$

where, for dimensional reasons, we have introduced Δu_{1max}. Then, we expand

$$P(u-\Delta u)\,W(u-\Delta u) \approx P(u)\,W(u) - \frac{\partial(P\,W)}{\partial u}\,\Delta u + \frac{1}{2}\frac{\partial^2(P\,W)}{\partial u^2}(\Delta u)^2 \,, \quad (4.9)$$

$$P(n+\Delta n) \approx P(n) + \frac{\partial P}{\partial n}\,\Delta n \,, \quad (4.10)$$

to obtain the Fokker–Planck equation

$$\frac{\partial P}{\partial n} = -\frac{\partial}{\partial u}(B\,P) + \frac{1}{2}\frac{\partial^2}{\partial u^2}(D\,P)\,. \quad (4.11)$$

Here, we have defined the frictional coefficient

$$B(u) = \frac{1}{\Delta n}\int \Delta u\, W(u, n, \Delta u, \Delta n)\, d(\Delta u)\,, \quad (4.12)$$

and the diffusion coefficient

$$D(u) = \frac{1}{\Delta n}\int (\Delta u)^2\, W(u, n, \Delta u, \Delta n)\, d(\Delta u)\,. \quad (4.13)$$

If the underlying dynamics are Hamiltonian, the relation

$$B = \frac{1}{2}\frac{dD}{du} \quad (4.14)$$

can be derived so that the Fokker–Planck equation appears in the form

$$\frac{\partial P}{\partial n} = \frac{\partial}{\partial u}\left(\frac{D}{2}\frac{\partial P}{\partial u}\right). \quad (4.15)$$

In the preceding calculation, we have ignored the dependence of W on the initial phase distribution. This was done since we expect that after a "correlation time" n_c (i.e., some collision times), any reasonable smooth initial phase distribution relaxes to a uniform phase distribution. Thus, the condition $\Delta n \gg 1$ actually means $\Delta n \gg n_c$.

The Fokker–Planck equation as derived here shows its generality for a diffusive scaling. The structure we have obtained is, in principle, quite satisfactory. However, we should note that the coefficients B and D, although written explicitly, are not yet specific enough for concrete systems. We shall come back to this point later.

4.1.2
The Boltzmann Equation

In the following, we review the Boltzmann equation for a dilute neutral gas in order to demonstrate fundamentals of kinetic theory. We consider the classical dynamics of a distribution function $f(\boldsymbol{r}, \boldsymbol{v}, t)$ which is normalized according to

$$\int d^3r\, d^3v\, f(\boldsymbol{r}, \boldsymbol{v}, t) = N\,. \quad (4.16)$$

For a collision, we can define two characteristic times: τ_c, the time of a collision, which is related to a finite range r_c of the interaction potential via $\tau_c = r_c/v_{th}$, and the time between two collisions $\tau = 1/nr_c^2 v_{th}$. The latter formula gives the collision frequency $\nu = 1/\tau$. Obviously, during one time unit, a particle will have $nr_c^2 v_{th}$ collisions. The inverse is the average time between collisions. For a neutral gas, r_c is well-defined, and in the region $\tau_c \ll \tau \rightarrow r_c \ll n^{-1/3}$, we can consider independent binary collisions. This situation will not occur in a plasma with long-range forces. Note that the condition means dilute gases with short-range interactions. Boltzmann derived an equation for the distribution function of neutral particles. Let us consider volume elements d^3r and d^3v, being large compared to microscopic volumes but small on macroscopic scales. Microscopic volume means that we have "enough" particles (e.g., 10^4) in the volume for the statistics; however, that number should be small compared to the total number of particles (e.g., 10^{21}). During the Hamiltonian dynamics in the presence of an external force $\boldsymbol{F}$ (but without interactions), motion occurs from d^3rd^3v to $d^3r'd^3v'$ with $d^3rd^3v = d^3r'd^3v'$. Without collisions (interactions), the particle numbers in the volume elements would be conserved. Collisions will cause changes according to

$$\left[f\left(\boldsymbol{r} + \boldsymbol{v}dt, \boldsymbol{v} + \frac{1}{m}\boldsymbol{F}dt, t + dt\right) - f(\boldsymbol{r}, \boldsymbol{v}, t)\right] d^3rd^3v$$
$$= \left.\frac{df}{dt}\right|_{coll} dtd^3rd^3v\,. \tag{4.17}$$

The left-hand side will be simplified by Taylor expansion. The right-hand side is evaluated via a "Stoßzahl-Ansatz,"

$$\left[\frac{\partial}{\partial t} + \boldsymbol{v}\cdot\nabla + \frac{1}{m}\boldsymbol{F}\cdot\frac{\partial}{\partial \boldsymbol{v}}\right] f(\boldsymbol{r}, \boldsymbol{v}, t)$$
$$= \int d^3v_2 \int d^3v_3 \int d^3v_4\, W(\boldsymbol{v}, \boldsymbol{v}_2; \boldsymbol{v}_3, \boldsymbol{v}_4)\left[f_3 f_4 - f_1 f_2\right], \tag{4.18}$$

where $f_i \equiv f(\boldsymbol{r}, \boldsymbol{v}_i, t)$ for $i = 1, 2, 3, 4$ with $\boldsymbol{v}_1 = \boldsymbol{v}$. We shall explain the ansatz in the following paragraph in more detail.

First, $W(\boldsymbol{v}_1, \boldsymbol{v}_2; \boldsymbol{v}_3, \boldsymbol{v}_4)$ is the probability that the collision of two particles with velocities $\boldsymbol{v}_1$ and $\boldsymbol{v}_2$ results in new particle velocities $\boldsymbol{v}_3$ and $\boldsymbol{v}_4$. Because of symmetry relations for W, the right-hand side of Eq. (4.18) can be written in the form as presented. Since we consider identical particles, the following symmetry should hold, that is,

$$W(\boldsymbol{v}, \boldsymbol{v}_2; \boldsymbol{v}_3, \boldsymbol{v}_4) = W(\boldsymbol{v}_2, \boldsymbol{v}; \boldsymbol{v}_4, \boldsymbol{v}_3)\,. \tag{4.19}$$

Furthermore, symmetry should also hold for rotations with an orthogonal matrix D, that is,

$$W(D\boldsymbol{v}, D\boldsymbol{v}_2; D\boldsymbol{v}_3, D\boldsymbol{v}_4) = W(\boldsymbol{v}, \boldsymbol{v}_2; \boldsymbol{v}_3, \boldsymbol{v}_4)\,, \tag{4.20}$$

with reflections as a special case:

$$W(-\boldsymbol{v}, -\boldsymbol{v}_2; -\boldsymbol{v}_3, -\boldsymbol{v}_4) = W(\boldsymbol{v}, \boldsymbol{v}_2; \boldsymbol{v}_3, \boldsymbol{v}_4)\,. \tag{4.21}$$

Time-inversion leads to

$$W(\boldsymbol{v}, \boldsymbol{v}_2; \boldsymbol{v}_3, \boldsymbol{v}_4) = W(-\boldsymbol{v}_3, -\boldsymbol{v}_4; -\boldsymbol{v}, -\boldsymbol{v}_2) . \tag{4.22}$$

The combination of reflections and time-reversion results in

$$W(\boldsymbol{v}, \boldsymbol{v}_2; \boldsymbol{v}_3, \boldsymbol{v}_4) = W(\boldsymbol{v}_3, \boldsymbol{v}_4; \boldsymbol{v}, \boldsymbol{v}_2) . \tag{4.23}$$

The latter has been already used when writing the "Stoßzahl-Ansatz" in the shown form.

Next, we note that during (elastic) collision, momentum and energy are conserved. When relating the transition probability W to the collision cross section σ, we explicitly take care of the latter by postulating

$$\begin{aligned} &W(\boldsymbol{v}_1, \boldsymbol{v}_2; \boldsymbol{v}_3, \boldsymbol{v}_4) \\ &= \sigma(\boldsymbol{v}_1, \boldsymbol{v}_2; \boldsymbol{v}_3, \boldsymbol{v}_4)\, \delta(\boldsymbol{p}_1 + \boldsymbol{p}_2 - \boldsymbol{p}_3 - \boldsymbol{p}_4)\, \delta\left(\frac{\boldsymbol{p}_1^2}{2m} + \frac{\boldsymbol{p}_2^2}{2m} - \frac{\boldsymbol{p}_3^2}{2m} - \frac{\boldsymbol{p}_4^2}{2m}\right) . \end{aligned} \tag{4.24}$$

The Dirac Delta-functions take care of the conservation laws. More details, that is, the explicit form of σ, require a specification of the interaction potential. We will assume that we have a short range interaction potential which only depends on the distance between the particles. In a dilute gas, collisions can then be considered as binary processes. When particles one and two collide, the velocities change from v_1, v_2 to v_1', v_2'. For identical particles, momentum end energy conservation lead to

$$\boldsymbol{v}_1 + \boldsymbol{v}_2 = \boldsymbol{v}_1' + \boldsymbol{v}_2' , \quad \boldsymbol{v}_1^2 + \boldsymbol{v}_2^2 = \boldsymbol{v}_1'^2 + \boldsymbol{v}_2'^2 . \tag{4.25}$$

The following lines repeat some of the basic relations known from two-body scattering. Introducing coordinates in a center of gravity system, we write

$$\boldsymbol{V} = \frac{1}{2}(\boldsymbol{v}_1 + \boldsymbol{v}_2) , \quad \boldsymbol{u} = \boldsymbol{v}_1 - \boldsymbol{v}_2 , \quad \boldsymbol{V}' = \frac{1}{2}(\boldsymbol{v}_1' + \boldsymbol{v}_2') , \quad \boldsymbol{u} = \boldsymbol{v}_1' - \boldsymbol{v}_2' . \tag{4.26}$$

Obviously,

$$\boldsymbol{V} = \boldsymbol{V}' \quad \text{and} \quad |\boldsymbol{u}| = |\boldsymbol{u}'| . \tag{4.27}$$

Both results follow from the conservation laws. The first one is rather straightforward; the second one follows when subtracting from both sides of the energy conservation (after multiplication by a factor 2) the squares of the respective sides of the momentum conservation. The transformations are such that the Jacobi determinant is one, that is,

$$d^3 v_1 d^3 v_2 = d^3 V d^3 u = d^3 V' d u' = d^3 v_1' d^3 v_2' . \tag{4.28}$$

When using the center-of-mass system and an effective one-particle description in the center-of-mass system, the reduced mass $\mu = m/2$ appears. Furthermore, in the center-of-mass system we have

$$v_{1s} = v_1 - V = \frac{1}{2}\boldsymbol{u}\,, \quad v_{2s} = v_2 - V = -\frac{1}{2}\boldsymbol{u}\,, \tag{4.29}$$

$$v'_{1s} = \frac{1}{2}\boldsymbol{u}'\,, \quad v'_{2s} = -\frac{1}{2}\boldsymbol{u}'\,. \tag{4.30}$$

Incident particles with intensity $I = n|\boldsymbol{u}|$ are scattered into space angle $(\Omega, d\Omega)$. The number of scattered particles (per unit time) depends on the collision parameter s and can be written as

$$\frac{dN(\Omega)}{dt} = \sigma(\Omega, \boldsymbol{u}) I d\Omega = I s d\varphi(-ds)\,. \tag{4.31}$$

Because of rotational symmetry, the φ-dependence becomes irrelevant. The minus sign has been introduced to take care of the fact that smaller collision parameters correspond to larger scattering angles. Since $d\Omega = \sin\vartheta\, d\vartheta\, d\varphi$, we find

$$\sigma(\Omega, \boldsymbol{u}) = -\frac{1}{\sin\vartheta} s \frac{ds}{d\vartheta} = -\frac{1}{\sin\vartheta}\frac{1}{2}\frac{ds^2}{d\vartheta}\,. \tag{4.32}$$

Clearly, there should be a unique relation between the collision parameter s and the scattering angle ϑ. When deriving the latter (in classical mechanics) for an interaction potential $w(\boldsymbol{r}) \equiv w(r)$, one introduces the so called asymptotic angle φ_a via $\vartheta = \pi - 2\varphi_a$, and from standard textbooks [33] we may take

$$\varphi_a = \int_{r_{min}}^{\infty} dr \frac{\mu s u}{r^2 \sqrt{2\mu\left(E - w(r) - \frac{\mu^2 s^2 u^2}{r^2}\right)}}\,. \tag{4.33}$$

Here,

$$E = \frac{\mu}{2}u^2\,, \quad l = \mu s u \tag{4.34}$$

and r_{min} follows from

$$w(r_{min}) + \frac{l^2}{2\mu r_{min}^2} = E\,. \tag{4.35}$$

The scattering angle ϑ is related to the asymptotic relative velocities via

$$\cos\vartheta = \frac{(v_1 - v_2)\cdot(v'_1 - v'_2)}{|v_1 - v_2||v'_1 - v'_2|}\,. \tag{4.36}$$

Note that the collision term contains $f'_1 f'_2$. The primed distribution functions have to be calculated for velocities v'_1 and v'_2 which, after collision, change to v_1 and v_2.

Coming back to the collision term in the Boltzmann equation, we can now proceed as follows. When binary and central collisions are considered, we relate the, for example, loss rate to the collision cross section σ. Consider a particle with velocity $\boldsymbol{v}_2$ approaching another particle with velocity $\boldsymbol{v}_1$ with a collision parameter s. The incoming (relative) flux is $f(\boldsymbol{r}, \boldsymbol{v}_2, t)|\boldsymbol{v}_2 - \boldsymbol{v}_1| d^3 v_2$, and we have the scattering relation (4.31)

$$\begin{aligned} & f(\boldsymbol{r}, \boldsymbol{v}_2, t)\, |\boldsymbol{v}_2 - \boldsymbol{v}_1| d^3 v_2 \,(-s d\varphi ds) \\ & = f(\boldsymbol{r}, \boldsymbol{v}_2, t)|\boldsymbol{v}_2 - \boldsymbol{v}_1| d^3 v_2 \sigma\,(\Omega, |\boldsymbol{v}_1 - \boldsymbol{v}_2|)\, d\Omega\,. \end{aligned} \tag{4.37}$$

To include all collisions during time dt with particles of a certain velocity $\boldsymbol{v}_1 = \boldsymbol{v}$ in $d^3 v_1$ (and volume element $d^3 r$), we have to multiply by $f(\boldsymbol{r}, \boldsymbol{v}_1, t) d^3 v_1 d^3 r dt$ to get the loss after integration over Ω and v_2, that is,

$$\begin{aligned} dN_- &\equiv l d^3 r d^3 v_1 dt \\ &= \int d^3 v_2 \int d\Omega\, |\boldsymbol{v}_2 - \boldsymbol{v}_1| \sigma\,(\Omega, |\boldsymbol{v}_2 - \boldsymbol{v}_1|)\, f_1 f_2 d^3 r d^3 v_1 dt\,, \end{aligned} \tag{4.38}$$

where $f_{1,2} \equiv f(\boldsymbol{r}, \boldsymbol{v}_{1,2}, t)$.

In a similar way, we can calculate the gain by particles with velocities $(\boldsymbol{v}_1', d^3 v_1')$ being scattered into $(\boldsymbol{v}_1, d^3 v_1)$:

$$\begin{aligned} dN_+ &\equiv g d^3 r d^3 v_1 dt \\ &= \int d\Omega \int d^3 v_2' |\boldsymbol{v}_1' - \boldsymbol{v}_2'| \sigma\,(\Omega, |\boldsymbol{v}_1' - \boldsymbol{v}_2'|)\, f_1' f_2' d^3 r d^3 v_1' dt\,, \end{aligned} \tag{4.39}$$

where $f_{1,2}' = f(\boldsymbol{r}, \boldsymbol{v}_{1,2}', t)$, and $\boldsymbol{v}_1', \boldsymbol{v}_2'$ are the two initial velocities which after the collision evolve into $\boldsymbol{v}_1$ and $\boldsymbol{v}_2$. If $\boldsymbol{v}_1$ is fixed, momentum conservation determines $\boldsymbol{v}_2$. Now, $d^3 v_1' d^3 v_2' = d^3 v_1 d^3 v_2$ (Jacobi determinant equal to one), and we may write

$$g d^3 r d^3 v_1 dt = \int d^3 v_2 \int d\Omega\, |\boldsymbol{v}_1 - \boldsymbol{v}_2| \sigma\,(\Omega, |\boldsymbol{v}_1 - \boldsymbol{v}_2|)\, f_1' f_2' d^3 r d^3 v_1 dt\,. \tag{4.40}$$

The collision term then follows as

$$\left.\frac{df}{dt}\right|_{coll} \equiv g - l = \int d^3 v_2 d\Omega\, |\boldsymbol{v}_2 - \boldsymbol{v}_1| \sigma\,(\Omega, |\boldsymbol{v}_2 - \boldsymbol{v}_1|) \left[f_1' f_2' - f_1 f_2\right]. \tag{4.41}$$

To relate this to the formulation with the transition probability W, we note that the integration over Ω means integration over all directions of $\boldsymbol{u}'$. Making use of

$$\begin{aligned} \boldsymbol{u}'^2 - \boldsymbol{u}^2 &= \boldsymbol{v}_1'^2 - 2\boldsymbol{v}_1' \cdot \boldsymbol{v}_2' + \boldsymbol{v}_2'^2 - \boldsymbol{v}_1^2 + 2\boldsymbol{v}_1 \cdot \boldsymbol{v}_2 - \boldsymbol{v}_2^2 \\ &= -4\boldsymbol{V}'^2 + 2\boldsymbol{v}_1'^2 + 2\boldsymbol{v}_2'^2 + 4\boldsymbol{V}^2 - 2\boldsymbol{v}_1^2 - 2\boldsymbol{v}_2^2 \\ &= 2\left[\boldsymbol{v}_1'^2 + 2\boldsymbol{v}_2'^2 - \boldsymbol{v}_1^2 - \boldsymbol{v}_2^2\right], \end{aligned} \tag{4.42}$$

we find for

$$\begin{aligned}\int d\Omega\,|v_2 - v_1| &= \int d\Omega\, u = \int du'\, d\Omega\, \delta(u' - u)u' \\ &= \int du' u'^2 d\Omega\, \delta\left[(u' + u)(u' - u)\frac{1}{2}\right] \\ &= \int d^3u' \delta\left(\frac{u'^2}{2} - \frac{u^2}{2}\right)\int d^3V' \delta(V' - V) \\ &= 4\int d^3v_1' d^3v_2' \delta\left(\frac{v_1'^2 + v_2'^2}{2} - \frac{v_1^2 + v_2^2}{2}\right)\delta\left(v_1' + v_2' - v_1 - v_2\right).\end{aligned} \tag{4.43}$$

Thus, we have the relation

$$\begin{aligned}W\left(v_1, v_2; v_1', v_2'\right) &= 4\sigma\left(\Omega, |v_2 - v_1|\right)\delta\left(\frac{v_1'^2 + v_2'^2}{2} - \frac{v_1^2 + v_2^2}{2}\right) \\ &\quad \times \delta\left(v_1' + v_2' - v_1 - v_2\right).\end{aligned}$$

That means

$$\sigma\left(v_1, v_2; v_1', v_2'\right) = 4m^4\sigma\left(\Omega, |v_2 - v_1|\right) \tag{4.44}$$

when comparing with the formulation presented at the beginning.

In conclusion, we may use two equivalent forms of the Boltzmann equation, either

$$\begin{aligned}&\left[\frac{\partial}{\partial t} + v\cdot\nabla + \frac{1}{m}F\cdot\frac{\partial}{\partial v}\right] f(r, v, t) \\ &= \int d^3v_2 \int d^3v_3 \int d^3v_4\, W\left(v, v_2; v_3, v_4\right)\left[f_3 f_4 - f_1 f_2\right]\end{aligned} \tag{4.45}$$

or

$$\begin{aligned}&\left[\frac{\partial}{\partial t} + v\cdot\nabla + \frac{1}{m}F\cdot\frac{\partial}{\partial v}\right] f(r, v, t) \\ &= \int d^3v_2 d\Omega\, |v_2 - v_1|\sigma\left(\Omega, |v_2 - v_1|\right)\left[f_1' f_2' - f_1 f_2\right].\end{aligned} \tag{4.46}$$

4.1.3
The Boltzmann H-Theorem

The Boltzmann equation is not time-symmetric, that is, it shows irreversibility. The latter has already been demonstrated by Boltzmann when deriving the so called H-theorem. Consider the functional

$$H(r, t) = \int d^3v\, f(r, v, t)\log f(r, v, t) \tag{4.47}$$

and its time-derivative $\dot{H} = dH/dt$. (The following arguments do not depend on whether we use log or ln in the definition of H.) Although more general conclusions are available, let us consider the case when no external force F is present.

Then, it is straightforward to prove $\dot{H} \leq 0$ in the following steps. When calculating $\dot{H}$ in the situation without an external force, we may assume a space-independent distribution function $f = f(\boldsymbol{v}, t)$ in which case the Boltzmann equation reads

$$\frac{\partial f}{\partial t} = \left.\frac{d f}{d t}\right|_{coll} . \tag{4.48}$$

Then,

$$\dot{H} = \int d^3 v (1 + \log f) \dot{f} = -I \tag{4.49}$$

with

$$I = \int d^3 v_1 \int d^3 v_2 \int d^3 v_3 \int d^3 v_4 \, W(\boldsymbol{v}_1, \boldsymbol{v}_2; \boldsymbol{v}_3, \boldsymbol{v}_4) \times (f_1 f_2 - f_3 f_4) \left[1 + \log f_1\right] . \tag{4.50}$$

Using the properties of W, especially the symmetries for the interchanges $1 \leftrightarrow 2$, $3 \leftrightarrow 4$ and $1 \leftrightarrow 3, 2 \leftrightarrow 4$, we find

$$I = \frac{1}{4} \int d^3 v_1 \int d^3 v_2 \int d^3 v_3 \int d^3 v_4 \, W\left(\boldsymbol{v}_1, \boldsymbol{v}_2; \boldsymbol{v}_3, \boldsymbol{v}_4\right) \times (f_1 f_2 - f_3 f_4) \log \frac{f_1 f_2}{f_3 f_4} . \tag{4.51}$$

Since for $x, y > 0$ generally $(x - y) \log x/y \geq 0$, we have $I \geq 0$, and therefore $\dot{H} \leq 0$ follows.

4.1.4
The Boltzmann Entropy

We would like to emphasize the relation of H to the negative entropy $-S$. We use a simple model to demonstrate the relationship. Let us discretize the velocity space by defining unit boxes $1, \ldots, M$ which can be filled with totally N particles. We denote by $f(k)$ the number of particles in the k-th box. Obviously, $\sum_{k=1}^{M} f(k) = N$. Each distribution of the particles over the boxes represents a microscopic state. The number of ways of putting all particles into the boxes not caring about the internal arrangements in the boxes is proportional to

$$\frac{N!}{\prod_{k=1}^{M} f(k)!} . \tag{4.52}$$

Following Boltzmann by defining the entropy S as the logarithm of that number, we obtain

$$S = \log N! - \sum_{k=1}^{M} \log f(k)! . \tag{4.53}$$

With the help of Stirling's formula for large numbers, $\log N! \approx N \log N - N$, we get

$$S \approx N \log N - N - \sum_{k=1}^{M} f(k) \log f(k) + \sum_{k=1}^{M} f(k) \, . \tag{4.54}$$

The constant $N \log N$ is often dropped. We interpret the boxes as equally spaced volume elements in velocity space. Within the continuum limit, we let $k \to v$ and

$$f(k) \equiv f(v)\Delta v \to f(v) dv \tag{4.55}$$

within a one-dimensional description. $f(v)$ becomes the velocity distribution function. The three-dimensional generalization is trivial. The formula

$$S \approx -\sum_{k=1}^{M} f(k) \log f(k) \tag{4.56}$$

becomes

$$S \approx -\sum_{j} f(v_j)\Delta v \log\left[f(v_j)\Delta v\right] = -\sum_{j} f(v_j)\Delta v \left[\log f(v_j) + \log \Delta v\right] . \tag{4.57}$$

We perform the limit $\Delta v \to dv$ and

$$\Delta v \log \Delta v = \frac{\log \Delta v}{\frac{1}{\Delta v}} \sim -\Delta v \to -dv \tag{4.58}$$

to obtain

$$S \approx \int dv f - \int dv f \log f \, . \tag{4.59}$$

Again, dropping the constant and generalizing to the three-dimensional situation,

$$S \approx -\int d^3 v f \log f \, , \tag{4.60}$$

will follow.

4.1.5 Transition to Equilibrium and Maxwellian

At the minimum with $\dot{H} = 0$,

$$f_1 f_2 = f_3 f_4 \tag{4.61}$$

should be satisfied. In other words,

$$\log f_1 + \log f_2 = \log f_3 + \log f_4 \, . \tag{4.62}$$

This requires $\log f$ to be constructed from constants of motion. Introducing coefficients α, β and $\boldsymbol{u}$, we make the ansatz

$$\log f = \alpha + \beta\left[\boldsymbol{u}\cdot m\boldsymbol{v} - \frac{m}{2}v^2\right], \tag{4.63}$$

leading to the Maxwellian

$$f(\boldsymbol{v}) = n\left(\frac{m}{2\pi k_B T}\right)^{3/2} \exp\left\{-\frac{m}{2k_B T}(\boldsymbol{v}-\boldsymbol{u})^2\right\}. \tag{4.64}$$

Generalization to space-dependent distribution functions $f(\boldsymbol{r}, \boldsymbol{v}; t)$ are possible, but not discussed herein.

In summary, the Boltzmann equation, in its original derivation as presented here, makes use of the "Stoßzahl-Ansatz" which in some respect, is not very systematic. In the following, we shall show how kinetic equations being more applicable to plasmas can be derived more systematically from the Liouville equation.

4.2 BBGKY Hierarchy

The Liouville equation

$$\frac{\partial \rho}{\partial t} = \{H, \rho\} \tag{4.65}$$

is the starting point of any classical statistical description in plasma physics. It is formulated in canonically conjugate space and momenta coordinates q and p, respectively. In thermodynamic equilibrium, the left-hand side of Eq. (4.65) is zero. Kinetic theory determines the explicit time-dependence (and perhaps the approach to equilibrium). Therefore, in the following, we keep the left-hand side of Eq. (4.65).

A kinetic equation is a closed equation for the one-particle distribution function. The latter follows by "reduction" from the Liouville density ρ. In the following, we designate the species (electrons or ions) by a superscript (in Greek labels). For N_α particles of species α, we define

$$f^\alpha(\boldsymbol{q}, \boldsymbol{p}; t) \equiv f_1^\alpha(\boldsymbol{q}_1, \boldsymbol{p}_1; t) := N_\alpha \int d^3q_2 \ldots d^3q_N d^3p_2 \ldots d^3p_N \rho\,, \tag{4.66}$$

where for indistinguishable particles, the index one for "particle number" one is written just for convenience. In a similar way, we can define the two-particle distribution function

$$f_2^{\alpha\alpha}(\boldsymbol{q}_1, \boldsymbol{p}_1, \boldsymbol{q}_2, \boldsymbol{p}_2; t) = N_\alpha(N_\alpha - 1)\int d^3q_3 \ldots d^3q_N d^3p_3 \ldots d^3p_N \rho\,, \tag{4.67}$$

$$f_2^{\alpha\beta}(\boldsymbol{q}_1, \boldsymbol{p}_1, \boldsymbol{q}_2, \boldsymbol{p}_2; t) = N_\alpha N_\beta \int d^3q_3 \ldots d^3q_N d^3p_3 \ldots d^3p_N \rho\,, \tag{4.68}$$

where slight differences appear, depending on the species of the two particles. For obvious reasons, we shall use $N_\alpha - 1 \approx N_\alpha$. What is quite important, is the normalization

$$\int f^\alpha(\boldsymbol{q}, \boldsymbol{p}; t) d^3 p = n_\alpha(\boldsymbol{q}, t) \,.$$

Without an (external) magnetic field, we have a simple relation between the canonical momentum $\boldsymbol{p}_j$ and the velocity $\boldsymbol{v}_j$ in the electrostatic limit. Kinetic theory is often formulated in terms of space and velocity coordinates; in the following, we demonstrate the procedure for the magnetic-field-free case $\boldsymbol{p}_j = m\boldsymbol{v}_j$. The transformation

$$f_1^\alpha\left(\boldsymbol{q}_1, \boldsymbol{v}_1; t\right) = m_\alpha^3 f_1^\alpha\left(\boldsymbol{q}_1, \boldsymbol{p}_1 = m\boldsymbol{v}_1; t\right) \tag{4.69}$$

obviously follows from the "normalization"

$$\int f^\alpha\left(\boldsymbol{q}, \boldsymbol{v}; t\right) d^3 v = n_\alpha(\boldsymbol{q}, t) \,. \tag{4.70}$$

The reduction of the Liouville equation (4.65) to a kinetic equation leads to a hierarchy problem. Let us rewrite Eq. (4.65) in the form

$$\frac{\partial \rho}{\partial t} = L\rho \tag{4.71}$$

with the Liouville operator

$$\begin{aligned} L &= \sum_{j=1}^{N} \left[\{H, \boldsymbol{q}_j\} \cdot \frac{\partial}{\partial \boldsymbol{q}_j} + \{H, \boldsymbol{p}_j\} \cdot \frac{\partial}{\partial \boldsymbol{p}_j} \right] \\ &= -\sum_{j=1}^{N} \left[\frac{\partial H}{\partial \boldsymbol{p}_j} \cdot \frac{\partial}{\partial \boldsymbol{q}_j} - \frac{\partial H}{\partial \boldsymbol{q}_j} \cdot \frac{\partial}{\partial \boldsymbol{p}_j} \right] . \end{aligned} \tag{4.72}$$

Here, the Hamiltonian enters. Having in mind $\boldsymbol{p}_j = m_j \boldsymbol{v}_j$, we write the Hamiltonian as

$$H = H(\boldsymbol{q}, \boldsymbol{v}) = \sum_{j=1}^{N} \left[\frac{m_j}{2} v_j^2 + e_j \phi(\boldsymbol{q}_j) \right] + \sum_{i<j} \phi_{ij}(\boldsymbol{q}_i, \boldsymbol{q}_j) \,. \tag{4.73}$$

We have denoted by $\phi(\boldsymbol{q}_j)$ the scalar potential of an external electric field $\boldsymbol{E}_0$ which arises at the position of the j-th particle with charge e_j. In addition, the interaction potential energy appears,

$$\phi_{ij} = \frac{1}{4\pi\varepsilon_0} \frac{e_i e_j}{|\boldsymbol{q}_i - \boldsymbol{q}_j|} \,. \tag{4.74}$$

Inserting the Hamiltonian equation (4.73) into the Liouville operator (4.72), we obtain

$$\begin{aligned} L &= -\sum_{j=1}^{N} \left[\boldsymbol{v}_j \cdot \frac{\partial}{\partial \boldsymbol{q}_j} - \frac{e_j}{m_j} \frac{\partial \phi}{\partial \boldsymbol{q}_j} \cdot \frac{\partial}{\partial \boldsymbol{v}_j} \right] + \sum_{j=1}^{N} \sum_{i=1}^{j-1} \frac{1}{m_j} \frac{\partial \phi_{ij}}{\partial \boldsymbol{q}_j} \cdot \frac{\partial}{\partial \boldsymbol{v}_j} \\ &\equiv L^{(1)} + L^{(2)} \,. \end{aligned} \tag{4.75}$$

We have split the Liouville operator into two parts, a first part which can be understood as a one-particle propagator, and a second which takes care of the interaction.

When reducing the Liouville equation (4.65) into an equation for the one-particle distribution function (4.66), we assume that the probability densities vanish for $|\boldsymbol{q}_j|, |\boldsymbol{p}_j| \to \infty$. Let us consider a particle of species $\alpha = e, i$ and integrate over the coordinates of all the other particles. Without any restriction, we choose $\boldsymbol{q}_1$, $\boldsymbol{v}_1$ and t as the remaining independent variables. The part

$$\left[\frac{\partial \rho}{\partial t}\right]^{(1)} = L^{(1)}\rho \tag{4.76}$$

of Eq. (4.75) immediately leads to

$$\left[\frac{\partial}{\partial t} f^{\alpha}(\boldsymbol{q}_1, \boldsymbol{v}_1; t)\right]^{(1)} = L_1^{\alpha} f^{\alpha}\left(\boldsymbol{q}_1, \boldsymbol{v}_1; t\right), \tag{4.77}$$

with

$$L_j^{\alpha} = -\boldsymbol{v}_j \cdot \frac{\partial}{\partial \boldsymbol{q}_j} - \frac{e_\alpha}{m_\alpha} \boldsymbol{E}_0(\boldsymbol{q}_j) \cdot \frac{\partial}{\partial \boldsymbol{v}_j} . \tag{4.78}$$

At this point, two remarks need to be made. First, we have

$$N_\alpha \int d^3 q_2 \ldots d^3 p_N L^{(1)} \rho = L_1^{\alpha} f^{\alpha} \tag{4.79}$$

since only the part L_1^{α} of $L^{(1)}$ (i.e., $j = 1$) contributes. Otherwise, integration by parts leads to

$$\begin{aligned} \int d^3 q_2 \ldots d^3 p_N L_j^{\alpha} \rho &= -\int \ldots d^3 q_j d^3 p_j \ldots \left[\frac{\partial H}{\partial \boldsymbol{p}_j} \cdot \frac{\partial \rho}{\partial \boldsymbol{q}_j} - \frac{\partial H}{\partial \boldsymbol{q}_j} \cdot \frac{\partial \rho}{\partial \boldsymbol{p}_j}\right] \\ &= 0 \text{, for } j \neq 1 . \end{aligned} \tag{4.80}$$

The other remark concerns the inclusion of an external magnetic field. In addition to the external electric field $\boldsymbol{E}_0$, a quite straightforward generalization leads to

$$L_j^{\alpha} = -\boldsymbol{v}_j \cdot \frac{\partial}{\partial \boldsymbol{q}_j} - \frac{e_\alpha}{m_\alpha} \left[\boldsymbol{v}_j \times \boldsymbol{B}(\boldsymbol{q}_j) + \boldsymbol{E}_0(\boldsymbol{q}_j)\right] \cdot \frac{\partial}{\partial \boldsymbol{v}_j} . \tag{4.81}$$

Next, we calculate the contribution of $L^{(2)}$. One easily recognizes that a reduction is again possible. However, because of the interaction potential, now the two-particle distribution function appears. We have to take care of the types of particles. The factors $1/(N_\alpha N_\beta)$ and $1/N_\alpha^2$, respectively, cancel because of the same numbers of contributions. Finally, we arrive at

$$\left[\frac{\partial}{\partial t} f^{\alpha}(\boldsymbol{q}_1, \boldsymbol{v}_1; t)\right]^{(2)} = \sum_{\beta=e,i} \int d^3 q_2 d^3 v_2 L_{12}^{\alpha\beta} f_2^{\alpha\beta}\left(\boldsymbol{q}_1, \boldsymbol{v}_1, \boldsymbol{q}_2, \boldsymbol{v}_2; t\right) \tag{4.82}$$

with the definition

$$L_{jk}^{\alpha\beta} = \frac{\partial \phi_{jk}}{\partial \boldsymbol{q}_j} \cdot \left[\frac{1}{m_\alpha} \frac{\partial}{\partial \boldsymbol{v}_j} - \frac{1}{m_\beta} \frac{\partial}{\partial \boldsymbol{v}_k} \right] . \tag{4.83}$$

A particle at position $\boldsymbol{q}_j$ with the velocity $\boldsymbol{v}_j$ belongs to species α, and the other particle (k) belongs to species β. Of course, $\alpha = \beta$ is allowed. Within the operator $L_{12}^{\alpha\beta}$, the indices 1 and 2 belong to different particles. Generally, for the operators L_1^α and $L_{12}^{\alpha\beta}$, the lower indices 1 and 2 designate the coordinates which are used for differentiation. Combining expressions (4.77) and (4.83), we obtain the first equation of the so called BBGKY (Bogoliubov, Born, Green, Kirkwood, Yvon) hierarchy,

$$\begin{aligned} \partial_t f^\alpha(\boldsymbol{q}_1, \boldsymbol{v}_1; t) &= L_1^\alpha f^\alpha (\boldsymbol{q}_1, \boldsymbol{v}_1; t) \\ &\quad + \sum_{\beta=e,i} \int d^3 q_2 d^3 v_2 L_{12}^{\alpha\beta} f^{\alpha\beta} (\boldsymbol{q}_1, \boldsymbol{v}_1, \boldsymbol{q}_2, \boldsymbol{v}_2; t) . \end{aligned} \tag{4.84}$$

We have omitted the lower indices at f^α and $f^{\alpha\beta}$ since no confusion is expected. Furthermore, we have abbreviated

$$\frac{\partial}{\partial t} = \partial_t , \quad \frac{\partial}{\partial \boldsymbol{q}} = \partial_q , \quad \frac{\partial}{\partial \boldsymbol{v}} = \partial_v \tag{4.85}$$

to improve readability. The other abbreviations, for example,

$$d^6 1 \equiv d^3 q_1 d^3 v_1 , \quad d^6 2 \equiv d^3 q_2 d^3 v_2 , \quad \text{etc.} \tag{4.86}$$

are shortcuts which should be self-explanatory.

Equation (4.84) is not closed; it is the first equation of a hierarchy of equations. In Eq. (4.84), the two-particle distribution function appears. When we derive an equation for the two-particle distribution function, the three-particle distribution function $f_3^{\alpha\beta\gamma}(\boldsymbol{q}_1, \boldsymbol{v}_1, \boldsymbol{q}_2, \boldsymbol{v}_2, \boldsymbol{q}_3, \boldsymbol{v}_3; t)$ will appear.

In order to gain more insight into a possible truncation procedure for the hierarchy, we define the two-particle correlation function $g^{\alpha\beta}$ via

$$f^{\alpha\beta}(\boldsymbol{q}_1, \boldsymbol{v}_1, \boldsymbol{q}_2, \boldsymbol{v}_2; t) = f^\alpha(\boldsymbol{q}_1, \boldsymbol{v}_1; t) f^\beta(\boldsymbol{q}_2, \boldsymbol{v}_2; t) + g^{\alpha\beta}(\boldsymbol{q}_1, \boldsymbol{v}_1, \boldsymbol{q}_2, \boldsymbol{v}_2; t) . \tag{4.87}$$

A good assumption is that $g^{\alpha\beta}$ only depends on the distance and vanishes for $|\boldsymbol{q}_1 - \boldsymbol{q}_2| \to \infty$. For $N \to \infty$, the normalization leads to

$$\int d^3 q_1 d^3 q_2 d^3 v_1 d^3 v_2 g^{\alpha\beta} = 0 . \tag{4.88}$$

The three-particle correlation function $g^{\alpha\beta\gamma}$ is defined via

$$f^{\alpha\beta\gamma} = f^\alpha f^\beta f^\gamma + f^\alpha g^{\beta\gamma} + f^\beta g^{\alpha\gamma} + f^\gamma g^{\alpha\beta} + g^{\alpha\beta\gamma} . \tag{4.89}$$

For simplicity, we did not write the arguments $\boldsymbol{q}_1, \boldsymbol{v}_1, \boldsymbol{q}_2, \ldots$; they follow straightforwardly form the upper indices α, β and γ, where particle 1 belongs to species α, particle 2 to β, and particle 3 to species γ. Next, following most derivations of kinetic equations, we make the ansatz

$$g^{\alpha\beta\gamma}\left(q_1, v_1, q_2, v_2, q_3, v_3; t\right) \approx 0 , \tag{4.90}$$

that is, we neglect the three-body correlation. That approximation is quite good in "dilute" plasmas. Then, the situations where three particles are very close to each other are rare. The approximation improves according to how larger the number of particles in the Debye sphere is. Remember that the number of particles in the Debye sphere is proportional to $n^{-1/2}$. For obvious reasons, the approximation Eq. (4.90) is called the "weak coupling approximation."

A short rearrangement of terms in the equations derived so far allows, under the assumption Eq. (4.90), one to write the following closed system for f^α and $g^{\alpha\beta}$, namely,

$$\begin{aligned}\partial_t f^\alpha(q_1, v_1; t) = L_1^\alpha f^\alpha(q_1, v_1; t) &+ \sum_{\beta=e,i} \int d^6 2 L_{12}^{\alpha\beta} f^\alpha(q_1, v_1; t) f^\beta(q_2, v_2; t) \\ &+ \sum_{\beta=e,i} \int d^6 2 L_{12}^{\alpha\beta} g^{\alpha\beta}(q_1, v_1, q_2, v_2; t) ,\end{aligned} \tag{4.91}$$

$$\begin{aligned}\partial_t g^{\alpha\beta}(q_1, v_1, q_2, v_2; t) = &\left(L_1^\alpha + L_2^\beta\right) g^{\alpha\beta}(q_1, v_1, q_2, v_2; t) \\ &+ L_{12}^{\alpha\beta} g^{\alpha\beta}(q_1, v_1, q_2, v_2; t) \\ &+ \sum_{\gamma=e,i} \int d^6 3 \Big[L_{13}^{\alpha\gamma} f^\alpha(q_1, v_1; t) g^{\beta\gamma}(q_2, v_2, q_3, v_3; t) \\ &+ L_{23}^{\beta\gamma} f^\beta(q_2, v_2; t) g^{\alpha\gamma}(q_1, v_1, q_3, v_3; t) \\ &+ \left(L_{13}^{\alpha\gamma} + L_{23}^{\beta\gamma}\right) f^\gamma(q_3, v_3; t) g^{\alpha\beta}(q_1, v_1, q_2, v_2; t) \Big] \\ &+ L_{12}^{\alpha\beta} f^\alpha(q_1, v_1; t) f^\beta(q_2, v_2; t) .\end{aligned} \tag{4.92}$$

Note that we have not arrived at a kinetic equation.

Before reducing the system Eqs. (4.91) and (4.92), let us briefly comment again on a systematic procedure in order to include an external magnetic field. As we have already mentioned, we are using the velocities instead of the canonical momenta as variables. That transformation from momenta to velocities is, in general, not a canonical transformation, but can be written as a Lie transformation. With the Lie brackets $[\ldots, \ldots]$, we can write the Liouville operator (4.72) as

$$L = \sum_{j=1}^{N} \left[[H, q_j] \cdot \frac{\partial}{\partial q_j} + [H, v_j] \cdot \frac{\partial}{\partial v_j} \right] . \tag{4.93}$$

First, if we only consider the one-particle contributions, we may use

$$[H, q_j] = -v_j \tag{4.94}$$

and

$$[H, \boldsymbol{v}_j] = -\frac{e_j}{m_j}\left[\boldsymbol{v}_j \times \boldsymbol{B}(\boldsymbol{q}_j) + \boldsymbol{E}_0(\boldsymbol{q}_j)\right]. \tag{4.95}$$

When we evaluate $L_{12}^{\alpha\beta}$, we only need the space-dependent interaction potential, and no changes occur compared to Eq. (4.83). The Lie formalism becomes very important when deriving drift- and gyro-kinetic equations since then we can transform via the Lie transformation technique to guiding center variables. For more details on the Lie transformation see Appendix D.

Let us now come back to Eqs. (4.91) and (4.92). In a first step, we simplify Eq. (4.92). Mainly, three approximations are in use. In the first one, the Vlasov approximation, we set

$$g^{\alpha\beta} = 0\,. \tag{4.96}$$

In the second approximation, due to Landau, we anticipate weak coupling,

$$\left(g^{\alpha\beta} \ll f^{\alpha} f^{\beta}\right) \tag{4.97}$$

and neglect in Eq. (4.92) all contributions from particle 3, that is, we approximate

$$\left[\partial_t - L_1^{\alpha} - L_2^{\beta}\right] g^{\alpha\beta}(\boldsymbol{q}_1, \boldsymbol{v}_1, \boldsymbol{q}_2, \boldsymbol{v}_2; t) \approx L_{12}^{\alpha\beta} f^{\alpha}(\boldsymbol{q}_1, \boldsymbol{v}_1; t) f^{\beta}(\boldsymbol{q}_2, \boldsymbol{v}_2; t)\,. \tag{4.98}$$

Often, that approximation is called the "Zweierstoß-Annahme" in analogy to the Boltzmann ansatz.

Balescu [51], Lenard [52], and Guernsey [53] have improved the Landau approximation by taking into account the shielding contributions of particle 3. This results in

$$\begin{aligned}\left[\partial_t - L_1^{\alpha} - L_2^{\beta}\right] g^{\alpha\beta}(\boldsymbol{q}_1, \boldsymbol{v}_1, \boldsymbol{q}_2, \boldsymbol{v}_2; t) &= L_{12}^{\alpha\beta} f^{\alpha}(\boldsymbol{q}_1, \boldsymbol{v}_1; t) f^{\beta}(\boldsymbol{q}_2, \boldsymbol{v}_2; t) \\ &+ \sum_{\gamma=e,i} \int d^6 3 \left[L_{13}^{\alpha\gamma} f^{\alpha}(\boldsymbol{q}_1, \boldsymbol{v}_1; t) g^{\beta\gamma}(\boldsymbol{q}_2, \boldsymbol{v}_2, \boldsymbol{q}_3, \boldsymbol{v}_3; t) \right. \\ &+ \left. L_{23}^{\beta\gamma} f^{\beta}(\boldsymbol{q}_2, \boldsymbol{v}_2; t) g^{\alpha\gamma}(\boldsymbol{q}_1, \boldsymbol{v}_1, \boldsymbol{q}_3, \boldsymbol{v}_3; t)\right]. \end{aligned} \tag{4.99}$$

In all three approximations, the terms

$$\left[L_{12}^{\alpha\beta} + \sum_{\gamma=e,i} \int d^6 3 \left[L_{13}^{\alpha\gamma} + L_{23}^{\beta\gamma}\right] f^{\gamma}(\boldsymbol{q}_3, \boldsymbol{v}_3; t)\right] g^{\alpha\beta}(\boldsymbol{q}_1, \boldsymbol{v}_1, \boldsymbol{q}_2, \boldsymbol{v}_2; t) \tag{4.100}$$

are being neglected in the dynamics of $g^{\alpha\beta}(\boldsymbol{q}_1, \boldsymbol{v}_1, \boldsymbol{q}_2, \boldsymbol{v}_2; t)$.

The equation for $g^{\alpha\beta}$ is, in any case except for the Vlasov approximation, a differential equation of first order in time. If we solve it exactly, the solution will depend on initial values, and thereby on the whole history in time. Such a behavior is called non-Markovian. In the so called kinetic regime, we assume that $g^{\alpha\beta}$ only depends on time functionally via f, that is,

$$g^{\alpha\beta}(t) \approx g^{\alpha\beta}[f(t)] . \qquad (4.101)$$

It is assumed, and can be shown by simple examples, that the influence of the initial values disappears after some time. Thus, with the functional dependence Eq. (4.101), we will get a kinetic equation of the form

$$\begin{aligned} \partial_t f^\alpha(\mathbf{q}_1, \mathbf{v}_1; t) = {} & L_1^\alpha f^\alpha(\mathbf{q}_1, \mathbf{v}_1; t) \\ & + \sum_{\beta=e,i} \int d^3 q_2 d^3 v_2 L_{12}^{\alpha\beta} f^\alpha(\mathbf{q}_1, \mathbf{v}_1; t) f^\beta(\mathbf{q}_2, \mathbf{v}_2; t) \\ & + K^\alpha\{f^\alpha(t)\} , \end{aligned} \qquad (4.102)$$

where of course the collision term K^α originates from the specific form of $g^{\alpha\beta}$.

The second term on the right-hand side of Eq. (4.102) takes care of the averaged field ("mean field approximation")

$$\begin{aligned} & \sum_{\beta=e,i} \int d^3 q_2 d^3 v_2 L_{12}^{\alpha\beta} f^\alpha f^\beta \\ & = \sum_{\beta=e,i} \int d^3 q_2 d^3 v_2 \frac{\partial \phi_{12}^{\alpha\beta}}{\partial \mathbf{q}_1} \cdot \left[\frac{1}{m_\alpha} \frac{\partial}{\partial \mathbf{v}_1} - \frac{1}{m_\beta} \frac{\partial}{\partial \mathbf{v}_2} \right] f^\alpha(\mathbf{q}_1, \mathbf{v}_1; t) f^\beta(\mathbf{q}_2, \mathbf{v}_2; t) \\ & = \left[\frac{\partial}{\partial \mathbf{q}_1} \sum_{\beta=e,i} \int d^3 q_2 d^3 v_2 \phi_{12}^{\alpha\beta} f^\beta(\mathbf{q}_2, \mathbf{v}_2; t) \right] \cdot \frac{1}{m_\alpha} \frac{\partial f^\alpha(\mathbf{q}_1, \mathbf{v}_1; t)}{\partial \mathbf{v}_1} \\ & \equiv -\frac{e_\alpha}{m_\alpha} \langle \mathbf{E}(\mathbf{q}_1; t) \rangle \cdot \frac{\partial f^\alpha}{\partial \mathbf{v}_1} . \end{aligned} \qquad (4.103)$$

The mean electric field $\langle \mathbf{E} \rangle$, caused by all the (other) particles at position $\mathbf{q}_1$, is here introduced via

$$\langle \mathbf{E}(\mathbf{q}_1; t) \rangle = -\frac{\partial}{\partial \mathbf{q}_1} \langle \phi(\mathbf{q}_1; t) \rangle , \qquad (4.104)$$

with

$$\langle \phi(\mathbf{q}_1; t) \rangle = \frac{1}{4\pi\varepsilon_0} \sum_{\beta=e,i} e_\beta \int d^3 q_2 d^3 v_2 \frac{1}{|\mathbf{q}_1 - \mathbf{q}_2|} f^\beta(\mathbf{q}_2, \mathbf{v}_2; t) . \qquad (4.105)$$

Thus, we can write the kinetic equation (4.102) for $f^\alpha \equiv f^\alpha(\boldsymbol{q}_1, \boldsymbol{v}_1; t)$ in the form

$$\frac{\partial f^\alpha}{\partial t} + \boldsymbol{v}_1 \cdot \frac{\partial f^\alpha}{\partial \boldsymbol{q}_1} + \frac{e_\alpha}{m_\alpha} \left[\boldsymbol{v}_\alpha \times \boldsymbol{B}(\boldsymbol{q}_1) + \boldsymbol{E}_0(\boldsymbol{q}_1) \right] \cdot \frac{\partial f^\alpha}{\partial \boldsymbol{v}_1} + \frac{e_\alpha}{m_\alpha} \langle \boldsymbol{E}(\boldsymbol{q}_1; t) \rangle \cdot \frac{\partial f^\alpha}{\partial \boldsymbol{v}_1}$$
$$= K^\alpha \{f^\alpha(t)\} \,. \tag{4.106}$$

The total electric field consists of the external field plus the average (mean) field,

$$\boldsymbol{E}(\boldsymbol{q}_1) = \boldsymbol{E}_0(\boldsymbol{q}_1) + \langle \boldsymbol{E}(\boldsymbol{q}_1; t) \rangle \,. \tag{4.107}$$

The collision term takes care of the close collisions.

4.3
Vlasov Equation and Landau Damping

In this section, we consider consequences following from the Vlasov equation. Thus, we neglect collisions. Furthermore, we linearize the Vlasov equation to find the oscillatory behavior of a collisionless and unmagnetized plasma. Neglecting collisions means that we use a kinetic equation without a binary collision term K^α, that is, the Vlasov equation has $K^\alpha = 0$. The Vlasov equation in an external electric field $\boldsymbol{E}_0$ and an external magnetic field $\boldsymbol{B}_0$ reads

$$\frac{\partial f^\alpha(\boldsymbol{q}_1, \boldsymbol{v}_1; t)}{\partial t} + \boldsymbol{v}_1 \cdot \frac{\partial f^\alpha(\boldsymbol{q}_1, \boldsymbol{v}_1; t)}{\partial \boldsymbol{q}_1} + \frac{e_\alpha}{m_\alpha} \left[\boldsymbol{v}_1 \times \boldsymbol{B}_0 + \boldsymbol{E}_0 + \boldsymbol{E}(\boldsymbol{q}_1; t) \right]$$
$$\cdot \frac{\partial f^\alpha(\boldsymbol{q}_1, \boldsymbol{v}_1; t)}{\partial \boldsymbol{v}_1} = 0 \,, \tag{4.108}$$

where in the electrostatic approximation ($\nabla \times \boldsymbol{E} = 0$), the self-consistent electric field $\boldsymbol{E}$ follows from the Poisson equation

$$\nabla \cdot \boldsymbol{E} = \frac{1}{\varepsilon_0} \sum_{\beta=e,i} e_\beta \int d^3 v_1 \, f^\beta(\boldsymbol{q}_1, \boldsymbol{v}_1; t) \,. \tag{4.109}$$

Anticipating the derivation from the BBGKY-hierarchy, we have denoted the space coordinate by $\boldsymbol{q}$. Later, we shall return to the more common nomenclature $\boldsymbol{r} \equiv \boldsymbol{q}$. Generally, and especially for relativistic plasmas, the electrostatic approximation will not be appropriate, and self-consistent electric as well as magnetic fields have to be taken into account. In addition to Eq. (4.109), we then have to use the full set of Maxwell equations, that is, to include

$$\nabla \cdot \boldsymbol{B} = 0, \quad \nabla \times \boldsymbol{E} = -\frac{\partial \boldsymbol{B}}{\partial t}, \quad \nabla \times \boldsymbol{B} = \mu_0 \boldsymbol{j} + \varepsilon_0 \mu_0 \frac{\partial \boldsymbol{E}}{\partial t} \tag{4.110}$$

with

$$j = \sum_{\beta=e,i} e_\beta \int d^3 v_1 \, f^\beta(\mathbf{q}_1, \mathbf{v}_1; t) \mathbf{v}_1 \,. \tag{4.111}$$

Because of the self-consistent (mean) fields $\mathbf{E}$ and $\mathbf{B}$, the Vlasov equation is a nonlinear integro-differential equation.

Before solving the linearized Vlasov equation, let us make a short remark about the applicability of the Vlasov equation. The neglect of binary collisions, that is, $K^\alpha \approx 0$, can be justified when the number of particles in the Debye sphere tends to infinity. The latter statement can be understood from the following argument based on the so called Klimontovich equations [4]. The distribution function in Γ-space can be written as

$$F = \sum_{j=1}^{N} F_j \equiv \sum_{j=1}^{N} \delta(\mathbf{q} - \mathbf{q}_j(t)) \delta(\mathbf{v} - \mathbf{v}_j(t)) \,. \tag{4.112}$$

Taking the time-derivative of F_j, we have to insert the exact dynamics of each particle j,

$$\dot{\mathbf{q}}_j(t) = \mathbf{v}_j(t) \,, \tag{4.113}$$

$$m_j \dot{\mathbf{v}}_j(t) = e_j \mathbf{E}\left[\mathbf{q}_j(t); t\right] + e_j \mathbf{v}_j \times \mathbf{B}\left[\mathbf{q}_j(t); t\right], \tag{4.114}$$

where $\mathbf{E}$ and $\mathbf{B}$ are the exact (total) electric and magnetic fields at the position of particle j. The properties of the Dirac delta function allow one to write

$$\partial_t F_j(\mathbf{q}, \mathbf{v}; t) + \mathbf{v} \cdot \partial_\mathbf{q} F_j(\mathbf{q}, \mathbf{v}; t) + \frac{e_j}{m_j}(\mathbf{E} + \mathbf{v} \times \mathbf{B}) \cdot \partial_\mathbf{v} F_j(\mathbf{q}, \mathbf{v}; t) = 0 \,. \tag{4.115}$$

Although Eq. (4.115) looks similar to the Vlasov equation (4.108), it is totally different. F_j contains the exact dynamics with all fluctuations. Only after averaging can we relate it to f. Thus, we identify

$$F_j(\mathbf{q}, \mathbf{v}; t) = \langle F_j \rangle + \delta F_j \equiv f^\alpha + \delta f^\alpha \,. \tag{4.116}$$

From Eq. (4.115), after averaging we get (for simplicity, no external magnetic field is assumed in the following calculation)

$$\frac{\partial f^\alpha(\mathbf{q}, \mathbf{v}; t)}{\partial t} + \mathbf{v} \cdot \frac{\partial f^\alpha(\mathbf{q}, \mathbf{v}; t)}{\partial \mathbf{q}} + \frac{e_\alpha}{m_\alpha}\left[\mathbf{v} \times \langle \mathbf{B}(\mathbf{q}; t)\rangle + \mathbf{E}_0 + \langle \mathbf{E}(\mathbf{q}; t)\rangle\right]$$
$$\cdot \frac{\partial f^\alpha(\mathbf{q}, \mathbf{v}; t)}{\partial \mathbf{v}}$$
$$= -\frac{e_\alpha}{m_\alpha}\left\langle (\delta \mathbf{E} + \mathbf{v} \times \delta \mathbf{B}) \cdot \frac{\partial \delta f^\alpha}{\partial \mathbf{v}}\right\rangle. \tag{4.117}$$

Comparing with the Vlasov equation (4.108), we recognize that the right-hand side is identical to K^α. The latter is obtained now as the average of the product of fluctuating quantities due to the single particle behavior,

$$K^\alpha = -\frac{e_\alpha}{m_\alpha}\left\langle(\delta \boldsymbol{E} + \boldsymbol{v} \times \delta \boldsymbol{B}) \cdot \frac{\partial \delta f^\alpha}{\partial \boldsymbol{v}}\right\rangle . \tag{4.118}$$

Now, we use the following "Gedankenexperiment." Let us "break" the N_α particles into "subparticles." We postulate for the number of subparticles $N_\alpha \to \infty$, $m_\alpha \to 0$, and $e_\alpha \to 0$, such that $N_\alpha e_\alpha = \text{const}$, $e_\alpha/m_\alpha = \text{const}$, and $v_\alpha = \text{const}$. By that procedure, the plasma frequencies ω_{pe} and ω_{pi} as well as the Debye lengths λ_{De} and λ_{Di} remain unchanged, provided $T_\alpha \to 0$. We have $N_D \sim n\lambda_D^3 \sim N_\alpha \to \infty$. According to the well-known predictions of statistical physics, the strength of the fluctuations is

$$\frac{\delta f^\alpha}{f^\alpha} \sim \frac{1}{\sqrt{N_\alpha}} \tag{4.119}$$

and therefore $\delta f^\alpha \sim N_\alpha^{1/2} \sim N_D^{1/2}$. On the other hand, Maxwell's equations lead to $\delta \boldsymbol{E}, \delta \boldsymbol{B} \sim e_\alpha \delta f^\alpha \sim N_\alpha^{-1} N_\alpha^{1/2} \sim N_\alpha^{-1/2} \sim N_D^{-1/2}$. As a consequence, the ordering of K^α does not change in this "Gedankenexperiment." On the other hand, the left-hand side of Eq. (4.117) is proportional to N_α. Since $N_\alpha \sim N_D$, the influence of the right-hand side of Eq. (4.117) vanishes for $N_D \to \infty$. In the latter limit, the Vlasov equation becomes exact.

Now, we come back to the Vlasov equation (4.108) in electrostatic approximation. For simplicity, we assume at present $\boldsymbol{E}_0 = \boldsymbol{B}_0 = 0$. Furthermore, we consider in the following only high-frequency phenomena which occur at the electron plasma frequency. Then, the ions form a (more or less) immobile background at high frequencies ($\omega \simeq \omega_{pe} \gg \omega_{pi}$). Later, we shall generalize to lower frequencies when the ions actively take part. When only one component, here the electrons, is important, we omit the index $\alpha = e$ ($e_e = -e$) and linearize with respect to an homogeneous equilibrium distribution, introducing the notation

$$f(\boldsymbol{q}, \boldsymbol{v}; t) = f_0(\boldsymbol{v}) + f_1(\boldsymbol{q}, \boldsymbol{v}; t) . \tag{4.120}$$

The linearized system of equations reads

$$\partial_t f_1 + \boldsymbol{v} \cdot \nabla f_1 - \frac{e}{m_e} \boldsymbol{E}_1 \cdot \partial_{\boldsymbol{v}} f_0 = 0 , \quad \nabla \cdot \boldsymbol{E}_1 = -\frac{1}{\varepsilon_0} e \int d^3v\, f_1 . \tag{4.121}$$

Note that, to zeroth order, the ion and electron densities are equal,

$$n_i = \int f^i(\boldsymbol{q}, \boldsymbol{v}; t) d^3v = \int f_0(\boldsymbol{v}) d^3v = n_e . \tag{4.122}$$

In the absence of external fields, we do not have any preferred direction. Then, we may choose the coordinate system such that $\boldsymbol{k} = k\hat{x}$ with $k > 0$. Within the corresponding one-dimensional consideration ($q_x = x$), we assume that all spatial variations are only in one direction x, and we start with a naive normal mode ansatz

$$f_1 \sim e^{ikx - i\omega t} , \tag{4.123}$$

to obtain

$$f_1 = \frac{ieE_{1x}}{m_e} \frac{\frac{\partial f_0}{\partial v_x}}{\omega - kv_x} , \qquad ik\varepsilon_0 E_{1x} = -e \int d^3v \, f_1 . \tag{4.124}$$

We combine both equations to one "dispersion relation"

$$1 = -\frac{\omega_{pe}^2}{k} \int_{-\infty}^{+\infty} du \frac{\frac{dg(u)}{du}}{\omega - ku} , \tag{4.125}$$

where

$$g(u) = \frac{1}{n_{e0}} \int dv_y dv_z \, f_0(u = v_x, v_y, v_z) . \tag{4.126}$$

The "parallel" velocity component is $u = \boldsymbol{k} \cdot \boldsymbol{v}/|\boldsymbol{k}| \mathrel{\hat{=}} v_x$. When interpreting Eq. (4.125) as a relation between frequency ω and wavenumber k, we face a problem. The integration contour produces divergences when $\omega - ku = 0$ at $dg(u)/du \neq 0$. The ambiguity reflects the fact that we did not properly handle the initial value problem in time. According to Landau, we may perform a Fourier analysis in space but have to use the Laplace transform in time. Some details about Fourier and Laplace transforms can be found in the Appendix B.

Without any other preferred direction (no external fields), we choose a coordinate system whose x direction points into the wave propagation direction $\hat{x}$ such that $\boldsymbol{k} = k\hat{x}$ with $k > 0$. In other words, we introduce a Fourier mode

$$f_1 \sim f_1(\boldsymbol{v}; t) e^{ikx} , \tag{4.127}$$

into the linearized Vlasov system. Then, we obtain

$$\partial_t f_1 + ikv_x f_1 - \frac{e}{m_e} E_{1x} \frac{\partial f_0}{\partial v_x} = 0 , \tag{4.128}$$

$$ikE_{1x} = -\frac{1}{\varepsilon_0} e \int f_1 d^3v . \tag{4.129}$$

Because of $\nabla \times \boldsymbol{E} \approx 0$, we have $E_{1y} = E_{1z} = 0$. Next, we perform the Laplace transform in time (for details see the Appendix B) for the distribution function

$$F(p, \boldsymbol{v}) = \int_0^\infty f_1(\boldsymbol{v}; t) e^{-pt} dt , \qquad \mathrm{Re}\, p > \gamma , \tag{4.130}$$

with the inverse

$$f_1(\boldsymbol{v}; t) = \frac{1}{2\pi i} \int_{\lambda - i\infty}^{\lambda + i\infty} F(p, \boldsymbol{v}) e^{pt} dp . \tag{4.131}$$

Similarly, for the electric field

$$E_{1x}(p) = \int_0^\infty E_{1x}(t)e^{-pt}dt\,, \quad \text{Re}\, p > \gamma\,, \tag{4.132}$$

$$E_{1x}(t) = \frac{1}{2\pi i}\int_{\sigma-i\infty}^{\sigma+i\infty} E_{1x}(p)e^{pt}dp\,. \tag{4.133}$$

The back-transformations in the complex p-plane are performed along lines parallel to the imaginary axis, being situated right to all singularities of $F(p, \boldsymbol{v})$ and $E_{1x}(p)$, respectively. That situation is depicted in the Appendix B. Multiplying both sides of Eq. (4.128) by e^{-pt} and integrating over t from 0 to ∞, we obtain

$$(p + ikv_x)F(p, \boldsymbol{v}) - \frac{e}{m_e}E_{1x}(p)\frac{\partial f_0}{\partial v_x} = f_1(\boldsymbol{v}; 0) \tag{4.134}$$

and similarly for Eq. (4.129)

$$ikE_{1x}(p) = -4\pi e\int F(p, \boldsymbol{v})d^3v\,. \tag{4.135}$$

Next, we introduce for the initial distribution of f_1 (note that the function g for the zeroth order distribution function remains as defined above)

$$G(u) = \int dv_y dv_z\, f_1\left(u = v_x, v_y, v_z; 0\right) \tag{4.136}$$

to combine Eqs. (4.134) and (4.135) in the form

$$E_{1x}(p) = -\frac{4\pi e}{ik}\frac{\int_{-\infty}^{+\infty}\frac{G(u)}{p+iku}du}{1 - \frac{\omega_{pe}^2}{k}\int\frac{dg}{du}\frac{du}{ku-ip}}\,, \quad \text{Re}\, p > 0\,. \tag{4.137}$$

Back-transformation leads to $E_{1x}(t)$,

$$E_{1x}(t) = \frac{1}{2\pi i}\int_{\sigma-i\infty}^{\sigma+i\infty} E_{1x}(p)e^{pt}dp\,. \tag{4.138}$$

As discussed already, the contour parallel to the imaginary axis has to be such that σ is bigger than any Re p_ν of the singularities p_ν of $E_{1x}(p)$. Singularities of $E_{1x}(p)$ occur at the zeros of the denominator on the right-hand side of Eq. (4.137).

In principle, the problem is solved. The remaining section deals with the practical evaluation of Eq. (4.138). For a stable situation with Re $p_\nu < 0$, any path with $\sigma > 0$ can be chosen. For unstable situations which occur for Re $p_\nu > 0$, the integration path parallel to the imaginary axis requires $\sigma > max_\nu$ Re p_ν. For the practical evaluation of Eq. (4.138), we will first generalize $E_{1x}(p)$ to the whole p-plane by analytic continuation. $E_{1x}(p)$ has been obtained so far for Re $p > 0$. In the latter area, we introduce the abbreviation (remember the choice $k > 0$)

$$H(p) := \frac{\omega_{pe}^2}{k^2} \int_{-\infty}^{+\infty} \frac{dg}{du} \frac{du}{u - \frac{ip}{k}} \quad \text{for } Rep > 0 \,. \tag{4.139}$$

Note that $g(u)$ is given along the real axis. Integration occurs along the (real) u-axis. Therefore, no singularity occurs for Re $p > 0$. We now write Eq. (4.137) as

$$E_{1x}(p) = -\frac{4\pi e}{ik} \frac{\int_{-\infty}^{+\infty} \frac{G(u)}{p+iku} du}{1 - H(p)} \,, \quad \text{Re } p > 0 \,. \tag{4.140}$$

For the practical evaluation of Eq. (4.138), we would like to have $E_{1x}(p)$ and thereby the right-hand side of Eq. (4.140) also in the left half-plane Re $p \leq 0$. We analytically continue numerator and denominator for Re $p \leq 0$ by considering the integrands in the complex u-plane. We will treat the numerator and denominator on the right-hand side of Eq. (4.140) in similar ways, but exemplify the details only on $H(p)$. For $H(p)$, the path of integration will be deformed when we analytically continue from Re $p > 0$ to Re $p \leq 0$. The singularity at $u = ip/k$ then moves from the upper u-half-plane to the lower u-half-plane. $H(p)$ remains continuous if we deform the u-integration contour appropriately. The pole at $u = ip/k$ should not cross the u-integration contour, as shown in Figure 4.1. The u-integration path should always pass under the pole $u = ip/k$ for analytic continuation. The corresponding evaluation of $H(p)$ in the different p-areas, especially during the crossing of the real u-axis, is easily performed by making use of the Plemelj formula

$$\frac{1}{x \pm i0} = \mathcal{P}\frac{1}{x} \mp i\pi\delta(x) \tag{4.141}$$

and the corresponding integration formulas,

$$\int_{-\infty}^{+\infty} \frac{f(x)}{x \pm i0} dx = \lim_{\varepsilon \to 0} \left[\int_{|x| \geq \varepsilon} \frac{f(x)}{x} dx + \int_{C_\pm} \frac{f(z)}{z} dz \right] , \tag{4.142}$$

where C_+ (bzw. C_-) are small half-circles with radii ε in the upper (lower) z-half-plane, respectively. The final result is ($k > 0$)

$$H(p) = \frac{\omega_{pe}^2}{k^2} \int_{-\infty}^{+\infty} \frac{\frac{dg}{du}}{u - \frac{ip}{k}} du \quad \text{for Re } p > 0 \,, \tag{4.143}$$

$$H(p) = \frac{\omega_{pe}^2}{k^2} \left[\mathcal{P} \int_{-\infty}^{+\infty} \frac{\frac{dg}{du}}{u - \frac{ip}{k}} du + i\pi \frac{dg}{du}\bigg|_{\frac{ip}{k}} \right] \quad \text{for Re } p = 0 \,, \tag{4.144}$$

$$H(p) = \frac{\omega_{pe}^2}{k^2} \left[\int_{-\infty}^{+\infty} \frac{\frac{dg}{du}}{u - \frac{ip}{k}} du + 2\pi i \frac{dg}{du}\bigg|_{\frac{ip}{k}} \right] \quad \text{for Re } p < 0 \,. \tag{4.145}$$

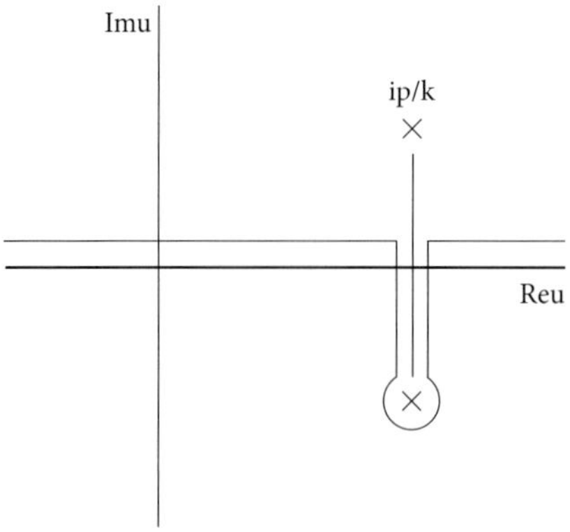

Figure 4.1 Deformation of the path in the complex u-plane when determining $H(p)$ for Re $p \leq 0$ according to Landau.

We assume that dg/du does not introduce singularities. The function g is a distribution with a finite integral value, and thus it should not contribute to a pole.

Now, we return to the back-transformation Eq. (4.138). Unstable distributions are of course allowed when choosing $\sigma > 0$ properly. An example is shown in Figure 4.2. Starting from the line going from $\sigma - i\infty$ to $\sigma + i\infty$, we shift by $\sigma + s$ into the direction of negative Re p, however only in the region $p_- \leq \text{Im}\, p \leq p_+$ for $|p_-|, |p_+| \to \infty$. The contribution from p_j with

$$H(p_j) = 1\,, \tag{4.146}$$

leads to

$$E_{1x}(t) = \sum_j \exp(p_j t)\text{Res}(p_j)\,. \tag{4.147}$$

In this context, one should remember that we have anticipated the standard sign conventions for Laplace transforms [54], meaning that $p \mathrel{\hat{=}} i\omega$ corresponds to a "mode" $\sim \exp(i\omega t)$. Note the difference in sign when compared to the previous normal mode ansatz $\sim \exp(ikx - i\omega t)$. For $t \to \infty$ the main contribution originates from the pole with the largest real part. Other contributions with the largest real part are negligible: The small sections parallel to the imaginary axis vanish since they are proportional to $\exp(-st)$. From the sections parallel to the real axis, effectively only the two most upward and most downward, respectively, do not cancel. However, for the latter in the limit $|p| \to \infty$, we have $|E_{1x}(p)| \sim 1/|p| \to 0$. Finally, along both the path sections being parallel to the imaginary axis and tending to $\pm\infty$, the integrands strongly oscillate and the integral vanishes for $|p_+|, |p_-| \to \infty$. Thus, in the limit $t \to \infty$, we obtain

$$E_{1x}(t) \approx \text{Res}(p)\exp(pt)\,, \tag{4.148}$$

where p is the pole with the largest real part.

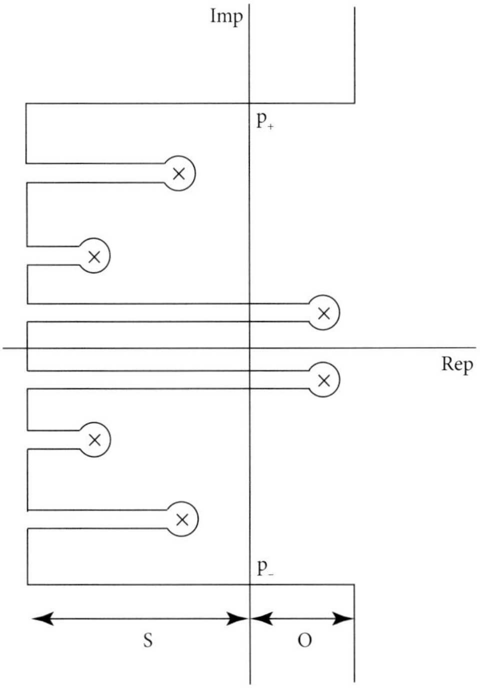

Figure 4.2 Deformation of the path Re $p = \sigma$ for evaluating Landau damping.

Summarizing the discussion, we may write the dispersion relation in the form

$$1 - H(p) \equiv \varepsilon(k, p) = 0\,, \tag{4.149}$$

which determines the solutions $p = p(k)$. In the notation of normal mode description $\sim \exp(ikx - i\omega t)$, the identification $\omega := ip$ is often used. (Note the sign change compared to the discussion of Fourier and Laplace transformations in the Appendix B.) Then, with $p \mathrel{\hat{=}} -i\omega$ and $\omega = \mathrm{Re}\,\omega + i\,\mathrm{Im}\,\omega \equiv \omega_r + i\gamma$ we formulate the dispersion relation for an electron plasma as

$$\varepsilon(k, \omega = ip) := 1 - \frac{\omega_{pe}^2}{k^2} \int \frac{dg}{du} \frac{du}{u - \frac{\omega}{k}} = 0\,, \quad \gamma > 0\,, \tag{4.150}$$

remembering the Landau contour for integration as discussed above. For unstable situations (Re $p \mathrel{\hat{=}} \gamma > 0$), we can use Eq. (4.150) directly. Then, there is a simple procedure to calculate the Landau damping rate for small damping,

$$\frac{|\gamma|}{\omega_r} \ll 1\,, \tag{4.151}$$

where $\gamma = p_r = \mathrm{Im}\,\omega$ and $\omega_r = \mathrm{Re}\,\omega$. Taylor expanding Eq. (4.150) around $\omega = \omega_r$ and having in mind $\mathrm{Im}\,\varepsilon \sim \gamma$ for small γ, we obtain to first order in γ

$$\mathrm{Re}\,\varepsilon(k, \omega_r) + i\,\mathrm{Im}\,\varepsilon(k, \omega_r) + i\gamma \left.\frac{\partial\,\mathrm{Re}\,\varepsilon(k, \omega)}{\partial \omega}\right|_{\omega_r} = 0\,. \tag{4.152}$$

This leads to

$$\mathrm{Re}\,\varepsilon(k,\omega_r) \approx 0 \tag{4.153}$$

and

$$\gamma = -\frac{\mathrm{Im}\,\varepsilon(k,\omega_r)}{\left.\frac{\partial\,\mathrm{Re}\,\varepsilon(k,\omega)}{\partial\omega}\right|_{\omega_r}} . \tag{4.154}$$

Let us start with the discussion of $\varepsilon(k,\omega_r)$, which according to the Landau description we should write in the form

$$\varepsilon(k,\omega_r) = 1 - \frac{\omega_{pe}^2}{k^2}\mathcal{P}\int_{-\infty}^{+\infty}\frac{\frac{dg}{du}}{u-\frac{\omega_r}{k}}du - i\pi\frac{\omega_{pe}^2}{k^2}\left.\frac{dg}{du}\right|_{u=\frac{\omega_r}{k}} . \tag{4.155}$$

From here, immediately

$$\mathrm{Re}\,\varepsilon(k,\omega_r) = 1 - \frac{\omega_{pe}^2}{k^2}\mathcal{P}\int_{-\infty}^{+\infty}\frac{\frac{dg}{du}}{u-\frac{\omega_r}{k}}du \tag{4.156}$$

and

$$\mathrm{Im}\,\varepsilon(k,\omega_r) = -\pi\frac{\omega_{pe}^2}{k^2}\left.\frac{dg}{du}\right|_{u=\frac{\omega_r}{k}} \tag{4.157}$$

follow. Next, we calculate

$$\left.\frac{\partial\,\mathrm{Re}\,\varepsilon(k,\omega)}{\partial\omega}\right|_{\omega_r} = \frac{\partial\,\mathrm{Re}\,\varepsilon(k,\omega_r)}{\partial\omega_r} = -\frac{\omega_{pe}^2}{k^2}\frac{\partial}{\partial\omega_r}\mathcal{P}\int_{-\infty}^{+\infty}\frac{\frac{dg}{du}}{u-\frac{\omega_r}{k}}du . \tag{4.158}$$

Before using a Maxwellian distribution, the following steps can be performed by just applying general properties of the zeroth order distribution function, provided we evaluate for large phase velocities ω_r/k. At large phase velocities $g(u=\omega_r/k)$ is small, such that

$$\begin{aligned}
\mathrm{Re}\,\varepsilon(k,\omega_r) &\approx 1 - \frac{\omega_{pe}^2}{k^2}\int_{-\infty}^{+\infty}\frac{g(u)}{\left(u-\frac{\omega_r}{k}\right)^2}du \\
&\approx 1 - \frac{\omega_{pe}^2}{\omega_r^2}\int_{-\infty}^{+\infty}du\,g(u)\left[1+\frac{2uk}{\omega_r}+\frac{3u^2k^2}{\omega_r^2}+\ldots\right] \\
&\approx 1 - \frac{\omega_{pe}^2}{\omega_r^2} - \frac{3k^2 v_{the}^2\omega_{pe}^2}{\omega_r^4} .
\end{aligned} \tag{4.159}$$

We have used

$$\int duug(u) = 0\,. \tag{4.160}$$

We obtain the approximate dispersion relation

$$\omega_r^2 \approx \omega_{pe}^2 + 3k^2 v_{the}^2 \tag{4.161}$$

which can be used to calculate

$$\left.\frac{\partial\,\mathrm{Re}\,\varepsilon}{\partial\omega}\right|_{\omega_r} \approx \frac{2}{\omega_{pe}} \tag{4.162}$$

such that

$$\gamma \approx \left.\frac{\pi\omega_{pe}^3}{2k^2}\frac{dg(u)}{du}\right|_{u=\frac{\omega_r}{k}} \tag{4.163}$$

follows.

For Maxwellians

$$f_0 = \frac{n_0}{(2\pi)^{3/2} v_{the}^3} \exp\left[-\frac{\left(v_x^2 + v_y^2 + v_z^2\right)}{2v_{the}^2}\right] \tag{4.164}$$

and

$$g(u) = \frac{1}{(2\pi)^{1/2} v_{the}} \exp\left[-\frac{u^2}{2v_{the}^2}\right], \tag{4.165}$$

we obtain

$$\gamma \approx -\omega_{pe}\left(\frac{\pi}{8}\right)^{1/2} (k\lambda_{De})^{-3} \exp\left[-\frac{1}{2}(k\lambda_{De})^{-2} - \frac{3}{2}\right]. \tag{4.166}$$

Figure 4.3 shows $|\gamma|/\omega_{pe}$ depending on $k\lambda_{De}$. A maximum at $k\lambda_{De} \approx 0.5$ occurs. However, one should be careful regarding its interpretation since the present derivation is only applicable for $|\gamma/\omega_{pe}| \ll 1$ and $k\lambda_{De} \ll 1$.

Landau damping is usually interpreted as a net damping effect: Particles moving slightly slower than the phase velocity of the wave will gain energy, while those moving a little bit larger than the phase velocity will transfer energy to the wave. A monotonically decreasing velocity distribution ($g'(u) < 0$) will be needed for a net energy loss (damping) of the wave.

As can be discussed on the basis of Figure 4.4, the situation is more complex. Trajectories are shown in the x,u-phase space. The potential energy is assumed proportional to $\cos(kx)$. The reference frame is moving with the phase-velocity v_φ. Then,

$$E = \frac{1}{2} m_e u^2 + a\cos(kx)\,, \tag{4.167}$$

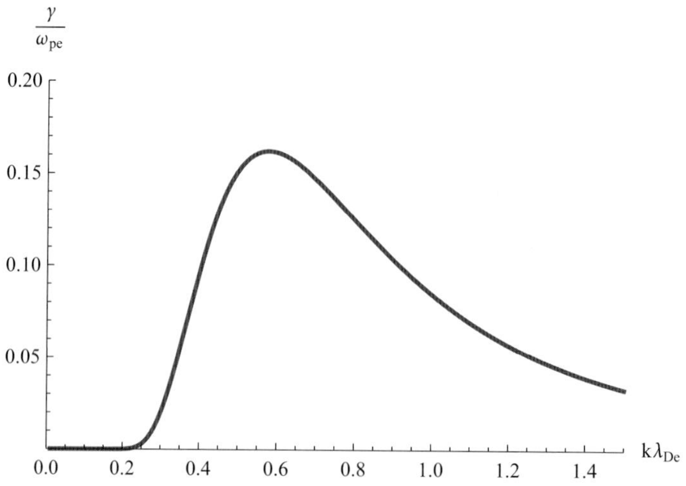

Figure 4.3 Landau damping rate γ (in ω_{pe}) as a function of $k\lambda_{De}$ as determined in Eq. (4.166).

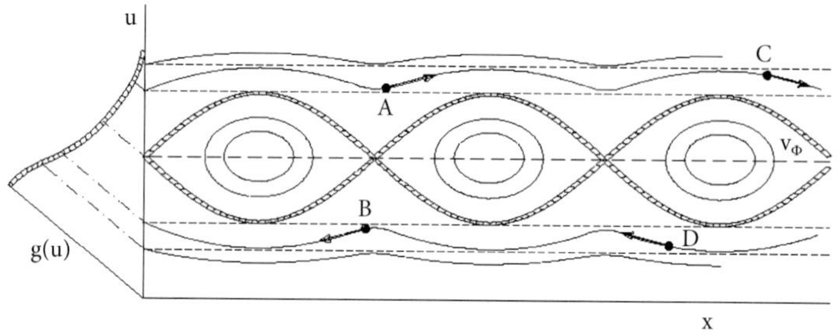

Figure 4.4 x,u-phase space plot for a discussion of the energy balance during Landau damping.

and different curves belong to different energies E. The wave moves with phase velocity

$$v_\phi = \frac{\omega}{k} \tag{4.168}$$

to the right. The symmetry axis in Figure 4.4 represents $u = v_\phi$. The third axis, $g(u)$, indicates the number of particles with

$$g(u_1) > g(v_\phi) > g(u_2) \quad \text{for} \quad u_1 < v_\phi < u_2 \,. \tag{4.169}$$

Now, let us have a look at typical situations characterized by A, B, C, and D. At locations A and D, particles gain energy (at least initially) from the wave, whereas at B and C, particles are decelerated. Because of $g(u_D) > g(u_B)$ and $g(u_A) > g(u_C)$, a net energy loss (damping) of the wave occurs.

For a quantitative (heuristic) estimate of Landau damping, one can use the following idea. We divide the particle system into "beams" with velocity u and densities n_u. The probability of a beam is given by $g(u)du$. When perturbed, the perturbed quantities u_1 and n_1 appear on top of u and n_u, respectively. We start from a linearized force balance

$$m_e\left(\frac{\partial u_1}{\partial t}+u\frac{\partial u_1}{\partial x}\right)=-eE_1\sin(kx-\omega t) \tag{4.170}$$

and a linearized particle continuity equation

$$\frac{\partial n_1}{\partial t}+u\frac{\partial n_1}{\partial x}=-n_u\frac{\partial u_1}{\partial x}\,. \tag{4.171}$$

The solutions are

$$u_1=-\frac{eE_1}{m_e}\,\frac{\cos(kx-\omega t)-\cos(kx-kut)}{\omega-ku} \tag{4.172}$$

and

$$n_1=-n_u\frac{eE_1k}{m_e}\,\frac{\cos(kx-\omega t)-\cos(kx-kut)-(\omega-ku)t\sin(kx-kut)}{(\omega-ku)^2}\,. \tag{4.173}$$

With the latter, the force density (per unit volume) at a beam is

$$F_u\approx-eE_1\sin(kx-\omega t)\,[n_u+n_1]\,. \tag{4.174}$$

The energy density change (per unit time) can be calculated approximately by

$$\frac{dW}{dt}\approx(u+u_1)F_u\,. \tag{4.175}$$

Averaging over space leads to

$$\begin{aligned}\left\langle\frac{dW}{dt}\right\rangle_u &\approx \frac{e^2E_1^2}{2m_e}n_u\\ &\times\left[\frac{\sin(\omega t-kut)}{\omega-ku}+ku\frac{\sin(\omega t-kut)-(\omega-ku)t\cos(\omega t-kut)}{(\omega-ku)^2}\right].\end{aligned} \tag{4.176}$$

Finally, we sum over all beam contributions,

$$\begin{aligned}\sum_u\left\langle\frac{dW}{dt}\right\rangle_u &\mathrel{\hat{=}} n_0\int\frac{g(u)}{n_u}\left\langle\frac{dW}{dt}\right\rangle_u du\\ &\approx\frac{\varepsilon_0}{2}E_1^2\omega_{pe}^2\int_{-\infty}^{+\infty}g(u)\frac{d}{du}\left[u\frac{\sin(\omega t-kut)}{\omega-ku}\right]du\,.\end{aligned} \tag{4.177}$$

The total energy should remain constant, $W+W_w=\text{const}$, where

$$W_w=\frac{\varepsilon_0}{2}E_1^2 \tag{4.178}$$

is the (electrostatic) wave energy density. Thus, we can rewrite Eq. (4.177) as

$$\frac{dW_w}{dt} = -W_w \omega_{pe}^2 \int_{-\infty}^{+\infty} g(u) \frac{d}{du}\left[u \frac{\sin(\omega t - kut)}{\omega - ku}\right] du . \tag{4.179}$$

Integration by parts and use of the asymptotic formula

$$\delta\left(u - \frac{\omega}{k}\right) = \frac{k}{\pi} \lim_{t\to\infty} \frac{\sin(\omega - ku)t}{\omega - ku} \tag{4.180}$$

lead to

$$\frac{dW_w}{dt} = W_w \pi \frac{\omega_{pe}^3}{k^2} g'\left(\frac{\omega}{k}\right) \approx 2\gamma W_w . \tag{4.181}$$

Note $W_w \sim E_1^2$ and $E_1 \sim \exp(\gamma t)$. By this heuristic argumentation, we can find a damping rate

$$\gamma \approx \frac{\pi}{2} \frac{\omega_{pe}^3}{k^2} g'\left(u = \frac{\omega}{k}\right) \tag{4.182}$$

which agrees with the previously derived formula.

It should be emphasized that the literature on the physics of Landau damping is huge and by far not covered by the present presentation. Especially the arguments of "phase mixing," for example, as discussed by van Kampen [55], are beyond the scope of the present overview.

4.4
Z-Function and Dispersive Properties of a Collisionless and Unmagnetized Plasma

Let us now formulate the linear dispersion relation of a Vlasov plasma in a mathematically tractable form. We assume a Maxwell distribution as the zeroth order distribution function. It would be straightforward to derive similar formulas for other distribution functions, for example, a Druyvestein distribution.

Let us anticipate a Maxwellian for $g(u)$ (being normalized to one) as the equilibrium distribution function. Since we expect that it corresponds to stable situations we evaluate $H(p)$ for Re $p < 0$ and start with the dispersion relation of an electron plasma with immobile ions in the form (we again substitute $k_B T_e \to T_e$)

$$\frac{k^2}{\omega_{pe}^2} = -\frac{\left(\frac{m_e}{T_e}\right)^{3/2}}{(2\pi)^{1/2}} \left\{ \int_{-\infty}^{+\infty} \frac{u \exp\left(-\frac{m_e u^2}{2T_e}\right)}{u - i\frac{p}{k}} du + 2\pi i \frac{ip}{k} \exp\left[-\frac{m_e}{2T_e}\left(\frac{ip}{k}\right)^2\right] \right\} . \tag{4.183}$$

It is appropriate to introduce dimensionless variables

$$t := u\left(\frac{m_e}{2T_e}\right)^{1/2}, \quad \zeta := \frac{ip}{k}\left(\frac{m_e}{2T_e}\right)^{1/2} \tag{4.184}$$

to obtain

$$-k^2\lambda_{De}^2 = \frac{1}{\sqrt{\pi}}\int_{-\infty}^{+\infty}\frac{te^{-t^2}}{t-\zeta}dt + 2i\sqrt{\pi}\zeta e^{-\zeta^2}. \tag{4.185}$$

Let us have a closer look at the integrand of the first term on the right-hand side. Since all odd contributions in t do not contribute, we perform the following algebraic transformation

$$\frac{t}{t-\zeta} = \frac{t(t+\zeta)}{t^2-\zeta^2} \to \frac{t^2}{t^2-\zeta^2} = \frac{t^2-\zeta^2+\zeta^2}{t^2-\zeta^2}, \tag{4.186}$$

leading to

$$-k^2\lambda_{De}^2 = 1 + \frac{1}{\sqrt{\pi}}\zeta^2 e^{-\zeta^2}\left\{\int_{-\infty}^{+\infty}\frac{e^{-(t^2-\zeta^2)}}{t^2-\zeta^2}dt + \frac{2\pi i}{\zeta}\right\} \equiv J(\zeta). \tag{4.187}$$

Next, we apply some more transformations, such as,

$$\frac{e^{-(t^2-\zeta^2)}}{t^2-\zeta^2} = -\int_0^1 e^{-(t^2-\zeta^2)s}ds + \frac{1}{t^2-\zeta^2}, \tag{4.188}$$

and use the mathematical results

$$\int_{-\infty}^{+\infty}\frac{1}{t^2-\zeta^2}dt = -\frac{i\pi}{\zeta}, \quad \operatorname{Im}\zeta < 0, \quad \int_{-\infty}^{+\infty}e^{-(t^2-\zeta^2)s}dt = \sqrt{\pi}\frac{1}{\sqrt{s}}e^{\zeta^2 s}. \tag{4.189}$$

By introducing $\nu = i\zeta\sqrt{s}$, we arrive at

$$J(\zeta) = 1 + 2i\zeta e^{-\zeta^2}\left\{\frac{1}{2}\sqrt{\pi} + \int_0^{i\zeta}e^{-\nu^2}d\nu\right\} = 1 + 2i\zeta e^{-\zeta^2}\int_{-\infty}^{i\zeta}e^{-\nu^2}d\nu. \tag{4.190}$$

In the literature [29], the abbreviation

$$Z(\zeta) := 2ie^{-\zeta^2}\int_{-\infty}^{i\zeta}e^{-\nu^2}d\nu \tag{4.191}$$

has become quite common. Using it, the dispersion relation reads

$$k^2\lambda_{De}^2 = -[1 + \zeta Z(\zeta)] \equiv \frac{1}{2} Z'(\zeta) \,. \tag{4.192}$$

For completeness, we also introduce the G-function which one can also find in the literature. We state the equivalence

$$G(\zeta) \equiv Z(-\zeta) = \frac{1}{\sqrt{\pi}} \int_{-\infty}^{+\infty} \frac{e^{-p^2}}{\zeta - p} dp \,. \tag{4.193}$$

On the right-hand side, the integration path is along the real axis which lies above the pole with $\operatorname{Im}\zeta < 0$. For $\operatorname{Im}\zeta > 0$, we have to proceed in the same way as discussed previously for the Landau contour. To comment on the equivalence Eq. (4.193), we note the following formulas:

$$G(z) = \frac{1}{\sqrt{\pi}} \int \frac{e^{-\xi^2}}{z - \xi} d\xi \,, \tag{4.194}$$

$$z G(z) - 1 = \frac{1}{\sqrt{\pi}} \int \frac{\xi e^{-\xi^2}}{z - \xi} d\xi \,, \tag{4.195}$$

$$G'(z) = -\frac{2}{\sqrt{\pi}} \int \frac{\xi e^{-\xi^2}}{z - \xi} d\xi \,. \tag{4.196}$$

Combining them, we obtain the differential equation

$$\frac{1}{2} G' + z G - 1 = 0 \,. \tag{4.197}$$

The solution of the homogeneous part of the differential equation is $G = ce^{-z^2}$; c is an arbitrary constant. By variation of parameters, we find a special solution of the original inhomogeneous differential equation (4.197) in the form

$$G(z) = 2e^{-z^2} \int_0^z e^{p^2} dp \,. \tag{4.198}$$

By combining the general solution of the homogeneous differential equation with the special solution of the inhomogeneous differential equation, we can construct the solution which fulfills $G(z = 0) = i\sqrt{\pi}$, that is,

$$G(z) = \left[2 \int_0^z e^{p^2} dp + i\sqrt{\pi} \right] e^{-z^2} = 2e^{-z^2} \int_{-i\infty}^{z} e^{p^2} dp \,. \tag{4.199}$$

Identifying $p = i\nu$ and $z = -\zeta$, we obtain

$$G(-\zeta) = 2ie^{-\zeta^2} \int_{-\infty}^{i\zeta} e^{-\nu^2} d\nu = Z(\zeta) \,, \tag{4.200}$$

that is, Eq. (4.193). The function $Z(-\zeta)$ is defined in Eq. (4.193) for $\operatorname{Im}\zeta < 0$. It can also be written as

$$Z(\zeta) = \frac{1}{\sqrt{\pi}} \int_{-\infty}^{+\infty} \frac{e^{-p^2}}{-\zeta - p} dp = \frac{1}{\sqrt{\pi}} \int_{-\infty}^{+\infty} \frac{e^{-\nu^2}}{\nu - \zeta} d\nu \quad \text{for } \operatorname{Im}\zeta > 0 \,. \tag{4.201}$$

Note that the definition Eq. (4.191) is more general.

In many cases, the knowledge of the asymptotic behaviors of the G- or Z-function is sufficient. For small arguments ($|\zeta| \ll 1$), we have

$$Z(\zeta) \approx i\sqrt{\pi} e^{-\zeta^2} - 2\zeta \left(1 - \frac{2}{3}\zeta^2 + \frac{4}{15}\zeta^4 - \cdots\right), \tag{4.202}$$

whereas for large arguments ($|\zeta| \gg 1$),

$$Z(\zeta) \approx i\sigma\sqrt{\pi} e^{-\zeta^2} - \frac{1}{\zeta}\left(1 + \frac{1}{2\zeta^2} + \frac{3}{4\zeta^4} + \cdots\right) \tag{4.203}$$

holds. Here, σ takes the values (with $\zeta = x + iy$) [29]

$$\sigma = \begin{cases} 0 & \text{for } y > \frac{1}{|x|} \,, \\ 1 & \text{for } |y| < \frac{1}{|x|} \,, \\ 2 & \text{for } y < -\frac{1}{|x|} \,. \end{cases} \tag{4.204}$$

These formulas become plausible when, first, expanding for $|\zeta| \gg 1$

$$\frac{1}{\nu - \zeta} \approx -\frac{1}{\zeta}\left[1 + \frac{\nu}{\zeta} + \left(\frac{\nu}{\zeta}\right)^2 + \left(\frac{\nu}{\zeta}\right)^3 + \cdots\right], \tag{4.205}$$

one remembers for a Gaussian

$$\frac{1}{\sqrt{\pi}} \int_{-\infty}^{+\infty} \nu^2 e^{-\nu^2} d\nu = \frac{1}{2}\,, \quad \frac{1}{\sqrt{\pi}} \int_{-\infty}^{+\infty} \nu^4 e^{-\nu^2} d\nu = \frac{3}{4}\,, \quad \ldots \tag{4.206}$$

to find the real part of Eq. (4.203). Secondly, for $|\zeta| \ll 1$,

$$\begin{aligned} \frac{1}{\sqrt{\pi}} \int_{-\infty}^{+\infty} \frac{e^{-\nu^2}}{\nu - \zeta} d\nu &= \frac{1}{\sqrt{\pi}} \int_{-\infty}^{+\infty} \frac{e^{-\mu^2 - 2\zeta\mu - \zeta^2}}{\mu} d\mu \\ &= \frac{e^{-\zeta^2}}{\sqrt{\pi}} \int_{-\infty}^{+\infty} \frac{e^{-\mu^2}}{\mu} \left(1 - 2\zeta\mu + 2\zeta^2\mu^2 - \frac{4}{3}\zeta^3\mu^3 + \cdots\right) d\mu \\ &\approx i\pi \frac{e^{-\zeta^2}}{\sqrt{\pi}} - 2\zeta \frac{e^{-\zeta^2}}{\sqrt{\pi}} \int_{-\infty}^{+\infty} e^{-\mu^2} \left(1 + \frac{2}{3}\zeta^2\mu^2 + \cdots\right) d\mu \,. \end{aligned} \tag{4.207}$$

Here, we have used

$$\int_{-\infty}^{+\infty} \frac{e^{-\mu^2}}{\mu} d\mu \to \mathcal{P} \int_{-\infty}^{+\infty} \frac{e^{-\mu^2}}{\mu} d\mu = i\pi \,. \tag{4.208}$$

Expanding $\exp(-\zeta^2)$, we find

$$\frac{1}{\sqrt{\pi}} \int_{-\infty}^{+\infty} \frac{e^{-\nu^2}}{\nu - \zeta} d\nu \approx i\sqrt{\pi} e^{-\zeta^2} - 2\zeta(1 - \zeta^2 + \cdots)\left(1 + \frac{\zeta^2}{3} + \cdots\right) . \tag{4.209}$$

Multiplication of the bracket terms and proper rearrangements lead to Eq. (4.202).

In the "olden days of plasma physics," tables containing data for the *Z*-function were published. Meanwhile, simple computer codes exist to calculate all needed values. So far, for the reason of a simple demonstration, we have only considered the dynamical response of one component (electrons). It is, however, quite straightforward for generalizing a two-component plasma consisting of electrons and (mobile) ions. We introduce the index $j = e, i$. From the linearized Vlasov equation, one quickly concludes that from

$$f_1 = \frac{i q_j E_{1x}}{m_j} \frac{\frac{\partial f_{0j}}{\partial v_x}}{\omega - k v_x} , \tag{4.210}$$

the density response of species j follows in the form

$$\delta n_j = \frac{i q_j E_{ix} n_{j0}}{2k m_j v_{thj}^2} Z'(\zeta_j) \,. \tag{4.211}$$

The prime (′) denotes differentiation of the *Z*-function with respect to its argument

$$\zeta_j = \frac{ip}{k} \left[\frac{m_j}{2T_j}\right]^{1/2} , \tag{4.212}$$

and $v_{thj}^2 = (T_j/m_j)^{1/2}$ is the thermal velocity of species j. Poisson's equation, which was used in deriving the *electrostatic* dispersion relation, now reads for an electron–ion plasma

$$\nabla \cdot \boldsymbol{E} = \frac{1}{\varepsilon_0} \sum_j q_j \delta n_j \,. \tag{4.213}$$

It becomes equivalent to

$$k^2 = \frac{\omega_{pe}^2}{2v_{the}^2} Z'(\zeta_e) + \frac{\omega_{pi}^2}{2v_{thi}^2} Z'(\zeta_i) , \tag{4.214}$$

which constitutes the linear dispersion relation for an electron-ion plasma in the electrostatic approximation. Obviously, for high-frequency oscillations ($\omega \geq \omega_{pe}$), the ion contribution is negligible which formally can be put in the form $\omega_{pi} \approx 0$. In that case, we should approximately solve

$$k^2\lambda_{De}^2 \approx \frac{1}{2}Z'(\zeta_e)\,. \tag{4.215}$$

The optical branch shows for $k \to 0$ large phase velocities, leading to

$$\frac{\omega}{k} \gg v_{the}\,. \tag{4.216}$$

Thus, we may approximate the Z-function for $|\zeta_e| \gg 1$. In that region, the real part of the Z-function is asymptotically

$$Z'(\zeta_e) \approx \frac{1}{\zeta_e^2} + \frac{3}{2\zeta_e^4} + \cdots\,. \tag{4.217}$$

For calculating the (real) frequency $\omega \approx \operatorname{Re}\omega$, the asymptotic expansion leads to

$$k^2\lambda_{De}^2 \approx \frac{k^2 v_{the}^2}{\omega^2} + 3\frac{k^4 v_{the}^4}{\omega^4}\,. \tag{4.218}$$

Its approximate solution is

$$\omega^2 \approx \omega_{pe}^2\left(1 + 3k^2\lambda_{De}^2\right)\,. \tag{4.219}$$

The solution shows the typical behavior of an optical branch and satisfies the assumptions which we made during its derivation. For small values of $k\lambda_{De}$, inclusion of the imaginary contributions in the dispersion relation ($\sigma = 1$) lead to

$$k^2\lambda_{De}^2 \approx \frac{k^2 v_{the}^2}{\omega^2} - i\sqrt{\pi}\frac{\omega}{k}\frac{1}{\sqrt{2}}\frac{1}{v_{the}}e^{-\frac{\omega^2}{2k^2 v_{the}^2}}\,. \tag{4.220}$$

We solve this equation by setting $\omega = \operatorname{Re}\omega + i\gamma$ and approximating ω^2 by Eq. (4.219) whenever appropriate. A short calculation leads to

$$\frac{\gamma}{\omega_{pe}} \approx -\left(\frac{\pi}{8}\right)^{1/2}(k\lambda_{De})^{-3}\exp\left[-\frac{1}{2}(k\lambda_{De})^{-2} - \frac{3}{2}\right]. \tag{4.221}$$

This is identical with the previously derived formula for the electron Landau damping rate. Note that this derivation is only valid for $k\lambda_{De} \ll 1$. When comparing with numerical solutions, it turns out that (practically) up to $k\lambda_{De} \leq 0.2$, it is not too bad. More exact expressions can be easily obtained from the full Eq. (4.215).

A more formal aspect should, however, be mentioned. Equation (4.215) is a nonlinear equation for the complex frequency. Thus, several branches should be expected as solutions. Mathematically, this is indeed the case. Physically, the higher branches are strongly damped such that these solutions do not show up in practice.

Let us now look for the low-frequency or acoustic branch. Then, we have to include the ion dynamics. It will turn out that $\zeta_e \ll 1$ is a good approximation for $\operatorname{Re}\omega \sim k \to 0$. Then, $Z'(\zeta_e) \approx -2$, and the dispersion relation reads

$$1 + k^2\lambda_{De}^2 \approx \lambda_{De}^2\frac{\omega_{pi}^2}{2v_{thi}^2}Z'(\zeta_i)\,. \tag{4.222}$$

Similar to the calculation of the electron plasma oscillations, we first assume $\zeta_i \gg 1$. The phase velocity of the low-frequency oscillations is expected to be of the order of the sound velocity, which in turn is much larger than the ion thermal velocity. Under that assumption (which will be checked ex post), from the asymptotic expansion, we approximately get

$$1 + k^2\lambda_{De}^2 \approx \lambda_{De}^2\omega_{pi}^2\left(\frac{k^2}{\omega^2} + 3\frac{k^4}{\omega^4}v_{thi}^2\right) . \tag{4.223}$$

Its approximate solution is

$$\omega^2 \approx \frac{k^2 c_s^2}{1 + k^2\lambda_{De}^2} . \tag{4.224}$$

For reasons which are obvious from the derivation, these modes are called ion-acoustic modes. The ion-acoustic velocity is

$$c_s = \left[\frac{k_B(T_e + 3T_i)}{m_i}\right]^{1/2} . \tag{4.225}$$

For $T_e \gg T_i$, one obtains the ion-acoustic velocity at electron temperature

$$c_s \approx \left(\frac{k_B T_e}{m_i}\right)^{1/2} . \tag{4.226}$$

We can also calculate the ion Landau damping by considering the imaginary parts. Starting from the asymptotically valid form ($\sigma = 1$)

$$\frac{k^2 v_{thi}^2}{\omega^2}\left(1 + 3\frac{T_i}{T_e}\right) - i\left(\frac{\pi}{2}\right)^{1/2}\frac{\omega}{k}\frac{1}{v_{thi}}e^{-\frac{\omega^2}{2k^2 v_{thi}^2}} \approx \frac{T_i}{T_e}\left(1 + k^2\lambda_{De}^2\right), \tag{4.227}$$

we set $\omega = \mathrm{Re}\omega + i\gamma$ and use Eq. (4.224) to obtain

$$\frac{\gamma}{\mathrm{Re}\omega} \approx -\left(\frac{\pi}{8}\right)^{1/2}\frac{T_e}{T_i}\left(3 + \frac{T_e}{T_i}\right)e^{-\frac{\left(3+\frac{T_e}{T_i}\right)}{2}} . \tag{4.228}$$

As mentioned, this formula is only true for $k\lambda_{De} \ll 1$ and $T_e/T_i \gg 1$. The last condition guarantees $\zeta_i \gg 1$. Obviously, for example, when $1 \le T_e/T_i \le 10$, we need a more exact evaluation of the Z-functions, which of course can be done. The evaluations then show that the damping is a monotonously increasing function of T_i/T_e.

Let us make a few more general remarks about the dispersion relation. It will be written in the form

$$\varepsilon \equiv \varepsilon(\boldsymbol{k}, \omega) = 0 , \quad \varepsilon = 1 - \sum_j \frac{\omega_{pj}^2}{k^2}\boldsymbol{k}\cdot\int\frac{\frac{\partial f_j}{\partial \boldsymbol{v}}}{\boldsymbol{k}\cdot\boldsymbol{v} - \omega}d^3v . \tag{4.229}$$

The notation ε is chosen deliberately to identify it with the dielectric constant. Take note that f_j is normalized to one for particles of species j. We may introduce susceptibilities via

$$\varepsilon = 1 + \sum_j \chi_j \,. \tag{4.230}$$

Obviously,

$$\chi_j = -\frac{\omega_{pj}^2}{k^2}\boldsymbol{k}\cdot\int \frac{\frac{\partial f_j}{\partial \boldsymbol{v}}}{\boldsymbol{k}\cdot\boldsymbol{v}-\omega} d^3v = -\frac{\omega_{pj}^2}{2v_{thj}^2 k^2} Z'(\zeta_j)\,, \tag{4.231}$$

with $\boldsymbol{k} = k\hat{x}$ and $v_x = u$. The result can be interpreted in a simple manner, for example, for the electron susceptibility. Let us consider the (nonrelativistic) electron oscillation in an electric field

$$\boldsymbol{E} = \boldsymbol{E}_0 e^{i(\boldsymbol{k}\cdot\boldsymbol{r}-\omega t)}\,. \tag{4.232}$$

The oscillation $\delta\boldsymbol{r}$ is governed by

$$\delta\ddot{\boldsymbol{r}} = \frac{q_e}{m_e}\boldsymbol{E}_0 e^{-i(\omega-\boldsymbol{k}\cdot\boldsymbol{v})t} e^{i\boldsymbol{k}\cdot(\boldsymbol{r}_0+\delta\boldsymbol{r})} \approx \frac{q_e}{m_e}\boldsymbol{E}_0 e^{i[\boldsymbol{k}\cdot\boldsymbol{r}_0-(\omega-\boldsymbol{k}\cdot\boldsymbol{v})t]}\,. \tag{4.233}$$

The Doppler shift $\omega - \boldsymbol{k}\cdot\boldsymbol{v}$ arises because of $\boldsymbol{r} = \boldsymbol{r}_0 + \boldsymbol{v}t + \delta\boldsymbol{r}$. The approximate solution

$$\delta\boldsymbol{r} \approx -\frac{q_e}{m_e}\frac{\boldsymbol{E}}{(\omega-\boldsymbol{k}\cdot\boldsymbol{v})^2} \tag{4.234}$$

determines the dipole moment

$$\boldsymbol{p} = q_e\delta\boldsymbol{r} \approx -\frac{q_e^2\boldsymbol{E}}{m_e(\omega-\boldsymbol{k}\cdot\boldsymbol{v})^2}\,, \tag{4.235}$$

which in turn determines the polarization

$$\boldsymbol{P} = -\frac{n_{e0}q_e^2}{m_e}\boldsymbol{E}\int \frac{f_e d^3v}{(\omega-\boldsymbol{k}\cdot\boldsymbol{v})^2}\,. \tag{4.236}$$

Because of $\boldsymbol{D} = \varepsilon_0\boldsymbol{E} + \boldsymbol{P} = \varepsilon\varepsilon_0\boldsymbol{E}$, $\boldsymbol{P} = \varepsilon_0\chi\boldsymbol{E}$, a comparison with $\varepsilon = 1+\chi$ makes the Eq. (4.231) plausible.

The total wave energy density of a mode $E = E(t)\sin(kx-\omega t)$ with frequency ω and slowly varying envelope $E(t)$ is

$$U = \frac{\varepsilon_0 E^2}{4}\frac{\partial}{\partial\omega}(\varepsilon\omega)\,. \tag{4.237}$$

For a justification, we go back to the Vlasov equation which we multiply by $m_e v$ and $1/2m_e v^2$, respectively, and average over space in the form

$$\langle\ldots\rangle := \frac{k}{2\pi}\int_0^{\frac{2\pi}{k}} dx\ldots\,. \tag{4.238}$$

A short calculation leads to

$$\partial_t\left\langle\int m_e v f dv\right\rangle \equiv \frac{dP}{dt} = q_e\left\langle E\int f dv\right\rangle , \tag{4.239}$$

$$\partial_t\left\langle\int \frac{1}{2} m_e v^2 f dv\right\rangle \equiv \frac{dK}{dt} = q_e\left\langle E\int v f dv\right\rangle . \tag{4.240}$$

The perturbed distribution function $f = f_0 + f_1$ will be calculated from

$$\frac{df_1}{dt} \approx -\frac{q_e}{m_e} f_0' E(t)\sin(kx-\omega t) . \tag{4.241}$$

Integration leads to

$$\frac{m_e}{q_e} f_1 = f_0' E(t)\frac{\cos(kx-\omega t)}{kv-\omega} - f_0'\frac{dE}{dt}\frac{\sin(kx-\omega t)}{(kv-\omega)^2} + \int_{-\infty}^{t} \frac{d^2E}{d\tau^2} f_0' \frac{\sin(kx_0+kv\tau-\omega\tau)}{(kv-\omega)^2} d\tau . \tag{4.242}$$

Inserting this on the right-hand sides of Eqs. (4.239) and (4.240), and by neglecting terms being proportional to the second time-derivative d^2E/dt^2, we obtain

$$P \approx -\frac{\omega_{pe}^2}{n_{e0}}\frac{\varepsilon_0 E^2}{4}\int\frac{f_0' dv}{(kv-\omega)^2} , \tag{4.243}$$

$$K \approx -\frac{\omega_{pe}^2}{n_{e0}}\frac{\varepsilon_0 E^2}{4}\int\frac{f_0' v dv}{(kv-\omega)^2} . \tag{4.244}$$

Combining the electrostatic wave energy density with the kinetic energy density K,

$$U = \frac{\varepsilon_0 E^2}{4} + K , \tag{4.245}$$

we finally obtain

$$\begin{aligned} U &\approx \frac{\varepsilon_0 E^2}{4}\left[1 - \frac{\omega_{pe}^2}{kn_{eo}}\int\frac{f_0'(kv-\omega)}{(kv-\omega)^2}dv - \frac{\omega}{k}\frac{\omega_{pe}^2}{n_{e0}}\int\frac{f_0' dv}{(kv-\omega)^2}\right] \\ &\approx \frac{\varepsilon_0 E^2}{4}\frac{\partial}{\partial\omega}(\varepsilon\omega) , \end{aligned} \tag{4.246}$$

thus proving the assertion.

Energy density U and momentum density are related via

$$\frac{U}{\omega} = \frac{P}{k} . \tag{4.247}$$

4.5
Landau–Fokker–Planck Equation

We now return to the more general description and include contributions from the two-particle distribution function g. The derivation of the Landau–Fokker–Planck equation starts from the approximation

$$\partial_t f^\alpha(1;t) = L_1^\alpha f^\alpha(1;t) + \sum_\beta \int d^6 2 L_{12}^{\alpha\beta} f^\alpha(1;t) f^\beta(2;t) + \sum_\beta \int d^6 2 L_{12}^{\alpha\beta} g^{\alpha\beta}(1,2;t) , \tag{4.248}$$

$$\partial_t g^{\alpha\beta}(1,2;t) - (L_1^\alpha + L_2^\beta) g^{\alpha\beta}(1,2;t) = L_{12}^{\alpha\beta} f^\alpha(1;t) f^\beta(2;t) . \tag{4.249}$$

The numbers one and two abbreviate the set of coordinates $\boldsymbol{q}_1, \boldsymbol{v}_1$ and $\boldsymbol{q}_2, \boldsymbol{v}_2$, respectively. When we derive from this set of equations a kinetic equation, that is, a closed equation for the one-particle distribution function f, we should solve Eq. (4.249) for $g^{\alpha\beta}$. Equation (4.249) is an inhomogeneous equation; for its solution, we introduce the propagator $U_{12}^{\alpha\beta}$ for the homogeneous part by

$$\partial_t U_{12}^{\alpha\beta}(t) - \left(L_1^\alpha + L_2^\beta\right) U_{12}^{\alpha\beta}(t) = 0 \tag{4.250}$$

with the initial condition $U_{12}^{\alpha\beta}(0) = 1$. The solution of the homogeneous part of Eq. (4.249) can then be written as

$$g^{\alpha\beta}(1,2;t) = U_{12}^{\alpha\beta}(t) g^{\alpha\beta}(1,2;0) . \tag{4.251}$$

It is easy to check that

$$g^{\alpha\beta}(1,2;t) = U_{12}^{\alpha\beta}(t) g^{\alpha\beta}(1,2;0) + \int_0^t d\tau\, U_{12}^{\alpha\beta}(t-\tau) L_{12}^{\alpha\beta} f^\alpha(1;\tau) f^\beta(2;\tau) \tag{4.252}$$

solves the full Eq. (4.249). The solution shows that the whole time history enters on the right-hand side of Eq. (4.252). In other words, when we insert $g^{\alpha\beta}$ from Eq. (4.252) into Eq. (4.248), the collision term

$$K^\alpha = \sum_\beta \int d^6 2 \int_0^t d\tau\, L_{12}^{\alpha\beta} U_{12}^{\alpha\beta}(t-\tau) L_{12}^{\alpha\beta} f^\alpha(1;\tau) f^\beta(2;\tau) + \sum_\beta \int d^6 2 L_{12}^{\alpha\beta} U_{12}^{\alpha\beta}(t) g^{\alpha\beta}(1,2;0) \tag{4.253}$$

appears which has some behavior which we cannot accept for a collision term in a kinetic equation for two reasons:

- It requires knowledge of the distribution function f at all previous times $0 \le \tau \le t$.
- The initial correlation $g(1,2;0)$ enters explicitly.

Although mathematically correct, we need some additional input from physics to simplify the collision term. For the following calculation, we first note that

$$U_{12}^{\alpha\beta}(t) = U_1^{\alpha}(t)\,U_2^{\beta}(t) \tag{4.254}$$

where U_1^{α} and U_2^{β} satisfy the equations

$$\partial_t U_1^{\alpha}(t) = L_1^{\alpha} U_1^{\alpha}(t)\,, \quad \partial_t U_2^{\beta}(t) = L_2^{\beta} U_2^{\beta}(t)\,, \tag{4.255}$$

respectively. Remembering the structure of L_j, we recognize that the action of the propagator U is equivalent to an integration along the *unperturbed orbits*. For example, we note

$$U_1^{\alpha}(t) f^{\alpha}(1;0) \approx f^{\alpha}(1;t) \approx f^{\alpha}(\boldsymbol{q}_1(-t), \boldsymbol{v}_1(-t); 0) \tag{4.256}$$

to lowest order in the orbit of a particle. Starting with the second term on the right-hand side of Eq. (4.253), we have

$$\begin{aligned}
&\sum_\beta \int d^6 2 L_{12}^{\alpha\beta} U_{12}^{\alpha\beta} g^{\alpha\beta}(1,2;0) \\
&= \sum_\beta \int d^3r\, d^3v_2\, L_{12}^{\alpha\beta} U_1^{\alpha} U_2^{\beta} g^{\alpha\beta}(\boldsymbol{q}_1, \boldsymbol{r}, \boldsymbol{v}_1, \boldsymbol{v}_2; 0) \\
&= \sum_\beta \int d^3r\, d^3v_2\, L_{12}^{\alpha\beta} g^{\alpha\beta}\left[\boldsymbol{q}_1(-t), \boldsymbol{r}(-t), \boldsymbol{v}_1(-t), \boldsymbol{v}_2(-t); 0\right].
\end{aligned} \tag{4.257}$$

We have introduced the difference $\boldsymbol{r} = \boldsymbol{q}_2 - \boldsymbol{q}_1$ of the two coordinates. The time shift in a system with, for example, a constant external magnetic field for a particle of species α leads to

$$\boldsymbol{q}(-t) = \begin{pmatrix} q_x - \Omega_\alpha^{-1} v_y(\cos\Omega_\alpha t - 1) - \Omega_\alpha^{-1} v_x \sin\Omega_\alpha t \\ q_y + \Omega_\alpha^{-1} v_x(\cos\Omega_\alpha t - 1) - \Omega_\alpha^{-1} v_y \sin\Omega_\alpha t \\ q_z - v_z t \end{pmatrix} \tag{4.258}$$

in terms of the initial values. The z-component, that is, the component parallel to the external magnetic field, is most important in the present context. (In the case of no external magnetic field, it represents the typical time evolution of any component.) Let us estimate the difference, that is, $r_z(-t) = q_{2z} - q_{1z} - (v_{2z} - v_{1z})t \sim t$. Obviously, the difference linearly grows with time. Now, we take into account that the correlation length is of the order of the Debye length. The correlation function $g^{\alpha\beta}$ becomes small at $r \ge \lambda_D$. Therefore, for times

$$t \ge \frac{\lambda_D}{v_{2z} - v_{1z}} \sim \frac{\lambda_D}{v_{th}} \sim \omega_p^{-1}, \tag{4.259}$$

the influence of the initial correlations $g^{\alpha\beta}(1,2;0)$ becomes negligible. In an electron-ion plasma, we therefore define the characteristic correlation time (corresponding to the effective duration of a collision) as

$$\tau_c = max(\omega_{pe}^{-1}, \omega_{pi}^{-1}) \tag{4.260}$$

and assume that for $t \geq \tau_c$, the initial configuration will be "forgotten." The regime $t \geq \tau_c$ is called the kinetic regime. Within the latter,

$$U_{12}(t) g^{\alpha\beta}(1,2;0) \approx 0 \tag{4.261}$$

holds. Let us now have a look at the first term on the right-hand side of Eq. (4.253). We change variables from τ to $t - \tau$ and obtain for that term

$$\sum_\beta \int d^6 2 \int_0^t d\tau L_{12}^{\alpha\beta} U_{12}^{\alpha\beta}(\tau) L_{12}^{\alpha\beta} f^\alpha(1;t-\tau) f^\beta(2;t-\tau) \, . \tag{4.262}$$

Next, we use Eqs. (4.254) and (4.256) and approximate

$$f^\alpha(1;t-\tau) \approx U_1^\alpha(-\tau) f^\alpha(1;t) \, . \tag{4.263}$$

A similar equation holds for the other particle two. The inverse of the propagator U follows straightforwardly because of $U_1^\alpha(\tau)U_1^\alpha(-\tau) = 1$. Summarizing, the collision term can be approximated in the Landau form

$$\begin{aligned} K^\alpha \approx & \sum_\beta \int d^3 v_2 \int d^3 r \int_0^t d\tau \frac{1}{m_\alpha} \frac{\partial}{\partial \boldsymbol{v}_1} \cdot [\nabla \phi_{\alpha\beta}(\boldsymbol{r}(\tau))] [\nabla \phi_{\alpha\beta}(\boldsymbol{r}(-\tau))] \\ & \cdot \left\{ U_1^\alpha(\tau) \frac{1}{m_\alpha} \frac{\partial}{\partial \boldsymbol{v}_1} U_1^\alpha(-\tau) - U_2^\beta(\tau) \frac{1}{m_\beta} \frac{\partial}{\partial \boldsymbol{v}_2} U_2^\beta(-\tau) \right\} f^\alpha(1;t) f^\beta(2;t) \, . \end{aligned} \tag{4.264}$$

The gradient operator ∇ is with respect to the difference $\boldsymbol{r}$. Now, note the two factors $\phi_{\alpha\beta}(\boldsymbol{r}(\tau))$ and $\phi_{\alpha\beta}(\boldsymbol{r}(-\tau))$ in the integrand on the right-hand side. They restrict the region of integration to $0 \leq r \leq r_c$ and $0 \leq \tau \leq \tau_c$. Using therefore in the integrals a cut-off at the correlation length $r_c \approx max(\lambda_{De}, \lambda_{Di})$ and the correlation time $\tau_c \approx \omega_{pi}^{-1}$, we may approximate in the integrand

$$U_1^\alpha(\tau) \frac{\partial}{\partial \boldsymbol{v}_1} U_1^\alpha(-\tau) f_1^\alpha(1;t) \approx \frac{\partial}{\partial \boldsymbol{v}_1} f_1^\alpha(1;t) \, . \tag{4.265}$$

That approximation (in the integrand) can be argued as follows. The propagator $U_1^\alpha(-\tau)$ causes a shift along the unperturbed orbits. Since Eq. (4.258) holds together with

$$\boldsymbol{v}(-t) = \begin{pmatrix} v_x \cos \Omega_\alpha t - v_y \sin \Omega_\alpha t \\ v_y \cos \Omega_\alpha t + v_x \sin \Omega_\alpha t \\ v_z \end{pmatrix} , \tag{4.266}$$

we may estimate

$$
\begin{aligned}
&U_1^\alpha(\tau)\frac{\partial}{\partial \boldsymbol{v}_1}U_1^\alpha(-\tau)f^\alpha(1;t)\\
&\approx \Bigg\{\hat{x}\left[\frac{1}{\Omega_\alpha}\sin\Omega_\alpha\tau\frac{\partial}{\partial x}+\frac{1}{\Omega_\alpha}(\cos\Omega_\alpha\tau-1)\frac{\partial}{\partial y}\right.\\
&\quad\left.+\cos\Omega_\alpha\tau\frac{\partial}{\partial v_x}-\sin\Omega_\alpha\tau\frac{\partial}{\partial v_y}\right]\\
&\quad+\hat{y}\left[-\frac{1}{\Omega_\alpha}(\cos\Omega_\alpha\tau-1)\frac{\partial}{\partial x}+\frac{1}{\Omega_\alpha}\sin\Omega_\alpha\tau\frac{\partial}{\partial y}\right.\\
&\quad\left.+\sin\Omega_\alpha\tau\frac{\partial}{\partial v_x}+\cos\Omega_\alpha\tau\frac{\partial}{\partial v_y}\right]+\hat{z}\left[\tau\frac{\partial}{\partial z}+\frac{\partial}{\partial v_z}\right]\Bigg\}f^\alpha(1;t)\\
&\approx\left[\frac{\partial}{\partial \boldsymbol{v}_1}+\tau\frac{\partial}{\partial \boldsymbol{q}_1}\right]f^\alpha(1;t)+\mathcal{O}(\Omega_\alpha\tau_c)\,. \qquad (4.267)
\end{aligned}
$$

The right-hand side simplifies for vanishing external magnetic fields ($\Omega_\alpha = 0$), and the estimate becomes obvious. Though also for

$$|\Omega_\alpha|\tau_c \ll 1 \qquad (4.268)$$

the estimate holds. Next, we make use of the fact that the τ-integration is effectively restricted to $\tau \le \tau_c$. Then, for the ratio we get

$$\tau\frac{\partial}{\partial q_1}f^\alpha : \frac{\partial}{\partial v_1}f^\alpha \sim \frac{\tau_c v_{th\alpha}}{L_H} \ll 1\,. \qquad (4.269)$$

Here, we have assumed that the space-dependence of the one-particle distribution function f occurs via the (hydrodynamic) moments: density, mean velocity, temperature, and so on. The latter varies on the hydrodynamic length scale L_H, that is,

$$L_H^{-1} \sim |\nabla \ln f^\alpha|\,. \qquad (4.270)$$

The hydrodynamic length scale satisfies

$$L_H \gg r_c \sim \lambda_{De}\,. \qquad (4.271)$$

The corresponding hydrodynamic time scale is estimated as

$$\tau_H^\alpha = \frac{L_H}{v_{th\alpha}}\,. \qquad (4.272)$$

Again, we assume $\tau_H^\alpha \gg \tau_c$, which has been used in Eq. (4.269). These arguments support Eq. (4.265). In a similar manner, we can estimate

$$
\begin{aligned}
f^\alpha(1;t)f^\beta(2;t) &= f^\alpha(\boldsymbol{q}_1,\boldsymbol{v}_1;t)f^\beta(\boldsymbol{q}_1+\boldsymbol{r},\boldsymbol{v}_2;t)\\
&\approx f^\alpha(\boldsymbol{q}_1,\boldsymbol{v}_1;t)\left[f^\beta(\boldsymbol{q}_1,\boldsymbol{v}_2;t)+\boldsymbol{r}\cdot\frac{\partial}{\partial \boldsymbol{q}_1}f^\beta(\boldsymbol{q}_1,\boldsymbol{v}_2;t)+\ldots\right]\\
&\approx f^\alpha(\boldsymbol{q}_1,\boldsymbol{v}_1;t)f^\beta(\boldsymbol{q}_1,\boldsymbol{v}_2;t)\,,
\end{aligned}
\qquad (4.273)
$$

neglecting terms of order r_c/L_H. For all estimates, please remember again that they are only used under the integral sign. Within the unperturbed-orbit approximation, neglecting terms of order $\Omega_\alpha \tau_c$, we also have

$$\phi_{\alpha\beta}(\boldsymbol{q}_2(-\tau) - \boldsymbol{q}_1(-\tau)) \approx \phi_{\alpha\beta}(\boldsymbol{r} - (\boldsymbol{v}_2 - \boldsymbol{v}_1)\tau) \,. \tag{4.274}$$

Putting everything together, the collision term Eq. (4.264) simplifies to

$$\begin{aligned} K^\alpha \approx & \sum_\beta \int d^3 v_2 \frac{1}{m_\alpha} \frac{\partial}{\partial v_{1\nu}} \mathcal{T}^{\alpha\beta}_{\nu\mu}(\boldsymbol{g}) \\ & \times \left(\frac{1}{m_\alpha} \frac{\partial}{\partial v_{1\mu}} - \frac{1}{m_\beta} \frac{\partial}{\partial v_{2\mu}} \right) f^\alpha(\boldsymbol{q}_1, \boldsymbol{v}_1; t) f^\beta(\boldsymbol{q}_1, \boldsymbol{v}_2; t) \,, \end{aligned} \tag{4.275}$$

with the difference velocity

$$\boldsymbol{g} = \boldsymbol{v}_2 - \boldsymbol{v}_1 \tag{4.276}$$

and the Einstein convention for summation over ν and μ. Although formally written correctly, in the tensor

$$\mathcal{T}^{\alpha\beta}(\boldsymbol{g}) := \int d^3 r \int_0^\infty d\tau \left[\nabla \phi_{\alpha\beta}(\boldsymbol{r}) \right] \left[\nabla \phi_{\alpha\beta}(\boldsymbol{r} - \boldsymbol{g}\tau) \right] , \tag{4.277}$$

the integration effectively is restricted to r_c and τ_c. First, we evaluate Eq. (4.277) for (unshielded) Coulomb interaction

$$\phi_{\alpha\beta}(\boldsymbol{r}) = \int d^3 k e^{i\boldsymbol{k}\cdot\boldsymbol{r}} \tilde{\phi}_{\alpha\beta}(\boldsymbol{k}) \tag{4.278}$$

with

$$\tilde{\phi}_{\alpha\beta} = \frac{1}{4\pi\varepsilon_0} \frac{1}{2\pi^2} \frac{e_\alpha e_\beta}{k^2} \,. \tag{4.279}$$

A short calculation leads to

$$\begin{aligned} \mathcal{T}^{\alpha\beta}(\boldsymbol{g}) &= \int d^3 r \int_0^\infty d\tau \int e^{i(\boldsymbol{k}+\boldsymbol{k}')\cdot\boldsymbol{r} - i\boldsymbol{k}'\cdot\boldsymbol{g}\tau} (i\boldsymbol{k})(i\boldsymbol{k}') \tilde{\phi}_{\alpha\beta}(\boldsymbol{k}) \tilde{\phi}_{\alpha\beta}(\boldsymbol{k}') d^3 k d^3 k' \\ &= 8\pi^4 \int d^3 k \boldsymbol{k}\boldsymbol{k} \tilde{\phi}^2_{\alpha\beta}(\boldsymbol{k}) \delta(\boldsymbol{k}\cdot\boldsymbol{g}) = 8\pi^4 \int_0^\infty dk k^4 \tilde{\phi}^2_{\alpha\beta}(k) \int_{-1}^1 d\cos\theta \\ & \times \int_0^{2\pi} d\varphi \delta(kg\cos\theta) \begin{pmatrix} \sin\theta\cos\varphi \\ \sin\theta\sin\varphi \\ \cos\theta \end{pmatrix} \begin{pmatrix} \sin\theta\cos\varphi \\ \sin\theta\sin\varphi \\ \cos\theta \end{pmatrix} \equiv [1 - \hat{z}\hat{z}]\, T^{\alpha\beta} \,, \end{aligned} \tag{4.280}$$

where

$$T^{\alpha\beta} = 8\pi^5 \int\limits_0^\infty dk k^4 \tilde{\phi}^2_{\alpha\beta}(k) \int\limits_{-1}^{1} d\mu(1-\mu^2)\delta(kg\mu) \, . \tag{4.281}$$

The μ-integration can easily be performed and results in

$$T^{\alpha\beta} = \frac{8\pi^5}{g} \int\limits_0^\infty dk k^3 \tilde{\phi}^2_{\alpha\beta}(k) \equiv \frac{A^{\alpha\beta}}{g} \, . \tag{4.282}$$

Finally, we choose the coordinate system such that $\mathbf{g} = g\hat{z}$. Then, the tensor $\mathcal{T}^{\alpha\beta}$ can be written as

$$\mathcal{T}^{\alpha\beta}_{\nu\mu}(\mathbf{g}) = \left(\delta_{\nu\mu} - \frac{g_\nu g_\mu}{g^2}\right)\frac{A^{\alpha\beta}}{g} \, , \tag{4.283}$$

where, for (unshielded) Coulomb interaction,

$$A^{\alpha\beta}_{Coulomb} = \frac{1}{\pi\varepsilon_0^2} e^2_\alpha e^2_\beta \int\limits_0^\infty dk \frac{1}{k} \, . \tag{4.284}$$

However, the latter expression diverges at $k = 0$ and $k \to \infty$. The limit $k \to \infty$ corresponds to small distances when, of course, the classical description breaks down. However, at small distances already in the classical regime, the weak-coupling approximation will also lose its validity. On the other hand, $k \to 0$ belongs to large distances where the Coulomb interaction will be shielded. The dynamical shielding (by third particles) was incorporated by Balescu, Lenard, and Guernsey [51–53], leading to the Balescu–Lenard–Guernsey kinetic equation. Here, we only argue with static shielding by replacing the thus far used unscreened Coulomb interaction by

$$\phi^{effektiv}_{\alpha\beta} = \frac{1}{4\pi\varepsilon_0} e_\alpha e_\beta \frac{e^{-r/\lambda_D}}{r} \, , \tag{4.285}$$

leading after Fourier transformation to

$$\tilde{\phi}^{effektiv}_{\alpha\beta} = \frac{1}{4\pi\varepsilon_0} \frac{e_\alpha e_\beta}{2\pi^2} \frac{1}{k^2 + \kappa_D^2} \, . \tag{4.286}$$

Introducing this into Eq. (4.282) will cure the small k divergence. On the other hand, for small distances (large k), we shall introduce a cut-off, such that

$$k < k_{max} \approx \frac{12\pi\varepsilon_0 k_B T_\alpha}{Ze^2} \tag{4.287}$$

is used. Here, $k_{max}^{-1} \sim Ze^2_\alpha/12\pi\varepsilon_0 k_B T_\alpha$ is the collision parameter for 90^0-deflections. In the two-body problem, the deflection angle θ for small-angle scattering is

$$\theta = \frac{Ze^2}{2\pi\varepsilon_0 b\mu v^2_{rel}} \, , \tag{4.288}$$

where b is the collision parameter, μ the reduced mass, and v_{rel} the relative speed. The collision parameter $b_{\pi/2}$ for $\theta = \pi/2$ depends on the type of collisions (ee, ei, ii); this dependence will not be discussed at the present stage (but later). Thus, when one does not use the screened Debye potential, that is, if one works with the Coulomb potential directly, one limits the integration on both sides, once from the 90° impact parameter $b_{\pi/2}$ (in both cases) and on the other hand by the Debye length (in the case of an unshielded Coulomb potential). The different procedures lead to slightly different factors (Coulomb logarithms) in the final results.

Introducing the mentioned *ad hoc* limitations for an integration over the Debye potential, we find

$$\begin{aligned} A^{\alpha\beta} &\approx 8\pi^5 \int_0^{k_{max}} dk k^3 \left[\tilde{\phi}_{\alpha\beta}^{effektiv}\right]^2 \\ &= \frac{1}{(4\pi\varepsilon_0)^2} \pi e_\alpha^2 e_\beta^2 \left[\frac{\kappa_D^2}{k_{max}^2 + \kappa_D^2} - 1 + \ln \frac{k_{max}^2 + \kappa_D^2}{\kappa_D^2}\right] \\ &\approx \frac{1}{(4\pi\varepsilon_0)^2} 2\pi e_\alpha^2 e_\beta^2 \ln \Lambda_B \end{aligned} \tag{4.289}$$

with

$$\ln \Lambda \approx \ln \frac{k_{max}}{k_{min}} \mathrel{\hat{=}} \ln \frac{k_{max}}{\kappa_D} \approx \ln \frac{6\pi\varepsilon_0 \lambda_D k_B (T_e + T_i)}{Z e^2} \equiv \ln \Lambda_B \,. \tag{4.290}$$

The definition (4.290) was used by Balescu [40], and we shall also use it whenever appropriate.

Summarizing, the Landau–Fokker–Planck collision term for weakly coupled plasma with effective Debye interaction turns out to be

$$\begin{aligned} K^\alpha \approx& \frac{1}{(4\pi\varepsilon_0)^2} \sum_{\beta=e,i} 2\pi e_\alpha^2 e_\beta^2 \ln \Lambda_B \int d^3 v_2 \frac{1}{m_\alpha} \frac{\partial}{\partial v_{1\nu}} G_{\nu\mu}(\boldsymbol{g}) \\ &\times \left[\frac{1}{m_\alpha} \frac{\partial}{\partial v_{1\mu}} - \frac{1}{m_\beta} \frac{\partial}{\partial v_{2\mu}}\right] f^\alpha(\boldsymbol{q}_1, \boldsymbol{v}_1; t) f^\beta(\boldsymbol{q}_1, \boldsymbol{v}_2; t) \,, \end{aligned} \tag{4.291}$$

with the Landau tensor

$$G_{\nu\mu}(\boldsymbol{g}) = \frac{g^2 \delta_{\nu\mu} - g_\nu g_\mu}{g^3} \,. \tag{4.292}$$

Equation (4.291) contains the average Coulomb logarithm $\ln \Lambda_B$. Because of the two assumptions made at the end of the present derivation, we should not overestimate the numerical values.

Next, we will write the kinetic equation in Fokker–Planck form, thereby justifying its name, that is, Landau–Fokker–Planck equation. For simplicity, we restrict ourselves to the dynamics of one component, let us say electrons (with immobile ions). Since, the Landau-tensor can be written in the form

$$\partial_{\boldsymbol{v}_1} \partial_{\boldsymbol{v}_1} g = \frac{g^2 I - \boldsymbol{g}\boldsymbol{g}}{g^3} , \tag{4.293}$$

we abbreviate the kinetic equation as

$$\frac{d f(\boldsymbol{r}_1, \boldsymbol{v}_1, t)}{dt} = \frac{1}{(4\pi\varepsilon_0)^2} \frac{2\pi e^4 \ln \Lambda_B}{m_e^2} \partial_{\boldsymbol{v}_1} \cdot \left[(\partial_{\boldsymbol{v}_1} f(\boldsymbol{v}_1)) \cdot \partial_{\boldsymbol{v}_1} \partial_{\boldsymbol{v}_1} \int d^3 v_2 g f(\boldsymbol{v}_2) \right.$$
$$\left. - f(\boldsymbol{v}_1) \int d^3 v_2 \partial_{\boldsymbol{v}_1} (\partial_{\boldsymbol{v}_1} \cdot \partial_{\boldsymbol{v}_1}) g f(\boldsymbol{v}_2) \right] . \tag{4.294}$$

This form is obtained after integration by parts, and $\partial_{\boldsymbol{v}_2} g = -\partial_{\boldsymbol{v}_1} g$ was used. Furthermore, the left-hand side of the kinetic equation is abbreviated by df/dt, and on the right-hand side not all variables of f are shown. Next, after a simple algebraic manipulation, we write Eq. (4.294) in the form

$$\frac{df}{dt} = \frac{1}{(4\pi\varepsilon_0)^2} \frac{2\pi e^4 \ln \Lambda_B}{m_e^2} \left\{ \partial_{\boldsymbol{v}_1} \partial_{\boldsymbol{v}_1} : \left[f(\boldsymbol{v}_1) \partial_{\boldsymbol{v}_1} \partial_{\boldsymbol{v}_1} \int d^3 v_2 g f(\boldsymbol{v}_2) \right] \right.$$
$$\left. - 2\partial_{\boldsymbol{v}_1} \cdot \left[f(\boldsymbol{v}_1) \int d^3 v_2 \partial_{\boldsymbol{v}_1} (\partial_{\boldsymbol{v}_1} \cdot \partial_{\boldsymbol{v}_1}) g f(\boldsymbol{v}_2) \right] \right\} \tag{4.295}$$

and make use of

$$(\partial_{\boldsymbol{v}_1} \cdot \partial_{\boldsymbol{v}_1})\, g = \frac{2}{g} . \tag{4.296}$$

As intended, we obtain the Landau equation in Fokker–Planck form

$$\frac{d f(\boldsymbol{v}_1)}{dt} = -\partial_{\boldsymbol{v}_1} \cdot [\boldsymbol{A} f(\boldsymbol{v}_1)] + \frac{1}{2} \partial_{\boldsymbol{v}_1} \partial_{\boldsymbol{v}_1} : [\mathcal{B}\, f(\boldsymbol{v}_1)] . \tag{4.297}$$

Two coefficients appear, namely, the dynamical friction coefficient

$$\boldsymbol{A} = \frac{1}{(4\pi\varepsilon_0)^2} \frac{8\pi e^4 \ln \Lambda_B}{m_e^2} \partial_{\boldsymbol{v}_1} \int d^3 v_2 \frac{f(\boldsymbol{v}_2)}{|\boldsymbol{v}_1 - \boldsymbol{v}_2|} \tag{4.298}$$

and the diffusion (in velocity space) coefficient

$$\mathcal{B} = \frac{1}{(4\pi\varepsilon_0)^2} \frac{4\pi e^4 \ln \Lambda_B}{m_e^2} \partial_{\boldsymbol{v}_1} \partial_{\boldsymbol{v}_1} \int d^3 v_2 |\boldsymbol{v}_1 - \boldsymbol{v}_2| f(\boldsymbol{v}_2) . \tag{4.299}$$

The friction term preferentially reduces the particle motion in its original direction of motion, whereas the diffusion term forces an increase in the transversal component. Furthermore, by estimating the order of magnitude of the friction term, we obtain the approximate inverse relaxation time $\tau_R^{-1} \approx \omega_{pe} \ln \Lambda_B / \Lambda_B$ which agrees with our expectations from heuristic arguments.

The kinetic equation (4.297) is of Fokker–Planck form. Let us relate the coefficients $\boldsymbol{A}$ and $\mathcal{B}$ to velocity moments. Coulomb interactions cause velocity changes. We introduce the probability $W(\boldsymbol{v}, \Delta\boldsymbol{v})$ for a velocity change $\Delta\boldsymbol{v}$ during time Δt. The distribution function changes during Δt; we calculate the new value caused by collisions by summing over all transition probabilities according to

$$f(\boldsymbol{q}, \boldsymbol{v}; t + \Delta t) = \int d^3 \Delta v\, f(\boldsymbol{q}, \boldsymbol{v} - \Delta\boldsymbol{v}; t)\, W(\boldsymbol{v} - \Delta\boldsymbol{v}, \Delta\boldsymbol{v}) . \tag{4.300}$$

The change of the distribution function due to collisions may be written as

$$\left.\frac{df}{dt}\right|_{coll} \Delta t \approx f(\boldsymbol{q}, \boldsymbol{v}; t + \Delta t) - f(\boldsymbol{q}, \boldsymbol{v}; t) \,. \tag{4.301}$$

When evaluating the right-hand side, we make use of a Taylor expansion in the integrand of Eq. (4.300),

$$f(\boldsymbol{q}, \boldsymbol{v} - \Delta\boldsymbol{v}; t)\, W(\boldsymbol{v} - \Delta\boldsymbol{v}, \Delta\boldsymbol{v}) \approx f(\boldsymbol{q}, \boldsymbol{v}; t)\, W(\boldsymbol{v}, \Delta\boldsymbol{v}) - \frac{\partial(f\,W)}{\partial \boldsymbol{v}} \cdot \Delta\boldsymbol{v} + \sum_{\nu,\mu} \frac{1}{2} \frac{\partial^2(f\,W)}{\partial v_\nu \partial v_\mu} \Delta v_\nu \Delta v_\mu - \cdots \tag{4.302}$$

and introduce the moments

$$\int d^3\Delta v\, W(\boldsymbol{v}, \Delta\boldsymbol{v}) = 1 \,, \qquad \int d^3\Delta v\, W(\boldsymbol{v}, \Delta\boldsymbol{v}) \Delta\boldsymbol{v} = \langle \Delta\boldsymbol{v} \rangle_t \Delta t \,, \tag{4.303}$$

$$\int d^3\Delta v\, W(\boldsymbol{v}, \Delta\boldsymbol{v}) \Delta v_\nu \Delta v_\mu = \langle \Delta v_\nu \Delta v_\mu \rangle_t \Delta t \,. \tag{4.304}$$

The first relation in Eq. (4.303) follows from the normalization of the probability. The next two relations in Eqs. (4.303) and (4.304) show expressions being proportional to Δt. This is consistent with the statistical independence of successive collisions. For example, instead of Eq. (4.304), consider the simpler discrete procedure $\langle \Delta v_\nu \Delta v_\mu \rangle = \sum_i \sum_j \langle \Delta v_\nu^i \Delta v_\mu^j \rangle \approx \sum_i \langle \Delta v_\nu^i \Delta v_\mu^i \rangle$ where the sum over collisions i and j becomes proportional to the sum of collisions and thereby to Δt and not $(\Delta t)^2$. The resulting form

$$\left.\frac{df}{dt}\right|_{coll} = -\frac{\partial}{\partial \boldsymbol{v}} \cdot \left[\langle \Delta\boldsymbol{v} \rangle_t f\right] + \frac{1}{2} \sum_{\nu,\mu} \frac{\partial^2}{\partial v_\nu \partial v_\mu} \left[\langle \Delta v_\nu \Delta v_\mu \rangle_t f\right] \tag{4.305}$$

is equivalent to Eq. (4.297).

The coefficients (4.298) and (4.299) can be evaluated after specifying the distribution function *f*. More or less as an exercise, let us assume a Maxwellian distribution

$$f = \frac{n_{e0} b^3}{\pi^{3/2}} \exp\left(-b^2 v_1^2\right) \,. \tag{4.306}$$

One finds

$$A = -\frac{1}{(4\pi\varepsilon_0)^2} \frac{8\pi e^4 n_{e0} \ln \Lambda_B}{m_e^2 v_1^3} v_1 \phi_1(b v_1) \tag{4.307}$$

and

$$B_{\nu\mu} = \frac{4\pi e^4 n_{e0} \ln \Lambda_B}{(4\pi\varepsilon_0)^2 m_e^2} \frac{\partial^2}{\partial v_{1\nu} \partial v_{1\mu}} \left[\left(1 + \frac{1}{2b^2 v_1^2}\right) \phi(b v_1) + \frac{1}{\pi^{1/2} b v_1} e^{-b^2 v_1^2}\right] , \tag{4.308}$$

where

$$\phi(x) = \frac{2}{\sqrt{\pi}} \int^{x} e^{-x^2} dx\,, \quad \phi_1(x) = \phi(x) - x \frac{d\phi}{dx}\,. \tag{4.309}$$

Putting everything together results in

$$\left.\frac{df}{dt}\right|_{coll} = 0\,. \tag{4.310}$$

This is interesting since it supports the conception that electron-electron collisions lead to a "Maxwellization" of the electron component.

Finally, let us generalize the obtained results to more than one component, that is, for an electron–ion plasma. Designating the species under consideration by the index α, we obtain a collision term K^α in the corresponding Landau–Fokker–Planck equation which consists of two parts,

$$K^\alpha = \sum_\beta K^{\alpha\beta} = K^{\alpha e} + K^{\alpha i}\,. \tag{4.311}$$

It is quite straightforward to prove the following relations:

$$\int d^3 v_1 K^\alpha = 0\,, \quad \sum_\alpha m_\alpha \int d^3 v_1 \boldsymbol{v}_1 K^\alpha = 0\,,$$
$$\sum_\alpha \frac{1}{2} m_\alpha \int d^3 v_1 v_1^2 K^\alpha = 0\,. \tag{4.312}$$

These relations express the global conservation laws for particle number, total momentum and total energy (for elastic collisions). The superscripts $\alpha\beta$ designate a collision of a particle of species α with a particle of species β. Also, within a component, one finds the following conserved quantities:

$$\int d^3 v_1 K^{\alpha\alpha} = 0\,, \quad m_\alpha \int d^3 v_1 \boldsymbol{v}_1 K^{\alpha\alpha} = 0\,, \quad \frac{1}{2} m_\alpha \int d^3 v_1 v_1^2 K^{\alpha\alpha} = 0 \tag{4.313}$$

for all $\alpha = e, i$. Detailed balances in momentum and energy transfers, respectively, between electrons and ions exist. Collisions between particles of different species lead to

$$\int d^3 v_1 K^{\alpha\beta} = 0 \quad \text{for } \alpha, \beta = e, i\,; \alpha \neq \beta\,, \tag{4.314}$$

$$m_i \int d^3 v_1 \boldsymbol{v}_1 K^{ie} + m_e \int d^3 v_1 \boldsymbol{v}_1 K^{ei} = 0\,, \tag{4.315}$$

$$\frac{1}{2} m_i \int d^3 v_1 v_1^2 K^{ie} + \frac{1}{2} m_e \int d^3 v_1 v_1^2 K^{ei} = 0\,. \tag{4.316}$$

We may define a friction force density and a collisional rate of heat exchange,

$$\boldsymbol{R}^{\alpha} = m_{\alpha} \int d^3 v \boldsymbol{v} K^{\alpha} , \tag{4.317}$$

$$Q^{\alpha} = \frac{1}{2} m_{\alpha} \int d^3 v |\boldsymbol{v} - \boldsymbol{u}^{\alpha}|^2 K^{\alpha} , \tag{4.318}$$

respectively, which satisfy

$$\boldsymbol{R}^{e} = \boldsymbol{R}^{ei} = -\boldsymbol{R}^{ie} = -\boldsymbol{R}^{i} , \tag{4.319}$$

$$Q^{e} = Q^{ei} = -Q^{ie} - \left(\boldsymbol{u}^{e} - \boldsymbol{u}^{i}\right) \cdot \boldsymbol{R}^{ei} , \tag{4.320}$$

$$Q^{i} = Q^{ie} \approx -3 n_i \frac{Z}{\tau_e} \frac{m_e}{m_i} (T_i - T_e) . \tag{4.321}$$

These relations express the conservation laws, though additionally, they show the weak energy exchange between electrons and ions. The latter fact is the reason for introducing electron and ion temperatures separately as plasma-dynamical variables. When we express the friction force in terms of collision frequencies $\nu \sim 1/\tau$, we may write

$$\boldsymbol{R}^{ei} = -\nu_{ei} m_e n_e (\boldsymbol{u}_e - \boldsymbol{u}_i) , \quad \boldsymbol{R}^{ie} = -\nu_{ie} m_i n_i (\boldsymbol{u}_i - \boldsymbol{u}_e) . \tag{4.322}$$

The collision frequencies can be found in the simplest way from the linearized collision term. Like-particle collisions conserve the number, momentum, and energy of each species; they efficiently redistribute momentum and energy among and are responsible for local (near) equilibrium states (after a time τ_{α}). Unlike-particle collisions exchange momentum and energy between species; they transfer energy between ions and electrons extremely slowly (characteristic time $(m_i/m_e)\tau_{\alpha}$). Since for the relaxation times ($\tau \sim 1/\nu$)

$$\tau_{ee} \sim \tau_{ei} \sim \tau_e \quad (Z = 1) , \quad \tau_{ie} \gg \tau_{ii} \sim \tau_i \tag{4.323}$$

holds, we are left with two characteristic times [29], that is,

$$\tau_e = \frac{3(4\pi\varepsilon_0)^2}{4\sqrt{2\pi}} \frac{m_e^{1/2} T_e^{3/2}}{n_i Z^2 e^4 \ln \Lambda} , \quad \tau_i = \frac{3(4\pi\varepsilon_0)^2}{4\sqrt{\pi}} \frac{m_i^{1/2} T_i^{3/2}}{n_i Z^4 e^4 \ln \Lambda} . \tag{4.324}$$

Here, $\ln \Lambda \equiv \ln(r_{\max}/r_{\min})$ is the Coulomb logarithm as defined in [29]. Typically, $\ln \Lambda \approx 10$–20. More exact values are given in [29]. Note that we use a slightly different definition for the Coulomb logarithm, namely,

$$\ln \Lambda_B = 4\pi\varepsilon_0 \frac{3(T_e + T_i)\lambda_D}{2Ze^2} , \quad \lambda_D = \left[\frac{Ze^2(n_e T_e + n_i T_i)}{\varepsilon_0 T_e T_i (1 + Z)}\right]^{-1/2} . \tag{4.325}$$

The characteristic times and collision frequencies can be found in [29]. From the orders of magnitude, we conclude that a fusion plasma is not necessarily collisional, especially in the hot core. Though because of confinement and trapping, although the mean free paths may be much longer than the machine dimensions, particles can stay long enough in the plasma before reaching the walls. Thus, it can be expected that even weak collisions contribute to transport.

The derivation of the relaxation times is one of the main tasks of kinetic theory. Looking at the parameter dependencies, we find that the relaxation times increase with temperature. Then, from a very simple model, we can obtain the estimate for the resistivity

$$\eta \equiv \frac{1}{\sigma} \sim \frac{m_e}{e^2 n \tau_e} \,. \tag{4.326}$$

Thus, the resistivity of a plasma will decrease with temperature. The question of how to obtain more exact results for the plasma (transport) coefficients belongs to transport theory. Let us elucidate the principle procedure in the following calculation.

Suppose the system is spatially homogeneous and stationary in the presence of a small electric field $\boldsymbol{E}$. We approximate the Landau–Fokker–Planck equation by

$$-\frac{e}{m_e}\boldsymbol{E}\cdot\frac{\partial f_{e0}}{\partial \boldsymbol{v}} \approx \frac{n_i Z^2 e^4 \ln \Lambda_B}{8\pi\varepsilon_0^2 m_e^2 v^3}\left[\frac{1}{\sin\theta}\frac{\partial}{\partial\theta}\left(\sin\theta\frac{\partial f_e}{\partial\theta}\right) + \frac{1}{\sin^2\theta}\frac{\partial^2 f_e}{\partial\phi^2}\right]. \tag{4.327}$$

Here, we have used a polar coordinate system in velocity space with a z-axis in the direction of the electric field. Azimuthal symmetry ($\partial/\partial\phi = 0$) is a reasonable assumption. Note also that we treat the electric field as a perturbation, allowing one to assume the zeroth order distribution function f_{e0} to be of Maxwellian form; $f_e = f_{e0} + f_{e1}$. This leads to

$$\frac{eEv}{T_e} f_{e0}\cos\theta \approx \frac{n_i Z^2 e^4 \ln \Lambda_B}{8\pi\varepsilon_0^2 m_e^2 v^3}\frac{1}{\sin\theta}\frac{\partial}{\partial\theta}\left(\sin\theta\frac{\partial f_{e1}}{\partial\theta}\right) \,. \tag{4.328}$$

The solution for f_{e1} is easy to obtain as

$$f_{e1} \approx -\frac{4\pi\varepsilon_0^2 m_e^2 E v^4 f_{e0}\cos\theta}{n_i Z^2 e^3 T_e \ln \Lambda_B} \,. \tag{4.329}$$

Calculating the z-component of the electric current density for $n_e = Z n_i$

$$\begin{aligned} j_z &= -e\int f_{e1} v\cos\theta\, d^3v \approx \frac{8\pi\varepsilon_0^2 m_e^2 E}{n_i Z^2 e^2 T_e \ln \Lambda_B}\int_0^\infty v^7 f_{e0}\, dv \int_0^\pi \cos^2\theta\sin\theta\, d\theta \\ &= \frac{32\sqrt{\pi}\varepsilon_0^2 E (2T_e)^{3/2}}{\sqrt{m_e} Z e^2 \ln \Lambda_B} \,, \end{aligned} \tag{4.330}$$

we can read off the resistivity

$$\eta = \frac{\sqrt{m_e} Z e^2 \ln \Lambda_B}{32\sqrt{\pi}\varepsilon_0^2 (2T_e)^{3/2}} \,. \tag{4.331}$$

The simple model already leads to quite an accurate result.

4.6
Kinetic Description of Strongly Magnetized Plasmas

The idea of the following considerations is to simplify the kinetic description for a magnetized plasma (with $\rho_i/L \ll 1$). Considering the particles, we can subdivide the motion into a fast gyration (with small gyro-radii ρ) and a drift in the direction perpendicular to the magnetic field. The motion parallel to the magnetic field is more or less not hindered at all, and therefore we keep $v_\parallel$ as a independent variable. From a simplified point of view, the exact particle positions are replaced by the gyrocenters, the perpendicular velocities are identified as the drift velocities, and $\boldsymbol{r}, v_\parallel, t$ remain as independent variables. (We should have in mind that the exact transformation of the so called drift-kinetic description has a few more technical aspects.) As we shall see in the next subsection, such a strategy leads to the drift-kinetic equation.

However, we should have in mind that by replacing the particle positions by their gyrocenters, we can no longer resolve space scales of the order of the gyroradius. Remember that we have a slow scale L (of the order of the toroidal minor radius or density gradient scale-length). Fast variations may occur for small, but not necessarily linear departures from the stationary state. If the wavelengths of the fast variations are of the order of the ion Larmor radius ρ_i, electric fields will also vary on the fast scale, and the particles see different field strengths during their gyromotion. Taking into account that effect is the purpose of the so called gyrokinetic description being discussed in the second subsection. However, with respect to the latter, we should mention that the full nonlinear theory causes several problems; see [56, 57] and references therein. The main reason is that when we subdivide f into

$$f = f_{slow} + \Delta f_{fast} , \tag{4.332}$$

$\Delta \ll 1$ is required for magnetized plasmas since otherwise for $\Delta \sim \mathcal{O}(1)$ and $\lambda_{fast} \sim \rho_i$, the plasma would be effectively demagnetized. In the following, we present only a short overview over the general procedure. For more details, we refer the reader to the literature [56, 57] and references therein.

4.6.1
The Drift-Kinetic Equation

As has just been mentioned, a kinetic equation for magnetized plasmas may be much simpler than, for example, the Landau–Fokker–Planck equation since it suppresses details on the gyroradius scale. The guiding centers have a velocity

$$\boldsymbol{v}_{gc} = \hat{b} v_\parallel + \boldsymbol{v}_d , \tag{4.333}$$

where the drift velocity to lowest orders is

$$\boldsymbol{v}_d = \boldsymbol{v}_E + \frac{1}{\Omega}\hat{b} \times \left(\frac{\mu}{m}\nabla B + v_\parallel^2 \boldsymbol{\kappa}\right) . \tag{4.334}$$

The latter consists of the $E \times B$-drift $\boldsymbol{v}_E$ and the curvature drift ($\boldsymbol{\kappa} = \hat{b} \cdot \nabla \hat{b}$). Note that according to the scaling

$$\Delta \equiv 0\,, \quad \frac{v_E}{v_{th}} \sim \delta\,, \quad \partial_t \sim \delta\Omega\,, \tag{4.335}$$

the polarization drift is of higher order. In (4.334), μ is the magnetic moment and for its magnitude, we have

$$\mu = \frac{m v_\perp^2}{2B}\,, \quad \frac{d\mu}{dt} = \mathcal{O}(\delta)\,. \tag{4.336}$$

In other words, μ is an adiabatic invariant.

The total guiding-center energy is the sum of kinetic and potential energy,

$$U = \frac{m v_\parallel^2}{2} + \mu B + e\phi\,. \tag{4.337}$$

Note, that the last two terms represent the potential energy of a guiding center. Calculating the time change of the energy, we make the ansatz

$$\frac{d}{dt}\left[\frac{m}{2} v_\parallel^2\right] \approx \boldsymbol{v}_{gc} \cdot \left[e\boldsymbol{E} - \nabla(\mu B)\right] \tag{4.338}$$

for the time change of the kinetic energy of a guiding center. The second term on the right-hand side takes care of the mirror force $\boldsymbol{F}_m = -\nabla(\mu B)$. One finds

$$\begin{aligned}\frac{dU}{dt} &\equiv \left(\frac{\partial}{\partial t} + \boldsymbol{v}_{gc} \cdot \nabla\right) U \approx e\frac{d\phi}{dt} + \mu\frac{dB}{dt} + \boldsymbol{v}_{gc} \cdot \left[e\boldsymbol{E} - \nabla(\mu B)\right] \\ &= e\frac{d\phi}{dt} + \mu\frac{dB}{dt} - \frac{e}{c}\boldsymbol{v}_{gc} \cdot \frac{\partial \boldsymbol{A}}{\partial t}\,.\end{aligned} \tag{4.339}$$

Without doing the exact calculation (see remarks below), we expect that the gyrophase-averaged distribution function $\bar{f} = \bar{f}(\boldsymbol{r}, U, t)$ obeys the drift-kinetic equation

$$\frac{\partial \bar{f}}{\partial t} + \boldsymbol{v}_{gc} \cdot \nabla \bar{f} + \frac{dU}{dt}\frac{\partial \bar{f}}{\partial U} = 0\,. \tag{4.340}$$

In that form of the drift-kinetic equation, we have to insert the expression for dU/dt shown in the previous formula.

As mentioned, a more satisfactory derivation of the drift-kinetic equation from the Landau–Fokker–Planck equation requires a more systematic procedure. Besides the transformation to new variables (from $\boldsymbol{r}, \boldsymbol{v}$ to $\boldsymbol{Y}$ (gyrocenter), U, μ, and φ (gyrophase)), the collision term has to be reconsidered. The basic idea is to rewrite the Landau–Fokker–Planck equation in the new variables and to recognize that the variation of the distribution function with respect to the gyrophase φ is fast. That behavior suggests that one make use of a multiple-time analysis. In addition, each part of the distribution function is being split into a gyrophase averaged part and a rapidly oscillating part. When doing that (for a systematic derivation, see Balescu [58]), we arrive at

$$\frac{\partial \bar{f}}{\partial t}+\left(v_{\parallel}\hat{b}+\boldsymbol{v}_d\right)\cdot\nabla_Y \bar{f}-ev_{\parallel}\hat{b}\cdot\frac{\partial \boldsymbol{A}}{\partial t}\frac{\partial \bar{f}}{\partial U}\approx \overline{K_0\{f_0,f_1\}}-\overline{K_1\{f_0,f_0\}}\,, \quad (4.341)$$

where the right-hand side represents the averaged linear collision term. Its evaluation is by no means trivial. Note that the space derivative is now with respect to the guiding-center position $\boldsymbol{Y}$. In addition, we have written the drift-kinetic equation for a species of electric charge e (including a sign). In general, we obtain drift kinetic equations for each species separately.

4.6.2 The Gyrokinetic Approach

Thus far, we have assumed that the fields are not varying rapidly (on the gyroradius scale). However, it is possible that instabilities develop with $k_\perp \rho_i \sim \mathcal{O}(1)$. Let us first, for demonstrative purposes, assume that the perturbations are electrostatic, with the electric field $\boldsymbol{E}_1$. During its gyration, the particle will experience varying electric field strengths, resulting in an additional averaged $E \times B$ velocity

$$\langle \boldsymbol{v}_1\rangle_{es} \approx \frac{c}{B_0}\langle \boldsymbol{E}_1(\boldsymbol{r})\rangle \times \hat{b}\,. \quad (4.342)$$

Let us name the actual position of the particle as $\boldsymbol{r} = \boldsymbol{Y} + \boldsymbol{\rho}$, with

$$\boldsymbol{\rho} = -\frac{v_\perp}{\Omega}\boldsymbol{n}_2 = -\frac{v_\perp}{\Omega}(\cos\varphi\,\hat{e}_1 - \sin\varphi\,\hat{e}_2)\,. \quad (4.343)$$

Let us further consider one (linear) mode

$$\boldsymbol{E}_1(\boldsymbol{r}) = \boldsymbol{E}_1 e^{i\boldsymbol{k}_\perp\cdot\boldsymbol{\rho}} \equiv \boldsymbol{E}_{1A}(\boldsymbol{Y})e^{i\boldsymbol{k}_\perp\cdot(\boldsymbol{Y}+\boldsymbol{\rho})}\,. \quad (4.344)$$

The index A characterizes the slowly varying part of the amplitude. We evaluate

$$\begin{aligned}\langle \boldsymbol{E}_1(\boldsymbol{r})\rangle &\equiv \boldsymbol{E}_{1A}\int \frac{d\varphi}{2\pi}e^{i\frac{v_\perp}{\Omega}(k_2\sin\varphi - k_1\cos\varphi)}\\ &= \boldsymbol{E}_{1A}J_0\left(\frac{v_\perp}{\Omega}k_\perp\right) \equiv -i\boldsymbol{k}_\perp\phi_A J_0(\rho k_\perp)\end{aligned} \quad (4.345)$$

when $k_1 = k_\perp\cos\psi$ and $k_2 = k_\perp\sin\psi$. Thus, for electrostatic perturbations, we obtain the linearized gyrokinetic equation

$$\begin{aligned}&\frac{\partial \bar{f}_1}{\partial t}+\left(v_\parallel\hat{b}+\boldsymbol{v}_d\right)\cdot\nabla\bar{f}_1 - ev_\parallel\hat{b}\cdot\frac{\partial\boldsymbol{A}}{\partial t}\frac{\partial\bar{f}_1}{\partial U}\\ &\approx i\frac{1}{B_0}J_0(\rho k_\perp)\,\phi_A(\boldsymbol{k}\times\hat{b})\cdot\nabla\bar{f}_0\end{aligned} \quad (4.346)$$

in the collisionless approximation. Note that $\bar{f} = \bar{f}_0 + \bar{f}_1$, and $\bar{f}_1$ is the (small) response to the (electrostatic) rapid field perturbation $\boldsymbol{E}_1(\boldsymbol{r})$.

In the presence of (additional) rapid magnetic perturbations $\boldsymbol{B}_1(\boldsymbol{r})$, more first order terms appear. The solution of the linearized equation of motion

$$m\frac{d\boldsymbol{v}}{dt} = e\left[\boldsymbol{v}_1 \times \boldsymbol{B}_0 + \boldsymbol{v} \times \boldsymbol{B}_1\right] \tag{4.347}$$

is straightforward if we remember the derivation of the $E \times B$ drift:

$$\boldsymbol{v}_1 \approx -\frac{1}{B_0}\,\hat{b} \times (\boldsymbol{v} \times \boldsymbol{B}_1) = -\boldsymbol{v}\frac{\hat{b}\cdot\boldsymbol{B}_1}{B_0} + \boldsymbol{B}_1\frac{\hat{b}\cdot\boldsymbol{v}}{B_0} = -\boldsymbol{v}_\perp\frac{B_{1\parallel}}{B_0} + \boldsymbol{B}_{1\perp}\frac{v_\parallel}{B_0}\,. \tag{4.348}$$

Thus, the additional averaged drift is

$$\begin{aligned}\langle \boldsymbol{v}_1\rangle_{em} &= -\frac{1}{B_0}\left\langle B_{1\parallel}\boldsymbol{v}_\perp\right\rangle + \frac{v_\parallel}{B_0}\left\langle \boldsymbol{B}_{1\perp}\right\rangle \\ &\approx -i\frac{v_\perp}{B_0}B_{A\parallel}J_1(\rho k_\perp)\hat{k}_\perp \times \hat{b} + i\frac{v_\parallel}{B_0}\boldsymbol{k}_\perp \times \hat{b}A_{A\parallel}J_0(\rho k_\perp)\,.\end{aligned} \tag{4.349}$$

Since the field will also be time-dependent $\sim e^{-i\omega t}$, the energy is no longer constant and we have to calculate

$$\begin{aligned}\left\langle\frac{dU}{dt}\right\rangle &= -i\omega e\left[\langle\phi_1\rangle - \langle\boldsymbol{v}\cdot\boldsymbol{A}_1\rangle\right] \\ &\approx -i\omega e\Big[J_0(k_\perp\rho)\phi_A - iv_\perp J_1(k_\perp\rho)(\hat{k}_\perp \times \hat{b})\cdot\boldsymbol{A}_{A\perp} \\ &\quad - v_\parallel J_0(k_\perp\rho)A_{A\parallel}\Big]\,.\end{aligned} \tag{4.350}$$

The following additional terms

$$-\langle\boldsymbol{v}_1\rangle_{em}\cdot\nabla\bar{f}_0 - \langle\boldsymbol{v}_1\rangle_{es}\cdot\nabla\bar{f}_0 - \left\langle\frac{dU}{dt}\right\rangle\frac{\partial\bar{f}_0}{\partial U} \tag{4.351}$$

appear on the right-hand side of the linearized gyrokinetic equation. They can be combined, finally leading to

$$\begin{aligned}&\frac{\partial\bar{f}_1}{\partial t} + \left(v_\parallel\hat{b} + \boldsymbol{v}_d\right)\cdot\nabla\bar{f}_1 - ev_\parallel\hat{b}\cdot\frac{\partial\boldsymbol{A}}{\partial t}\frac{\partial\bar{f}_1}{\partial U} \\ &\approx i\left[J_0(k_\perp\rho)\left\{\phi_A - v_\parallel A_{A\parallel}\right\} + \frac{v_\perp}{k_\perp}J_1(k_\perp\rho)B_{A\parallel}\right] \\ &\quad\times\left\{e\omega\frac{\partial\bar{f}_0}{\partial U} + \frac{1}{B_0}\left(\boldsymbol{k}_\perp \times \hat{b}\right)\cdot\nabla\bar{f}_0\right\}.\end{aligned} \tag{4.352}$$

This is the (collisionless) linear gyrokinetic equation for rapid perturbations (on the gyroradius scale). As mentioned, when large disturbances take place, the linearization breaks down and we have to return to the (unreduced) description by the Landau–Fokker–Planck or Balescu–Lenard–Guernsey equations.

5
Fluid Description

In this chapter, we proceed to the so called macroscopic description of plasmas which contains less information than the kinetic formulation. Knowledge of a distribution function f provides us with the statistical information in the $\boldsymbol{r}, \boldsymbol{p}$-phase space at any time t. (Here, we do not elaborate on the difference between the descriptions using $\boldsymbol{p}$ or $\boldsymbol{v}$, as previously mentioned.) When we are not interested in the details in velocity space, we may integrate over velocities in order to obtain equations for the so called moments of the distribution function. Of course, mathematically a finite number of moments contains less information than the distribution function itself. We should mention that the physical consequences of the reduced, macroscopic description also depends on time- and space-scales. We assume that the moments (slowly) vary on the so called hydrodynamic time- and space-scales which are generally different from the characteristic microscopic variations.

We shall develop the macroscopic theory for plasmas within the following sections. First, we define moments. When aiming for the equations for the moments, we face a hierarchy problem, similar to the difficulties when deriving a (closed) kinetic equation. To illustrate which steps are needed for reaching a closed system of equations for the first three moments, we must next summarize, for pedagogical reasons, the steps known from neutral gas theory. When doing so, we do not present all of the details of the calculations since they will appear later when we develop the plasma pendant.

5.1
Moments and Hierarchy of Moment Equations

Averaging over velocity space,

$$\langle \cdots \rangle_\alpha = \frac{\int d^3v \cdots f^\alpha}{\int d^3v\, f} , \tag{5.1}$$

we may introduce plasmadynamic (hydrodynamic in the case of neutral fluids) variables, for example, particle density, mean velocity, and temperature, respective-

High Temperature Plasmas, Theory and Mathematical Tools for Laser and Fusion Plasmas, First Edition.
Karl-Heinz Spatschek.

ly. For the species α, we define the particle density

$$n_\alpha(\boldsymbol{r}, t) = \int d^3v\, f^\alpha \tag{5.2}$$

(note that the normalization of the distribution function varies), the mean velocity

$$\boldsymbol{u}_\alpha = \langle \boldsymbol{v} \rangle , \tag{5.3}$$

and the temperature (again measured in eV; the Boltzmann constant k_B is suppressed)

$$T_\alpha = \frac{1}{3} m_\alpha \langle |\boldsymbol{v} - \boldsymbol{u}_\alpha|^2 \rangle , \tag{5.4}$$

just to name a few.

For a derivation of the dynamical equations for the moments, we start from the kinetic equation in general form

$$\frac{\partial f^\alpha}{\partial t} + \boldsymbol{v} \cdot \frac{\partial f^\alpha}{\partial \boldsymbol{r}} + \frac{1}{m_\alpha} \boldsymbol{F}_\alpha \cdot \frac{\partial f^\alpha}{\partial \boldsymbol{v}} = K^\alpha \{ f^\alpha(t) \} , \tag{5.5}$$

where $\boldsymbol{F}$ is the force and K^α is the collision term. When building the equations for the moments, we keep in mind that $\boldsymbol{r}$, $\boldsymbol{v}$, and t are independent. Furthermore, $f^\alpha \to 0$ as $v \to \infty$.

We start with the zeroth moment n_α. Its dynamical equation is immediately obtained by integrating both sides of Eq. (5.5) over velocity space. Independence from variables as well as Gauss theorem lead to

$$\frac{\partial n_\alpha}{\partial t} + \nabla \cdot (n_\alpha \boldsymbol{u}_\alpha) = 0 . \tag{5.6}$$

Next, we multiply Eq. (5.5) by $\boldsymbol{v}$ and integrate over velocity space,

$$\int d^3v\boldsymbol{v} \left[\frac{\partial f^\alpha}{\partial t} + \boldsymbol{v} \cdot \frac{\partial f^\alpha}{\partial \boldsymbol{r}} + \frac{1}{m_\alpha} \boldsymbol{F}_\alpha \cdot \frac{\partial f^\alpha}{\partial \boldsymbol{v}} \right] = \int d^3v\boldsymbol{v} K^\alpha \{ f^\alpha(t) \} . \tag{5.7}$$

Before proceeding with the evaluation of the second integral, we introduce the variable

$$\boldsymbol{v}' = \boldsymbol{v} - \boldsymbol{u}_\alpha . \tag{5.8}$$

Since $\boldsymbol{u}_\alpha \equiv \boldsymbol{u}_\alpha(\boldsymbol{r}, t)$ is space-dependent, so will $\boldsymbol{v}'$. Therefore, in the second term on the right-hand side, we should take out the operator $\partial/\partial\boldsymbol{r}$ first (since it does not act on $\boldsymbol{v}$, but on $\boldsymbol{v}'$). Furthermore, we assume the force as only $\boldsymbol{r}$-dependent, or in the more general case of the Lorentz form

$$\boldsymbol{F}_\alpha = q_\alpha \left[\boldsymbol{E}(\boldsymbol{r}, t) + \boldsymbol{v} \times \boldsymbol{B}(\boldsymbol{r}, t) \right] , \tag{5.9}$$

such that the third term leads to

$$\int d^3v\boldsymbol{v} q_\alpha \left[\boldsymbol{E}(\boldsymbol{r}, t) + \boldsymbol{v} \times \boldsymbol{B}(\boldsymbol{r}, t) \right] \cdot \frac{\partial f^\alpha}{\partial \boldsymbol{v}} = n_\alpha q_\alpha \left[\boldsymbol{E}(\boldsymbol{r}, t) + \boldsymbol{u}_\alpha \times \boldsymbol{B}(\boldsymbol{r}, t) \right] . \tag{5.10}$$

The collisional contribution from the right-hand side will be abbreviated as

$$\int d^3 v \boldsymbol{v} K^\alpha \{f^\alpha(t)\} = \frac{1}{m_\alpha} \boldsymbol{R}_{\alpha\alpha'} , \tag{5.11}$$

with

$$\boldsymbol{R}_{\alpha\alpha'} = -\nu_{\alpha\alpha'} m_\alpha n_\alpha (\boldsymbol{u}_\alpha - \boldsymbol{u}_{\alpha'}) , \quad \alpha, \alpha' = e, i . \tag{5.12}$$

By combining all terms, we end up with

$$\begin{aligned} \frac{\partial (n_\alpha \boldsymbol{u}_\alpha)}{\partial t} + \frac{\partial}{\partial \boldsymbol{r}} \cdot (n_\alpha \boldsymbol{u}_\alpha \boldsymbol{u}_\alpha) = n_\alpha \frac{q_\alpha}{m_\alpha} \left[\boldsymbol{E}(\boldsymbol{r}, t) + \boldsymbol{u}_\alpha \times \boldsymbol{B}(\boldsymbol{r}, t) \right] \\ - \frac{1}{m_\alpha} \frac{\partial}{\partial \boldsymbol{r}} \cdot \overleftrightarrow{\boldsymbol{P}}_\alpha + \frac{1}{m_\alpha} \boldsymbol{R}_{\alpha\alpha'} , \end{aligned} \tag{5.13}$$

with the pressure tensor

$$\overleftrightarrow{P}_\alpha \equiv \overleftrightarrow{\boldsymbol{P}}_\alpha = m_\alpha \int d^3 v' \boldsymbol{v}' \boldsymbol{v}' f^\alpha , \quad P_{\alpha i j} = m_\alpha \int d^3 v' v'_i v'_j f^\alpha . \tag{5.14}$$

Without the shift in the velocity by $\boldsymbol{u}_\alpha$, we would have the stress tensor

$$\overleftrightarrow{\mathcal{P}}_\alpha = m_\alpha \int d^3 v \boldsymbol{v} \boldsymbol{v} f^\alpha , \quad \mathcal{P}_{\alpha i j} = m_\alpha \int d^3 v v_i v_j f^\alpha . \tag{5.15}$$

The relation between both is

$$\overleftrightarrow{\mathcal{P}}_\alpha = \overleftrightarrow{P}_\alpha + m_\alpha \boldsymbol{u}_\alpha \boldsymbol{u}_\alpha . \tag{5.16}$$

We have defined

$$\frac{\partial}{\partial \boldsymbol{r}} \cdot \overleftrightarrow{P}_\alpha \equiv \nabla \cdot \overleftrightarrow{P}_\alpha , \quad \left[\nabla \cdot \overleftrightarrow{P}_\alpha \right]_j = m_\alpha \sum_{i=1}^{3} \frac{\partial}{\partial x_i} \int d^3 v v'_i v'_j f^\alpha . \tag{5.17}$$

Also,

$$\left[\frac{\partial}{\partial \boldsymbol{r}} \cdot (n_\alpha \boldsymbol{u}_\alpha \boldsymbol{u}_\alpha) \right]_j \equiv [\nabla \cdot (n_\alpha \boldsymbol{u}_\alpha \boldsymbol{u}_\alpha)]_j = \sum_{i=1}^{3} \frac{\partial}{\partial x_i} (n_\alpha u_{\alpha i} u_{\alpha j}) . \tag{5.18}$$

The pressure tensor $\overleftrightarrow{\boldsymbol{P}}_\alpha$ as a higher moment is unknown. For an isotropic distribution functions in $\boldsymbol{v}'$, the diagonal elements are equal ($n_\alpha k_B T_\alpha$ for a Maxwellian), and the off-diagonal terms vanish. When assuming an isotropic distribution, we introduce the scalar pressure

$$p_\alpha = \frac{m_\alpha}{3} \int d^3 v \boldsymbol{v}' \cdot \boldsymbol{v}' f^\alpha . \tag{5.19}$$

We write

$$\overleftrightarrow{P}_\alpha = p_\alpha \overleftrightarrow{I} + \overleftrightarrow{\pi}_\alpha \equiv p_\alpha I + \overleftrightarrow{\pi}_\alpha , \tag{5.20}$$

when no preferred direction occurs; $\overleftrightarrow{\pi}_\alpha$ is called the dissipative part of the pressure tensor. Furthermore, using

$$\frac{\partial(n_\alpha \boldsymbol{u}_\alpha)}{\partial t} + \frac{\partial}{\partial \boldsymbol{r}} \cdot (n_\alpha \boldsymbol{u}_\alpha \boldsymbol{u}_\alpha) = \boldsymbol{u}_\alpha \frac{\partial n_\alpha}{\partial t} + \boldsymbol{u}_\alpha \nabla \cdot (n_\alpha \boldsymbol{u}_\alpha) + n_\alpha \frac{\partial \boldsymbol{u}_\alpha}{\partial t} + n_\alpha (\boldsymbol{u}_\alpha \cdot \nabla) \boldsymbol{u}_\alpha \tag{5.21}$$

and the continuity equation, we finally end up with the momentum equation

$$\frac{\partial \boldsymbol{u}_\alpha}{\partial t} + (\boldsymbol{u}_\alpha \cdot \nabla) \boldsymbol{u}_\alpha = \frac{q_\alpha}{m_\alpha} [\boldsymbol{E} + \boldsymbol{u}_\alpha \times \boldsymbol{B}] - \frac{1}{m_\alpha n_\alpha} \nabla P_\alpha + \frac{1}{m_\alpha n_\alpha} \boldsymbol{R}_{\alpha\alpha'} \,. \tag{5.22}$$

Next, we multiply Eq. (5.5) by $1/2 m_\alpha v^2$ and integrate over velocity space,

$$\frac{m_\alpha}{2} \int d^3 v v^2 \left[\frac{\partial f^\alpha}{\partial t} + \boldsymbol{v} \cdot \frac{\partial f^\alpha}{\partial \boldsymbol{r}} + \frac{1}{m_\alpha} \boldsymbol{F}_\alpha \cdot \frac{\partial f^\alpha}{\partial \boldsymbol{v}} \right] = \frac{m_\alpha}{2} \int d^3 v v^2 K^\alpha \{ f^\alpha(t) \} \,. \tag{5.23}$$

Again, we use the transformation Eq. (5.8), and we evaluate term by term. The first term on the left-hand side leads immediately to

$$\frac{m_\alpha}{2} \int d^3 v v^2 \frac{\partial f^\alpha}{\partial t} = \frac{\partial}{\partial t} \left(\frac{3}{2} p_\alpha + \frac{m_\alpha}{2} n_\alpha u_\alpha^2 \right) . \tag{5.24}$$

The second term may be written as

$$\frac{m_\alpha}{2} \nabla \cdot \int d^3 v v^2 \boldsymbol{v} f^\alpha = \nabla \cdot \left(\mathcal{Q}_\alpha + \overleftrightarrow{P}_\alpha \cdot \boldsymbol{u}_\alpha + \frac{3}{2} p_\alpha \boldsymbol{u}_\alpha + \frac{m_\alpha}{2} n_\alpha u_\alpha^2 \boldsymbol{u}_\alpha \right) , \tag{5.25}$$

where the heat flux density

$$\boldsymbol{q}_\alpha \equiv \frac{m_\alpha}{2} \int d^3 v' v'^2 \boldsymbol{v}' f^\alpha \tag{5.26}$$

has been defined. The latter is related to the energy flux density

$$\mathcal{Q}_\alpha \equiv \frac{m_\alpha}{2} \int d^3 v v^2 \boldsymbol{v} f^\alpha \tag{5.27}$$

via

$$\mathcal{Q}_\alpha = \boldsymbol{q}_\alpha + \overleftrightarrow{P}_\alpha \cdot \boldsymbol{u}_\alpha + \frac{3}{2} p_\alpha \boldsymbol{u}_\alpha + \frac{1}{2} m_\alpha n_\alpha u_\alpha^2 \boldsymbol{u}_\alpha \,. \tag{5.28}$$

The third term containing the force is straightforward to obtain. First, we rearrange

$$\frac{q_\alpha}{2} \int d^3 v v^2 [\boldsymbol{E} + \boldsymbol{v} \times \boldsymbol{B}] \cdot \frac{\partial f^\alpha}{\partial \boldsymbol{v}} = -\frac{q_\alpha}{2} \int d^3 v v^2 \frac{\partial}{\partial \boldsymbol{v}} \cdot [\boldsymbol{E} + \boldsymbol{v} \times \boldsymbol{B}] f^\alpha \tag{5.29}$$

in order to obtain after integration by parts

$$\frac{q_\alpha}{2} \int d^3 v v^2 \left[\boldsymbol{E} + \boldsymbol{v} \times \boldsymbol{B}\right] \cdot \frac{\partial f^\alpha}{\partial \boldsymbol{v}} = -q_\alpha n_\alpha \boldsymbol{E} \cdot \boldsymbol{u}_\alpha \,. \tag{5.30}$$

The collisional term contributes in the abbreviated form

$$\frac{m_\alpha}{2} \int d^3 v v^2 K^\alpha \left\{f^\alpha(t)\right\} \equiv -\left.\frac{dW}{dt}\right|_\alpha \,. \tag{5.31}$$

Putting everything together, we arrive at

$$\begin{aligned} &\frac{\partial}{\partial t}\left(\frac{3}{2} p_\alpha + \underbrace{\frac{m_\alpha}{2} n_\alpha u_\alpha^2}_{*}\right) + \nabla \cdot \left(\boldsymbol{q}_\alpha + \frac{5}{2} p_\alpha \boldsymbol{u}_\alpha + \underbrace{\frac{m_\alpha}{2} n_\alpha u_\alpha^2 \boldsymbol{u}_\alpha}_{**}\right) \\ &= q_\alpha n_\alpha \boldsymbol{E} \cdot \boldsymbol{u}_\alpha - \left.\frac{dW}{dt}\right|_\alpha \,. \end{aligned} \tag{5.32}$$

The terms marked by $*$ and $**$ can be further simplified by using the continuity Eq. (5.6) and the momentum Eq. (5.22). We get

$$* + ** = n_\alpha \left(\frac{\partial}{\partial t} + \boldsymbol{u}_\alpha \cdot \nabla\right) \frac{m_\alpha u_\alpha^2}{2} \,. \tag{5.33}$$

The right-hand side can be further simplified by using an additional identity from dotting Eq. (5.22) with $\boldsymbol{u}_\alpha$. We also substitute

$$\boldsymbol{u}_\alpha \cdot \nabla \boldsymbol{u}_\alpha = \nabla \left(\frac{u_\alpha^2}{2}\right) - \boldsymbol{u}_\alpha \times \nabla \times \boldsymbol{u}_\alpha \tag{5.34}$$

into

$$\begin{aligned} &\boldsymbol{u}_\alpha \cdot \left\{\frac{\partial \boldsymbol{u}_\alpha}{\partial t} + (\boldsymbol{u}_\alpha \cdot \nabla)\, \boldsymbol{u}_\alpha\right\} \\ &= \boldsymbol{u}_\alpha \cdot \left\{\frac{q_\alpha}{m_\alpha}\left[\boldsymbol{E} + \boldsymbol{u}_\alpha \times \boldsymbol{B}\right] - \frac{1}{m_\alpha n_\alpha} \nabla p_\alpha + \frac{1}{m_\alpha n_\alpha} \boldsymbol{R}_{\alpha\alpha'}\right\} . \end{aligned} \tag{5.35}$$

Then, we obtain

$$* + ** = \boldsymbol{u}_\alpha \cdot \left\{q_\alpha n_\alpha \boldsymbol{E} - \nabla p_\alpha + \boldsymbol{R}_{\alpha\alpha'}\right\} . \tag{5.36}$$

Then, the energy equation appears in the form

$$\left(\frac{\partial}{\partial t} + \boldsymbol{u}_\alpha \cdot \nabla\right) \frac{3 p_\alpha}{2} + \frac{5}{2} p_\alpha \nabla \cdot \boldsymbol{u}_\alpha = -\nabla \cdot \boldsymbol{q}_\alpha - \boldsymbol{R}_{\alpha\alpha'} \cdot \boldsymbol{u}_\alpha - \left.\frac{dW}{dt}\right|_\alpha \,. \tag{5.37}$$

Here, we have replaced

$$\frac{3}{2} p_\alpha \nabla \cdot \boldsymbol{u}_\alpha + \overleftrightarrow{P}_\alpha : \nabla \boldsymbol{u}_\alpha = \frac{5}{2} p_\alpha \nabla \cdot \boldsymbol{u}_\alpha \,. \tag{5.38}$$

5.2
Truncation of the Corresponding Hierarchy in the Case of the Boltzmann Equation

The statistical description of neutral particles in the classical regime is known as the celebrated Boltzmann equation

$$\frac{\partial f_1}{\partial t} + \boldsymbol{v}_1 \cdot \frac{\partial f_1}{\partial \boldsymbol{r}_1} + \frac{\boldsymbol{F}}{m} \cdot \frac{\partial f_1}{\partial \boldsymbol{v}_1} = \int d^3 v_2 \int d\Omega\, \sigma(\Omega) |\boldsymbol{v}_2 - \boldsymbol{v}_1| \left(f_2' f_1' - f_2 f_1 \right) . \tag{5.39}$$

The Boltzmann equation is a closed integro-differential equation for the one-particle distribution function f, where here and in the following – whenever appropriate – the index points at the variables, that is, $f_1 = f(\boldsymbol{r}_1, \boldsymbol{v}_1, t)$. The left-hand side of Eq. (5.39) is the free streaming term, whereas the right-hand side is the binary collision term. The purpose of plasma kinetic theory is to derive the corresponding equation(s) for a system of charged particles. Let us now demonstrate how the closed hydrodynamic description is linked to kinetic theory regarding the example of the Boltzmann equation.

5.2.1
Hierarchical Form of the First Moment Equations for the Boltzmann Equation

By straightforward integrations, as discussed in the previous section, we obtain the generally valid equations

$$\frac{\partial n}{\partial t} + \nabla \cdot (n \boldsymbol{u}) = 0 , \tag{5.40}$$

$$\left(\frac{\partial}{\partial t} + \boldsymbol{u} \cdot \nabla \right) \boldsymbol{u} = \frac{1}{m} \boldsymbol{F} - \frac{1}{mn} \nabla \cdot \overleftrightarrow{P} , \tag{5.41}$$

$$\left(\frac{\partial}{\partial t} + \boldsymbol{u} \cdot \nabla \right) T = -\frac{2}{3n} \nabla \cdot \boldsymbol{q} - \frac{2}{3mn} \overleftrightarrow{P} : \overleftrightarrow{\Lambda} . \tag{5.42}$$

The moments appear. The components of the pressure tensor $\overleftrightarrow{P}$ are defined as

$$P_{ij} = mn \langle (v_i - u_i)(v_j - u_j) \rangle . \tag{5.43}$$

Also, the heat flux appears as a higher moment, namely,

$$\boldsymbol{q} = \frac{1}{2} mn \langle (\boldsymbol{v} - \boldsymbol{u}) |\boldsymbol{v} - \boldsymbol{u}|^2 \rangle . \tag{5.44}$$

Finally, we have abbreviated

$$\Lambda_{ij} = \frac{1}{2} m \left(\frac{\partial u_i}{\partial x_j} + \frac{\partial u_j}{\partial x_i} \right) \tag{5.45}$$

as the components of a tensor $\overleftrightarrow{\Lambda}$. It is very important to emphasize that we have not yet obtained the hydrodynamic equations for density, mean velocity and temperature since the higher moments are not known (yet). The evaluation of, for example, the pressure tensor and the heat current, is the highly nontrivial task of transport theory. The latter uses the kinetic theory quite effectively, as will now be outlined for our introductory example of a neutral fluid.

5.2.2
Truncation of the Hierarchy, Transport Coefficients, and the Euler Equation

Let us abbreviate the kinetic equation by

$$\frac{\partial f}{\partial t} + D f = \frac{1}{\varepsilon} J(f|f) \,. \tag{5.46}$$

Here, a very important assumption has been introduced. We assume the (neutral) system to be collision-dominated by introducing the (huge but only symbolic) factor $1/\varepsilon$ in front of the collision term. Having an equation of that type, the multiple-scale analysis is the appropriate method of solution. In neutral fluid theory, it is known as the Chapman–Enskog method. We expand

$$f = f^{(0)} + \varepsilon f^{(1)} + \varepsilon^2 f^{(2)} + \cdots \,, \qquad f^{(\nu)} = f^{(\nu)}(\boldsymbol{r}, \boldsymbol{v}; t_0, t_1, t_2, \ldots) \,, \tag{5.47}$$

$$\frac{\partial}{\partial t} \rightarrow \frac{\partial}{\partial t_0} + \varepsilon \frac{\partial}{\partial t_1} + \varepsilon^2 \frac{\partial}{\partial t_2} + \cdots \tag{5.48}$$

to get to lowest order

$$J^{(0)}\left(f^{(0)} | f^{(0)}\right) = 0 \,. \tag{5.49}$$

Its solution is the celebrated Maxwell distribution

$$f^{(0)}(\boldsymbol{r}, \boldsymbol{v}; t) = n \left(\frac{m}{2\pi T}\right)^{3/2} \exp\left[-\frac{m}{2T}(\boldsymbol{v} - \boldsymbol{u})^2\right] ; \tag{5.50}$$

the hydrodynamic variables appear as parameters. (It is important to note that the parameters appearing in the lowest order distribution function are by definition the exact hydrodynamic quantities.) With the help of the Maxwell distribution, we can evaluate the higher moments. Introducing the results into the general equations for the first three moments leads to the famous Euler equation for the momentum balance,

$$\frac{\partial n}{\partial t} + \nabla \cdot (n \boldsymbol{u}) = 0 \,, \tag{5.51}$$

$$\left(\frac{\partial}{\partial t} + \boldsymbol{u} \cdot \nabla\right) \boldsymbol{u} = \frac{\boldsymbol{F}}{m} - \frac{1}{mn} \nabla p \,, \tag{5.52}$$

$$\left(\frac{\partial}{\partial t} + \boldsymbol{u} \cdot \nabla\right) T = -\frac{2}{3} T \nabla \cdot \boldsymbol{u} \,. \tag{5.53}$$

The factor 3/2 originates from the specific heat at constant volume

$$c_V = \frac{3}{2} \,. \tag{5.54}$$

The three equations consisting of the mass continuity equation, the Euler equation, and the temperature equation are the hydrodynamic equations in dissipationless fluids. Sometimes, the total system is also called Euler equations. Viscosity does not appear.

5.2.3 Equation of State

In the present approximation, the temperature equation and the continuity equation may be combined to

$$\frac{dn}{dt} - c_V \frac{n}{T}\frac{dT}{dt} = 0\,, \tag{5.55}$$

leading to

$$\frac{d}{dt}\left(nT^{-c_V}\right) = 0\,. \tag{5.56}$$

Using $p = nT$, we may write

$$pn^{-5/3} = \text{const}\,. \tag{5.57}$$

5.2.4 Sound Wave Dispersion from the Euler Equations

Linearizing the Euler equations about a state with constant density (n_0) and temperature (T_0), and assuming no flow velocity to zeroth order ($\boldsymbol{u}_0 = 0$), the linearized Euler equations are

$$\frac{\partial \delta n}{\partial t} = -n_0 \nabla \cdot \delta \boldsymbol{u}\,, \quad m n_0 \frac{\partial \delta \boldsymbol{u}}{\partial t} = -\nabla \delta p\,, \quad \frac{\delta n}{n_0} = \frac{3}{2}\frac{\delta T}{T_0}\,. \tag{5.58}$$

The first two equations can be combined to

$$\frac{\partial^2 \delta n}{\partial t^2} = \frac{1}{m}\nabla^2 \delta p\,. \tag{5.59}$$

Having in mind $\delta p = n_0 \delta T + T_0 \delta n$, we immediately find

$$\frac{\partial^2 \delta p}{\partial t^2} = \frac{5 T_0}{3m}\nabla^2 \delta p\,. \tag{5.60}$$

Thus, we obtain compressional waves (longitudinal oscillations) with the phase velocity

$$c_s = \sqrt{\frac{5 T_0}{3m}}\,. \tag{5.61}$$

These are the well-known sound waves.

5.2.5
Next Order Approximation and Navier–Stokes Equations

A better description is obtained when higher order corrections are included. The equation for $f^{(1)}$ is

$$\left(\frac{\partial}{\partial t_0} + D\right) f^{(0)} = J^{(1)}\left(f^{(0)} | f^{(1)}\right), \tag{5.62}$$

with the constraint

$$\int d^3 v\, f^{(1)} \begin{pmatrix} 1 \\ \boldsymbol{v} \\ v^2 \end{pmatrix} = 0\,. \tag{5.63}$$

Its solution is a quite sophisticated task. We write in an abbreviated form (with summation over repeated indices)

$$f^{(1)} = -\left[\frac{1}{T}\frac{\partial T}{\partial x_i} g_i \mathcal{F}(g) + \frac{1}{T}\Lambda_{ij}\left(g_i g_j - \frac{1}{3}\delta_{ij} g^2\right)\mathcal{G}(g)\right] f^{(0)}\,, \tag{5.64}$$

where $\boldsymbol{g} := \boldsymbol{v} - \boldsymbol{u}$. The functions $\mathcal{F}$ and $\mathcal{G}$ can be found by, for example, series solutions. We do not discuss details of that calculation here. Instead, in the next subsection, we present a simplified solution based on the Krook collision term. Incorporating the first order, the basic equations are the continuity equation, the Navier–Stokes equation, and the heat conduction equation:

$$\frac{\partial n}{\partial t} + \nabla\cdot(n\boldsymbol{u}) = 0\,, \tag{5.65}$$

$$\left(\frac{\partial}{\partial t} + \boldsymbol{u}\cdot\nabla\right)\boldsymbol{u} = \frac{\boldsymbol{F}}{m} - \frac{1}{mn}\nabla\left(p - \frac{\mu}{3}\nabla\cdot\boldsymbol{u}\right) + \frac{\mu}{mn}\nabla^2\boldsymbol{u}\,, \tag{5.66}$$

$$\left(\frac{\partial}{\partial t} + \boldsymbol{u}\cdot\nabla\right) T = -\frac{2}{3}(\nabla\cdot\boldsymbol{u})\, T + \frac{2K}{3nm}\nabla^2 T\,. \tag{5.67}$$

The pressure tensor and the heat current are in this approximation

$$P_{ij} = p\delta_{ij} - \frac{2\mu}{m}\left(\Lambda_{ij} - \frac{m}{3}\delta_{ij}\nabla\cdot\boldsymbol{u}\right), \tag{5.68}$$

$$\boldsymbol{q} = -K\nabla T\,, \tag{5.69}$$

where the viscosity and heat conductivity are calculated as

$$\mu = \frac{m^2}{15T}\int d^3 g g^4 f^{(0)}\mathcal{G}(g)\,, \tag{5.70}$$

$$K = \frac{m}{6T}\int d^3 g g^4 f^{(0)}\mathcal{F}(g)\,, \tag{5.71}$$

respectively.

5.2.6 Simplified Solution with a Krook Collision Term

In this subsection, we first estimate the collision term of the Boltzmann equation

$$\left(\frac{\partial f}{\partial t}\right)_{coll} \equiv \int d^3 v_2 \int d\Omega\, \sigma(\Omega)|\boldsymbol{v}_2 - \boldsymbol{v}_1| \left(f_2' f_1' - f_2 f_1\right), \tag{5.72}$$

and then use the simplified version for calculating $f^{(1)}$. Introducing

$$\delta f_\nu \equiv f_\nu - f_\nu^{(0)}, \tag{5.73}$$

where the index ν characterizes the set of variables $\boldsymbol{r}, \boldsymbol{v}_\nu, t$, we may write the linearized collision term as

$$\left(\frac{\partial f}{\partial t}\right)_{coll} \approx \int d^3 v_2 \int d\Omega\, \sigma(\Omega)|\boldsymbol{v}_2 - \boldsymbol{v}_1| \times \left(f_2^{(0)\prime}\delta f_1' - f_2^{(0)}\delta f_1 + f_1^{(0)\prime}\delta f_2' - f_1^{(0)}\delta f_2\right). \tag{5.74}$$

Expecting each term on the right-hand side of similar order, we pick out the second one for the estimate

$$-\delta f_1 \int d^3 v_2 \int d\Omega\, \sigma(\Omega)|\boldsymbol{v}_2 - \boldsymbol{v}_1| f_2^{(0)} \equiv -\frac{\delta f_1}{\tau}. \tag{5.75}$$

This suggests the formal replacement

$$\left(\frac{\partial f}{\partial t}\right)_{coll} \approx -\frac{f - f^{(0)}}{\tau}, \tag{5.76}$$

which is known as a Krook model.

The zeroth order $f^{(0)}$ is a well-known Maxwellian which – besides the explicit velocity dependence – depends on $\boldsymbol{r}$ and t through n, T, and $\boldsymbol{u}$. The first-order distribution function now directly follows from

$$\left(\frac{\partial}{\partial t} + \boldsymbol{v}\cdot\nabla + \frac{\boldsymbol{F}}{m}\right) f^{(0)} \approx \frac{f^{(1)}}{\tau}. \tag{5.77}$$

When calculating the left-hand side, we note the following derivatives

$$\frac{\partial f^{(0)}}{\partial n} = \frac{f^{(0)}}{n}, \quad \frac{\partial f^{(0)}}{\partial T} = \frac{1}{T}\left(\frac{m}{2T}g^2 - \frac{3}{2}\right) f^{(0)}, \tag{5.78}$$

$$\frac{\partial f^{(0)}}{\partial u_i} = \frac{m}{T} g_i f^{(0)}, \quad \frac{\partial f^{(0)}}{\partial v_i} = -\frac{m}{T} g_i f^{(0)}. \tag{5.79}$$

Using these derivatives, the first order solution is obtained as

$$f^{(1)} \approx -\tau f^{(0)} \left[\frac{1}{n}\hat{D} n + \frac{1}{T}\left(\frac{m}{2T}g^2 - \frac{3}{2}\right)\hat{D} T + \frac{m}{T}\sum_j g_j \hat{D} u_j - \frac{1}{T}\boldsymbol{F}\cdot\boldsymbol{g}\right], \tag{5.80}$$

where

$$\hat{D} \equiv \frac{\partial}{\partial t} + \boldsymbol{v} \cdot \nabla \,. \tag{5.81}$$

We have

$$\hat{D} n = -n \nabla \cdot \boldsymbol{u} + \boldsymbol{g} \cdot \nabla n \,, \quad \hat{D} T = -\frac{2}{3} T \nabla \cdot \boldsymbol{u} + \boldsymbol{g} \cdot \nabla T \,, \tag{5.82}$$

$$\hat{D} u_j = \frac{1}{nm} \frac{\partial p}{\partial x_j} + \frac{F_j}{m} + \sum_i g_i \frac{\partial u_j}{\partial x_i} \,, \quad p = nT \,. \tag{5.83}$$

Now, we can write

$$f^{(1)} \approx -\tau \left[\frac{1}{T} \sum_i \frac{\partial T}{\partial x_i} g_i \left(\frac{m}{2T} g^2 - \frac{5}{2} \right) + \frac{1}{T} \sum_{i,j} \Lambda_{ij} \left(g_i g_j - \frac{1}{3} \delta_{ij} g^2 \right) \right] f^{(0)} \,. \tag{5.84}$$

The previously introduced functions $\mathcal{F}$ and $\mathcal{G}$ follow as

$$\mathcal{F} \approx \tau \left(\frac{m}{2T} g^2 - \frac{5}{2} \right) \,, \quad \mathcal{G} \approx \tau \,. \tag{5.85}$$

Next, we calculate the heat flux density $\boldsymbol{q}$. The first order distribution function $f^{(1)}$ contributes to $\boldsymbol{q}$ in the form $-K \nabla T$, where

$$K = \frac{m^2 \tau}{6T} \int d^3 g g^4 \left(\frac{m}{2T} g^2 - \frac{5}{2} \right) f^{(0)} = \frac{5}{2} \tau T n \,. \tag{5.86}$$

In a similar way, we calculate the pressure tensor $\overleftrightarrow{P}$ to obtain the representation $P_{ij} = p \delta_{ij} + \pi_{ij}$ with

$$\pi_{ij} = -\frac{\tau m}{T} \sum_{k,l} \Lambda_{kl} \int d^3 g g_i g_j \left(\underbrace{g_k g_l}_{(1)} - \frac{1}{3} \delta_{kl} g^2 \right) f^{(0)} \,. \tag{5.87}$$

When determining the components in detail, we note that π_{ij} is a symmetric tensor with zero trace. Its first part originating from the underbraced terms (1) will be proportional to Λ_{ij},

$$\pi_{ij}^{(1)} \sim \Lambda_{ij} \,. \tag{5.88}$$

The previously defined tensor Λ_{ij} is also symmetric, but has the nonvanishing trace

$$\sum_i \Lambda_{ii} = m \sum_i \frac{\partial u_i}{\partial x_i} = m \nabla \cdot \boldsymbol{u} \,. \tag{5.89}$$

This suggests the ansatz

$$\pi_{ij} \sim \left(\Lambda_{ij} - \frac{m}{3} \delta_{ij} \nabla \cdot \boldsymbol{u} \right) . \tag{5.90}$$

To determine the proportionality factor, it is sufficient to calculate one component, that is,

$$\begin{aligned} \pi_{12} &= -\frac{\tau m}{T} \sum_{k,l} \Lambda_{kl} \int d^3 g g_1 g_2 \left(g_k g_l - \frac{1}{3} \delta_{kl} g^2 \right) f^{(0)} \\ &= -2 \frac{\tau m}{T} \Lambda_{12} \int d^3 g g_1^2 g_2^2 f^{(0)} \\ &= -2 \frac{\tau n T}{m} \Lambda_{12} \equiv -2 \frac{\mu}{m} \Lambda_{12} . \end{aligned} \tag{5.91}$$

Thus, we obtain

$$\pi_{ij} = p \delta_{ij} - 2 \frac{\mu}{m} \left(\Lambda_{ij} - \frac{m}{3} \delta_{ij} \nabla \cdot \boldsymbol{u} \right) . \tag{5.92}$$

The dissipative part of the pressure tensor contains the viscosity

$$\mu = \tau n T . \tag{5.93}$$

For the ratio,

$$\frac{K}{\mu} = \frac{5}{2} = \frac{5}{3} c_V \tag{5.94}$$

follows.

5.3 General Outline and Models for Plasmas

In principle, plasma dynamical equations are derived in a similar way as outlined for neutral gases starting from the Boltzmann equation. However, in plasmas, the long-range Coulomb forces cause significant differences. We outline the procedure when starting from the Landau–Fokker–Planck equation. This is the basis for classical plasma transport theory. Neoclassical transport theory starts from the drift-kinetic equation in curved magnetic fields which however is not discussed here. For that fusion plasma physics application, we refer to, for example, [14, 58, 59].

5.3.1 General Starting Point

We start with the reminder that transport theory is based on kinetic equations. One reduces the information by deriving equations for the moments of the distribution function(s). After defining hydrodynamic variables, we can immediately derive exact – although not closed – equations by integrating the kinetic equation. Let us do it here for the plasmadynamical variables $n_\alpha(\boldsymbol{r},t) = \int d^3v\, f^\alpha(\boldsymbol{v},\boldsymbol{r},t)$, $n_\alpha(\boldsymbol{r},t)\boldsymbol{u}^\alpha(\boldsymbol{r},t) = \int d^3v\boldsymbol{v}\, f^\alpha(\boldsymbol{v},\boldsymbol{r},t)$, $n_\alpha T_\alpha = 1/3 m_\alpha \int d^3v|\boldsymbol{v}-\boldsymbol{u}^\alpha|^2 f^\alpha(\boldsymbol{v},\boldsymbol{r},t)$. The scalar pressure is $p_\alpha = n_\alpha T_\alpha$ when temperature is measured in eV. To avoid confusion with vector components, we occasionally write the particle species index α as an upper index.

Straightforward integration of the Landau–Fokker–Planck equation leads to

$$\partial_t n_\alpha + \nabla \cdot (n_\alpha \boldsymbol{u}^\alpha) = 0\,, \tag{5.95}$$

$$\begin{aligned}&\partial_t\left(m_\alpha n_\alpha u_r^\alpha\right) + \nabla_m\left(m_\alpha n_\alpha u_r^\alpha u_m^\alpha + \delta_{rm} n_\alpha T_\alpha + \Pi_{rm}^\alpha\right)\\ &\quad - e_\alpha n_\alpha\left(E_r + \varepsilon_{rmn} u_m^\alpha B_n\right) = R_r^\alpha\,,\end{aligned} \tag{5.96}$$

$$n_\alpha \partial_t T_\alpha = -n_\alpha\left(\boldsymbol{u}^\alpha\cdot\nabla\right)T_\alpha - \frac{2}{3}n_\alpha T_\alpha \nabla\cdot\boldsymbol{u}^\alpha - \frac{2}{3}\Pi_{mn}^\alpha\nabla_m u_n^\alpha - \frac{2}{3}\nabla_m q_m^\alpha + \frac{2}{3}Q^\alpha\,, \tag{5.97}$$

where

$$R_r^\alpha \equiv m_\alpha \int K^\alpha v_r d^3v\,, \quad Q^\alpha \equiv \frac{1}{2}m_\alpha \int K^\alpha|\boldsymbol{v}-\boldsymbol{u}^\alpha|^2 d^3v\,. \tag{5.98}$$

Also, the dissipative part of the pressure tensor

$$\begin{aligned}\Pi_{rs}^\alpha(\boldsymbol{r},t) &= m_\alpha \int d^3v\left[v_r - u_r^\alpha\right]\left[v_s - u_s^\alpha\right] f^\alpha(\boldsymbol{v},\boldsymbol{r},t)\\ &\quad - \frac{1}{3}m_\alpha \delta_{rs}\int d^3v|\boldsymbol{v}-\boldsymbol{u}^\alpha|^2 f^\alpha(\boldsymbol{v},\boldsymbol{r},t)\end{aligned} \tag{5.99}$$

and the heat flux density

$$q_r^\alpha(\boldsymbol{r},t) = \frac{1}{2}m_\alpha \int d^3v\left[v_r - u_r^\alpha\right]|\boldsymbol{v}-\boldsymbol{u}^\alpha|^2 f^\alpha(\boldsymbol{v},\boldsymbol{r},t) \tag{5.100}$$

are still undetermined. Dynamic equations for the mass density ρ, the charge density σ, the mean mass velocity $\boldsymbol{u}$, the electric current density $\boldsymbol{j}$ follow from here by taking sums or differences. Note that all of these equations still contain the exact distribution function(s) $f^\alpha(\boldsymbol{v},\boldsymbol{r},t)$. For a further simplification, one only considers phenomena taking place on a hydrodynamic scale being much slower than the collisional time scale. Then, the distribution function(s) $f^\alpha(\boldsymbol{v},\boldsymbol{r},t)$ can be assumed to be close to a Maxwellian

$$f_0^\alpha = n_\alpha\left(\frac{m_\alpha}{2\pi T_\alpha}\right)^{3/2}\exp\left[-\frac{m_\alpha|\boldsymbol{v}-\boldsymbol{u}^\alpha|^2}{2T_\alpha}\right], \tag{5.101}$$

with varying coefficients n_α, $\boldsymbol{u}^\alpha$, and T_α. The small deviations χ^α, where

$$f^\alpha = f_0^\alpha\left[1 + \chi^\alpha\right], \tag{5.102}$$

are expanded in a certain basis, for example, Hermite polynomials. With specific truncation assumptions (which will be discussed in the chapter on linear transport theory), the transport equations can be calculated up to a desired degree of accuracy, and one obtains a closed set of so called two-fluid equations. For more details see, for example, the excellent book by Balescu [40, 58].

At this stage, let us only make one important remark. When looking within transport theory for (iterated) approximate solutions of the kinetic equations(s), we find as the dominating part a Maxwellian. For a Maxwellian, the collision term vanishes, which is a requirement to zeroth order within the collision dominated regime. When we consider wave-kinetic equations, we also consider to lowest order vanishing collision terms, however, in the opposite limit of negligible particle collisions.

5.3.2
Simple Two-Fluid Model in Unmagnetized Plasma

Thus far, we have presented the first three moment equations for plasmas in the two-fluid approximation. However, they have one disadvantage, namely, they do not form a closed set of equations. Truncations are necessary. Transport theory will provide the systematic procedure to do that. However, for a first insight into standard problems of plasma dynamics, simple models may be helpful. We shall provide a simple two-fluid model in the following using ad hoc arguments. We will consider a situation with approximately an isotropic Maxwellian distribution, and apply the scalar pressure assumption. As the result, we expect a model similar to the Euler equations.

First, we take a look at the energy balance and ignore the two collisional contributions describing energy exchange between the components. This might be adequate for so called collisionless plasmas. Still, the unknown heat flux term is present. We expect that when the heat conductivity becomes large, the term with the heat flux density will dominate unless the temperature becomes spatially uniform. Considering phenomena with low phase-velocity compared to the thermal velocity, heat transport will equalize the temperature(s) and we shall postulate constant temperatures, that is,

$$\frac{v_{phase}}{v_{th\alpha}} \ll 1 \Rightarrow T_\alpha = \text{const} \quad \text{(isothermal)} . \tag{5.103}$$

The opposite limit is the adiabatic limit when the heat flux will be ignored,

$$\frac{v_{phase}}{v_{th\alpha}} \gg 1 \Rightarrow \left(\frac{\partial}{\partial t} + \boldsymbol{u}_\alpha \cdot \nabla\right) \frac{N p_\alpha}{2} + \frac{N+2}{2} p_\alpha \nabla \cdot \boldsymbol{u}_\alpha \approx 0 \quad \text{(adiabatic)} . \tag{5.104}$$

The adiabatic equation of state can be reformulated after introducing

$$\gamma = \frac{N+2}{N} , \qquad N = 1, 2, 3 \quad \text{dimension of the system} , \tag{5.105}$$

and noting

$$\left(\frac{\partial}{\partial t} + \boldsymbol{u}_\alpha \cdot \nabla\right) p_\alpha + \gamma p_\alpha \nabla \cdot \boldsymbol{u}_\alpha = \frac{d p_\alpha}{dt} - \gamma p_\alpha \frac{1}{n_\alpha} \frac{d n_\alpha}{dt} . \tag{5.106}$$

Integration leads to

$$p_\alpha \sim n_\alpha^\gamma \,, \quad p_\alpha = \frac{p_{\alpha 0}}{n_{\alpha 0}^\gamma} n_\alpha^\gamma \quad \text{(adiabatic)} \,. \tag{5.107}$$

In the opposite case, we have

$$p_\alpha = n_\alpha T_\alpha \,, \quad T_\alpha = \text{const} \quad \text{(isothermal)} \,. \tag{5.108}$$

In a collisionless, unmagnetized, isotropic plasma, we may use as the simplest set of two-fluid equations

$$\frac{\partial n_\alpha}{\partial t} + \nabla \cdot (n_\alpha \boldsymbol{u}_\alpha) = 0 \,, \tag{5.109}$$

$$m_\alpha n_\alpha \left(\frac{\partial \boldsymbol{u}_\alpha}{\partial t} + \boldsymbol{u}_\alpha \cdot \nabla \boldsymbol{u}_\alpha \right) = e_\alpha n_\alpha \boldsymbol{E} - \nabla p_\alpha \,, \tag{5.110}$$

closing the system by using one of the above mentioned equations of state $p_\alpha = p_\alpha(n_\alpha)$. Of course, in addition, Maxwell equations should be used for the fields.

5.3.2.1 Electron Plasma Oscillations

We present a short derivation of electron plasma oscillations which we already detected within the kinetic description. From kinetic theory, we already know that the real frequency obeys in the long-wavelength limit

$$\omega^2 \approx \omega_{pe}^2 + 3k^2 v_{the}^2 \,. \tag{5.111}$$

Thus, when we aim for a rederivation within the two-fluid model, we use the fact that the oscillations are of high frequency $\omega \geq \omega_{pe}$, and in the long-wavelength limit the phase velocity becomes very large, $\omega/k \gg v_{the}$ for $k \to 0$. As a consequence of the latter, we may use an adiabatic equation of state for the electrons,

$$p_e \sim n_e^\gamma \,, \quad \text{with} \quad \gamma = 3 \quad \text{for } N = 1 \,. \tag{5.112}$$

The high frequency nature of electron plasma oscillations allows one to consider the ions as an immobile background. Thus, linearizing the simple two-fluid equations about $n_{e0} = n_{i0} \equiv n_0 = \text{const}$ and $\boldsymbol{u}_{e0} = 0$, we obtain for $\delta n_e \equiv \delta n$ and $\delta \boldsymbol{u}_e \equiv \delta \boldsymbol{u}$

$$\frac{\partial \delta n}{\partial t} = -\nabla \cdot (n_0 \delta \boldsymbol{u}) \,, \tag{5.113}$$

$$m_e n_0 \frac{\partial \delta \boldsymbol{u}}{\partial t} = -3 T_e \nabla \delta n - e n_0 \delta \boldsymbol{E} \,, \tag{5.114}$$

$$\nabla \cdot \delta \boldsymbol{E} = -\frac{1}{\varepsilon_0} e \delta n \,. \tag{5.115}$$

We consider in the one-dimensional model a Fourier mode $\delta n \sim \delta u_x \sim \delta E_x \sim e^{ikx - i\omega t}$ and omit the index x in the following. The linear homogeneous system of equations for the Fourier amplitudes

$$-\omega\delta n = -kn_0\delta u\,, \quad -mn_0 i\omega\delta u = -3T_e ik\delta n - en_0\delta E\,,$$

$$ik\delta E = -\frac{1}{\varepsilon_0}e\delta n\,, \tag{5.116}$$

has a nontrivial solution provided

$$\omega^2 = \omega_{pe}^2 + 3\frac{T_e}{m_e}k^2\,. \tag{5.117}$$

That condition is easily found by, for example, eliminating δn and δE from the above equations. The dispersion relation is the expected result which we have already derived from kinetic theory.

5.3.2.2 **Ion-Sound Waves**

In an unmagnetized plasma, a second branch also exists which generally is called the acoustic branch. In an electron-ion plasma, it is known as the ion-acoustic oscillation. For its derivation, we may assume the case $T_e \gg T_i$ which will lead to

$$v_{thi} \ll \frac{\omega}{k} \ll v_{the}\,. \tag{5.118}$$

At the end, we shall find that this is a consistent assumption. Obviously, because of the low frequencies under consideration, we now have to consider the ion dynamics in the equations of motion. The electron inertia can be neglected since the frequencies are well-below the electron plasma frequency. Electrons are approximately Boltzmann distributed. After linearizing and Fourier transforming the simple two-fluid model with adiabatic ions and isothermal electrons in a one-dimensional description, we obtain

$$-\omega\delta n_i = -kn_0\delta u_i\,, \quad -m_i n_0 i\omega\delta u_i = en_0\delta E\,, \tag{5.119}$$

$$0 \approx -T_e ik\delta n_e - en_0\delta E\,, \quad ik\delta E = \frac{1}{\varepsilon_0}e(\delta n_i - \delta n_e)\,. \tag{5.120}$$

The last equations yield the electron oscillations as

$$\delta n_e = \delta n_i - \frac{ik\varepsilon_0}{e}\delta E\,. \tag{5.121}$$

Using that in the momentum balance of electrons, we find the electric field oscillations in terms of the ion oscillations,

$$\delta E = -\frac{ikT_e}{en_o\left(1 + k^2\lambda_{De}^2\right)}\delta n_i\,. \tag{5.122}$$

With the ion momentum balance, we express the ion fluid velocity in terms of the ion density oscillations,

$$-m_i n_0\omega\delta u_i = -\frac{kT_e}{\left(1 + k^2\lambda_{De}^2\right)}\delta n_i\,. \tag{5.123}$$

When inserting these results into the ion continuity equation, a short rearrangement leads for nonvanishing ion oscillations to

$$\omega^2 \approx \frac{k^2 c_s^2}{1 + k^2 \lambda_{De}^2} , \tag{5.124}$$

where the ion-sound velocity has been defined as

$$c_s = \sqrt{\frac{T_e}{m_i}} . \tag{5.125}$$

Clearly, for $T_e \gg T_i$ the assumption made at the beginning applies.

5.3.2.3 Electromagnetic Waves

In an unmagnetized plasma, electromagnetic waves may propagate provided $\omega > \omega_{pe}$. To show this, we couple the Maxwell equations for a pure electromagnetic wave

$$\nabla \times \boldsymbol{E} = -\frac{\partial \boldsymbol{B}}{\partial t} , \quad \nabla \times \boldsymbol{B} = \mu_0 \boldsymbol{j} + \varepsilon_0 \mu_0 \frac{\partial \boldsymbol{E}}{\partial t} \tag{5.126}$$

to the electron plasma motion in the plasma. By first combining the Maxwell equations, we immediately obtain

$$\nabla \times (\nabla \times \boldsymbol{E}) \equiv \nabla(\nabla \cdot \boldsymbol{E}) - \nabla^2 \boldsymbol{E} = -\mu_0 \frac{\partial \boldsymbol{j}}{\partial t} - \varepsilon_0 \mu_0 \frac{\partial^2 \boldsymbol{E}}{\partial t^2} . \tag{5.127}$$

The electric current density $\boldsymbol{j}$ will be calculated from

$$\boldsymbol{j} \equiv \delta \boldsymbol{j} \approx \sum_\alpha n_0 e_\alpha \delta \boldsymbol{u}_\alpha . \tag{5.128}$$

Linearizing the momentum equations in the zero pressure approximation $p_\alpha \approx 0$, we obtain

$$m_\alpha \frac{\partial \delta \boldsymbol{u}_\alpha}{\partial t} \approx e_\alpha \delta \boldsymbol{E} \tag{5.129}$$

or

$$\frac{\partial \boldsymbol{j}}{\partial t} \approx \underbrace{\sum_\alpha \frac{e_\alpha^2 n_0}{m_\alpha}}_{\equiv \varepsilon_0 \omega_p^2} \delta \boldsymbol{E} . \tag{5.130}$$

The wave equation becomes

$$\nabla(\nabla \cdot \boldsymbol{E}) - \nabla^2 \boldsymbol{E} = -\varepsilon_0 \mu_0 \omega_p^2 \boldsymbol{E} - \varepsilon_0 \mu_0 \frac{\partial^2 \boldsymbol{E}}{\partial t^2} . \tag{5.131}$$

After Fourier transformation, we get, with $c^2 = 1/\varepsilon_0 \mu_0$,

$$-\boldsymbol{k}(\boldsymbol{k} \cdot \boldsymbol{E}) + k^2 \boldsymbol{E} = \frac{\omega^2 - \omega_p^2}{c^2} \boldsymbol{E} . \tag{5.132}$$

Transverse electromagnetic waves have $\boldsymbol{k} \perp \boldsymbol{E}$ such that

$$\omega^2 \approx \omega_p^2 + k^2 c^2 \tag{5.133}$$

appears as the electromagnetic dispersion relation. In the nonrelativistic limit $c \gg v_{the}$, the zero pressure assumption is justified.

On the other hand, the zero pressure limit yields for electrostatic $\boldsymbol{k} \parallel \boldsymbol{E}$ oscillations $\omega^2 \approx \omega_p^2$ in agreement with the previous calculation.

5.3.3
Drift Model

Now, let us briefly discuss the drift truncation when the $E \times B$ velocity is small compared to the thermal velocity. The assumed distribution function deviates from a Maxwellian

$$f_M(\boldsymbol{v}) = \frac{n m^{3/2}}{(2\pi T)^{3/2}} e^{-\frac{mv^2}{2T}} \tag{5.134}$$

by two terms, that is,

$$f \approx f_M(\boldsymbol{v}) \left[1 + \frac{u_\| v_\|}{v_{th}^2}\right] - \boldsymbol{\rho} \cdot \nabla f_M + \mathcal{O}(\delta^2)\,. \tag{5.135}$$

The first term in this ansatz yields the parallel velocity through

$$\int d^3 v\, f_M(\boldsymbol{v}) \frac{m v_\| u_\|}{T} v_\| = n u_\|\,, \tag{5.136}$$

while the second one leads to the drift velocities

$$\int d^3 v\, (-\boldsymbol{\rho} \cdot \nabla f_M)\, \boldsymbol{v}_\perp = \frac{p}{m\Omega} \boldsymbol{b} \times \left(\nabla \ln p + \frac{e \nabla \phi}{T}\right) \equiv n \boldsymbol{u}_\perp\,. \tag{5.137}$$

The other moments can also be evaluated, although the algebra is quite tedious. Therefore, we do not present the details here. In summary, using the drift ordering in all equations, we finally arrive in a quasineutral case for $P = p_e + p_i$, $p_e \approx p_i \approx p$, and $T_e \approx T_i \approx T$ at the following (drift) model [59]

$$\frac{dn}{dt} + n \nabla \cdot \boldsymbol{u} = 0\,, \tag{5.138}$$

$$\frac{dP}{dt} - \boldsymbol{v}_{pi} \cdot \nabla P + \frac{5}{3} P \nabla \cdot \left(\boldsymbol{v}_{E\times B} + \boldsymbol{b} u_\|\right) = 0\,, \tag{5.139}$$

$$m_i n \left[\frac{d \boldsymbol{v}_{E\times B}}{dt} + \frac{d}{dt}(\boldsymbol{b} u_\|) - \boldsymbol{v}_{pi} \cdot \nabla(\boldsymbol{b} u_\|)\right] + \nabla P - \boldsymbol{j} \times \boldsymbol{B} = 0\,, \tag{5.140}$$

$$\boldsymbol{E} + \boldsymbol{u} \times \boldsymbol{B} + \frac{1}{en}\left(\frac{1}{2} \nabla P - \boldsymbol{j} \times \boldsymbol{B}\right) - \eta \left[\boldsymbol{j} - \frac{3}{4} \frac{n}{B} \boldsymbol{b} \times \nabla \frac{P}{n}\right] = 0\,. \tag{5.141}$$

The relations

$$\nabla \times \boldsymbol{B} \approx \mu_0 \boldsymbol{j} , \quad \nabla \times \boldsymbol{E} = -\frac{\partial \boldsymbol{B}}{\partial t} , \tag{5.142}$$

$$\frac{d}{dt} = \partial_t + \boldsymbol{u} \cdot \nabla , \quad \boldsymbol{v}_{E\times B} = \frac{\boldsymbol{E} \times \boldsymbol{B}}{B^2} \tag{5.143}$$

have to be added. The ion polarization drift is defined as

$$\boldsymbol{v}_{pi} = \frac{1}{2\Omega_i m_i n} \boldsymbol{b} \times \nabla P . \tag{5.144}$$

We again simplified the problem by changing from the detailed description in configuration and velocity space (via the distribution functions f^α) to a reduced one only in configuration space (via a finite number of moments $n, \boldsymbol{u}, P, \boldsymbol{B}, \ldots$). When we are not interested in detailed wave-particle resonances, and when slow (hydrodynamic) processes are of interest, such a reduction is advisable. The momentum equations still contain important informations, for example, about nonlinear processes, collective effects, and so on [60–63]. To demonstrate this, let us recall, for example, the electrostatic drift waves.

5.3.3.1 Electrostatic Drift Waves

Within the drift model, it is easy to derive the dispersion of low frequency drift waves which occur in inhomogeneous plasmas. Let us discuss the electrostatic limit with $\boldsymbol{E} = -\nabla\phi$ in a constant magnetic field (index 0: $\boldsymbol{B} = \boldsymbol{B}_0$) such that $\nabla_\parallel = \boldsymbol{b}_0 \cdot \nabla$. We shall finally discuss the nonresistive case $\eta \approx 0$ and consider the ion polarization drift v_{pi} as small. Because of $\nabla \cdot \boldsymbol{v}_{E\times B} = 0$, $\nabla \cdot \boldsymbol{v}_{pi} = -\boldsymbol{v}_{pi} \cdot \nabla \ln n$, and $\nabla \cdot (\boldsymbol{b} u_\parallel) = \boldsymbol{b}_0 \cdot \nabla u_\parallel = \nabla_\parallel u_\parallel$, the basic Eqs. (5.138)–(5.141) can be considerably simplified. Equation (5.138) becomes

$$\frac{\partial n}{\partial t} + (\boldsymbol{u} - \boldsymbol{v}_{pi}) \cdot \nabla n \approx -n \nabla_\parallel u_\parallel . \tag{5.145}$$

The parallel component of Eq. (5.140) becomes

$$m_i n \left(\frac{\partial}{\partial t} + (\boldsymbol{u} - \boldsymbol{v}_{pi}) \cdot \nabla \right) u_\parallel + \nabla_\parallel u_\parallel \approx 0 . \tag{5.146}$$

Again, the parallel component, now of Eq. (5.141), will be used to find

$$-\nabla_\parallel \phi + \frac{1}{2en} \nabla_\parallel P \approx \eta j_\parallel \approx 0 . \tag{5.147}$$

When linearizing in the slab approximation (the zeroth order quantities get the index 0, and the first order is designated by the index 1), we use $u_{\parallel 0} = 0$ and, for example,

$$\nabla n_0 = \hat{x} n_0' , \quad \nabla n_1 = i\boldsymbol{k} n_1 , \quad \frac{\partial n_1}{\partial t} = -i\omega n_1 , \tag{5.148}$$

and so on. Furthermore, we introduce $\boldsymbol{k}_\perp = (\boldsymbol{b} \times \hat{x}) \cdot \boldsymbol{k} = k_y$, and

$$\omega_* \equiv -k_y \frac{T}{e B_0} \frac{P_0'}{P_0}, \qquad \omega_E \equiv k_y \frac{\phi_0'}{B_0}. \tag{5.149}$$

Equations (5.145) and (5.147) then lead to

$$-i(\omega - \omega_E)\frac{P_1}{P_0} + i\omega_* \frac{e\phi_1}{T} \approx -i\frac{5}{3} k_\parallel u_{\parallel 1}, \tag{5.150}$$

$$-i(\omega - \omega_E) u_{\parallel 1} \approx -i k_\parallel \frac{P_1}{m_i n_0}, \tag{5.151}$$

$$\frac{P_1}{P_0} \approx \frac{e\phi_1}{T}. \tag{5.152}$$

The solvability condition for nontrivial solutions of the linear homogeneous system is

$$1 - \frac{5}{3} \frac{2k_\parallel^2 T}{m_i(\omega - \omega_E)^2} = \frac{\omega_*}{\omega - \omega_E}, \tag{5.153}$$

with the approximate solution

$$\omega \approx \omega_E + \omega_* \left[1 + \frac{5}{3} \frac{2k_\parallel^2 T}{m_i \omega_*^2} \right]. \tag{5.154}$$

We observe a Doppler shift (ω_E). Besides that, for small $k_\parallel$, the characteristic drift wave frequency is ω_*.

5.3.4
Braginskii Equations

Thus far, we have considered simple fluid models which, for example, correspond to the Euler equations for neutral fluids. The derivation of the general two-fluid model is the core mission of transport theory. The systematic procedure will be outlined in a separate chapter on linear transport theory. Here, we present the result which was first obtained by Braginsii [64] as a summary. For the electrons, the balances for particle number density, momentum, and pressure are

$$\frac{\partial n_e}{\partial t} + \nabla \cdot (n_e \boldsymbol{u}_e) = 0, \tag{5.155}$$

$$m_e n_e \left(\frac{\partial \boldsymbol{u}_e}{\partial t} + \boldsymbol{u}_e \cdot \nabla \boldsymbol{u}_e \right) + \nabla p_e + \nabla \cdot \overleftrightarrow{\pi}_e + e n_e (\boldsymbol{E} + \boldsymbol{u}_e \times \boldsymbol{B}) = \boldsymbol{F}, \tag{5.156}$$

$$\frac{3}{2} \left(\frac{\partial p_e}{\partial t} + \boldsymbol{u}_e \cdot \nabla p_e \right) + \frac{5}{2} p_e \nabla \cdot \boldsymbol{u}_e + \overleftrightarrow{\pi}_e : \nabla \boldsymbol{u}_e + \nabla \cdot \boldsymbol{q}_e = W_e. \tag{5.157}$$

A similar set of equations occurs for the ions ($Z = 1$)

$$\frac{\partial n_i}{\partial t} + \nabla \cdot (n_i \boldsymbol{u}_i) = 0 \,, \tag{5.158}$$

$$m_i n_i \left(\frac{\partial \boldsymbol{u}_i}{\partial t} + \boldsymbol{u}_i \cdot \nabla \boldsymbol{u}_i \right) + \nabla p_i + \nabla \cdot \overleftrightarrow{\pi}_i - e n_i \left(\boldsymbol{E} + \boldsymbol{u}_i \times \boldsymbol{B} \right) = -\boldsymbol{F} \,, \tag{5.159}$$

$$\frac{3}{2} \left(\frac{\partial p_i}{\partial t} + \boldsymbol{u}_i \cdot \nabla p_i \right) + \frac{5}{2} p_i \nabla \cdot \boldsymbol{u}_i + \overleftrightarrow{\pi}_i : \nabla \boldsymbol{u}_i + \nabla \cdot \boldsymbol{q}_i = W_i \,. \tag{5.160}$$

We should have $n_e \approx n_i \equiv n$ for an electron-proton plasma. The important point is that the transport coefficients have been calculated.

For an unmagnetized plasma with $\Omega_e \tau_e$, $\Omega_i \tau_i \ll 1$, we have [64]

$$\boldsymbol{F} = \frac{n e \boldsymbol{j}}{\sigma_\parallel} - 0.71 n \nabla T_e \,, \quad W_i = \frac{3 m_e}{m_i} \frac{n (T_e - T_i)}{\tau_e} \,, \quad W_e = -W_i + \frac{\boldsymbol{j} \cdot \boldsymbol{F}}{n e} \,, \tag{5.161}$$

$$\boldsymbol{j} = -n e (\boldsymbol{u}_e - \boldsymbol{u}_i) \,, \quad \sigma_\parallel = 1.96 \frac{n e^2 \tau_e}{m_e} \,, \tag{5.162}$$

$$\tau_e = \frac{6 \sqrt{2} \pi^{3/2} \varepsilon_0^2 \sqrt{m_e} T_e^{3/2}}{\ln \Lambda n e^4} \,, \quad \tau_i = \frac{12 \pi^{3/2} \varepsilon_0^2 \sqrt{m_i} T_i^{3/2}}{\ln \Lambda n e^4} \,, \tag{5.163}$$

$$\boldsymbol{q}_e = -\kappa_\parallel^e \nabla T_e - 0.71 \frac{T_e}{e} \boldsymbol{j} \,, \quad \boldsymbol{q}_i = -\kappa_\parallel^i \nabla T_i \,, \tag{5.164}$$

$$\kappa_\parallel^e = 3.2 \frac{n \tau_e T_e}{m_e} \,, \quad \kappa_\parallel^i = 3.9 \frac{n \tau_i T_i}{m_i} \,, \tag{5.165}$$

$$(\pi_e)_{\alpha\beta} = -\eta_0^e \left(\frac{\partial u_\alpha}{\partial x_\beta} + \frac{\partial u_\beta}{\partial x_\alpha} - \frac{2}{3} \nabla \cdot \boldsymbol{u}_e \delta_{\alpha\beta} \right) \,, \tag{5.166}$$

$$(\pi_i)_{\alpha\beta} = -\eta_0^i \left(\frac{\partial u_\alpha}{\partial x_\beta} + \frac{\partial u_\beta}{\partial x_\alpha} - \frac{2}{3} \nabla \cdot \boldsymbol{u}_i \delta_{\alpha\beta} \right) \,, \tag{5.167}$$

$$\eta_0^e = 0.73 n \tau_e T_e \,, \quad \eta_0^i = 0.96 n \tau_i T_i \,. \tag{5.168}$$

Note that $\ln \Lambda$ depends on the type of collisions, and is typically between 10 and 20 in a fusion plasma.

For a magnetized plasma with $\Omega_e \tau_e$, $\Omega_i \tau_i \geq 1$, the corresponding values are also known [64], they can be found, for example, in [29].

5.4 MHD Model

Thus far, we have defined plasmadynamical moments for each species. We remember the particle density

$$n_\alpha(\boldsymbol{r}, t) = \int d^3 v \; f^\alpha(\boldsymbol{r}, \boldsymbol{v}; t) \tag{5.169}$$

of species $\alpha = e, i$. The particle current density

$$\boldsymbol{\Gamma}_\alpha(\boldsymbol{r}, t) \equiv n_\alpha(\boldsymbol{r}, t)\boldsymbol{u}_\alpha(\boldsymbol{r}, t) = \int d^3v \boldsymbol{v} f^\alpha(\boldsymbol{r}, \boldsymbol{v}; t) \tag{5.170}$$

contains the mean velocity $\boldsymbol{u}_\alpha$. The energy density is first defined with the total velocity in the form

$$n_\alpha \mathcal{E}_\alpha = \frac{1}{2} m_\alpha \int d^3v v^2 f^\alpha(\boldsymbol{r}, \boldsymbol{v}; t) . \tag{5.171}$$

Taking into account the mean velocity of component α, the thermal energy density will be defined with the (random) relative velocity such that

$$n_\alpha \varepsilon_\alpha = \frac{1}{2} m_\alpha \int d^3v |\boldsymbol{v} - \boldsymbol{u}_\alpha|^2 f^\alpha(\boldsymbol{r}, \boldsymbol{v}; t) . \tag{5.172}$$

Obviously, we have the relation

$$n_\alpha \mathcal{E}_\alpha = \frac{1}{2} m_\alpha n_\alpha |\boldsymbol{u}_\alpha|^2 + n_\alpha \varepsilon_\alpha . \tag{5.173}$$

Within thermodynamic equilibrium and for $\Lambda \gg 1$, we expect

$$p_\alpha = \frac{2}{3} n_\alpha \varepsilon_\alpha \tag{5.174}$$

for an ideal gas. This suggests the general definition

$$p_\alpha(\boldsymbol{r}, t) = \frac{1}{3} m_\alpha \int d^3v |\boldsymbol{v} - \boldsymbol{u}_\alpha|^2 f^\alpha(\boldsymbol{r}, \boldsymbol{v}; t) \tag{5.175}$$

for the scalar pressure. We introduce the pressure tensor via

$$\overleftrightarrow{P}_\alpha = m_\alpha \int d^3v (\boldsymbol{v} - \boldsymbol{u}_\alpha)(\boldsymbol{v} - \boldsymbol{u}_\alpha) f^\alpha(\boldsymbol{r}, \boldsymbol{v}; t) . \tag{5.176}$$

Having in mind the equilibrium relation

$$\varepsilon_\alpha = \frac{3}{2} k_B T_\alpha , \tag{5.177}$$

we define the temperatures T_α via

$$n_\alpha k_B T_\alpha(\boldsymbol{r}, t) = \frac{1}{3} m_\alpha \int d^3v |\boldsymbol{v} - \boldsymbol{u}_\alpha|^2 f^\alpha(\boldsymbol{r}, \boldsymbol{v}; t) . \tag{5.178}$$

The variables $n_\alpha, \boldsymbol{u}_\alpha, T_\alpha$ (for $\alpha = e, i$) are the plasmadynamical variables in a two-fluid description of a plasma.

However, by adding and subtracting appropriately, we can introduce new "variables" like the mass density

$$\rho(\boldsymbol{r}, t) = \sum_\alpha m_\alpha n_\alpha , \tag{5.179}$$

the charge density

$$\Sigma(\boldsymbol{r}, t) = \sum_{\alpha} e_{\alpha} n_{\alpha} , \tag{5.180}$$

the total momentum density, and thereby the center of mass velocity $\boldsymbol{u}$,

$$\rho \boldsymbol{u}(\boldsymbol{r}, t) = \sum_{\alpha} m_{\alpha} n_{\alpha} \boldsymbol{u}_{\alpha} , \tag{5.181}$$

the electric current density

$$\boldsymbol{j}(\boldsymbol{r}, t) = \sum_{\alpha} e_{\alpha} n_{\alpha} \boldsymbol{u}_{\alpha} , \tag{5.182}$$

and the total energy density

$$\rho \mathcal{E} = \sum_{\alpha} n_{\alpha} \mathcal{E}_{\alpha} = \frac{1}{2} \rho u^2 + \rho \varepsilon . \tag{5.183}$$

Note that, in general, ε does not agree with $\sum_{\alpha} \varepsilon_{\alpha}$. We keep the two temperatures T_e and T_i, assuming a rapid thermalization within each component, but only a small energy exchange between the components. The set of hydrodynamical variables ρ, $\boldsymbol{u}$, T_e, T_i plus electrodynamical variables Σ and $\boldsymbol{j}$ is, of course, equivalent to the total set of plasmadynamical variables $n_{\alpha}, \boldsymbol{u}_{\alpha}, T_{\alpha} (\alpha = e, i)$. The latter may be obtained from the hydrodynamic plus electrodynamic variables $\rho, \boldsymbol{u}, T_e, T_i, \Sigma, \boldsymbol{j}$. Setting $e_i = Ze$ and $e_e = -e$, we have

$$n_e = \frac{Ze\rho - m_i \Sigma}{e(m_i + Z m_e)} , \quad n_i = \frac{e\rho + m_e \Sigma}{e(m_i + Z m_e)} , \tag{5.184}$$

$$\boldsymbol{u}_e = \frac{Ze\rho \boldsymbol{u} - m_i \boldsymbol{j}}{Ze\rho - m_i \Sigma} , \quad \boldsymbol{u}_i = \frac{\rho e \boldsymbol{u} + m_e \boldsymbol{j}}{e\rho + m_e \Sigma} . \tag{5.185}$$

Thus, we may easily translate the equations for $n_{\alpha}, \boldsymbol{u}_{\alpha}, T_{\alpha} (\alpha = e, i)$ into those for $\rho, \boldsymbol{u}, T_e, T_i, \Sigma, \boldsymbol{j}$ without gaining any simplification. However, in certain parameter regimes, some of the latter variables are more important than the others, and a significant simplification may become available.

5.4.1 MHD Ordering

We first introduce the popular MHD (magnetohydrodynamic) truncation in some ad hoc manner. We assume that the model should be valid on large space scales L. Large, that is, compared to the Debye length and the (ion) Larmor radius. Then, $\Sigma \approx 0$ will be used. Let us further assume temperatures to be comparable, $T_e \approx T_i \sim T$. The one-fluid description further anticipates that electrons and ions more or less respond together. The time scales τ should be long compared to, for example, the inverse of the ion gyrofrequency. To better understand the synchronized response of electrons and ions, let us consider the momentum balance of species α which may be reformulated as

$$n_\alpha \boldsymbol{u}_{\alpha\perp} = \frac{1}{m_\alpha \Omega_\alpha} \boldsymbol{b} \times \left[\partial_t (m_\alpha n_\alpha \boldsymbol{u}_\alpha) + \nabla \cdot \overleftrightarrow{\mathcal{P}}_\alpha - e_\alpha n_\alpha \boldsymbol{E} - \boldsymbol{R}_\alpha \right] , \tag{5.186}$$

where $\overleftrightarrow{\mathcal{P}}_\alpha$ is the stress tensor and $\boldsymbol{b}$ is the unit magnetic field vector. Electrons and ions respond together when on the right-hand side, the $E \times B$ velocity

$$\boldsymbol{u}_\perp \approx \frac{\boldsymbol{E} \times \boldsymbol{B}}{B^2} \equiv \boldsymbol{u}_{E\times B} \tag{5.187}$$

dominates. The diamagnetic drift should be smaller than the $E \times B$-drift. Thus, one should allow for large (perpendicular to the magnetic field) electric fields. The resulting $E \times B$-velocity $\boldsymbol{u}_{E\times B}$ can be of the order of the thermal velocities. The corresponding MHD approximation thus allows for the possibility of fast and most violent motion. Then, in general, the electromagnetic contribution dominates over the electrostatic one. This can be seen as follows. Having in mind $\boldsymbol{E} = -\nabla\phi - \partial_t \boldsymbol{A}$, we may estimate the various contributions for

$$\frac{e\phi}{T_e} \sim \mathcal{O}(1) , \quad A \sim \mathcal{O}(BL) , \quad \frac{L}{\tau} \sim \mathcal{O}(v_{th}) \tag{5.188}$$

with $v_{th} \sim v_{the} \sim v_{thi}$. We further introduce

$$\delta_\alpha \equiv \frac{\rho_\alpha}{L} \ll 1 , \quad \rho_\alpha = \frac{v_{th\alpha}}{|\Omega_\alpha|} . \tag{5.189}$$

Within this ordering,

$$\frac{|\frac{e_\alpha}{m_\alpha} \nabla\phi|}{|\Omega_\alpha|} \sim \frac{T_\alpha}{L m_\alpha |\Omega_\alpha|} \sim \frac{\rho_\alpha}{L} v_{th\alpha} \sim \delta_\alpha v_{th\alpha} , \qquad \frac{|\frac{e_\alpha}{m_\alpha} \partial_t \boldsymbol{A}|}{|\Omega_\alpha|} \sim \frac{L}{\tau} \sim v_{th} \tag{5.190}$$

follows. Note that $A \sim \mathcal{O}(BL)$ and $E \sim |\partial \boldsymbol{A}/\partial t| \sim v_{th} B$ are essential. The truncation of the hierarchy of moments is then performed with Maxwellians (index M) for each component,

$$f^\alpha \approx f_M^\alpha (\boldsymbol{v} - \boldsymbol{u}) + \mathcal{O}(\delta_\alpha) , \tag{5.191}$$

where $\boldsymbol{u} \approx \boldsymbol{u}_{E\times B} + \boldsymbol{b} u_\parallel$. Thus far, the parallel flow $u_\parallel$ is still unspecified [25]. The assumption of Maxwellians implies that the pressure tensor is isotropic and that the heat flux is negligible. We shall show in the following that these assumptions will lead to

$$\frac{d\rho}{dt} + \rho \nabla \cdot \boldsymbol{u} = 0 , \tag{5.192}$$

$$\rho \frac{d\boldsymbol{u}}{dt} - \boldsymbol{j} \times \boldsymbol{B} + \nabla p = 0 , \tag{5.193}$$

$$\boldsymbol{E} + \boldsymbol{u} \times \boldsymbol{B} = \eta \boldsymbol{j} , \tag{5.194}$$

$$\nabla \times \boldsymbol{E} + \partial_t \boldsymbol{B} = 0 , \tag{5.195}$$

$$\nabla \times \boldsymbol{B} - \mu_0 \boldsymbol{j} = 0 , \tag{5.196}$$

$$\frac{dp}{dt} + \frac{5}{3} p \nabla \cdot \boldsymbol{u} = 0 . \tag{5.197}$$

Here,

$$\frac{d}{dt} = \frac{\partial}{\partial t} + \boldsymbol{u} \cdot \nabla , \tag{5.198}$$

and p is the total pressure. In ideal MHD, the resistivity is ignored, that is, $\eta \approx 0$. Note that two equations can immediately be used to eliminate $\boldsymbol{E}$ and $\boldsymbol{j}$. Then, we are left with eight scalar equations for eight components of ρ, p, $\boldsymbol{u}$, $\boldsymbol{B}$.

To "derive" the MHD equations systematically [65], we start from the kinetic equation, for example, the Landau–Fokker–Planck equation without particle sources and sinks. Binary collisions will be included. We shall take care of the appropriate conservation laws during (elastic) collisions. Let us abbreviate the kinetic equation in the form

$$\partial_t f^\alpha + \boldsymbol{v} \cdot \nabla f^\alpha + \frac{e_\alpha}{m_\alpha} (\boldsymbol{E} + \boldsymbol{v} \times \boldsymbol{B}) \cdot \partial_{\boldsymbol{v}} f^\alpha = K^\alpha \equiv \sum_\beta K^{\alpha\beta} , \tag{5.199}$$

where $f^\alpha \equiv f^\alpha(\boldsymbol{r}, \boldsymbol{v}; t)$. We multiply by powers of velocity $\boldsymbol{v}$, sum over species α, and finally integrate over velocity space. Of course, the resulting moment equations have to be supplemented by the Maxwell equations

$$\nabla \cdot \boldsymbol{E} = \frac{1}{\varepsilon_0} \sum_\alpha e_\alpha \int d^3 v \, f^\alpha , \tag{5.200}$$

$$\nabla \times \boldsymbol{E} = -\frac{\partial \boldsymbol{B}}{\partial t} , \tag{5.201}$$

$$\nabla \cdot \boldsymbol{B} = 0 , \tag{5.202}$$

$$\nabla \times \boldsymbol{B} = \mu_0 \varepsilon_0 \frac{\partial \boldsymbol{E}}{\partial t} + \mu_0 \sum_\alpha e_\alpha \int d^3 v \boldsymbol{v} f^\alpha . \tag{5.203}$$

For the zeroth moment, the calculation is simple. Taking care of the conservation properties of the collision term, we immediately obtain

$$\frac{\partial \rho}{\partial t} + \nabla \cdot (\rho \boldsymbol{u}) = 0 . \tag{5.204}$$

Next, we consider

$$\int d^3 v \sum_\alpha m_\alpha \boldsymbol{v} \left[\partial_t f^\alpha + \boldsymbol{v} \cdot \nabla f^\alpha + \frac{e_\alpha}{m_\alpha} (\boldsymbol{E} + \boldsymbol{v} \times \boldsymbol{B}) \cdot \partial_{\boldsymbol{v}} f^\alpha \right]$$
$$= \sum_{\alpha\beta} \int d^3 v \boldsymbol{v} K^{\alpha\beta} = 0 . \tag{5.205}$$

Before introducing the random velocity $\boldsymbol{v}' = \boldsymbol{v} - \boldsymbol{u}$ with respect to the mean mass velocity $\boldsymbol{u} = \boldsymbol{u}(\boldsymbol{r}, t)$, we make use of the independence of $t, \boldsymbol{r}, \boldsymbol{v}$. We introduce the pressure tensor

$$P_{rs} = \sum_\alpha m_\alpha \int d^3v (v_r - u_r)(v_s - u_s) f^\alpha(\boldsymbol{r}, \boldsymbol{v}; t) , \tag{5.206}$$

which does not agree, in principle, with the sum over the pressure tensors of the components,

$$P_{rs} \neq \sum_\alpha P^\alpha_{rs} . \tag{5.207}$$

The total scalar pressure is defined by

$$p = \frac{1}{3} \mathrm{Tr} \overleftrightarrow{P} , \tag{5.208}$$

such that $P_{rs} = p\delta_{rs} + \pi_{rs}$ when the situation is, to lowest order, isotropic. The dissipative part of the pressure tensor,

$$\begin{aligned} \pi_{rs} = &\sum_\alpha m_\alpha \int d^3v (v_r - u_r)(v_s - u_s) f^\alpha(\boldsymbol{r}, \boldsymbol{v}; t) \\ &- \frac{1}{3} \sum_\alpha m_\alpha \int d^3v |\boldsymbol{v} - \boldsymbol{u}|^2 f^\alpha(\boldsymbol{r}, \boldsymbol{v}; t) \delta_{rs} , \end{aligned} \tag{5.209}$$

is assumed to be of higher order in the MHD approximation. In a similar way, we define the heat flux vector

$$\boldsymbol{q}(\boldsymbol{r}, t) = \frac{1}{2} \sum_\alpha m_\alpha \int d^3v (\boldsymbol{v} - \boldsymbol{u}) |\boldsymbol{v} - \boldsymbol{u}|^2 f^\alpha(\boldsymbol{r}, \boldsymbol{v}; t) . \tag{5.210}$$

A short calculation leads to

$$\frac{\partial \boldsymbol{u}}{\partial t} + \nabla \cdot (\rho \boldsymbol{u} \boldsymbol{u}) = \left[\sum_\alpha n_\alpha e_\alpha \right] \boldsymbol{E} + \boldsymbol{j} \times \boldsymbol{B} - \nabla \cdot \overleftrightarrow{P} . \tag{5.211}$$

First, by $\Sigma(\boldsymbol{r}, t) = \sum_\alpha e_\alpha n_\alpha \approx 0$, the plasma is assumed to be essentially neutral. Note that we will apply MHD to large-scale phenomena. Second, in the same way as for the two-fluid equation, we may use Eq. (5.204) to rearrange terms with the result

$$\rho \frac{d\boldsymbol{u}}{dt} - \boldsymbol{j} \times \boldsymbol{B} + \nabla p = 0 . \tag{5.212}$$

A simple form of Ohm's law for the current density $\boldsymbol{j}$ follows from the electron momentum equation

$$m_e \left(\frac{\partial \boldsymbol{u}_e}{\partial t} + \boldsymbol{u}_e \cdot \nabla \boldsymbol{u}_e \right) \approx -e(\boldsymbol{E} + \boldsymbol{u}_e \times \boldsymbol{B}) - \frac{1}{n_e} \nabla p_e - \nu_{ei} m_e (\boldsymbol{u}_e - \boldsymbol{u}_i) . \tag{5.213}$$

Within the MHD scaling, we can drop the terms on the left-hand side. They are small compared to the magnetic force term on the right-hand side,

$$\left|\frac{\partial \boldsymbol{u}_e}{\partial t} + \boldsymbol{u}_e \cdot \nabla \boldsymbol{u}_e\right| \sim \frac{v_{the}}{\tau} \ll v_{the}|\Omega_e| \sim \frac{e}{m_e}|\boldsymbol{u}_e \times \boldsymbol{B}| \, . \tag{5.214}$$

Furthermore, we replace

$$\boldsymbol{u}_e = \boldsymbol{u}_i - \frac{1}{en_e}\boldsymbol{j} \approx \boldsymbol{u} - \frac{1}{en_e}\boldsymbol{j} \, , \tag{5.215}$$

and rewrite Eq. (5.213) as

$$\boldsymbol{E} + \boldsymbol{u} \times \boldsymbol{B} - \frac{1}{en_e}\boldsymbol{j} \times \boldsymbol{B} + \frac{1}{en_e}\nabla p_e \approx \eta \boldsymbol{j} \, , \tag{5.216}$$

where

$$\eta = \frac{m_e \nu_{ei}}{e^2 n_e} \tag{5.217}$$

is the resistivity. Except in the so called Hall-MHD model, the Hall term $1/(en_e)\boldsymbol{j} \times \boldsymbol{B}$ is usually dropped. We can find arguments for that assumption by comparing the Hall term with either the resistive term (right-hand side) or the pressure term. In resistive MHD, the ratio of the Hall term to the resistive term is of order of magnitude

$$\frac{|\Omega_e|}{\nu_{ei}} \ll 1 \, . \tag{5.218}$$

Otheqrwise, when Eq. (5.212) leads to $j \sim \rho v_{th}/(\tau B)$, we compare the Hall term with $\boldsymbol{u} \times \boldsymbol{B}$ to obtain for the ratio

$$\frac{1}{|\Omega_i|\tau} \ll 1 \tag{5.219}$$

within the MHD scaling. Thus, the Hall term is negligible when the pressure term is small in the MHD equation of motion Eq. (5.212). In the latter case, we write Ohm's law as

$$\boldsymbol{E} + \boldsymbol{u} \times \boldsymbol{B} = \eta \boldsymbol{j} \, . \tag{5.220}$$

Let us compare in Eq. (5.212) the pressure term with the magnetic force term. We find

$$\frac{|\nabla p|}{|\boldsymbol{j} \times \boldsymbol{B}|} \sim \frac{p}{\frac{B^2}{2\mu_0}} \equiv \beta \, . \tag{5.221}$$

The ratio of plasma pressure to magnetic pressure is known as plasma$-\beta$. For $\beta \ll 1$, the Hall term will also be small. When we add Maxwell's induction equation (Faraday's law)

$$\nabla \times \boldsymbol{E} + \partial_t \boldsymbol{B} = 0 \tag{5.222}$$

together with Ampère's law

$$\nabla \times \boldsymbol{B} - \mu_0 \boldsymbol{j} = 0\,, \tag{5.223}$$

ignoring the displacement current for sufficiently slow processes (phase velocities much smaller than the velocity of light), we have almost finished deriving the closed set of MHD equations. For the pressure, we may use the adiabatic law

$$\frac{dp}{dt} + \frac{5}{3} p \nabla \cdot \boldsymbol{u} \equiv \frac{dp}{dt} + \frac{5}{3}\frac{p}{\rho}\frac{d\rho}{dt} = 0\,. \tag{5.224}$$

We remind the reader that

$$\frac{d}{dt} = \frac{\partial}{\partial t} + \boldsymbol{u} \cdot \nabla \tag{5.225}$$

is the total derivative in the one-fluid system. Finally, we may assume dominating diagonal terms in the pressure tensor such that the MHD equations (5.192)–(5.197) follow. We should note that instead of Eq. (5.197), which expresses the adiabatic equation of state,

$$\frac{d}{dt}\left(\frac{p}{\rho^{5/3}}\right) = 0\,, \tag{5.226}$$

also other closures are in use. The isothermal equation of state is

$$\frac{d}{dt}\left(\frac{p}{\rho}\right) = 0\,. \tag{5.227}$$

Also, in some situations, incompressibility may be the better choice. When

$$\nabla \cdot \boldsymbol{u} = 0 \tag{5.228}$$

applies, Eq. (5.197) leads to

$$\frac{d}{dt} p = 0\,. \tag{5.229}$$

We will conclude with several additional remarks. The charge density continuity equation

$$\partial_t \Sigma + \nabla(\boldsymbol{j}) = 0 \tag{5.230}$$

is consistent with the derived set of MHD equations. The neglection of the dissipative part of the pressure tensor is related to

$$\pi_{rs} \sim \mathcal{O}(\delta)\,. \tag{5.231}$$

More detailed models are sometimes used, for example, the so called electron MHD.

The literature on MHD is huge. Important problems fill entire books. We refer to [66, 67] and references therein for a comprehensive overview over MHD.

5.5
Simple MHD Applications

In general, we have to supplement the MHD equations (5.192)–(5.197) by appropriate boundary conditions. In principle, boundary conditions are already discussed in classical electrodynamics so that we waive a general discussion here.

MHD is the basic theory for fusion plasmas. Thus, fusion physicists will consult more specific books, for example, [14, 59, 68] just to mention a few of them, for special aspects. Basic aspects of MHD are broadly discussed, for example, in [67, 69]. In the following, we will illustrate the power of MHD with a few examples.

5.5.1
Frozen-in Magnetic Field Lines

The evolution of a magnetic field in a plasma with constant electric conductivity is governed by

$$\frac{\partial \boldsymbol{B}}{\partial t} - \nabla \times (\boldsymbol{u} \times \boldsymbol{B}) \approx \frac{1}{\mu_0 \sigma} \nabla^2 \boldsymbol{B} . \tag{5.232}$$

The right-hand side represents a diffusion term. For $\boldsymbol{u} \approx 0$, we may estimate the characteristic time for diffusion by

$$\tau \sim \mu_0 \sigma L^2 , \tag{5.233}$$

where L is the characteristic length for variations of the magnetic field $\boldsymbol{B}$. When τ is large (e.g., in the limit $\sigma \to \infty$), we may neglect the diffusion term. Large τ means large compared to the characteristic time of the phenomenon under consideration. Within ideal MHD ($\sigma \to \infty$), we have

$$\frac{\partial \boldsymbol{B}}{\partial t} \approx \nabla \times (\boldsymbol{u} \times \boldsymbol{B}) . \tag{5.234}$$

The latter case describes frozen-in magnetic field lines. To see this, we integrate over a surface moving with the plasma (magneto-fluid) to obtain

$$\int \frac{\partial \boldsymbol{B}}{\partial t} \cdot d\boldsymbol{F} - \int (\boldsymbol{u} \times \boldsymbol{B}) \cdot d\boldsymbol{l} = \int \frac{\partial \boldsymbol{B}}{\partial t} \cdot d\boldsymbol{F} + \int \boldsymbol{B} \cdot (\boldsymbol{u} \times d\boldsymbol{l}) = \frac{d}{dt} \int \boldsymbol{B} \cdot d\boldsymbol{F} = 0 . \tag{5.235}$$

5.5.2
MHD Equilibria

Let us consider the ideal MHD equations static case ($\partial_t = 0$, $\boldsymbol{u} = 0$)

$$\nabla p = (\boldsymbol{j} \times \boldsymbol{B}) , \quad \nabla \times \boldsymbol{B} = \mu_0 \boldsymbol{j} , \quad \nabla \cdot \boldsymbol{B} = 0 . \tag{5.236}$$

For obvious reasons, the case is called ideal magneto-hydrostatics. Eliminating $\boldsymbol{j}$ leads to

$$(\nabla \times \boldsymbol{B}) \times \boldsymbol{B} = \mu_0 \nabla p \; . \tag{5.237}$$

Writing the last equation as

$$\nabla \left(p + \frac{B^2}{2\mu_0} \right) = \boldsymbol{B} \cdot \nabla \boldsymbol{B} \; , \tag{5.238}$$

we can easily interpret the relation as a pressure balance. Besides the kinetic pressure p, a magnetic pressure also appears. The right-hand side tells us that a magnetic tension is present. The case $\nabla p = 0$ is called a force-free equilibrium, which satisfies

$$(\nabla \times \boldsymbol{B}) \times \boldsymbol{B} = 0 \; . \tag{5.239}$$

In a force-free equilibrium, $\nabla \times \boldsymbol{B} \sim \boldsymbol{B}$ should hold, that is,

$$\nabla \times \boldsymbol{B} = \alpha(\boldsymbol{r}) \boldsymbol{B} \; . \tag{5.240}$$

Since the magnetic filed is divergence-free, we obtain

$$\boldsymbol{B} \cdot \nabla \alpha = 0 \; . \tag{5.241}$$

That means that magnetic field lines should be located in the surfaces $\alpha =$ const, which are called magnetic surfaces.

When $\nabla p \neq 0$, similar considerations as above lead to

$$\boldsymbol{j} \cdot \nabla p = 0 \; , \quad \boldsymbol{B} \cdot \nabla p = 0 \; . \tag{5.242}$$

This means that the surfaces $p(\boldsymbol{r}) =$ const are both magnetic as well as current surfaces. Because of $\nabla \cdot \boldsymbol{B} = 0$ and $\nabla \cdot \boldsymbol{j} = 0$, both fields $\boldsymbol{B}(\boldsymbol{r})$ and $\boldsymbol{j}(\boldsymbol{r})$ are either closed or extend to infinity. Thus, the magnetic surfaces $p(\boldsymbol{r}) =$ const are either tubes or tori.

As an example, we will investigate the possible magnetic field configuration under the assumption of axisymmetry configuration in the scalar pressure approximation. Introducing a cylindrical coordinate system r, ϕ, z, we assume azimuthal symmetry regarding the z-axis; see Figure 5.1. As we shall discuss in later sections, because of $\nabla \cdot \boldsymbol{B} = 0$, the magnetic field $\boldsymbol{B}$ can be written in terms of two Euler potentials. The most general form for an axisymmetric magnetic field is

$$\boldsymbol{B} = \frac{1}{2\pi} (\nabla \psi \times \nabla \phi + \mu_0 I \nabla \phi) \equiv \boldsymbol{B}_{pol} + \boldsymbol{B}_{tor} \; , \quad \boldsymbol{B}_{tor} = \frac{1}{2\pi} \mu_0 I \nabla \phi \; . \tag{5.243}$$

Here, $I = I(r, z)$ is a current flowing in z-direction and being linked to a circle of radius r with the center at the axis at axial location z. Then, integrating over the area of the circle,

$$\int \nabla \times \boldsymbol{B}_{tor} \cdot d\boldsymbol{F} = \oint_C \boldsymbol{B}_{tor} \cdot d\boldsymbol{l} = \mu_0 I \; . \tag{5.244}$$

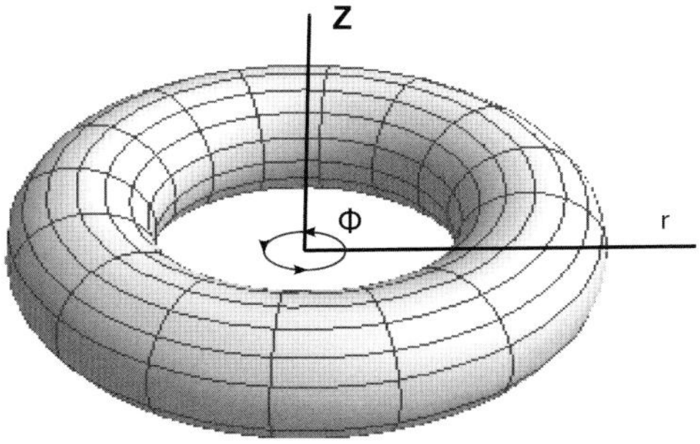

Figure 5.1 Cylindrical coordinate system for an axisymmetric configuration.

Next, we can determine the poloidal flux which we can identify with ψ. Again, we will integrate over a surface perpendicular to the symmetry axis. In cylindrical coordinates, the azimuthal component of the poloidal magnetic field vanishes ($\boldsymbol{B}_{pol} \cdot \hat{\phi} = 0$). Integration leads to

$$\int \boldsymbol{B}_{pol} \cdot d\boldsymbol{F} = \frac{1}{2\pi} \int_0^r (\nabla\psi \times \nabla\phi) \cdot \hat{z} 2\pi r' dr' = \int_0^r \frac{d\psi}{dr'} dr' = \psi \,. \tag{5.245}$$

Thus, the Euler potentials allow the simple identification of the poloidal flux with ψ. In cylindrical coordinates, $\nabla\phi = r^{-1}\hat{\phi}$ holds (whereas, for later use, we would have $\nabla\phi = (r \sin\theta)^{-1}\hat{\phi}$ in spherical coordinates).

Having defined the poloidal and toroidal magnetic fields in an axisymmetric situation, we may calculate the corresponding current densities via $\mu_0 \boldsymbol{j} = \nabla \times \boldsymbol{B}$, that is,

$$\boldsymbol{j}_{pol} \equiv \frac{1}{\mu_0} \nabla \times \boldsymbol{B}_{pol} = \frac{1}{2\pi} \nabla I \times \nabla\phi \,, \tag{5.246}$$

$$\boldsymbol{j}_{tor} \equiv \frac{1}{\mu_0} \nabla \times \boldsymbol{B}_{tor} = -\frac{r^2}{2\pi\mu_0} \nabla \cdot \left(\frac{1}{r^2} \nabla\psi \right) \nabla\phi \,. \tag{5.247}$$

As expected, the toroidal current density is in azimuthal direction, and the poloidal current density is in the r, z-plane. Next, we introduce the results into $\nabla p = \boldsymbol{j} \times \boldsymbol{B}$, leading to

$$\nabla p = \underbrace{\boldsymbol{j}_{pol} \times \boldsymbol{B}_{pol}}_{*} + \underbrace{\boldsymbol{j}_{tor} \times \boldsymbol{B}_{tor}}_{=0} + \boldsymbol{j}_{pol} \times \boldsymbol{B}_{tor} + \boldsymbol{j}_{tor} \times \boldsymbol{B}_{pol} \,. \tag{5.248}$$

In axisymmetric situations ($\partial/\partial\phi = 0$), we should have $(\nabla p)_\phi = 0$. Now, let us have a look at the term

$$* \sim (\nabla I \times \nabla\phi) \times (\nabla\psi \times \nabla\phi) \sim \nabla\phi \,, \tag{5.249}$$

which is the only term on the right-hand side of Eq. (5.248) in the direction of $\hat{\phi}$. Thus, axisymmetry requires "$* = 0$". This will be satisfied if I is only a function of

ψ, that is, $I = I(\psi)$ with $\nabla I = I'\nabla\psi$. The poloidal current then becomes

$$j_{pol} = \frac{I'}{2\pi}\nabla\psi \times \nabla\phi \,. \tag{5.250}$$

The evaluation of Eq. (5.248) is now straightforward,

$$\nabla p = -\frac{1}{(2\pi)^2}\left[\frac{\mu_0 I I'}{r^2} + \frac{1}{\mu_0}\nabla\cdot\left(\frac{1}{r^2}\nabla\psi\right)\right]\nabla\psi \,. \tag{5.251}$$

Thus, $\nabla p \parallel \nabla\psi$, and for isotropic pressure p the latter is only a function of ψ, that is, $p = p(\psi)$. Introducing $\nabla p = p'\nabla\psi$, from Eq. (5.251), we get

$$\nabla\cdot\left(\frac{1}{r^2}\nabla\psi\right) + \frac{\mu_0^2 I I'}{r^2} + 4\pi^2\mu_0 p' = 0 \,, \tag{5.252}$$

which is called the Grad–Shafranov equation [70, 71] (in cylindrical coordinates). The Grad–Shafranov equation contains the unknown functions $I = I(\psi)$ and $p = p(\psi)$ which should be specified for its solution. In general, the Grad–Shafranov equation is a nonlinear partial differential equation. It has been shown by Solovev [72] that the linear Grad–Shafranov equation already possesses quite useful solutions for tokamak physics. Suppose the plasma pressure is a linear function of ψ, for example, $p = p_0 + \lambda\psi$. In addition, the current I is modeled as a line current along the z-axis such that $I' = 0$ within the plasma. Then, for $r > 0$, the (linear) Grad–Shafranov equation is

$$\frac{\partial^2\psi}{\partial r^2} - \frac{1}{r}\frac{\partial\psi}{\partial r} + \frac{\partial^2\psi}{\partial z^2} + 4\pi^2 r^2\mu_0\lambda = 0 \,. \tag{5.253}$$

It has the exact solution

$$\psi(r, z) = \psi_0\frac{r^2}{r_0^4}\left(2r_0^2 - r^2 - 4\alpha^2 z^2\right) \,, \tag{5.254}$$

with constants ψ_0, r_0, and α which are related according to

$$\lambda = \frac{2\psi_0}{\pi^2 r_0^4\mu_0}(1 + \alpha^2) \,. \tag{5.255}$$

A typical contour plot in the r, z-plane is shown in Figure 5.2. Closed contours, that is, closed magnetic surfaces $\psi = \text{const}$, do exist. The closed contours are separated from contours going to infinity by a separatrix $r^2 + 4\alpha^2 z^2 = 2r_0^2$. The common center of all closed surfaces is called the magnetic axis.

For $I = 0$, an analogue to Eq. (5.253) exists in spherical coordinates,

$$\frac{\partial^2\psi}{\partial r^2} + \frac{1}{r^2}\frac{\partial^2\psi}{\partial\theta^2} - \frac{\operatorname{cotg}\theta}{r^2}\frac{\partial\psi}{\partial\theta} = -4\pi^2\mu_0 r^2\sin^2\theta\frac{dp}{d\psi} \,. \tag{5.256}$$

The homogeneous ($p = 0$) equation possesses a solution with a dipole structure $\psi \sim -r^{-1}\sin^2\theta$.

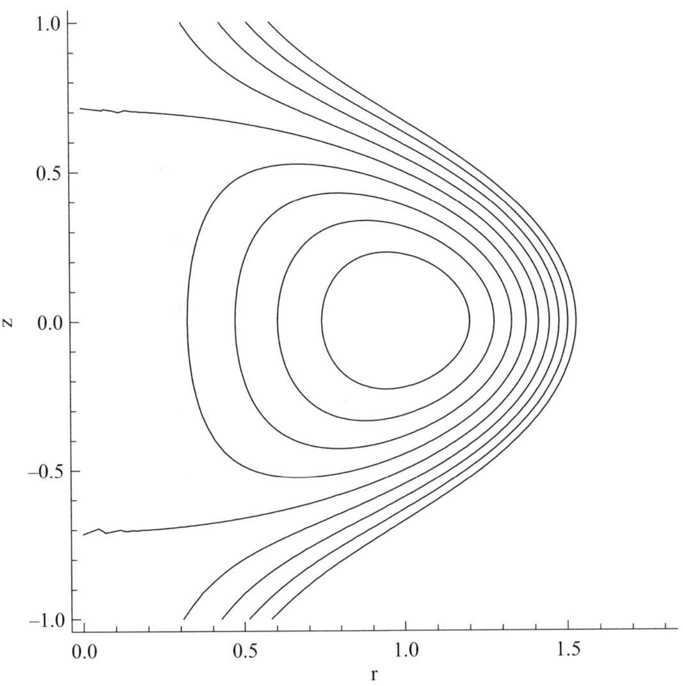

Figure 5.2 Contour plot of the Solovev solution [72] of the Grad–Shafranov equation (5.252).

Next, we consider a straight linear pinch discharge configuration with infinite extension in z-direction. We assume cylindrical symmetry with an electric current in z-direction, that is, $\boldsymbol{j} = j(r)\hat{z}$. The current produces an azimuthal magnetic field $\boldsymbol{B} = B(r)\hat{\varphi}$. Again we consider the stationary case ($\partial_t = 0$) without mean flow ($\boldsymbol{u} = 0$). The pressure balance equation (5.238) simplifies (remember $[(\boldsymbol{B} \cdot \nabla)\boldsymbol{B}]_r = -B^2(r)/r$),

$$\frac{dp}{dr} = -\frac{1}{2\mu_0 r^2}\frac{d}{dr}(r^2 B^2) \,. \tag{5.257}$$

The azimuthal component of the magnetic field is determined by

$$B(r) = \frac{\mu_0}{r}\int_0^r r j(r) dr \,. \tag{5.258}$$

In a simple application we specify $j(r)$ as constant for $r < R$ and zero for $r > R$. Then

$$B(r) = \frac{\mu_0}{2\pi}\frac{Ir}{R^2} \quad \text{for } r < R \,, \quad B(r) = \frac{\mu_0}{2\pi}\frac{I}{r} \quad \text{for } r > R \,, \tag{5.259}$$

where I is the total pinch current. The pressure $p(r)$ follows by integration,

$$p(r) = \frac{\mu_0 I^2}{4\pi^2 R^2}\left(1 - \frac{r^2}{R^2}\right) \quad \text{for } r < R \,, \tag{5.260}$$

when we apply the obvious boundary condition $p(r) = 0$ for $r > R$. The number N of confined particles will be related to the pinch current I. At a given constant temperature and with $p = nk_B T$, we have

$$N = \int_0^R n(r) 2\pi r dr \,, \tag{5.261}$$

such that

$$N k_B T = \frac{\mu_0 I^2}{8\pi} \,. \tag{5.262}$$

The magnetic surfaces $\psi = \text{const}$ are cylinders around the z-axis.

Summarizing, we have found exact equilibrium solutions, although the terminology is not quite correct from the thermodynamic point of view. The characterization as stationary solutions is quite better. The reason is that the solutions are not required to be stable. We shall come back to this aspect in the chapter on stability.

5.5.3 Alfvén Waves

In a magnetized plasma, the equation of motion contains the additional $\boldsymbol{j} \times \boldsymbol{B}$-force which tries to bring the magnetic filed lines in straight direction. Thus, also at zero pressure some oscillatory modes are possible. The additional force density can be written in the form

$$\boldsymbol{j} \times \boldsymbol{B} = -\nabla \left(\frac{B^2}{2\mu_0} \right) + \frac{1}{\mu_0} (\boldsymbol{B} \cdot \nabla) \boldsymbol{B} \,. \tag{5.263}$$

Let us now calculate some of its consequences within the ideal MHD approximation (conductivity $\sigma \to \infty$). The zeroth order, unperturbed equilibrium situation may have constant mass density, an homogeneous magnetic field, and zero flow velocity. The electric field $\boldsymbol{E}$ follows as

$$\boldsymbol{E} = -\boldsymbol{u} \times \boldsymbol{B} \,; \tag{5.264}$$

coupled equations for ρ, $\boldsymbol{u}$, and $\boldsymbol{B}$ remain to be solved in the perturbed situation. The scalar pressure p will be determined via $p = p(\rho)$ from the equation of state. Small deviations (index 1) from the equilibrium (Index 0) are introduced as

$$\boldsymbol{B} = \boldsymbol{B}_0 + \boldsymbol{B}_1(\boldsymbol{r}, t) \,, \quad \rho = \rho_0 + \rho_1(\boldsymbol{r}, t) \,, \quad \boldsymbol{u} = \boldsymbol{u}_1(\boldsymbol{r}, t) \,. \tag{5.265}$$

The linearized MHD equations read

$$\partial_t \rho_1 + \rho_0 \nabla \cdot \boldsymbol{u}_1 = 0 \,, \tag{5.266}$$

$$\rho_0 \partial_t \boldsymbol{u}_1 + \left. \frac{\partial p}{\partial \rho} \right|_0 \nabla \rho_1 + \frac{\boldsymbol{B}_0}{\mu_0} \times (\nabla \times \boldsymbol{B}_1) = 0 \,, \tag{5.267}$$

$$\partial_t \boldsymbol{B}_1 - \nabla \times (\boldsymbol{u}_1 \times \boldsymbol{B}_0) = 0 \,. \tag{5.268}$$

After differentiating Eq. (5.267) with respect to time, we may eliminate $\partial_t \boldsymbol{B}_1$ and $\partial_t \rho_1$. Then, we can make use of the following vector identity, that is,

$$\boldsymbol{B}_0 \times (\nabla \times \boldsymbol{B}_1) = (\nabla \boldsymbol{B}_1) \cdot \boldsymbol{B}_0 - \boldsymbol{B}_0 \cdot \nabla \boldsymbol{B}_1 = \nabla (\boldsymbol{B}_1 \cdot \boldsymbol{B}_0) - \boldsymbol{B}_0 \cdot \nabla \boldsymbol{B}_1 \,. \quad (5.269)$$

Next,

$$\frac{\partial}{\partial t} \boldsymbol{B}_0 \times (\nabla \times \boldsymbol{B}_1) = \nabla \left[\boldsymbol{B}_0 \cdot \nabla (\boldsymbol{u}_1 \times \boldsymbol{B}_0) \right] - (\boldsymbol{B}_0 \cdot \nabla) \nabla \times (\boldsymbol{u}_1 \times \boldsymbol{B}_0) \,. \quad (5.270)$$

Finally,

$$\nabla \left[\boldsymbol{B}_0 \cdot \nabla (\boldsymbol{u}_1 \times \boldsymbol{B}_0) \right] = \nabla \left\{ \nabla \cdot \left[(\boldsymbol{u}_1 \times \boldsymbol{B}_0) \times \boldsymbol{B}_0 \right] \right\} = \nabla \left\{ -B_0^2 \nabla \cdot \boldsymbol{u}_{1\perp} \right\} \,. \quad (5.271)$$

We make use of a coordinate system with

$$\boldsymbol{B}_0 = B_0 \hat{z} \,, \quad (5.272)$$

and determine the pressure variations via

$$\nabla p_1 = \left. \frac{\partial p}{\partial \rho} \right|_0 \nabla \rho_1 \equiv c_s^2 \nabla \rho_1 \,. \quad (5.273)$$

Putting everything together, we obtain

$$\frac{\partial^2 \boldsymbol{u}_1}{\partial t^2} - c_s^2 \nabla (\nabla \cdot \boldsymbol{u}_1) - v_A \frac{\partial}{\partial z} \left[\nabla \times (\boldsymbol{u}_1 \times \boldsymbol{v}_A) \right] - v_A^2 \nabla \left(\nabla \cdot \boldsymbol{u}_{1\perp} \right) = 0 \,, \quad (5.274)$$

with the Alfvén velocity

$$\boldsymbol{v}_A = \frac{\boldsymbol{B}_0}{\sqrt{\mu_0 \rho_0}} \,. \quad (5.275)$$

Let us exemplify the new type of solution for the simple case $k_\perp = 0$. We consider a Fourier mode

$$\boldsymbol{u}_1(\boldsymbol{r}, t) = \boldsymbol{u}_1 e^{i \boldsymbol{k} \cdot \boldsymbol{r} - i \omega t} \quad (5.276)$$

with $k_\perp = 0$. Thus, the propagation is along the zeroth order magnetic field. For $\boldsymbol{k} \parallel \boldsymbol{v}_A$, we obtain

$$\left(k^2 v_A^2 - \omega^2 \right) \boldsymbol{u}_1 + \left(\left. \frac{\partial p}{\partial \rho} \right|_0 \frac{1}{v_A^2} - 1 \right) k^2 (\boldsymbol{v}_A \cdot \boldsymbol{u}_1) \boldsymbol{v}_A = 0 \,. \quad (5.277)$$

We may separate the solutions into two branches. First, for longitudinal oscillations $\boldsymbol{u}_1 \parallel \boldsymbol{v}_A \parallel \boldsymbol{k}$, the phase velocity ω / k agrees with the sound velocity $\sqrt{\partial p / \partial \rho |_0}$. Second, when $\boldsymbol{u}_1$ is perpendicular to $\boldsymbol{v}_A \parallel \boldsymbol{k}$, that is, for $\boldsymbol{u}_1 \cdot \boldsymbol{v}_A = 0$, we find the Alfvén waves with the frequency

$$\omega = k v_A \,. \quad (5.278)$$

The transverse waves are accompanied by lateral oscillations of the magnetic field lines ($\boldsymbol{k} \cdot \boldsymbol{B}_1 = 0$).

For the general dispersion properties of MHD waves, we refer to the overview over Alfvén waves (slow, fast, shear, and compressional modes) in [15, 73].

5.5.4
Energy Conservation in Ideal MHD

We conclude this section with a short discussion on the energy balance in ideal MHD. Multiplying Eq. (5.193) with $\boldsymbol{u}$ leads to

$$\frac{1}{2}\rho\left[\partial_t u^2 + \boldsymbol{u}\cdot\nabla u^2\right] = -\boldsymbol{u}\cdot\nabla p + \frac{1}{\mu_0}\boldsymbol{u}\cdot(\nabla\times\boldsymbol{B})\times\boldsymbol{B}\,. \tag{5.279}$$

Using Eq. (5.192), we can reformulate the left-hand side,

$$\partial_t\left(\frac{1}{2}\rho u^2\right) + \nabla\cdot\left(\frac{u^2}{2}\rho\boldsymbol{u}\right) = -\boldsymbol{u}\cdot\nabla p + \frac{1}{\mu_0}\boldsymbol{u}\cdot(\nabla\times\boldsymbol{B})\times\boldsymbol{B}\,. \tag{5.280}$$

To evaluate the first term on the right-hand side, we make use of Eq. (5.197) with $\gamma = 5/3$ and find for $\boldsymbol{u}\cdot\nabla p$,

$$\boldsymbol{u}\cdot\nabla p = \frac{1}{\gamma-1}\partial_t p + \frac{\gamma}{\gamma-1}\nabla\cdot(p\boldsymbol{u})\,. \tag{5.281}$$

Finally,

$$\begin{aligned}\boldsymbol{u}\cdot(\nabla\times\boldsymbol{B})\times\boldsymbol{B} &= -(\boldsymbol{u}\times\boldsymbol{B})\cdot(\nabla\times\boldsymbol{B})\\ &= \boldsymbol{E}\cdot(\nabla\times\boldsymbol{B}) = \boldsymbol{B}\cdot\nabla\times\boldsymbol{E} - \nabla\cdot(\boldsymbol{E}\times\boldsymbol{B})\\ &= -\frac{1}{2}\partial_t B^2 - \nabla\cdot(\boldsymbol{E}\times\boldsymbol{B})\,.\end{aligned} \tag{5.282}$$

Combining everything, the local energy equation appears as

$$\partial_t\left[\frac{1}{2}\rho u^2 + \frac{p}{\gamma-1} + \frac{B^2}{2\mu_0}\right] + \nabla\cdot\left[\frac{u^2}{2}\rho\boldsymbol{u} + \frac{\gamma}{\gamma-1}p\boldsymbol{u} + \frac{1}{\mu_0}\boldsymbol{E}\times\boldsymbol{B}\right] = 0\,. \tag{5.283}$$

The first three terms represent the kinetic energy, the thermal free energy, and the magnetic energy, respectively. Because of the presence of electromagnetic fields, the Poynting flux $1/\mu_0\boldsymbol{E}\times\boldsymbol{B}$ arises among the fluxes. Appropriate boundary conditions (vanishing of the motion and the fields at infinity) allow one to identify, after space-integration, the conserved energy,

$$\int dV\left[\frac{1}{2}\rho u^2 + \frac{p}{\gamma-1} + \frac{B^2}{2\mu_0}\right] = E, \tag{5.284}$$

for ideal MHD in the scalar pressure approximation.

6
Principles of Linear and Stochastic Transport

Transport theory is a key issue in theoretical plasma physics. We distinguish between linear transport theory (being described in the first part of this chapter for straight magnetic field lines) and nonlinear transport which becomes necessary to consider because of the failure of linear transport in the presence of strong fluctuations (that will be considered in the second part of this chapter for the so called stochastic transport). Magnetic field topologies as, for example, being realized in tokamaks and stellarators, require even in the linear regime, a transport theory which goes beyond the straight magnetic field line approximation considered here. When only presenting the basic ideas of transport theory, we abandon a presentation of the neoclassical theory which is a topic of its own; see, for example, [58].

In the last chapter, we presented the fluid approach which becomes a closed description after making some ad hoc assumptions on the distribution function(s). Linear transport theory provides us with a systematic procedure to evaluate transport coefficients by taking into account the linear kinetic plasma response. Thus, the starting point of a linear transport theory is a kinetic equation. Here, we shall use the Landau–Fokker–Planck equation; the aim then is to derive the coefficients which have been summarized in the section on the Braginskii equations. Since the evaluation is quite elaborate and requires the evaluation of many details, we shall only present an overview. Also, the more fusion-applied directions like neoclassical transport, which has been developed from the drift-kinetic equation for tokamak- and stellarator-like magnetic field configurations, cannot be discussed here because of the space limitations already mentioned.

In principle, we distinguish in transport theory between the hydrodynamic scales (time scale T_H and space scale L_H on which variations are slow and weak, respectively) and collisional scales (which are fast).

A kinetic equation generally contains a convective term $(\boldsymbol{v} \cdot \nabla f)$ which is mainly responsible for the nonlinear terms in the momentum equations, and collisional terms K^{α}. Momenta vary on the hydrodynamic scales T_H and L_H, whereas the collisional term may be approximated by

$$K^{\alpha\beta} \sim \frac{1}{\tau_{R\alpha\beta}} f^{\alpha} , \tag{6.1}$$

High Temperature Plasmas, Theory and Mathematical Tools for Laser and Fusion Plasmas, First Edition.
Karl-Heinz Spatschek.

where $\tau_{R\alpha\beta}$ is the collisional relaxation time. As previously discussed, we should distinguish between fast relaxations within one component and slow exchanges between components. For estimating orders of magnitudes, we may use the relation $T_H \sim L_H/v_{th}$ between the hydrodynamic scales. Thus, linear plasma transport theory starts with a zeroth order ansatz satisfying the fast exchange within the components, and then determines the (slow and weak) linear response to external forces with the help of a kinetic equation.

6.1
Moments in Linear Transport Theory

Starting with the outlined scheme, we first investigate the contributions from K^{ee} and K^{ii}. They are responsible for the approach of the local distribution function to an equilibrium-like form. To zeroth order,

$$K^{\alpha\alpha} \approx 0 \quad \text{for } \alpha = e, i \tag{6.2}$$

is the starting point. These equations are satisfied for Maxwellians

$$f_0^\alpha = n_\alpha \left(\frac{m_\alpha}{2\pi k_B T_\alpha}\right)^{3/2} \exp\left[-\frac{m_\alpha|\boldsymbol{v} - \boldsymbol{u}^\alpha|^2}{2 k_B T_\alpha}\right], \tag{6.3}$$

which still contain fields (momenta or plasmadynamical variables). It is easy to validate

$$\frac{1}{m_\alpha}\left(\frac{\partial}{\partial v_{1\nu}} - \frac{\partial}{\partial v_{2\nu}}\right) f_0^\alpha(\boldsymbol{v}_1) f_0^\alpha(\boldsymbol{v}_2) = -\frac{1}{k_B T_\alpha} g_\nu f_0^\alpha(\boldsymbol{v}_1) f_0^\alpha(\boldsymbol{v}_2), \tag{6.4}$$

and the collision term vanishes because of

$$G_{\mu\nu} g_\nu = \frac{g^2 \delta_{\mu\nu} - g_\mu g_\nu}{g^3} g_\nu = 0. \tag{6.5}$$

Here, exactly the plasmadynamical variables n_α, T_α, and $\boldsymbol{u}^\alpha$ appear. The general forms of the dynamical equations for the latter have been presented. Let us assume that the time- and space-variations of the plasmadynamical variables are on the hydrodynamical scales,

$$n_\alpha = n_\alpha(\boldsymbol{x}, t), \quad T_\alpha = T_\alpha(\boldsymbol{x}, t), \quad \boldsymbol{u}^\alpha = \boldsymbol{u}^\alpha(\boldsymbol{x}, t). \tag{6.6}$$

Writing the general solution of the kinetic equation as

$$f^\alpha = f_0^\alpha [1 + \chi^\alpha], \tag{6.7}$$

we introduce the deviation $\chi^\alpha = \chi^\alpha(\boldsymbol{x}, \boldsymbol{v}, t)$ as the deviation from the local Maxwell distribution f_0^α. When, in the following, determining χ^α, we demand that the moments of f^α exactly agree with the moments appearing in f_0^α. In other words,

$$\int d^3v\, f^\alpha = \int d^3v\, f_0^\alpha = n_\alpha(\boldsymbol{x}, t) , \tag{6.8}$$

$$\int d^3v \boldsymbol{v}\, f^\alpha = \int d^3v \boldsymbol{v}\, f_0^\alpha = n_\alpha(\boldsymbol{x}, t)\boldsymbol{u}^\alpha(\boldsymbol{x}, t) , \tag{6.9}$$

$$\frac{1}{3}m_\alpha \int d^3v |\boldsymbol{v} - \boldsymbol{u}^\alpha|^2 f^\alpha = \frac{1}{3}m_\alpha \int d^3v |\boldsymbol{v} - \boldsymbol{u}^\alpha|^2 f_0^\alpha = n_\alpha(\boldsymbol{x}, t) k_B T_\alpha(\boldsymbol{x}, t) . \tag{6.10}$$

Before discussing the consequences for $f_0^\alpha \chi^\alpha$ and χ^α, let us introduce the relative velocity $\boldsymbol{v} - \boldsymbol{u}^\alpha$, which we shall normalize with the thermal velocity $v_{th\alpha} = (k_B T_\alpha)^{1/2}/m_\alpha^{1/2}$ for each component α. In the following, we omit the index α whenever no big confusion can be expected. Then,

$$\boldsymbol{c} = \boldsymbol{c}^\alpha = \left(\frac{m_\alpha}{k_B T_\alpha}\right)^{1/2} (\boldsymbol{v} - \boldsymbol{u}^\alpha) . \tag{6.11}$$

Similarly, we introduce, instead of f_0^α, the normalized function ϕ_0^α via

$$f_0^\alpha = n_\alpha \left(\frac{m_\alpha}{k_B T_\alpha}\right)^{1/2} \phi_0^\alpha , \quad \phi_0 = \phi_0^\alpha = \frac{1}{(2\pi)^{3/2}} e^{\frac{c^2}{2}} . \tag{6.12}$$

Again, not all indices are shown all the time.

We shall expand χ^α in terms of Gauss–Hermite polynomials. The latter are defined by

$$\tilde{H}^{(m)}_{r_1 \ldots r_m}(\boldsymbol{c}) = (-1)^m e^{\frac{c^2}{2}} \frac{\partial}{\partial c_{r1}} \cdots \frac{\partial}{\partial c_{rm}} e^{-\frac{c^2}{2}} . \tag{6.13}$$

The upper index designates the order $m = 0, 1, \ldots$; the lower indices show which m components are being used in the differentiations with respect to the velocity components $r_1, \ldots, r_m$. The Hermite polynomials are appropriate since we have a Gaussian (Maxwellian) as weight function. The expansion is formally written as

$$\chi^\alpha(\boldsymbol{x}, \boldsymbol{c}, t) = \sum_{m=0}^{\infty} \frac{1}{m!} \tilde{h}^{\alpha(m)}_{r_1 \ldots r_m}(\boldsymbol{x}, t) \tilde{H}^{(m)}_{r_1 \ldots r_m}(\boldsymbol{c}) \tag{6.14}$$

with the expansion coefficients

$$\tilde{h}^{\alpha(m)}_{r_1 \ldots r_m}(\boldsymbol{x}, t) = \int d^3c \phi_0(\boldsymbol{c}) \chi^\alpha(\boldsymbol{x}, \boldsymbol{c}, t) \tilde{H}^{(m)}_{r_1 \ldots r_m}(\boldsymbol{c}) . \tag{6.15}$$

The Einstein convention for summation over repeated velocity indices $r_\nu = x, y, z$ has been applied. By contracting indices, we obtain, for example,

$$\hat{H}^{(2n)}(c) = \sum_{s_1,\ldots,s_n} \tilde{H}^{(2n)}_{s_1 s_1 s_2 s_2 \ldots s_n s_n}(c) , \tag{6.16}$$

$$\hat{H}^{(2n+1)}_r(c) = \sum_{s_1,\ldots,s_n} \tilde{H}^{(2n+1)}_{r s_1 s_1 \ldots s_n s_n}(c) , \tag{6.17}$$

$$\hat{H}^{(2n)}_{r_1 r_2}(c) = \sum_{s_1,\ldots,s_{n-1}} \tilde{H}_{r_1 r_2 s_1 s_1 \ldots s_{n-1} s_{n-1}}(c) - \frac{1}{3}\delta_{r_1 r_2} \hat{H}^{(2n)}(c) , \tag{6.18}$$

and so on. In the so called irreducible representation, we make use of

$$H^{(0)} = 1 , \quad H^{(2)} = \frac{1}{\sqrt{6}}(c^2-3) , \quad H^{(4)} = \frac{1}{2\sqrt{30}}(c^4-10c^2+15) , \ldots \tag{6.19}$$

$$H^{(1)}_r = c_r , \quad H^{(3)}_r = \frac{1}{\sqrt{10}} c_r(c^2-5) , \quad H^{(5)}_r = \frac{1}{2\sqrt{70}} c_r(c^4-14c^2+35) , \ldots \tag{6.20}$$

$$H^{(2)}_{rs} = \frac{1}{\sqrt{2}}\left(c_r c_s - \frac{1}{3}\delta_{rs} c^2\right) , \quad H^{(4)}_{rs} = \frac{1}{2\sqrt{7}}\left(c_r c_s - \frac{1}{3}\delta_{rs} c^2\right)(c^2-7) , \ldots \tag{6.21}$$

$$H^{(3)}_{rsp} = c_r c_s c_p - \frac{1}{5} c^2 \left(c_r \delta_{sp} + c_s \delta_{rp} + c_p \delta_{rs}\right) , \ldots \tag{6.22}$$

and so on. These are scalar, vectorial, and tensorial quantities, respectively. Since every symmetric tensor of rank ≥ 2 (here $\tilde{H}^{(m)}_{r_1\ldots r_m}$) can be uniquely decomposed into irreducible components, we may express $\tilde{H}^{(m)}_{r_1\ldots r_m}$ in terms of the just defined $H^{(\nu+\mu)}_{r_1 r_2 \ldots r_\nu}$, $\nu = 0, 1, 2, \ldots, \mu = 0, 2, 4$, for example,

$$\tilde{H}^{(2)}_{rs} = \sqrt{2}\left(H^{(2)}_{rs} + \frac{1}{\sqrt{3}} H^{(2)} \delta_{rs}\right) . \tag{6.23}$$

This leads to the expansion

$$\chi^\alpha = \sum_{m=0}^{\infty} h^{\alpha(2m)}(x, t) H^{(2m)}(c) + \sum_{m=0}^{\infty} h^{\alpha(2m+1)}_r(x, t) H^{(2m+1)}_r(c) + \sum_{m=1}^{\infty} h^{\alpha(2m)}_{rs}(x, t) H^{(2m)}_{rs}(c) + \ldots \tag{6.24}$$

The coefficients satisfy h

$$h^{\alpha(m)}_{r_1\ldots r_n} = \int d^3 c \phi_0(c) \chi^\alpha(x, c, t) H^{(m)}_{r_1\ldots r_n}(c) . \tag{6.25}$$

Because of the constraints

$$\int d^3 c \phi_0 \chi^\alpha = \int d^3 c \phi_0 c_r \chi^\alpha = \int d^3 c \phi_0 c^2 \chi^\alpha = 0, \tag{6.26}$$

we have

$$h^{\alpha(0)} = h_r^{\alpha(1)} = h^{\alpha(2)} = 0\,. \tag{6.27}$$

Since

$$\int d^3c\, H_{pq}^{(2)} H_{rs}^{(2)} \phi_0 = \frac{1}{2}\left(\delta_{rp}\delta_{sq} + \delta_{rq}\delta_{sp} - \frac{2}{3}\delta_{rs}\delta_{pq}\right), \tag{6.28}$$

the dissipative part of the pressure tensor π_{rs}^{α} may be expressed as

$$\pi_{rs}^{\alpha} = \sqrt{2} n_\alpha k_B T_\alpha h_{rs}^{\alpha(2)}\,. \tag{6.29}$$

In a similar way, the heat flux is

$$q_r^{\alpha} = \left(\frac{5}{2}\right)^{1/2} m_\alpha \left(\frac{k_B T_\alpha}{m_\alpha}\right)^{3/2} n_\alpha h_r^{\alpha(3)}\,. \tag{6.30}$$

Thus far, the calculation is exact. However, the coefficients on the right-hand sides are still undetermined. To proceed, we truncate the expansion, assuming that the vectorial and traceless polynomials of second order are sufficient to describe weakly unisotropic situations. In other words, we shall only use the polynomials $H^{(\ldots)}$, $H_r^{(\ldots)}$, and $H_{rs}^{(\ldots)}$ as basis functions. Then, the finite expansion reads

$$\chi^{\alpha}(\boldsymbol{x}, \boldsymbol{c}, t) \approx \sum_r c_r B_r^{\alpha}(c; \boldsymbol{x}, t) + \sum_{r,s}\left(c_r c_s - \frac{1}{3}c^2\delta_{rs}\right) C_{rs}^{\alpha}(c; \boldsymbol{x}, t)\,. \tag{6.31}$$

The coefficients B and C still contain the summation over m. For practical reasons, we also have to truncate that summation. Within the *13 moment approximation*, we only use

$$c_r B_r^{\alpha} \approx h_r^{\alpha(3)} H_r^{(3)}\,, \quad \left(c_r c_s - \frac{1}{3}c^2\delta_{rs}\right) C_{rs}^{\alpha} \approx h_{rs}^{\alpha(2)} H_{rs}^{(2)}\,. \tag{6.32}$$

The next better approximation is the *21 moment approximation* which uses

$$\begin{aligned} c_r B_r^{\alpha} &\approx h_r^{\alpha(3)} H_r^{(3)} + h_r^{\alpha(5)} H_r^{(5)}\,, \\ \left(c_r c_s - \frac{1}{3}c^2\delta_{rs}\right) C_{rs}^{\alpha} &\approx h_{rs}^{\alpha(2)} H_{rs}^{\alpha(2)} + h_{rs}^{\alpha(4)} H_{rs}^{(4)}\,. \end{aligned} \tag{6.33}$$

The degree of accuracy can be further improved, for example, the *29 moment approximation*

$$c_r B_r^{\alpha} \approx h_r^{\alpha(3)} H_r^{(3)} + h_r^{\alpha(5)} H_r^{(5)} + h_r^{\alpha(7)} H_r^{(7)}\,, \tag{6.34}$$

$$\left(c_r c_s - \frac{1}{3}c^2\delta_{rs}\right) C_{rs}^{\alpha} \approx h_{rs}^{\alpha(2)} H_{rs}^{\alpha(2)} + h_{rs}^{\alpha(4)} H_{rs}^{\alpha(4)} + h_{rs}^{\alpha(6)} H_{rs}^{\alpha(6)}\,, \tag{6.35}$$

contains eight more moments. How do we count the moments? The 13 moment approximation consists of the five plasmadynamical variables n_α, u_r^{α}, and T_α, the three components $h_r^{\alpha(3)}$, and the five independent entries of the symmetric and traceless tensor $h_{rs}^{\alpha(2)}$. Obviously, the plurality because of the various species α is not counted. The next approximation additionally contains three vector and five tensor components. The series of approximations would contain 13, 21, 29, 37, ... moments. Increasing the number of variables by eight gives a better approximation; the performance of the approximation can be easily checked. The differences between the 13 moment approximation and the 21 moment approximation are generally significant, whereas the 29 moment approximation usually does not improve very much.

Summarizing, we perform two steps. The deviations from anisotropy are taken into account via the vectorial and rank-2 tensors only. The contributions of these two nonscalar parts are calculated successively in either the 13, 21 or the 29 moment approximation. For example, besides the plasmadynamical variables n_e, n_i, T_e, T_i, $\boldsymbol{u}^e$, and $\boldsymbol{u}^i$, we have in the 13 moment approximation in addition to $h_r^{\alpha(3)}$ and $h_{rs}^{\alpha(2)}$. In the 21 moment approximation, new additions are $h_r^{\alpha(3)}$, $h_r^{\alpha(5)}$, $h_{rs}^{\alpha(2)}$, $h_{rs}^{\alpha(4)}$. In the 29-moment-approximation $h_r^{\alpha(3)}$, $h_r^{\alpha(5)}$, $h_r^{\alpha(7)}$, $h_{rs}^{\alpha(2)}$, $h_{rs}^{\alpha(4)}$, and $h_{rs}^{\alpha(6)}$ are being added. In all cases, we have $h^{\alpha(0)} = h^{\alpha(2)} = h_r^{\alpha(1)} = 0$.

Some of the moments are privileged [40]. When we transform to the hydrodynamical variables ρ, $\sigma \approx 0$, $\boldsymbol{u}$, T_e, T_i, and $\boldsymbol{j}$, we introduce $j_r \equiv e n_e (T_e/m_e)^{1/2} h_r^{(1)}$ (new definition of $h_r^{(1)}$), and find $\pi_{rs}^\alpha \sim h_{rs}^{\alpha(2)}$ and $q_r^\alpha \sim h_r^{\alpha(3)}$. These are the privileged moments compared to $h_r^{\alpha(5)}$, $h_{rs}^{\alpha(4)}$, ... The reason for this nomenclature will become obvious soon.

Let us advert to the 21 moment approximation in the quasineutral case $\sigma \approx 0$. The displacement current will be neglected in the Maxwell equations such that $\nabla \cdot \boldsymbol{j} \approx 0$ and $\partial_t \sigma \approx 0$ follow. We start from

$$\partial_t \rho + \nabla \cdot (\rho \boldsymbol{u}) = 0 \,, \tag{6.36}$$

$$\begin{aligned}\partial_t(\rho u_r) = &-\nabla_s(\rho u_r u_s) - \frac{Z k_B}{m_i}\nabla_r\left[\rho\left(T_e + \frac{1}{Z}T_i\right)\right] \\ &- \sqrt{2}\frac{Z k_B}{m_i}\nabla_s\left[\rho\left(T_e h_{rs}^{e(2)} + \frac{1}{Z}T_i h_{rs}^{i(2)}\right)\right] \\ &+ e\frac{Z\rho}{m_i}\left(\frac{k_B T_e}{m_e}\right)^{1/2}\varepsilon_{rsm} h_s^{(1)} B_m \,,\end{aligned} \tag{6.37}$$

$$\begin{aligned}\partial_t T_e = &-\boldsymbol{u}\cdot\nabla T_e - \frac{2}{3}T_e \nabla\cdot\boldsymbol{u} + \left(\frac{k_B T_e}{m_e}\right)^{1/2} h_r^{(1)}\nabla_r T_e \\ &+ \frac{2}{3}T_e\nabla_r\left[\left(\frac{k_B T_e}{m_e}\right)^{1/2} h_r^{(1)}\right] \\ &- \frac{2\sqrt{2}}{3}T_e h_{rs}^{e(2)}\nabla_r u_s + \frac{2\sqrt{2}}{3}T_e h_{rs}^{e(2)}\nabla_r\left[\left(\frac{k_B T_e}{m_e}\right)^{1/2} h_s^{(1)}\right] \\ &- \frac{\sqrt{10}}{3}\frac{1}{\rho}\nabla_r\left[\rho T_e\left(\frac{k_B T_e}{m_e}\right)^{1/2} h_r^{e(3)}\right] - \frac{1}{Z k_B}Q^{(2)} - \frac{2}{3}T_e h_n^{(1)} Q_n^{(1)} \,,\end{aligned} \tag{6.38}$$

$$\partial_t T_i = -\boldsymbol{u} \cdot \nabla T_i - \frac{2}{3} T_i \nabla \cdot \boldsymbol{u} - \frac{2\sqrt{2}}{3} T_i h^{i(2)}_{rs} \nabla_r u_s - \frac{\sqrt{10}}{3} \frac{1}{\rho} \nabla_r \left[\rho T_i \left(\frac{k_B T_i}{m_i} \right)^{1/2} h^{i(3)}_r \right] + \frac{1}{k_B} Q^{(2)} . \tag{6.39}$$

Here, we have defined

$$Q^{(2)} = \frac{2}{3} \frac{m_i}{\rho} Q^{ie} , \quad Q^{(1)}_r = -\frac{m_i}{Z \rho m_e} \left(\frac{m_e}{k_B T_e} \right)^{1/2} R^{ei}_r . \tag{6.40}$$

In addition, we have the dynamical equations for the coefficients, namely,

$$\partial_t h^{(1)}_r = \left(\frac{m_e}{k_B T_e} \right)^{1/2} \left[\frac{e}{m_e} E_r + \frac{1}{m_e n_e} \nabla_r (n_e k_B T_e) + \frac{e}{m_e} \varepsilon_{rmn} u_m B_n \right] - \frac{e}{m_e} \varepsilon_{rmn} h^{(1)}_m B_n + Q^{(1)}_r + L^{(1)}_r + C^{(1)}_r + N^{(1)}_r , \tag{6.41}$$

where

$$L^{(1)}_r = \sqrt{2} \left(\frac{k_B T_e}{m_e} \right)^{1/2} \frac{1}{\rho k_B T_e} \nabla_s \left[\rho \left(k_B T_e h^{e(2)}_{rs} - Z \mu k_B T_i h^{i(2)}_{rs} \right) \right], \tag{6.42}$$

$$C^{(1)}_r = -\boldsymbol{u} \cdot \nabla h^{(1)}_r - h^{(1)}_m \nabla_m u_r + \frac{1}{3} h^{(1)}_r \nabla \cdot \boldsymbol{u} + \frac{\sqrt{2}}{3} h^{(1)}_r h^{e(2)}_{mn} \nabla_m \left[u_n - \left(\frac{k_B T_e}{m_e} \right)^{1/2} h^{(1)}_n \right], \tag{6.43}$$

$$N^{(1)}_r = \left(\frac{k_B T_e}{m_e} \right)^{1/2} \left[\frac{\sqrt{10}}{6} h^{(1)}_r \frac{1}{\rho (k_B T_e)^{3/2}} \nabla_m \left(\rho (k_B T_e)^{3/2} h^{e(3)}_m \right) + h^{(1)}_m \nabla_m h^{(1)}_r - \frac{1}{3} h^{(1)}_r \frac{1}{(k_B T_e)^{1/2}} \nabla_m \left((k_B T_e)^{1/2} h^{(1)}_m \right) \right] - \frac{1}{3} h^{(1)}_r h^{(1)}_n Q^{(1)}_n . \tag{6.44}$$

The other nonprivileged moment equations are

$$\partial_t h^{\alpha(3)}_r = -\left(\frac{5}{2} \right)^{1/2} \left(\frac{k_B T_\alpha}{m_\alpha} \right)^{1/2} \frac{1}{T_\alpha} \nabla_r T_\alpha + \frac{e_\alpha}{m_\alpha} \varepsilon_{rmn} h^{\alpha(3)}_m B_n + Q^{\alpha(3)}_r + L^{\alpha(3)}_r + M^{(3)}_r + C^{\alpha(3)}_r + N^{\alpha(3)}_r , \tag{6.45}$$

$$\partial_t h^{\alpha(5)}_r = \frac{e_\alpha}{m_\alpha c} \varepsilon_{rmn} h^{\alpha(5)}_m B_n + Q^{\alpha(5)}_r + L^{\alpha(5)}_r + M^{(5)}_r + C^{\alpha(5)}_r + N^{\alpha(5)}_r , \tag{6.46}$$

$$\partial_t h_{rs}^{\alpha(2)} = -\frac{1}{\sqrt{2}}\tau_{rs|pq}\nabla_p u_q + \frac{2e_\alpha}{m_\alpha c}\tau_{rs|pq}\varepsilon_{pmn}h_{qm}^{\alpha(2)}B_n + Q_{rs}^{\alpha(2)} + L_{rs}^{\alpha(2)} + M_{rs}^{\alpha(2)} + C_{rs}^{\alpha(2)} + N_{rs}^{\alpha(2)} , \tag{6.47}$$

$$\partial_t h_{rs}^{\alpha(4)} = \frac{2e_\alpha}{m_\alpha c}\tau_{rs|pq}\varepsilon_{pmn}h_{qm}^{\alpha(4)}B_n + Q_{rs}^{\alpha(4)} + L_{rs}^{\alpha(4)} + M_{rs}^{\alpha(4)} + C_{rs}^{\alpha(4)} + N_{rs}^{\alpha(4)} . \tag{6.48}$$

The operator $\tau_{rs|pq}$ causes symmetrization,

$$\tau_{rs|pq} = \frac{1}{2}\left(\delta_{rp}\delta_{sq} + \delta_{rq}\delta_{sp} - \frac{1}{3}\delta_{rs}\delta_{pq}\right) , \tag{6.49}$$

and the collisional contributions are contained in

$$Q_{r_1 r_2 \dots}^{\alpha(m)} = \frac{1}{n_\alpha}\int d^3 v\, H_{r_1 r_2 \dots}^{(m)}\left[\left(\frac{m_\alpha}{k_B T_\alpha}\right)^{1/2}(v - u^\alpha)\right] K^\alpha . \tag{6.50}$$

Next, we recognize

$$Q_r^{(1)}, Q^{(2)} \sim \mathcal{O}\left(\frac{m_e}{m_i}\right), \tag{6.51}$$

and to lowest order we shall neglect these contributions. Furthermore, since we developed linear transport theory, nonlinear terms like $-2/3 T_e h_n^{(1)} Q_n^{(1)}$, or $N_r^{(1)}$ will be first ignored. This implicates that the dynamical equations for ρ, $\boldsymbol{u}$, T_e, and T_i do not contain collisional terms explicitly (ρ, $\boldsymbol{u}$, T_e, and T_i are quasiconserved hydrodynamic moments).

The dynamical equations for ρ, $\boldsymbol{u}$, T_e, and T_i form a *first group* – the so called *hydrodynamic* plasma equations. These equations contain, besides ρ, $\boldsymbol{u}$, T_e, and T_i, a *second group* of variables, namely, $h_r^{(1)}$, $h_{rs}^{\alpha(2)}$, $h_r^{\alpha(3)}$, which are called *privileged nonhydrodynamic* moments.

The dynamical equations for $j_r \sim h_r^{(1)}$ (generalized Ohm's law), and $q_r^\alpha \sim h_r^{\alpha(3)}$ (heat conduction equation), and for the dissipative part of the pressure tensor $\pi_{rs}^\alpha \sim h_{rs}^{\alpha(2)}$ form the *second group* of equations. Compared to the rest, the second group is privileged since it contains source terms. The latter are gradients of ρ, $\boldsymbol{u}$, T_e, T_i as well as the electrodynamical fields. In a more general terminology, they are called thermodynamic forces. As we shall see, the source terms in the equations for $h^{(1)}$, $h_{rs}^{\alpha(2)}$, and $h_r^{\alpha(3)}$ are the reason for the functional dependence of the privileged nonhydrodynamic variables on the hydrodynamic variables which will suggest a meaningful truncation of the hierarchy.

Before going into more details, let us qualify the terms Q, L, M, C, and N. Let us start with the *collisional* contributions $Q^{(2)}$, $Q_r^{(1)}$, …, $Q_{r_1 r_2}^{\alpha(m)}$ … They are clearly defined in terms of the distribution functions f^α which appear in the collision term K^α of the kinetic equation. Inserting the expansion of the distribution function in terms of Hermite polynomials, we face a straightforward, but algebraically overloaded, problem which we do not present in a detailed manner. The final results [40] start with

$$Q^{(2)} = -\frac{2Z}{\tau_e}\frac{m_e}{m_i}k_B(T_i - T_e)\,, \tag{6.52}$$

which is of order m_e/m_i since

$$\tau_e = \frac{3(4\pi\varepsilon_0)^2}{4\sqrt{2\pi}}\frac{m_e^{1/2}T_e^{3/2}}{n_i Z^2 e^4 \ln\Lambda_B}\,, \tag{6.53}$$

with

$$\ln\Lambda_B = 4\pi\varepsilon_0\frac{3(T_e+T_i)\lambda_D}{2Ze^2}\,, \qquad \lambda_D = \left[\frac{Ze^2(n_e T_e + n_i T_i)}{\varepsilon_0 T_e T_i(1+Z)}\right]^{-1/2}\,. \tag{6.54}$$

The other terms will be written in standard notation [40] by introducing $Q_r^{e(1)} = Q_r^{(1)}$ and $h_r^{e(1)} = h_r^{(1)}$. Within a linear approximation

$$Q_r^{e(2n+1)} = \sum_{m=0}^{3} C_{2n+1,2m+1}h_r^{e(2m+1)}\,,$$
$$Q_r^{i(2n+1)} = \sum_{m=1}^{3} \overline{C}_{2n+1,2m+1}h_r^{i(2m+1)}\,, \tag{6.55}$$

$$Q_{rs}^{e(2n)} = \sum_{m=1}^{3} C_{2n,2m}h_{rs}^{e(2m)}\,, \quad Q_{rs}^{i(2n)} = \sum_{m=1}^{3} \overline{C}_{2n,2m}h_{rs}^{i(2m)}\,, \tag{6.56}$$

for $n = 0, 1, 2, 3$. Here, $C_{\nu,\mu}$ and $\overline{C}_{\nu,\mu}$ are given [40] as products of numerical constants with τ_e^{-1} (for electrons) or τ_i^{-1} (for ions). Note that

$$\tau_i = \left(\frac{m_i}{m_e}\right)^{1/2} Z^{-2}\left(\frac{T_i}{T_e}\right)^{3/2}\tau_e\,. \tag{6.57}$$

The *hierarchial* terms L contain higher moments. Later, we shall consider the terms

$$L_{rs}^{\alpha(2)} = -\frac{2}{\sqrt{5}}\tau_{rs|pq}\left(\frac{k_B T_\alpha}{m_\alpha}\right)^{1/2}\frac{1}{n_\alpha T_\alpha^{3/2}}\nabla_p\left(n_\alpha T_\alpha^{3/2}h_q^{\alpha(3)}\right)\,, \tag{6.58}$$

$$L_r^{\alpha(3)} = -\left(\frac{14}{5}\frac{k_B T_\alpha}{m_\alpha}\right)^{1/2}\frac{1}{n_\alpha T_\alpha^2}\nabla_m\left(n_\alpha T_\alpha^2 h_{rm}^{\alpha(4)}\right) \tag{6.59}$$

in more detail. The terms M contain derivatives of lower moments. Later, we shall especially consider the terms

$$M_{rs}^{\alpha(2)} = \begin{cases} \sqrt{2}\tau_{rs|pq}\left(\frac{k_B T_e}{m_e}\right)^{1/2}\nabla_p h_q^{(1)} & \text{for } \alpha = e\,, \\ 0 & \text{for } \alpha = i\,, \end{cases} \tag{6.60}$$

and

$$M_r^{\alpha(3)} = -\frac{2}{\sqrt{5}} \left(\frac{k_B T_\alpha}{m_\alpha} \right)^{1/2} T_\alpha^{-7/2} \nabla_m \left(T_\alpha^{7/2} h_{rm}^{\alpha(2)} \right) . \tag{6.61}$$

The *convective* terms C contain moments of the same or lower order as well as $\boldsymbol{u}^\alpha$ and its derivatives. In the following discussion,

$$C_{rs}^{\alpha(2)} = -\boldsymbol{u}^\alpha \cdot \nabla h_{rs}^{\alpha(2)} + \frac{2}{3} h_{rs}^{\alpha(2)} \nabla \cdot \boldsymbol{u}^\alpha - 2\tau_{rs|pq} h_{pm}^{\alpha(2)} \nabla_m u_q^\alpha + \frac{2\sqrt{2}}{3} h_{rs}^{\alpha(2)} h_{mn}^{\alpha(2)} \nabla_n u_m^\alpha \tag{6.62}$$

and

$$C_r^{\alpha(3)} = -\boldsymbol{u}^\alpha \cdot \nabla h_r^{\alpha(3)} - \frac{7}{5} h_m^{\alpha(3)} \nabla_m u_r^\alpha - \frac{2}{5} h_m^{\alpha(3)} \nabla_r u_m^\alpha + \frac{3}{5} h_r^{\alpha(3)} \nabla \cdot \boldsymbol{u}^\alpha + \sqrt{2} h_r^{\alpha(3)} h_{ns}^{\alpha(2)} \nabla_n u_s^\alpha \tag{6.63}$$

will be most relevant. As mentioned, the *nonlinear* terms N will be ignored.

6.2
The Hydrodynamic Regime in Linear Transport Theory

We now start with a reduction of the basic equations. We note that the hydrodynamic equations for ρ, $\boldsymbol{u}$, T_e, and T_i do not contain collisional contributions, provided $Q^{(2)}$ will be neglected for $T_i \approx T_e$ and $m_e \ll m_i$. The rest of the equations contain two characteristic time scales, the collisional one $\tau_{coll} = max(\tau_e, \tau_i)$, where usually $\tau_{coll} \approx \tau_i$, and the hydrodynamic time $\tau_H \equiv T_H$. The latter enters the equations for the privileged nonhydrodynamic moments via the thermodynamic forces (e.g., gradients of the hydrodynamical variables). Let us introduce

$$\frac{1}{\tau_{T_\alpha}} = v_{th\alpha} |\nabla \ln T_\alpha| \, , \quad \frac{1}{\tau_\rho} = v_{the} |\nabla \ln \rho| \, ,$$
$$\frac{1}{\tau_u} = |\nabla u| \, , \quad \frac{1}{\tau_E} = \frac{e}{m_e v_{the}} E \, . \tag{6.64}$$

The hydrodynamic time τ_H is then of the order

$$\tau_H \sim min(\tau_\rho, \tau_{T_\alpha}, \tau_u, \tau_E) \, . \tag{6.65}$$

The ratio of τ_{coll} and τ_H forms the smallness parameter

$$\lambda_H = \frac{\tau_{coll}}{\tau_H} \ll 1 \, , \tag{6.66}$$

which will be used in linear transport theory.

To analyze how collisional effects contribute to the development of the hydrodynamical variables, we formally write, for example,

$$\partial_t h^{(1)} = \frac{\tilde{H}}{\tau_H} - \frac{\tilde{B} h^{(1)}}{\tau_B} + \frac{\tilde{Q}^{(1)}}{\tau_{coll}} + L^{(1)} + C^{(1)} + N^{(1)} . \tag{6.67}$$

The term $\tilde{H}$ stands for the source term for the hydrodynamical variables, $\tilde{B}$ is the contribution from the magnetic fields, $\tilde{Q}^{(1)}$ is the collisional contribution, $L^{(1)}$ the hierarchial, $C^{(1)}$ the convective, and $N^{(1)}$ is the nonlinear contribution (if considered at all). A new time, τ_B, originating from the gyromotion in the external magnetic field appears, which will be estimated by $\tau_B \approx |\Omega_e^{-1}|$ or $\tau_B \approx \Omega_i^{-1}$ depending on the species under consideration. In principle, we could order τ_B with respect to τ_{coll}. We refrain from doing that and keep the term $\tilde{B}$ without an additional ordering (let us say of order one). The balance of the $\tilde{Q}^{(1)}$-term with the $\tilde{H}$-term suggests the scaling

$$h^{(1)} \sim \mathcal{O}(\lambda_H) \tag{6.68}$$

since $\tilde{H}$, in contrast to $\tilde{Q}$, does not contain any contribution from $h^{(1)}$. To first order in λ_H, we obtain

$$\partial_t h^{(1)} + \frac{\tilde{B} h^{(1)}}{\tau_B} \approx \frac{\tilde{H}}{\tau_H} + \frac{\tilde{Q}^{(1)}}{\tau_{coll}} . \tag{6.69}$$

The terms $L^{(1)}$, $C^{(1)}$, and $N^{(1)}$ are of higher order (as can be seen from Eq. (6.67) after multiplication with τ_{coll}: $h^{(1)} \sim \lambda_H$, $N^{(1)}\tau_{coll} \sim \lambda_H^2$, and so on).

The equations for the nonhydrodynamical moments now are in the 21 moment approximation

$$\begin{aligned} \partial_t h_r^{(1)} - \Omega_e \varepsilon_{rmn} h_m^{(1)} b_n &= -\frac{1}{\tau_e}\left[c_{11} h_r^{(1)} + c_{13} h_r^{e(3)} + c_{15} h_r^{e(5)}\right] \\ &+ \left(\frac{m_e}{k_B T_e}\right)^{1/2} \left[\frac{e}{m_e} E_r + \frac{1}{m_e \rho}\nabla_r(\rho k_B T_e) + \frac{e}{m_e c}\varepsilon_{rmn} u_m B_n\right] , \end{aligned} \tag{6.70}$$

$$\begin{aligned} \partial_t h_r^{e(3)} - \Omega_e \varepsilon_{rmn} h_m^{e(3)} b_n &= -\frac{1}{\tau_e}\left[c_{31} h_r^{(1)} + c_{33} h_r^{e(3)} + c_{35} h_r^{e(5)}\right] \\ &\quad - \left(\frac{5}{2}\right)^{1/2} \left(\frac{k_B T_e}{m_e}\right)^{1/2} \frac{1}{T_e}\nabla_r T_e , \end{aligned} \tag{6.71}$$

$$\partial_t h_r^{e(5)} - \Omega_e \varepsilon_{rmn} h_m^{e(5)} b_n = -\frac{1}{\tau_e}\left[c_{51} h_r^{(1)} + c_{53} h_r^{e(3)} + c_{55} h_r^{e(5)}\right], \tag{6.72}$$

$$\begin{aligned} &\partial_t h_{rs}^{e(2)} - \Omega_e \left(\varepsilon_{rmn} h_{sm}^{e(2)} + \varepsilon_{smn} h_{rm}^{e(2)}\right) b_n \\ &= -\frac{1}{\tau_e}\left[c_{22} h_{rs}^{e(2)} + c_{24} h_{rs}^{e(4)}\right] - \sqrt{2}\tau_{rs|pq}\nabla_p u_q , \end{aligned} \tag{6.73}$$

$$\partial_t h_{rs}^{e(4)} - \Omega_e \left(\varepsilon_{rmn} h_{sm}^{e(4)} + \varepsilon_{smn} h_{rm}^{e(4)}\right) b_n = -\frac{1}{\tau_e}\left[c_{42} h_{rs}^{e(2)} + c_{44} h_{rs}^{e(4)}\right], \tag{6.74}$$

$$\partial_t h_r^{i(3)} - \Omega_i \varepsilon_{rmn} h_m^{i(3)} b_n = -\frac{1}{\tau_i}\left[\overline{c}_{33} h_r^{i(3)} + \overline{c}_{35} h_r^{i(5)}\right] - \left(\frac{5}{2}\right)^{1/2}\left(\frac{k_B T_i}{m_i}\right)^{1/2}\frac{1}{T_i}\nabla_r T_i \,, \quad (6.75)$$

$$\partial_t h_r^{i(5)} - \Omega_i \varepsilon_{rmn} h_m^{i(5)} b_n = -\frac{1}{\tau_i}\left[\overline{c}_{53} h_r^{i(3)} + \overline{c}_{55} h_r^{i(5)}\right], \quad (6.76)$$

$$\partial_t h_{rs}^{i(2)} - \Omega_i\left(\varepsilon_{rmn} h_{sm}^{i(2)} + \varepsilon_{smn} h_{rm}^{i(2)}\right) b_n = -\frac{1}{\tau_i}\left[\overline{c}_{22} h_{rs}^{i(2)} + \overline{c}_{24} h_{rs}^{i(4)}\right] - \sqrt{2}\tau_{rs|pq}\nabla_p u_q \,, \quad (6.77)$$

$$\partial_t h_{rs}^{i(4)} - \Omega_i\left(\varepsilon_{rmn} h_{sm}^{i(4)} + \varepsilon_{smn} h_{rm}^{i(4)}\right)_n = -\frac{1}{\tau_i}\left[\overline{c}_{42} h_{rs}^{i(2)} + \overline{c}_{44} h_{rs}^{i(4)}\right]. \quad (6.78)$$

Here, we have abbreviated

$$b_n = \frac{B_n}{B}\,, \quad C_{\dots} = -\frac{1}{\tau_e} c_{\dots}\,, \quad \overline{C}_{\dots} = -\frac{1}{\tau_i}\overline{c}_{\dots}\,. \quad (6.79)$$

The gyrofrequencies contain a sign, that is, $\Omega_e = -(eB/m_e)$ and $\Omega_i = (ZeB/m_i)$. We did not list all of the known constants, for example, $c_{11} = 1$, $c_{13} = 3/\sqrt{10}, \dots$, $\overline{c}_{33} = 2\sqrt{2}/5$, and so on.

The vectorial moment equations decouple from the tensorial moment equations. Next, we shall simplify, making use of the already mentioned ordering $\mu = m_e/m_i \ll 1$, $\sigma \approx 0$, $\tau_{coll}/\tau_H \ll 1$, $h_{\dots}^{\alpha} \sim \lambda_H$ and so on. It would be practically impossible, and fortunately physically meaningless, if transport theory would require one to solve the initial value problem of the just posted differential equations. The physical reason is rather simple. When the nonhydrodynamic variables only depend on time functionally through the hydrodynamic variables, the structure of the equations can be abbreviated as

$$\tau_{coll}\partial_t h_{\dots}^{\alpha(m)} \sim \frac{\tau_{coll}}{\tau_H}\lambda_H = \lambda_H^2 \ll \lambda_H \,, \quad (6.80)$$

meaning that we can effectively solve algebraic equations after ignoring the left-hand sides

$$\partial_t h_{\dots}^{\alpha(m)} \approx 0 \,. \quad (6.81)$$

The mathematical argument is as follows. We abbreviate the just mentioned structure of equations (with a single equation and not for a system of equations; the latter can be discussed with similar arguments and conclusions) as

$$\partial_t y = -\frac{|c|}{\tau_{coll}} y + q \,, \quad (6.82)$$

where q is the source term containing the thermodynamic forces. The solution of the homogeneous part of the equation is

$$y = \alpha \exp\left(-\frac{|c|t}{\tau_{coll}}\right). \quad (6.83)$$

The variation of parameters leads to

$$y(t) = y(0) \exp\left(-\frac{|c|t}{\tau_{coll}}\right) + \int_0^t d\tau q(\tau) e^{-|c|\frac{(t-\tau)}{\tau_{coll}}} . \tag{6.84}$$

Since $|c| > 0$ is of order one, the influence of the initial values disappears after $t \gg \tau_{coll}$. The time variation of the source term $q(t)$ is on the $\tau_H \gg \tau_{coll}$ scale. Therefore, we may approximate

$$\begin{aligned} \int_0^t d\tau q(\tau) e^{-|c|\frac{(t-\tau)}{\tau_{coll}}} &= \int_0^t d\sigma q(t-\sigma) e^{-|c|\frac{\sigma}{\tau_{coll}}} \\ &\approx \int_0^\infty d\sigma q(t-\sigma) e^{-|c|\frac{\sigma}{\tau_{coll}}} , \end{aligned} \tag{6.85}$$

for $t \gg \tau_{coll}$. By expanding q for $t \gg \tau_{coll}$ as

$$q(t-\sigma) \approx q(t) + \mathcal{O}\left(\frac{\tau_{coll}}{\tau_H}\right), \tag{6.86}$$

with the result

$$y(t) \approx \frac{\tau_{coll}}{|c|} q(t) \quad \text{for} \quad t \gg \tau_{coll}. \tag{6.87}$$

This approximate solution follows for Eq. (6.82) when we approximate Eq. (6.82) by an algebraic equation which is formally obtained by setting $\partial_t y \approx 0$.

Summarizing, in the so called hydrodynamic regime

$$\tau_{coll} \ll t \sim \mathcal{O}(\tau_H), \tag{6.88}$$

the equations for the nonhydrodynamic variables become simple, that is, (linear) algebraic equations which can be solved straightforwardly. Inserting the solutions into the equations for the hydrodynamic variables leads to the (linear) transport equations.

6.3 Summary of Linear Transport Coefficients

Let us summarize the results of the transport calculations outlined in the two previous sections. We start with the equations for the hydrodynamic variables, assuming

$\sigma \approx 0$,

$$\partial_t \rho + \nabla \cdot (\rho \boldsymbol{u}) = 0 \,, \tag{6.89}$$

$$\begin{aligned}\rho \partial_t u_r = &-\rho u_s \nabla_s u_r - \frac{Z}{m_i} \nabla_r \left[\rho \left(k_B T_e + \frac{1}{Z} k_B T_i \right) \right] \\ &- \sqrt{2} \frac{Z}{m_i} \nabla_s \left[\rho \left(k_B T_e h_{rs}^{e(2)} + \frac{1}{Z} k_B T_i h_{rs}^{i(2)} \right) \right] \\ &+ e \frac{Z\rho}{m_i} \left(\frac{k_B T_e}{m_e} \right)^{1/2} \varepsilon_{rsm} h_s^{(1)} B_m \,,\end{aligned} \tag{6.90}$$

$$\begin{aligned}\partial_t T_e = &-\boldsymbol{u} \cdot \nabla T_e - \frac{2}{3} T_e \nabla \cdot \boldsymbol{u} \\ &+ \left(\frac{k_B T_e}{m_e} \right)^{1/2} h_r^{(1)} \nabla_r T_e + \frac{2}{3} T_e \nabla_r \left[\left(\frac{k_B T_e}{m_e} \right)^{1/2} h_r^{(1)} \right] \\ &- \frac{2\sqrt{2}}{3} T_e h_{rs}^{e(2)} \nabla_r u_s + \frac{2\sqrt{2}}{3} T_e h_{rs}^{e(2)} \nabla_r \left[\left(\frac{k_B T_e}{m_e} \right)^{1/2} h_s^{(1)} \right] \\ &- \frac{\sqrt{10}}{3} \frac{1}{\rho} \nabla_r \left[\rho T_e \left(\frac{k_B T_e}{m_e} \right)^{1/2} h_r^{e(3)} \right] ,\end{aligned} \tag{6.91}$$

$$\begin{aligned}\partial_t T_i = &-\boldsymbol{u} \cdot \nabla T_i - \frac{2}{3} T_i \nabla \cdot \boldsymbol{u} - \frac{2\sqrt{2}}{3} T_i \, h_{rs}^{i(2)} \nabla_r u_s \\ &- \frac{\sqrt{10}}{3} \frac{1}{\rho} \nabla_r \left[\rho T_i \left(\frac{k_B T_i}{m_i} \right)^{1/2} h_r^{i(3)} \right] .\end{aligned} \tag{6.92}$$

They contain the privileged nonhydrodynamic coefficients $h_r^{(1)}$, $h_{rs}^{(2)}$, and $h_r^{\alpha(3)}$ which are related to the electric current density $\boldsymbol{j}$, the dissipative part of the pressure tensor π_{rs}^{α}, and the heat flux $\boldsymbol{q}^{\alpha}$ through

$$\begin{aligned} j_r &= e n_e \left(\frac{k_B T_e}{m_e} \right)^{1/2} h_r^{(1)} \,, \quad \pi_{rs}^{\alpha} = \sqrt{2} n_\alpha k_B T_\alpha h_{rs}^{\alpha(2)} \,, \\ q_r^{\alpha} &= \left(\frac{5}{2} \right)^{1/2} m_\alpha \left(\frac{k_B T_\alpha}{m_\alpha} \right)^{3/2} n_\alpha h_r^{\alpha(3)} \,. \end{aligned} \tag{6.93}$$

The particle densities are approximately $n_e \approx Z\rho/m_i$ and $n_i \approx \rho/m_i$. The quantities $\boldsymbol{j}$, π_{rs}^{e}, q_r^{e}, π_{rs}^{i}, and q_r^{i} can be expressed in terms of the thermodynamic forces. Depending on the degree of approximation, the numerical factors change. We present the result for the electric current density in the form

$$\begin{aligned}\boldsymbol{j} = &\boldsymbol{b} \left\{ \sigma_\parallel \hat{\boldsymbol{E}} \cdot \boldsymbol{b} + \alpha_\parallel (-\nabla k_B T_e) \cdot \boldsymbol{b} \right\} + \boldsymbol{b} \times \left\{ \sigma_\wedge \hat{\boldsymbol{E}} + \alpha_\wedge (-\nabla k_B T_e) \right\} \\ &+ \boldsymbol{b} \times \left\{ \sigma_\perp (\hat{\boldsymbol{E}} \times \boldsymbol{b}) + \alpha_\perp (-\nabla k_B T_e) \times \boldsymbol{b} \right\} .\end{aligned} \tag{6.94}$$

Here,

$$\hat{\boldsymbol{E}} = \boldsymbol{E} + (\boldsymbol{u} \times \boldsymbol{B}) + \frac{1}{e\rho} \nabla (\rho k_B T_e) \,. \tag{6.95}$$

The coefficients are defined as

$$\sigma_A = \frac{e^2 Z\rho}{m_e m_i}\tau_e\tilde{\sigma}_A\,, \qquad \alpha_A = \left(\frac{5}{2}\right)^{1/2}\frac{eZ\rho}{m_e m_i}\tau_e\tilde{\alpha}_A\,, \tag{6.96}$$

with index A characterizing the direction with respect to the external magnetic field, $A = \|$, that is, parallel to $\boldsymbol{b}$, $A = \wedge$, that is, along $\boldsymbol{b}\times\nabla\cdots$, $A = \perp$, that is, along $\boldsymbol{b}\times[(\nabla\cdots)\times\boldsymbol{b}]$. Numerical values of $\tilde{\sigma}_A, \tilde{\alpha}_A, \tilde{\kappa}^e_A$ can be calculated (see below). The heat fluxes are

$$\begin{aligned}\boldsymbol{q}^e = \boldsymbol{b}&\left\{\alpha_\| k_B T_e(\hat{\boldsymbol{E}}\cdot\boldsymbol{b}) + \kappa^e_\|(-\nabla k_B T_e)\cdot\boldsymbol{b}\right\}\\ &+\boldsymbol{b}\times\left\{\alpha_\wedge k_B T_e\hat{\boldsymbol{E}} + \kappa^e_\wedge(-\nabla k_B T_e)\right\}\\ &+\boldsymbol{b}\times\left\{\alpha_\perp k_B T_e(\hat{\boldsymbol{E}}\times\boldsymbol{b}) + \kappa^e_\perp(-\nabla k_B T_e)\times\boldsymbol{b}\right\},\end{aligned} \tag{6.97}$$

$$\boldsymbol{q}^i = -\kappa^i_\|\boldsymbol{b}(\boldsymbol{b}\cdot\nabla k_B T_i) - \kappa^i_\wedge(\boldsymbol{b}\times\nabla k_B T_i) - \kappa^i_\perp\left\{\boldsymbol{b}\times[(\nabla k_B T_i)\times\boldsymbol{b}]\right\}. \tag{6.98}$$

The additional coefficients

$$\kappa^e_A = \frac{5}{2}\frac{Z\rho k_B T_e}{m_e m_i}\tau_e\tilde{\kappa}^e_A\,, \qquad \kappa^i_A = \frac{5}{2}\frac{\rho k_B T_i}{m_i^2}\tau_i\tilde{\kappa}^i_A\,, \tag{6.99}$$

appear. Defining

$$v_{rs} = 2\tau_{rs|pq}\nabla_p u_q\,, \tag{6.100}$$

one can write the components of the dissipative part of the pressure tensor as

$$\pi^\alpha_{zz} = -\eta^\alpha_\| v_{zz}\,, \tag{6.101}$$

$$\pi^\alpha_{xz} = -\eta^\alpha_2 v_{xz} + \eta^\alpha_1 v_{yz}\,, \tag{6.102}$$

$$\pi^\alpha_{yz} = -\eta^\alpha_1 v_{xz} - \eta^\alpha_2 v_{yz}\,, \tag{6.103}$$

$$\pi^\alpha_{xx} = -\frac{1}{2}\left(\eta^\alpha_\| + \eta^\alpha_4\right)v_{xx} - \frac{1}{2}\left(\eta^\alpha_\| - \eta^\alpha_4\right)v_{yy} + \eta^\alpha_3 v_{xy}\,, \tag{6.104}$$

$$\pi^\alpha_{yy} = -\frac{1}{2}\left(\eta^\alpha_\| - \eta^\alpha_4\right)v_{xx} - \frac{1}{2}\left(\eta^\alpha_\| + \eta^\alpha_4\right)v_{yy} - \eta^\alpha_3 v_{xy}\,, \tag{6.105}$$

$$\pi^\alpha_{xy} = -\frac{1}{2}\eta^\alpha_3 v_{xx} + \frac{1}{2}\eta^\alpha_3 v_{yy} - \eta^\alpha_4 v_{xy}\,, \tag{6.106}$$

where

$$\eta^e_A = \frac{Z\rho}{m_i}k_B T_e\tau_e\tilde{\eta}^e_A\,, \quad \eta^i_A = \frac{\rho}{m_i}k_B T_i\tau_i\tilde{\eta}^i_A\,, \quad A = \|, 1, 2, 3, 4\,. \tag{6.107}$$

Traditionally, the following quantities are called transport coefficients for the components $\alpha = e, i$: Three coefficients for the electric conductivity σ (or resistivity η), three coefficients for the thermal conductivity κ^α, three thermoelectric coefficients α, and five viscosity coefficients η^α. The dimensionless quantities $\tilde{\sigma}$, $\tilde{\alpha}$, $\tilde{\kappa}^e$, $\tilde{\kappa}^i$, $\tilde{\eta}^e$, $\tilde{\eta}^i$ depend on Z and $\Omega_\alpha \tau_\alpha$ (except $\|$).

In the following, we summarize for $Z = 1$ the characteristic values. Let us start with $A = \|$. The following numbers have been obtained (going up to the 29 moment expansion), that is,

$$\begin{aligned} \tilde{\sigma}_\| &\approx 1.9\,, \quad \tilde{\alpha}_\| \approx -0.8\,, \quad \tilde{\kappa}^e_\| \approx 1.6\,, \\ \tilde{\kappa}^i_\| &\approx 2.2\,, \quad \tilde{\eta}^e_\| \approx 0.7\,, \quad \tilde{\eta}^i_\| \approx 1.3\,. \end{aligned} \tag{6.108}$$

The perpendicular coefficients $(\wedge, \perp)$ show some nontrivial behavior with variations in $\Omega_\alpha \tau_\alpha$. $\tilde{\sigma}_\perp$ is monotonically decreasing with a characteristic value 0.4 at $|\Omega_e|\tau_e = 1$. $\tilde{\sigma}_\wedge$ has a maximum at $|\Omega_e|\tau_e \approx 0.3$ and the value 0.7 at $|\Omega_e|\tau_e = 1$. $\tilde{\alpha}_\perp$ is monotonically increasing with the value 0.02 at $|\Omega_e|\tau_e = 1$. $\tilde{\alpha}_\wedge$ has a minimum at $|\Omega_e|\tau_e \approx 0.3$ and the value -0.3 at $|\Omega_e|\tau_e = 1$. $\tilde{\kappa}^e_\perp$ is monotonically decreasing with the value 0.3 at $|\Omega_e|\tau_e = 1$. $\tilde{\kappa}^e_\wedge$ has a maximum at $|\Omega_e|\tau_e \approx 0.3$ and the value 0.5 at $|\Omega_e|\tau_e = 1$. $\tilde{\eta}^e_2$ is monotonically decreasing with the value 0.2 at $|\Omega_e|\tau_e = 1$. A similar behavior is seen for $\tilde{\eta}^e_4$ with the value 0.4 at $|\Omega_e|\tau_e = 1$. $\tilde{\eta}^e_1$ has a maximum at $|\Omega_e|\tau_e \approx 0.3$ and the value 0.3 at $|\Omega_e|\tau_e = 1$. The maximum of $\tilde{\eta}^e_3$ is not very pronounced; a characteristic value is 0.3 at $|\Omega_e|\tau_e = 1$.

The behavior of $\tilde{\kappa}^i$ and $\tilde{\eta}^i$ for ions is similar to that for electrons; characteristic values at $\Omega_i \tau_i = 1$ are $\tilde{\kappa}^i_\perp \approx 0.4$, $\tilde{\kappa}^i_\wedge \approx -0.7$, $\tilde{\eta}^i_2 \approx 0.2$, $\tilde{\eta}^i_4 \approx 0.4$, $\tilde{\eta}^i_1 \approx -0.4$, and $\tilde{\eta}^i_3 \approx -0.6$.

For strong magnetic fields ($|\Omega_\alpha|\tau_\alpha \gg 1$), the following formulas are asymptotically correct:

$$\tilde{\sigma}_\perp \approx \frac{1}{(\Omega_e \tau_e)^2}\,, \quad \tilde{\sigma}_\wedge \approx -\frac{1}{\Omega_e \tau_e}\,, \tag{6.109}$$

$$\tilde{\alpha}_\perp \approx \frac{\frac{3}{\sqrt{10}}}{(\Omega_e \tau_e)^2}\,, \quad \tilde{\alpha}_\wedge \approx 0\,, \tag{6.110}$$

$$\tilde{\kappa}^\alpha_\perp \approx \frac{c^\alpha_{33}}{(\Omega_\alpha \tau_\alpha)^2}\,, \quad \tilde{\kappa}^\alpha_\wedge \approx -\frac{1}{\Omega_\alpha \tau_\alpha}\,, \tag{6.111}$$

$$\tilde{\eta}^\alpha_2 \approx \frac{c^\alpha_{22}}{4(\Omega_\alpha \tau_\alpha)^2}\,, \quad \tilde{\eta}^\alpha_1 \approx -\frac{1}{\Omega_\alpha \tau_\alpha}\,, \tag{6.112}$$

$$\tilde{\eta}^\alpha_4 \approx \frac{c^\alpha_{22}}{(\Omega_\alpha \tau_\alpha)^2}\,, \quad \tilde{\eta}^\alpha_3 \approx -\frac{1}{2\Omega_\alpha \tau_\alpha}\,. \tag{6.113}$$

Corrections are of order $(\Omega_\alpha \tau_\alpha)^{-3}$. The following abbreviations are

$$c^\alpha_{22} = \begin{cases} \frac{3}{5}\left(2 + \sqrt{2}\frac{1}{Z}\right) & \text{for } \alpha = e\,, \\ \frac{3\sqrt{2}}{5} & \text{for } \alpha = i\,, \end{cases} \tag{6.114}$$

$$c^\alpha_{33} = \begin{cases} \frac{1}{10}\left(13 + 4\sqrt{2}\frac{1}{Z}\right) & \text{for } \alpha = e\,, \\ \frac{2\sqrt{2}}{5} & \text{for } \alpha = i\,. \end{cases} \tag{6.115}$$

For more details, we refer the reader to the more specialized literature, for example, [29, 40].

Finally, we discuss transport in the absence of an external magnetic field, that is, $\boldsymbol{B} = 0$. In the case $\boldsymbol{B} = 0$, the system is isotropic, and the transport coefficients are scalars. We have

$$\boldsymbol{j} = \sigma_{\parallel} \hat{\boldsymbol{E}} + \alpha_{\parallel}(-\nabla k_B T_e) \,, \tag{6.116}$$

$$\boldsymbol{q}^e = \alpha_{\parallel} k_B T_e \hat{\boldsymbol{E}} + \kappa_{\parallel}^e(-\nabla k_B T_e) \,, \tag{6.117}$$

$$\boldsymbol{q}^i = \kappa_{\parallel}^i(-\nabla k_B T_i) \,, \tag{6.118}$$

$$\pi_{rs}^{\alpha} = -\eta_{\parallel}^{\alpha} \nu_{rs} \,, \tag{6.119}$$

where

$$\hat{\boldsymbol{E}} = \boldsymbol{E} + \frac{1}{e n_e} \nabla p_e \,. \tag{6.120}$$

Formally, these equations agree with the previous ones for $\sigma_{\perp} = \sigma_{\parallel}$, $\sigma_{\wedge} = 0$, $\eta_2^{\alpha} = \eta_4^{\alpha} = \eta_{\parallel}^{\alpha}$, $\eta_1^{\alpha} = \eta_3^{\alpha} = 0$. Clearly, a temperature gradient can cause an electric current (thermoelectric current), and an electric field will stimulate a heat current. Heat conduction is mainly accomplished by electrons,

$$\kappa_{\parallel}^i \ll \kappa_{\parallel}^e \,, \tag{6.121}$$

for $T_e \geq T_i$. On the other hand, for momentum transport, and thereby the viscosity, ions are dominating, which is reflected in

$$\eta_{\parallel}^i \gg \eta_{\parallel}^e \,. \tag{6.122}$$

We conclude this section by mentioning another transport coefficient, namely, diffusion. The simplest estimate of a diffusion coefficient D in a quiescent, unmagnetized plasma can be obtained via the random walk formula

$$D \equiv \chi_{\parallel} = \frac{(\Delta x)^2}{2\Delta t} \approx \nu \lambda_{\text{mfp}}^2 \approx \frac{v_{th}^2}{\nu} \,, \tag{6.123}$$

where λ_{mfp} is the mean free path, v_{th} the thermal velocity, and ν the collision frequency. Here, we do not indicate the particle species which could be easily done by introduced appropriate indices (e for electrons and i for ions). It is clear how to generalize that procedure to the transport perpendicular to an external magnetic field. The characteristic step size is the (thermal) Larmor radius ρ_L, leading to

$$D \equiv \chi_{\perp} = \frac{(\Delta x)^2}{2\Delta t} \approx \nu \rho_L^2 \approx v_{th}^2 \frac{\nu}{\Omega_L^2} \,, \tag{6.124}$$

where $\Omega_L = eB/m$ is the gyrofrequency. Obviously, one predicts from here (the temperature T is measured in electron volt)

$$\chi_{\perp} = \frac{v m T}{e^2 B^2} \sim \frac{T}{B^2} \,, \tag{6.125}$$

as has been done for other transport coefficients in the limit of strong magnetic fields. In most fusion devices, magnitude and scaling of the transport do not agree with the B^{-2}-prediction. The ambipolar diffusion coefficient may have an anomalous factor of 10^2. Also, often the electron thermal conductivity deviates from theoretical predictions by a factor $10-10^3$, and the ion thermal conductivity may be different by a factor of 10.

6.4
Nonlinear Transport Phenomenology

Nonlinear signatures of particle and heat transport are generic in, for example, fluid dynamics, plasma physics, and astrophysics. Transport anomalies reach from Bohm-like diffusion in gas discharges up to low-energy cosmic ray penetration into the heliosphere. In the classical picture, large magnetic guiding fields reduce the perpendicular diffusion decisively. Collisions appear as an obstacle for the free motion along the field lines on the one hand, and increase, on the other hand, the transport in perpendicular direction. The specific reason for the extraordinary interest in the mechanisms of diffusion in stochastic fields lies in the unexpected large losses caused by anomalous transport [11, 68]. Anomalous particle and heat losses have enormous consequences for practical applications. For example, in magnetic fusion devices, heat fluxes must be relatively low and well-distributed over the wall to reach tolerable local power depositions. Therefore, it is desirable to identify the basic processes which rule anomalous losses from open nonlinear (chaotic) systems.

Anomalous transport analysis is one of the key problems in plasma research and nuclear fusion applications [25]. During the last decades, considerable progress was reported in understanding the basic features [74]. First, linear transport theory has been modified to take care of the geometrical effects and large mean free paths. The neoclassical theory (see [58] and references therein) is a great intellectual and practical success. Nevertheless, it is not able to resolve all the problems. As it is well-known [75], in many cases, the strong deviation of the diffusion rate from classical or neoclassical predictions is due to nonlinear effects caused by (electrostatic as well as electromagnetic) fluctuations. In the past, several attempts have been made for a self-consistent theory of nonlinear transport; see [25, 76] and references therein. Quasilinear theory and transport estimates based on the weak-turbulence description are by far the most successful analytical approaches. However, it is well-known [75] that they are of limited applicability. Strong plasma turbulence is a very complex and complicated problem. Physical kinetics of plasma turbulence through a focus on quasiparticle models has recently been summarized [77]. Analytical evaluations are generally difficult, and therefore numerical simulations become more and more important. They lead to a rich data base with many hints for fundamental transport scalings. In Appendix E, we briefly comment on low-dimensional basis systems for a reduced description.

Anomalous charged particle transport is also a long-standing problem in astrophysical issues [78–91]. A variety of problems, such as, low-energy cosmic ray penetration into the heliosphere, the transport of galactic cosmic rays in and out of the interstellar magnetic field, the Fermi acceleration mechanism, and so on, are on the top of the astrophysical agenda [92]. In most cases, astrophysical transport is collisionless. Often, no dominant guiding fields exist.

Nonlinear effects are responsible for the differences between linear theoretical predictions and experimental facts. A very simple argument sheds light on the appearance of nonlinearities in transport theory. The continuity equation for the particle current density $\boldsymbol{\Gamma}$ may be formally written as

$$\partial_t \boldsymbol{\Gamma} = -\nabla \cdot (\boldsymbol{u}\boldsymbol{\Gamma}) - \nabla P - \nabla \cdot \Pi + \frac{e}{m}\boldsymbol{\Gamma} \times \boldsymbol{B} + \frac{e}{m}\boldsymbol{E} + \frac{1}{m}\boldsymbol{R} . \tag{6.126}$$

On the right-hand side of the convective term, pressure and field contributions as well as the resistive term appear. We decompose each variable into an averaged and a fluctuating part, respectively, for example,

$$X := \langle X \rangle + \delta X . \tag{6.127}$$

A short calculation shows the averaged x-component of the current density in the form

$$\langle \Gamma_x \rangle = \underbrace{\frac{1}{m\Omega}\langle R_y \rangle}_{classical} + \underbrace{\frac{1}{\Omega}\left\langle \left(\hat{b} \times \nabla \cdot \Pi\right) \cdot \hat{x} \right\rangle}_{neoclassical} + \Gamma_x^{anomalous} + \Gamma_x^{time\text{-}dependent} . \tag{6.128}$$

The first part, caused by the friction term, is the classical one. The second one, from the pressure tensor, is mainly discussed in neoclassical theory. The third one on the right-hand side is the term we are mainly interested in within the next section,

$$\Gamma_x^{anomalous} = \frac{1}{B}\langle \delta n \delta E_y \rangle + \frac{1}{B}\left\{ \langle \delta \Gamma_\parallel \delta B_x \rangle - \langle \delta \Gamma_x \delta B_\parallel \rangle \right\} . \tag{6.129}$$

The fourth, time-dependent term is usually neglected.

We should emphasize that this simple argument does not cover the important aspect of gyroaveraging. The latter is the central part of gyrokinetic models.

6.4.1
Fluctuation Spectra and Transport

Weak turbulence theory turned out to be a very successful approach, being specifically developed for plasma fluctuations with a characteristic frequency. Essentially, in its simplest form, it results from perturbation theory. It is thus restricted to weak fluctuation levels. Introducing, for example, for the second moment of fluctuating electrostatic potentials, the spectral density $n(\boldsymbol{k}, t)$,

$$\left\langle \delta\varphi_{\boldsymbol{k}} \delta\varphi_{\boldsymbol{k}'}^{*} \right\rangle = n(\boldsymbol{k}, t)\delta\left(\boldsymbol{k} - \boldsymbol{k}'\right) , \tag{6.130}$$

we obtain a wave-kinetic equation in the form [77]

$$\frac{\partial n(\boldsymbol{k},t)}{\partial t} = \mathcal{I}(\boldsymbol{k},t) + \Gamma(\boldsymbol{k}) n(\boldsymbol{k},t) \,, \tag{6.131}$$

where on the right-hand side a nonlinear term ($\mathcal{I}$) and a damping term (Γ) appear. When damping can be neglected in a certain (inertial) region of $\boldsymbol{k}$–space, we may write for the spectral energy density $E(k)$ in isotropic turbulence

$$\frac{\partial E(k)}{\partial t} + \frac{\partial P(k)}{\partial k} = 0 \,. \tag{6.132}$$

This equation can be used for analyzing wave number spectra.

Dimensional arguments are often useful. Let us repeat the simple dimensional analysis known from fluid theory. In three-dimensional space, we can write the dimensions of energy E and spectral energy density $E(k)$ as

$$\begin{aligned} [E] &= \left[\frac{gL^2}{T^2}\right] = \left[\frac{\rho L^5}{T^2}\right] \quad \Rightarrow \\ \left[E(k)\right] &= \left[\frac{\rho L^3}{T^2}\right] \quad \text{and} \quad \left[P(k)\right] = \left[\frac{\rho L^2}{T^3}\right] \end{aligned} \tag{6.133}$$

since we define

$$\frac{E}{V} = \int E(k) dk \tag{6.134}$$

independent of the dimension of the system. The famous argument of Kolmogorov [93] (with application to fluid turbulence) states that the time T can be eliminated via the constant transfer function P leading to

$$\left[E(k)\right] = \left[\frac{\rho L^3}{T^2}\right] \sim \rho^{1/3} P^{2/3} k^{-3+4/3} \sim k^{-5/3} \,, \tag{6.135}$$

that is, the famous Kolmogorov spectrum.

When a characteristic frequency appears in plasmas, then the situation is different. The characteristic time follows from the finite frequency $\omega = \omega(k)$, and we can formulate a nondimensional variable ξ in the form

$$\xi = \frac{P k^2}{\rho \omega^3} \,. \tag{6.136}$$

Now, more freedom appears since we may introduce an arbitrary function f of the dimensionless quantity to obtain for the spectral density

$$E(k) \sim \rho \omega^2(k) k^{-3} f(\xi). \tag{6.137}$$

When $f(\xi) \sim \xi^{1/3}$, one recovers some fundamental aspects of drift-wave turbulence. However, in the strict sense, the latter functional form is not uniquely prescribed.

There have been many attempts to generalize weak-turbulence theory to strong plasma turbulence via renormalization procedures [94–97]. Although intellectually convincing, for practical applications, the evaluation is in most cases technically very ambitious (and perhaps more time-consuming than solving the original gyrokinetic problem).

Especially for drift-wave turbulence, semiempirical models exist which are published under different names, for example, mixing length theory, the marginal stabilization scenarios, and so on [12].

For electrostatic turbulence, the nonlinear current density in radial direction r is

$$\Gamma_r \approx \langle \delta n \delta u_r \rangle \hat{=} - D \frac{\langle n \rangle}{L_n} , \tag{6.138}$$

where we expressed on the right-hand side the current density as an effective diffusion current with a diffusion coefficient D and an inhomogeneity length L_n. For drift waves, the linear growth rate, frequency, density fluctuation, and $E \times B$-velocity are

$$\frac{\gamma_k}{\omega_k} \sim \delta_k , \quad \omega_k \approx \frac{kT}{eBL_n} ,$$
$$\delta n_k \approx \langle n \rangle \frac{e\phi_k(1 - i\delta_k)}{T} , \quad \delta u_r \approx ik \frac{\phi_k}{B} , \tag{6.139}$$

respectively. Putting these terms together, we arrive at

$$\Gamma_r \approx \langle n \rangle \frac{T}{eB} \sum_k k \delta_k \left| \frac{\delta n_k}{\langle n \rangle} \right|^2 . \tag{6.140}$$

The latter expression contains the still unknown fluctuation spectrum. For the level of the latter, we assume that it increases until the corresponding local density change eliminates the instability driving density gradient:

$$\nabla \delta n \hat{=} k \delta n_k \sim \frac{\langle n \rangle}{L_n} . \tag{6.141}$$

We finally obtain

$$D \approx \frac{\gamma_{max}}{k_{max}^2} , \tag{6.142}$$

when we approximate the sum by its maximum term. This expression is often used for estimates of diffusion processes. It is related to the mixing length expression. According to Prandtl, the turbulent diffusion coefficient can be written as

$$D \approx \frac{L_r^2}{\tau_c} , \tag{6.143}$$

where L_r is the characteristic length (mode radial correlation length), and τ_c is the eddy turnover time (autocorrelation time for the turbulent field).

Now, we may have two extremely different scenarios. First, small-scale turbulence with $L_r \approx \rho_s \sim \rho_i$ and the diamagnetic frequency

$$\tau_c^{-1} \approx \frac{v_{thi}}{a} , \quad k \approx k_\theta \approx \rho_i^{-1} , \quad L_n \approx a . \tag{6.144}$$

Combining these characteristic scales, we obtain

$$D \sim \rho_i^2 \tau_c^{-1} \sim D_{Bohm} \frac{\rho_i}{a} \sim T^{3/2} B^{-2} , \tag{6.145}$$

which is known as the gyro-Bohm scaling.

On the other hand, for large-scale turbulence, for example, with $L_r \approx \sqrt{a\rho_s}$, where a is the dimension of the system (minor radius of the tokamak), we estimate

$$D \sim \rho_i v_{thi} \sim D_{Bohm} \left[\frac{m^2}{s} \right] = \frac{T \text{ [eV]}}{16B \text{ [T]}} \sim T B^{-1} . \tag{6.146}$$

This is the celebrated Bohm scaling.

Other scalings have also been proposed. For example, the Goldstone regime $D \sim \sqrt{\rho_i}$ in the L-regime.

The main question remains: How can we predict the characteristic scales? Several procedures have been proposed during the last decades. A quite convincing one is the following self-organization hypothesis:

> "Let a system be modeled by a nonlinear partial differential equation, with dissipation, for the field u. The system contains (at least) two quadratic or higher-order conserved quantities in the absence of dissipation. One of the conserved quantities, let us say $A(u)$, decays faster than the other(s), e.g. $B(u)$. The modal cascade is expected to reach a quasistationary state which minimizes A for constant B, i.e. $\delta A - \lambda \delta B = 0$."

This formulation, as well as arguments for it, can be found in the review by Hasegawa [98].

Besides the evident progress in analytical theory, one should have in mind that (really) self-consistent models are generally very difficult to solve and often show their power only qualitatively, that is, in scaling predictions. Heuristic, quasi-self-consistent models are often ad hoc, are good for rough interpretations, but are, strictly speaking, not predictive. Nonlocality, intermittency, interaction with coherent structures, and many other open problems which could not be solved thus far in full generality exist. Many new insights rely on numerical simulations. The progress in numerical codes is enormous; unfortunately, analytical theory is behind, but has remained indispensable.

6.5 Simple Models in Stochastic Transport Theory

A possible approach to the problem of anomalous plasma transport is to consider the (passive) motion of (test) particles under the influence of given perturbations; see [99] and references therein. Such a treatment is quite common in fluid turbulence [100] where passive motion of scalars, vectors, particles, and so on has been investigated extensively. In magnetic fusion research, there exists an additional, qualitatively important reason to investigate particle motion in given stochastic fields. Perturbations in the magnetic field structure are more or less unavoidable because of errors in the coil arrangements of the devices. In addition, and recently that aspect became very important, additional coils are being installed in tokamaks to control the particle and heat loads on the walls via magnetic stochastization of the edge [101–105]. By title stochastic transport, we mean transport in stochastic fields produced by some external means.

The strength of the magnetic field fluctuations may be quite small, for example, less than one-tenth of a percent of the zeroth order confining field (which defines the *parallel* direction). Nevertheless, magnetic fluctuations have a strong influence on *perpendicular* transport. Particles will propagate in the perpendicular direction significantly faster than predicted by classical theory, as has been shown in classical papers [78, 80, 81, 106, 107]. According to these findings, the perpendicular diffusion coefficient shows a decisive dependency on the parallel diffusion coefficient. This is caused by the fact that perpendicular fluctuations in the magnetic field are small channels of parallel diffusivity in perpendicular direction.

The kinetics of test particle diffusion with special emphasis on the collisional limit was discussed within a modified perturbation theory [108]. Myra *et al.* [109] showed, both analytically and by Monte Carlo simulation, that the diffusion coefficient can be sensitive to the spectral width of the magnetic turbulence. A general variational principle for particle diffusion across braided magnetic surfaces was established by Laval [110]. Models for magnetic field line configurations were discussed by several authors, for example, in [65, 111–113]. Statistical plasma transport theory [55, 114, 115] prompted a strong interest [116, 117]. The Langevin equation model [115] was used for analyzing the subdiffusive behavior of charged particles in stochastic magnetic fields [118, 119]. The relation to continuous time random walks was elucidated [120]. The reliable transition from Euler to Liouville correlation functions turned out to be a very important step [119, 121–125]. Effects of time-dependence of the magnetic fluctuations [126] and plasma flows [127] on particle diffusion were investigated. Trajectory structures and transport became a main topic [128–140]. Larmor radius effects have been shown to be important in the percolative regime [141–143]. New processes, like a ratchet current in plasma devices, were discussed; see, for example, [144–146] and references therein.

For the calculation of effective perpendicular transport, one needs the parallel diffusion coefficient. In the collisionless case, classical theory predicts a superdiffusive transport in parallel direction. In fact, that transport will not occur as long as magnetic perturbation fields act as an additional source for collision-like interactions. Even without binary collisions, the parallel motion should show diffusive behavior. That behavior was recently discussed for plasmas in the presence of magnetic fluctuations.

The total area of "nonlinear transport" is huge. Any global overview would go beyond the scope of a single chapter. Fortunately, several specialized books, for example, [12, 24, 76, 77], already exist which should be consulted when all aspects of nonlinear transport are of interest. In the present chapter, we concentrate on one single subtopic, namely, "stochastic transport theory" in fluctuating magnetic fields. This topic has several applications in nuclear fusion and astrophysics. It has the additional advantage that "simple" analytical models can be developed which allow a lot of insight into nonlinear transport processes. Thus, the following may be considered as a tutorial into elementary aspects of nonlinear plasma transport. For the more advanced aspects of nonlinear transport, we refer to the still growing literature, including the numerical simulations which have become of increasing importance.

A first view of anomalous transport in stochastic magnetic fields may be based on the following fluid picture. Suppose that magnetic fluctuations are present in the plasma such that the total field consists of (strong) zeroth order contributions $\boldsymbol{B}_0$ and fluctuations $\delta \boldsymbol{B}$,

$$\boldsymbol{B} = \boldsymbol{B}_0 + \delta \boldsymbol{B} \,, \quad \text{with} \quad \boldsymbol{b} = \frac{\delta \boldsymbol{B}}{B} \,. \tag{6.147}$$

We assume that components of the fluctuating magnetic field $\delta \boldsymbol{B}$ being perpendicular to the zeroth order (main) magnetic field $\boldsymbol{B}_0$ do exist locally. In order to avoid problems of ambipolarity, let us consider heat transport. To zeroth order, heat transport is mainly along the (zeroth order) magnetic field; $\kappa_\perp \ll \kappa_\parallel$ holds for the linear heat conductivities. Including the magnetic perturbations, we may formulate the dominating terms in the heat conduction equation as

$$\frac{\partial T}{\partial t} \approx \kappa_\parallel \left(\boldsymbol{n} \cdot \nabla\right)^2 T \,. \tag{6.148}$$

The normal vector $\boldsymbol{n}$ is along the total magnetic field. We have

$$\boldsymbol{n} \equiv \frac{\boldsymbol{B}}{B} \,, \qquad \boldsymbol{n}_0 \equiv \frac{\boldsymbol{B}_0}{B_0} \,, \qquad \kappa_\parallel \approx \frac{v_{th}^2}{\nu_{coll}} \gg \kappa_\perp \,, \tag{6.149}$$

where ν_{coll} is the binary collision frequency. Introducing fluctuations, we obtain an additional heat transport in the perpendicular direction with respect to the zeroth order magnetic field. Assuming that on the (zeroth order) magnetic surfaces the temperature is uniform, we may write for the change of the averaged (over the zeroth order magnetic surfaces) temperature $\langle T \rangle$ up to second order

$$\frac{\partial \langle T \rangle}{\partial t} = \kappa_\parallel \langle b^2 \rangle \frac{\partial^2 \langle T \rangle}{\partial x_\perp^2} + 2\kappa_\parallel \left\langle b \frac{\partial}{\partial x_\perp} \left(\boldsymbol{n}_0 \cdot \nabla\right) \delta T \right\rangle \,. \tag{6.150}$$

The first term on the right-hand side already shows that due to the (perpendicular) fluctuations, a quite effective perpendicular transport may appear which is caused by an effective parallel heat conduction.

The phenomena summarized so far gave rise to the often used paradigm: Plasma stochasticity predominantly originating from plasma instabilities in high-dimensional phase space is counterproductive to good confinement, that is, enhanced losses appear (due to microinstabilities, MHD instabilities, ripple loss, etc.).

6.5.1
Description of Stochastic (Magnetic) Fields

Nonlinear dynamics already taught us that a system with a few degrees of freedom may become chaotic. In plasma, sources of magnetic stochasticity are, for example, imperfect coils for confinement, heating, shaping as well as correction coils, MHD control coils and boundary layer coils, ergodic divertor coils, and so on. In a "new" type of view, stochastization may have positive aspects since it is manageable in some respect, and may help for edge localized mode (ELM) mitigation [102, 147, 148], particle exhaust, zonal flows [149], radial electric fields and transport barriers [150–153], understanding of fast particle escape rates [154–156], heat flow patterns [157, 158], and so on. Specific coil arrangements have been already tested in Tore-Supra, W7-AS (island divertor), DIII-D, TEXTOR, JET, and ASDEX-U, and are being planned for ITER. In the following, we present the basic techniques to describe and investigate stochastic magnetic fields.

For the magnetic field, we use the Clebsch representation [159]

$$\boldsymbol{B} = \nabla\psi \times \nabla\theta - \nabla H \times \nabla\varphi \,. \tag{6.151}$$

This representation follows, for example, from the vector potential in cylindrical coordinates $\varphi \hat{=} z$ (toroidal "height"), θ (azimuthal poloidal coordinate), and ρ (radial coordinate in the poloidal cross section)

$$\boldsymbol{A} = A_\rho \nabla\rho + A_\theta \nabla\theta + A_\varphi \nabla\varphi \,. \tag{6.152}$$

Introducing

$$A_\rho = \frac{\partial G(\rho,\theta,\varphi)}{\partial\rho} \,, \tag{6.153}$$

and defining

$$\psi = A_\theta - \frac{\partial G(\rho,\theta,\varphi)}{\partial\theta} \,, \quad -H = A_\varphi - \frac{\partial G(\rho,\theta,\varphi)}{\partial\varphi} \,, \tag{6.154}$$

we arrive at the Clebsch form. To show this, we insert the vector potential components as defined via Eqs. (6.153) and (6.154) into $\boldsymbol{B} = \nabla \times \boldsymbol{A}$ and use in cylindrical coordinates

$$(\nabla \times \boldsymbol{A})_\rho = \frac{1}{\rho}\frac{\partial A_z}{\partial\theta} - \frac{\partial A_\theta}{\partial z} \,, \tag{6.155}$$

$$(\nabla \times \boldsymbol{A})_\theta = \frac{\partial A_\rho}{\partial z} - \frac{\partial A_z}{\partial\rho} \,, \tag{6.156}$$

$$(\nabla \times \boldsymbol{A})_z = \frac{1}{\rho}\frac{\partial(\rho A_\theta)}{\partial\rho} - \frac{1}{\rho}\frac{\partial A_\rho}{\partial\theta} \,. \tag{6.157}$$

Using the Clebsch representation, it is quite obvious that the magnetic field line system is of Hamiltonian form. Starting from the differentials

$$d\psi = \boldsymbol{B} \cdot \nabla\psi = -(\nabla H \times \nabla\varphi) \cdot \nabla\psi, \tag{6.158}$$

$$= -\left(\frac{\partial H}{\partial \theta}\nabla\theta \times \nabla\varphi\right) \cdot \nabla\psi = -\frac{\partial H}{\partial \theta}(\nabla\psi \times \nabla\theta) \cdot \nabla\varphi\,, \tag{6.159}$$

$$d\theta = \boldsymbol{B} \cdot \nabla\theta = -(\nabla H \times \nabla\varphi) \cdot \nabla\theta, \tag{6.160}$$

$$= -\left(\frac{\partial H}{\partial \psi}\nabla\psi \times \nabla\varphi\right) \cdot \nabla\theta = \frac{\partial H}{\partial \psi}(\nabla\psi \times \nabla\theta) \cdot \nabla\varphi\,, \tag{6.161}$$

$$d\varphi = \boldsymbol{B} \cdot \nabla\varphi = (\nabla\psi \times \nabla\theta) \cdot \nabla\varphi\,, \tag{6.162}$$

we immediately get the symplectic structure

$$\frac{d\psi}{d\varphi} = -\frac{\partial H}{\partial \theta}\,, \quad \frac{d\theta}{d\varphi} = \frac{\partial H}{\partial \psi}\,. \tag{6.163}$$

Compared to classical mechanics, φ corresponds to "time," and ψ is the conjugate "momentum" to the "coordinate" θ. Within the Hamiltonian formalism, one can introduce action (in the following, the set of action variables is denoted by $\boldsymbol{J}$) and angle (in the following the set of angle variables is denoted by $\boldsymbol{\theta}$) coordinates. Generally speaking, when the Hamiltonian consists of an integrable part H_0 (leading to closed magnetic flux surfaces) and a perturbation

$$H(\boldsymbol{J}, \boldsymbol{\theta}) = H_0(\boldsymbol{J}) + \delta H(\boldsymbol{J}, \boldsymbol{\theta})\,, \tag{6.164}$$

the canonical equations read

$$\frac{\partial H}{\partial \boldsymbol{J}} = \dot{\boldsymbol{\theta}} = \boldsymbol{\Omega}(\boldsymbol{J}) + \frac{\partial \delta H}{\partial \boldsymbol{J}}\,, \quad \frac{\partial H}{\partial \boldsymbol{\theta}} = -\dot{\boldsymbol{J}} = \frac{\partial \delta H}{\partial \boldsymbol{\theta}}\,. \tag{6.165}$$

Within a Fourier decomposition

$$\delta H(\boldsymbol{J}, \boldsymbol{\theta}) = \sum_{\boldsymbol{m}} e^{i\boldsymbol{m}\cdot\boldsymbol{\theta}} H_{\boldsymbol{m}}, \tag{6.166}$$

we get the following approximate solution

$$\boldsymbol{\theta} \approx \boldsymbol{\theta}_0 + \boldsymbol{\Omega} t\,, \quad \Delta \boldsymbol{J} = -\int dt \frac{\partial \delta H}{\partial \boldsymbol{\theta}} \approx -i \sum_{\boldsymbol{m}} \frac{\boldsymbol{m} H_{\boldsymbol{m}} e^{i\boldsymbol{m}\cdot\boldsymbol{\theta}_0}\left(e^{i\boldsymbol{m}\cdot\boldsymbol{\Omega} t} - 1\right)}{\boldsymbol{m}\cdot\boldsymbol{\Omega}} \tag{6.167}$$

with resonances for $\boldsymbol{m} \cdot \boldsymbol{\Omega}(\boldsymbol{J}) = 0$. As it is well-known, the resonances result in the breaking of magnetic surfaces. Stochastic field line motion can occur.

For tokamak applications, the perturbation terms may be written with toroidal and poloidal mode numbers in the form $\delta H \sim \varepsilon H_{mn} \cos(m\theta + n\varphi + \chi_{m0})$. Resonances occur at the rational magnetic surface ψ_{mn} with $q(\psi_{mn}) = m/n$, creating chains of islands. The island width is [26]

$$W_{mn} = 4 \left| \frac{\varepsilon H_{mn}(\psi_{mn})}{\frac{dq^{-1}}{d\psi}} \right|^{1/2} , \tag{6.168}$$

and chaos sets in when

$$\sigma_{Chir} = \frac{W_{mn} + W_{m+1n}}{|\psi_{m+1n} - \psi_{mn}|} \geq 1 . \tag{6.169}$$

This (qualitative) theoretical prediction has been successfully tested in many numerical applications and is also very helpful for interpreting experimental results.

Now, consider the cylindrical coordinate system (R, φ, Z), where R is a major radius, φ is a toroidal angle, and Z is a vertical coordinate. The field line equations in this coordinate system are given by

$$\frac{1}{R}\frac{dZ}{d\varphi} = \frac{B_Z}{B_\varphi} , \quad \frac{1}{R}\frac{dR}{d\varphi} = \frac{B_R}{B_\varphi} . \tag{6.170}$$

The magnetic field components B_R, B_φ, B_Z can be determined by the vector potential $\boldsymbol{A}$. The coordinates (R, φ, Z) are related to the previously defined coordinates (r, θ, φ) via

$$R = R_0 + r\cos\theta , \quad Z = r\sin\theta . \tag{6.171}$$

We choose the radial component of the vector potential A_R to be zero, that is, $A_R = 0$. The reason for this choice is the gauge invariance of the vector potential. We suppose that the Z-component of the vector potential (A_Z), which determines the toroidal field $B_\varphi = -\partial A_Z/\partial R$, to be in form $A_Z = -B_0 R_0/R$, that is, $B_\varphi = B_0 R_0/R$, where B_0 is the strength of the magnetic field at the major radius of the torus $R = R_0$. We also assume that the equilibrium poloidal magnetic field of the plasma (B_R, B_Z) and the perturbed magnetic field are created by external coils such that they may completely be described by the toroidal component of the vector potential $A_\varphi(R, \varphi, Z)$ in the form

$$B_Z = \frac{1}{R}\frac{\partial(RA_\varphi)}{\partial R} , \quad B_R = -\frac{\partial A_\varphi}{\partial Z} . \tag{6.172}$$

Introducing the normalized coordinate z and the canonical momentum p_z as

$$z = \frac{Z}{R_0} , \quad p_z = \ln\frac{R}{R_0} ,$$

one obtains Eq. (6.170) in Hamiltonian form

$$\frac{dz}{d\varphi} = \frac{\partial H}{\partial p_z} , \quad \frac{dp_z}{d\varphi} = -\frac{\partial H}{\partial z} , \tag{6.173}$$

where the Hamiltonian function $H = H(z, p_z, \varphi)$ is the normalized φ-component of the vector potential, that is,

$$H \equiv H(z, p_z, \varphi) = \frac{R(p_z) A_\varphi(R(p_z), \varphi, z R_0)}{B_0 R_0^2}, \tag{6.174}$$

with $R = R_0 \exp(p_z)$.

In the axisymmetric case, the magnetic field does not depend on the toroidal angle φ, $A_\varphi = A_\varphi(R, Z)$, and thus $H = H(z, p_z)$. In this case, the Hamiltonian system (6.173) is completely integrable. The field lines lie on nested toroidal surfaces determined by the surface function $H(z, p_z) = f(Z, R) = \text{const}$. The section of a toroidal surface with the plane $\varphi = \text{const}$ is shown in Figure 6.1. One can introduce the action-angle variables (I, ϑ) via

$$I = \frac{1}{2\pi} \oint_C p_z dz, \quad \vartheta = \frac{\partial}{\partial I} \int^z p_z(z', I) dz', \tag{6.175}$$

where the integration is taken along the closed contour C following from the cross section of the surface function $f(R, Z) = \text{const}$ with the poloidal plane $\varphi = \text{const}$ (see Figure 6.1). The action variable I coincides with the normalized toroidal flux ψ because of

$$I = \frac{1}{2\pi} \oint_C p_z dz = \frac{1}{2\pi} \int_S dp_z dz = \frac{1}{2\pi R_0^2 B_0} \int_S B_\varphi(R, Z) dR dZ = \psi, \tag{6.176}$$

which has the meaning of the normalized flux of the toroidal field B_φ through the area S enclosed by the closed contour C on the poloidal plane $\varphi = \text{const}$. The angle variable ϑ is exactly the *intrinsic* poloidal angle.

In action-angle variables (ψ, ϑ), the Hamiltonian $H = H(\psi)$ and the field lines obey Eq. (6.163). The inverse safety factor $q(\psi)$ is determined by $dH(\psi)/d\psi$. It can also be found from the equation of field lines Eq. (6.173). According to its definition, q is equal to the number of toroidal turns per one poloidal turn, that is, $q = \Delta\varphi/2\pi$, where $\Delta\varphi$ is the increment of the toroidal angle φ when a field line makes one full poloidal turn. Then, from the first equation (6.173), it follows that

$$q(\psi) = \frac{\Delta\varphi}{2\pi} = \int_C \frac{dz}{\frac{\partial H}{\partial p_z}}, \tag{6.177}$$

where the integral is taken along the closed contour C of $H = H_0(z, p_z) = \text{const}$.

The geometrical coordinates R, Z (or r, θ) of field lines are periodic functions of the angle variable ϑ with $R(\vartheta, \psi) = R(\vartheta + 2\pi, \psi)$, $Z(\vartheta, \psi) = Z(\vartheta + 2\pi, \psi)$. Therefore, they can be presented in Fourier series

$$Z(\vartheta, \psi) = \sum_m Z_m(\psi) e^{im\vartheta}, \quad R(\vartheta, \psi) = \sum_m R_m(\psi) e^{im\vartheta}, \tag{6.178}$$

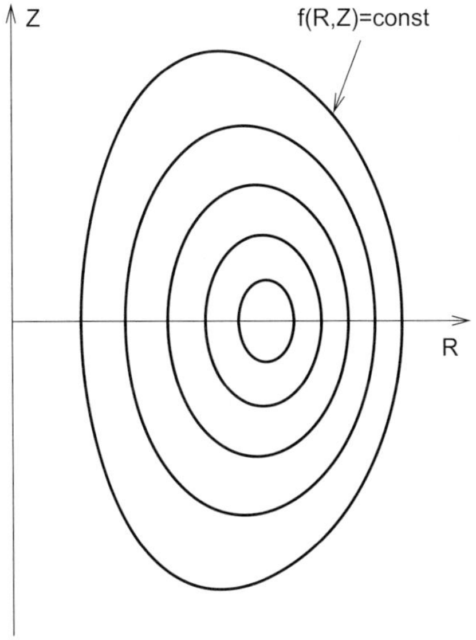

Figure 6.1 Magnetic flux surfaces $H(z, p_z) = f(Z, R) = \text{const.}$

or

$$r(\vartheta, \psi) = \sum_{m=0}^{\infty} \left(r_m^{(c)}(\psi) \cos m\vartheta + r_m^{(s)}(\psi) \sin m\vartheta \right), \tag{6.179}$$

$$\theta(\vartheta, \psi) = \vartheta + \sum_{m=0}^{\infty} \alpha_m(\psi) \sin m\vartheta . \tag{6.180}$$

The coefficients $R_m(\psi)$, $Z_m(\psi)$ (or $r_m^{(c,s)}(\psi)$, $\alpha_m(\psi)$), which depend on the toroidal flux ψ, can be found by integrating the Hamiltonian field line equations (6.173). The field lines $R(\varphi)$, $Z(\varphi)$ (or $r(\varphi)$, $\theta(\varphi)$) in real geometrical space (R, Z) or (r, θ) are determined by Eqs. (6.178) and (6.179), respectively, when taking the intrinsic poloidal angle ϑ as $\vartheta = \varphi/q(\psi) + \vartheta_0$.

In the presence of nonaxisymmetric magnetic perturbations, the toroidal component of the vector potential $A_\varphi(R, Z, \varphi)$ can be presented as a sum

$$A_\varphi = A_\varphi^{(0)}(R, Z) + A_\varphi^{(per)}(R, Z, \varphi) . \tag{6.181}$$

In typical situations, the perturbed part of the component A_Z is small and can be neglected in comparison with the unperturbed part $A_Z^{(0)}(R)$ which determines the toroidal magnetic field B_φ. Then, the magnetic perturbation $\delta \boldsymbol{B}(R, Z, \varphi)$ has only two nonzero components, namely, δB_R and δB_Z. The latter, according to (6.172), are expressed via the perturbed part of the vector potential $A_\varphi^{(per)}(R, Z, \varphi)$ in the form

$$\delta \boldsymbol{B}(R, Z, \varphi) = \left(-\hat{e}_R \frac{1}{R} \frac{\partial}{\partial Z} + \hat{e}_Z \frac{1}{R} \frac{\partial}{\partial R} \right) R A_\varphi^{(per)}(R, Z, \varphi) . \tag{6.182}$$

Introducing the toroidal flux ψ and the intrinsic poloidal angle ϑ according to Eq. (6.175) for the equilibrium magnetic field, the Hamiltonian equations for perturbed field lines follow as

$$\frac{d\vartheta}{d\varphi} = \frac{1}{q(\psi)} + \varepsilon \frac{\partial H_1}{\partial \psi}, \qquad \frac{d\psi}{d\varphi} = -\varepsilon \frac{\partial H_1}{\partial \vartheta}. \tag{6.183}$$

The perturbed Hamiltonian $\varepsilon H_1 \equiv H_1(\psi, \vartheta, \varphi)$ is given by

$$H_1(\psi, \vartheta, \varphi) = \frac{R(\psi, \vartheta)}{R_0^2 B_0} A_\varphi^{(per)}(R(\psi, \vartheta), Z(\psi, \vartheta), \varphi), \tag{6.184}$$

where $R(\psi, \vartheta) = R_0 + r(\psi, \vartheta)\cos\theta(\psi, \vartheta)$. For

$$A_\varphi^{(per)}(r, \theta, \varphi) = \varepsilon B_0 R_0 a(r, \theta, \varphi), \tag{6.185}$$

$$a(r, \theta, \varphi) = \sum_m a_{mn}(r, \theta)\cos(m\theta + n\varphi + \chi_{m0}), \tag{6.186}$$

it has the Fourier expansion

$$H_1(\psi, \vartheta, \varphi) = \varepsilon \sum_m H_{mn}(\psi)\cos(m\vartheta + n\varphi + \chi_{m0}). \tag{6.187}$$

The dimensionless perturbation parameter ε characterizes the strength of the perturbations. According to Eq. (6.185), the Fourier components $H_{m,n}(\psi)$ are given by the integrals

$$\begin{aligned} H_{mn}(\psi) &= \mathrm{Re} \iint_0^{2\pi} \frac{R(\psi, \vartheta)}{(2\pi)^2 R_0^2 B_0 \varepsilon} A_\varphi^{(per)}(R(\psi, \vartheta), Z(\psi, \vartheta), \varphi)\, e^{-im\vartheta - in\varphi} d\vartheta\, d\varphi \\ &= \mathrm{Re} \iint_0^{2\pi} \frac{R(\psi, \vartheta)}{(2\pi)^2 R_0} a(r(\psi, \vartheta), \theta(\psi, \vartheta), \varphi)\, e^{-im\vartheta - in\varphi} d\vartheta\, d\varphi. \end{aligned} \tag{6.188}$$

One can also expand the perturbed magnetic field, $\delta \boldsymbol{B}$, in a Fourier series with respect to the intrinsic poloidal angle ϑ and the toroidal angle φ similar to the expansion of the perturbation Hamiltonian H_1 given by Eq. (6.187), that is,

$$\delta \boldsymbol{B}(\psi, \vartheta, \varphi) = \sum_m \boldsymbol{B}_{mn}(\psi)\sin(m\vartheta + n\varphi + \chi_{m0}). \tag{6.189}$$

Often, one uses the Fourier components $\boldsymbol{B}_{mn}(\psi)$ to estimate the width of magnetic islands formed in the presence of magnetic perturbations.

The relation between the Fourier components of the perturbation Hamiltonian $H_{mn}(\psi)$ and the perturbation field $\boldsymbol{B}_{mn}(\psi)$ can be found using Eqs. (6.182) and (6.184). Suppose that $|dH_{mn}/d\psi| \ll m|H_{mn}|$; then, one obtains

$$|\boldsymbol{B}_{mn}(\psi)| \approx \varepsilon m H_{mn}(\psi) B_0 \equiv m H_{mn}(\psi) B_c, \quad \text{with} \quad \varepsilon = \frac{B_c}{B_0}. \tag{6.190}$$

6.5.2
Symplectic Mappings

The conventional method for determining the magnetic field topology is numerical field line tracing and the subsequent Poincaré display. In the mapping approach, one wishes to avoid small time-step integration, replacing the integration by return mappings of variables (θ, ψ) (at certain poloidal sections of torus at $\varphi = \varphi_k$). In general, one considers the case when a toroidal mode number n takes values proportional to a certain number N ($N \geq 1$) which corresponds to the N-fold symmetry of magnetic perturbations along the toroidal angle φ, that is, $n = sN$, $s = \pm 1, \pm 2, \dots$ Let $\theta = \vartheta_k, \psi = \psi_k$ be the poloidal angle and toroidal flux at the poloidal sections $\varphi = \varphi_k = 2\pi k/N$, $k = 0, \pm 1, \pm 2, \dots$, respectively. One may designate the map by

$$(\vartheta_{k+1}, \psi_{k+1}) = \hat{M}(\vartheta_k, \psi_k), \tag{6.191}$$

which relates the variables (θ, ψ) at the successive sections φ_k and φ_{k+1}. Then, the map $(\hat{M})^N$ defines the Poincaré return map. The flux-preserving property of the mapping Eq. (6.191) is expressed by

$$\left| \frac{\partial(\vartheta_{k+1}, \psi_{k+1})}{\partial(\vartheta_k, \psi_k)} \right| = 1\,. \tag{6.192}$$

The construction of "realistic" mappings is an art. Symplectic mappings for integrating Hamiltonian systems based on the Hamilton–Jacobi method and perturbation theory have been developed in [160–164].

Following the method of canonical transformation for constructing mappings, the general form of the mapping Eq. (6.191) can be cast into the flux-preserving form (see [162]):

$$\Psi_k = \psi_k - \varepsilon \frac{\partial S^{(k)}}{\partial \vartheta_k}\,, \qquad \Theta_k = \vartheta_k + \varepsilon \frac{\partial S^{(k)}}{\partial \Psi_k}\,, \tag{6.193}$$

$$\Psi_{k+1} = \Psi_k\,, \quad \bar{\Theta}_k = \Theta_k + w(\Psi_k, \varepsilon)(\varphi_{k+1} - \varphi_k)\,, \tag{6.194}$$

$$\psi_{k+1} = \Psi_{k+1} + \varepsilon \frac{\partial S^{(k+1)}}{\partial \vartheta_{k+1}}\,, \qquad \vartheta_{k+1} = \bar{\Theta}_k - \varepsilon \frac{\partial S^{(k+1)}}{\partial \Psi_{k+1}}\,, \tag{6.195}$$

where $w(\Psi, \varepsilon) = \partial \mathcal{H}/\partial \Psi$ is the frequency of perturbed motion, and $S^{(k)} \equiv S(\vartheta_k, \Psi_k)$ is the value of the generating function $G(\vartheta, \Psi, \varphi, \varphi_0; \varepsilon)$ taken at sections $\varphi = \varphi_k$, that is, $S(\vartheta_k, \Psi_k) = G(\vartheta_k, \Psi_k, \varphi_k, \varphi_0; \varepsilon)$. The generating function obeys the Hamilton–Jacobi equation,

$$H\left(\Psi + \varepsilon \frac{\partial G}{\partial \vartheta}, \vartheta, \varphi\right) + \varepsilon \frac{\partial G}{\partial \varphi} = \mathcal{H}(\Psi, \varepsilon)\,, \tag{6.196}$$

in the finite interval $\varphi_k < \varphi < \varphi_{k+1}$, satisfying the condition $G|_{\varphi=\varphi_0} = 0$ at the initial value φ_0 ($\varphi_k < \varphi_0 < \varphi_{k+1}$).

In the first order of ε, the frequency $w(\Psi, \varepsilon)$ is determined by the inverse safety factor $q(\Psi)$:

$$w(\Psi, \varepsilon) = \frac{1}{q(\Psi)}, \tag{6.197}$$

and the generating function $G(\vartheta, \Psi, \varphi, \varphi_0)$ in the finite interval $\varphi_{k+1} < \varphi < \varphi_k$ is given by [162]

$$G(\vartheta, \Psi, \varphi, \varphi_0) = -(\varphi - \varphi_0) \sum_{m,n} H_{mn}(\Psi) \left[a(x_{mn}) \sin(m\vartheta - n\varphi + \chi_{mn}) + b(x_{mn}) \cos(m\vartheta - n\varphi + \chi_{mn}) \right], \tag{6.198}$$

where

$$a(x) = \frac{1 - \cos x}{x}, \quad b(x) = \frac{\sin x}{x}, \quad x_{mn} = \left(\frac{m}{q(\Psi)} - n \right) (\varphi - \varphi_0). \tag{6.199}$$

The free parameter φ_0 lies in the interval $\varphi_k \le \varphi_0 \le \varphi_{k+1}$.

The mapping Eqs. (6.193)–(6.195) can be also considered as the alternative method of the symplectic integration of Hamiltonian systems. Having this in mind, we can use it for other Hamiltonian systems, for example, the drift motion of particles in stochastic magnetic fields.

We conclude this subsection by summarizing some results which have been obtained by the symplectic mapping procedure for the onset of chaotic drift motion due to resonant magnetic field perturbations [155]. The symplectic mapping is now the drift Hamiltonian [159, 165] including magnetic perturbations. In the example, the latter are taken from from TEXTOR-DED. The chaotic region at the plasma edge is formed due to interactions of several resonant drift islands. Their radial positions with respect to the divertor coils, and the distances between them, are important factors in formation of particle chaos. These factors are determined not only by the plasma parameters, as in the case of magnetic field line stochasticity, but also by the particle parameters. It was shown that due to a toroidal drift, the positions of resonant drift surfaces are shifted with respect to resonant magnetic surfaces. The shift depends not only on the energy of particles, but also on their directions of motion. For copassing particles, the resonant surfaces are shifted inwardly, while for counterpassing particles, they shift outwardly. Figure 6.2 shows an example.

An equilibrium electric field and a rotating magnetic field also significantly vary the positions of resonant drift surfaces, acting selectively on the copassing and counterpassing particles, respectively. For instance, a rotating field of frequency $f = 10^4$ Hz shifts the resonant drift surface outward (inward) up to 2–3 cm for the copassing (counterpassing) particles. The shift of resonant drift surfaces significantly affects the formation of the stochastic zone of drift motion since the magnetic field perturbation strongly decays in the inward radial direction. Particularly,

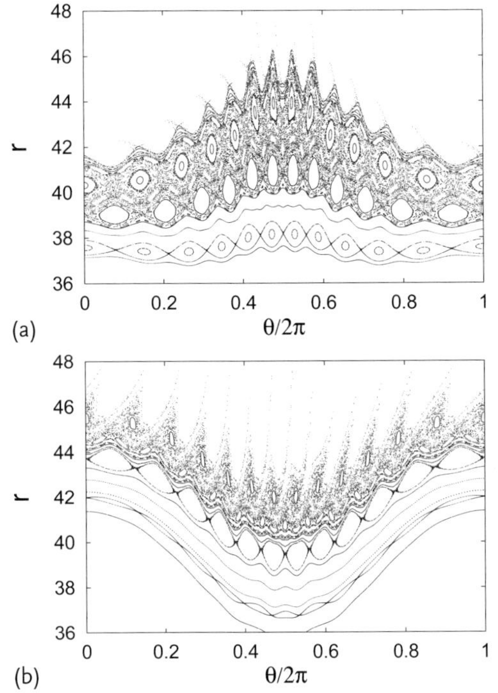

Figure 6.2 Drift surfaces and chaotic drift motion for (a) co- and (b) counter-passing particles, respectively [155].

it has been shown that the ergodic regions for copassing and counterpassing particles are different. This difference becomes more pronounced with an increase in particle energy. The calculations have direct applications for runaway losses in tokamaks [154].

6.5.3
The Standard Map as a Simple Example for Stochastic Field Line Dynamics

We now leave the systematic approach to construct "realistic" mappings and turn to more intuitive procedures. Let us present a simple-minded approach to a map which allows one to do some analytical estimates. Of course, the price for simplicity is less reliability. We start with a magnetic field configuration possessing small deviations in the perpendicular direction compared to a guiding zeroth order field,

$$\boldsymbol{B} = \boldsymbol{B}_0 + \delta \boldsymbol{B}_\perp \,, \qquad B_0 \gg |\delta \boldsymbol{B}_\perp| \,. \tag{6.200}$$

As before, $\boldsymbol{B}_0$ is a constant magnetic field in z-direction and $\delta \boldsymbol{B}_\perp$ is the perpendicular perturbation field having components in x- and y-direction. The relative components are defined via $\boldsymbol{b}_\perp = \delta \boldsymbol{B}_\perp / B_0$ as b_x and b_y. When we propagate a step along a magnetic field line, the components dx, dy, and dz of the infinitesimal path are approximately related by

$$\frac{dy}{\delta B_{\perp y}} \approx \frac{dz}{B_0} \approx \frac{dx}{\delta B_{\perp x}} , \tag{6.201}$$

defining the geometry of the line. We may introduce a parameter t (which will later be identified with time for particle motion), for example, in the simple form

$$\frac{dx}{dt} = b_x \frac{dz}{dt} , \tag{6.202}$$

$$\frac{dy}{dt} = b_y \frac{dz}{dt} , \tag{6.203}$$

$$\frac{dz}{dt} = \eta_{\parallel} \equiv \text{const} . \tag{6.204}$$

These equations already show some formal similarity to the Langevin equations. At this stage, however, no stochastic source has been introduced. For tokamak-like configurations, several simple forms of the perturbation field have been discussed [166–169]. Starting from the Fourier decomposition of the x-component,

$$b_x = \sum_{m,n} b_{mn}(x) e^{i(2\pi m y - 2\pi n z)} + c.c. , \tag{6.205}$$

we may identify x with a normalized radial coordinate, y with the normalized poloidal, and z with the normalized toroidal (angle) coordinate. The radial coordinate will be measured in the shear length L_s, whereas y and z change by one after a poloidal or toroidal turn, respectively. The y-component is assumed to take care of the shear,

$$\delta B_{\perp y} = B_0 \frac{x}{L_s} \quad \to \quad b_y = x . \tag{6.206}$$

Next, we simplify to

$$m = 1 , \quad b_{1n} = \varepsilon \frac{e^{i2\pi\varphi_{10}}}{2i} \equiv \frac{\varepsilon}{2i} , \quad b_{mn} = 0 \quad \text{for} \quad m \neq 1 , \tag{6.207}$$

with constant phase $\varphi_{10} \equiv 0$ (see below). The strength of the perturbation is determined by ε. Using

$$\sum_n b_{1n} e^{i(2\pi y - 2\pi n z)} + c.c. = \varepsilon \sum_{n \in \mathbb{N}} \sin(2\pi y - 2\pi n z) \tag{6.208}$$

and $\sin(2\pi y - 2\pi n z) = \sin(2\pi y)\cos(2\pi n z) - \cos(2\pi y)\sin(2\pi n z)$, we end up with

$$b_x = \varepsilon \sin(2\pi y) \left[1 + 2 \sum_{n \in \mathbb{N}+} \cos(2\pi n z) \right] . \tag{6.209}$$

Noting the identity (Poisson summation formula)

$$\sum_{n \in \mathbb{Z}} \cos(2\pi n z) = \sum_{k \in \mathbb{Z}} \delta(z - k) , \tag{6.210}$$

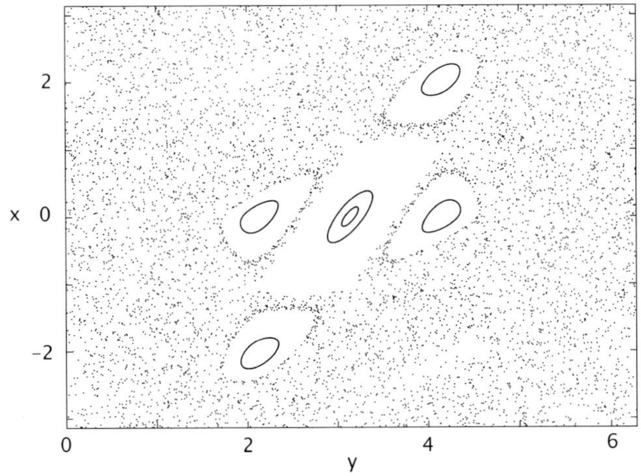

Figure 6.3 Poincaré plot of the standard map Eqs. (6.213) and (6.214) for $\varepsilon = 2.4$.

we may write

$$b_x = \varepsilon \sin(2\pi y) \sum_{k \in \mathbb{Z}} \delta(z - k) . \tag{6.211}$$

Physically, the Dirac delta function δ produces "kicks" every "time" z changes by $\Delta z \equiv 1$. Because of Eqs. (6.202) and (6.203), a "kick" results in changes, that is,

$$\Delta x = \varepsilon \sin(2\pi y) , \quad \Delta y = x . \tag{6.212}$$

In other words, we may interpret the result in a poloidal cross section (Poincaré section) as the changes of the penetration points of magnetic field lines which occur every toroidal turn. Equation (6.212) is equivalent to the standard map written for the $(t+1)$-th iterations y_{t+1} and x_{t+1} in terms of the t-th iteration. For convenience, we may transform $2\pi x \to x$, $2\pi y \to y$ and $2\pi\varepsilon \to \varepsilon$ to get the form [27]

$$x_{t+1} = x_t + \varepsilon \sin(y_t) , \tag{6.213}$$

$$y_{t+1} = y_t + x_{t+1} . \tag{6.214}$$

Figure 6.3 shows a Poincaré-plot of the standard map for $\varepsilon = 2.4$ in the map equations Eqs. (6.213) and (6.214). We have chosen 10 initial points; the number of iterations was 1000. One can recognize that for this relatively large control parameter ε, already a quite large stochastic area exists. Embedded into the stochastic sea are islands, a central one at $(0, \pi)$, and four significant ones surrounding the central one. Some surfaces with irrational winding numbers are designated by solid lines. Several interesting features of nonlinear dynamics hidden in the standard map are discussed in the literature; see, for example, [27, 170].

In the present context, the divergence of field lines, the Lyapunov exponent, and the Kolmogorov length are of interest. The general theory of the latter tools can be found in standard text books on nonlinear dynamics [171].

Looking within the standard map model for the deviations of two, initially adjacent field lines, the deviation

$$d\boldsymbol{I}_k = \begin{pmatrix} dx_k \\ dy_k \end{pmatrix}, \quad d\boldsymbol{I}_{k+1} = J_k d\boldsymbol{I}_k \,, \quad J_k = \begin{pmatrix} \frac{\partial x_{k+1}}{\partial x_k} & \frac{\partial x_{k+1}}{\partial y_k} \\ \frac{\partial y_{k+1}}{\partial x_k} & \frac{\partial y_{k+1}}{\partial y_k} \end{pmatrix} \tag{6.215}$$

follows via the Jacobian matrix J_k. Note that det $J_k = 1$. The eigenvalues $\lambda_1^{(k)}$ and $\lambda_2^{(k)}$ of the Jacobian

$$\begin{vmatrix} \frac{\partial x_{k+1}}{\partial x_k} - \lambda_{1,2}^{(k)} & \frac{\partial x_{k+1}}{\partial y_k} \\ \frac{\partial y_{k+1}}{\partial x_k} & \frac{\partial y_{k+1}}{\partial y_k} - \lambda_{1,2}^{(k)} \end{vmatrix} = 0 \,, \quad \lambda^{(k)} = max\left(\lambda_1^{(k)}, \lambda_2^{(k)}\right) \,, \tag{6.216}$$

determine the Lyapunov exponent. We may define a global Lyapunov exponent for unstable orbits

$$\lambda = \lim_{N\to\infty} \frac{1}{N} \ln \prod_{k=1}^{N} \lambda^{(k)} > 0 \,. \tag{6.217}$$

From the latter, we can define a local e-folding length (Kolmogorov length) as

$$L_K = \frac{1}{\lambda} \,. \tag{6.218}$$

In general, the Kolmogorov length will be determined numerically, starting with the Lyapunov exponent λ. Various numerical procedures have been proposed [172–174]. For the standard map, the Chirikov method [172] is very effective. Defining the differences between field lines (x_t, y_t) and (x_t', y_t') as

$$\eta_t = x_t' - x_t \,, \quad \xi_t = y_t' - y_t \,, \tag{6.219}$$

we determine the differences at each step by iterating Eqs. (6.213) and (6.214). For neighboring field lines, we use

$$\lim_{y_t'\to y_t} \left(\frac{\sin(y_t') - \sin(y_t)}{y_t' - y_t} \right) = \cos(y_t) \tag{6.220}$$

to obtain

$$\eta_{t+1} \approx \eta_t + \varepsilon \xi_t \cos(y_t) \,, \tag{6.221}$$

$$\xi_{t+1} \approx \xi_t + \eta_{t+1} \,. \tag{6.222}$$

The distance

$$\Delta r(t) = \sqrt{\eta_t^2 + \xi_t^2} \quad \text{with initial value} \quad \Delta r(0) = \Delta r_0 \tag{6.223}$$

determines the Lyapunov exponent via

$$\lambda = \lim_{t\to\infty} \frac{1}{t} \ln \left(\frac{|\Delta r(t)|}{|\Delta r_0|} \right) \tag{6.224}$$

after a reasonable amount of iterations of Eqs. (6.221) and (6.222). The exponential divergence may be written as

$$\left\langle [\Delta r(t)]^2 \right\rangle \sim [\Delta r(0)]^2 \exp\left\{\frac{2t}{L_K}\right\} . \tag{6.225}$$

However, Eqs. (6.221) and (6.222) are only valid for small η and ξ. Therefore, from time to time, a rescaling becomes necessary during the iteration. As a typical number, in total, 10^5 iterations are in most cases sufficient for the determination of the Lyapunov exponent. For very large control parameters ε, Chirikov [172] has calculated the analytical limit

$$L_K = \frac{1}{\ln\left(\frac{\varepsilon}{2}\right)} . \tag{6.226}$$

We present some details on the evaluation of the Kolmogorov lengths in other systems later.

One may generalize the standard map by including collisions. For that, we interpret Eqs. (6.202)–(6.204) as the equations of motion for particles moving along the magnetic fields. We generalize these equations to stochastic differential equations

$$\frac{dx}{dt} = b_{\perp x}\frac{dz}{dt} + \eta_{\perp x} , \tag{6.227}$$

$$\frac{dy}{dt} = b_{\perp y}\frac{dz}{dt} + \eta_{\perp y} , \tag{6.228}$$

$$\frac{dz}{dt} = \eta_{\parallel} , \tag{6.229}$$

by including collisional disturbances. The velocity changes due to collisions are characterized by the statistical properties

$$\langle \eta_x(t+\tau)\eta_x(t)\rangle_{\perp} = \left\langle \eta_y(t+\tau)\eta_y(t)\right\rangle_{\perp} = \frac{v_{th}^2}{2} e^{-\nu_{coll}|\tau|}\cos(\Omega\tau) , \tag{6.230}$$

$$\langle \eta_z(t+\tau)\eta_z(t)\rangle_{\parallel} = \left\langle \eta_{\parallel}(t+\tau)\eta_{\parallel}(t)\right\rangle = \frac{v_{th}^2}{2} e^{-\nu_{coll}|\tau|} , \tag{6.231}$$

where ν_{coll} is the collision frequency and $\Omega = qB_0/m$ is the gyrofrequency. With respect to the map Eqs. (6.213) and (6.214), collisions will introduce additional dislocations. Let us assume that every collision shifts the radial position of a particle in the x, y-plane by the Larmor radius $\rho = v_{th}/\Omega$. The shift could be inwards or outwards which will be reflected in the sign $\pm\rho$. A collision will also change the sign of the parallel velocity $\eta_{\parallel}$. We take care of this by changing notation $\eta_{\parallel} \to v_{th}$; the latter may change sign during a collision. Thus, the following picture emerges. After one iteration, with probability P, a collision occurs. In case of a collision, the x-coordinate (and indirectly also the y-coordinate) changes by $\rho_t = \pm\rho$. Without collision, $\rho_t = 0$. The sign $\pm$ is chosen randomly. In case of a collision, $\eta_{\parallel} = v_{th}$ changes sign. For example, for $v_{th} = \pm 1$, that is, $|v_{th}| = 1$, the collision frequency is $\nu_{coll} = P$. Summarizing, we postulate the following collisional map [169]

$$x_{t+1} = x_t + \varepsilon \sin(y_t) + \rho_t \,, \tag{6.232}$$

$$y_{t+1} = y_t + x_{t+1} \,, \tag{6.233}$$

$$z_{t+1} = z_t + v_{th}, \tag{6.234}$$

where $v_{th} = -1$ means that Eqs. (6.232) and (6.233) should be inverted.

The mean free path λ_{coll} is the mean value of $t\Delta z$ under the condition that no collision occurs. The probability of no collision after t iterations is $(1-P)^t$, leading to

$$\lambda_{coll} = \langle z \rangle = \frac{\sum_{t=0}^{\infty} t(1-P)^t}{\sum_{t=0}^{\infty} (1-P)^t} \Delta z \,. \tag{6.235}$$

Denominator

$$\sum_{t=0}^{\infty} (1-P)^t = \frac{1}{P} \tag{6.236}$$

and numerator

$$\sum_{t=0}^{\infty} t(1-P)^t = -(1-P)\frac{\partial}{\partial P} \sum_{t=0}^{\infty} (1-P)^t = \frac{1-P}{P^2} \tag{6.237}$$

are straightforward to calculate, resulting in

$$\lambda_{coll} = \frac{1-P}{P} \tag{6.238}$$

for $|v| = \Delta z = 1$. The classical (collisional) parallel diffusion coefficient is therefore

$$\chi_{\parallel} = \frac{v \lambda_{coll}}{2} = \frac{1-P}{2P} \,. \tag{6.239}$$

Classical (collisional) perpendicular diffusion can be calculated as

$$\chi_{\perp} = \frac{\nu_{coll} \rho^2}{2} = \frac{P}{2\Omega_0^2} \,. \tag{6.240}$$

Figure 6.4 shows a Poincaré plot of Eqs. (6.232)–(6.234) for the same parameters as in Figure 6.3. The Larmor radius was chosen as $\rho = 10^{-8}$; the collision frequency is $\nu_{coll} = P = 0.1$. Comparing with the (collisionless) standard map Eqs. (6.232) and (6.233), the rough impression remains unchanged, mainly because of the small value for ρ. For larger values $\rho \gg 10^{-8}$, one clearly recognizes the smearing of the islands. For $\rho = 0.01$ and $\rho = 0.02$, details of the islands are shown in Figure 6.5.

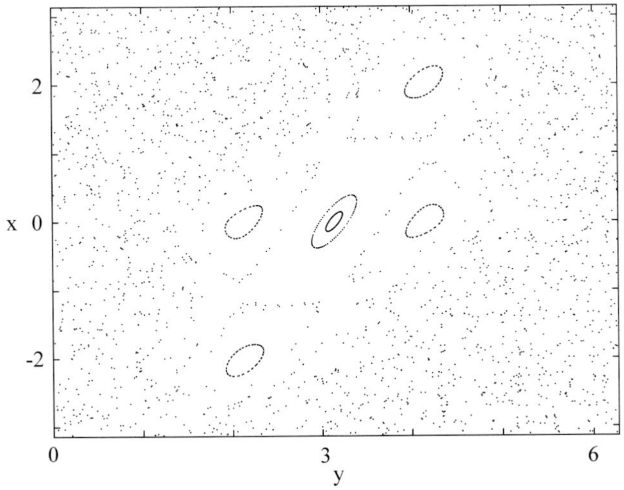

Figure 6.4 Poincaré plot of the weakly collisional generalized standard map Eqs. (6.232)–(6.234) for $\varepsilon = 2.4$ with $\rho = 10^{-8}$, $|v_{th}| = 1$, and $\nu_{coll} = 0.1$.

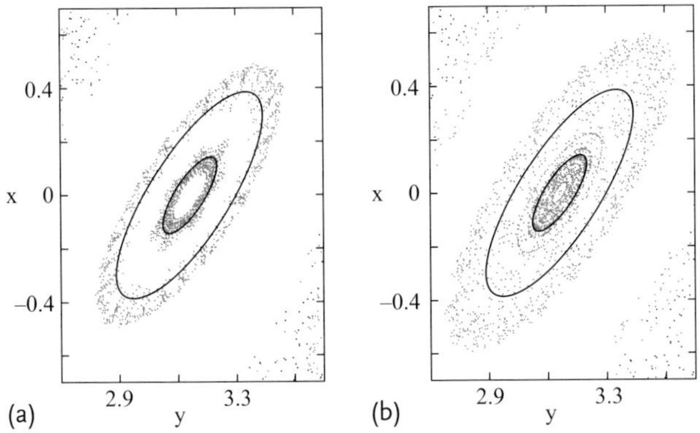

Figure 6.5 Comparison of Poincaré plots near the islands of the standard map. The solid lines are from the (collisionless) standard map whereas the dots originate from the collisional generalized standard map. The common parameters are $\varepsilon = 2.4$ and $\nu_{coll} = 0.1$. (a) is for $\rho = 0.01$; (b) has $\rho = 0.02$.

As a matter of course, the previous generalized collisional standard map is only applicable for rare collisions, that is, less than one collision per toroidal turn. The correct statistical evaluation requires $P \ll 1$ since otherwise the expressions for λ_{coll} and ν_{coll} would become inconsistent. Note that for $v_{th} \equiv 1$ the relation $\lambda_{coll} \approx v_{th}\nu_{coll} = \nu_{coll}$ would break down. It is not justified to use $P \to 1$ apart from the fact that $P > 1$ is not allowed at all. When we want to include higher collision frequencies, we should subdivide the map. Instead of a step size of $\Delta z = 1$, we might use $\Delta z = 1/N$ with $N \in \mathbb{N}^+$. For $N = 8$, that subdivision is visualized in Figure 6.6. Let us start for $m = 1$ with the Fourier decomposition of the x-

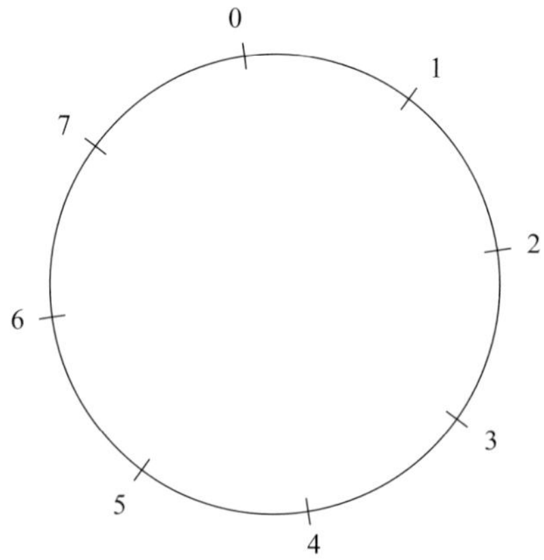

Figure 6.6 Sketch of the idea to generalize the standard map for strong collisions. Instead of one Poincaré section per toroidal turn, we may have $N = 8$ sections along the torus. In each plane, an (additional) collisional displacement may be introduced.

component of the magnetic field. We choose the form of [166], that is,

$$b_x = \sum_{n\in\mathbb{Z}} b_{1n}(x) e^{i(2\pi y - 2\pi n z)} + c.c., \quad b_{1n} = \varepsilon \frac{e^{i2\pi\varphi_{1p}}}{2i} \tag{6.241}$$

where p has been introduced according to $p = n \mod N$ for a subdivision into N sections. Rechester *et al.* [166] have introduced random phases φ_{1p} between zero and one since in their case, the magnetic perturbations originate from short-wavelength plasma microinstabilities. When the magnetic perturbations are deterministic, we might choose $\varphi_{1p} = 0$, as we shall do at the end. For possible generalizations, we keep the phases during the next steps. Note that

$$a \mod m = a - \left[\frac{a}{m}\right] m\,, \tag{6.242}$$

where $[\ldots]$ is the largest integer being smaller or equal to a/m, such that

$$p = n \mod N: \quad p = 0, \pm 1, \pm 2, \ldots, \pm(N-1)\,. \tag{6.243}$$

The generalized weakly collisional map corresponds to $N = 1$ and $p = 0$. Considering

$$b_x = \varepsilon \sum_{n\in\mathbb{Z}} \sin(2\pi y - 2\pi n z + 2\pi\varphi_{1p}), \tag{6.244}$$

we can change the summation index according to $n = jN + p$, when introducing $j \in \mathbb{Z}$ and $p \in \{0, 1, \ldots, N-1\}$ leading to

$$\sum_{n\in\mathbb{Z}} \sin(2\pi y - 2\pi n z + 2\pi\varphi_{1p}) = \sum_{j\in\mathbb{Z}} \sum_{p=0}^{N-1} \sin(2\pi y - 2\pi j N z - 2\pi p z + 2\pi\varphi_{1p})\,. \tag{6.245}$$

Because of

$$\begin{aligned}
&\sin\left(2\pi y - 2\pi j Nz - 2\pi pz + 2\pi\varphi_{1p}\right) \\
&= \sin\left(2\pi y - 2\pi pz + 2\pi\varphi_{1p}\right)\cos(2\pi j Nz) \\
&\quad - \cos\left(2\pi y - 2\pi pz + 2\pi\varphi_{1p}\right)\sin(2\pi j Nz),
\end{aligned} \tag{6.246}$$

we obtain

$$b_x = \varepsilon \sum_{j\in\mathbb{Z}} \sum_{p=0}^{N-1} \sin\left(2\pi y - 2\pi pz + 2\pi\varphi_{1p}\right)\cos(2\pi j Nz)\,. \tag{6.247}$$

Note that the terms proportional to $\sin(2\pi j Nz)$ will cancel when we combine negative and positive j-values in the summation. Making use of the Poisson summation formula analogous to (6.210), we find

$$b_x = \varepsilon \sum_{p=0}^{N-1} \sin\left(2\pi y - 2\pi pz + 2\pi\varphi_{1p}\right) \sum_{k\in\mathbb{Z}} \delta(Nz - k)\,. \tag{6.248}$$

Compared to the result (6.211), “kicks” now occur after $\Delta z = 1/N$. When we remember from the collisionless case

$$\frac{\Delta x}{\Delta z} \sim \frac{\Delta x}{N^{-1}} \sim b_x\,, \quad \frac{\Delta y}{\Delta z} \sim \frac{\Delta y}{N^{-1}} \sim x\,, \tag{6.249}$$

it becomes natural to define a “small step standard map” in the form

$$x_{t+1} = x_t + \frac{\varepsilon}{N}\sum_{p=0}^{N-1} \sin\left(2\pi y_t - 2\pi pz_t + 2\pi\varphi_{1p}\right)\,, \tag{6.250}$$

$$y_{t+1} = y_t + \frac{1}{N}x_{t+1}\,. \tag{6.251}$$

Note that by increasing the index t, we wander around the steps marked in Figure 6.6.

One can again rescale $2\pi x \to x$, $2\pi y \to y$ and $2\pi\varepsilon \to \varepsilon$ as before. Inclusion of collisions with probability P in each Poincaré section then leads for $\varphi_{1p} = 0$ to

$$x_{t+1} = x_t + \frac{\varepsilon}{N}\sum_{p=0}^{N-1} \sin(y_t - 2\pi pz_t) + \rho_t\,, \tag{6.252}$$

$$y_{t+1} = y_t + \frac{1}{N}x_{t+1}\,, \tag{6.253}$$

$$z_{t+1} = z_t + v_{th}\,, \tag{6.254}$$

with $v_{th} = \pm 1/N$.

Now, the collision frequency is $\nu_{coll} = P/|v_{th}| = PN$.

6.5.4
Tokamap as a Twist Map with Polar Axis

A Hamiltonian twist map was introduced by Balescu [175] as a representation of the stroboscopic plot of magnetic field lines in a toroidal confinement device as used in fusion physics. To indicate its relevance for general tokamak considerations (see also the so called Wobig map [176]), it was called the tokamap. The tokamap is compatible with minimal toroidal geometry requirements (in particular, the polar axis cannot be crossed upon iteration). It depends on two parameters: the stochasticity parameter and the winding number on axis. With increasing values of the stochasticity parameter, chaotic regions appear mostly near the edge of the torus, while the zone near the magnetic axis remains very robust. The tokamap has the following form

$$\psi_{k+1} = \psi_k - L\frac{\psi_{k+1}}{1+\psi_{k+1}}\sin(\vartheta_k)\,, \tag{6.255}$$

$$\vartheta_{k+1} = \vartheta_k + 2\pi\Omega\left(\psi_{k+1}\right) - L\frac{1}{(1+\psi_{k+1})^2}\cos(\vartheta_k)\,. \tag{6.256}$$

The variable ψ is related in a poloidal cross section of a tokamak to the radial coordinate while ϑ is the poloidal angle coordinate. Compared to the formulation in [177], $\vartheta = 2\pi\theta$ and $\Omega \equiv W = (2-\psi)(2-2\psi+\psi^2)/4$ holds.

It should be emphasized that the tokamap has *not* been rigorously derived from a continuous Hamiltonian system. It was supposed [178] that the tokamap originates from the Hamiltonian [179]

$$\begin{aligned} H &= H_0(\psi) + \varepsilon H_1(\psi,\vartheta;\varphi) \\ &= \int\frac{d\psi}{q(\psi)} - 2\pi L_c\frac{\psi}{1+\psi}\cos(\vartheta)\sum_{k=-\infty}^{\infty}\delta(\varphi-2\pi k)\,. \end{aligned} \tag{6.257}$$

Performing a "trivial" one-step integration in φ over a distance 2π, relates the canonical equations based on Eq. (6.257) "in some sense" to the tokamap with the identification

$$2\pi L_c \mathrel{\hat{=}} L\,. \tag{6.258}$$

However, a derivation of the tokamap from the Hamiltonian equation (6.257) encounters a difficulty related with the presence of delta functions (see [180] and references therein). A regularization procedure has been proposed in [180]. Instead of the Hamiltonian equation (6.257), one might use

$$H = H_0(\psi) + \varepsilon H_1(\psi,\vartheta;\varphi) = \int\frac{d\psi}{q(\psi)} - L_c\frac{\psi}{1+\psi}\cos(\vartheta)\sum_{s=-M}^{M}\cos(s\varphi)\,, \tag{6.259}$$

containing the sum of a finite number M of trigonometric functions. Using the Poisson summation rule, it is easy to see that the Hamiltonian equation (6.257) follows from the regularized Hamiltonian equation (6.259) in the limit $M \to \infty$. Applying the construction of canonical mappings developed in [164] to the Hamiltonian equation (6.259), and then performing the limit $M \to \infty$, one can derive the *symmetric tokamap*. The (standard nonsymmetric) tokamap Eq. (6.255) cannot be derived from the Hamiltonian equation (6.259) using the regularization method.

In general, the tokamap is an implicit map, but the first equation (6.255) can be explicitly resolved with respect to ψ_{k+1},

$$\psi_{k+1} = \frac{1}{2}\left[\sqrt{P^2(\psi_k, \vartheta_k) + 4\psi_k} - P(\psi_k, \vartheta_k)\right],$$
$$P(\psi_k, \vartheta_k) = 1 - \psi_k + L\sin(\vartheta_k)\,. \tag{6.260}$$

The implicit equation for ϑ_{k+1} cannot be solved analytically. For this purpose, one can use numerical methods to solve algebraic equations, for example, the Newton method.

In the following, we shall use the tokamap for explicit evaluations. We interpret the dynamics evolving from the tokamap as magnetic filed line motion. From Poincaré sections (which qualitatively look similar to Figure 6.2), we can recognize different areas, some of them are stochastic, others belonging to still integrable regions. In the (completely) stochastic regions, standard theories can be applied to investigate the field line wandering. Here, we elucidate the quasilinear description.

In the following, we concentrate on the continuous model and investigate the motion in ψ-direction (radial direction) which is expected to be diffusive [181]. For large control parameters L, we first reckon with quasilinear diffusion in a Fokker–Planck description. We shall compute $\psi_m(\varphi) \equiv \langle \psi(\varphi) \rangle$ and $\langle [\Delta\psi(\varphi)]^2 \rangle \equiv \langle [\psi(\varphi) - \psi_m(\varphi)]^2 \rangle$ and start with the latter.

We consider the Hamiltonian system in action-angle variables ψ, ϑ and "time" φ,

$$\frac{d\psi}{d\varphi} \equiv \dot{\psi} = -\frac{\partial H}{\partial \vartheta}\quad , \quad \frac{d\vartheta}{d\varphi} \equiv \dot{\vartheta} = \frac{\partial H}{\partial \psi}\,, \tag{6.261}$$

where the Hamiltonian H may be written as $H = H_0(\psi) + \varepsilon H_1(\psi, \vartheta; \varphi)$. Let us briefly review the derivation of the quasilinear diffusion equation [182, 183] which appears in Fokker–Planck form. In this case, the Liouville equation for the probability (phase space) density $\mathcal{P}(\psi, \vartheta; \varphi)$ reads

$$\frac{d\mathcal{P}}{d\varphi} \equiv \frac{\partial \mathcal{P}}{\partial \varphi} + \dot{\vartheta}\frac{\partial \mathcal{P}}{\partial \vartheta} + \dot{\psi}\frac{\partial \mathcal{P}}{\partial \psi} = \frac{\partial \mathcal{P}}{\partial \varphi} + \frac{\partial H}{\partial \psi}\frac{\partial \mathcal{P}}{\partial \vartheta} - \frac{\partial H}{\partial \vartheta}\frac{\partial \mathcal{P}}{\partial \psi} = 0\,. \tag{6.262}$$

We shall make use of the smallness parameter ε when defining multiple time scales $\varphi_0 = \varphi, \varphi_1 = \varepsilon\varphi, \varphi_2 = \varepsilon^2\varphi, \ldots$ (i.e., the physical dynamics evolves on different time scales, the fastest being identified as φ_0, the next one as φ_1, and so on). We also expand the distribution function in the form

$$\mathcal{P} = \mathcal{P}_0(\psi, \vartheta; \varphi_0, \varphi_1, \varphi_2, \ldots) + \varepsilon\mathcal{P}_1(\psi, \vartheta; \varphi_0, \varphi_1, \varphi_2, \ldots)$$
$$+ \varepsilon^2\mathcal{P}_2(\psi, \vartheta; \varphi_0, \varphi_1, \varphi_2, \ldots) + \ldots \tag{6.263}$$

Each contribution on the right-hand side (r.h.s.) will be decomposed in the form $\mathcal{P}_\nu = \bar{\mathcal{P}}_\nu + \tilde{\mathcal{P}}_\nu$, where $\bar{\mathcal{P}}_\nu \equiv 1/(2\pi) \int d\vartheta \mathcal{P}_\nu = \langle \mathcal{P}_\nu \rangle_\vartheta$. Introducing the above ansatz into the (kinetic) equation (6.262), we obtain in successive orders

$$\frac{\partial \mathcal{P}_0}{\partial \varphi_0} + \frac{\partial H_0}{\partial \psi}\frac{\partial \mathcal{P}_0}{\partial \vartheta} = 0\,, \tag{6.264}$$

$$\frac{\partial \mathcal{P}_1}{\partial \varphi_0} + \frac{\partial \mathcal{P}_0}{\partial \varphi_1} + \frac{\partial H_0}{\partial \psi}\frac{\partial \mathcal{P}_1}{\partial \vartheta} + \frac{\partial H_1}{\partial \psi}\frac{\partial \mathcal{P}_0}{\partial \vartheta} - \frac{\partial H_1}{\partial \vartheta}\frac{\partial \mathcal{P}_0}{\partial \psi} = 0\,, \tag{6.265}$$

$$\frac{\partial \mathcal{P}_2}{\partial \varphi_0} + \frac{\partial \mathcal{P}_1}{\partial \varphi_1} + \frac{\partial \mathcal{P}_0}{\partial \varphi_2} + \frac{\partial H_0}{\partial \psi}\frac{\partial \mathcal{P}_2}{\partial \vartheta} + \frac{\partial H_1}{\partial \psi}\frac{\partial \mathcal{P}_1}{\partial \vartheta} - \frac{\partial H_1}{\partial \vartheta}\frac{\partial \mathcal{P}_1}{\partial \psi} = 0\,, \tag{6.266}$$

and so on. This ordering is based on the assumption that an unperturbed path with $\Omega \equiv dH_0/d\psi =$ const exists and that the correlation time is larger than the time to proceed along this path between steps. We shall come back to this point later. To lowest order in ε, Eq. (6.264) can be simplified, and the two projections

$$\frac{\partial \bar{\mathcal{P}}_0}{\partial \varphi_0} = 0\,, \quad \frac{\partial \tilde{\mathcal{P}}_0}{\partial \varphi_0} + \Omega \frac{\partial \tilde{\mathcal{P}}_0}{\partial \vartheta} = 0\,, \tag{6.267}$$

will be solved in the forms $\bar{\mathcal{P}}_0 = \bar{\mathcal{P}}_0(\psi; \varphi_1, \varphi_2, \ldots)$, $\tilde{\mathcal{P}}_0 = \tilde{\mathcal{P}}_0(\psi, \vartheta - \Omega\varphi_0; \varphi_0 = 0, \varphi_1, \varphi_2, \ldots) \equiv 0$. $\tilde{\mathcal{P}}_0 \equiv 0$ is a reasonable choice for the initial condition. Otherwise, we would produce large "drift velocities", and we are not interested in such situations. Substituting this value into Eq. (6.265), and decomposing the resulting equation into the averaged and fluctuation contributions, respectively, we obtain

$$\frac{\partial \tilde{\mathcal{P}}_1}{\partial \varphi_0} + \Omega \frac{\partial \tilde{\mathcal{P}}_1}{\partial \vartheta} = \left[\frac{\partial H_1}{\partial \vartheta}\frac{\partial \mathcal{P}_0}{\partial \psi}\right]^{\sim}\,, \quad \frac{\partial \bar{\mathcal{P}}_1}{\partial \varphi_0} = -\frac{\partial \bar{\mathcal{P}}_0}{\partial \varphi_1}\,. \tag{6.268}$$

Let us first consider the last equation which can be integrated since the r.h.s. is independent of φ_0. However, the secular term ($\sim \varphi_0$) would soon violate the ordering, and for a consistent procedure $\bar{\mathcal{P}}_0$ should not depend on φ_1. To summarize the results obtained so far, we write $\mathcal{P}_0 = \bar{\mathcal{P}}_0(\psi; \varphi_2, \ldots)$, $\mathcal{P}_1 = \bar{\mathcal{P}}_1(\psi; \varphi_1, \varphi_2, \ldots) + \tilde{\mathcal{P}}_1(\psi, \vartheta; \varphi_0, \varphi_1, \varphi_2, \ldots)$. Introducing the notation $V_1 \equiv -\partial \varepsilon H_1/\partial\vartheta$, the solution of Eq. (6.268) is

$$\tilde{\mathcal{P}}_1 = -\int_0^{\varphi_0} d\varphi_0' V_1\left[\psi, \vartheta - \Omega\varphi_0'; \varphi_0 - \varphi_0'\right] \frac{\partial \bar{\mathcal{P}}_0}{\partial \psi} + \tilde{\mathcal{P}}_1(\psi, \vartheta - \Omega\varphi_0; \varphi_0 = 0)\,, \tag{6.269}$$

and we neglect the contribution from the initial value in the following. Note that per definition, H_1 only depends on φ_0.

The explicit solution shows that $\tilde{\mathcal{P}}_1$ does not depend on φ_1. Next, we have to investigate the order ε^2. We have for the ϑ-averaged part (after partial integration)

$$\frac{\partial \bar{\mathcal{P}}_2}{\partial \varphi_0} = -\frac{\partial \bar{\mathcal{P}}_1}{\partial \varphi_1} - \frac{\partial \bar{\mathcal{P}}_0}{\partial \varphi_2} + \frac{\partial}{\partial \psi}\left\langle \frac{\partial H_1}{\partial \vartheta}\tilde{\mathcal{P}}_1 \right\rangle_\vartheta\,. \tag{6.270}$$

From here, we first conclude that $\bar{\mathcal{P}}_2$ is a function of φ_0 only via the φ_0-dependence of H_1 (since also $\tilde{\mathcal{P}}_1$ depends on φ_0 only via H_1). Thus, assuming H_1 perturbations to be periodic with period $T_0 = 2\pi$, we introduce the average $\langle \cdots \rangle_{\varphi_0} = 1/(2\pi) \int d\varphi_0 \cdots$, and obtain from (6.270),

$$\frac{\partial \langle \mathcal{P}_1 \rangle_{\varphi_0}}{\partial \varphi_1} = -\frac{\partial \bar{\mathcal{P}}_0(\psi; \varphi_2, \ldots)}{\partial \varphi_2} + \frac{\partial}{\partial \psi} \left\langle \frac{\partial H_1}{\partial \vartheta} \tilde{\mathcal{P}}_1 \right\rangle_{\vartheta, \varphi_0} . \tag{6.271}$$

The right-hand side is not a function of φ_1 and therefore, to avoid secular terms,

$$\frac{\partial \bar{\mathcal{P}}_0}{\partial \varphi_2} = \frac{\partial}{\partial \psi} \left\langle \frac{\partial H_1}{\partial \vartheta} \tilde{\mathcal{P}}_1 \right\rangle_{\vartheta, \varphi_0} \tag{6.272}$$

should hold.

Introducing the solution $\mathcal{P}_1$, we finally arrive at the quasilinear equation

$$\frac{\partial \bar{\mathcal{P}}_0}{\partial \varphi_2} = \frac{\partial}{\partial \psi} \left(\mathcal{D} \frac{\partial \bar{\mathcal{P}}_0}{\partial \psi} \right), \tag{6.273}$$

where

$$\mathcal{D} \equiv \mathcal{D}(\psi) = \left\langle \int_0^{\varphi_0} d\varphi_0' \, V_1 \left[\psi, \vartheta - \Omega \varphi_0'; \varphi_0 - \varphi_0' \right] V_1 \left[\psi, \vartheta; \varphi_0 \right] \right\rangle_{\vartheta, \varphi_0} . \tag{6.274}$$

Thus, we finally arrived at the standard diffusion equation with the quasilinear diffusion coefficient by a systematic multiple scale formalism.

Before proceeding, let us review some standard estimates. Using the correlation length L_{corr} for field line fluctuations, we write

$$\mathcal{D} \sim D_{m(agnetic)} = \int_0^\infty d\zeta \left\langle b_x \left[\boldsymbol{x}_\perp(\zeta), \zeta \right] b_x \left[\boldsymbol{x}_\perp(0), 0 \right] \right\rangle \sim b^2 L_{corr} . \tag{6.275}$$

Introducing the Eulerian perpendicular ($\lambda_\perp$) and parallel ($\lambda_\parallel$) correlation lengths, respectively, we can distinguish the two cases:

$$\Delta x < \lambda_\perp : \; D_m \lambda_\parallel < \lambda_\perp^2 \to L_{corr} \approx \lambda_\parallel \to D_m \approx b^2 \lambda_\parallel , \tag{6.276}$$

$$\Delta x > \lambda_\perp : \; D_m \lambda_\parallel > \lambda_\perp^2 \to D_m L_{corr} \approx \lambda_\perp^2 \to D_m \approx \frac{b^2 \lambda_\perp^2}{D_m} \to D_m \approx b \lambda_\perp . \tag{6.277}$$

Summarizing, we have

$$D_m \approx \begin{cases} b^2 \lambda_\parallel & \text{for} \quad b < \frac{\lambda_\perp}{\lambda_\parallel} , \\ b \lambda_\perp & \text{for} \quad b > \frac{\lambda_\perp}{\lambda_\parallel} . \end{cases} \tag{6.278}$$

The first result $D_m \approx b^2\lambda_{\|}$ for $\lambda_\perp \to \infty$ is known in the literature as the quasilinear result, whereas the second one, $D_m \approx b\lambda_\perp$ for $\lambda_{\|} \to \infty$, $\lambda_\perp$ finite, is due to Kadomtsev and Pogutse [107]. As we shall see, there is a problem with the second result since it does not correctly take trapping into account.

For the Hamiltonian equation (6.257), the contribution $V_1(\psi, \vartheta; \varphi)$ is given by

$$V_1(\psi, \vartheta, \varphi) = -2\pi L_c \frac{\psi}{1+\psi} \sin(\vartheta) \sum_{k=-\infty}^{\infty} \delta(\varphi - 2\pi k) . \tag{6.279}$$

Using it in Eq. (6.274) leads to

$$\begin{aligned} \mathcal{D}(\psi) = {} & \frac{L_c^2 \psi^2}{(1+\psi)^2} \int_0^{2\pi} d\varphi \sum_{n=-\infty}^{\infty} \delta(\varphi - 2\pi n) \int_0^{2\pi} d\vartheta \sin\vartheta \\ & \times \int_0^{\varphi} d\widetilde{\varphi} \sin(\vartheta - \Omega\widetilde{\varphi}) \sum_{m=-\infty}^{\infty} \delta(\varphi - \widetilde{\varphi} - 2\pi m) . \end{aligned} \tag{6.280}$$

First, we integrate over φ, leading to

$$\begin{aligned} \mathcal{D}(\psi) = {} & \frac{L_c^2 \psi^2}{(1+\psi)^2} \int_0^{2\pi} d\vartheta \sin\vartheta \int_0^{2\pi} d\widetilde{\varphi} \sin(\vartheta - \Omega\widetilde{\varphi}) \\ & \times \frac{1}{2} \sum_{m=-\infty}^{\infty} \delta(-\widetilde{\varphi} - 2\pi(m-1)) . \end{aligned} \tag{6.281}$$

The factor 1/2 appears because of the half-integration over the Delta function at $\varphi = 2\pi$.

Next, we integrate over φ. Here, the ambiguity because of the Delta function (whether we take $0-\varepsilon$ and $2\pi-\varepsilon$ or zero and 2π as integration limits, respectively) appears. In the quasilinear limit, after one turn in φ over 2π, the system is already decorrelated from its initial value, and therefore we replace the integration limits by

$$\begin{aligned} & \int_{0-\varepsilon}^{2\pi-\varepsilon} d\widetilde{\varphi} \sin(\vartheta - \Omega\widetilde{\varphi}) \sum_{m=-\infty}^{\infty} \delta(-\widetilde{\varphi} - 2\pi(m-1)) \\ & = \begin{cases} \sin\vartheta , & \varepsilon \to +0 ; \\ \frac{1}{2}\sin\vartheta + \frac{1}{2}\sin(\vartheta - 2\pi\Omega) , & \varepsilon \equiv 0 . \end{cases} \end{aligned} \tag{6.282}$$

In the quasilinear limit $\varepsilon \to +0$, we obtain

$$\mathcal{D}_{ql}(\psi) \approx \frac{L_c^2}{2} \frac{\psi^2}{(1+\psi)^2} \int_0^{2\pi} d\vartheta \sin^2\vartheta \approx \frac{\pi L_c^2}{2} \frac{\psi^2}{(1+\psi^2)^2} . \tag{6.283}$$

This is a space-(ψ-)dependent diffusion coefficient. On the other hand, if correlations exist for a longer time, we take the case $\varepsilon \equiv 0$ and because of $\cos(2\pi\Omega) = 2\cos^2(\pi\Omega) - 1$, we obtain

$$\mathcal{D}_{lc}(\psi) \approx \frac{\pi L_c^2}{2} \frac{\psi^2}{(1+\psi)^2} \cos^2(\pi\Omega) . \tag{6.284}$$

We emphasize that *without the space-dependence*, the just calculated expression $\mathcal{D}_{ql}$ agrees with the expected quasilinear diffusion coefficient for the corresponding map. It is easy to show that

$$\mathcal{D}_{ql} = \frac{\pi}{2} L_c^2 \tag{6.285}$$

is equivalent to

$$D_{ql} = \frac{1}{4} L^2 . \tag{6.286}$$

The latter follows from estimating the MSD from the map via

$$\langle(\psi_k - \psi_0)^2\rangle \approx L^2 \sum_{\nu,\mu=0}^{k-1} \langle \sin\vartheta_\nu \sin\vartheta_\mu \rangle \approx \frac{1}{2} L^2 k , \tag{6.287}$$

where we have ignored the factor $\psi/(1+\psi)$. It should be emphasized that the evaluation of this standard form of a quasilinear diffusion coefficient results from the recipe of only considering the contributions $\nu = \mu$ in the sum on the right-hand side. If we would also consider correlations over longer times by using

$$\vartheta_{\mu+1} = \vartheta_\mu + 2\pi\Omega , \quad \sin\vartheta_{\mu+1} = \sin\vartheta_\mu \cos(2\pi\Omega) + \sin(2\pi\Omega)\cos\vartheta_\mu , \tag{6.288}$$

we would also get the cos terms because of $\cos(2\pi\Omega) = 2\cos^2(\pi\Omega) - 1$.

Let us come back to the quasilinear case. When we use the definition of a (running) diffusion coefficient from the MSD, we define

$$\langle(\psi_k - \psi_0)^2\rangle \approx 2Dk . \tag{6.289}$$

Let us assume for the moment that there is a simple relation of the Fokker–Planck diffusion coefficient $\mathcal{D}_{ql}$ to the MSD development, then we would postulate for the continuous system

$$\langle(\psi_\varphi - \psi_0)^2\rangle \mathrel{\hat{=}} 2\mathcal{D}_{ql}\varphi \mathrel{\hat{=}} 2\mathcal{D}_{ql} 2\pi k \mathrel{\hat{=}} 2\frac{\pi}{2} L_c^2 2\pi k \mathrel{\hat{=}} \frac{1}{2} L^2 k , \tag{6.290}$$

which shows the equivalence. It was found numerically [179] that an additional factor appears,

$$D \approx 0.36 \mathcal{D}_{ql} . \tag{6.291}$$

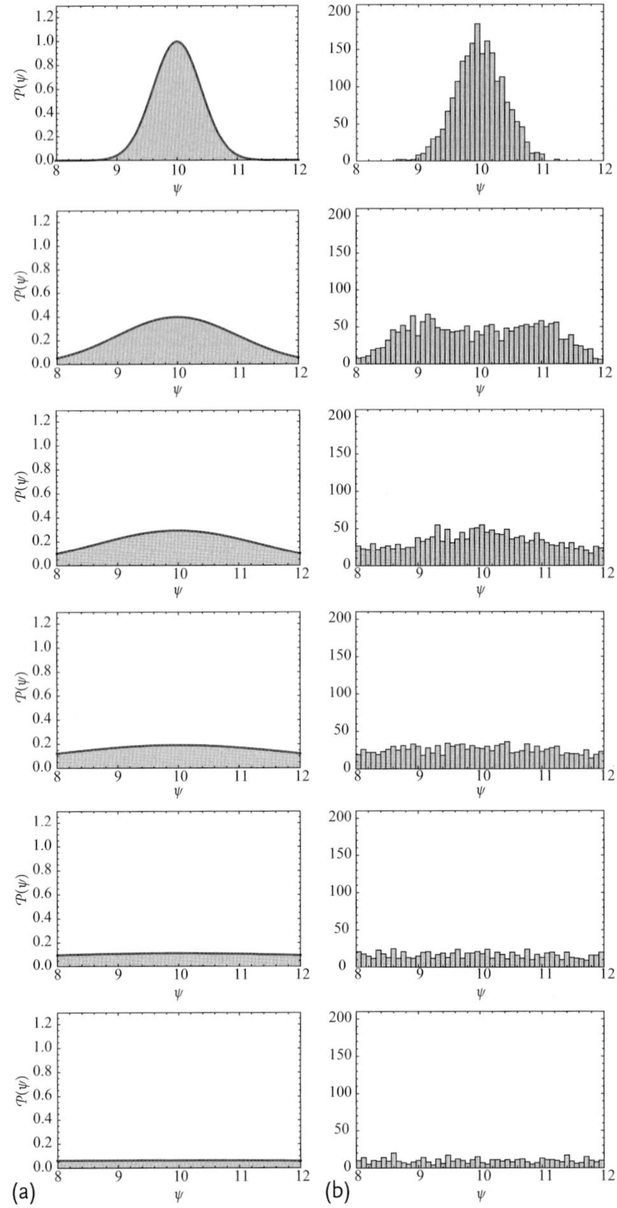

Figure 6.7 Evaluation of a Gaussian initial distribution for the probability distribution $\mathcal{P}(\psi)$ at $L = 9/2\pi$. (a) Direct solution of the Fokker–Planck equation with the quasilinear diffusion coefficient $\mathcal{D}_{ql}(\psi)$ at $L_c = 9$. (b) Iteration of the tokamap. The various figures correspond to "times" $\varphi = 2\pi k$ with $k = 0, 1, 2, 5, 15$, and 50, respectively, from top to bottom [181].

The latter scaling also agrees with the numerical values given explicitly in [177], although the discussion in [177] does not point out that discrepancy. The factor was explained in [181].

After solving the Fokker–Planck equation with the space-dependent quasilinear diffusion coefficient $\mathcal{D}_{ql}(\psi) = (\pi/2)L_c^2\psi^2/(1+\psi)^2$, one may compare the results with a direct iteration of the tokamap. The comparison is shown in Figure 6.7. The agreement is convincing. In Figure 6.7a, the solution of the Fokker–Planck equation is shown at "times" $\varphi = 2\pi k$ for $k = 0, 1, 2, 5, 15$, and 50. Figure 6.7b results from the statistics based on simulations with the tokamap. The "time" evolution of the probability density is shown at the corresponding k-values.

6.6 Basic Statistics for Magnetic Field Lines and Perpendicular Particle Diffusion

Basic statistical assumptions for fluctuating fields, which are often used in theory for qualitative predictions as well as quantitative estimates, will be summarized in the present section. The models look simple, and everybody should be aware that they cannot cover all the details appearing in complex plasma systems. The situation is ambivalent. Models should not deviate too much from realistic situations, though on the other hand, they should be simple enough to allow analytical calculations as far as possible. In this subsection, the preference is on analytical manageability.

Let us start with the current density for a three-dimensional diffusive process

$$\boldsymbol{\Gamma} = -\overleftrightarrow{D}\nabla n\,, \tag{6.292}$$

with the diffusion tensor $\overleftrightarrow{D}$, which in the gyrotropic case may be written as

$$\overleftrightarrow{D} = \chi_\perp(\hat{e}_x\hat{e}_x + \hat{e}_y\hat{e}_y) + \chi_{zz}\hat{e}_z\hat{e}_z\,. \tag{6.293}$$

The particle continuity equation leads to

$$\frac{\partial n}{\partial t} = \left[\chi_\perp(\nabla_x^2 + \nabla_y^2) + \chi_{zz}\nabla_z^2\right] n\,. \tag{6.294}$$

Associating with the particle density n, being normalized to the total particle number N, the probability to find a particle at a certain position $\boldsymbol{r}$ at time t, we determine mean values via

$$\langle\cdots\rangle = \frac{1}{N}\int d^3r\cdots n\,. \tag{6.295}$$

For purely diffusive processes, we find

$$\frac{d}{dt}\langle\boldsymbol{r}\rangle = 0\,. \tag{6.296}$$

Next, defining

$$\delta \boldsymbol{r} = \boldsymbol{r}(t) - \langle \boldsymbol{r}(t) \rangle \,, \tag{6.297}$$

we introduce the mean square displacement

$$\left\langle (\delta \boldsymbol{r})^2 \right\rangle = \left\langle (\boldsymbol{r})^2 \right\rangle - \langle \boldsymbol{r} \rangle^2 \,. \tag{6.298}$$

One immediately gets for purely diffusive processes

$$\frac{d}{dt} \left\langle (\delta \boldsymbol{r})^2 \right\rangle = \frac{d}{dt} \left[\left\langle (\boldsymbol{r})^2 \right\rangle - \langle \boldsymbol{r} \rangle^2 \right] = \frac{d}{dt} \left\langle (\boldsymbol{r})^2 \right\rangle . \tag{6.299}$$

The right-hand side will be evaluated according to

$$\frac{d}{dt} \left\langle (\boldsymbol{r})^2 \right\rangle = \frac{1}{N} \int d^3 r \boldsymbol{r}^2 \frac{\partial n}{\partial t} = \frac{1}{N} \int d^3 r \boldsymbol{r}^2 \left[\chi_\perp \left(\nabla_x^2 + \nabla_y^2 \right) + \chi_{zz} \nabla_z^2 \right] n \,. \tag{6.300}$$

For space-independent diffusion coefficients, we integrate by parts and obtain after a short calculation for the displacements in the different directions

$$\chi_\perp(t) = \frac{1}{2} \frac{d}{dt} \langle x^2 \rangle = \frac{1}{2} \frac{d}{dt} \langle y^2 \rangle \,, \quad \chi_{zz}(t) = \frac{1}{2} \frac{d}{dt} \langle z^2 \rangle \,, \tag{6.301}$$

that is, the running diffusion coefficients

$$\chi_\perp(t) = \frac{1}{2} \frac{d}{dt} \left\langle \delta x^2(t) \right\rangle = \frac{1}{2} \frac{d}{dt} \left\langle \delta y^2(t) \right\rangle , \quad \chi_{zz}(t) = \frac{1}{2} \frac{d}{dt} \left\langle \delta z^2(t) \right\rangle \,. \tag{6.302}$$

The latter equations (6.302) also apply for advection-diffusion equations

$$\frac{\partial n}{\partial t} = -\boldsymbol{V}(t) \cdot \nabla n(\boldsymbol{r}, t) + \left[\chi_\perp \left(\nabla_x^2 + \nabla_y^2 \right) + \chi_{zz} \nabla_z^2 \right] n(\boldsymbol{r}, t) \tag{6.303}$$

when

$$\frac{d}{dt} \langle \boldsymbol{r} \rangle = \boldsymbol{V}(t) \neq 0 \,. \tag{6.304}$$

We have found the famous Einstein relations (6.302) between the mean square displacements and the running diffusion coefficients. From here, we can immediately derive the Green–Kubo formula.

Let us have a closer look at the mean square displacement Eq. (6.298). We may evaluate by integration. Let us demonstrate that on the z-component of a purely diffusive process,

$$\delta z(t) = z(t) - \langle z(t) \rangle \,, \quad \langle z(t) \rangle = \text{const} \,, \quad \frac{dz(t)}{dt} = v_z(t) \tag{6.305}$$

in the form

$$z(t) = \underbrace{z(t_0 = 0)}_{z_0} + \int\limits_0^t dt' v_z(t') = z_0 + \delta z(t) \,. \tag{6.306}$$

Obviously, $\langle \delta z(t) \rangle = 0$ and

$$\langle [\delta z(t)]^2 \rangle = \int_0^t dt_1 \int_0^t dt_2 \underbrace{\langle v_z(t_1) v_z(t_2) \rangle}_{R_{zz}(t_1,t_2)} . \tag{6.307}$$

For homogeneous (in time) situations, the velocity correlation function R_{zz} is only a function of the time difference,

$$\langle v_z(t_1) v_z(t_2) \rangle \equiv R_{zz}(t_1, t_2) = R_{zz}(t_2 - t_1) . \tag{6.308}$$

In that case, we may evaluate

$$\begin{aligned}
\int_0^t dt_1 \int_0^t dt_2 R_{zz}(t_1, t_2) &= 2 \int_0^t ds \int_0^s dt_2 R_{zz}(t_2 - s) = 2 \int_0^t ds \int_{-s}^0 d\tau R_{zz}(\tau) \\
&= 2 \int_0^t ds \int_0^s d\tau R_{zz}(\tau) = 2 \left[t \int_0^t d\tau R_{zz}(\tau) - \int_0^t ds s R_{zz}(s) \right] \\
&= 2 \int_0^t ds (t - s) R_{zz}(s) , \quad t > 0 .
\end{aligned} \tag{6.309}$$

Combining with Eq. (6.302), and differentiating with respect to t, we obtain

$$\chi_{zz}(t) = \int_0^t ds R_{zz}(s) , \quad \chi_{zz} \equiv \chi_{zz}(t \to \infty) = \int_0^\infty ds R_{zz}(s) , \tag{6.310}$$

that is, the famous Green–Kubo formula expressing the diffusion coefficient as a time-integral of the velocity autocorrelation function. Of course, the present result can be generalized to the other diffusion coefficients, for example,

$$\chi_\perp = \int_0^\infty ds R_{xx}(s) , \quad R_{xx}(t_1, t_2) \equiv R_{xx}(t_2 - t_1) = \langle v_x(t_1) v_x(t_2) \rangle . \tag{6.311}$$

Obviously, the Green–Kubo formulas constitute direct access to the diffusion coefficient χ_{zz}. We may use it to determine the parallel diffusion coefficient perturbatively. Also, an alternate formulation (see Eq. (6.519) which will be derived later) has some advantages.

A very simple form of the velocity autocorrelation function can be derived from the Langevin equations for the motion of particles in an external magnetic field $B_0 \hat{z}$ in the presence of collisions with collision frequency ν, but without magnetic fluctuations. The corresponding "zeroth order" velocity is denoted as $\boldsymbol{v} \equiv \boldsymbol{\eta}$. The

Langevin equations are

$$\frac{d\boldsymbol{r}}{dt} = \boldsymbol{\eta}\,, \tag{6.312}$$

$$\frac{d\eta_\perp}{dt} = \Omega\,\boldsymbol{\eta}_\perp \times \hat{z} - \nu\eta_\perp + \boldsymbol{a}_\perp(t)\,, \tag{6.313}$$

$$\frac{d\eta_\parallel}{dt} = -\nu\eta_\parallel + a_z(t)\,. \tag{6.314}$$

We already solved for the velocity correlation functions

$$\overline{R_{xx}^{(0)}} = \frac{1}{2}v_{th}^2 \exp\left(-\nu\,|\tau|\right)\cos(\Omega\,\tau)\,, \quad \overline{R_{yy}^{(0)}} = \frac{1}{2}v_{th}^2 \exp\left(-\nu\,|\tau|\right)\cos(\Omega\,\tau)\,, \tag{6.315}$$

$$\overline{R_{xy}^{(0)}} = \frac{1}{2}v_{th}^2 \exp\left(-\nu\,|\tau|\right)\sin(\Omega\,\tau)\,, \quad \overline{R_{yx}^{(0)}} = -\frac{1}{2}v_{th}^2 \exp\left(-\nu\,|\tau|\right)\sin(\Omega\,\tau)\,, \tag{6.316}$$

$$\overline{R_{zz}^{(0)}} = \frac{1}{2}v_{th}^2 \exp\left(-\nu\,|\tau|\right)\,, \tag{6.317}$$

for $\tau = t_1 - t_2$.

Using $R_{xx} = \overline{R_{xx}^{(0)}}$ in the Green–Kubo formula (6.311), we obtain

$$\chi_\perp(t) = \frac{v_{th}^2}{2}\frac{\nu + [\Omega\,\sin(\Omega\,t) - \nu\cos(\Omega\,t)]\exp(-\nu t)}{\nu^2 + \Omega^2}\,, \tag{6.318}$$

with the asymptotic value

$$\chi_\perp = \frac{v_{th}^2}{2}\frac{\nu}{\nu^2 + \Omega^2}\,. \tag{6.319}$$

In a similar manner, using $R_{zz} = \overline{R_{zz}^{(0)}}$ in the Green–Kubo formula (6.310), we obtain

$$\chi_\parallel(t) \equiv \chi_{zz}(t) = \frac{v_{th}^2}{2}\frac{1 - \exp(-\nu t)}{\nu}\,, \tag{6.320}$$

with the asymptotic value

$$\chi_\parallel \equiv \chi_{zz} = \frac{v_{th}^2}{2\nu}\,. \tag{6.321}$$

6.6.1 Correlation Functions for Magnetic Field Fluctuations

In the present subsection, we investigate possible forms for the magnetic fluctuation autocorrelation functions. In contrast to the constant magnetic field considered thus far, we assume that the magnetic field has a fluctuating component $\boldsymbol{b}$.

Let us introduce

$$B = B_0 \left(b_0 \hat{e}_z + b \right) , \tag{6.322}$$

where $B_0 = \text{const}$ covers the dimension and $b_0 \equiv 1$ represents the constant external magnetic field. The challenge is to formulate magnetic autocorrelation functions which are not in conflict with the constraint

$$\nabla \cdot B = B_0 \nabla \cdot b = 0 \tag{6.323}$$

imposed by Maxwell's equations. We introduce the perpendicular and parallel correlation lengths $\lambda_\perp$ and $\lambda_\parallel$, respectively. We normalize the space coordinates by these lengths,

$$\frac{x}{\lambda_\perp} \to x \,, \quad \frac{y}{\lambda_\perp} \to y \,, \quad \frac{z}{\lambda_\parallel} \to z \,, \tag{6.324}$$

(in the following x, y, z are the dimensionless values, where x, y will be combined to $x_\perp$), and introduce the vector potential

$$A = A_0 + \delta A \,, \tag{6.325}$$

with

$$\delta A = \lambda_\perp B_0 \psi(x, y, z) \hat{e}_z \,, \quad \langle \psi \rangle = 0 \,. \tag{6.326}$$

Because of $B = \nabla \times A$, we immediately find

$$b_x = \frac{\partial \psi}{\partial y} \,, \quad b_y = -\frac{\partial \psi}{\partial x} \,. \tag{6.327}$$

Note that in the present formulation, we use $b_z = 0$ because of $b_z \ll b_0$. By prescribing the statistics of the vector potential, we avoid the problems with the zero divergence of the magnetic field.

Let us introduce the notation for the correlation function in real space,

$$\langle \psi(x_{\perp 0}, z_0) \psi(x_{\perp 0} + x_\perp, z_0 + z) \rangle = \beta^2 C_\perp(x_\perp) C_\parallel(z) \,. \tag{6.328}$$

Making use of the Fourier transform

$$\psi(x_\perp, z) = \int d^2 k_\perp \int d k_\parallel e^{i k_\perp \cdot x_\perp + i k_\parallel z} \hat{\psi}(k_\perp, k_\parallel) \,, \tag{6.329}$$

we have the correlation function in Fourier space

$$\left\langle \hat{\psi}(k_\perp, k_\parallel) \hat{\psi}\left(k'_\perp, k'_\parallel\right) \right\rangle = \beta^2 \hat{C}_\perp(k_\perp) \hat{C}_\parallel(k_\parallel) \delta\left(k_\perp + k'_\perp\right) \delta\left(k_\parallel + k'_\parallel\right) \,. \tag{6.330}$$

When no other information is available, we may introduce the exponential ansatz

$$C_\perp(x_\perp) = e^{-\frac{1}{2}(x^2 + y^2)} \,, \quad C_\parallel(z) = e^{-\frac{1}{2} z^2} \,, \tag{6.331}$$

leading to

$$\hat{C}_\perp(\boldsymbol{k}_\perp) = \frac{1}{2\pi} e^{-\frac{1}{2}k_\perp^2} , \quad \hat{C}_\parallel(k_\parallel) = \frac{1}{\sqrt{2\pi}} e^{-\frac{1}{2}k_\parallel^2} . \tag{6.332}$$

Next, we calculate

$$\langle b_i(\mathbf{0}, 0) b_j(\boldsymbol{x}_\perp, z)\rangle \equiv \beta^2 E_{\perp ij}(\boldsymbol{x}_\perp) C_\parallel(z) \equiv E_{ij}(\boldsymbol{r}) , \quad i, j = x, y . \tag{6.333}$$

Making use of the Fourier forms, it is straightforward to show

$$E_{\perp xx}(\boldsymbol{x}_\perp) = -\frac{\partial^2}{\partial y^2} C_\perp(\boldsymbol{x}_\perp) , \quad E_{\perp yy}(\boldsymbol{x}_\perp) = -\frac{\partial^2}{\partial x^2} C_\perp(\boldsymbol{x}_\perp) ,$$
$$E_{\perp xy}(\boldsymbol{x}_\perp) = E_{\perp yx}(\boldsymbol{x}_\perp) = \frac{\partial^2}{\partial x \partial y} C_\perp(\boldsymbol{x}_\perp) . \tag{6.334}$$

The ansatz (6.331) immediately leads to

$$E_{\perp xx}(\boldsymbol{x}_\perp) = (1 - y^2)\, e^{-\frac{1}{2}(x^2+y^2)} , \quad E_{\perp yy}(\boldsymbol{x}_\perp) = (1 - x^2)\, e^{-\frac{1}{2}(x^2+y^2)} ,$$
$$E_{\perp xy}(\boldsymbol{x}_\perp) = E_{\perp yx}(\boldsymbol{x}_\perp) = x y e^{-\frac{1}{2}(x^2+y^2)} . \tag{6.335}$$

Thus, back to dimensional variables, we arrive at the form

$$\begin{aligned} E_{ij}(\boldsymbol{r}) &\equiv \langle b_i(\boldsymbol{r}) b_j(0)\rangle \\ &= \beta^2 \begin{pmatrix} 1 - \frac{y^2}{\lambda_\perp^2} & \frac{xy}{\lambda_\perp^2} \\ \frac{xy}{\lambda_\perp^2} & 1 - \frac{x^2}{\lambda_\perp^2} \end{pmatrix} \exp\left(-\frac{x^2 + y^2}{2\lambda_\perp^2} - \frac{z^2}{2\lambda_\parallel^2} \right) , \quad i, j = 1, 2 , \end{aligned} \tag{6.336}$$

where β designates the strength, $\lambda_\perp$ the perpendicular and $\lambda_\parallel$ the parallel correlation length. This form of the autocorrelation matrix is consistent with the Maxwell equation $\nabla \cdot \boldsymbol{B} = 0$. One should have in mind that the correlation tensor is an ensemble average. For homogenous statistical systems we may perform the average for example as a space average,

$$\langle \ldots \rangle = \frac{1}{V} \int d^3 x_0 \ldots \tag{6.337}$$

Fourier transformation of Eq. (6.336),

$$E_{ij}(\boldsymbol{k}) = \frac{1}{(2\pi)^3} \int E_{ij}(\boldsymbol{r}) \exp(-i\boldsymbol{k} \cdot \boldsymbol{r}) d^3 r \tag{6.338}$$

with the inverse (note the asymmetric form of the used Fourier transform)

$$E_{ij}(\boldsymbol{r}) = \int d^3 k\, E_{ij}(\boldsymbol{k}) \exp(i\boldsymbol{k} \cdot \boldsymbol{r}) \tag{6.339}$$

is straightforward. The final result is

$$E_{ij}(\boldsymbol{k}) = \left(k_\perp^2 \delta_{ij} - k_i k_j\right) A(\boldsymbol{k}) \,, \tag{6.340}$$

$$A(\boldsymbol{k}) = \frac{\beta^2}{(2\pi)^{3/2}} \lambda_\| \lambda_\perp^4 \exp\left(-\frac{1}{2}\lambda_\|^2 k_\|^2 - \frac{1}{2}\lambda_\perp^2 k_\perp^2\right) . \tag{6.341}$$

The Lagrangian correlator for magnetic field lines is defined as

$$L_{mn}(\zeta) = \langle b_m(\boldsymbol{x}_\perp(\zeta), \zeta) b_n(\boldsymbol{x}_\perp(0), 0)\rangle \,, \quad m, n = x, y \,, \tag{6.342}$$

where $\boldsymbol{x}_\perp(\zeta)$ is the perpendicular position of the magnetic field line starting at $\boldsymbol{x}_\perp(0)$ for $\zeta = 0$. Balescu and coworkers [123] introduced the Corrsin approximation be rewriting Eq. (6.342) as

$$L_{mn}(\zeta) = \int d\rho_x d\rho_y \, \langle b_m(\boldsymbol{\rho}_\perp, \zeta)\, b_n(\boldsymbol{x}_\perp(0), 0) \delta\left(\boldsymbol{\rho}_\perp - \boldsymbol{x}_\perp(\zeta)\right)\rangle \,,$$
$$m, n = x, y \,. \tag{6.343}$$

Then, the exact propagator $\delta(\boldsymbol{\rho}_\perp - \boldsymbol{x}_\perp(\zeta))$ will be replaced by its ensemble average, leading to

$$L_{mn}(\zeta) \approx \int dr_x dr_y \, \langle b_m(\boldsymbol{r}_\perp, \zeta)\, b_n(\boldsymbol{0}, 0)\rangle \langle \delta\left(\boldsymbol{r}_\perp - \delta \boldsymbol{x}_\perp(\zeta)\right)\rangle \,,$$
$$m, n = x, y \,, \tag{6.344}$$

where $\boldsymbol{r}_\perp = \boldsymbol{\rho}_\perp - \boldsymbol{x}_\perp(0)$ and $\delta \boldsymbol{x}_\perp(\zeta) = \boldsymbol{x}_\perp(\zeta) - \boldsymbol{x}_\perp(0)$. Thus, on the right-hand side, the Eulerian correlator of magnetic field fluctuations appears. Together with the averaged propagator

$$\gamma(\boldsymbol{r}_\perp, \zeta) \equiv \langle \delta\left(\boldsymbol{r}_\perp - \delta \boldsymbol{x}_\perp(\zeta)\right)\rangle \,, \quad m, n = x, y \,, \tag{6.345}$$

the Lagrange correlation will be written in the form

$$L_{mn}(\zeta) \approx \int dr_x dr_y \, E_{mn}(\boldsymbol{r}_\perp, \zeta)\gamma(\boldsymbol{r}_\perp, \zeta) \,, \quad m, n = x, y \,. \tag{6.346}$$

Homogeneity and gyrotropy (i.e., isotropy with respect to the $\boldsymbol{B}_0$-axis) imply the properties [123]

$$\gamma(\boldsymbol{r}_\perp, \zeta) = \gamma(r_\perp, \zeta) \,, \quad L_{mn} = L\delta_{mn} \,, \quad m, n = x, y \,. \tag{6.347}$$

For calculating the averaged propagator γ, we make use of the Fourier representation

$$\gamma(\boldsymbol{r}_\perp, \zeta) = \int dk_x dk_y e^{i\boldsymbol{k}_\perp \cdot \boldsymbol{r}_\perp} \left\langle e^{-i\boldsymbol{k}_\perp \cdot \delta \boldsymbol{x}_\perp(\zeta)}\right\rangle \,, \tag{6.348}$$

and the cumulant expansion

$$\left\langle e^{-i\boldsymbol{k}_\perp \cdot \delta \boldsymbol{x}_\perp(\zeta)}\right\rangle \approx e^{-\frac{1}{2}\sum_{r,s} k_r k_s \Gamma_{rs}(\zeta)} \,, \tag{6.349}$$

with

$$\Gamma_{rs}(\zeta) = \langle \delta x_r(\zeta) \delta x_s(\zeta) \rangle \ , \quad r, s = x, y \ . \tag{6.350}$$

For the proof of $L_{mn} = L\delta_{mn}$, we go back to the Green–Kubo formula, that is,

$$\Gamma_{mn} = \int_0^{\zeta} d\zeta_1 \int_0^{\zeta} d\zeta_2 \underbrace{L_{mn}(\zeta_1, \zeta_2)}_{L_{mn}(\zeta_2 - \zeta_1)} = 2 \int_0^{\zeta} d\zeta' \, (\zeta - \zeta') \, L_{mn}(\zeta') \ . \tag{6.351}$$

Obviously, Γ_{mn} satisfies the differential equation

$$\frac{d^2 \Gamma_{mn}}{d\zeta^2} = 2 L_{mn} \ . \tag{6.352}$$

Inserting the expression (6.346) on the right-hand side leads to

$$\frac{d^2 \Gamma_{mn}}{d\zeta^2} = 2 \int d^2 k_\perp \int d k_\parallel e^{i k_\parallel \zeta} E_{mn}(\boldsymbol{k}_\perp, k_\parallel) e^{-\frac{1}{2} \sum_{r,s} k_r k_s \Gamma_{rs}(\zeta)} \ . \tag{6.353}$$

The integrations can be performed when Gaussian Euler correlation functions are used. When introducing the abbreviations

$$\zeta = \lambda_\parallel \tau \ , \quad g_{mn} = \frac{\Gamma_{mn}(\lambda_\parallel \tau)}{\lambda_\perp^2} \ , \qquad \alpha = \beta \frac{\lambda_\parallel}{\lambda_\perp} , \tag{6.354}$$

the result is

$$\frac{d^2 g_{mn}}{d\tau^2} = \frac{2\alpha^2 e^{-\frac{\tau^2}{2}}}{\left\{ [1 + g_{xx}(\tau)] [1 + g_{yy}(\tau)] - g_{xy}^2(\tau) \right\}^{3/2}} \left[\delta_{mn} + g_{mn}(\tau) \right] \ . \tag{6.355}$$

From Eq. (6.350), we deduce the initial values

$$g_{mn}(0) = 0 \ , \quad \left. \frac{d g_{mn}}{d\tau} \right|_{\tau=0} = 0 \ . \tag{6.356}$$

For $m \neq n$, this is a homogeneous differential equation. When $g_{mn} = 0$ initially, it will vanish at all times. Therefore, we conclude that

$$g_{mn}(\tau) = 0 \quad \text{for} \quad m \neq n \ . \tag{6.357}$$

On the other hand, for the diagonal elements, we introduce $g(\tau) \equiv g_{xx} - g_{yy}$ to again obtain a homogeneous differential equation for g. The same arguments as above lead to $g_{xx} = g_{yy}$ and therefore finally to Eq. (6.347).

Using these properties, we can simplify Eq. (6.348) with the result

$$\gamma(r_\perp, \zeta) = \frac{1}{4\pi \int_0^\zeta d\zeta' (\zeta - \zeta') L(\zeta')} \exp\left\{ -\frac{r_\perp^2}{4 \int_0^\zeta d\zeta' (\zeta - \zeta') L(\zeta')} \right\}, \tag{6.358}$$

which will be inserted into Eq. (6.346). Using a Gaussian Euler correlator, we finally obtain for L the integral equation

$$L(\zeta) = \beta^2 \frac{\lambda_\perp^4 e^{-\frac{\zeta^2}{2\lambda_\parallel^2}}}{\left[\lambda_\perp^2 + 2 \int_0^\zeta d\zeta' (\zeta - \zeta') L(\zeta')\right]^2} . \tag{6.359}$$

In the case of $\lambda_\perp \to \infty$, the approximate Lagrange correlation function

$$L(\zeta) = \beta^2 e^{-\frac{\zeta^2}{2\lambda_\parallel^2}} \tag{6.360}$$

coincides with the Eulerian one.

In context with magnetic field line divergence, derivatives of magnetic fluctuations become of interest. Let us introduce for the derivatives of magnetic field fluctuations

$$b_{m,\alpha} \equiv \frac{d}{dx_\alpha} b_m , \quad \alpha = 1, 2 \hat{=} x, y , \quad \boldsymbol{r} = \begin{pmatrix} x_1 \\ x_2 \\ \zeta \end{pmatrix} \equiv \begin{pmatrix} x \\ y \\ \zeta \end{pmatrix} \equiv \begin{pmatrix} \boldsymbol{x}_\perp \\ \zeta \end{pmatrix} . \tag{6.361}$$

The Euler correlator for derivatives will be calculated from the Euler correlator for fields

$$E_{mn} = \langle b_m(\boldsymbol{r}) b_n(\boldsymbol{0}) \rangle \tag{6.362}$$

by using the Fourier transform

$$E_{mn}^{\alpha\beta} = \int d^3k e^{i\boldsymbol{k}\cdot\boldsymbol{r}} k_\alpha k_\beta E_{mn}(\boldsymbol{k}) . \tag{6.363}$$

In the integrand, we use Eqs. (6.341) and (6.340). We immediately notice the symmetries

$$E_{mn}^{\alpha\beta} = E_{mn}^{\beta\alpha} = E_{nm}^{\alpha\beta} = E_{nm}^{\beta\alpha} . \tag{6.364}$$

Five independent components remain:

$$E_{xx}^{xx} = E_{yy}^{yy} = -E_{xy}^{xy} = \frac{\beta^2}{\lambda_\perp^2}\left(1 - \frac{x^2}{\lambda_\perp^2}\right)\left(1 - \frac{y^2}{\lambda_\perp^2}\right)\mathcal{E}\,, \tag{6.365}$$

$$E_{xx}^{yy} = \frac{\beta^2}{\lambda_\perp^2}\left(3 - 6\frac{y^2}{\lambda_\perp^2} + \frac{y^4}{\lambda_\perp^4}\right)\mathcal{E}\,, \quad E_{yy}^{xx} = \frac{\beta^2}{\lambda_\perp^2}\left(3 - 6\frac{x^2}{\lambda_\perp^2} + \frac{x^4}{\lambda_\perp^4}\right)\mathcal{E}\,, \tag{6.366}$$

$$E_{yy}^{xy} = -E_{yx}^{xx} = -\frac{\beta^2 xy}{\lambda_\perp^4}\left(3 - \frac{x^2}{\lambda_\perp^2}\right)\mathcal{E}\,,$$

$$E_{xx}^{xy} = -E_{yx}^{yy} = -\frac{\beta^2 xy}{\lambda_\perp^4}\left(3 - \frac{y^2}{\lambda_\perp^2}\right)\mathcal{E}\,, \tag{6.367}$$

where

$$\mathcal{E} \equiv \exp\left(-\frac{x^2 + y^2}{2\lambda_\perp^2} - \frac{\zeta^2}{2\lambda_\parallel^2}\right)\,. \tag{6.368}$$

One may calculate the Lagrangian correlator

$$E_{mn}^{\alpha\beta} = \langle b_{m,\alpha}(\boldsymbol{x}_\perp(\zeta), \zeta) b_{n,\beta}(\boldsymbol{0}, 0)\rangle \tag{6.369}$$

in a similar way as we have estimated the Lagrangian correlator for magnetic field fluctuations. The averaged propagator has the same form as before and

$$L_{mn}^{\alpha\beta}(\zeta) \approx \int dr_x dr_y\, E_{mn}^{\alpha\beta}(\boldsymbol{r}_\perp, \zeta)\gamma(\boldsymbol{r}_\perp, \zeta)\,, \quad m, n = x, y\,. \tag{6.370}$$

All nonvanishing components can be expressed in terms of a single function $K(\zeta)$ in the form

$$L_{xx}^{xx}(\zeta) = L_{yy}^{yy}(\zeta) = -L_{xy}^{xy}(\zeta) = -L_{xy}^{yx}(\zeta) \equiv K(\zeta)\,, \tag{6.371}$$

$$L_{xx}^{yy}(\zeta) = L_{yy}^{xx}(\zeta) = 3K(\zeta)\,, \quad L_{xx}^{xy}(\zeta) = L_{xy}^{yy}(\zeta) = L_{yy}^{xy}(\zeta) = L_{yx}^{xx}(\zeta) = 0\,. \tag{6.372}$$

Thus, we only have to explicitly calculate one element. Let us choose $L_{xx}^{xx}(\zeta)$, which after integration over the angles is

$$K(\zeta) = \beta^2 \frac{2\pi}{\lambda_\perp^2} \frac{e^{-\frac{\zeta^2}{2\lambda_\parallel^2}}}{4\int_0^\zeta d\zeta'(\zeta - \zeta')L(\zeta')} \times \int dr_\perp r_\perp \left[1 - \frac{r_\perp^2}{\lambda_\perp^2} + \frac{r_\perp^4}{8\lambda_\perp^4}\right] e^{-\frac{r_\perp^2}{4\int_0^\zeta d\zeta'(\zeta-\zeta')L(\zeta')}} e^{-\frac{r_\perp^2}{2\lambda_\perp^2}}\,. \tag{6.373}$$

Here, L is the Lagrangian correlator for magnetic field lines. We perform the $r_\perp$-integration to obtain

$$K(\zeta) = \beta^2 e^{-\frac{\zeta^2}{2\lambda_\parallel^2}} \frac{\lambda_\perp^4}{\left[\lambda_\perp^2 + 2\int_0^\zeta d\zeta'(\zeta - \zeta')L(\zeta')\right]^3} . \tag{6.374}$$

As an application, we calculate the diffusion associate with magnetic field line wandering. The equation for a magnetic field line Eq. (6.202) may be written as

$$\frac{dx}{dz} = b_x . \tag{6.375}$$

This equation has to be supplemented by the statistical properties for the magnetic fluctuations. It is a Langevin-type equation. The solution for the mean square displacement follows the lines presented for the Green–Kubo formula, resulting in

$$\langle \delta x^2(z) \rangle = \int_0^z dz' \int_0^z dz'' \langle b_x(z') \, b_x(z'') \rangle = 2 \int_0^z dz' \, (z - z') \, E_{xx}(z') . \tag{6.376}$$

In the quasilinear limit $\lambda_\perp \to \infty$, we use

$$E_{xx} = E_{xx}(z) = \beta^2 \exp\left(-\frac{z^2}{2\lambda_\parallel^2}\right) \tag{6.377}$$

to obtain

$$\langle \delta x^2(z) \rangle = 2\lambda_\parallel^2 \beta^2 \left\{ \sqrt{\frac{\pi}{2}} \frac{z}{\lambda_\parallel} \operatorname{erf}\left(\frac{z}{\sqrt{2}\lambda_\parallel}\right) + \left[\exp\left(-\frac{z^2}{2\lambda_\parallel^2}\right) - 1\right] \right\} . \tag{6.378}$$

For large z, the right-hand side behaves as

$$\langle \delta x^2(z) \rangle \sim 2\sqrt{\frac{\pi}{2}} \lambda_\parallel \beta^2 z , \tag{6.379}$$

leading to the magnetic diffusion coefficient

$$D = \sqrt{\frac{\pi}{2}} \beta^2 \lambda_\parallel . \tag{6.380}$$

The factor $\sqrt{\pi/2}$ originates from the assumed exponential form of the Euler correlation function. Therefore, when being less specific in the form of the correlator, the form of the magnetic diffusion coefficient is assumed as

$$D_{m(agnetic)} \approx \beta^2 \lambda_\parallel . \tag{6.381}$$

6.6.2
Elementary Estimates of the Kolmogorov Length L_K

The Kolmogorov length characterizes the exponential divergence of neighboring magnetic field lines. Standard books on nonlinear dynamical systems, see, for example, [27], present overviews over expressions for L_K. We have already mentioned the Kolmogorov length Eq. (6.226) for the standard map in the completely stochastic parameter regime. In this subsection, we review a result which will be used later (see Eq. (6.444)) following [123].

We determine the separation of magnetic field lines in the quasilinear regime. Two distinct field lines start at different positions. The (longitudinal) coordinate ζ is used as the parameter to determine the (lateral) positions $\boldsymbol{x}_{\perp 1}$ and $\boldsymbol{x}_{\perp 2}$. The components Δx and Δy of the distance $\Delta \boldsymbol{x}_\perp = \boldsymbol{x}_{\perp 2} - \boldsymbol{x}_{\perp 1}$ obey the equations

$$\frac{d\Delta x}{d\zeta} = b_x(\boldsymbol{x}_{\perp 2}(\zeta), \zeta) - b_x(\boldsymbol{x}_{\perp 1}(\zeta), \zeta) , \tag{6.382}$$

$$\frac{d\Delta y}{d\zeta} = b_y(\boldsymbol{x}_{\perp 2}(\zeta), \zeta) - b_y(\boldsymbol{x}_{\perp 1}(\zeta), \zeta) . \tag{6.383}$$

After linearization, we obtain

$$\frac{d\Delta x}{d\zeta} \approx b_{x,x}(\boldsymbol{x}_\perp(\zeta), \zeta)\Delta x + b_{x,y}(\boldsymbol{x}_\perp(\zeta), \zeta)\Delta y , \tag{6.384}$$

$$\frac{d\Delta y}{d\zeta} \approx b_{y,x} y(\boldsymbol{x}_\perp(\zeta), \zeta)\Delta x + b_{y,y}(\boldsymbol{x}_\perp(\zeta), \zeta)\Delta y . \tag{6.385}$$

The derivatives $b_{m,n}$ are functions of the coordinates. Because of the ζ-dependence of the coefficients, even the linearized equations for the distance do not posses *simple* exponential solutions. As has been discussed, the derivatives $b_{m,n}$ are also random fields, and a statistical description is appropriate. From now on, we use Eqs. (6.384) and (6.385) as the starting equations (replacing $\approx$ by $=$). Straightforward algebraic manipulations lead (with the notation $(\Delta x)^2 \equiv \Delta x^2$ and so on) to

$$\frac{d\Delta x^2}{d\zeta} = 2b_{x,x}(\boldsymbol{x}_\perp(\zeta), \zeta)\Delta x^2 + 2b_{x,y}(\boldsymbol{x}_\perp(\zeta), \zeta)\Delta x\Delta y , \tag{6.386}$$

$$\frac{d\Delta y^2}{d\zeta} = 2b_{y,x}(\boldsymbol{x}_\perp(\zeta), \zeta)\Delta x\Delta y + 2b_{y,y}(\boldsymbol{x}_\perp(\zeta), \zeta)\Delta y^2 , \tag{6.387}$$

$$\begin{aligned}\frac{d\Delta x\Delta y}{d\zeta} &= b_{x,x}(\boldsymbol{x}_\perp(\zeta), \zeta)\Delta x\Delta y + b_{x,y}(\boldsymbol{x}_\perp(\zeta), \zeta)\Delta y^2 \\ &\quad + b_{y,x}(\boldsymbol{x}_\perp(\zeta), \zeta)\Delta x\Delta y + b_{y,y}(\boldsymbol{x}_\perp(\zeta), \zeta)\Delta y^2 .\end{aligned} \tag{6.388}$$

For an iterative, quasilinear solution, we formally integrate these equations over ζ to replace the squares on the right-hand side by the formal solutions. After averaging, we find

$$\frac{d\langle\Delta x^2\rangle}{d\zeta} \approx 4\bar{L}^{xx}_{xx}\langle\Delta x^2\rangle + 4\bar{L}^{xy}_{xx}\langle\Delta x\Delta y\rangle + 2\bar{L}^{yx}_{xx}\langle\Delta x\Delta y\rangle + 2\bar{L}^{yy}_{xx}\langle\Delta y^2\rangle + 2\bar{L}^{yx}_{xy}\langle\Delta x^2\rangle + 2\bar{L}^{yy}_{xy}\langle\Delta x\Delta y\rangle\,, \tag{6.389}$$

and similar equations for $d\langle\Delta y^2\rangle/d\zeta$ and $d\langle\Delta x\Delta y\rangle/d\zeta$. The coefficients

$$\bar{L}^{\alpha\beta}_{mn} = \int_0^\infty d\zeta\, L^{\alpha\beta}_{mn}(\zeta) \tag{6.390}$$

follow by integrating the correlation function for derivatives of magnetic field fluctuations. Implicitly, we have assumed

$$\lambda_{\|} \ll L_K\,, \tag{6.391}$$

where L_K is the characteristic (exponentiation) length for the mean squares. Making use of the properties of the correlation functions $L^{\alpha\beta}_{mn}(\zeta)$ and introducing

$$\mathcal{K} = \int_0^\infty K(\zeta)d\zeta \equiv \frac{1}{4L_K}\,, \tag{6.392}$$

we obtain

$$\frac{d\langle\Delta x^2\rangle}{d\zeta} = 2\mathcal{K}\langle\Delta x^2\rangle + 6\mathcal{K}\langle\Delta y^2\rangle\,, \tag{6.393}$$

$$\frac{d\langle\Delta y^2\rangle}{d\zeta} = 6\mathcal{K}\langle\Delta x^2\rangle + 2\mathcal{K}\langle\Delta y^2\rangle\,, \tag{6.394}$$

$$\frac{d\langle\Delta x\Delta y\rangle}{d\zeta} = -4\mathcal{K}\langle\Delta x\Delta y\rangle\,. \tag{6.395}$$

The solution of this linear system of first order differential equations with constant coefficients is straightforward. The analysis shows that two eigenvalues $-L_K^{-1}$ are negative (and degenerate) and one is positive, leading to the exponential growth

$$\langle\Delta x^2\rangle \sim \langle\Delta y^2\rangle \sim \exp\left[2\frac{\zeta}{L_K}\right]; \tag{6.396}$$

L_K is the exponentiation (Kolmogorov) length. In general, it is not easy to determine. However, in the limit of very large perpendicular correlation lengths $\lambda_\perp$, with

$$K(\zeta) = \beta^2 e^{-\frac{\zeta^2}{2\lambda_\|^2}} \frac{\lambda_\perp^4}{\left[\lambda_\perp^2 + 2\int_0^\zeta d\zeta'(\zeta-\zeta')\,L(\zeta')\right]^3} \approx \frac{\beta^2}{\lambda_\perp^2} e^{-\frac{\zeta^2}{2\lambda_\|^2}}\,, \tag{6.397}$$

the integration over ζ leads to

$$L_K \approx \sqrt{\frac{2}{\pi}}\frac{\lambda_\perp^2}{4\beta^2\lambda_\|}\,. \tag{6.398}$$

The condition (6.391) determines the region of validity

$$4\sqrt{\frac{\pi}{2}}\beta^2\frac{\lambda_{\parallel}^2}{\lambda_{\perp}^2} \ll 1\,, \tag{6.399}$$

which is true for small Kubo numbers.

6.7
Phenomenology of Stochastic Particle Diffusion Theory in Perpendicular Direction

The following subsections present heuristic theories of test particle diffusion in specified stochastic magnetic fields. We distinguish between the spatial diffusion of magnetic lines characterized by a magnetic diffusion coefficient, and the motion of particles along as well as decorrelation of particles from a given line by, for example, collisions. We start with negligible binary collisions, and subsequently, discuss models in subregimes with increasing collisional effects. The results can be used to estimate the electron thermal conductivity in specified, chaotic tokamak boundaries. Ion orbits and ambipolarity are not discussed. That is the reason for the applicability of the models for only electron thermal conductivity, although we name the expressions as (test) particle diffusion coefficients.

Jokipii [78] and Rosenbluth *et al.* [183] showed the diffusive motion of magnetic field lines in disturbed systems. Stix [184] was probably the first to apply the ideas of magnetic turbulence to tokamaks [185]. Jokipii and Parker [81] noted the importance of magnetic stochasticity in astrophysical transport problems. Rechester and Rosenbluth [106] calculated electron heat transport in a tokamak with destroyed magnetic surfaces. At the same time, Kadomtsev and Pogutse [107] derived a diffusion coefficient for stochastic plasmas with strong collisions. After these pioneering works, many authors, for example, [108, 110, 122, 128–130, 186, 187] and many others, rounded the theory off, although today no complete self-consistent theory of transport in stochastic plasmas is available.

The starting point for a simple but systematic stochastic theory for perpendicular transport of electrons and ions are stochastic differential equations. The details of that method has been extensively worked out in the excellent monograph [76] by Balescu so that we do not repeat these calculations here. The following remarks should only give an impression as to how that type of transport theory works.

Balescu and coworkers [76] advanced the field of stochastic transport considerably by using the so called V-Langevin equations

$$\frac{dx_p(t)}{dt} = b_x\left[x_p(t), y_p(t), z_p(t)\right]\frac{dz_p(t)}{dt} + \eta_{\perp x}(t)\,, \tag{6.400}$$

$$\frac{dy_p(t)}{dt} = b_y\left[x_p(t), y_p(t), z_p(t)\right]\frac{dz_p(t)}{dt} + \eta_{\perp y}(t)\,, \tag{6.401}$$

$$\frac{dz_p(t)}{dt} = \eta_{\parallel}(t)\,. \tag{6.402}$$

The V-Langevin equations make use of the guiding center approach for small Larmor radii of the particles. Without collisions ($\eta_\perp = \eta_\parallel = 0$), the particle positions $\boldsymbol{r}_p(t)$ follow the magnetic field lines. Collisions cause diffusive motion along the zeroth order magnetic field (parallel direction) as well as deviations from the perturbed field lines in perpendicular direction.

The V-Langevin equation is a simplification of the A-Langevin equation (Acceleration Langevin equation)

$$\frac{d\boldsymbol{v}}{dt} = \frac{Ze}{m}\boldsymbol{v} \times \boldsymbol{B} - \nu \boldsymbol{v} + \boldsymbol{a}\,, \tag{6.403}$$

which has to be used when finite Larmor radii become important.

The V-Langevin equations allow straightforward estimates of the particle diffusion. Making use of the Taylor–Green–Kubo formula [188–190] in three-dimensional geometry

$$D = \frac{1}{3}\int_0^\infty d\tau \left\langle \boldsymbol{u}\left[\boldsymbol{x}(\tau), \tau\right] \cdot \boldsymbol{u}[0,0]\right\rangle, \tag{6.404}$$

we obtain for the magnetic diffusion coefficient in x-direction the estimate already used in the evaluation of the quasilinear expression, namely,

$$D_{m(agnetic)} \sim \int_0^\infty d\zeta \left\langle b_x\left[\boldsymbol{x}_\perp(\zeta), \zeta\right] b_x\left[\boldsymbol{x}_\perp(0), 0\right]\right\rangle \sim b^2 L_{corr}\,. \tag{6.405}$$

However, one should note that the Taylor–Green–Kubo formula contains the Lagrangian correlation function

$$L_{rs}[\zeta] = \left\langle b_r\left[\boldsymbol{x}_\perp(\zeta), \zeta\right] b_s\left[\boldsymbol{x}_\perp(0), 0\right]\right\rangle, \tag{6.406}$$

and not the (simpler) Eulerian correlation function. Reasonable assumptions for the latter have been discussed in the last section.

The relation between the Lagrange correlation function and the Euler correlation function is a difficult task which, in principle, requires the knowledge of the exact dynamics. All discussions of the following pages are centered around that problem. Taking arguments from fluid dynamics, the Corrsin approximation [121] is often applied when the exact propagator is replaced by the averaged propagator. The averaged propagator can be further evaluated by making use of a cumulant expansion [100]. Within the Corrsin approximation, Balescu and coworkers were able to derive an integral equation for the Lagrangian form of the correlation function. Its solution, for example, for magnetic field line wandering, yields

$$D_m \approx \begin{cases} b^2\lambda_\parallel & \text{for } \lambda_\perp \to \infty \\ b\lambda_\perp & \text{for } \lambda_\parallel \to \infty \end{cases}\,. \tag{6.407}$$

These expressions as well as more general results on particle diffusion perpendicular to a strong external magnetic field will be discussed in the following on the basis of heuristic arguments.

6.7.1
Perpendicular Particle Diffusion

Using the quasilinear magnetic diffusion coefficient, we can use random walk estimates for the particle diffusion coefficient. In the collisionless case, we find the particle diffusion coefficient in terms of the magnetic (field line) diffusion coefficient $D_m \equiv D_{m(agnetic)}$ in the form

$$D_{\perp particle} \sim \frac{\langle(\Delta x)^2\rangle}{\Delta\tau} \approx \frac{D_{m(agnetic)} l}{\frac{l}{v_{th}}} = D_{m(agnetic)} v_{th} \approx b^2 \lambda_{\parallel} v_{th} \,. \quad (6.408)$$

This collisionless quasilinear result, which occurs for large (infinite) perpendicular correlation lengths ($\lambda_\perp \to \infty$), is based on the following picture. For a particle propagating a distance $l \approx \lambda_\parallel$ along a magnetic field line with small perpendicular disturbance δB compared to the zeroth order field B_0, we obtain from the field line equations of motion for the transverse displacement Δr

$$\frac{\Delta r}{\delta B} \approx \frac{\lambda_\parallel}{B_0} \,. \quad (6.409)$$

When the particle approximately has constant speed $v \approx v_{th}$, the time step will be estimated by

$$\Delta\tau \approx \frac{\lambda_\parallel}{v_{th}} \,. \quad (6.410)$$

The random walk formula leads to

$$D_{\perp particle} \sim \frac{(\Delta r)^2}{\Delta\tau} \approx v_{th} \left(\frac{\delta B}{B_0}\right)^2 \lambda_\parallel \sim v_{th} D_m , \quad (6.411)$$

as stated in Eq. (6.408).

Collisions will influence the quasilinear result. For weak collisions ($\lambda_{coll} \gg \lambda_\parallel$) and large perpendicular correlation lengths $\lambda_\perp \to \infty$, the result will be approximately the same as in the collisionless case, that is, Eq. (6.408).

Next, we explicitly take into account finite perpendicular correlation lengths and small collision rates. We start with diffusion caused by magnetic field line stochasticity, as given by the formula for perpendicular magnetic diffusion after a distance $L_\parallel$,

$$(\Delta r)^2 \approx 2 D_m L_\parallel \,. \quad (6.412)$$

Along $L_\parallel$, decorrelation will be achieved. The decorrelation length $L_\parallel$ should be larger than the Kolmogorov length L_K. Then, a small initial cross-field displacement will be significantly amplified. Collisional motion along the magnetic field lines relates the characteristic time $\Delta\tau$ to the (parallel) length $L_\parallel$ in the form

$$L_\parallel^2 \approx 2\chi_\parallel \Delta\tau \,. \quad (6.413)$$

The time-step $\Delta\tau$ is the decorrelation time during which the perpendicular deviation grows from zero to $\lambda_\perp$. Combining Eq. (6.412) with Eq. (6.413), we obtain the perpendicular particle diffusion coefficient

$$D_\perp = \frac{1}{2}\frac{(\Delta r)^2}{\Delta\tau} \approx \frac{D_m\sqrt{2\chi_\parallel}}{\sqrt{\Delta\tau}}\,, \tag{6.414}$$

which will be the starting point for the following estimates.

When we apply the magnetic diffusion coefficient D_m, we are in the limit

$$\frac{\lambda_\parallel}{\lambda_\perp} \ll 1\,. \tag{6.415}$$

Furthermore, we assume that the (classical) parallel diffusion coefficient is much larger than the perpendicular one,

$$\frac{\chi_\perp}{\chi_\parallel} \ll 1\,, \quad \leftrightarrow \quad \frac{\nu_{coll}}{\Omega} \ll 1\,. \tag{6.416}$$

The characteristic lengths will be ordered as follows [76], that is,

$$\lambda_{coll} \ll L_K \lesssim L_\parallel\,. \tag{6.417}$$

Detailed theory [123] shows that

$$\lambda_\parallel < L_K \tag{6.418}$$

is required for the Markovian approximation when evaluating integrals over correlation functions.

Within the Kadomtsev–Pogutse model [107], the time-step $\Delta\tau \equiv \Delta\tau_{KP}$ is determined by

$$\lambda_\perp^2 \approx 2\chi_\perp\Delta\tau_{KP}\,. \tag{6.419}$$

In Figure 6.8, this situation is shown in an illustration. Collisional broadening occurs around a magnetic field line. When the lateral displacement becomes of the order of the perpendicular correlation length, the particle will become decorrelated from the magnetic field line. This defines $L_\parallel$. Because of the strong field line wandering, the mean displacement Δr may be larger than the perpendicular correlation length. Making use of this ordering leads to the perpendicular particle diffusion coefficient

$$D_{KP} \approx \frac{2D_m\sqrt{\chi_\parallel\chi_\perp}}{\lambda_\perp} \approx \frac{D_m v_{th}\rho}{\lambda_\perp}\,, \tag{6.420}$$

that is, the so called Kadomtsev–Pogutse diffusion coefficient which, in the present form, is independent of the collision frequency ν_{coll}. As we shall see, it can be considered as the limit of the Rechester–Rosenbluth diffusion coefficient D_{RR} (see below) for large collision frequencies.

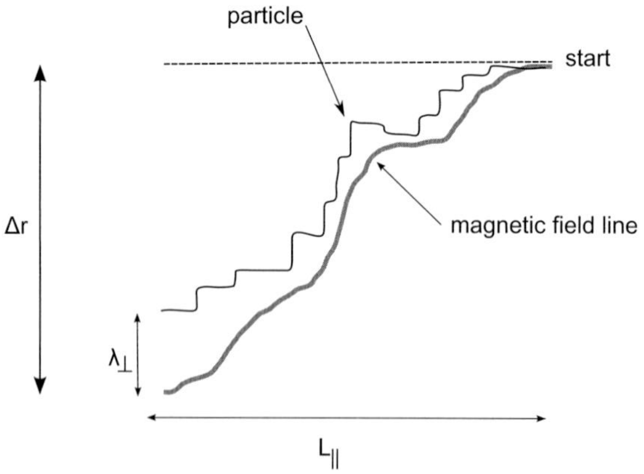

Figure 6.8 Sketch of the physical conditions for the validity of the Kadomtsev–Pogutse diffusion coefficient.

The present form of the Kadomtsev–Pogutse diffusion coefficient should not be mistaken for $D_{KP}^{(II)} \approx b\lambda_\perp v_{th}$, originating from the second form of Eq. (6.278) which sometimes is also named the Kadomtsev–Pogutse diffusion coefficient. The form $D_{KP}^{(II)}$ was derived in the percolation limit without taking care of trapping. Therefore, it is not correct.

In the present case, the decorrelation length will be of the form

$$L_\| \approx \lambda_\perp \sqrt{\frac{\chi_\|}{\chi_\perp}} \equiv L_{KP}\;; \tag{6.421}$$

L_{KP} is called the Kadomtsev–Pogutse characteristic length. We note the order of magnitude

$$L_{KP} \approx \frac{\Omega}{\nu_{coll}} \lambda_\perp \gg \lambda_\perp\,. \tag{6.422}$$

The determination of the time-step $\Delta\tau \equiv \Delta\tau_{RR}$ within the Rechester–Rosenbluth model [106] is a little bit more complex since now it is assumed that the main decorrelation occurs because of the exponential divergence of neighboring field lines, as it is sketched in Figure 6.9. The condition for ensuring that the decorrelation mechanism due to the chaotic field prevails over the collisional cross-field displacement is

$$L_\| < L_{KP}\,. \tag{6.423}$$

Decorrelation after exponential field line separation (with the Kolmogorov length L_K) stops when the perpendicular distance approaches the perpendicular correlation length,

$$\lambda_\perp \approx \delta r_0 \exp\left\{\frac{L_\|}{L_K}\right\}. \tag{6.424}$$

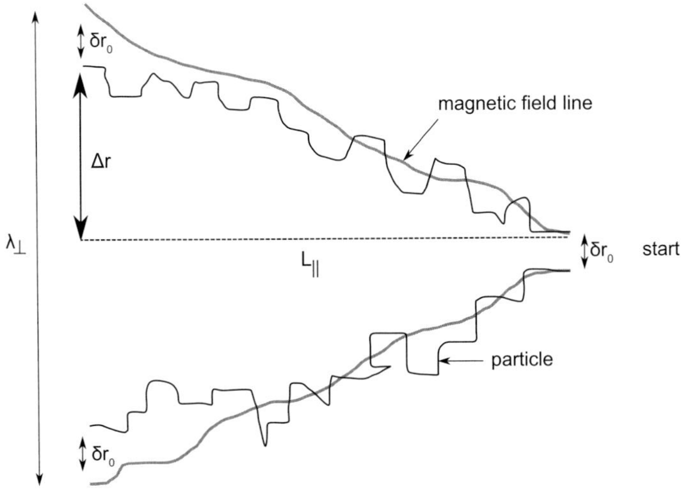

Figure 6.9 Sketch of the scenario for the Rechester–Rosenbluth diffusion coefficient.

Then, the particle has traveled a longitudinal distance $L_{\|}$ during time $\Delta\tau_{RR}$; both quantities are approximately related via Eq. (6.413). The starting width δr_0 will be determined self-consistently when noting that the repeating process (see Figure 6.9) requires

$$\delta r_0^2 \approx 2\chi_{\perp}\Delta\tau_{RR} . \tag{6.425}$$

Here, it is assumed that collisional broadening puts different particles on different filed lines. Combining the last two equations together with Eq. (6.413), we find an equation for $\Delta\tau_{RR}$

$$\lambda_{\perp} \approx \sqrt{2\chi_{\perp}}\sqrt{\Delta\tau_{RR}}\exp\left\{\frac{\sqrt{2\chi_{\|}}\sqrt{\Delta\tau_{RR}}}{L_K}\right\} \tag{6.426}$$

which, on the one hand guarantees self-consistency, but on the other hand cannot be solved explicitly for $\Delta\tau_{RR}$. We act on the assumption that $\lambda_{\perp}$, $\chi_{\perp}$, $\chi_{\|}$, and L_K are either known or can be easily determined for a specific situation.

Introducing the new variable μ via [169]

$$\sqrt{\frac{\chi_{\|}}{\Delta\tau_{RR}}} = \frac{1}{\sqrt{2}}\frac{v_{th}}{\sqrt{\mu}} , \tag{6.427}$$

we may rewrite Eq. (6.426) in the form

$$L_K \ln\left[\frac{\lambda_{\perp}}{\rho\sqrt{\mu}}\right] \approx \lambda_{coll}\sqrt{\mu} . \tag{6.428}$$

This is an equation for the variable μ. Its solution determines the Rechester–Rosenbluth diffusion coefficient in the form

$$D_{RR} \approx \frac{D_m v_{th}}{\sqrt{\mu}} . \tag{6.429}$$

Other reformulations of the Rechester–Rosenbluth formula may be useful. In its original form [106], it was written as

$$D_{RR} \approx \frac{2D_m\chi_{\|}}{L_K \ln\left(\frac{\lambda_\perp}{L_K}\sqrt{\frac{\chi_{\|}}{\chi_\perp}}\right)} \sim \frac{D_m\chi_{\|}}{L_K}\,, \tag{6.430}$$

which follows from Eq. (6.414) and the iterative solution

$$\sqrt{\Delta\tau_{RR}} \approx \frac{L_K}{\sqrt{2\chi_{\|}}} \ln\sqrt{\frac{\lambda_\perp}{\sqrt{2\chi_\perp \Delta\tau_{RR}}}} \approx \frac{L_K}{\sqrt{2\chi_{\|}}} \ln\sqrt{\frac{\lambda_\perp\sqrt{2\chi_{\|}}}{\sqrt{2\chi_\perp} L_K}}\,. \tag{6.431}$$

With Eq. (6.413), we obtain

$$L_{\|} \gg L_K\,; \tag{6.432}$$

however, for estimates, one often uses $L_{\|} \sim \mathcal{O}(L_K)$.

Let us introduce the Kubo number. In general, the Kubo number is the ratio between the distance ℓ traveled during an autocorrelation time and the correlation distance ℓ_{corr},

$$K \approx \frac{\ell}{\ell_{corr}}\,. \tag{6.433}$$

Let us apply this definition to the deviations in perpendicular direction. Propagating in parallel direction the distance $\lambda_{\|}$, for $b \ll 1$, the perpendicular excursion is $b\lambda_{\|}$, which we should compare with $\lambda_\perp$. Thus, setting $\ell \approx b\lambda_{\|}$ and $\ell_{corr} \approx \lambda_\perp$, we obtain

$$K \approx \frac{b\lambda_{\|}}{\lambda_\perp}\,. \tag{6.434}$$

This definition leads to the scaling

$$D_{RR} \sim \chi_{\|} b^2 K^2\,, \tag{6.435}$$

that is, the diffusion coefficient increases with the square of the Kubo number.

When a field line (or particle) is trapped, it sticks to an island in Poincaré section. With island-like structures in phase space, the system is not completely stochastic. Nevertheless, for motion along a trapped field line, the covered distance along the field line (unless it propagates into the (perpendicular) uncorrelated area at perpendicular distance $\lambda_\perp$) becomes very large, such that $K \gg 1$. On the other hand, when a field line is not exploring an uncorrelated perpendicular region in moving a parallel distance $\lambda_{\|}$, then $K \approx \ell/\lambda_\perp \approx b\lambda_{\|}/\lambda_\perp \ll 1$.

For the Kadomtsev–Pogutse formula, we will get the scaling

$$D_{KP} \approx \frac{b^2\lambda_{\|}}{\lambda_\perp}\sqrt{\chi_{\|}\chi_\perp} \sim bKD_{Bohm}\,, \tag{6.436}$$

where the Bohm diffusion coefficient

$$D_{Bohm} \approx \frac{1}{16}\sqrt{\chi_{\|}\chi_{\perp}} \sim \frac{T}{eB} \tag{6.437}$$

has been introduced. We recognize the linear dependence on the Kubo number. We shall come back to the role of the Kubo number after a short comparison of the two diffusion coefficients.

Introducing from Eq. (6.419) the time-step within the Kadomtsev–Pogutse model

$$\Delta\tau_{KP} \approx \frac{\lambda_{\perp}^2}{2\chi_{\perp}}, \tag{6.438}$$

we find from Eq. (6.426) for the decorrelation time $\Delta\tau_{RR}$ the implicit equation

$$\sqrt{\Delta\tau_{KP}} \approx \sqrt{\Delta\tau_{RR}}\exp\left[\frac{\sqrt{\Delta\tau_{RR}}\sqrt{2\chi_{\|}}}{L_K}\right] \approx \sqrt{\Delta\tau_{RR}}\exp\left[\frac{\sqrt{\Delta\tau_{RR}}v_{th}}{\sqrt{\nu_{coll}}L_K}\right]. \tag{6.439}$$

From here, two conclusions follow. First,

$$\Delta\tau_{RR} \leq \Delta\tau_{KP}, \tag{6.440}$$

and therefore

$$D_{RR} \geq D_{KP}, \tag{6.441}$$

that is, the Rechester–Rosenbluth coefficient is always (however, as long as it model assumptions apply) the dominating transport process. Second, we have the limit

$$D_{RR} \to D_{KP} \quad \text{for} \quad \nu_{coll} \to \infty\,; \tag{6.442}$$

that is, D_{KP} is the lower limit for D_{RR} for large collision frequencies. In Figure 6.10, we compare the Kadomtsev–Pogutse and Rechester–Rosenbluth diffusion formulas with each other. Several parameters of the stochastic system are needed for the evaluations. When not known otherwise, we take them from the standard map. That means, we identify D_m with the diffusion coefficient of the standard map Eqs. (6.213) and (6.214) for $\varepsilon = 10$. Also, the Kolmogogrov length is numerically evaluated from the standard map for $\varepsilon = 10$. The other parameters are $\lambda_{\perp} = 2\pi$ and $v_{th} = 1$. Two different Larmor radii, $\rho = 0.1$ and $\rho = 0.01$, are chosen for demonstration. For smaller Larmor radii, higher collision frequencies are needed for $D_{KP} \sim \mathcal{O}(D_{RR})$. Also, the smaller ρ is, the smaller are the (perpendicular) diffusion coefficients. The numerics confirms Eqs. (6.441) and (6.442).

The parameter values where the Kadomtsev–Pogutse regime begins can be estimated as follows. For D_{KP}, we require for the decorrelation time

$$\Delta\tau_{KP} \approx \Delta\tau_{RR} \quad \leftrightarrow \quad \frac{\lambda_{\perp}^2}{2\chi_{\perp}} \gtrsim \frac{L_K^2}{2\chi_{\|}}\,. \tag{6.443}$$

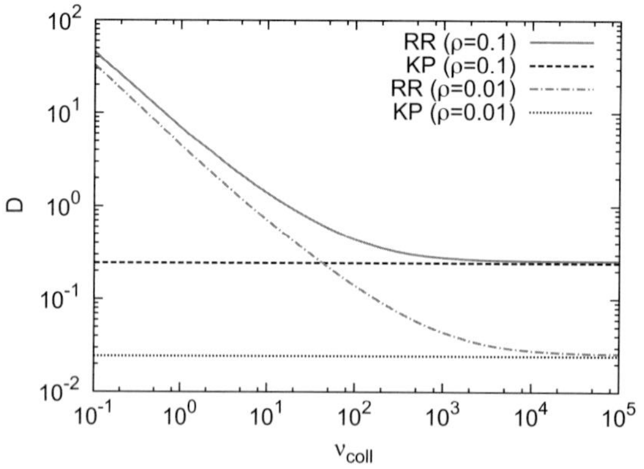

Figure 6.10 Diffusion coefficients as a function of collision frequency for stochastic parameter $\varepsilon = 10$. Two different Larmor radii, $\rho = 0.1$ and $\rho = 0.01$, are chosen for demonstration. The limit process $D_{RR} \rightarrow D_{KP}$ for $\nu_{coll} \rightarrow \infty$ is shown.

When using [123]

$$L_K \approx \sqrt{\frac{2}{\pi} \frac{\lambda_\perp^2}{4\beta^2 \lambda_\parallel}}, \tag{6.444}$$

we may cast this in the form

$$\frac{\nu_{coll}}{\Omega} \gtrsim \beta K , \tag{6.445}$$

where K is the Kubo number.

Another collisional case has been discussed in the literature [108, 184–187]. In the so called collisional fluid limit for strong collisions, one assumes that a fluid element is convected along the actual magnetic field line, resulting, from purely geometrical arguments, in

$$\Delta x \sim b \Delta z . \tag{6.446}$$

Therefore,

$$D_{\perp particle} \sim \frac{\langle (\Delta x)^2 \rangle}{\Delta \tau} \sim b^2 \frac{\langle (\Delta z)^2 \rangle}{\Delta \tau} \sim b^2 \frac{D^{classical}_{\parallel particle} \Delta \tau}{\Delta \tau} \sim b^2 \chi_\parallel . \tag{6.447}$$

Here, we have introduced the classical (collisional) diffusion coefficients

$$D^{classical}_{\perp particle} \equiv \chi_\perp = \frac{1}{2} \rho^2 \nu_{coll} , \quad D^{classical}_{\parallel particle} \equiv \chi_\parallel = \frac{1}{2} \lambda^2_{coll} \nu_{coll} , \tag{6.448}$$

with Larmor radius $\rho = v_{th}/\Omega$ and collisional mean free path $\lambda_{coll} \approx v_{th}/\nu_{coll}$. The result Eq. (6.448) also follows from the point of view that for strong collisions, particles moving along stochastic field lines will effectively be decorrelated from the field lines after $\lambda_{coll} \approx v_{th}/\nu_{coll}$,

$$D_{\perp particle} \sim \frac{\langle (\Delta x)^2 \rangle}{\Delta \tau} \sim \frac{b^2 \lambda_{coll} \lambda_{coll}}{\tau_{coll}} \sim b^2 \lambda^2_{coll} \nu_{coll} \sim b^2 D^{classical}_{\parallel particle} . \tag{6.449}$$

However, for strong collisions $\lambda_{coll} \ll \lambda_{\|}$ and negligible Larmor radii (when no significant perpendicular excursions occur except magnetic field line divergence), the (collisional) diffusive motion along the main magnetic field (z-direction) leads to the broadening

$$z^2 \approx 2\chi_{\|} t \tag{6.450}$$

during time $t \hat{=} \Delta\tau$. Neglecting collisional broadening in the perpendicular direction, we get the perpendicular movement only due to field line wandering, that is, for a (parallel) length z, the (square of the) perpendicular deviation is

$$(\Delta r)^2 \approx 2D_m z \approx 2D_m\sqrt{2\chi_{\|}}\sqrt{t}\,. \tag{6.451}$$

An estimate based on the random walk model will lead to

$$\frac{(\Delta r)^2}{t} \sim D_{RR}^{sub}(t) \sim \frac{D_m\sqrt{\chi_{\|}}}{\sqrt{t}} \to 0 \quad \text{for} \quad t \to \infty\,. \tag{6.452}$$

This is the famous subdiffusive behavior in the infinitely magnetized case ($\rho = 0$) first formulated by Rechester and Rosenbluth [106]. However, general deviations of particles from the magnetic field lines are important. The prediction of subdiffusive nature is well-grounded. However, the value 1/2 for the exponent ν in $(\Delta r)^2 \sim t^\nu$ is not free of doubt when based on the rough arguments presented above.

6.7.2
Test of the Diffusion Predictions with the Standard Map

For testing the predictions for anomalous transport, for example, the Rechester–Rosenbluth prediction D_{RR}, we apply the standard map with collisions Eqs. (6.252)–(6.254) as discussed before. We thus take

$$x_{t+1} = x_t + \frac{\varepsilon}{N}\sum_{p=0}^{N-1}\sin(y_t - 2\pi p z_t) + \rho_t\,, \tag{6.453}$$

$$y_{t+1} = y_t + \frac{1}{N}x_{t+1}\,, \tag{6.454}$$

$$z_{t+1} = z_t + v_t\,, \tag{6.455}$$

as the model for the stochastic system. The map allows for strong collisions when we choose $N \gg 1$ by subdividing the torus into N toroidal steps (with poloidal cross sections acting as Poincaré sections). The parameter $v_t = \pm 1/N$ changes sign each time when a collision occurs (the thermal velocity v_{th} is normalized to one). After a collision, the map is inverted for backward iteration. The collision frequency is $\nu_{coll} = PN$ when we allow for a collision with probability P in each cross section. In the case of a collision, a shift $\rho_t = \pm\rho$ with random sign occurs within the x, y-plane. The other parameters are as follows. The classical (collisional) parallel and

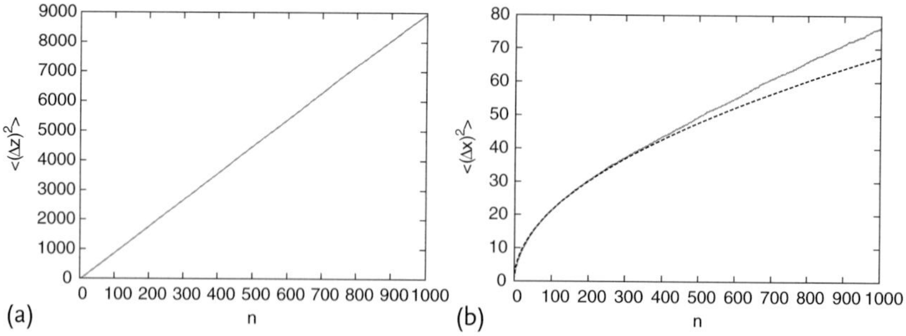

Figure 6.11 Mean square displacements versus number of iterations n for $\varepsilon = 2.4$, $P = 0.1$, $N = 1$ and $\rho = 0$ in the map Eqs. (6.252)–(6.254). (a) Parallel displacement $\langle(\Delta z)^2\rangle$; (b) perpendicular displacements $\langle(\Delta x)^2\rangle$. The broken line is a fitting curve $\langle(\Delta x)^2\rangle \sim \sqrt{n}$.

perpendicular diffusion coefficients are $\chi_\parallel = 0.5(1 - P)/PN$ and $\chi_\perp = 0.5P\rho^2 N$, respectively. For the parallel mean free path, we get $\lambda_{coll} = (1 - P)/PN$. The transverse autocorrelation length of the standard map is $\lambda_\perp = 2\pi$. The Kolmogorov length L_K can by determined numerically by standard methods [171–173].

We begin the comparison with the case $\rho \equiv 0$. In that situation, no perpendicular displacements from the magnetic field lines occur. The particles follow the field lines (when interpreting the map Eqs. (6.252)–(6.254) as the Poincaré map for field lines), moving back and force along z due to collisions. The motion along z should be diffusive. Figure 6.11a shows this behavior.

$$\langle(\Delta z)^2\rangle \approx \frac{1 - P}{P} n = 9n \quad \text{for} \quad P = 0.1 \,. \tag{6.456}$$

On the other hand, the perpendicular transport should be subdiffusive. Figure 6.11b shows this behavior for small $n < 1000$. We have fitted the curve with $\langle(\Delta x)^2\rangle \approx 2.2\sqrt{n}$ for $\varepsilon = 2.4$, $P = 0.1$, $N = 1$. Note that for these parameter values, the factor 2.2 quite well agrees with $2D_m\sqrt{2\chi_\parallel}$, as predicted by Eq. (6.451), when (numerically) $D_m \approx 0.4$ for the standard map at $\varepsilon = 2.4$ and $\chi_\parallel = 4.5$ at $P = 0.1$.

For larger n-values, the subdiffusive behavior is less clearly seen. The reasons are numerical inaccuracies. For the back-iteration after a collision, we invert the map, but quite high numerical accuracy is needed for an exact backtracking along a field line. The mapping shows visible numerical errors after approximately 1000 iterations.

The Rechester–Rosenbluth formula (6.429) was derived from random walk estimates. It would therefore not be astonishing when besides the correct qualitative prediction, the numerical values of the diffusion coefficient would be error-prone. As has been tested by Rax and White [169], surprisingly one constant factor k is sufficient for good quantitative predictions over broad parameter ranges. Following Rax and White [169], we introduce a fitting parameter $k \approx 0.5$ into the formula (6.429), thus replacing D_{RR} from Eq. (6.429) with kD_{RR}.

Figure 6.12 shows, similar to Figure 2 in [169], the Rechester–Rosenbluth coefficient kD_{RR} and the numerically determined diffusion coefficient D_m from the

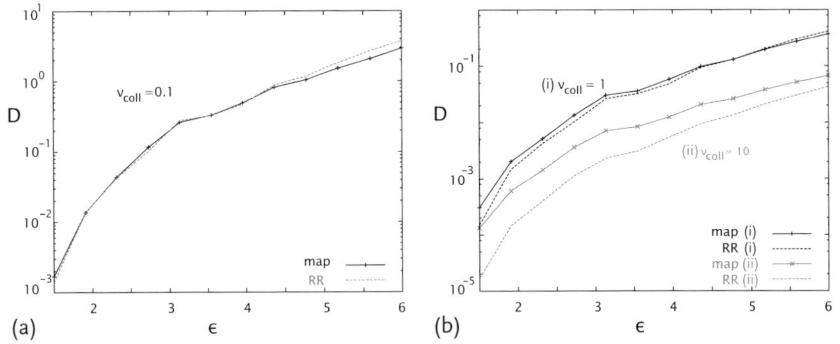

Figure 6.12 Rechester–Rosenbluth coefficients $D \mathrel{\hat{=}} k D_{RR}$ compared to numerical diffusion coefficients $D \mathrel{\hat{=}} D_m$ from the standard map for $\nu_{coll} = 0.1$ (a) and for (i) $\nu_{coll} = 1$ and (ii) $\nu_{coll} = 10$ (b) as functions of the control parameter ε in the (collisional) standard map. The Larmor radius is $\rho = 10^{-8}$. The solid lines show the numerical results and the predications of Rechester–Rosenbluth are given by the dashed lines.

standard map for $\nu_{coll} = 0.1$ as functions of the control parameter ε in the (collisional) standard map Eqs. (6.252)–(6.254) for $N = 1$. The Larmor radius is $\rho = 10^{-8}$. The solid line shows the numerical results and the predications of Rechester–Rosenbluth are given by the broken line. The agreement is formidable. Next, one may be interested in larger collision frequencies. In Figure 6.12, the Rechester–Rosenbluth coefficient $k D_{RR}$ is shown together with the numerically determined diffusion coefficients D_m of the generalized collisional standard map for $\nu_{coll} = 1$ and $\nu_{coll} = 10$, respectively, as functions of the control parameter ε. In both cases, the Larmor radius is $\rho = 10^{-8}$. The solid lines show the numerical results and the predications of Rechester–Rosenbluth are given by the broken lines. One finds a good approximation to the numerical values by Rechester–Rosenbluth (with $k = 0.5$) for $\nu_{coll} < 1$; see Figure 6.12. The reason for the discrepancy (see Figure 6.13) at large collision frequencies $\nu_{coll} > 1$ is actually quite simple, and does not contradict the statement that also for large collision frequencies the Rechester–Rosenbluth formula $k D_{RR}$ should be used for the diffusion process due to stochastic (magnetic) fields. For higher collision frequencies, the classical perpendicular diffusion coefficient becomes more important and we have to take care of it. One finds that for the whole space, the numerical results fit by a function which is the addition of the classical perpendicular diffusion coefficient and Rechester–Rosenbluth coefficient (see Figure 6.14).

Finally, we may estimate the transition to the Kadomtsev–Pogutse regime. Let us consider the case $\varepsilon = 10$, $\rho = 10^{-2} \to \Omega = 100$, and $\lambda_\perp = 2\pi$, as shown in Figure 6.13a. When using Eq. (6.443) in the form

$$\nu_{coll} \gtrsim \Omega \frac{\lambda_\perp}{L_K} , \qquad (6.457)$$

we estimate the Kolmogorov length either by Eqs. (6.226) or (6.444). Interestingly, both formulas predict the same order of magnitude $L_K \sim 0.5$, such that at $\nu_{coll} \approx$

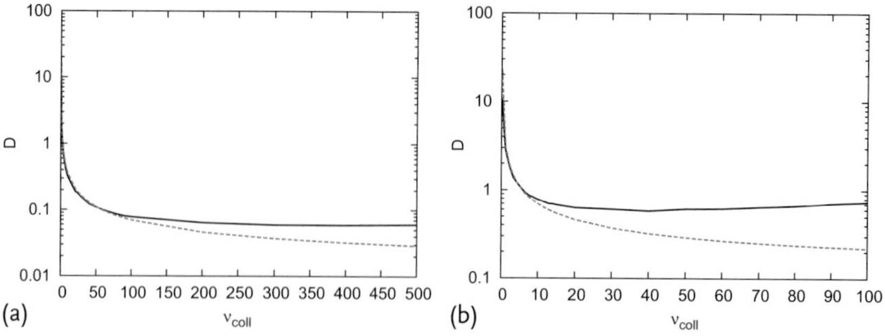

Figure 6.13 Rechester–Rosenbluth coefficients $D \hat{=} D_{RR}$ compared to numerical diffusion coefficients $D \hat{=} D_{eff}$ from the standard map as functions of ν_{coll}. The control parameter $\varepsilon = 10$ in the (collisional) standard map. The Larmor radius $\rho = 10^{-2}$ (a) and $\rho = 10^{-1}$ (b), respectively, is now much larger than in Figure 6.12. The solid lines show the numerical results and the predications of Rechester–Rosenbluth are given by the broken line.

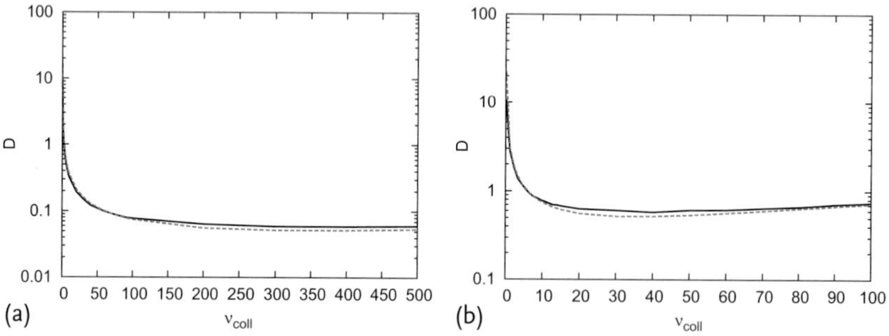

Figure 6.14 Same as in Figure 6.13, but the broken lines now show the addition of the Rechester–Rosenbluth coefficient and classical perpendicular diffusion coefficient.

1000, the diffusion coefficient is $D \approx D_{KP} \approx 0.04$, in agreement with the value shown in Figure 6.13a.

6.7.3
Trapping and Percolation ($K > 1$)

The theories mentioned so far work very well (in their respective regimes) as long as no trapping occurs. Whether trapping becomes significant or not can be measured by the so called Kubo number. The Kubo number is the ratio between the distance traveled during an autocorrelation time and the correlation distance. If the Kubo number is large, the system spirals around for some time within the highly correlated area, that is, it is being trapped. For magnetic field lines, we define the Kubo number as

$$K = \frac{b\lambda_{\parallel}}{\lambda_{\perp}} . \tag{6.458}$$

From here, we clearly recognize that for $K \ll 1$, no trapping occurs. On the other hand, the Kadomtsev–Pogutse formula $D_{KP}^{(II)} \sim b\lambda_\perp \sim K^0$ is problematic since it has been derived for $\lambda_\parallel \to \infty$ when $K \to \infty$. The actual diffusion coefficient should decrease with $K \sim \lambda_\parallel$ and not stay constant $\sim K^0$ as above. Before presenting any detailed calculation, let us give a simple argument [91].

Let us consider a model field which additively consists of a zeroth order field $B_0\hat{z}$, a random part, and a trapping component, such that the field lines follow

$$\frac{dx}{dz} = b_x^{trapping}(x,y) + b_x^{random}(z)\,, \quad \frac{dy}{dz} = b_y^{trapping}(x,y) + b_y^{random}(z)\,. \tag{6.459}$$

Next, we model the trapping contribution by

$$\delta \boldsymbol{B}^{trapping} = B_0 \boldsymbol{b}^{trapping} = \nabla \times a\hat{z}\,, \quad \text{with} \quad a \approx a_0 \exp\left\{-\frac{x^2+y^2}{2\sigma^2}\right\}. \tag{6.460}$$

The averaged motion in the x,y-plane near the center of an island at $x, y = 0$ is approximately given by

$$\frac{d^2x}{dz^2} \sim \frac{db_x^{trapping}}{dz} \sim \frac{1}{B_0}\frac{d}{dz}\frac{\partial a}{\partial y} \sim -\frac{a_0}{B_0\sigma^2}\frac{dy}{dz} \sim \frac{a_0}{B_0^2\sigma^2}\frac{\partial a}{\partial x} \sim -\frac{a_0^2}{B_0^2\sigma^4}x\,. \tag{6.461}$$

In other words, to lowest order, the x-coordinate of a field line follows from

$$\frac{d^2x}{dz^2} \approx -\kappa^2 x\,, \quad \kappa = \frac{a_0}{B_0\sigma^2}\,. \tag{6.462}$$

As expected for trapping, the field line winds around the z-direction with "frequency" κ. Looking for diffusion in radial (perpendicular) direction, we calculate with the random part of the magnetic field the mean radial displacement $\langle \Delta r^2 \rangle$. The radial ($r$) mean field displacement follows as

$$\langle \Delta r^2 \rangle \approx \iint dz' dz'' \left\langle b_r^{random}(z') b_r^{random}(z'') \right\rangle. \tag{6.463}$$

We introduce the relative distance $\Delta z = z' - z''$ and assume as a result of field line circulation

$$\left\langle b_r^{random}(z') b_r^{random}(z'') \right\rangle \approx \underbrace{\left\langle b_x^{random}(\Delta z) b_x^{random}(0) \right\rangle}_{B_{xx}(\Delta z)} \cos(\kappa \Delta z)\,. \tag{6.464}$$

Note the geometrical factor $\cos(\kappa\Delta z)$ which arises because of the trapping of the field line. Changing from z'- and z''-integration to integration over $z = z'$ and Δz, homogeneity leads to (compare Eq. (6.309))

$$\langle \Delta r^2 \rangle = \iint dz' dz'' \left\langle b_r^{random}(z') b_r^{random}(z'') \right\rangle \sim 2z\, P_{xx}(\kappa) \tag{6.465}$$

with the Fourier transform

$$P_{xx}(\kappa) = \int B_{xx}(\Delta z) \exp\{-i\kappa \Delta z\} d\Delta z \tag{6.466}$$

of the correlation function B_{xx} with $B_{xx}(\Delta z) = B_{xx}(-\Delta z)$. Assuming, for example, for the fluctuation spectrum a Kolmogorov spectrum

$$P_{xx}(\kappa) \approx \frac{C}{\left[1 + (\kappa\lambda_\parallel)^2\right]^{5/6}}, \tag{6.467}$$

we can estimate the radial diffusion coefficient in the form

$$D_r \approx D_{quasilinear} \frac{P_{xx}(\kappa)}{P_{xx}(0)} \sim \begin{cases} D_{quasilinear} & \text{for} \quad \kappa\lambda_\parallel \ll 1\,, \\ \kappa^{-5/3} \to 0 & \text{for} \quad \kappa\lambda_\parallel \gg 1\,. \end{cases} \tag{6.468}$$

Here, the quasilinear magnetic diffusion coefficient is

$$D_{quasilinear} = P_{xx}(0) = \iint d\Delta z \left\langle b_x^{random}(\Delta z) b_x^{random}(0) \right\rangle \sim b^{random2} \lambda_\parallel\,. \tag{6.469}$$

The interpretation of κ in terms of the Kubo number is simple:

$$\kappa\lambda_\parallel = \frac{a_0}{B_0 \sigma^2} \lambda_\parallel \sim \frac{a_0}{B_0 \sigma} \frac{\lambda_\parallel}{\sigma} \sim b^{trapping} \frac{\lambda_\parallel}{\lambda_\perp} \sim K. \tag{6.470}$$

From here, we conclude that the diffusion coefficient decreases with $K \to \infty$ as it should. However, the order of decrease depends on the spectrum; different forms can be used [191]. A systematic method and the principal behavior will be discussed next.

In a series of papers Vlad *et al.* [132, 134, 136, 141] developed a new and systematic procedure for the Lagrangian correlation functions called the decorrelation trajectory method. It was designed for large Kubo numbers and can be regarded as a substitute for the Corrsin approximation. The method has been used within the guiding center approximation (when the so called V-Langevin equation applies). It was first developed for test particle motion in a constant strong magnetic field B_0 in z-direction under the action of a fluctuating electrostatic potential in the absence of collisions, and later generalized to magnetic fluctuations. In the present subsection, we follow the review [137].

The statistical properties of the fluctuating electrostatic potential $\phi(\boldsymbol{r}_\perp, t)$, leading to the filed $\boldsymbol{E} = -\nabla\phi$ being perpendicular to the strong magnetic field, may be given in Euler formulation. In the following, we normalize the components x and y of $\boldsymbol{r}_\perp$ by the perpendicular correlation length $\lambda_\perp$, time t by $\lambda_\perp / V$ where V is the typical $E \times B$ velocity, and potential by the characteristic amplitude ϕ_0. Then, $V = \phi_0/(\lambda_\perp B_0)$. The equation of motion

$$\frac{d\boldsymbol{r}_\perp}{dt} = \hat{z} \times \nabla\phi \tag{6.471}$$

transforms to nondimensional form

$$\frac{dx}{dt} = -\frac{\partial \phi}{\partial y}\,, \quad \frac{dy}{dt} = \frac{\partial \phi}{\partial x}\,. \tag{6.472}$$

These equations are of canonical form with the Hamiltonian $H(x \hat{=} p, y \hat{=} q, t) \equiv \phi(x, y, t)$.

Thus far, we allow for time-dependent fluctuations (without time-dependence, we denominate the system as frozen). The Eulerian correlation function for homogeneous, isotropic, and stationary stochastic processes will be written as

$$\langle \phi(0,0,0)\phi(x,y,t)\rangle = E(x,y)T(t)\,, \tag{6.473}$$

where the functions E and T should have a maximum at the origin $x = y = t = 0$, for example, in *nonnormalized form*,

$$E(x,y) = \exp\left(-\frac{x^2+y^2}{2\lambda_\perp^2}\right), \quad T(t) = \exp\left(-\frac{t}{K}\right). \tag{6.474}$$

Here, $K = V\tau_c/\lambda_\perp$ is the Kubo number constructed with the correlation length $\lambda_\perp$ and the correlation time τ_c. For the DCT method, we first define subensembles S containing all realizations of the potential fluctuation $\phi(x, y, t)$ which have a common fixed value ϕ^0 at the origin, together with fixed values of the x- and y-derivatives. The latter determine via Eq. (6.472) the initial velocity components $v_x^0 = v_x(0,0,t)$ and $v_y^0 = v_y(0,0,t)$. We define $P_0(\phi^0, v_x^0, v_y^0)$ as the probability distribution of the initial values in the global ensemble. With these definitions, the Lagrangian velocity correlation function can be reformulated as

$$\begin{aligned} L_{xx} &= \int d\phi^0 dv_x^0 dv_y^0 P_0\left(\phi^0, v_x^0, v_y^0\right) \langle v_x(0,0,0)v_x(x(t),y(t),t)\rangle^S \\ &= \int d\phi^0 dv_x^0 dv_y^0 P_0\left(\phi^0, v_x^0, v_y^0\right) v_x^0 \langle v_x(x(t),y(t),t)\rangle^S\,. \end{aligned} \tag{6.475}$$

Since the initial values are fixed in the subensembles, the velocity autocorrelation in the subensemble reduces to a product of the initial value with the average velocity in the same subensemble S. Whereas the global average of the velocity component would vanish, its average in the subensemble is nonvanishing. For evaluating the integrals, we first assume a Gaussian probability distribution [76]

$$P_0\left(\phi^0, v_x^0, v_y^0\right) = \frac{1}{(2\pi)^{3/2}} \exp\left(-\frac{\phi^{02} + v_x^{02} + v_y^{02}}{2}\right). \tag{6.476}$$

The conditional probability $\langle v_x(x(t), y(t), t)\rangle^S$ is being calculated via

$$\begin{aligned} \langle v_x(x(t),y(t),t)\rangle^S &= \frac{\langle v_x(x,y,t)\delta\left[\phi(0,0,0)-\phi^0\right]\delta\left[\boldsymbol{v}(0,0,0)-\boldsymbol{v}^0\right]\rangle}{\langle \delta\left[\phi(0,0,0)-\phi^0\right]\delta\left[\boldsymbol{v}(0,0,0)-\boldsymbol{v}^0\right]\rangle} \\ &= \frac{1}{P_0\left(\phi^0, v_x^0, v_y^0\right)} \\ &\quad\times \int ds \int d^2q\, e^{is\phi^0 + i\boldsymbol{q}\cdot\boldsymbol{v}^0} \left\langle v_x(x,y,t)e^{-is\phi(0,0,0) - i\boldsymbol{q}\cdot\boldsymbol{v}(0,0,0)}\right\rangle\,. \end{aligned} \tag{6.477}$$

The last bracket on the right-hand side will be evaluated by using a "trick" and subsequent cumulant expansion for Gaussian processes,

$$\left\langle v_x(x,y,t) e^{-is\phi(0,0,0)-i\boldsymbol{q}\cdot\boldsymbol{v}(0,0,0)}\right\rangle = i\frac{\partial}{\partial a}\left\langle e^{-is\phi(0,0,0)-i\boldsymbol{q}\cdot\boldsymbol{v}(0,0,0)-iav_x}\right\rangle$$
$$\approx i\frac{\partial}{\partial a}\exp\left\{-\frac{1}{2}\left\langle\left[s\phi(0,0,0)+\boldsymbol{q}\cdot\boldsymbol{v}(0,0,0)+av_x\right]^2\right\rangle\right\}\Bigg|_{a=0}$$
$$=\left\{-isE_{\phi|x}-iq_xE_{x|x}-iq_yE_{y|x}\right\}Te^{-\frac{1}{2}(s^2+q^2)}\,. \tag{6.478}$$

We have defined

$$\langle\phi(0,0,0)v_x(x,y,t)\rangle = E_{\phi|x}(x,y)T\left(\frac{t}{K}\right)\,, \tag{6.479}$$

$$\langle v_x(0,0,0)\phi(x,y,t)\rangle = E_{x|\phi}(x,y)T\left(\frac{t}{K}\right)\,, \tag{6.480}$$

$$\langle v_j(0,0,0)v_x(x,y,t)\rangle = E_{j|x}(x,y)T\left(\frac{t}{K}\right)\,. \tag{6.481}$$

Their values can be calculated from the Euler correlators,

$$E_{\phi|x} = -E_{x|\phi} = -\frac{\partial}{\partial y}E(x,y)\,,\quad E_{\phi|y} = -E_{y|\phi} = \frac{\partial}{\partial x}E(x,y)\,, \tag{6.482}$$

$$E_{x|x} = -\frac{\partial^2}{\partial y^2}E(x,y)\,,\quad E_{y|y} = -\frac{\partial^2}{\partial x^2}E(x,y)\,, \tag{6.483}$$

$$E_{x|y} = E_{y|x} = \frac{\partial^2}{\partial xy}E(x,y)\,. \tag{6.484}$$

The values at the origin (index 0) are

$$E^0_{\phi|x} = E^0_{x|\phi} = E^0_{\phi|y} = E^0_{y|\phi} = E^0_{x|y} = 0\,, \tag{6.485}$$

$$E^0_{x|x} = E^0_{y|y} = 1\,,\quad T(0) = 1\,. \tag{6.486}$$

Because of these properties, the simple form Eq. (6.478) appears. Inserting Eq. (6.478) into (6.477), after integration, we obtain

$$\langle v_x(x(t),y(t),t)\rangle^S = \left[\phi^0E_{\phi|x}+v^0_xE_{x|x}+v^0_yE_{y|x}\right]T\left(\frac{t}{K}\right)$$
$$\equiv v^S_x(x,y)T\left(\frac{t}{K}\right)\,. \tag{6.487}$$

This averaged velocity in the subensemble can be derived from the averaged potential. The latter is obtained by a similar procedure as above and may be written in the form

$$\langle\phi(x(t),y(t),t)\rangle^S \equiv \phi^S(x,y)T\left(\frac{t}{K}\right)\,, \tag{6.488}$$

with

$$\phi^S(x, y) = \left[\phi^0 E_{x|x} + v_x^0 E_{x|\phi} + v_y^0 E_{y|\phi}\right]. \tag{6.489}$$

We have

$$v_x^S = -\frac{\partial \phi^S(x, y)}{\partial y}, \quad v_y^S = \frac{\partial \phi^S(x, y)}{\partial x}. \tag{6.490}$$

Balescu *et al.* defined the so called decorrelation trajectory (DCT) as the deterministic trajectory $x^S(t), y^S(t)$ solving the Hamiltonian equations

$$v_x^S(x^S, y^S) T\left(\frac{t}{K}\right) \equiv \frac{dx^S(t)}{dt} = -\frac{\partial \langle \phi\left(x^S, y^S, t\right)\rangle^S}{\partial y^S}, \tag{6.491}$$

$$v_y^S\left(x^S, y^S\right) T\left(\frac{t}{K}\right) \equiv \frac{dy^S(t)}{dt} = \frac{\partial \langle \phi\left(x^S, y^S, t\right)\rangle^S}{\partial x^S}. \tag{6.492}$$

The solution of the nonlinear DCT equations of motion can be obtained numerically. The DCT is not the trajectory of a physical particle. Because of

$$\langle v_x(0,0,0) v_x\left[x(t), y(t), t\right]\rangle^S = v_x^0 \langle v_x\left[x(t), y(t), t\right]\rangle^S, \tag{6.493}$$

it can be understood as an indicator of correlations. When $\boldsymbol{v}^S = 0$, the fluctuating field is no longer correlated to its initial value.

Once the DCT orbits have been determined, the Lagrangian velocity autocorrelation will be calculated via

$$\begin{aligned} L_{xx}^{\mathrm{DCT}} &\approx \int d\phi^0 dv_x^0 dv_y^0 P_0\left(\phi^0, v_x^0, v_y^0\right) v_x^0 \langle v_x\left(x^S(t), y^S(t), t\right)\rangle^S \\ &= \int d\phi^0 dv_x^0 dv_y^0 P_0\left(\phi^0, v_x^0, v_y^0\right) v_x^0 v_x^S\left(x^S(t), y^S(t), t\right) T\left(\frac{t}{K}\right). \end{aligned} \tag{6.494}$$

The substitution

$$v_x\left[x(t), y(t), t\right] \to \langle v_x\left[x^S(t), y^S(t), t\right]\rangle^S \tag{6.495}$$

is the main assumption in the DCT method. Effectively, the Lagrangian average is replaced by an Eulerian average, being evaluated along the deterministic DCT. The diffusion coefficient is obtained from the time-integral after postulating an expression for the spatial part of the Euler correlation. The operation requires a rather nontrivial numerical code. Nevertheless, the procedure has been successfully applied by several authors [132, 134, 136, 141, 192].

We conclude this section by a few remarks on how to apply the method to transport in stochastic magnetic fields. The perturbation field $\boldsymbol{b}$ is assumed to only have a two-dimensional structure. The stochastic field $\boldsymbol{b}$ is generated by the scalar magnetic potential $\phi(\boldsymbol{x}, z)$,

$$\boldsymbol{b}(\boldsymbol{x}, z) = \nabla\phi(\boldsymbol{x}, z) \times \boldsymbol{e}_z, \tag{6.496}$$

which we will call the flux function. The vector $\boldsymbol{x} = (x, y)$ refers to the perpendicular coordinates, whereas z can be regarded as the parallel coordinate. Magnetic field lines follow from the flux function by the relation $d\boldsymbol{x}/dz = \boldsymbol{b}$. Thus, for magnetic field line motion, the situation is formally similar to the previous electrostatic case.

The situation becomes more complicated for particle motion in stochastic magnetic fields. Let us elucidate this on the example of the V-Langevin equations (6.400)–(6.402) without collisions. The Lagrangian correlation for $v_y \approx \eta_y$ may then be written as

$$\begin{aligned} L &= \langle v_y(x(t), y(t), z(t), t) v_y(0,0,0,0)\rangle \\ &= \langle v_z(t) b_y(x(t), y(t), z(t), t) v_z(0) b_y(0,0,0,0)\rangle \\ &\approx \langle \eta_z(t) b_y(x(t), y(t), z(t), t) \eta_z(0) b_y(0,0,0,0)\rangle \,, \end{aligned} \tag{6.497}$$

when η_i is the ith velocity component in the absence of magnetic field fluctuations. Within the DCT formulation, we start from

$$L^{\mathrm{DCT}} = \int_{-\infty}^{\infty} P\left(\eta_z^0\right) P\left(\boldsymbol{b}^0\right) P\left(\phi^0\right) \eta^0 b_y^0 \left\langle \eta_z(t) b_y\left[\boldsymbol{x}(t), z(t), t\right]\right\rangle^S d\eta_z^0 d^2 b^0 d\phi^0 \,, \tag{6.498}$$

where the notation is analogous to Eq. (6.475). Again, making a Gaussian ansatz for the probability distributions, we have to evaluate $\langle \eta_z(t) b_y[\boldsymbol{x}(t), z(t), t]\rangle^S$ by similar methods as discussed above for electrostatic fluctuations. For more details, the reader is referred to the actual literature, for example, [143].

As mentioned, quasilinear diffusion can only be expected in the fully stochastic region. In an (open) chaotic system, (laminar) regions with small connection lengths of magnetic field lines exist. There, the motion is nondiffusive. Experimental hints exist for nondiffusive areas. For example, characteristic spatial heat flux patterns (see the cover picture) have been observed in bounded stochastic plasmas [193, 194]. They show a characteristic number of stripes on the target plates

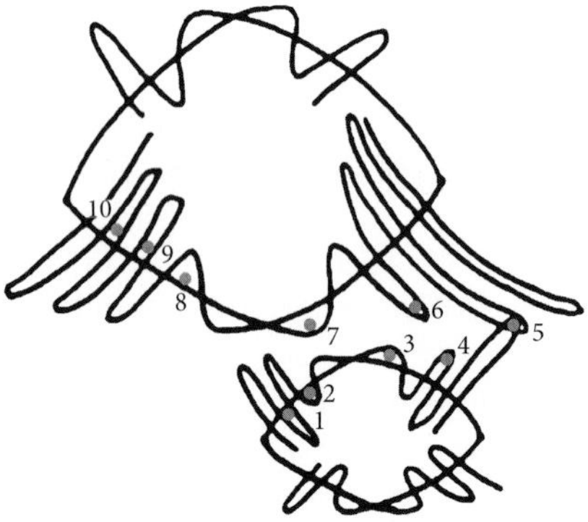

Figure 6.15 Sketch of the stable and unstable manifolds of periodic hyperbolic fixed points and their relation to field line wandering 1, 2, 3, ... in a Poincaré plot.

when plotted in the (φ, θ)-plane, where φ is the toroidal angle and θ is the poloidal angle of the torus. In addition, when varying the edge safety factor q_a, and projecting on the (q_a, θ)-plane, generic butterfly-like structures appear. In order to analyze and classify the spatial structures of heat flux patterns [194], one can use the concept of magnetic footprints [195, 196] together with an analysis of the stable and unstable manifolds of hyperbolic periodic points [170, 179, 197, 198] of magnetic resonances present in the stochastic plasma edge region. Field line wandering occurs along the stable and unstable manifolds of hyperbolic fixed points, as indicated in Figure 6.15. This is a very efficient process near the plasma boundary and has been analyzed in several recent publications on that specific subject.

6.8 Stochastic Theory of the Parallel Test Particle Diffusion Coefficient

The motion of particles along a zeroth order magnetic field will also be influenced by perpendicular magnetic field fluctuations. The estimate of the longitudinal diffusion coefficient (and the corresponding parallel mean free path) is the topic of the following presentation.

6.8.1 Fundamental Relations for the Parallel Diffusion Coefficient

We now calculate the particle diffusion in the direction of a strong external magnetic field. That physical process will be abbreviated by "parallel diffusion." In principle, the same methods as in the previous section can be applied to that problem. However, it turns out that more elegant procedures are available for parallel diffusion. They will be used in the following.

We start from the one-particle distribution function $f = f(z, \mu, t)$ where μ is the cosine of the pitch angle and z is the space coordinate along the external magnetic field. Isotropy is assumed in the plane perpendicular to the external magnetic field. Because of the symmetry of the problem, the distribution function f will only depend on the coordinate z, the time t, the speed v, and the pitch angle μ. The variable v may be hidden since it does not change during interaction with magnetic fluctuations. Therefore, we formally write $f = f(z, \mu, t)$. The kinetic equation for f is a two-dimensional Fokker–Planck equation (see below) which can be derived from the relativistic Vlasov equation [92] by neglecting momentum diffusion due to the assumption of purely magnetic fluctuations. From f, we obtain the pitch-angle averaged particle density

$$M(z, t) = \frac{1}{2} \int_{-1}^{+1} d\mu \, f \tag{6.499}$$

with the normalization

$$\int_{-\infty}^{+\infty} dz\, M(z,t) = 1\,. \tag{6.500}$$

Then, a particle current density may be defined as

$$j_{\|} = \frac{1}{2}\int_{-1}^{+1} d\mu\, v\mu\, f\,, \tag{6.501}$$

where $\|$ denotes the direction of the external magnetic field $\boldsymbol{B} \approx B_0\hat{z}$. Here, $\boldsymbol{v}$ denotes the velocity of the particle, the parallel velocity component is $v_{\|}$. Obviously, we have for the parallel velocity component

$$v_{\|} = v\mu\,, \quad \mu = \cos(\vartheta)\,, \tag{6.502}$$

where $\vartheta = \sphericalangle(\boldsymbol{v}, \boldsymbol{B})$. When only magnetic fields act on the particle, the speed v will be unchanged, although the direction of $\boldsymbol{v}$ varies.

The total magnetic field consists of the large background field $\boldsymbol{B}_0$ plus fluctuations,

$$\boldsymbol{B} = B_0\hat{z} + \delta\boldsymbol{B} \mathrel{\hat{=}} B_0(b_0\hat{z} + \boldsymbol{b})\,, \quad b_0 \equiv 1\,. \tag{6.503}$$

A simple indexing in the form b_ν, $\nu = 0, 1, 2, 3$ can be used with $b_x = b_1$, $b_y = b_2$, $b_z = b_3$. For strong background fields, we may neglect $b_3 \ll b_0$. In principle, we may use $b_0 = 1$; however, sometimes we keep b_0 to make some formulas more transparent. The dimension of the magnetic field is always included in B_0, and we should keep in mind $\delta\boldsymbol{B} = B_0\boldsymbol{b}$. The nonrelativistic equations of motion are

$$\dot{v}_x = \Omega\, v_y + \Omega\left(v_y\frac{\delta B_z}{B_0} - v_z\frac{\delta B_y}{B_0}\right), \tag{6.504}$$

$$\dot{v}_y = -\Omega\, v_x + \Omega\left(v_z\frac{\delta B_x}{B_0} - v_x\frac{\delta B_z}{B_0}\right), \tag{6.505}$$

$$\dot{v}_z = \Omega\left(v_x\frac{\delta B_y}{B_0} - v_y\frac{\delta B_x}{B_0}\right), \tag{6.506}$$

where the gyrofrequency $\Omega = qB_0/m$ has been introduced. q is the charge and m is the (rest) mass of the particle. Equation (6.506) can be written in the form

$$\dot{\mu} = \frac{1}{\rho_L}\left(v_x\frac{\delta B_y}{B_0} - v_y\frac{\delta B_x}{B_0}\right) \tag{6.507}$$

when the Larmor radius $\rho_L = v/\Omega$ is used.

Next, we shall calculate the pitch angle diffusion coefficient

$$D_{\mu\mu} = \lim_{t\to\infty}\int_0^t dt'\,\langle\dot{\mu}(t')\dot{\mu}(0)\rangle\,. \tag{6.508}$$

Within a Fokker–Planck description, the change of the distribution function $f(z, \mu, t)$ follows from

$$\frac{\partial f}{\partial t} + v\mu\frac{\partial f}{\partial z} = \frac{\partial}{\partial \mu}\left(D_{\mu\mu}\frac{\partial f}{\partial \mu}\right). \tag{6.509}$$

We shall come back to the Fokker–Planck description later. The following derivation will show that (quasilinear) Fokker–Planck scattering in velocity space will lead to the diffusion approximation in configuration space. We have already mentioned that the magnetic fluctuations do not change the magnitude of the particle velocity. However, they may lead to a total isotropization on the long times

$$t \to \infty: \quad f \to M(z, t) = \frac{1}{2}\int_{-1}^{+1} d\mu\, f\,. \tag{6.510}$$

From the Fokker–Planck equation (6.509), we obtain

$$\frac{\partial M}{\partial t} + \frac{\partial j_{\|}}{\partial z} = 0\,. \tag{6.511}$$

We may also use Eq. (6.509) to express

$$\frac{\partial f}{\partial \mu} = \frac{1}{D_{\mu\mu}}\int_{-1}^{\mu} d\nu\left[\frac{\partial f}{\partial t} + v\nu\frac{\partial f}{\partial z}\right], \tag{6.512}$$

which can be incorporated into the equation for the current density

$$j_{\|} \equiv -\frac{v}{4}\int_{-1}^{+1} d\mu\frac{\partial(1-\mu^2)}{\partial \mu} f = \frac{v}{4}\int_{-1}^{+1} d\mu(1-\mu^2)\frac{\partial f}{\partial \mu}\,. \tag{6.513}$$

Note $D_{\mu\mu}(\mu = \pm 1) = 0$ because of Eq. (6.508) and $\dot{\mu} \sim v_x$ (or v_y) $\sim v\sqrt{1-\mu^2}$. Inserting Eq. (6.512) into Eq. (6.513), we obtain

$$j_{\|} = \frac{v}{4}\int_{-1}^{+1} d\mu\frac{1-\mu^2}{D_{\mu\mu}}\int_{-1}^{\mu} d\nu\frac{\partial f}{\partial t} + \frac{v^2}{4}\int_{-1}^{+1} d\mu\frac{1-\mu^2}{D_{\mu\mu}}\int_{-1}^{\mu} d\nu\nu\frac{\partial f}{\partial z}\,. \tag{6.514}$$

For large times t, we assume $f \to M$, and therefore by replacing f by M, we get asymptotically

$$\begin{aligned} j_{\|} &\approx \frac{v}{4}\frac{\partial M}{\partial t}\int_{-1}^{+1} d\mu\frac{(1-\mu^2)(1+\mu)}{D_{\mu\mu}} - \frac{v^2}{8}\frac{\partial M}{\partial z}\int_{-1}^{+1} d\mu\frac{(1-\mu^2)^2}{D_{\mu\mu}} \\ &\equiv \chi_{zt}\frac{\partial M}{\partial t} - \chi_{zz}\frac{\partial M}{\partial z}\,. \end{aligned} \tag{6.515}$$

We have defined two coefficients χ_{zt} and χ_{zz}. Comparing the two terms on the right-hand side of Eq. (6.515), we can argue that the second one dominates. For justification, we reformulate Eq. (6.515) as

$$j_{\|} \approx \chi_{zt}\frac{\partial M}{\partial t} - \chi_{zz}\frac{\partial M}{\partial z} = -\frac{\partial}{\partial z}\left[\chi_{zt} j_{\|} + \chi_{zz} M\right] . \tag{6.516}$$

Because of the tendency to isotropy ($f \to M$) also for large times,

$$t \to \infty: \quad j_{\|} = \frac{v}{2}\int_{-1}^{+1} d\mu \mu f \to \frac{Mv}{2}\int_{-1}^{+1} \mu d\mu = 0 . \tag{6.517}$$

The smaller $j_{\|}$ is, the less important the first term in Eq. (6.515) becomes. Thus, we may approximate (after waiting for reasonable long times)

$$j(z,t) \approx -\chi_{zz}\frac{\partial M(z,t)}{\partial z} . \tag{6.518}$$

We arrived at the famous diffusion equation. The diffusion coefficient is

$$\chi_{zz} = \frac{v^2}{8}\int_{-1}^{+1} \frac{(1-\mu^2)^2}{D_{\mu\mu}} d\mu . \tag{6.519}$$

That is an important relation between the diffusion coefficient in parallel direction and the pitch angle diffusion coefficient $D_{\mu\mu}$. Inserting Eq. (6.507) into the definition of $D_{\mu\mu}$, we may compose the total result into the following contributions

$$\begin{aligned} D_{\mu\mu} &= \int_0^\infty dt \langle \dot\mu(t)\dot\mu(0)\rangle \\ &= \frac{1}{\rho_L^2}\int_0^\infty dt \left\{\langle v_x(t)v_x(0)b_y(t)b_y(0)\rangle + \langle v_y(t)v_y(0)b_x(t)b_x(0)\rangle \right. \\ &\quad \left. - \langle v_x(t)v_y(0)b_y(t)b_x(0)\rangle - \langle v_y(t)v_x(0)b_x(t)b_y(0)\rangle\right\} \\ &\equiv D_{\mu\mu}^{(I)} + D_{\mu\mu}^{(II)} + D_{\mu\mu}^{(III)} + D_{\mu\mu}^{(IV)} . \end{aligned} \tag{6.520}$$

The four contributions on the right-hand side will be calculated separately in the following.

By evaluating the parallel diffusion coefficient χ_{zz}, we shall have direct access to the parallel mean free path $\lambda_{\|}^{mfp}$. The relation

$$\chi_{zz} = \frac{1}{3}\langle v\rangle \lambda_{\|}^{mfp} \tag{6.521}$$

holds and will be justified in the following.

Diffusion is the physical process of particle spreading from a region of higher concentration to a region of lower concentration. The average distance a particle travels between collisions is called the mean free path. The relation

$$\frac{\partial M}{\partial t} = \chi_{zz} \frac{\partial^2 M}{\partial z^2} \tag{6.522}$$

is known as Fick's second law of diffusion. The diffusion coefficient χ_{zz} has the unit $m^2 s^{-1}$ and provides a measure of the number of particles moving through a particular cross section per unit of time.

The diffusion coefficient is related to the mean free path via the Einstein–Smoluchowski equation

$$\chi_{zz} \sim \frac{\lambda_{\parallel}^{mfp2}}{\tau} , \tag{6.523}$$

where τ is the average time between collisions. Assuming the latter can be calculated from the average speed $\langle v \rangle$ and the mean free path $\lambda_{\parallel}^{mfp}$, we note

$$\tau \sim \frac{\lambda_{\parallel}^{mfp}}{\langle v \rangle} \tag{6.524}$$

and obtain

$$\chi_{zz} \sim \lambda_{\parallel}^{mfp} \langle v \rangle . \tag{6.525}$$

To accurately determine the constant of proportionality in Eq. (6.524), we shall discuss the physical statistics behind the relations just mentioned in more detail. We consider self-diffusion in a system of identical particles (gas) assuming that the collisions occur randomly in gas. The probability that a particle can travel a distance z without collisions can be assumed in the form

$$P_0(z) \approx \frac{1}{\lambda} e^{-\frac{z}{\lambda}} , \tag{6.526}$$

with $\lambda \equiv \lambda_{\parallel}^{mfp}$. From here, we calculate the mean distance a particle travels in the mean without any collision,

$$\langle z \rangle = \int_0^\infty z P_0(z) dz = \lambda , \tag{6.527}$$

which is called the mean free path. The distribution Eq. (6.526) follows from the independency of collisions along the following argument. The average number of collisions per unit length is $1/\lambda$, and the probability that a collision occurs in an interval dz is dz/λ. Then, the probability that no collision occurs in the interval $z + dz$ is

$$P_0(z + dz) \approx P_0(z) + dz \frac{dP_0}{dz} \approx \underbrace{P_0(z)}_{\text{no collision in } z} \underbrace{\left(1 - \frac{dz}{\lambda}\right)}_{\text{no collision in } dz} , \tag{6.528}$$

leading to the differential equation

$$\frac{dP_0}{dz} \approx -\frac{1}{\lambda} P_0 \,, \tag{6.529}$$

which has the (normalized) solution Eq. (6.526).

Next, we introduce the average speed $\langle v \rangle$ and the average relative speed $\langle v_r \rangle$ for a gas of identical particles. The two speeds are defined as

$$\langle v \rangle = \int d^3 v \, f_M(\boldsymbol{v}) |\boldsymbol{v} - \boldsymbol{V}| \,, \tag{6.530}$$

$$\langle v_r \rangle = \int d^3 v_A d^3 v_B \, f_M(\boldsymbol{v}_A) f_M(\boldsymbol{v}_B) |\boldsymbol{v}_A - \boldsymbol{v}_B| \,, \tag{6.531}$$

where $\boldsymbol{V}$ is the average directed velocity and f_M is, for example, a Maxwellian

$$f_M(\boldsymbol{v}) = \left(\frac{m}{2\pi k_B T}\right)^{3/2} \exp\left(-\frac{m(\boldsymbol{v} - \boldsymbol{V})^2}{2k_B T}\right) \tag{6.532}$$

with temperature T. The integrations lead to

$$\langle v_r \rangle = \sqrt{\frac{16 k_B T}{\pi m}} = \sqrt{2} \langle v \rangle . \tag{6.533}$$

In the center of mass system, the reduced mass $m_r = m/2$ appears for identical particles. Therefore, when calculating $\langle v_r \rangle$, we may use the formula for $\langle v \rangle$ by replacing $m \to m_r$, leading to the factor $\sqrt{2}$ in the above formula. Assuming a collision cross section σ for binary collisions, the collision frequency is

$$\nu_{\text{coll}} \equiv \tau_{\text{coll}}^{-1} \equiv n\sigma \langle v_r \rangle , \tag{6.534}$$

leading to the mean free path

$$\lambda = \langle v \rangle \tau_{\text{coll}} = \frac{1}{\sqrt{2} n\sigma} \,. \tag{6.535}$$

For the relation to the self-diffusion coefficient χ_{zz}, we identify some of the (otherwise identical) particles as "tracer particles." Self-diffusion describes the transport of "tracer particles" through a gas of otherwise identical particles. The rate at which spatial inhomogeneities of "tracer particles" become smoothed out is determined by the self-diffusion coefficient χ. Let us assume that the density of "tracer particles" $n^T(z)$ varies in z-direction while the total particle density n is held constant. We determine the net flux through an imaginary wall at $z = 0$. Let dS be a segment of the wall. The origin of the spatial coordinates should be at the center of dS. Consider a volume element dV containing "tracer particles" hitting the wall from above. It should be centered at r, ϑ, φ in polar coordinates. The average number of "tracer particles" undergoing collisions in dV per unit time is

$$\frac{n^T(z) dV}{\tau_{\text{coll}}} = \frac{\langle v \rangle n^T}{\lambda} dV \,. \tag{6.536}$$

After collisions, the "tracer particles" leave the volume element in random directions. The fraction moving to dS is determined by the space angle under which dS is seen from dV, that is,

$$\frac{d\Omega}{4\pi} = \frac{dS|\cos\vartheta|}{4\pi r^2} . \tag{6.537}$$

The probability to reach dS before any new collision is $P_0(r)$. Thus, the number of "tracer particles" which collide in dV and reach dS per unit time is

$$dn^T = \frac{\langle v\rangle n^T dV}{\lambda} \frac{dS|\cos\vartheta|}{4\pi r^2} e^{-\frac{r}{\lambda}} . \tag{6.538}$$

The total number of "tracer particles" hitting a unit area of wall per unit time from above is

$$\dot{N}_+ = \frac{\langle v\rangle}{4\pi\lambda} \int_0^\infty dr r^2 \int_0^{\frac{\pi}{2}} d\vartheta \sin\vartheta \int_0^{2\pi} d\varphi\, n^T(z) \cos\vartheta \frac{e^{-\frac{r}{\lambda}}}{r^2} . \tag{6.539}$$

Note that for constant n^T, we would obtain the known result

$$\dot{N}_+ = \frac{n\langle v\rangle}{4} \quad \text{for} \quad n^T = n = \text{const} . \tag{6.540}$$

Considering particles moving to dS from below, we get a similar expression, except that we have to integrate over ϑ from $\pi/2$ to π, and $|\cos\vartheta| = -\cos\vartheta$, leading to

$$\dot{N}_+ = -\frac{\langle v\rangle}{4\pi\lambda} \int_0^\infty dr r^2 \int_{\frac{\pi}{2}}^{\pi} d\vartheta \sin\vartheta \int_0^{2\pi} d\varphi\, n^T(z) \cos\vartheta \frac{e^{-\frac{r}{\lambda}}}{r^2} . \tag{6.541}$$

Therefore, the net rate is

$$\dot{N}_+ - \dot{N}_- = \frac{\langle v\rangle}{4\pi\lambda} \int_0^\infty dr r^2 \int_0^{\pi} d\vartheta \sin\vartheta \int_0^{2\pi} d\varphi\, n^T(z) \cos\vartheta \frac{e^{-\frac{r}{\lambda}}}{r^2} . \tag{6.542}$$

When evaluating the right-hand side, we Taylor-expand up to the second order,

$$n^T(z) \approx n^T(0) + z\frac{\partial n^T}{\partial z} + \frac{z^2}{2}\frac{\partial^2 n^T}{\partial z^2} . \tag{6.543}$$

Only the second term of the Taylor expansion Eq. (6.543) contributes, leading to

$$\dot{N}_+ - \dot{N}_- \approx \frac{\langle v\rangle\lambda}{3}\frac{\partial n^T}{\partial z} . \tag{6.544}$$

If the density increases in the z-direction, $\dot{N}_+ - \dot{N}_- > 0$ and more particles will propagate in the negative z-direction (from above). We therefore can use the particle current density

$$\Gamma(z) \approx -\left(\dot{N}_+ - \dot{N}_-\right) \equiv -\chi_{zz}\frac{\partial n^T(z)}{\partial z} , \tag{6.545}$$

and we have proven Eq. (6.521). The continuity equation

$$\frac{\partial n^T}{\partial t} + \frac{\partial \Gamma}{\partial z} = 0 \tag{6.546}$$

leads to

$$\frac{\partial n^T}{\partial t} = \chi_{zz} \frac{\partial^2 n^T}{\partial z^2} = 0 \,, \tag{6.547}$$

that is, Eq. (6.522).

6.8.2
Pitch Angle Diffusion

Now, we are ready to calculate the pitch angle diffusion coefficient. According to Eq. (6.520), we have four contributions; we start with the first one. We evaluate the latter under the following assumption: Only the lowest order corrections due to magnetic field fluctuations are taken into account. That means we approximate

$$D_{\mu\mu}^{(I)} \equiv \frac{1}{\rho_L^2} \int_0^\infty dt \left\langle v_x(t) v_x(0) b_y(t) b_y(0) \right\rangle \approx \frac{1}{\rho_L^2} \int_0^\infty dt \left\langle \eta_x(t) \eta_x(0) b_y(t) b_y(0) \right\rangle . \tag{6.548}$$

Since the integrand already explicitly contains the magnetic field fluctuations up to a second order, for the velocity correlations, only collisions are taken into account. The velocity component η_x has to be calculated at time t following the path of the particle under the action of the random collisional force $\boldsymbol{a}$. As mentioned earlier, the force has a perpendicular ($\perp$) and a parallel ($\|$) component. In addition, the total correlation expresses an average over magnetic field (b) fluctuations. Thus, we may speak of a threefold stochastic process, indicating it by the notation

$$\left\langle \eta_x(t) \eta_x(0) b_y(t) b_y(0) \right\rangle \equiv \left\langle \left\langle \left\langle \eta_x(t) \eta_x(0) b_y(t) b_y(0) \right\rangle_\perp \right\rangle_\| \right\rangle_b . \tag{6.549}$$

Part of the total averaging may be performed separately. For example, the averaging over the magnetic field fluctuations does not affect the lowest order velocity component η_x. Denoting the path of the particle by $\boldsymbol{R}(t)$, which in the present approximation is only influenced by collisions, we may first consider

$$\langle b_i(t) b_j(0) \rangle_b = \left\langle \int_\infty^\infty E_{ij}(\boldsymbol{r}) \delta\left[\boldsymbol{r} - \boldsymbol{R}(t)\right] d^3 r \right\rangle_b . \tag{6.550}$$

We evaluate the latter within the Corrsin approximation

$$\left\langle \int_\infty^\infty E_{ij}(\boldsymbol{r}) \delta\left[\boldsymbol{r} - \boldsymbol{R}(t)\right] d^3 r \right\rangle_b \approx \int_\infty^\infty E_{ij}(\boldsymbol{r}) \left\langle \delta\left[\boldsymbol{r} - \boldsymbol{R}(t)\right] \right\rangle_b d^3 r . \tag{6.551}$$

The averaged propagator $\langle\delta[\boldsymbol{r}-\boldsymbol{R}(t)]\rangle_b$ will be reformulated by making use of the Fourier representation of the Delta function

$$\delta\left[\boldsymbol{r}-\boldsymbol{R}(t)\right]=\frac{1}{(2\pi)^3}\int e^{-ik\cdot[r-R(t)]}d^3k\,. \tag{6.552}$$

Introducing the Fourier transform of $E_{ij}(r)$, we obtain

$$\langle b_i(t)b_j(0)\rangle_b=\int E_{ij}(\boldsymbol{k})\left\langle\exp\left[i\boldsymbol{k}\cdot\boldsymbol{R}(t)\right]\right\rangle_b d^3k\,. \tag{6.553}$$

To lowest order, the trajectory is

$$\boldsymbol{R}(t)\approx\int_0^t\eta(t')dt'\,. \tag{6.554}$$

Making use of that approximate representation, we find

$$D_{\mu\mu}^{(I)}\approx\frac{1}{\rho_L^2}\int E_{yy}(\boldsymbol{k})\left\langle\left\langle\eta_y(t)\eta_y(0)\mathcal{E}\right\rangle_\perp\right\rangle_\parallel d^3k\,dt\,. \tag{6.555}$$

Similarly, we obtain

$$D_{\mu\mu}^{(II)}\approx\frac{1}{\rho_L^2}\int E_{xx}(\boldsymbol{k})\left\langle\left\langle\eta_y(t)\eta_y(0)\mathcal{E}\right\rangle_\perp\right\rangle_\parallel d^3k\,dt\,, \tag{6.556}$$

$$D_{\mu\mu}^{(III)}\approx-\frac{1}{\rho_L^2}\int E_{yx}(\boldsymbol{k})\left\langle\left\langle\eta_x(t)\eta_y(0)\mathcal{E}\right\rangle_\perp\right\rangle_\parallel d^3k\,dt\,, \tag{6.557}$$

$$D_{\mu\mu}^{(IV)}\approx-\frac{1}{\rho_L^2}\int E_{xy}(\boldsymbol{k})\langle\langle\eta_y(t)\eta_x(0)\ \mathcal{E}\rangle_\perp\rangle_\parallel d^3k\,dt\,, \tag{6.558}$$

where we have defined

$$\mathcal{E}\equiv\exp\left[ik_x\int_0^t\eta_x(t')dt'+ik_y\int_0^t\eta_y(t')dt'+ik_z\int_0^t\eta_z(t')dt'\right]. \tag{6.559}$$

Before discussing more general cases, let us first concentrate on the much simpler collisionless case for $\nu=0$. In that case, the velocities in the external magnetic field are straightforward to obtain in the form

$$\eta_x=v_\perp\cos(\phi_0-\Omega t)=\frac{v_\perp}{2}\left[e^{i(\phi_0-\Omega t)}+e^{-i(\phi_0-\Omega t)}\right], \tag{6.560}$$

$$\eta_y=v_\perp\sin(\phi_0-\Omega t)=\frac{v_\perp}{2i}\left[e^{i(\phi_0-\Omega t)}-e^{-i(\phi_0-\Omega t)}\right], \tag{6.561}$$

$$\eta_z=v_\parallel\,. \tag{6.562}$$

Here, ϕ_0 is an arbitrary phase which can be applied to perform an ensemble average. It is easy to see that by the averaging

$$\langle\cdots\rangle=\frac{1}{2\pi}\int_0^{2\pi}d\phi_0\cdots\,, \tag{6.563}$$

the correlation functions (6.315)–(6.317) for $\nu = 0$ can be recovered. In the following, we shall use $v_\perp = v\sqrt{1-\mu^2}$ and $v_\parallel = v\mu$ since in the absence of collisions, the velocity v is unchanged. The factor $\mathcal{E}$ as defined in Eq. (6.559) will be evaluated with

$$\int_0^t \eta_x(t')dt' = x_0 - \frac{v_\perp}{\Omega}\{\sin(\phi_0 - \Omega t) - \sin\phi_0\}, \tag{6.564}$$

$$\int_0^t \eta_y(t')dt' = y_0 + \frac{v_\perp}{\Omega}\{\cos(\phi_0 - \Omega t) - \cos\phi_0\}, \tag{6.565}$$

$$\int_0^t \eta_z(t')dt' = z_0 + v_\parallel t, \tag{6.566}$$

such that

$$\mathcal{E} = \exp\left\{ i\frac{k_\perp v_\perp}{\Omega}\left[\sin(\psi + \omega t - \phi_0) - \sin(\psi - \phi_0)\right] + ik_\parallel v_\parallel t\right\}. \tag{6.567}$$

Here, we have introduced the azimuthal angle ψ in k-space according to $k_x = k_\perp\cos\psi$, $k_y = k_\perp\sin\psi$, $k_z \equiv k_\parallel$. The initial position is assumed as $\mathbf{r}_0 = (x_0, y_0, z_0)$. Next, we combine the trigonometric functions. In order to perform the averaging, we shall use the representation

$$e^{iU\sin a} = \sum_{n=-\infty}^{\infty} J_n(U)e^{ina} \tag{6.568}$$

in terms of Bessel functions J_n of order n. A short calculation leads to the form

$$\mathcal{E} = \sum_{n=-\infty}^{\infty}\sum_{m=-\infty}^{\infty} J_n(W)J_m(W)e^{i(n-m)(\psi-\phi_0)+in\Omega t+ik_\parallel v_\parallel t} \tag{6.569}$$

where $W := k_\perp v_\perp/\Omega$. Now, we are ready to simplify

$$D_{\mu\mu}^{(I)} = \frac{1}{\rho_L^2}\int_0^\infty dt \int_{-\infty}^{\infty} d^3k\, E_{yy}(\mathbf{k})\left\langle\langle\eta_x(t)\eta_x(0)\mathcal{E}\rangle_\perp\right\rangle_\parallel, \tag{6.570}$$

where we have to introduce

$$\eta_x(t)\eta_x(0) = \frac{v_\perp^2}{4}\left[e^{i(2\phi_0-\Omega t)} + e^{-i\Omega t} + e^{i\Omega t} + e^{-i(2\phi_0-\Omega t)}\right]. \tag{6.571}$$

Averaging over the phase by operating with $1/(2\pi)\int_0^{2\pi} d\phi_0$, we obtain

$$D_{\mu\mu}^{(I)} = \frac{v_\perp^2}{4\rho_L^2}\int_0^\infty dt \int_{-\infty}^{\infty} d^3k\, E_{yy}(\mathbf{k}) \sum_{n=-\infty}^{\infty} e^{-in\Omega t+ik_\parallel v_\parallel t}$$
$$\times\left[J_{n+1}^2(W) + J_{n-1}^2(W) + J_{n+1}(W)J_{n-1}(W)\left(e^{i2\psi} + e^{-i2\psi}\right)\right]. \tag{6.572}$$

Note that the t-integration can easily be performed, provided that the magnetic fluctuation spectrum does not depend on time. We will use

$$\int_0^\infty dt e^{i(k_\parallel v_\parallel - n\Omega)t} = \pi\delta(k_\parallel v_\parallel - n\Omega) \equiv R_n(\mathbf{k}) = R_n(k_\parallel) \,. \tag{6.573}$$

For possible generalizations, we abbreviated the $k_\parallel$-resonance function by R_n. With this definition, we obtain

$$D_{\mu\mu}^{(I)} = \frac{v_\perp^2}{4\rho_L^2} \int_{-\infty}^{\infty} d^3k\, E_{yy}(\boldsymbol{k}) \times \sum_{n=-\infty}^{\infty} R_n(k_\parallel)\left[J_{n+1}^2 + J_{n-1}^2 + J_{n+1}J_{n-1}(e^{i2\psi} + e^{-i2\psi})\right]. \tag{6.574}$$

Obviously, the other contributions $D_{\mu\mu}^{(II)}$, $D_{\mu\mu}^{(III)}$ and $D_{\mu\mu}^{(VI)}$ can be calculated similarly when using

$$\eta_y(t)\eta_y(0) = -\frac{v_\perp^2}{4}\left[e^{i(2\phi_0-\Omega t)} - e^{-i\Omega t} - e^{i\Omega t} + e^{-i(2\phi_0-\Omega t)}\right], \tag{6.575}$$

$$\eta_x(t)\eta_y(0) = \frac{v_\perp^2}{4i}\left[e^{i(2\phi_0-\Omega t)} - e^{-i\Omega t} + e^{i\Omega t} - e^{-i(2\phi_0-\Omega t)}\right], \tag{6.576}$$

$$\eta_y(t)\eta_x(0) = \frac{v_\perp^2}{4i}\left[e^{i(2\phi_0-\Omega t)} + e^{-i\Omega t} - e^{i\Omega t} - e^{-i(2\phi_0-\Omega t)}\right]. \tag{6.577}$$

The final results are

$$D_{\mu\mu}^{I} = \frac{v_\perp^2}{4\rho_L^2} \int_{-\infty}^{\infty} d^3k\, E_{yy}(\boldsymbol{k}) \sum_{n=-\infty}^{\infty} R_n(k_\parallel) \times \left[J_{n+1}^2(W) + J_{n-1}^2(W) + J_{n+1}(W)J_{n-1}(W)\left(e^{2i\psi} + e^{-2i\psi}\right)\right], \tag{6.578}$$

$$D_{\mu\mu}^{II} = \frac{v_\perp^2}{4\rho_L^2} \int_{-\infty}^{\infty} d^3k\, E_{xx}(\boldsymbol{k}) \sum_{n=-\infty}^{\infty} R_n(k_\parallel) \times \left[J_{n+1}^2(W) + J_{n-1}^2(W) - J_{n+1}(W)J_{n-1}(W)\left(e^{2i\psi} + e^{-2i\psi}\right)\right], \tag{6.579}$$

$$D_{\mu\mu}^{III} = \frac{v_\perp^2}{i4\rho_L^2} \int_{-\infty}^{\infty} d^3k\, E_{yx}(\boldsymbol{k}) \sum_{n=-\infty}^{\infty} R_n(k_\parallel) \times \left[-J_{n+1}^2(W) + J_{n-1}^2(W) + J_{n+1}(W)J_{n-1}(W)\left(e^{2i\psi} - e^{-2i\psi}\right)\right], \tag{6.580}$$

$$D_{\mu\mu}^{IV} = \frac{v_\perp^2}{i4\rho_L^2} \int_{-\infty}^{\infty} d^3k\, E_{xy}(\boldsymbol{k}) \sum_{n=-\infty}^{\infty} R_n(k_\parallel)$$
$$\times \left[J_{n+1}^2(W) - J_{n-1}^2(W) + J_{n+1}(W) J_{n-1}(W)\left(e^{2i\psi} - e^{-2i\psi}\right)\right]. \tag{6.581}$$

Let us reconsider the collisionless case $\nu = 0$ without making explicit use of the Corrsin approximation. For the velocities, we use Eqs. (6.560) and (6.561). Then, Eq. (6.507) may be written as

$$\dot{\mu} = \frac{\Omega\sqrt{1-\mu^2}}{B_0}\left[\delta B_y \cos\phi - \delta B_x \sin\phi\right], \tag{6.582}$$

where $\phi = \phi_0 - \Omega t$. The magnetic field fluctuations will be represented in the form

$$\delta B_L(\mathbf{r}, t) := \frac{1}{\sqrt{2}}\left(\delta B_x(\mathbf{r}, t) + i\delta B_y(\mathbf{r}, t)\right), \tag{6.583}$$

$$\delta B_R(\mathbf{r}, t) := \frac{1}{\sqrt{2}}\left(\delta B_x(\mathbf{r}, t) - i\delta B_y(\mathbf{r}, t)\right). \tag{6.584}$$

Then, Eq. (6.507) appears as

$$\dot{\mu} = \frac{i\Omega}{\sqrt{2}B_0}\sqrt{1-\mu^2}\left[\delta B_R e^{i\phi} - \delta B_L e^{-i\phi}\right]. \tag{6.585}$$

We obtain the pitch angle diffusion coefficient Eq. (6.508) as

$$D_{\mu\mu} = \frac{\Omega^2(1-\mu^2)}{2B_0^2}\,\mathrm{Re}$$
$$\times \int_0^\infty dt \left[\mathcal{E}_{RR}e^{-i\Omega t} - \mathcal{E}_{RL}e^{2i\phi_0 - i\Omega t} - \mathcal{E}_{LR}e^{-2i\phi_0 + i\Omega t} + \mathcal{E}_{LL}e^{i\Omega t}\right], \tag{6.586}$$

with

$$\mathcal{E}_{XY}(\mathbf{r}, t) = \left\langle \delta B_X(\mathbf{r}, t)\delta B_Y^*(\mathbf{r}_0, 0)\right\rangle, \quad \text{where} \quad X, Y = R, L. \tag{6.587}$$

Next, we apply a Fourier decomposition of the magnetic fluctuations,

$$\delta B_{L,R}(\mathbf{r}, t) = \int d^3k\,\delta B_{L,R}(\mathbf{k}, t)e^{i\mathbf{k}\cdot\mathbf{r}(t)}, \tag{6.588}$$

leading to

$$\mathcal{E}_{XY} = \mathcal{E}_{XY}(\mathbf{r}, t, \mathbf{r}_0, 0) = \int d^3k \int d^3k' \left\langle \delta B_X(\mathbf{k}, t)\delta B_Y\left(\mathbf{k}', 0\right)\right\rangle e^{i\left[\mathbf{k}\cdot\mathbf{r}(t) - \mathbf{k}'\cdot\mathbf{r}_0\right]}. \tag{6.589}$$

Cylindrical coordinates in k-space

$$k_x = k_\perp \cos\psi\ , \quad k_y = k_\perp \sin\psi\ , \quad k_z = k_\parallel \tag{6.590}$$

lead to

$$\mathbf{k}\cdot\mathbf{r}(t) = \mathbf{k}\cdot\mathbf{r}_0 + \frac{k_\perp v_\perp}{\Omega}\left[\sin(\psi - \phi_0 + \Omega t) - \sin(\psi - \phi_0)\right] + k_\parallel v_\parallel t\ . \tag{6.591}$$

Putting everything together, we obtain in real space

$$\mathcal{E}_{XY}(\mathbf{r}, t, \mathbf{r}_0, 0) = \int d^3k \mathcal{E}_{XY}(\mathbf{k}, t) e^{iW\left[\sin(\psi-\phi_0+\Omega t)-\sin(\psi-\phi_0)\right]+iv_\parallel k_\parallel t}\ , \tag{6.592}$$

with

$$W = \frac{k_\perp v_\perp}{\Omega} = k_\perp \rho_L \sqrt{1-\mu^2}\ , \tag{6.593}$$

and

$$\mathcal{E}_{XY}(\mathbf{k}, t) = \frac{1}{(2\pi)^3}\int d^3r \mathcal{E}_{XY}(\mathbf{r}, t) e^{-i\mathbf{k}\cdot\mathbf{r}}\ . \tag{6.594}$$

Making use of Eq. (6.568), we may write

$$\begin{aligned} D_{\mu\mu} = &\frac{\Omega^2(1-\mu^2)}{2B_0^2}\,\mathrm{Re}\int_0^\infty dt \int d^3k \\ &\times \sum_{n,m=-\infty}^{\infty} J_n(W) J_m(W) e^{in(\psi-\phi_0+\Omega t)-im(\psi-\phi_0)+iv_\parallel k_\parallel t} \\ &\times \left[\mathcal{E}_{RR}(\mathbf{k}) e^{-i\Omega t} - \mathcal{E}_{RL}(\mathbf{k}) e^{2i\phi_0 - i\Omega t} - \mathcal{E}_{LR}(\mathbf{k}) e^{-2i\phi_0 + i\Omega t} + \mathcal{E}_{LL}(\mathbf{k}) e^{i\Omega t}\right]. \end{aligned} \tag{6.595}$$

Next, we average over all initial angles by applying $1/(2\pi)\int_0^{2\pi} d\phi_0$. It leads to

$$\begin{aligned} D_{\mu\mu} = &\frac{\Omega^2(1-\mu^2)}{2B_0^2}\,\mathrm{Re}\sum_{n=-\infty}^{\infty}\int d^3k\, R_n(k_\parallel)\left[J_{n+1}^2(W)\mathcal{E}_{RR}(\mathbf{k})\right. \\ &\left. + J_{n-1}^2(W)\mathcal{E}_{LL}(\mathbf{k}) - J_{n+1}(W)J_{n-1}(W)\left(\mathcal{E}_{RL}(\mathbf{k})e^{2i\psi} + \mathcal{E}_{LR}(\mathbf{k})e^{-2i\psi}\right)\right], \end{aligned} \tag{6.596}$$

where, as in Eq. (6.573),

$$R_n(k_\parallel) = \int_0^\infty dt e^{i(k_\parallel v_\parallel - n\Omega)t} = \pi\delta\left(k_\parallel v_\parallel - n\Omega\right). \tag{6.597}$$

Short calculations, using Eq. (6.594), lead to

$$\mathcal{E}_{RR}(\mathbf{k},t) = \frac{B_0^2}{2}\left[E_{xx}(\mathbf{k}) - i\,E_{yx}(\mathbf{k}) + i\,E_{xy}(\mathbf{k}) + E_{yy}(\mathbf{k})\right], \tag{6.598}$$

where $E_{ij}(\mathbf{k})$ has been defined in Eq. (6.336); furthermore, $b_i = \delta B_i/B_0$. In a similar manner,

$$\mathcal{E}_{LL}(\mathbf{k},t) = \frac{B_0^2}{2}\left[E_{xx}(\mathbf{k}) + i\,E_{yx}(\mathbf{k}) - i\,E_{xy}(\mathbf{k}) + E_{yy}(\mathbf{k})\right], \tag{6.599}$$

$$\mathcal{E}_{RL}(\mathbf{k},t) = \frac{B_0^2}{2}\left[E_{xx}(\mathbf{k}) - i\,E_{yx}(\mathbf{k}) - i\,E_{xy}(\mathbf{k}) - E_{yy}(\mathbf{k})\right], \tag{6.600}$$

$$\mathcal{E}_{LR}(\mathbf{k},t) = \frac{B_0^2}{2}\left[E_{xx}(\mathbf{k}) + i\,E_{yx}(\mathbf{k}) + i\,E_{xy}(\mathbf{k}) - E_{yy}(\mathbf{k})\right], \tag{6.601}$$

follow. Introducing the results into $D_{\mu\mu}$, we may split with respect to contributions from E_{ij} to obtain

$$\begin{aligned} D_{\mu\mu}^{(I)} = \frac{v_\perp^2}{4\rho_L^2}\int d^3k\,E_{yy}(\mathbf{k})\sum_{n=-\infty}^{\infty} R_n(k_\parallel) \\ \times\left[J_{n+1}^2(W) + J_{n-1}^2(W) + J_{n+1}(W)J_{n-1}(W)\left(e^{2i\psi} + e^{-2i\psi}\right)\right], \end{aligned} \tag{6.602}$$

$$\begin{aligned} D_{\mu\mu}^{(II)} = \frac{v_\perp^2}{4\rho_L^2}\int d^3k\,E_{xx}(\mathbf{k})\sum_{n=-\infty}^{\infty} R_n(k_\parallel) \\ \times\left[J_{n+1}^2(W) + J_{n-1}^2(W) - J_{n+1}(W)J_{n-1}(W)\left(e^{2i\psi} + e^{-2i\psi}\right)\right], \end{aligned} \tag{6.603}$$

$$\begin{aligned} -D_{\mu\mu}^{(III)} = \frac{i v_\perp^2}{4\rho_L^2}\int d^3k\,E_{yx}(\mathbf{k})\sum_{n=-\infty}^{\infty} R_n(k_\parallel) \\ \times\left[-J_{n+1}^2(W) + J_{n-1}^2(W) + J_{n+1}(W)J_{n-1}(W)\left(e^{2i\psi} - e^{-2i\psi}\right)\right], \end{aligned} \tag{6.604}$$

$$\begin{aligned} -D_{\mu\mu}^{(IV)} = \frac{i v_\perp^2}{4\rho_L^2}\int d^3k\,E_{xy}(\mathbf{k})\sum_{n=-\infty}^{\infty} R_n(k_\parallel) \\ \times\left[J_{n+1}^2(W) - J_{n-1}^2(W) + J_{n+1}(W)J_{n-1}(W)\left(e^{2i\psi} - e^{-2i\psi}\right)\right], \end{aligned} \tag{6.605}$$

where $(1-\mu^2) = v_\perp/v$ and $\rho_L = v/\Omega$. These formulas agree with the previously derived expressions.

6.8.2.1 Parallel Diffusion in the Collisionless, Quasilinear Limit $\lambda_\perp \to \infty$

The so called quasilinear case with a slab model for the magnetic fluctuation spectrum is frequently considered in anomalous transport theory. It applies in the limit

$\lambda_\perp \to \infty$. In that case, the magnetic fluctuation simplifies considerably. For an exponential dependence in parallel direction (note that, especially in astrophysics, different forms are in use for the parallel spectrum), we we make the ansatz

$$E_{ij}(\mathbf{k}) \equiv E_{ij}^{slab}(\mathbf{k}) = \frac{1}{\sqrt{2\pi}}\beta^2\lambda_\parallel e^{-\frac{1}{2}\lambda_\parallel^2 k_\parallel^2}\delta(k_x)\delta(k_y)\delta_{ij}. \tag{6.606}$$

The different formulation

$$E_{ij}(\mathbf{k}) = g^{slab}(k_\parallel)\frac{1}{k_\perp}\delta(k_\perp)\delta_{ij} \tag{6.607}$$

with appropriate expressions for $g^{slab}(k_\parallel)$ can be found in the literature [83, 88, 199]. We should note that then the cylindrical coordinates are used in k-space, and we should remember $dk_x dk_y \to k_\perp dk_\perp d\psi$ as well as

$$\int_{-\infty}^{\infty} dk_x \int_{-\infty}^{\infty} dk_y \delta(k_x)\delta(k_y) = 1 \to \int_0^{2\pi} d\psi \int_0^{\infty} dk_\perp \frac{1}{k_\perp \pi}\delta(k_\perp)k_\perp d\psi = 1\,. \tag{6.608}$$

The spectrum should be normalized according to

$$E_{ij}(\mathbf{r}=0) = \beta^2 = \int d^3k\, E_{ij}(\mathbf{k})\,. \tag{6.609}$$

First, we shall use Eqs. (6.602)–(6.605), although in the present special case $\nu = 0$ and $\lambda_\perp \to \infty$, a more straightforward way is available (see below). Inserting Eq. (6.606) into Eq. (6.602), we obtain

$$D_{\mu\mu}^{(I)} = \frac{\sqrt{\pi}\lambda_\parallel\beta^2 v_\perp^2}{4\sqrt{2}\rho_L^2}\sum_{n=-\infty}^{\infty}\int_{-\infty}^{\infty} d^3k e^{-\frac{1}{2}\lambda_\parallel^2 k_z^2}\delta(k_x)\delta(k_y)\delta(k_z v_\parallel - n\Omega)$$
$$\times\left[J_{n-1}^2(W) + 2\cos(2\psi)J_{n-1}(W)J_{n+1}(W) + J_{n+1}^2(W)\right], \tag{6.610}$$

where $k_\perp = \sqrt{k_x^2 + k_y^2}$ and $k_\parallel = k_z$. Integration over k_z leads to

$$D_{\mu\mu}^{(I)} = \frac{\sqrt{\pi}\lambda_\parallel\beta^2 v_\perp^2}{4\sqrt{2}\rho_L^2|v_\parallel|}\sum_{n=-\infty}^{\infty}\int_{-\infty}^{\infty} dk_x \int_{-\infty}^{\infty} dk_y e^{-\frac{1}{2}\lambda_\parallel^2\left(\frac{n\Omega}{v_\parallel}\right)^2}\delta(k_x)\delta(k_y)$$
$$\times\left[J_{n-1}^2(W) + 2\cos(2\psi)J_{n-1}(W)J_{n+1}(W) + J_{n+1}^2(W)\right]. \tag{6.611}$$

The property of the Bessel function

$$J_n(0) = 0 \quad \text{for} \quad n \neq 0\,, \quad J_0(0) = 1\,, \tag{6.612}$$

finally leads to

$$D_{\mu\mu}^{(I)} = \frac{\sqrt{\pi}}{2\sqrt{2}} \frac{\beta^2 \lambda_\parallel v_\perp^2}{\rho_L^2 |v_\parallel|} e^{-\frac{1}{2}\lambda_\parallel^2 \frac{\Omega^2}{v_\parallel^2}} . \tag{6.613}$$

Similar calculations show

$$D_{\mu\mu}^{(II)} = D_{\mu\mu}^{(I)} , \quad D_{\mu\mu}^{(III)} = D_{\mu\mu}^{(IV)} = 0 , \tag{6.614}$$

such that

$$D_{\mu\mu} = \frac{\sqrt{\pi}\beta^2 \lambda_\parallel v_\perp^2}{\sqrt{2}\rho_L^2 |v_\parallel|} e^{-\frac{1}{2}\lambda_\parallel^2 \frac{\Omega^2}{v_\parallel^2}} . \tag{6.615}$$

Before discussing the result, we come back to the point that under the present assumptions, the diffusion coefficient can be directly evaluated from Eqs. (6.555)–(6.558). For example, by inserting Eq. (6.606) into Eq. (6.555), we obtain

$$D_{\mu\mu}^{(I)} \approx \frac{1}{\rho_L^2} \beta^2 \lambda_\parallel \frac{v_\perp^2}{2\sqrt{2}\sqrt{\pi}} \int dt \int dk_\parallel \frac{1}{2} \left[e^{-i\Omega t} + e^{i\Omega t} \right] e^{ik_\parallel v_\parallel t} e^{-i\frac{1}{2}\lambda_\parallel^2 k_\parallel^2} . \tag{6.616}$$

Because of

$$\int_0^\infty dt e^{i(k_\parallel v_\parallel \pm n\Omega)t)} = \pi \delta \left(k_\parallel v_\parallel \pm n\Omega \right) \tag{6.617}$$

and

$$\int_{-\infty}^{\infty} dk_\parallel \delta \left(k_\parallel v_\parallel \pm n\Omega \right) \cdots = \frac{1}{|v_\parallel|} \int_{-\infty}^{\infty} dk_\parallel \delta \left(k_\parallel \pm \frac{n\Omega}{v_\parallel} \right) \cdots , \tag{6.618}$$

the integrations are straightforward to perform and immediately lead to

$$D_{\mu\mu}^{(I)} = \frac{\sqrt{\pi}}{2\sqrt{2}} \frac{\beta^2 \lambda_\parallel v_\perp^2}{\rho_L^2 |v_\parallel|} e^{-\frac{1}{2}\lambda_\parallel^2 \frac{\Omega^2}{v_\parallel^2}} , \tag{6.619}$$

in agreement with the previous result. The other expressions (6.556)–(6.558) will lead to the same results as derived before.

Following the notations in astrophysics, we next introduce the rigidity, that is,

$$R = \frac{\rho_L}{l_{slab}} \sim \frac{\rho_L}{\lambda_\parallel} , \tag{6.620}$$

where the so called slab bendover scale l_{slab} is defined so that the area $l_{slab}\beta^2$ is equal to the area under the correlation function, that is,

$$l_{slab} = \int_0^\infty dz e^{-\frac{1}{2}\frac{z^2}{\lambda_\parallel^2}} = \sqrt{\frac{\pi}{2}} \lambda_\parallel . \tag{6.621}$$

The rigidity provides a measure for the momentum of a particle. For large rigidities, we approximate

$$D_{\mu\mu} \approx \frac{\sqrt{\pi}\beta^2\lambda_{\parallel} v_{\perp}^2}{\sqrt{2}\rho_L^2 |v_{\parallel}|} = \frac{\sqrt{\pi}\beta^2\lambda_{\parallel} v}{\sqrt{2}\rho_L^2} \frac{1-\mu^2}{|\mu|} . \tag{6.622}$$

Inserting into Eq. (6.519), we can integrate analytically and find

$$\chi_{zz} = \frac{v^2}{8} \int_{-1}^{+1} \frac{(1-\mu^2)^2}{D_{\mu\mu}} d\mu \approx \frac{1}{8\sqrt{2}\sqrt{\pi}} \frac{v^3}{\Omega^2\beta^2\lambda_{\parallel}} . \tag{6.623}$$

Besides a numerical factor, this agrees with the result which has been derived on the basis of a small gyrocenter expansion [192]. In the latter paper, the "magnetic collision frequency"

$$\nu_{\text{mag}} = \sqrt{2}\Omega^2\beta^2 \frac{\lambda_{\parallel}}{v_{th}} \tag{6.624}$$

was introduced, and thus the quasilinear parallel and perpendicular diffusion coefficients, respectively, can be written in forms analogous to the classical ones, namely,

$$D_{\parallel} = \frac{v_{th}^2}{2\nu_{\text{mag}}} , \quad D_{\perp} = \frac{v_{th}^2 \nu_{\text{mag}}}{2\Omega^2} . \tag{6.625}$$

Equation (6.521) allows one to calculate the parallel mean free path

$$\lambda_{\parallel}^{mfp} = 3\frac{\chi_{zz}}{v} \approx \frac{3}{8\sqrt{2}\sqrt{\pi}} \frac{v^2}{\Omega^2\beta^2\lambda_{\parallel}} \sim R^2 . \tag{6.626}$$

The mean free path is proportional to the square of the rigidity in the quasilinear approximation.

Starting from the relation (6.519) between the parallel diffusion coefficient and the pitch-angle diffusion coefficient as well as the expressions (6.555)–(6.558) for the latter, more general cases can be considered, for example, $\nu \neq 0$ and finite $\lambda_{\perp}$. We will not show the details here.

7
Linear Waves and Instabilities

Nonequilibrium plasmas may develop in time in different ways, depending on boundary conditions. For example, they may simply relax to equilibrium, stay in a stable stationary state, or develop instabilities. As in any other system, instabilities are driven by sources of free energy. In plasmas, usual sources are pressure gradients and electric currents, accompanied by bad curvature, and so on. Depending on the time- and space-scales involved, the dynamical behavior of a plasma can be analyzed within different approximations, such as, the kinetic Vlasov equation, the two-fluid approximation, or the MHD model, just to name the most common ones. We have already presented examples of linear waves by drawing and solving dispersion relations. However, thus far in most cases, we only focused on the real frequency response, ignoring imaginary parts in the presence of free energy sources. In the present section, we shall develop more accurate solutions of dispersion relations, including unstable configurations.

The plasma universe breeds a whole zoo of plasma instabilities. Thus, it will not be the intention of the present chapter to cover all possible instabilities; we shall concentrate on a few typical examples and we make no claims regarding completeness. For more details, the reader is referred to more specialized monographs, for example, [17, 200, 201].

7.1
Waves and Instabilities in the Homogeneous Vlasov Description

7.1.1
The Penrose Criterion and Its Cognate Formulations

Let us start with the Vlasov description. When discussing Landau damping, we already started from the dispersion relation

$$k^2 = G\left(\frac{\omega}{k}\right) \equiv \omega_{pe}^2 \int \frac{dg}{du} \frac{du}{u - \frac{\omega}{k}} \tag{7.1}$$

for an electron system with fixed ion background. The integration contour for the integral on the right-hand side is given by the Landau prescription. The possible

High Temperature Plasmas, Theory and Mathematical Tools for Laser and Fusion Plasmas, First Edition.
Karl-Heinz Spatschek.

solutions for $\omega = ip$ depend on the zeroth order distribution function f_0 which in the present 1D description enters via g; see Eq. (4.126). Modes $\sim \exp(pt) \sim \exp(-i\omega t)$ are growing, that is, unstable, when

$$\operatorname{Im}\omega > 0\,. \tag{7.2}$$

Since condition Eq. (7.2) corresponds to Re $p > 0$, for that case we can can perform in Eq. (7.1) the integration along the real u-axis from $-\infty$ to $+\infty$. Defining the real phase velocity

$$u_p = \frac{\operatorname{Re}\omega}{k} = \operatorname{Re}\zeta\,, \quad \zeta = \frac{\omega}{k}\,, \tag{7.3}$$

where ζ is a complex variable, we obtain for small growth rates

$$G(u_p + i0) = \omega_{pe}^2 \left\{ \mathcal{P} \int_{-\infty}^{+\infty} \frac{g'(u)}{u - u_p} du + i\pi g'(u_p) \right\}. \tag{7.4}$$

The right-hand side of Eq. (7.1) can be interpreted as a mapping $G = G(\omega/k)$ of the upper half of the complex ω-plane (or ω/k-plane where $k > 0$ is real) into an area $\mathcal{G}$ of the complex G-plane. The curve Eq. (7.4) is the borderline of the area $\mathcal{G}$ in G-space. Note that $G(R + i0)$ is finite and continuous and $G(\omega/k)$ is holomorphic in the upper ω/k-plane. The directed curve $G(u_p + i0)$ starts and ends at $G = 0$ because of $G(\pm\infty) = 0$. The upper ω/k-plane is mapped into the interior of Eq. (7.4) (lying to the left of the curve). Suppose $G_0 \in \mathcal{G}$ is a point not on $G(R + i0)$, then by the argument principle, the curve Eq. (7.4) winds around G_0 (anticlockwise) as many times as $G(\omega/k)$ takes the value G_0 in the upper ω/k-plane.

Obviously, unstable solutions with $\operatorname{Im}\omega > 0$ exist if the curve Eq. (7.4) cuts the positive ($G = k^2$)-axis. Cutting the positive G-axis from below requires in Eq. (7.4) a sign change of $g'(u_p)$ with $g'(u_p) = 0$ at the intersection point. The sign change of g at $u = u_p$ should occur when the curve Eq. (7.4) (in the G-plane) approaches the real G-axis from below. Then, the positive values of $\operatorname{Im}\omega$ are to the left. Thus, with increasing u, the derivative g' changes from negative to positive values, meaning that $g(u_p)$ is a minimum value.

For a given distribution function, it is now quite easy to decide whether instability occurs or not. Integrating by parts, we may rewrite the real part of Eq. (7.4) as

$$\begin{aligned}
&\operatorname{Re} G(u_p + i0) \\
&= \lim_{\varepsilon \to 0} \left[\left\{ \int_{-\infty}^{u_p - \varepsilon} + \int_{u_p + \varepsilon}^{\infty} \right\} \left\{ \frac{g(u)}{(u - u_p)^2} du \right\} - \frac{g(u_p - \varepsilon)}{\varepsilon} - \frac{g(u_p + \varepsilon)}{\varepsilon} \right] \\
&= \mathcal{P} \int_{-\infty}^{+\infty} \frac{g(u) - g(u_p)}{(u - u_p)^2} du + \lim_{\varepsilon \to 0} \left[\frac{g(u_p) - g(u_p - \varepsilon)}{\varepsilon} - \frac{g(u_p + \varepsilon) - g(u_p)}{\varepsilon} \right] \\
&= \mathcal{P} \int_{-\infty}^{+\infty} \frac{g(u) - g(u_p)}{(u - u_p)^2} du
\end{aligned} \tag{7.5}$$

since

$$g(u_p) \lim_{\varepsilon \to 0} \left\{ \int_{-\infty}^{u_p - \varepsilon} + \int_{u_p + \varepsilon}^{\infty} \right\} \left\{ \frac{du}{(u - u_p)^2} \right\} = 2 \lim_{\varepsilon \to 0} \frac{g(u_p)}{\varepsilon} . \tag{7.6}$$

We may omit $\mathcal{P}$ in the result when the function g possesses a minimum at $u_p \equiv u_p^{min}$. Following Penrose [202], we can formulate the following instability criterion:
Sufficient and necessary for (exponential) instability are

- The distribution function g possesses at least one minimum at $u = u_p = u_p^{min}$.
- At the minimum,

$$\int_{-\infty}^{+\infty} \frac{g(u) - g(u_p^{min})}{(u - u_p^{min})^2} du > 0 . \tag{7.7}$$

In two examples, we show how the relevant integrals will be evaluated. Let us first assume a Maxwellian

$$g(u) = \frac{1}{(2\pi)^{1/2} v_{th}} \exp\left[-\frac{u^2}{2v_{th}^2} \right] \tag{7.8}$$

which, as we know already, belongs to a stable situation. Nevertheless, the evaluation of the image of the real ζ-axis according to Eq. (7.4) gives

$$G(u_p + i0) = -\frac{\omega_{pe}^2 v_{th}^{-3}}{(2\pi)^{1/2}}$$
$$\times \left\{ \mathcal{P} \int_{-\infty}^{+\infty} \frac{u \exp\left(-\frac{u^2}{2v_{th}^2}\right)}{u - u_p} du + i\pi u_p \exp\left[-\frac{u_p^2}{2v_{th}^2} \right] \right\} . \tag{7.9}$$

Obviously, it has no cut with the positive real G-axis. The real value at $u_p = 0$ is negative. The corresponding plot is shown in Figure 7.1.

The second example assumes a distribution function with a minimum. Instability becomes possible. A possible image of the real ζ-axis is shown in Figure 7.2. For simplicity, let us assume a superposition of two Maxwellians. For some analytical calculations, we approximate the Maxwellians with narrow widths such that

$$g \approx \left[\delta(u - U) + \delta(u + U) \right] \tag{7.10}$$

may be assumed. The symmetry suggests to assume the minimum at $u = 0$ with $g(0) \approx 0$. For the Penrose criterion Eq. (7.7), we have to evaluate the integral

$$\int_{-\infty}^{+\infty} \frac{g(u)}{u^2} du = \frac{2}{U^2} > 0 . \tag{7.11}$$

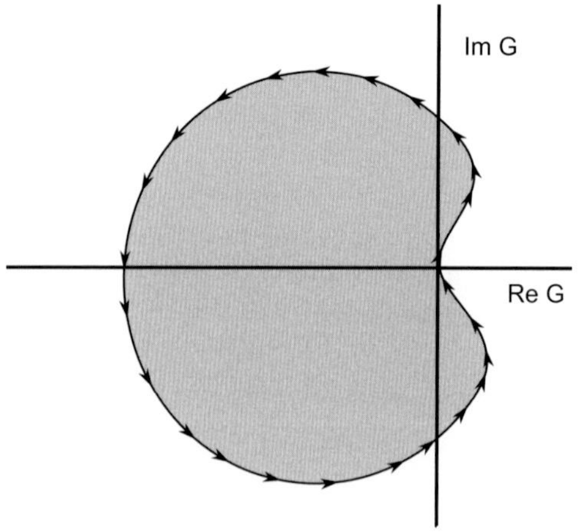

Figure 7.1 The curve $G(u_p)$ for a Maxwell distribution. The image of the upper ζ-half-plane is tinged with gray.

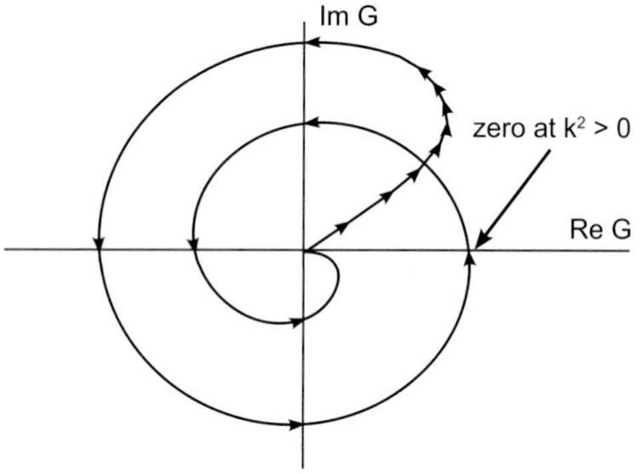

Figure 7.2 Typical example for the image of the real ζ-axis in the G-plane for a distribution function with a minimum.

Of course, now a cut with the positive real G-axis occurs for $U \neq 0$, implying instability.

We should note that for the distribution function Eq. (7.10), the dispersion relation can be solved directly. For narrow widths, we can ignore the imaginary terms, and the approximate dispersion relation is

$$\omega_{pe}^2 \left[\frac{1}{(kU - \omega)^2} + \frac{1}{(kU + \omega)^2} \right] = 1 \,. \tag{7.12}$$

For

$$|k| < \sqrt{2} \frac{\omega_{pe}}{U} \,, \tag{7.13}$$

only two purely real ω-solutions exist. Since the dispersion relation is of fourth order, two additional complex solutions for ω do exist. They correspond to exponentially growing and decaying solutions, respectively. The maximum growth rate of the two-beam instability is

$$\gamma = \frac{\omega_{pe}}{2} . \tag{7.14}$$

Let us have a look at the (in)stability criterion from another point of view. We start from a homogeneous situation in the presence of an external magnetic field $\boldsymbol{B}_0$; there should be no zeroth order electric field. The distribution function

$$g = g\left(\frac{m_e}{2} v^2\right) \tag{7.15}$$

is an exact solution of the electron Vlasov equation to zeroth order. Now, let us linearize about the stationary g, that is, introduce

$$f = g + \delta f . \tag{7.16}$$

The perturbation δf induces first-order electric and magnetic fields $\boldsymbol{E}_1$ and $\boldsymbol{B}_1$, respectively. Since g is chosen as an isotropic distribution function,

$$\frac{q_e}{m_e} \boldsymbol{v} \times \boldsymbol{B}_1 \cdot \partial_{\boldsymbol{v}} g = 0 \tag{7.17}$$

follows. Thus, the linearized Vlasov equation for electrons is

$$\partial_t \delta f + \boldsymbol{v} \cdot \partial_{\boldsymbol{r}} \delta f + \frac{q_e}{m_e} \boldsymbol{E}_1 \cdot \partial_{\boldsymbol{v}} g + \frac{q_e}{m_e} (\boldsymbol{v} \times \boldsymbol{B}_0) \cdot \partial_{\boldsymbol{v}} \delta f = 0 . \tag{7.18}$$

Next, we multiply both sides with δf and divide by

$$g' = \frac{d}{d\varepsilon} g(\varepsilon) , \quad \varepsilon \equiv \frac{1}{2} m_e v^2 . \tag{7.19}$$

Integration over velocity space finally leads to

$$\partial_t \int d^3 v \frac{(\delta f)^2}{g'} + 2 q_e \int d^3 v \, \boldsymbol{E}_1 \cdot \boldsymbol{v} \delta f = 0 . \tag{7.20}$$

Performing a similar calculation for ions, and introducing the index $\alpha = e, i$, summation leads to

$$\partial_t \sum_\alpha \int \frac{(\delta f_\alpha)^2}{g'_\alpha} d^3 v + 2 \boldsymbol{E}_1 \cdot \boldsymbol{j}_1 = 0 , \tag{7.21}$$

with the first-order electric current density $\boldsymbol{j}_1$. The Maxwell equations

$$\mu_0 \boldsymbol{j}_1 = \nabla \times \boldsymbol{B}_1 - \frac{1}{c^2} \partial_t \boldsymbol{E}_1 , \tag{7.22}$$

$$\partial_t \boldsymbol{B}_1 = -\nabla \times \boldsymbol{E}_1 , \tag{7.23}$$

together with

$$\int B_1 \cdot (\nabla \times E_1) d^3 r = \int E_1 \cdot (\nabla \times B_1) d^3 r, \tag{7.24}$$

finally lead to

$$\partial_t \left[\sum_\alpha \int d^3 v d^3 r \frac{(\delta f_\alpha)^2}{g'_\alpha} - \varepsilon_0 \int d^3 r \left(E_1^2 + B_1^2 \right) \right] = 0 \, . \tag{7.25}$$

Thus, the term in the angular bracket turns out to be conserved; it is negative for $g'_\alpha < 0$. The definiteness of

$$\sum_\alpha \int d^3 v d^3 r \frac{(\delta f_\alpha)^2}{-g'_\alpha} + \varepsilon_0 \int d^3 r \left(E_1^2 + B_1^2 \right) = \text{const} \tag{7.26}$$

for $g'_\alpha < 0$ forbids exponentially growing modes $\sim \exp(\gamma t)$. Distribution functions without minima ("single-humped distributions") do not become unstable, in agreement with the Penrose criterion.

The above consideration is strongly related to the minimum principle for the free energy. Assuming for simplicity $E_0 = 0$ and $B_0 = 0$, we define the intrinsic energy U as

$$U = \sum_\alpha \int d^3 v d^3 r \frac{m_\alpha}{2} v^2 f_\alpha + \frac{\varepsilon_0}{2} \int d^3 r \left(E^2 + B^2 \right) \tag{7.27}$$

and the entropy S as

$$S = -k_B \sum_\alpha \int d^3 v d^3 r f_\alpha \ln f_\alpha \, . \tag{7.28}$$

Variation of the free energy $F = U - TS$ leads in first order ($\delta F = 0$) to $f_{\alpha 0} = g_\alpha$, that is, a Maxwell distribution. The following corresponds therefore to the rather trivial investigation of the stability of a Maxwellian. The second-order variation is

$$\delta^2 F = \sum_\alpha \int d^3 v d^3 r k_B T_\alpha \frac{(\delta f_\alpha)^2}{g_\alpha} + \frac{\varepsilon_0}{2} \int d^3 r \left[(\delta E)^2 + (\delta B)^2 \right] . \tag{7.29}$$

Obviously, the (Maxwellian) system is stable.

To generalize to non-Maxwellian zeroth order distributions, we have to generalize the entropy to

$$S = -k_B \sum_\alpha \int d^3 v d^3 r F_\alpha(f_\alpha) \, . \tag{7.30}$$

For the functional F_α, from the vanishing of the first variation of F, one obtains

$$F'_\alpha(g_\alpha) = -\frac{1}{2} m_\alpha v^2 \, , \tag{7.31}$$

where the prime (′) denotes the differentiation with respect to the argument. We demand that Eq. (7.31) can be inverted to find the distribution function g_α being finite at large v. This requires $F''_\alpha > 0$. We shall get an isotropic distribution $g_\alpha = g_\alpha(+1/2 m_\alpha v^2)$. Differentiating Eq. (7.31) with respect to $1/2 m_\alpha v^2$ leads to

$$F''_\alpha g'_\alpha = -1 \,. \tag{7.32}$$

The second variation of the free energy then leads to

$$\delta^2 F = \sum_\alpha \int d^3 v d^3 r \frac{1}{2} \frac{(\delta f_\alpha)^2}{-g'_\alpha} + \frac{\varepsilon_0}{2} \int d^3 r \left[(\delta E)^2 + (\delta B)^2\right] \geq 0 \,. \tag{7.33}$$

Clearly, a distribution function without minima is stable.

7.1.2
Dispersion in Homogeneous, Magnetized Vlasov Systems

Let us now consider in more detail a collisionless plasma embedded in an homogeneous magnetic field $\boldsymbol{B}_0 = B_0 \hat{z}$; there should be no zeroth order electric filed ($\boldsymbol{E}_0 = 0$). The stationary Vlasov equation has solutions g_α which are well-behaved functions of the velocity componentsv_z and $v_\perp^2$, where $\perp$ designates the direction perpendicular to z. Thus, we may write $g_\alpha = g_\alpha(v_\perp^2, v_z)$. Linearizing in the form $f_\alpha = g_\alpha + \delta f_\alpha$, and considering normal modes $\sim \exp(-i\omega t + i\boldsymbol{k}\cdot\boldsymbol{r})$, we obtain for the perturbation

$$i(\boldsymbol{k}\cdot\boldsymbol{v} - \omega)\delta f_\alpha + (\boldsymbol{v}\times\boldsymbol{\Omega}_\alpha)\cdot\partial_{\boldsymbol{v}}\delta f_\alpha = -\frac{q_\alpha}{m_\alpha}(\boldsymbol{E}_1 + \boldsymbol{v}\times\boldsymbol{B}_1)\cdot\partial_{\boldsymbol{v}} g_\alpha \,. \tag{7.34}$$

Here, $\boldsymbol{\Omega}_\alpha = q_\alpha \boldsymbol{B}_0/m_\alpha = \Omega_\alpha \hat{z}$. Without loss of generality, we introduce $\boldsymbol{k} = k_x \hat{x} + k_z \hat{z}$ and $v_x = v_\perp \cos\varphi$, $v_y = v_\perp \sin\varphi$. With the gyrophase φ, a short calculation leads to

$$(\boldsymbol{v}\times\boldsymbol{\Omega}_\alpha)\cdot\partial_{\boldsymbol{v}}\delta f_\alpha = -\Omega_\alpha \frac{\partial \delta f_\alpha}{\partial\varphi} \,, \tag{7.35}$$

such that the homogeneous part of the linearized Vlasov equation is

$$i\left(\omega - k_x v_\perp \cos\varphi - k_z v_z\right)\delta f_\alpha^h + \Omega_\alpha \frac{\partial}{\partial\varphi}\delta f_\alpha^h = 0 \,. \tag{7.36}$$

Its solution

$$\delta f_\alpha^h = h e^{i k_x v_\perp \sin\frac{\varphi}{\Omega_\alpha}} e^{-i(\omega - k_z v_z)\frac{\varphi}{\Omega_\alpha}} \tag{7.37}$$

allows one to apply the method of variation of parameters with $h = h(v_\perp, v_z)$. Abbreviating the inhomogeneous term in the linearized Vlasov equation by $I_\alpha(v_\perp \cos\varphi, v_\perp \sin\varphi, v_z)$, we find for the solution of the linearized Vlasov equation

$$\delta f_\alpha = \frac{1}{\Omega_\alpha} \int^{\varphi} d\varphi' I_\alpha\left(v_\perp \cos\varphi', v_\perp \sin\varphi', v_z\right) \times e^{i\left(k_x \frac{v_\perp}{\Omega_\alpha}\right)\left(\sin\varphi - \sin\varphi'\right)} e^{-i(\omega - k_z v_z)\frac{\varphi-\varphi'}{\Omega_\alpha}} \,. \tag{7.38}$$

Exponentially unstable modes will have $\operatorname{Im}\omega > 0$. Let us start the special solution at $\varphi' = \infty$ and substitute $\tau := (\varphi - \varphi')/\Omega_\alpha$ such that

$$\delta f_\alpha = -\int_{-\infty}^{0} d\tau I_\alpha \left[v_\perp \cos(\varphi - \Omega_\alpha \tau), v_\perp \sin(\varphi - \Omega_\alpha \tau), v_z\right] \times e^{i\left(k_x \frac{v_\perp}{\Omega_\alpha}\right)[\sin\varphi - \sin(\varphi - \Omega_\alpha \tau)]} e^{i(k_z v_z - \omega)\tau} . \tag{7.39}$$

Next, we rewrite the inhomogeneity term $I_\alpha(v_x, v_y, v_z) = -q_\alpha/m_\alpha(\boldsymbol{E}_1 + \boldsymbol{v} \times \boldsymbol{B}_1) \cdot \partial_{\boldsymbol{v}} g_\alpha$ in the form

$$\begin{aligned} I_\alpha &= -\frac{q_\alpha}{m_\alpha}\left\{\frac{1}{v_\perp}\left[(\boldsymbol{E}_1 \cdot \boldsymbol{v}_\perp) + \frac{1}{\omega}(\boldsymbol{k} \cdot \boldsymbol{v}_\perp)(\boldsymbol{E}_1 \cdot \boldsymbol{v}) - \frac{1}{\omega}(\boldsymbol{v} \cdot \boldsymbol{k})(\boldsymbol{E}_1 \cdot \boldsymbol{v}_\perp)\right]\frac{\partial g_\alpha}{\partial v_\perp}\right. \\ &\quad \left. + \left[E_{1z} + \frac{1}{\omega}k_z(\boldsymbol{E}_1 \cdot \boldsymbol{v}) - \frac{1}{\omega}E_{1z}(\boldsymbol{v} \cdot \boldsymbol{k})\right]\frac{\partial g_\alpha}{\partial v_z}\right\} \\ &= -\frac{q_\alpha}{m_\alpha}\left\{\frac{1}{v_\perp}\boldsymbol{E}_1 \cdot \boldsymbol{v}_\perp\left(1 - \frac{k_z v_z}{\omega}\right)\frac{\partial g_\alpha}{\partial v_\perp} + \boldsymbol{E}_1 \cdot \boldsymbol{v}_\perp \frac{k_z}{\omega}\frac{\partial g_\alpha}{\partial v_z}\right. \\ &\quad \left. + E_{1z}\left[\left(1 - \frac{\boldsymbol{k}_\perp \cdot \boldsymbol{v}_\perp}{\omega}\right)\frac{\partial g_\alpha}{\partial v_z} + \frac{1}{\omega}\boldsymbol{k}_\perp \cdot \boldsymbol{v}_\perp \frac{v_z}{v_\perp}\frac{\partial g_\alpha}{\partial v_\perp}\right]\right\} . \end{aligned} \tag{7.40}$$

We have used $\boldsymbol{B}_1 = \omega^{-1}\boldsymbol{k} \times \boldsymbol{E}_1$. On the right-hand side with the terms $\sim \boldsymbol{v}_\perp$, we keep in mind the shift $\varphi \to \varphi - \Omega_\alpha \tau$. Also the relation

$$e^{i\kappa[\sin\varphi - \sin(\varphi - \chi)]} = \sum_{m,n=-\infty}^{+\infty} J_m(\kappa) J_n(\kappa) e^{i(m-n)\varphi + in\chi} \tag{7.41}$$

will be used. Inserting everything, we find a quite complicated expression for δf_α. The latter will be introduced into

$$\boldsymbol{j}_1 = \sum_\alpha q_\alpha \int \boldsymbol{v}\,\delta f_\alpha d^3 v . \tag{7.42}$$

The current density is the source term in the wave equation

$$\boldsymbol{k} \times (\boldsymbol{k} \times \boldsymbol{E}_1) + \frac{\omega^2}{c^2}\boldsymbol{E}_1 + i\omega\mu_0 \boldsymbol{j}_1 = 0 . \tag{7.43}$$

From here, we get an homogeneous algebraic system of equations which will be abbreviated as

$$\mathcal{D} \cdot \boldsymbol{E}_1 = 0 . \tag{7.44}$$

Nontrivial solutions $\boldsymbol{E}_1$ exist provided

$$\det \mathcal{D} = 0 . \tag{7.45}$$

Introducing $g_\alpha = n_\alpha F_\alpha$, we have the entries of $\mathcal{D}$ in the form

$$D_{xx}(\boldsymbol{k},\omega) = 1 - \frac{c^2 k_z^2}{\omega^2} + \sum_\alpha \frac{\omega_{p\alpha}^2}{\omega} \sum_{n=-\infty}^{\infty} \int d^3v v_\perp \frac{\frac{n^2}{b_\alpha^2} J_n^2(b_\alpha)}{(\omega - n\Omega_\alpha - k_z v_z)} \times \left[\frac{\partial F_\alpha}{\partial v_\perp} - \frac{k_z v_z}{\omega}\left(\frac{\partial F_\alpha}{\partial v_\perp} - \frac{v_\perp}{v_z}\frac{\partial F_\alpha}{\partial v_z}\right)\right], \tag{7.46}$$

$$D_{xy}(\boldsymbol{k},\omega) = i\sum_\alpha \frac{\omega_{p\alpha}^2}{\omega} \sum_{n=-\infty}^{\infty} \int d^3v v_\perp \frac{\frac{n}{b_\alpha} J_n(b_\alpha) J_n'(b_\alpha)}{(\omega - n\Omega_\alpha - k_z v_z)} \times \left[\frac{\partial F_\alpha}{\partial v_\perp} - \frac{k_z v_z}{\omega}\left(\frac{\partial F_\alpha}{\partial v_\perp} - \frac{v_\perp}{v_z}\frac{\partial F_\alpha}{\partial v_z}\right)\right] = -D_{yx}(\boldsymbol{k},\omega), \tag{7.47}$$

$$D_{xz}(\boldsymbol{k},\omega) = \frac{c^2 k_z k_\perp}{\omega^2} + \sum_\alpha \frac{\omega_{p\alpha}^2}{\omega} \sum_{n=-\infty}^{\infty} \int d^3v v_\perp \frac{\frac{n}{b_\alpha} J_n^2(b_\alpha)}{(\omega - n\Omega_\alpha - k_z v_z)} \times \left[\frac{\partial F_\alpha}{\partial v_z} + \frac{n\Omega_\alpha}{\omega}\left(\frac{v_z}{v_\perp}\frac{\partial F_\alpha}{\partial v_\perp} - \frac{\partial F_\alpha}{\partial v_z}\right)\right], \tag{7.48}$$

$$D_{yy}(\boldsymbol{k},\omega) = 1 - \frac{c^2(k_z^2 + k_\perp^2)}{\omega^2} + \sum_\alpha \frac{\omega_{p\alpha}^2}{\omega} \sum_{n=-\infty}^{\infty} \int d^3v v_\perp \frac{[J_n'(b_\alpha)]^2}{(\omega - n\Omega_\alpha - k_z v_z)} \times \left[\frac{\partial F_\alpha}{\partial v_\perp} - \frac{k_z v_z}{\omega}\left(\frac{\partial F_\alpha}{\partial v_\perp} - \frac{v_\perp}{v_z}\frac{\partial F_\alpha}{\partial v_z}\right)\right], \tag{7.49}$$

$$D_{yz}(\boldsymbol{k},\omega) = -i\sum_\alpha \frac{\omega_{p\alpha}^2}{\omega} \sum_{n=-\infty}^{\infty} \int d^3v v_\perp \frac{J_n(b_\alpha) J_n'(b_\alpha)}{(\omega - n\Omega_\alpha - k_z v_z)} \times \left[\frac{\partial F_\alpha}{\partial v_z} + \frac{n\Omega_\alpha}{\omega}\left(\frac{v_z}{v_\perp}\frac{\partial F_\alpha}{\partial v_\perp} - \frac{\partial F_\alpha}{\partial v_z}\right)\right], \tag{7.50}$$

$$D_{zx}(\boldsymbol{k},\omega) = \frac{c^2 k_z k_\perp}{\omega^2} + \sum_\alpha \frac{\omega_{p\alpha}^2}{\omega} \sum_{n=-\infty}^{\infty} \int d^3v v_z \frac{\frac{n}{b_\alpha} J_n^2(b_\alpha)}{(\omega - n\Omega_\alpha - k_z v_z)} \times \left[\frac{\partial F_\alpha}{\partial v_\perp} - \frac{k_z v_z}{\omega}\left(\frac{\partial F_\alpha}{\partial v_\perp} - \frac{v_\perp}{v_z}\frac{\partial F_\alpha}{\partial v_z}\right)\right], \tag{7.51}$$

$$D_{zy}(\boldsymbol{k},\omega) = i\sum_\alpha \frac{\omega_{p\alpha}^2}{\omega} \sum_{n=-\infty}^{\infty} \int d^3 v v_z \frac{J_n(b_\alpha) J_n'(b_\alpha)}{(\omega - n\Omega_\alpha - k_z v_z)} \times \left[\frac{\partial F_\alpha}{\partial v_\perp} - \frac{k_z v_z}{\omega} \left(\frac{\partial F_\alpha}{\partial v_\perp} - \frac{v_\perp}{v_z} \frac{\partial F_\alpha}{\partial v_z} \right) \right], \tag{7.52}$$

$$D_{zz}(\boldsymbol{k},\omega) = 1 - \frac{c^2 k_\perp^2}{\omega^2} + \sum_\alpha \frac{\omega_{p\alpha}^2}{\omega} \sum_{n=-\infty}^{\infty} \int d^3 v v_z \frac{J_n^2(b_\alpha)}{(\omega - n\Omega_\alpha - k_z v_z)} \times \left[\frac{\partial F_\alpha}{\partial v_z} + \frac{n\Omega_\alpha}{\omega} \left(\frac{v_z}{v_\perp} \frac{\partial F_\alpha}{\partial v_\perp} - \frac{\partial F_\alpha}{\partial v_z} \right) \right]. \tag{7.53}$$

We have set $b_\alpha = k_\perp v_\perp / \Omega_\alpha$ and $k_x = k_\perp$. The prime (′) at the Bessel functions J_n means differentiation with respect to the argument.

In the special case of wave propagation along $\boldsymbol{B}_0$, we let $k_\perp \to 0$ and get the much simpler expressions

$$D_{xz} = D_{zx} = D_{yz} = D_{zy} = 0 \tag{7.54}$$

as well as

$$\begin{aligned} D_{xx} &= D_{yy} \\ &= 1 - \frac{c^2 k_z^2}{\omega^2} + \frac{1}{4} \sum_\alpha \frac{\omega_{p\alpha}^2}{\omega} \\ &\quad \times \int d^3 v v_\perp \left[\frac{\partial F_\alpha}{\partial v_\perp} - \frac{k_z v_z}{\omega} \left(\frac{\partial F_\alpha}{\partial v_\perp} - \frac{v_\perp}{v_z} \frac{\partial F_\alpha}{\partial v_z} \right) \right] \\ &\quad \times \left[\frac{1}{\omega - \Omega_\alpha - k_z v_z} + \frac{1}{\omega + \Omega_\alpha - k_z v_z} \right], \end{aligned} \tag{7.55}$$

$$\begin{aligned} D_{xy} &= -D_{yx} \\ &= i \sum_\alpha \frac{\omega_{p\alpha}^2}{\omega} \int d^3 v v_\perp \left[\frac{\partial F_\alpha}{\partial v_\perp} - \frac{k_z v_z}{\omega} \left(\frac{\partial F_\alpha}{\partial v_\perp} - \frac{v_\perp}{v_z} \frac{\partial F_\alpha}{\partial v_z} \right) \right] \\ &\quad \times \left[\frac{1}{\omega - \Omega_\alpha - k_z v_z} - \frac{1}{\omega + \Omega_\alpha - k_z v_z} \right], \end{aligned} \tag{7.56}$$

$$D_{zz} = 1 + \sum_\alpha \frac{\omega_{p\alpha}^2}{\omega} \int d^3 v \frac{v_z \frac{\partial F_\alpha}{\partial v_z}}{\omega - k_z v_z} . \tag{7.57}$$

The last term leads to the well-known electrostatic dispersion relation

$$D_{zz} = 0 . \tag{7.58}$$

In addition, we have the electromagnetic part

$$D_{xx} D_{yy} + D_{xy}^2 = 0 , \tag{7.59}$$

which can be written in the form

$$(D_{xx} + i D_{xy})(D_{xx} - i D_{xy}) \equiv D^+ D^- = 0 , \tag{7.60}$$

leading to

$$1 - \frac{c^2 k_z^2}{\omega^2} + \sum_\alpha \frac{\omega_{p\alpha}^2}{\omega} \int d^3 v \frac{v_\perp}{2} \left[\frac{\partial F_\alpha}{\partial v_\perp} - \frac{k_z v_z}{\omega} \left(\frac{\partial F_\alpha}{\partial v_\perp} - \frac{v_\perp}{v_z} \frac{\partial F_\alpha}{\partial v_z} \right) \right]$$
$$\times \frac{1}{\omega \pm \Omega_\alpha - k_z v_z} = 0 \,. \tag{7.61}$$

Strictly perpendicular propagation to $\boldsymbol{B}_0$, that is, the limit $k_z \to 0$, is easy to handle when no current propagates parallel to $\boldsymbol{B}_0$,

$$\int_{-\infty}^{+\infty} dv_z v_z F_\alpha(v_\perp, v_z) = 0 \,. \tag{7.62}$$

Then,

$$D_{xz} = D_{zx} = D_{yz} = D_{zy} = 0 \,, \tag{7.63}$$

and the other elements are

$$D_{xx} = 1 + \sum_\alpha \frac{\omega_{p\alpha}^2}{\omega} \sum_{n=-\infty}^{+\infty} \int d^3 v v_\perp \frac{n^2 \frac{J_n^2(b_\alpha)}{b_\alpha^2}}{\omega - n\Omega_\alpha} \frac{\partial F_\alpha}{\partial v_\perp} \,, \tag{7.64}$$

$$D_{xy} = -D_{yx} = i \sum_\alpha \frac{\omega_{p\alpha}^2}{\omega} \sum_{n=-\infty}^{+\infty} d^3 v v_\perp \frac{n J_n(b_\alpha) \frac{J_n'(b_\alpha)}{b_\alpha}}{\omega - n\Omega_\alpha} \frac{\partial F_\alpha}{\partial v_\perp} \,, \tag{7.65}$$

$$D_{yy} = 1 - \frac{c^2 k_\perp^2}{\omega^2} + \sum_\alpha \frac{\omega_{p\alpha}^2}{\omega} \sum_{n=-\infty}^{+\infty} \int d^3 v v_\perp \frac{[J_n'(b_\alpha)]^2}{\omega - n\Omega_\alpha} \frac{\partial F_\alpha}{\partial v_\perp} \,, \tag{7.66}$$

$$D_{zz} = 1 - \frac{c^2 k_\perp^2}{\omega^2} - \sum_\alpha \frac{\omega_{p\alpha}^2}{\omega^2} + \sum_\alpha \frac{\omega_{p\alpha}^2}{\omega^2} \sum_{n=-\infty}^{+\infty} \int d^3 v \frac{n\Omega_\alpha J_n^2(b_\alpha)}{\omega - n\Omega_\alpha} \frac{v_z^2}{v_\perp} \frac{\partial F_\alpha}{\partial v_\perp} \,. \tag{7.67}$$

Again, we get two branches, namely,

$$D_{zz} = 0 \tag{7.68}$$

and

$$D_{xx} D_{yy} + D_{xy}^2 = 0 \,. \tag{7.69}$$

The first one leads to the "ordinary" mode with transversal polarization; the second one corresponds to the "extraordinary" mode with mixed polarization.

Two more simple, limiting cases are of interest. First, in case of purely electrostatic perturbations, we obtain the dispersion properties from

$$D_{elst}(\boldsymbol{k}, \omega) = 1 + \sum_\alpha \frac{\omega_{p\alpha}^2}{k^2} \sum_{n=-\infty}^{+\infty} \int d^3 v \frac{J_n^2(b_\alpha)}{\omega - n\Omega_\alpha - k_z v_z}$$
$$\times \left[k_z \frac{\partial F_\alpha}{\partial v_z} + \frac{n\Omega_\alpha}{v_\perp} \frac{\partial F_\alpha}{\partial v_\perp} \right] = 0 \,. \tag{7.70}$$

The electrostatic limit follows from the general dispersion relation for $\omega^2 \ll k^2c^2$ and $\omega_{p\alpha}^2 \ll k^2c^2$. The other simple case occurs for vanishing external magnetic fields $\boldsymbol{B}_0$. For isotropic distributions $F_\alpha = F_\alpha(v_\perp^2, v_z)$ in v_x and v_y, one obtains

$$D_{xx} = D_{yy} = 1 - \frac{c^2k_z^2}{\omega^2} + \sum_\alpha \frac{\omega_{p\alpha}^2}{\omega} \int d^3v \frac{\frac{v_\perp^2}{2}}{\omega - k_z v_z} \times \left[\left(1 - \frac{k_z v_z}{\omega}\right) \frac{\partial F_\alpha}{\partial v_\perp^2} + \frac{k_z}{\omega} \frac{\partial F_\alpha}{\partial v_z} \right], \tag{7.71}$$

whereas for more general distributions $F_\alpha = F_\alpha(v_x, v_y, v_z)$,

$$D_{xx} = 1 - \frac{c^2k_z^2}{\omega^2} + \sum_\alpha \frac{\omega_{p\alpha}^2}{\omega} \int d^3v \frac{v_x}{\omega - k_z v_z} \times \left[\left(1 - \frac{k_z v_z}{\omega}\right) \frac{\partial F_\alpha}{\partial v_x} + \frac{k_z v_x}{\omega} \frac{\partial F_\alpha}{\partial v_z} \right] \tag{7.72}$$

and

$$D_{yy} = 1 - \frac{c^2k_z^2}{\omega^2} + \sum_\alpha \frac{\omega_{p\alpha}^2}{\omega} \int d^3v \frac{v_y}{\omega - k_z v_z} \times \left[\left(1 - \frac{k_z v_z}{\omega}\right) \frac{\partial F_\alpha}{\partial v_y} + \frac{k_z v_y}{\omega} \frac{\partial F_\alpha}{\partial v_z} \right] \tag{7.73}$$

follow. The nondiagonal elements vanish, for example, $D_{xy} = D_{yx} = 0$, and so on. D_{zz} is of the same form as for propagation along a magnetic field $\boldsymbol{B}_0$.

Summarizing, for $B_0 = 0$, the electrostatic dispersion relation $D_{zz} = 0$ leads to high-frequency Langmuir oscillations

$$\omega_r^2 \approx \omega_{pe}^2(1 + 3k^2\lambda_{De}^2) \tag{7.74}$$

for the real part $\omega_r \equiv \operatorname{Re}\omega$ of ω in case of $\omega_r/k \gg (k_B T_e/m_e)^{1/2}$. The imaginary part $\omega_i \equiv \operatorname{Im}\omega$ describes the famous Landau damping

$$\omega_i \approx -\left(\frac{\pi}{8}\right)^{1/2} \frac{\omega_{pe}}{k^3\lambda_{De}^3} \exp\left(-\frac{1}{2k^2\lambda_{De}^2} - \frac{3}{2}\right). \tag{7.75}$$

For $T_e \gg T_i$ and $(k_B T_i/m_i)^{1/2} \ll \omega_r/k \ll (k_B T_e/m_e)^{1/2}$, ion-acoustic modes appear which obey

$$\omega_r^2 \approx \frac{k^2c_s^2}{1 + k^2\lambda_{De}^2}. \tag{7.76}$$

The electron Landau damping of these modes is

$$\omega_i \approx -\left(\frac{\pi}{8}\right)^{1/2} \left(\frac{m_e}{m_i}\right)^{1/2} (1 + k^2\lambda_{De}^2)^{-3/2} |\omega_r|. \tag{7.77}$$

Electromagnetic modes follow from

$$D_{xx}(k_z,\omega) = 1 - \frac{c^2 k_z^2}{\omega^2} - \sum_\alpha \frac{\omega_{p\alpha}^2}{\omega^2} + \sum_\alpha \frac{\omega_{p\alpha}^2}{\omega^2} \int d^3 v \frac{v_x^2}{\omega - k_z v_z} k_z \frac{\partial F_\alpha}{\partial v_z} = 0\,. \tag{7.78}$$

Introducing $\omega_p^2 = \sum_\alpha \omega_{p\alpha}^2$ and

$$F = \sum_\alpha \frac{\omega_{p\alpha}^2}{\omega_p^2} \int d v_x d v_y \frac{v_x^2}{W^2} F_\alpha(\boldsymbol{v})\,, \tag{7.79}$$

where W is the mean perpendicular velocity, one obtains

$$\omega^2 = \omega_p^2 + k_z^2 c^2 + \omega_p^2 \hat{G}\left(\frac{\omega}{k_z}\right), \tag{7.80}$$

with

$$\hat{G}(u) = W^2 \int_{-\infty}^{+\infty} dv \frac{F'(v)}{v-u} = W^2 \int_{-\infty}^{+\infty} dv \frac{F(v)}{(v-u)^2}\,. \tag{7.81}$$

Because of $\omega/k_z \geq c$, we may neglect $\hat{G}$ so that

$$\omega^2 \approx \omega_{pe}^2 + c^2 k_z^2\,. \tag{7.82}$$

The case of perpendicular propagation to an external magnetic field $\boldsymbol{B}_0 = B_0 \hat{z} \neq 0$ with $k_z = 0$ can be treated on the basis of Eq. (7.68) for $\boldsymbol{E}$ being parallel to $\boldsymbol{B}_0$; then, $\boldsymbol{k} = k_x \hat{x}$ is perpendicular to $\boldsymbol{B}_0$. The general solution of Eq. (7.68) is rather complicated. However, when the wavelength is much larger than the gyroradii, the approximate evaluation simplifies. Different branches occur which can easily be detected when ω is close to $n\Omega_\alpha$. Also, $k^2c^2 \gg \omega_{pe}^2$ may be assumed. For $n = 0$, we obtain the ordinary mode

$$\omega^2 \approx c^2 k^2 + \omega_{pe}^2\,, \tag{7.83}$$

whereas for $n \neq 0$, the frequency branches

$$\omega \approx n\Omega_\alpha \tag{7.84}$$

follow. In contrast to the ordinary mode, the solutions of Eq. (7.69) are strongly influenced by the magnetic field. For $\omega_{pe}^2 \ll k^2c^2$, we may neglect D_{xy} and should solve $D_{xx} \approx 0$ and $D_{yy} \approx 0$, respectively. Obviously, the first choice is nearly electrostatic (note $\boldsymbol{k} = k_x \hat{x}$) and is close to Eq. (7.70). In any case,

$$k^2 + \sum_\alpha \sum_{n=1}^{\infty} \frac{\omega_{p\alpha}^2 n^2 \Omega_\alpha^2}{\omega^2 - n^2\Omega_\alpha^2} \int J_n^2(b_\alpha) \frac{\partial F_\alpha}{\partial v_\perp^2} d^3 v = 0\,. \tag{7.85}$$

The solutions are known as Bernstein modes. They propagate in the frequency regions between the harmonics of Ω_α. To further evaluate the integral appearing in the second term on the left-hand side, we use a Maxwellian distribution and

$$\int_0^\infty \exp(-a^2x^2)x J_n(px) J_n(qx) dx = \frac{1}{2a^2} \exp\left(-\frac{p^2+q^2}{4a^2}\right) I_n\left(\frac{pq}{2a^2}\right); \tag{7.86}$$

I_n is the modified Bessel function of order n,

$$I_n(z) = e^{-i\pi\frac{n}{2}} J_n\left(e^{i\frac{\pi}{2}} z\right) . \tag{7.87}$$

For small k, the frequency of the lowest mode is

$$\omega \approx \omega_H \approx \sqrt{\omega_{pe}^2 + \Omega_e^2} , \tag{7.88}$$

that is, close to the upper (electron) hybrid frequency. The others have

$$\omega \approx \omega_n \approx n\Omega_e , \quad \text{for} \quad n \geq 2 . \tag{7.89}$$

In the case $\omega_H > 2|\Omega_e|$, that is, at large plasma densities, the situation is a little bit different. Then, the lowest frequency is slightly below the second harmonic, the others are still close to the higher harmonics of the gyrofrequency.

Electromagnetic cyclotron waves or extraordinary modes follow with Eq. (7.66) from $D_{yy} \approx 0$, that is,

$$k^2c^2 - \omega^2 - 2\pi\omega \sum_\alpha \sum_n \omega_{p\alpha}^2 \int dv_z dv_\perp^2 \frac{v_\perp^2 \left[J_n'(b_\alpha)\right]^2}{\omega - n\Omega_\alpha} \frac{\partial F_\alpha}{\partial v_\perp^2} = 0 . \tag{7.90}$$

The transversal oscillations ($\boldsymbol{k} = k_x\hat{x}$, $\boldsymbol{E} \approx E_y\hat{y}$) for $n = \pm 1$ approximately satisfy

$$k^2c^2 - \omega^2 + \sum_\alpha \frac{\omega^2\omega_{p\alpha}^2}{\omega^2 - \Omega_\alpha^2} \approx 0 , \tag{7.91}$$

that is, in the low-frequency limit ($\omega < \Omega_i$),

$$\omega^2 = \frac{k^2 v_A^2}{1 + \frac{v_A^2}{c^2}} . \tag{7.92}$$

These are magnetosonic waves with the Alfvén velocity

$$v_A = \frac{B_0}{\sqrt{\mu_0 n_{e0} m_i}} . \tag{7.93}$$

At higher frequencies ($\Omega_i < \omega < |\Omega_e| < \omega_{pe}$), the characteristic frequency is approximately

$$\omega \approx (\Omega_i |\Omega_e|)^{1/2} . \tag{7.94}$$

Electrostatic modes propagating parallel to an external magnetic field $\boldsymbol{B}_0$ (having $k_\perp = 0$) are governed by Eq. (7.57) with $D_{zz} = 0$. We have already found Langmuir waves and ion-acoustic oscillations. For electromagnetic waves along B_0, we should have a look at $D^\pm = 0$ which is identical to

$$\omega^2 = k^2 c^2 + 2\pi\omega \sum_\alpha \omega_{p\alpha}^2 \int dv_\perp dv_z v_\perp^3 \left(k_z v_z - \omega \pm \Omega_\alpha\right)^{-1} \times \left[\frac{\partial F_\alpha}{\partial v_\perp^2}\left(1 - \frac{k_z v_z}{\omega}\right) + \frac{k_z v_z}{\omega}\frac{\partial F_\alpha}{\partial v_z^2} \right] . \tag{7.95}$$

For cold plasmas [$F_\alpha \sim \delta(v_z)$], the complicated kinetic relation reduces to the hydrodynamic form

$$\frac{c^2 k_z^2}{\omega^2} = \varepsilon_\pm , \quad \varepsilon_+ = \varepsilon_1 + \varepsilon_2 , \quad \varepsilon_- = \varepsilon_1 - \varepsilon_2 , \tag{7.96}$$

with

$$\varepsilon_1 = 1 + \frac{\omega_{pe}^2}{\Omega_e^2 - \omega^2} + \frac{\omega_{pi}^2}{\Omega_i^2 - \omega^2} , \quad \varepsilon_2 = -\frac{\Omega_e}{\omega}\frac{\omega_{pe}^2}{\Omega_e^2 - \omega^2} - \frac{\Omega_i}{\omega}\frac{\omega_{pi}^2}{\Omega_i^2 - \omega^2} . \tag{7.97}$$

For frequencies $\omega \gg \Omega_i$, one finds the two approximate solutions, that is,

$$k = \frac{\omega}{c}\left(1 - \frac{\omega_{pe}^2}{\omega(\omega \mp |\Omega_e|)}\right)^{1/2} , \tag{7.98}$$

corresponding to circularly polarized electromagnetic waves propagating along a magnetic field. The upper sign belongs to right circular polarization (rotation in the same direction as the electron gyration), the lower sign belongs to left circular polarization. Right circularly polarized waves experience a resonance at $\omega \approx |\Omega_e|$ which requires a more exact kinetic treatment. (Remember that we made the cold plasma approximation.)

In the frequency regime $\Omega_i < \omega \ll |\Omega_e|$, only right circularly polarized waves can propagate with the wavenumber

$$k \approx \frac{\omega_{pe}}{c}\left(\frac{\omega}{|\Omega_e|}\right)^{1/2} . \tag{7.99}$$

They are called whistler waves since the propagation velocity depends on the frequency. A broadband packet of whistler modes shows the typical whistling sound at the position of an observer.

For lower frequencies, Eq. (7.98) has to be generalized by including the ion dynamics,

$$k^2 = \frac{\omega^2}{c^2}\left[1 - \frac{\omega_{pe}^2 + \omega_{pi}^2}{(\omega \pm \Omega_i)(\omega \mp |\Omega_e|)}\right] . \tag{7.100}$$

For $\omega \ll \Omega_i$, the new branch

$$k^2 \approx \frac{\omega^2}{c^2}\left[1 + \frac{n_{e0}(m_e + m_i)}{\epsilon_0 B_0^2}\right] \tag{7.101}$$

appears which can be rewritten as

$$\frac{\omega}{k} = \frac{v_A}{\left(1 + \frac{v_A^2}{c^2}\right)^{1/2}} . \tag{7.102}$$

We recognize the already found Alfvén waves which are transversal low-frequency modes, being independent of polarization. Next order corrections in the dispersion are of the order ω/Ω_i.

Finally, we briefly comment on the damping of the modes. For electrostatic modes propagating along $\boldsymbol{B}_0$, we can refer the reader to the previous discussion of electron and ion Landau damping. Waves with $k_z = 0$ are not significantly damped since the wave motion is not synchronized to the particle motion along B_0. On the other hand, for wave motion with $k_z \neq 0$, cyclotron damping can occur. When performing, for example, for $\omega_i \ll \omega_r$, similar calculations as before for Landau damping, we get similar results, however after replacing ω by $\omega \mp \Omega_\alpha$. Resonances between particles with velocity v_z and waves occur at

$$v_z = \frac{\omega_r \mp \Omega_\alpha}{k_z} . \tag{7.103}$$

We conclude with two examples. Assuming Maxwellian velocity distributions, for Alfvén waves, the damping rate $\gamma = \omega_i = \operatorname{Im}\omega$ is

$$\gamma \approx -\sqrt{\frac{\pi}{8}}\frac{\omega_{pi}^2}{|k_z|}\frac{1}{1 + \frac{c^2}{v_A^2}}\sqrt{\frac{m_i}{k_B T_i}}\exp\left(-\frac{B_0^2}{2\mu_0 n_{eo} k_B T_i}\frac{\Omega_i^2}{\omega_r^2}\right) . \tag{7.104}$$

The other example concerns whistler waves in the frequency regime $\Omega_i \ll \omega \ll |\Omega_e|$. The approximate damping rate is

$$\gamma \approx -\frac{\omega_{pe}^2}{|k_z|}\sqrt{\frac{m_e}{2k_B T_e}}\frac{1}{1 + k^2\frac{c^2}{\omega_r^2}}\exp\left(-\frac{\Omega_e^2}{k_z^2}\frac{m_e}{2k_B T_e}\right) . \tag{7.105}$$

7.1.3 Instabilities in Homogeneous Vlasov Systems

Now, let us turn to the explicit calculation of some exemplary plasma instabilities in homogeneous Vlasov systems. We start with a magnetic-field-free case in the electrostatic approximation. The distribution function consists of a superposition of two Maxwell distributions which has a minimum. According to the Penrose criterion, it may lead to instability. In the following, we may interpret F as a superposition of distribution functions belonging to different components α, β or as a "bump on the tail" distribution within one component. In any case, F is assumed as

$$F(v) = \frac{1}{\sqrt{\pi}}\left[\frac{1-\varepsilon}{v_T}e^{-\frac{v^2}{v_T^2}} + \frac{\varepsilon}{v_B}e^{-\frac{(v-v_D)^2}{v_B^2}}\right], \tag{7.106}$$

where the coefficients ε, v_B, v_T, and v_D are still to be fixed. Using the normalizations

$$\xi_1 = \frac{\omega}{k v_T}, \quad \xi_2 = \frac{\frac{\omega}{k} - v_D}{v_B}, \tag{7.107}$$

the dispersion relation is

$$1 - \frac{\omega_p^2}{k^2}\left[\frac{1-\varepsilon}{v_T^2}Z'(\xi_1) + \frac{\varepsilon}{v_B^2}Z'(\xi_2)\right] = 0\,. \tag{7.108}$$

We have further introduced $\omega_p^2 = \omega_{p\alpha}^2 + \omega_{p\beta}^2$ with $\omega_{p\alpha}^2 = (e^2/\varepsilon_0 m_\alpha)n_\alpha$ and $n_\alpha = (1-\varepsilon)\,n_{total}$. A similar definition has been used for $\omega_{p\beta}^2$. Asymptotic cases in which the Z-function can be expanded will be shown next.

In the first case, we assume $|\xi_1| \gg 1$, $|\xi_2| \ll 1$ such that the expansion yields

$$1 + 2\frac{\omega_p^2}{k^2}\left\{\frac{1-\varepsilon}{v_T^2}\left[-\frac{1}{2}\xi_1^{-2} - \frac{3}{4}\xi_1^{-4} - \dots\right] + \frac{\varepsilon}{v_B^2}\left[1 - 2\xi_2^2 + i\sqrt{\pi}\xi_2 + \dots\right]\right\}$$
$$= 0\,. \tag{7.109}$$

Now, we interpret F as the superposition of a Maxwellian for ions and a shifted Maxwellian for electrons such that

$$\varepsilon = \frac{\omega_{pe}^2}{\omega_p^2} \approx 1\,, \quad 1-\varepsilon = \frac{\omega_{pi}^2}{\omega_p^2} \approx \frac{m_e}{m_i}\,, \quad v_T = \sqrt{2}v_{thi}\,, \quad v_B = \sqrt{2}v_{the}\,. \tag{7.110}$$

For $v_D = 0$, we would recover the ion-acoustic modes. With $v_D \neq 0$, an ion-acoustic instability can take place. Introducing $u = \omega/k$, we obtain to lowest order

$$0 \approx 1 - \frac{\omega_{pi}^2}{u^2k^2} + \frac{\omega_{pe}^2}{v_{the}^2k^2} \quad \Rightarrow \quad u^2 \approx u_0^2 = \frac{c_s^2}{1+k^2\lambda_{De}^2}\,. \tag{7.111}$$

Corrections to $u \approx u_0$ in the form $u \approx u_0 + u_1$ follow from

$$0 \approx \frac{2\omega_{pi}^2}{u_0^3k^2}u_1 + i\sqrt{\frac{\pi}{2}}\frac{\omega_{pe}^2}{v_{the}^3k^2}(u_0 - v_D) \tag{7.112}$$

with the result

$$u_1 = -i\sqrt{\frac{\pi}{8}}\frac{m_i}{m_e}\frac{u_0^3}{v_{the}^3}\frac{u_0 - v_D}{u_0}u_0\,. \tag{7.113}$$

Clearly, for $v_D > u_0$, we have $\operatorname{Im}\omega > 0$, meaning instability, provided $|\xi_2| \approx |c_s - v_D|/v_{the} \ll 1$ is still satisfied.

The next example will be the *Buneman instability* [203] ("electron-ion two-stream instability") which will be discussed for $|\xi_1| \gg 1$, $|\xi_2| \gg 1$. The interpretation of F is the same as in the previous case. We further investigate the region $v_{ti} \ll u \ll v_D$ and $v_{te} \ll v_D$. Under these assumptions, the dispersion relation is

$$0 \approx 1 + \frac{2}{k^2}\left\{\frac{\omega_{pi}^2}{v_{thi}^2}\left[-\frac{1}{2}\left(\frac{v_{thi}}{u}\right)^2 - \frac{3}{2}\left(\frac{v_{thi}}{u}\right)^4 + \ldots\right]\right.$$
$$\left. + \frac{\omega_{pe}^2}{v_{the}^2}\left[-\frac{1}{2}\frac{v_{the}^2}{(u-v_D)^2} - \frac{3}{2}\frac{v_{the}^4}{(u-v_D)^4} + \ldots\right]\right\}. \tag{7.114}$$

The main conclusions can already be drawn from the lowest order form

$$k^2 \approx \frac{\omega_{pi}^2}{u^2} + \frac{\omega_{pe}^2}{(u-v_D)^2}. \tag{7.115}$$

Without solving the fourth-order equation explicitly, we approximate

$$k^2 \approx \frac{\omega_{pe}^2}{v_D^2} \tag{7.116}$$

as the value where maximum growth will occur. This can be justified *a posteriori* or by more accurate solutions. In that case, Eq. (7.115) simplifies to

$$\frac{m_e}{m_i} \approx \frac{u^2}{v_D^2}\left[1 - \left(1 - \frac{u}{v_D}\right)^{-2}\right]. \tag{7.117}$$

Introducing

$$x := \frac{-u}{v_D} \ll 1\,, \tag{7.118}$$

we may further simplify the dispersion relation to

$$x^3 \approx \frac{m_e}{2m_i}\,, \tag{7.119}$$

leading to the solution

$$x \approx \left(-\frac{1}{2} - i\frac{1}{2}\sqrt{3}\right)\left(\frac{m_e}{2m_i}\right)^{1/3}. \tag{7.120}$$

The growth rate $\gamma \approx 1/2\sqrt{3}(m_e/(2m_i))^{1/3} k v_D$ belongs to an oscillatory instability.

Within another interpretation of F, we can investigate beam instabilities ("bump on tail instability"). We consider an electron plasma with a smeared-out ion background. Then, ε designates the part of electrons with drift v_D within a beam of width v_B. To lowest order in ε, we again expand the Z-function such that

$$1 \approx \frac{\omega_{pe}^2}{k^2}\left(\frac{1}{u^2} + 3\frac{v_{the}^2}{u^4}\right) \tag{7.121}$$

follows with the approximate solution $\omega \approx \omega_{pe}$, or

$$u^2 \approx u_0^2 = \frac{\omega_{pe}^2}{k^2}\left(1 + 3k^2\lambda_{De}^2\right) . \tag{7.122}$$

For $k\lambda_{De} \ll 1$, the condition $\xi_1 \gg 1$ is satisfied. Therefore, the expansion leads to first order in ε

$$0 \approx -\varepsilon\left[-\frac{1}{2u_0^2} - \frac{3v_{the}^2}{2u_0^4}\right] + \frac{u_1}{u_0^3} + 6u_1\frac{v_{the}^2}{u_0^5} + i\sqrt{\pi}\varepsilon\frac{u_0 - v_D}{v_B^3} , \tag{7.123}$$

from where

$$\operatorname{Im}\omega = -\sqrt{\pi}k\varepsilon\left(\frac{u_0}{v_B}\right)^3 \frac{u_0 - v_D}{1 + 6\frac{v_{the}^2}{u_0^2}} \tag{7.124}$$

follows. Instability occurs for $v_D > u_0$. However, we should have in mind $|\xi_2| \ll 1$, that is,

$$\left|\frac{\frac{\omega_{pe}}{k} - v_D}{v_B}\right| \ll 1 , \tag{7.125}$$

which gives the region of applicability of the above calculation for the bump on tail instability.

Thus far, the streaming of plasma particles was the source of free energy leading to instability. Let us now briefly comment on deviations from isotropy in velocity space by considering the nonisotropic distribution functions

$$F_\alpha = \pi^{-3/2}(w_{\alpha\perp}^2 w_{\alpha\|})^{-1} \exp\left[-\frac{v_z^2}{w_{\alpha\|}^2} - \frac{v_y^2}{w_{\alpha\perp}^2}\right] \times \frac{1}{2}\left\{\exp\left[-\frac{(v_x + v_\alpha)^2}{w_{\alpha\perp}^2}\right] + \exp\left[-\frac{(v_x - v_\alpha)^2}{w_{\alpha\perp}^2}\right]\right\} . \tag{7.126}$$

Here, $w_{\alpha\|,\perp}^2 = 2(k_B T_{\alpha\|,\perp}/m_\alpha)$. The electromagnetic dispersion relation for $B_0 = 0$ and $k = k_\|$ is

$$-\omega^2 + k^2c^2 + \omega_p^2 - \sum_\alpha \omega_{p\alpha}^2 \frac{w_{\alpha\perp}^2 + 2v_\alpha^2}{w_{\alpha\|}^2}\left[1 + \xi_\alpha Z(\xi_\alpha)\right] = 0 . \tag{7.127}$$

Analyzing the dispersion relation shows that two types of solutions occur: Stable electromagnetic waves with phase velocities larger than c, and purely growing solutions $\omega \approx i\gamma$. For purely imaginary arguments ξ_α, we have $\xi_\alpha Z(\xi_\alpha) < 0$ and instability will occur in the region

$$0 < k^2 < k_0^2 = \sum_\alpha \frac{\omega_{p\alpha}^2}{c^2}\left(\frac{w_{\alpha\perp}^2 + 2v_\alpha^2}{w_{\alpha\|}^2} - 1\right) . \tag{7.128}$$

The condition for instability can be formulated as

$$\sum_{\alpha} \frac{\omega_{p\alpha}^2}{c^2} \left(\frac{w_{\alpha\perp}^2 + 2v_\alpha^2}{w_{\alpha\|}^2} - 1 \right) > 0\,. \tag{7.129}$$

Assuming an electron–proton plasma, the condition is

$$\frac{k_B T_{e\perp} + m_e v_e^2}{k_B T_{e\|}} + \frac{m_e}{m_i} \frac{k_B T_{i\perp} + m_i v_i^2}{k_B T_{i\|}} > 1 + \frac{m_e}{m_i}\,, \tag{7.130}$$

which can be further simplified for the Weibel instability [204] when

$$v_e \approx v_i \approx 0 \tag{7.131}$$

and the ion distribution is isotropic ($T_{i\perp} = T_{i\|}$). In that limit, the unstable region is

$$k^2 < \frac{\omega_{pe}^2}{c^2} \left(\frac{T_{e\perp}}{T_{e\|}} - 1 \right) \tag{7.132}$$

under the condition $T_{e\perp} > T_{e\|}$. Let us estimate the growth rate for $\varepsilon^2 = |T_{e\perp} - T_{e\|}|/T_{e\|} \ll 1$. Then,

$$\gamma^2 + k^2 c^2 - \omega_{pe}^2 \varepsilon^2 - \omega_{pe}^2 \frac{T_{e\perp}}{T_{e\|}} \xi_e Z(\xi_e) - \omega_{pi}^2 \xi_i Z(\xi_i) = 0\,. \tag{7.133}$$

Since $\xi_i = i\gamma/(\sqrt{2} k v_{thi})$ and $|\xi_i| \gg 1$, we obtain for $|\xi_e| \ll 1$ and $\varepsilon^2 \omega_{pe}^2 \gg \omega_{pi}^2$

$$k^2 c^2 - \omega_{pe}^2 \varepsilon^2 + \sqrt{\pi} \frac{\gamma}{k w_{e\|}} \omega_{pe}^2 \frac{T_{e\perp}}{T_{e\|}} \approx 0\,. \tag{7.134}$$

Its approximate solution is

$$\gamma \approx \frac{1}{\sqrt{\pi}} \frac{T_{e\|}}{T_{e\perp}} \left(\frac{k_0^2}{k^2} - 1 \right) \frac{c^2 k^2}{\omega_{pe}^2} k w_{e\|}\,. \tag{7.135}$$

Maximum growth occurs at $k^2 = k_0^2/2$ with

$$\gamma_{max} \approx \frac{1}{2} k_0 v_{the} \left(\frac{T_{e\|}}{T_{e\perp}} \right)^{1/2} \left| \frac{T_{e\perp}}{T_{e\|}} - 1 \right|^{1/2}\,. \tag{7.136}$$

Now, we turn to magnetized plasmas. Of course, when $\boldsymbol{k}$ is parallel to $B_0 \hat{z}$, nothing new will happen compared to the magnetic-field-free case ($B_0 = 0$) when we investigate $D_{zz}(k, \omega) = 0$. An interesting case is Eq. (7.61) for $k_\perp = 0$. After an integration by parts, we have

$$1 - \frac{c^2 k^2}{\omega^2} - \sum_{\alpha} \frac{\omega_{p\alpha}^2}{\omega^2} \int d^3 v \frac{(\omega - k_z v_z) F_\alpha - k_z \frac{v_x^2 + v_y^2}{2} \frac{\partial F_\alpha}{\partial v_z}}{\omega \pm \Omega_\alpha - k v_z} = 0\,. \tag{7.137}$$

Obviously, the detailed dependence of F_α on $v_\perp$ is not very important. When we introduce

$$F_\alpha = \pi^{-1/2} w_{\alpha\|}^{-1} \exp\left(-\frac{v_z^2}{w_{\alpha\|}^2}\right) G_\alpha\left(v_\perp^2\right) \tag{7.138}$$

with the normalized function G_α,

$$\int d^2 v_\perp G_\alpha\left(v_\perp^2\right) = 1 , \tag{7.139}$$

only the moment (effective "perpendicular" temperature)

$$k_B T_{\alpha\perp} = \frac{m_\alpha}{2} w_{\alpha\perp}^2 = \frac{m_\alpha}{2} \int d^2 v_\perp v_\perp^2 G_\alpha\left(v_\perp^2\right) \tag{7.140}$$

appears. Thus, we obtain (note $k = k_z$)

$$0 = 1 - \frac{k^2 c^2}{\omega^2} + \sum_\alpha \frac{\omega_{p\alpha}^2}{\omega^2} \left\{ \frac{\omega}{k w_{\alpha\|}} Z(\xi_\alpha^\pm) - \left(1 - \frac{T_{\alpha\perp}}{T_{\alpha\|}}\right) \left[1 + \xi_\alpha^\pm Z(\xi_\alpha^\pm)\right] \right\} , \tag{7.141}$$

with $\xi^\pm = (\omega \pm \Omega_\alpha)/k w_{\alpha\|} \equiv \omega_\alpha^\pm / k w_{\alpha\|}$. Depending on the parameters ($T_{\alpha\|} > T_{\alpha\perp}$ or $T_{\alpha\|} < T_{\alpha\perp}$), different instabilities are known. For $|\xi_\alpha| \gg 1$, the so called firehose instability appears. In that case, we approximate the dispersion relation as

$$0 \approx 1 - \frac{k^2 c^2}{\omega^2} - \sum_\alpha \frac{\omega_{p\alpha}^2}{\omega \omega_\alpha^\pm} + \sum_\alpha \frac{\omega_{p\alpha}^2}{\omega^2} \left(1 - \frac{T_{\alpha\perp}}{T_{\alpha\|}}\right) \frac{k^2 T_{\alpha\|}}{(\omega_\alpha^\pm)^2 m_\alpha} . \tag{7.142}$$

Assuming $|\omega| \ll |\Omega_\alpha|$ for all α, we use

$$\sum_\alpha \frac{\omega_{p\alpha}^2}{\Omega_\alpha} = 0 , \quad \sum_\alpha \frac{\omega_{p\alpha}^2}{\Omega_\alpha^2} = \frac{c^2}{v_A^2} ,$$
$$\sum_\alpha \omega_{p\alpha}^2 \frac{k_B (T_{\alpha\|} - T_{\alpha\perp})}{m_\alpha \Omega_\alpha^2} = \frac{1}{2} c^2 (\beta_\| - \beta_\perp) . \tag{7.143}$$

Here, β is the ratio of the kinetic pressure to the magnetic pressure, that is,

$$\beta \equiv 2\mu_0 \sum_\alpha n_\alpha \frac{k_B T_\alpha}{B_0^2} \equiv 2\mu_0 \frac{p}{B_0^2} . \tag{7.144}$$

The dispersion relation takes the simple form

$$0 \approx 1 - \frac{c^2 k^2}{\omega^2} + \frac{c^2}{v_A^2} + \frac{c^2 k^2}{\omega^2} \frac{1}{2} (\beta_\| - \beta_\perp) . \tag{7.145}$$

When the solution is written as $\omega = i\gamma$, we get

$$\gamma^2 = \frac{1}{2} \frac{k^2 v_A^2}{1 + \frac{v_A^2}{c^2}} (\beta_\| - \beta_\perp - 2) . \tag{7.146}$$

It shows instability for

$$\beta_\parallel > \beta_\perp + 2\,. \tag{7.147}$$

It is a purely growing mode. The result is not correct for very large k since then the used expansion of the Z-functions with respect to ξ_i will break down. Note that at $\beta_\parallel = \beta_\perp$, the real Alfvén-frequency appears for ω.

Finally, as a further example, we present the ordinary-mode electromagnetic instability. The wave vector is perpendicular to the external magnetic field. In this case, Eqs. (7.67) and (7.68) lead to

$$1 - \frac{c^2k^2}{\omega^2} - \sum_\alpha \frac{\omega_{p\alpha}^2}{\omega^2} + \sum_\alpha \frac{\omega_{p\alpha}^2}{\omega^2} \sum_{n=-\infty}^{+\infty} \frac{n\Omega_\alpha}{\omega - n\Omega_\alpha} \int d^3v\, J_n^2(b_\alpha) \frac{v_z^2}{v_\perp} \frac{\partial F_\alpha}{\partial v_\perp} = 0\,, \tag{7.148}$$

where $b_\alpha = kv_\perp/\Omega_\alpha$, $\mathbf{k} = k\hat{x}$, and $\mathbf{B}_0 = B_0\hat{z}$. The anisotropic distribution

$$F_\alpha(v_\perp^2, v_z) = \pi^{-3/2} w_{\alpha\parallel}^{-1} w_{\alpha\perp}^{-2} \exp\left(-\frac{v_\perp^2}{w_{\alpha\perp}^2}\right) \exp\left(\frac{-v_z^2}{w_{\alpha\parallel}^2}\right) \tag{7.149}$$

leads together with

$$\int_0^\infty dx\, x e^{-a^2x^2} J_n(px) J_n(qx) = \frac{1}{2a^2} \exp\left[-\frac{p^2+q^2}{4a^2}\right] I_n\left(\frac{pq}{2a^2}\right) \tag{7.150}$$

to

$$\omega^2 = k^2c^2 + \sum_\alpha \omega_{p\alpha}^2 \left\{1 + \sum_{n=-\infty}^{+\infty} \frac{T_{\alpha\parallel}}{T_{\alpha\perp}} \frac{n\Omega_\alpha}{\omega - n\Omega_\alpha} \exp(-\lambda_\alpha) I_n(\lambda_\alpha)\right\}. \tag{7.151}$$

The parameter

$$\lambda_\alpha = \frac{k^2 k_B T_{\alpha\perp}}{m_\alpha \Omega_\alpha^2} = \frac{1}{2} \frac{c^2k^2}{\omega_{p\alpha}^2} \beta_\perp \tag{7.152}$$

appears as the argument of the modified Bessel function I_n of first kind and order n. Besides the real solutions in the high-frequency regime, purely growing modes with $\mathrm{Re}\,\omega = 0$ exist. Using

$$I_n(x) = I_{-n}(x)\,, \quad \frac{n^2\Omega_\alpha^2}{\gamma^2 + n^2\Omega_\alpha^2} = 1 - \frac{\gamma^2}{\gamma^2 + n^2\Omega_\alpha^2}\,, \quad \sum_n I_n(\lambda_\alpha) = e^{\lambda_\alpha} \tag{7.153}$$

for $\omega = i\gamma$ and $\gamma \ll |\Omega_\alpha|$, we get the approximate dispersion relation

$$\gamma^2 L(k, \gamma^2) = R(k)\,. \tag{7.154}$$

Here,

$$L(k,\gamma^2) = 1 + \sum_\alpha \sum_{n\neq 0} \frac{T_{\alpha\|}}{T_{\alpha\perp}} \frac{\omega_{p\alpha}^2}{\gamma^2 + n^2\Omega_\alpha^2} \exp(-\lambda_\alpha) I_n(\lambda_\alpha)\,, \tag{7.155}$$

$$R(k) = \sum_\alpha \omega_{p\alpha}^2 \left[\frac{T_{\alpha\|}}{T_{\alpha\perp}} - 1 - \frac{T_{\alpha\|}}{T_{\alpha\perp}} \exp(-\lambda_\alpha) I_0(\lambda_\alpha)\right] - c^2k^2\,. \tag{7.156}$$

Since $I_n(x) > 0$, also $L(k,\gamma^2) > 0$, and instability requires $R(k) > 0$. Thus, there is an unstable region in k-space

$$\gamma^2 \geq \frac{R(k)}{L(k,0)} > 0\,. \tag{7.157}$$

For $T_i = T_{i\|} = T_{i\perp}$ and $m_e/m_i \to 0$, the expressions for L and R simplify considerably, for example,

$$R(k) \approx \omega_{pe}^2 \frac{T_{e\|}}{T_{e\perp}}\left[1 - \frac{T_{e\perp}}{T_{e\|}} - G(\lambda_e)\right], \quad G(\lambda_e) = \frac{2}{\beta_{e\|}}\lambda_e + \exp(-\lambda_e) I_0(\lambda_e)\,, \tag{7.158}$$

with $\beta_{e\|} = 2\mu_0 n_{eo} k_B T_{e\|}/B_0^2$. Analyzing the condition $R(k) > 0$ for instability, one finds growing modes for $T_{e\perp}/T_{e\|} \leq 1$ and $\beta_{e\|} \geq 2$.

7.2 Waves and Instabilities in Inhomogeneous Vlasov Systems

We generalize the previous section by allowing for spatial inhomogeneity. Still, we will mainly consider Vlasov systems.

7.2.1 Stationary Solutions and a Liapunov Stability Criterion

Stationary solutions g_α of the Vlasov equation have to obey

$$\left[\boldsymbol{v}\cdot\nabla + \frac{e_\alpha}{m_\alpha}(\boldsymbol{E} + \boldsymbol{v}\times\boldsymbol{B})\cdot\partial_{\boldsymbol{v}}\right] g_\alpha(\boldsymbol{q},\boldsymbol{v}) = 0\,. \tag{7.159}$$

We shall take advantage of the fact that solutions can be constructed from constants of motion. Generally, any function $f_\alpha(\boldsymbol{q},\boldsymbol{v};t) = f_\alpha(c_1,c_2,\ldots)$, where $c_1,c_2,\ldots$ are constants of motion, are solutions of the (time-dependent) Vlasov equations. Obviously, $df_\alpha/dt = \sum_i(\partial f_\alpha/\partial c_i)(dc_i/dt) = 0$. When the constants do not explicitly depend on time, $c_i = c_i(\boldsymbol{q},\boldsymbol{v})$, then $f_\alpha = g_\alpha(c_1,c_2,\ldots)$ is a solution of the stationary Vlasov equation. When using the integration along trajectories

$$\dot{\boldsymbol{q}}' = \boldsymbol{v}'\,,\quad \dot{\boldsymbol{v}}' = \frac{e_\alpha}{m_\alpha}\left[\boldsymbol{E} + \boldsymbol{v}'\times\boldsymbol{B}\right], \tag{7.160}$$

we require that at $t = t'$, the orbit $\boldsymbol{q}'(t')$ and $\boldsymbol{v}'(t')$ coincides with $\boldsymbol{q}$ and $\boldsymbol{v}$. Constants of motion $a = a(\boldsymbol{q}', \boldsymbol{v}')$ and $b = b(\boldsymbol{q}', \boldsymbol{v}')$ of Eq. (7.160) lead to the solution

$$g_\alpha = g_\alpha \left[a\left(\boldsymbol{q}', \boldsymbol{v}'\right), b\left(\boldsymbol{q}', \boldsymbol{v}'\right)\right]. \tag{7.161}$$

A simple conserved quantity of the Hamiltonian system with time-independent fields is the Hamiltonian

$$H = \frac{1}{2m_\alpha}\left(\boldsymbol{p}' - e_\alpha \boldsymbol{A}\right)^2 + q_\alpha \phi \tag{7.162}$$

itself. When we have a cyclic coordinate q'_k, then the conjugate momentum p'_k is also a conserved quantity. In the absence of external fields, $(E_0 = B_0 = 0)$, g_α may be an arbitrary function of v'_x, v'_y, and v'_z. For $E_0 = 0$ and $\boldsymbol{B}_0(\boldsymbol{q})\hat{z}$, the following conserved quantities are known

$$W_\alpha = \frac{1}{2} m_\alpha \left(v_x'^2 + v_y'^2 + v_z'^2\right), \quad p'_{\alpha\|} = m_\alpha v'_z, \quad L_z = (\boldsymbol{q}' \times \boldsymbol{p}') \cdot \hat{z}, \tag{7.163}$$

$$\xi_y = v'_y + \frac{e_\alpha}{m_\alpha} \int B_0(\boldsymbol{q}') d q'_x, \quad \xi_x = v'_x - \frac{e_\alpha}{m_\alpha} \int B_0(\boldsymbol{q}') d q'_y. \tag{7.164}$$

Besides these quantities, also

$$H_\alpha = \int G_\alpha(g_\alpha) d^3 q d^3 v, \tag{7.165}$$

where G_α is a bounded functional of g_α with finite derivative, will be conserved. Remember, for example, the entropy formula.

Now, let us derive a stability criterion for stationary inhomogeneous Vlasov systems. We specify to time-independent self-consistent electric and magnetic fields and assume that one coordinate is cyclic. For the self-consistent potentials ϕ^0 and $\boldsymbol{A}^0$, we assume

$$\frac{\partial}{\partial q_k}\phi^0 = \frac{\partial}{\partial q_k}\boldsymbol{A}^0 = 0 \quad \text{for} \quad k = y. \tag{7.166}$$

Writing g_α as $g_\alpha = g_\alpha(\varepsilon^0_\alpha, p^0_{\alpha k})$ with $\varepsilon^0_\alpha = m_\alpha/2v^2 + e_\alpha \phi^0$ and $p^0_{\alpha k} = m_\alpha v_k + e_\alpha A^0_k$, by integration, we arrive at

$$\rho^0 = \sum_\alpha \int d^3 v e_\alpha g_\alpha \left(\varepsilon^0_\alpha, p^0_{\alpha k}\right), \quad j^0_k = \sum_\alpha \int d^3 v e_\alpha v_k g_\alpha \left(\varepsilon^0_\alpha, p^0_{\alpha k}\right). \tag{7.167}$$

The charge density ρ^0 as well as the electric current density $\boldsymbol{j}^0$ become functionally dependent on ϕ^0 and A^0_k, that is, $\rho^0 = \rho^0(\phi^0, A^0_k)$, $j^0_k = j^0_k(\phi^0, A^0_k)$. They obey the differential equations $\nabla^2\phi^0 = -\rho^0/\varepsilon_0$, $\nabla^2 A^0_k = \mu_0 j^0_k$.

For a sufficient stability criterion, we introduce perturbations of the distribution functions in the form $f_\alpha = g_\alpha + \delta f_\alpha$. For reason of simplicity, we only keep the space variable x and ignore the variables z (i.e., we formulate the two-dimensional version of the stability criterion). We require $\partial_y = 0$ also for the perturbed (time-dependent) variables. A Liapunov functional for stability will be constructed out of the constants of motion

$$E = \sum_\alpha \int d^3q d^3v \frac{m_\alpha}{2} v^2 f_\alpha + \frac{1}{2} \int d^3q \left(\varepsilon_0 \mathbf{E}^2 + \frac{1}{\mu_0} \mathbf{B}^2 \right) \tag{7.168}$$

and

$$H = \sum_\alpha \int d^3q d^3v G_\alpha(f_\alpha, p_{\alpha y}) , \tag{7.169}$$

where the functional form of G_α will be determined later. We shall investigate the definiteness properties of $L := E + H$ in the vicinity of the stationary solutions g_α. Stability will be concluded when g_α corresponds to a minimum of L.

The first variation δL of L should vanish when a minimum occurs. The first variation is

$$\delta L = \sum_\alpha \int d^3q d^3v \left\{ \frac{\partial G_\alpha}{\partial g_\alpha} \delta f_\alpha + \frac{m_\alpha}{2} v^2 \delta f_\alpha + \frac{\partial G_\alpha}{\partial p_{\alpha y}} e_\alpha \delta A_y \right\}$$
$$+ \int d^3q \left\{ \varepsilon_0 \mathbf{E}^0 \cdot \delta \mathbf{E} + \frac{1}{\mu_0} \mathbf{B}^0 \cdot \delta \mathbf{B} \right\} . \tag{7.170}$$

The coefficients in the integrand have to be determined at the zeroth order values $\phi^0(x)$, $\mathbf{A}^0(x)$, ε^0_α, and $p^0_{\alpha y}$. Integrating by parts and using

$$m_\alpha \frac{\partial}{\partial p^0_{\alpha y}} G_\alpha \left(g_\alpha, p^0_{\alpha y} \right) = \frac{\partial}{\partial v_y} G_\alpha \left(g_\alpha, m_\alpha v_y + q_\alpha A^0_y \right)$$
$$- \frac{\partial g_\alpha}{\partial v_y} \frac{\partial}{\partial g_\alpha} G_\alpha \left(g_\alpha, p^0_{\alpha y} \right) , \tag{7.171}$$

one gets

$$\delta L = \sum_\alpha \int d^3q d^3v \left\{ \left[\frac{\partial G_\alpha}{\partial g_\alpha} + \frac{m_\alpha}{2} v^2 \right] \delta f_\alpha - \frac{e_\alpha}{m_\alpha} \delta A_y \frac{\partial G_\alpha}{\partial g_\alpha} \frac{\partial g_\alpha}{\partial v_y} \right\}$$
$$+ \int d^3q \left\{ \varepsilon_0 \phi^0 \nabla \cdot \delta \mathbf{E} + \frac{1}{\mu_0} \delta \mathbf{A} \cdot (\nabla \times \mathbf{B}_0) \right\}$$
$$= \sum_\alpha \int d^3q d^3v \left\{ \left[\frac{\partial G_\alpha}{\partial g_\alpha} + \varepsilon^0_\alpha \right] \delta f_\alpha - \frac{e_\alpha}{m_\alpha} \delta A_y \frac{\partial G_\alpha}{\partial g_\alpha} \frac{\partial g_\alpha}{\partial v_y} \right\}$$
$$+ \int d^3q \delta \mathbf{A} \cdot \mathbf{j}^0 . \tag{7.172}$$

This equation can be further simplified to

$$\delta L = \sum_\alpha \int d^3q d^3v \left\{ \left[\frac{\partial G_\alpha}{\partial g_\alpha} + \varepsilon_\alpha^0 \right] \delta f_\alpha \right.$$
$$\left. + e_\alpha \delta A_y \left[\frac{1}{m_\alpha} \frac{\partial}{\partial v_y} \left(\frac{\partial G_\alpha}{\partial g_\alpha} \right) + v_y \right] g_\alpha \right\} . \tag{7.173}$$

At this stage, we postulate the form of G_α such that $\delta L = 0$. The choice

$$\frac{\partial G_\alpha}{\partial g_\alpha} = -\varepsilon_\alpha^0 \tag{7.174}$$

does it. Defining the inverse function $\tilde{g}_\alpha$ of g_α with respect to $-\varepsilon_\alpha^0$, that is,

$$-\varepsilon_\alpha^0 = \tilde{g}_\alpha \left(g_\alpha, p_{\alpha y}^0 \right) , \tag{7.175}$$

the solution for G_α is

$$G_\alpha = G_\alpha \left(g_\alpha, p_{\alpha y}^0 \right) = \int_0^{g_\alpha} d\xi \tilde{g}_\alpha \left(\xi, p_{\alpha y}^0 \right) . \tag{7.176}$$

Now, the second variation

$$\delta^2 L = \frac{1}{2} \sum_\alpha \int d^3q d^3v \left\{ \frac{\partial^2 G_\alpha}{\partial g_\alpha^2} (\delta f_\alpha)^2 + 2 e_\alpha \frac{\partial^2 G_\alpha}{\partial g_\alpha \partial p_{\alpha y}} \delta f_\alpha \delta A_y \right.$$
$$\left. + e_\alpha^2 \frac{\partial^2 G_\alpha}{\partial p_{\alpha y}^2} (\delta A_y)^2 \right\} + \frac{1}{2} \int d^3q \left[\varepsilon_0 (\delta \boldsymbol{E})^2 + \frac{1}{\mu_0} (\delta \boldsymbol{B})^2 \right] . \tag{7.177}$$

Introducing the potentials,

$$\delta \boldsymbol{E} = -\nabla \phi - \frac{\partial}{\partial t} \delta \boldsymbol{A} , \tag{7.178}$$

together with the gauge $\nabla \cdot \delta \boldsymbol{A} = 0$, and using $(\delta \boldsymbol{B})^2 = (\delta B_y)^2 + (\nabla \delta A_y)^2$, we obtain

$$\delta^2 L = \frac{1}{2} \sum_\alpha \int d^3q d^3v \left\{ \frac{\partial^2 G_\alpha}{\partial g_\alpha^2} \left[\delta f_\alpha + \frac{\left(\frac{\partial^2 G_\alpha}{\partial g_\alpha \partial p_{\alpha y}} \right)}{\left(\frac{\partial^2 G_\alpha}{\partial g_\alpha^2} \right)} e_\alpha \delta A_y \right]^2 \right.$$
$$\left. + \left[\frac{\partial^2 G_\alpha}{\partial p_{\alpha y}^2} - \frac{\left(\frac{\partial^2 G_\alpha}{\partial g_\alpha \partial p_{\alpha y}} \right)^2}{\left(\frac{\partial^2 G_\alpha}{\partial g_\alpha^2} \right)} \right] (e_\alpha \delta A_y)^2 \right\}$$
$$+ \frac{1}{2} \int d^3q \left\{ \varepsilon_0 (\nabla \delta \phi)^2 + \frac{1}{\mu_0} (\nabla \delta A_y)^2 + \varepsilon_0 \left(\frac{\partial}{\partial t} \delta \boldsymbol{A} \right)^2 + \frac{1}{\mu_0} (\delta B_y)^2 \right\} . \tag{7.179}$$

We further have from Eqs. (7.174) and (7.175)

$$\frac{\partial^2 G_\alpha}{\partial g_\alpha^2} = \frac{\partial \tilde{g}_\alpha}{\partial g_\alpha} = -\left(\frac{\partial g_\alpha}{\partial \varepsilon^0}\right)^{-1} . \tag{7.180}$$

Next, differentiating both sides of Eq. (7.174) with respect to $p^0_{\alpha y}$ leads to

$$0 = \frac{\partial^2 G_\alpha}{\partial g_\alpha^2}\frac{\partial g_\alpha}{\partial p^0_{\alpha y}} + \frac{\partial^2 G_\alpha}{\partial g_\alpha \partial p_{\alpha y}} . \tag{7.181}$$

Therefore,

$$\frac{\partial^2 G_\alpha}{\partial g_\alpha \partial p_{\alpha y}} = -\frac{1}{e_\alpha}\frac{\partial^2 G_\alpha}{\partial g_\alpha^2}\frac{\partial g_\alpha}{\partial A^0_y} . \tag{7.182}$$

From

$$\sum_\alpha e_\alpha \int d^3v \frac{\partial G_\alpha}{\partial p_{\alpha y}} = -\sum_\alpha \int d^3v \frac{e_\alpha}{m_\alpha}\frac{\partial G_\alpha}{\partial g_\alpha}\frac{\partial g_\alpha}{\partial v_y} = -j^0_y , \tag{7.183}$$

we find

$$\frac{\partial j^0_y}{\partial A^0_y} = -\sum_\alpha \int d^3v \left[e_\alpha^2 \frac{\partial^2 G_\alpha}{\partial p^2_{\alpha y}} + e_\alpha \frac{\partial^2 G_\alpha}{\partial g_\alpha \partial p_{\alpha y}}\frac{\partial g_\alpha}{\partial A^0_y}\right] \tag{7.184}$$

or

$$\frac{\partial j^0_y}{\partial A^0_y} = -\sum_\alpha \int d^3v e_\alpha^2 \left[\frac{\partial^2 G_\alpha}{\partial p^2_{\alpha y}} - \frac{\left(\frac{\partial^2 G_\alpha}{\partial g_\alpha}\partial p_{\alpha y}\right)^2}{\left(\frac{\partial^2 G_\alpha}{\partial g_\alpha^2}\right)}\right] . \tag{7.185}$$

Putting everything together, we may write the second variation $\delta^2 L$ in a form which allows one to discuss its sign,

$$\begin{aligned}\delta^2 L = {} & \frac{1}{2}\sum_\alpha \int d^3q d^3v \left\{ -\left(\frac{\partial g_\alpha}{\partial \varepsilon^0_\alpha}\right)^{-1}\left(\delta f_\alpha - \frac{\partial g_\alpha}{\partial A^0_y}\delta A_y\right)^2\right\} \\ & + \frac{1}{2}\int d^3q \left\{ \varepsilon_0 (\nabla\delta\phi)^2 + \frac{1}{\mu_0}(\nabla\delta A_y)^2 \right. \\ & \left. - \frac{\partial j^0_y}{\partial A^0_y}(\delta A_y)^2 + \varepsilon_0(\partial_t \delta \mathbf{A})^2 + \frac{1}{\mu_0}(\delta B_y)^2 \right\} . \end{aligned} \tag{7.186}$$

It makes sense to assume for g_α a monotonic decrease with energy,

$$\frac{\partial g_\alpha}{\partial \varepsilon^0_\alpha} < 0 . \tag{7.187}$$

Most terms contributing to the second variation of L are positive (semidefinite). To investigate the remaining ones, we formulate the Schrödinger eigenvalue problem

$$-\nabla^2 \delta A_y - \mu_0 \frac{\partial j^0_y}{\partial A^0_y}\delta A_y = \lambda \delta A_y . \tag{7.188}$$

The "potential" is given by $\partial j^0_y/\partial A^0_y$. The eigenvalue condition $\lambda \geq 0$ represents a sufficient criterion for stability.

7.2.2 Instabilities in Inhomogeneous Vlasov Systems

The opposite case of instability can, in principle, be analyzed within a normal mode analysis. For weakly inhomogeneous systems, a local analysis leading to a local dispersion relation $\det \mathcal{D}(\boldsymbol{k}, \omega) = 0$ may be a reasonable approximation. However, the local approximation is in many cases too restrictive, and more general methods based on integral equations may be the matter of choice. In the following, we present examples of the normal mode analysis for weakly inhomogeneous systems.

Let us start with a stationary situation with an zeroth order magnetic field $\boldsymbol{B}_0 = B_0(x)\hat{z}$, depending only on the coordinate x (slab approximation) such that the vector potential can be written in the form $\boldsymbol{A}_0 \equiv \boldsymbol{A}^0 = A_0(x)\hat{y}$. There should be no zeroth order electric field, $\phi^0 \equiv \varphi_0 = 0$. In addition, we only investigate stability with respect to electrostatic perturbations $\boldsymbol{E}_1 = -\nabla\delta\varphi$.

The starting point is the linearized Vlasov equation

$$\partial_t \delta f_a + \boldsymbol{v} \cdot \partial_{\boldsymbol{q}} \delta f_a + \frac{e_a}{m_a}(\boldsymbol{v} \times \boldsymbol{B}_0) \cdot \partial_{\boldsymbol{v}} \delta f_a = \frac{e_a}{m_a} \nabla \delta\varphi \cdot \partial_{\boldsymbol{v}} g_a \,, \tag{7.189}$$

where the zeroth order distribution function g_a depends on $\varepsilon_{a\perp} = 1/2 m_a (v_x^2 + v_y^2)$, $p_{ay} = m_a v_y + e_a A_0(x)$, and v_z, that is, $g_a = g_a(\varepsilon_{a\perp}, p_{ay}, v_z)$. Through p_{ay}, the distribution function is space-dependent. Integrating along the trajectories $\boldsymbol{q}'(t'), \boldsymbol{v}'(t')$ with the initial conditions $\boldsymbol{q}'(t' = t) = \boldsymbol{q}$ and $\boldsymbol{v}'(t' = t) = \boldsymbol{v}$, we obtain the solution

$$\delta f_a = \frac{e_a}{m_a} \int_{-\infty}^{t} dt' \nabla' \delta\varphi \left(\boldsymbol{q}', \boldsymbol{v}', t'\right) \cdot \partial_{\boldsymbol{v}'} g_a \,. \tag{7.190}$$

Since x is the only relevant space-coordinate, we may Fourier-transform in y and z an. The normal mode ansatz

$$\delta f_a(x, y, z, \boldsymbol{v}, t) = \delta f_a\left(x, k_y, k_z, \boldsymbol{v}, t\right) \exp\left[i k_y y + i k_z z - i\omega t\right] \tag{7.191}$$

will be used. With that ansatz,

$$\delta f_a = \frac{e_a}{m_a} \int_{-\infty}^{t} dt' \nabla' \delta\varphi \cdot \partial_{\boldsymbol{v}'} g_a e^{i k_y (y' - y) + i k_z (z' - z) - i\omega(t' - t)} \tag{7.192}$$

follows. Weakly inhomogeneity will result in a weak space-dependence of g_a on x entering via the weakly varying magnetic field strength $B_0(x)$. To lowest order, we approximate $\hat{B}$ by $B_0(x) \approx \hat{B} = \text{const}$, taking the value close to the chosen position x. Under this assumption, the trajectories are

$$v_z' = v_z \,, \quad z' - z = v_z(t' - t) \,, \quad v_\perp^2 \equiv v_x^2 + v_y^2 = v_x'^2 + v_y'^2 \,, \tag{7.193}$$

and

$$v'_x = v_\perp \cos(\Omega_a \tau - \chi) \,, \quad v'_y = -v_\perp \sin(\Omega_a \tau - \chi) \,, \tag{7.194}$$

$$x' = x + \frac{v_\perp}{\Omega_a}\left[\sin(\Omega_a \tau - \chi) + \sin\chi\right] \,,$$
$$y' = y + \frac{v_\perp}{\Omega_a}\left[\cos(\Omega_a \tau - \chi) - \cos\chi\right] . \tag{7.195}$$

Here, $\tau = t' - t$.

When proceeding, we formally characterize the weak space-dependence by the smallness parameter $\delta_a \ll 1$. For example, when designating the weak dependence of g_a on p_{ay} (and not showing the dependence on $\varepsilon_{a\perp}$ and v_z), we first normalize and then formally write, with $\delta_a \ll 1$,

$$g_a = g_a\left[\ldots, \delta_a\left(\frac{v_y}{\hat{\Omega}_a L} + \frac{A_0}{\hat{B}L}\right)\right] \tag{7.196}$$

where $\hat{\Omega}_a = e_a \hat{B}/m_a$, and L is the characteristic inhomogeneity length. At the very end, we shall return to the previous variables by replacing δ_a by one. For small δ_a, we expand

$$g_a \approx g_a^{(0)}(\ldots) + \delta_a\left(\frac{v_y}{\hat{\Omega}_a L} + \frac{A_0}{\hat{B}L}\right) g_a^{(1)}(\ldots) \,. \tag{7.197}$$

Since $g_a^{(0)}$ as well as $g_a^{(1)}$ are isotropic in v_y and v_x, and the total system is neutral, we set $\delta_i = \delta_e \equiv \delta$. Furthermore, $A_0(x)$ follows from the Maxwell equation

$$\partial_x^2 A_0 = \delta\mu_0 \sum_a \int d^3v \frac{e_a}{\hat{\Omega}_a L} v_y^2 g_a^{(1)} \,. \tag{7.198}$$

Because of

$$\beta = \frac{\mu_0 p}{\hat{B}^2} \,, \tag{7.199}$$

the right-hand side will be estimated by $\delta\beta\hat{B}/L$. Since this constant is x-independent, we find the solution

$$A_0(x) = \frac{1}{2L}\delta\beta\hat{B}x^2 + \hat{B}x \,. \tag{7.200}$$

Thus, $\beta \ll 1$ is consistent with the adopted approximation. For g_a, we now approximately have

$$g_a = g_a^{(0)}(\ldots) + \delta\left(\frac{v_y}{\hat{\Omega}_j L} + \frac{x}{L} + \frac{1}{2}\delta\beta\frac{x^2}{L^2}\right) g_a^{(1)}(\ldots) \,. \tag{7.201}$$

To further evaluate Eq. (7.192), we additionally assume

$$|\partial_x \delta\varphi| \ll |k_y \delta\varphi|, |k_z \delta\varphi| \,. \tag{7.202}$$

The solution of the linearized Vlasov is now written in the form

$$\delta f_\alpha = i\frac{e_\alpha}{m_\alpha}\delta\varphi \int_{-\infty}^{0} d\tau \left[k_y \frac{\partial g_\alpha}{\partial v'_y} + k_z \frac{\partial g_\alpha}{\partial v'_z} \right] \times \exp\left\{ i\frac{v_\perp}{\Omega_\alpha}\left[\cos(\Omega_\alpha \tau - \chi) - \cos\chi\right] k_y \right\} \times \exp\{i(k_z v_z - \omega)\tau\}, \tag{7.203}$$

where we introduce

$$\frac{1}{m_\alpha}\frac{\partial g_\alpha}{\partial v'_y} = v'_y \frac{\partial g_\alpha}{\partial \varepsilon_{\alpha\perp}} + \frac{\partial g_\alpha}{\partial p_{\alpha y}} = -v_\perp \sin(\Omega_\alpha \tau - \chi)\frac{\partial g_\alpha}{\partial \varepsilon_{\alpha\perp}} + \frac{\partial g_\alpha}{\partial p_{\alpha y}} \tag{7.204}$$

and

$$\frac{\partial g_\alpha}{\partial v'_z} = \frac{\partial g_\alpha}{\partial v_z}\,. \tag{7.205}$$

Using

$$e^{i\xi \sin\delta} = \sum_{n=-\infty}^{\infty} J_n(\xi) e^{in\delta}\,, \quad \xi_\alpha = k_y \frac{v_\perp}{\Omega_\alpha}, \tag{7.206}$$

a short calculation leads to

$$\delta f_\alpha = \frac{e_\alpha}{m_\alpha}\delta\varphi \sum_{n=-\infty}^{\infty} \left[n\Omega_\alpha m_\alpha \frac{\partial g_\alpha}{\partial \varepsilon_{\alpha\perp}} + m_\alpha \frac{\partial g_\alpha}{\partial p_{\alpha y}} k_y + \frac{\partial g_\alpha}{\partial v_z} k_z \right] \times \frac{\exp\left\{in\left(\frac{\pi}{2} - \chi\right) - i\xi_\alpha \cos\chi\right\}}{k_z v_z - \omega + n\Omega_\alpha} J_n(\xi_\alpha)\,. \tag{7.207}$$

Next, we rewrite the right-hand side as

$$\begin{aligned}\delta f_\alpha &= \frac{e_\alpha}{m_\alpha}\delta\varphi \sum_{n=-\infty}^{\infty} \left[m_\alpha \frac{\partial g_\alpha}{\partial \varepsilon_{\alpha\perp}} J_n(\xi_\alpha) - G_\alpha \frac{J_n(\xi_\alpha)}{\omega - k_z v_z - n\Omega_\alpha} \right] \times e^{in\left(\frac{\pi}{2} - \chi\right) - i\xi_\alpha \cos\chi} \\ &= \frac{e_\alpha}{m_\alpha}\delta\varphi \left\{ m_\alpha \frac{\partial g_\alpha}{\partial \varepsilon_{\alpha\perp}} - G_\alpha \sum_{n=-\infty}^{+\infty} \frac{J_n(\xi_\alpha)}{\omega - k_z v_z - n\Omega_\alpha} e^{in\left(\frac{\pi}{2} - \chi\right) - i\xi_\alpha \cos\chi} \right\},\end{aligned} \tag{7.208}$$

where

$$G_\alpha := \omega m_\alpha \frac{\partial g_\alpha}{\partial \varepsilon_{\alpha\perp}} + k_z \left(\frac{\partial g_\alpha}{\partial v_z} - v_z m_\alpha \frac{\partial g_\alpha}{\partial \varepsilon_{\alpha\perp}} \right) + k_y m_\alpha \frac{\partial g_\alpha}{\partial p_{\alpha y}}\,. \tag{7.209}$$

Note

$$e^{i\xi_\alpha \cos\chi} = \sum_{n=-\infty}^{\infty} J_n(\xi_\alpha) e^{in\left(\frac{\pi}{2} - \chi\right)}\,. \tag{7.210}$$

When G_α only weakly depends on $p_{\alpha y}$, we may approximate

$$G_\alpha \approx G_\alpha\left(\varepsilon_{\alpha\perp}, v_z, \frac{e_\alpha}{m_\alpha}A_0\right) + m_\alpha v_y \frac{\partial G_\alpha}{\partial p_{\alpha y}}$$
$$= \hat{G}_\alpha + m_\alpha v_y \frac{\partial \hat{G}_\alpha}{\partial p_{\alpha y}} = \hat{G}_\alpha + m_\alpha v_\perp \sin\chi \frac{\partial \hat{G}_\alpha}{\partial p_{\alpha y}} . \tag{7.211}$$

Averaging δf_α over χ leads to

$$\frac{1}{2\pi}\int_0^{2\pi} d\chi \delta f_\alpha = \frac{e_\alpha}{m_\alpha}\delta\varphi \left\{ m_\alpha \frac{\partial \hat{g}_\alpha}{\partial \varepsilon_{\alpha\perp}} - \sum_{n=-\infty}^{+\infty} \frac{J_n^2(\xi_\alpha)}{\omega - k_z v_z - n\Omega_\alpha} G_{1\alpha} \right\} , \tag{7.212}$$

where

$$\hat{g}_\alpha = g_\alpha\left(\varepsilon_{\alpha\perp}, v_z, \frac{e_\alpha}{m_\alpha}A_0\right) , \tag{7.213}$$

$$G_{1\alpha} = \hat{G}_\alpha + \frac{m_\alpha \Omega_\alpha}{k_y} n \frac{\partial \hat{G}_\alpha}{\partial p_{\alpha y}} \approx \hat{G}_\alpha + \frac{n}{k_y}\frac{\partial \hat{G}_\alpha}{\partial x} , \tag{7.214}$$

and

$$J_{n-1}(\zeta) + J_{n+1}(\zeta) = \frac{2n}{\zeta} J_n(\zeta) \tag{7.215}$$

has been used.

For electrostatic fluctuations, we use the Poisson equation which, after introducing the just calculated density perturbations, leads to

$$1 - \sum_\alpha \frac{e_\alpha^2}{\varepsilon_0 m_\alpha k^2} \int d^3 v \left\{ \frac{\partial \hat{g}_\alpha}{\partial \varepsilon_\perp} - \sum_{n=-\infty}^{\infty} \frac{J_n^2(\xi_\alpha)}{\omega - n\Omega_\alpha - k_z v_z} G_{1\alpha} \right\} = 0 , \tag{7.216}$$

that is, the electrostatic dispersion relation. The notation has been simplified by introducing $\varepsilon_\perp = \varepsilon_{\alpha\perp}/m_\alpha$ (wherever no confusion is expected). The dispersion relation will be evaluated with $\hat{g}_\alpha = \hat{g}_\alpha(\varepsilon, e_\alpha A_0/m_\alpha)$ and $A_0 \approx \hat{B}_0 x$ such that for weak spatial variations,

$$\hat{g} \approx \hat{g}(\varepsilon, 0) + x\partial_x \hat{g}(\varepsilon, 0) \tag{7.217}$$

follows. Here, $\varepsilon = \varepsilon_\perp + v_z^2/2$, indicating that we have a joined dependence on $\varepsilon_\perp$ and v_z. Therefore,

$$k_z\left(\frac{\partial g_\alpha}{\partial v_z} - v_z \frac{\partial g_\alpha}{\partial \varepsilon_\perp}\right) = 0 \tag{7.218}$$

and

$$\hat{G}_\alpha = \omega \frac{\partial}{\partial \varepsilon} \left(\hat{g}_\alpha + x \hat{g}'_\alpha\right) + \frac{k_y}{\Omega_\alpha} \hat{g}'_\alpha \, ; \tag{7.219}$$

the dash (′) means differentiation with respect to x. The function $G_{1\alpha}$ will be approximated by

$$\begin{aligned} G_{1\alpha} &\approx \omega \frac{\partial}{\partial \varepsilon} \left(\hat{g}_\alpha + x \hat{g}'_\alpha\right) + \frac{k_y}{\Omega_\alpha} \hat{g}'_\alpha + \frac{n}{k_y} \omega \frac{\partial}{\partial \varepsilon} \hat{g}'_\alpha \\ &\approx \omega \frac{\partial}{\partial \varepsilon} \hat{g}_\alpha + \frac{k_y}{\Omega_\alpha} \hat{g}'_\alpha + \frac{n}{k_y} \omega \frac{\partial}{\partial \varepsilon} \hat{g}'_\alpha \, . \end{aligned} \tag{7.220}$$

We recognize that the local approximation effectively lets us evaluate the variables at $x \approx 0$. For specific applications, we have to choose the form of $\hat{g}_\alpha$, for example, as a local Maxwellian

$$\hat{g}_\alpha = n_\alpha(x) \left[\frac{2\pi k_B T_\alpha}{m_\alpha}\right]^{-\frac{3}{2}} \exp\left[-\frac{m_\alpha v^2}{2 k_B T_\alpha}\right] . \tag{7.221}$$

Then,

$$\frac{\partial}{\partial \varepsilon} \hat{g}_\alpha = -\frac{m_\alpha}{k_B T_\alpha} \hat{g}_\alpha \, , \quad \hat{g}'_\alpha = \frac{\partial \hat{g}_\alpha}{\partial x} = n'_\alpha \frac{\hat{g}_\alpha}{n_\alpha(x)} + T'_\alpha \frac{\partial \hat{g}_\alpha}{\partial T_\alpha} \, . \tag{7.222}$$

In the following, we shall further assume $T'_\alpha = 0$. In this case, the function $G_{1\alpha}$ can be written as

$$G_{1\alpha} \approx -\frac{\omega m_\alpha}{k_B T_\alpha} \left[1 + \left(\frac{n}{k_y} - \frac{k_y k_B T_\alpha}{\Omega_\alpha m_\alpha \omega}\right) \kappa\right] \hat{g}_\alpha \, , \tag{7.223}$$

with the inhomogeneity parameter

$$\kappa = \partial_x \left[\ln n_\alpha(x)\right] \, . \tag{7.224}$$

We will measure the wavelength of the perturbations in gyroradii by introducing

$$z_\alpha := \frac{k_y^2 k_B T_\alpha}{m_\alpha \Omega_\alpha^2} = k_y^2 \rho_\alpha^2 \, . \tag{7.225}$$

When restricting to low-frequency perturbations, even for $z_\alpha < 1$, we will consider the region

$$\frac{\Omega_\alpha z_\alpha}{\omega} \gg 1 \, . \tag{7.226}$$

Under this assumption, the term proportional to n on the right-hand side of Eq. (7.223) can be neglected. For the Bessel functions, the relation

$$\int_0^\infty d\xi \, \xi \, e^{-a^2 \xi^2} J_n^2(p\xi) = \frac{1}{2a^2} e^{-\frac{p^2}{2a^2}} I_n\left(\frac{p^2}{2a^2}\right) \tag{7.227}$$

further allows one to simplify the dispersion relation. In conclusion, for a local Maxwellian, the electrostatic dispersion relation is

$$1+\sum_\alpha \frac{1}{k^2\lambda_{D\alpha}^2}\left\{1+\frac{\omega-\omega_\alpha^*}{\sqrt{2}k_z v_{th\alpha}}\sum_{n=-\infty}^{\infty} Z\left(\frac{\omega-n\Omega_\alpha}{\sqrt{2}k_z v_{th\alpha}}\right) I_n(z_\alpha)e^{-z_\alpha}\right\}=0 \tag{7.228}$$

with $\omega_\alpha^* = k_y\rho_\alpha^2\kappa\Omega_\alpha$.

In the following, we present two examples. The first one applies to the low-frequency drift instability in the region $\omega \ll \Omega_i \ll |\Omega_e|$, $k_z v_{th\alpha} \ll |\Omega_\alpha|$ for $\alpha = e, i$, $k_z v_{thi} \ll \omega \ll k_z v_{the}$, $k^2\lambda_{De}^2 \ll z_i, z_e \ll 1$, and $z_i < 1$. In that case, we use the appropriate asymptotic expansions of the Z function. To lowest order, we obtain for the real part of the dispersion relation

$$\omega^2\left(1+\frac{c_s^2k_y^2}{\Omega_i^2}\right)-\omega\omega_e^*-c_s^2k_z^2\approx 0\,. \tag{7.229}$$

Two branches appear, namely, the already know ion-sound and the drift branch. The solutions can be written in the form

$$\omega\approx\frac{1}{2\delta}\left\{k_y v_{de}\pm\left[(k_y v_{de})^2+4\delta c_s^2k_z^2\right]^{1/2}\right\},\quad \delta=1+\frac{c_s^2k_y^2}{\Omega_i^2}\,. \tag{7.230}$$

The drift frequency $\omega_e^* = k_y v_{de}$ obeys the relation

$$v_{de}=\frac{\omega_e^*}{k_y}\,. \tag{7.231}$$

From the imaginary part of the dispersion relation, we determine the growth rate (as long as it exists and is small) from

$$\gamma\approx-\frac{\operatorname{Im}\varepsilon(\omega_0,k)}{\left.\frac{\partial\varepsilon}{\partial\omega}\right|_{\omega_0}} \tag{7.232}$$

where ω_0 satisfies $\operatorname{Re}\varepsilon(\omega_0,k)=0$. We have for

$$\begin{aligned}\operatorname{Im}\varepsilon\approx{}&\frac{\pi e}{\varepsilon_0 k_z k^2 B_0}n_{e0}(x)\left(\frac{m_e}{2\pi k_B T_e}\right)^{1/2}\\&\times\exp\left[-\frac{m_e\omega_0^2}{2k_B T_e k_z^2}\right]\left(k_y\kappa-\frac{m_e\omega_0\Omega_e}{k_B T_e}\right).\end{aligned} \tag{7.233}$$

Note that in the last factor on the right-hand side, the two terms have different signs because of $\omega_0\approx\omega_e^*$. From the real part of the dispersion relation, we get

$$\left.\frac{\partial\varepsilon}{\partial\omega}\right|_{\omega_0}\approx\left(\frac{\delta}{k\lambda_{De}}\right)^2\frac{1}{k_y v_{de}} \tag{7.234}$$

such that for $k^2 v_{de}^2 \gg \delta c_s^2 k_z^2$, growth occurs with the exponential rate

$$\gamma \approx (\pi\tilde{\beta})^{1/2} k_z^{-1} \left(\frac{k_y v_{de}}{\delta}\right)^2 \left(1 - \frac{1}{\delta}\right) \exp\left[-\tilde{\beta}\left(\frac{\omega}{k_z}\right)^2\right] . \tag{7.235}$$

Here, $\tilde{\beta} = m_e/2k_B T_e$. More rigorous calculations require the full knowledge of the full Z-function. The maximum of the growth rate occurs at $k_y \rho_s \sim 1$, and it becomes of the order of magnitude ω_0.

The second example considers the drift-cyclotron instability which occurs in the region $\omega \sim \ell\Omega_i \sim \omega_i^*$, $|\omega - \ell\Omega_i| \gg k_z v_{thi}$, $\omega \gg k_z v_{the}$, $z_i \gg 1$, and $z_e < 1$.

The appropriate expansions lead to a dispersion relation $\varepsilon \equiv 1 + \chi_e + \chi_i = 0$ with

$$\chi_e \approx -(k_y \lambda_{Di})^{-2} \left(\frac{\omega_i^*}{\omega}\right) + \frac{\omega_{pe}^2}{\Omega_e^2} , \tag{7.236}$$

$$\chi_i \approx (k_y \lambda_{Di})^{-2} \left\{1 + \frac{\omega - \omega_i^*}{\omega - \ell\Omega_i} I_l(z_i) e^{-z_i}\right\} . \tag{7.237}$$

Its analysis leads to the growth rate [201, 205]

$$\gamma \approx \frac{\ell^{3/2}\Omega_i}{(2\pi)^{1/4}} \left(\frac{m_e}{m_i} + \frac{\Omega_i}{\omega_{pi}}\right)^{1/2} \left(\frac{\kappa}{\rho_i}\right)^{1/2} . \tag{7.238}$$

7.3 Waves and Instabilities in the Magnetohydrodynamic Description

With less rigor, though far more efficient than in the more exact kinetic description, the macroscopic MHD model can be used to investigate large-scale instabilities. Actually, some of the most dangerous instabilities in a tokamak will occur on MHD scales. The strategy of tokamak physics is to avoid them with top priority. The present section is devoted to MHD stability in the ideal ($\eta = 0$) limit.

7.3.1 Hydromagnetic Variational Principle

Let us first develop criteria when an equilibrium (or better, stationary state) ($\partial_t = 0$) is stable [206–208]. We consider stationary states without flow $u_0 = 0$ in the ideal limit. In this case, the MHD model requires

$$\nabla p_0 = \frac{1}{\mu_0} (\nabla \times \boldsymbol{B}_0) \times \boldsymbol{B}_0 . \tag{7.239}$$

Small perturbations will be governed by

$$\rho_0 \partial_t \boldsymbol{u} = -\nabla p + \frac{1}{\mu_0} (\nabla \times \boldsymbol{B}_0) \times \boldsymbol{B} + \frac{1}{\mu_0} (\nabla \times \boldsymbol{B}) \times \boldsymbol{B}_0 \tag{7.240}$$

in the linear approximation. Quantities without any index are perturbed quantities; the index zero characterizes the zeroth order quantities. Differentiating with respect to time,

$$\rho_0 \partial_t^2 \boldsymbol{u} = -\nabla \dot{p} + \frac{1}{\mu_0}(\nabla \times \boldsymbol{B}_0) \times \nabla \times (\boldsymbol{u} \times \boldsymbol{B}_0) + \frac{1}{\mu_0}[\nabla \times \nabla \times (\boldsymbol{u} \times \boldsymbol{B}_0)] \times \boldsymbol{B}_0 \tag{7.241}$$

follows. This equation has to be supplemented by an equation of state,

$$\frac{d}{dt}\left(\frac{p}{\rho^\gamma}\right) = 0 , \tag{7.242}$$

which can be rewritten by making use of the continuity equation for ρ. To lowest order, that is, in the linearized version, we find for $\dot{p}$

$$\dot{p} = -\boldsymbol{u} \cdot \nabla p_0 - \gamma p_0 \nabla \cdot \boldsymbol{u} . \tag{7.243}$$

which can be used to formulate a closed equation for the time-development of $\boldsymbol{u}$.

Very often, the Lagrangian variable $\boldsymbol{\xi}$ is used instead of $\boldsymbol{u}$. Within the linearization for the perturbations, we may use $\boldsymbol{u} = \dot{\boldsymbol{\xi}}$ as will be briefly commented now. The Eulerian velocity $\boldsymbol{u}(\boldsymbol{r}, t)$ is the velocity of fluid elements at time t at a prescribed position $\boldsymbol{r}$. On the other hand, the Lagrangian velocity $\dot{\boldsymbol{\xi}}(\boldsymbol{r}_0, t)$ is the velocity of a fluid element which was at time t_0 at position $\boldsymbol{r}_0$. For small perturbations, we Taylor-expand

$$\dot{\boldsymbol{\xi}}(\boldsymbol{r}_0, t) = \boldsymbol{u}(\boldsymbol{r}, t) \approx \boldsymbol{u}(\boldsymbol{r}_0, t) + \underbrace{\boldsymbol{\xi} \cdot \nabla \boldsymbol{u}}_{\text{second order}} + \ldots \approx \boldsymbol{u}(\boldsymbol{r}_0, t) \tag{7.244}$$

and only use the first order contribution. Note, $\boldsymbol{u}_0 = 0$. When using $\boldsymbol{u} = \dot{\boldsymbol{\xi}}$, we may subsequently integrate over time. For stationary states, $\boldsymbol{\xi}(\boldsymbol{r}_0, 0) = \ddot{\boldsymbol{\xi}}(\boldsymbol{r}_0, 0) = 0$ should hold. We thus obtain

$$\rho_0 \ddot{\boldsymbol{\xi}} = \nabla (\boldsymbol{\xi} \cdot \nabla p_0 + \gamma p_0 \nabla \cdot \boldsymbol{\xi}) + \frac{1}{\mu_0}\{\nabla \times [\nabla \times (\boldsymbol{\xi} \times \boldsymbol{B}_0)]\} \times \boldsymbol{B}_0 + \frac{1}{\mu_0}(\nabla \times \boldsymbol{B}_0) \times [\nabla \times (\boldsymbol{\xi} \times \boldsymbol{B}_0)] . \tag{7.245}$$

Introducing the (quite complicated) operator $\boldsymbol{F}$, this equation will be abbreviated by

$$\rho_0 \ddot{\boldsymbol{\xi}} = \boldsymbol{F}(\boldsymbol{\xi}) . \tag{7.246}$$

A current, but mathematically not rigorous method separates space- and time-variables through the ansatz $\boldsymbol{\xi}(\boldsymbol{r}_0, t) = \boldsymbol{\xi}_k(\boldsymbol{r}_0)\tau_k(t)$. Because of

$$\rho_0 \boldsymbol{\xi}_k \ddot{\tau}_k = \boldsymbol{F}(\boldsymbol{\xi}_k)\tau_k , \tag{7.247}$$

the time-variation can be written in the form $\tau_k \sim \exp[i(\omega_k t + \varphi_k)]$, where ω_k^2 is the eigenvalue in

$$-\rho_0 \omega_k^2 \boldsymbol{\xi}_k = \boldsymbol{F}(\boldsymbol{\xi}_k) . \tag{7.248}$$

Since we are dealing with linear equations, the (more) general solution follows by superposition (summation over k)

$$\boldsymbol{\xi}(\boldsymbol{r}_0, t) = \sum_k a_k \boldsymbol{\xi}_k(\boldsymbol{r}_0) \exp\left[i\left(\omega_k t + \varphi_k\right)\right] . \tag{7.249}$$

However, this method is only rigorous if the operator $\boldsymbol{F}$ is complete. We shall now discuss the line of action when completeness cannot be assumed.

Summarizing, in the case of completeness, we can formulate the stability theorem excluding exponential growth: *A hydrodynamic equilibrium is stable if and only if all eigenvalues* $\rho_0^{-1}\boldsymbol{F}$ *are negative.* However, we advise caution also because of the reason that the equation $\boldsymbol{F}(\boldsymbol{\xi}_0) = 0$ has any number of eigenfunctions $\boldsymbol{\xi}_0 = \alpha(p_0)\boldsymbol{B}_0 + \beta(p_0)\nabla \times \boldsymbol{B}_0$, and because of degeneracy, algebraically growing modes are still possible.

Because of missing dissipation, it may be expected that the operator $\boldsymbol{F}$ is self-adjoint. Let us first present a plausibility argument for that. We interpret $\boldsymbol{F}$ as the force density in a Newtonian system such that

$$-\int_0^{\boldsymbol{\xi}} \boldsymbol{F}(\boldsymbol{\eta}) \cdot d\boldsymbol{\eta} = -\frac{1}{2}\boldsymbol{\xi} \cdot \boldsymbol{F}(\boldsymbol{\xi}) \tag{7.250}$$

is the change of potential energy density. In a conservative system, the total energy should be time-independent,

$$\frac{\partial}{\partial t} \int \left[\frac{1}{2}\rho_0 \dot{\boldsymbol{\xi}}^2 - \frac{1}{2}\boldsymbol{\xi} \cdot \boldsymbol{F}(\boldsymbol{\xi})\right] d^3 r = 0 . \tag{7.251}$$

Performing the time-derivative, we conclude

$$\int \dot{\boldsymbol{\xi}} \cdot \boldsymbol{F}(\boldsymbol{\xi}) d^3 r = \int \boldsymbol{\xi} \cdot \boldsymbol{F}(\dot{\boldsymbol{\xi}}) d^3 r . \tag{7.252}$$

Equation (7.248) is an equation of second order which can be replaced by a system of two first-order equations for the independent variables $\boldsymbol{\xi}$ and $\dot{\boldsymbol{\xi}}$. Thus, in Eq. (7.252), we may consider $\dot{\boldsymbol{\xi}}$ and $\boldsymbol{\xi}$ as independent such that the self-adjointness of $\boldsymbol{F}$ becomes plausible. The stability conclusion will be the same as in the previously considered case Eq. (7.247). Expanding in the potential energy density

$$\delta W = -\frac{1}{2}\int \boldsymbol{\xi} \cdot \boldsymbol{F}(\boldsymbol{\xi}) d^3 r \tag{7.253}$$

$\boldsymbol{\xi}$ in terms of eigenfunctions Eq. (7.249) of $\boldsymbol{F}$, and taking into account that the normalized eigenfunctions of a self-adjoint operator are orthogonal, we obtain

$$\delta W = \sum_k a_k^2 \cos^2(\omega_k t + \varphi_k)\omega_k^2 . \tag{7.254}$$

For instability, δW should become negative, which only occurs if at least one ω_k^2 is negative, validating instability.

Under the assumption that $\boldsymbol{F}$ is self-adjoint, the following theorem can be proven. *Necessary and sufficient for stability is the positive (semi-)definiteness of the potential energy density* δW. The proof of the self-adjointness of $\boldsymbol{F}$ is presented in the Appendix J.

The argumentation that positive δW imply stability is simple. We may use

$$K = \int \frac{1}{2}\rho_0 \dot{\xi}^2 d^3 r = E - \delta W \tag{7.255}$$

as a positive (semi-)definite functional, implying a norm for the perturbations. When $\delta W > 0$ because of $K < E$, the norm of the perturbation is bounded, thus meaning stability. For the opposite case, when δW can become negative, we show that

$$I := \frac{1}{2}\int \rho_0 \xi^2 d^3 r \tag{7.256}$$

will be exponentially growing with time. To show that [208], we differentiate twice with respect to time, $\ddot{I} = 2K - 2\delta W$. A short calculation leads to

$$\frac{d^2}{dt^2}\ln I = \frac{1}{I}\left[2K - 2\delta W - \frac{1}{I}\left(\int \rho_0 \xi \cdot \dot{\xi}\, d^3 r\right)^2\right]. \tag{7.257}$$

The last term on the right-hand side can be estimated with the help of the Schwarz inequality

$$\left(\int \rho_0 \xi \cdot \dot{\xi}\, d^3 r\right)^2 \le 4 I K\,. \tag{7.258}$$

When using this, we conclude

$$\frac{d^2}{dt^2}\ln I \ge -\frac{2E}{I}\,. \tag{7.259}$$

Provided there exists a function $\xi := \boldsymbol{v}_0(\boldsymbol{r})$ which realizes $\delta W < 0$, the initial value problem will be solved with $\xi(t=0) = \boldsymbol{v}_0(\boldsymbol{r})$ and $\dot{\xi}(t=0) = 0$. Obviously, then $E < 0$. Let us introduce new variables $E := -2\nu^2 I(0)$ and $y := \ln[I(t)/I(0)]$ such that $\ddot{y} \ge 2\nu^2 e^{-y}$ with $y(0) = \dot{y}(0) = 0$. To avoid a detailed discussion of the solutions of the inequality, we introduce the function $Y(t)$ which satisfies the same initial conditions as $y(t)$, but obeys the differential *equation*

$$\ddot{Y} = 2\nu^2 e^{-Y}\,. \tag{7.260}$$

Obviously, for all times, $y \ge Y$ holds. The solution of Eq. (7.260) is $Y(t) = \ln(\cosh \nu t)^2$. Going back to the original variables, the condition $y \ge Y$ transforms into

$$I(t) \ge I(0)\frac{e^{2\nu t} + 2 + e^{-2\nu t}}{4} \to \infty \quad \text{for} \quad t \to \infty\,. \tag{7.261}$$

The unstable behavior (for $\delta W < 0$) now follows from the definition Eq. (7.256) of I.

Having proven the stability theorem (assisted by the self-adjointness property of $\boldsymbol{F}$, as shown in the Appendix J), we will now evaluate the stability properties of some MHD equilibria. Hitherto we shall reformulate δW which contains $\boldsymbol{F}$,

$$\boldsymbol{F} = \underbrace{\nabla(\boldsymbol{\xi}\cdot\nabla p_0 + \gamma p_0 \nabla\cdot\boldsymbol{\xi})}_{\mathrm{I}} + \underbrace{\frac{1}{\mu_0}\left\{\nabla\times\left[\nabla\times(\boldsymbol{\xi}\times\boldsymbol{B}_0)\right]\right\}\times\boldsymbol{B}_0}_{\mathrm{II}}$$
$$+ \underbrace{\frac{1}{\mu_0}(\nabla\times\boldsymbol{B}_0)\times\left[\nabla\times(\boldsymbol{\xi}\times\boldsymbol{B}_0)\right]}_{\mathrm{III}} . \tag{7.262}$$

For the contributions from I, we shall use

$$\nabla\cdot\left[(\boldsymbol{\xi}\cdot\nabla p_0 + \gamma p_0\nabla\cdot\boldsymbol{\xi})\,\boldsymbol{\xi}\right] = \boldsymbol{\xi}\cdot\nabla(\boldsymbol{\xi}\cdot\nabla p_0 + \gamma p_0\nabla\cdot\boldsymbol{\xi})$$
$$+ (\nabla\cdot\boldsymbol{\xi})(\boldsymbol{\xi}\cdot\nabla p_0 + \gamma p_0\nabla\cdot\boldsymbol{\xi}) . \tag{7.263}$$

The contribution from the second term II, containing $\boldsymbol{Q} = \nabla\times(\boldsymbol{\xi}\times\boldsymbol{B}_0)$, will be calculated by making use of

$$\nabla\cdot\left[(\boldsymbol{\xi}\times\boldsymbol{B}_0)\times\boldsymbol{Q}\right] = \boldsymbol{Q}\cdot\nabla\times(\boldsymbol{\xi}\times\boldsymbol{B}_0) - (\boldsymbol{\xi}\times\boldsymbol{B}_0)\cdot\nabla\times\boldsymbol{Q}$$
$$= \boldsymbol{Q}^2 + \boldsymbol{\xi}\cdot\left[(\nabla\times\boldsymbol{Q})\times\boldsymbol{B}_0\right] . \tag{7.264}$$

Note that $\boldsymbol{Q}$ is the magnetic field perturbation in plasma, that is, $\boldsymbol{Q} \equiv \delta\boldsymbol{B}_{plasma}$. This interpretation follows from

$$\dot{\boldsymbol{B}} = -\nabla\times\boldsymbol{E} = \nabla\times(\boldsymbol{u}\times\boldsymbol{B}) \approx \nabla\times\left(\dot{\boldsymbol{\xi}}\times\boldsymbol{B}_0\right) \tag{7.265}$$

within ideal MHD. Finally, the contribution from III will be rewritten with the help of

$$\boldsymbol{\xi}\left[(\nabla\times\boldsymbol{B}_0)\times\boldsymbol{Q}\right] = -(\nabla\times\boldsymbol{B}_0)\cdot(\boldsymbol{\xi}\times\boldsymbol{Q}) . \tag{7.266}$$

When integrating over the volume of the plasma (the system consists of plasma with a variable boundary S to vacuum which is finally limited by a fixed surface S'), the divergence terms lead via the Gauss theorem to surface contributions. Therefore, we split the total δW into $\delta W = \delta W_{volume} + \delta W_{surface}$, where the volume term is

$$\delta W_{volume} =$$
$$\frac{1}{2}\int_V\left[\frac{Q^2}{\mu_0} + \frac{1}{\mu_0}(\nabla\times\boldsymbol{B}_0)\cdot(\boldsymbol{\xi}\times\boldsymbol{Q}) + (\nabla\cdot\boldsymbol{\xi})\boldsymbol{\xi}\cdot\nabla p_0 + \gamma p_0(\nabla\cdot\boldsymbol{\xi})^2\right] d^3r , \tag{7.267}$$

and the surface term is

$$\delta W_{surface} = -\frac{1}{2}\int\limits_S \left[\frac{1}{\mu_0} \underbrace{(\xi \times B_0) \times Q}_{-\xi(Q \cdot B_0) + B_0(\xi \cdot Q)} + \xi\,(\xi \cdot \nabla p_0 + \gamma\, p_0 \nabla \cdot \xi) \right] \cdot d^2 r\,. \tag{7.268}$$

Since we assume $B_0 \cdot n \sim B_0 \cdot d^2 r = 0$ at the surface, the surface term reduces to

$$\delta W_{surface} = \frac{1}{2}\int\limits_S \left[\frac{1}{\mu_0} \xi (Q \cdot B_0) - \xi\,(\xi \cdot \nabla p_0 + \gamma\, p_0 \nabla \cdot \xi) \right] \cdot d^2 r\,. \tag{7.269}$$

The surface term is zero if the normal displacement at the surface vanishes, as would be the case at a perfect conductor. When there is a vacuum region outside the plasma and no surface currents in the equilibrium configuration, the surface term represents the transfer of energy to the vacuum region [68]. This energy is $\int \delta B^2_{vacuum}/2\mu_0 d^3 r$ where δB_{vacuum} is the first order vacuum magnetic field contribution and the integration is over the vacuum region, provided that an additional contribution $\delta W_{interface}$ from the vacuum-plasma interface is ignored by assuming pressure continuity. This can be shown by the relations derived in the Appendix J. Summarizing, we may substitute $\delta W_{surface} \to \delta W_{vacuum} + \delta W_{interface}$ and $\delta W_{volume} \equiv \delta W_{plasma}$, resulting in

$$\begin{aligned} \delta W = {} & \\ & \frac{1}{2}\int\limits_{plasma} \left[\frac{Q^2}{\mu_0} + \frac{1}{\mu_0}(\nabla \times B_0) \cdot (\xi \times Q) + (\nabla \cdot \xi)\xi \cdot \nabla p_0 + \gamma\, p_0 (\nabla \cdot \xi)^2 \right] d^3 r \\ & + \frac{1}{2\mu_0} \int\limits_{vacuum} \delta B^2_{vacuum} d^3 r + \delta W_{interface} \\ & \equiv \delta W_{plasma} + \delta W_{vacuum} + \delta W_{interface}\,. \end{aligned} \tag{7.270}$$

The interface term is

$$\delta W_{interface} = -\frac{1}{2\mu_0} \int\limits_S d^2 r\, (\xi \cdot n)^2\, n \cdot \nabla \left(\mu_0 p_0 + \frac{B_0^2}{2} - \frac{B^2_{vacuum}}{2} \right). \tag{7.271}$$

We should remind the reader that the variational principle is based on ideal MHD. There is no reason to believe that resistivity will not be important for the development of instabilities. The instabilities being picked up by the variational principle are ideal in the sense that instability would even occur if the plasma were perfectly conducting. Still, there is the possibility of unstable resistive modes. Typical tokamak MHD instabilities [68] are

- the kink instability, which is an ideal MHD instability driven by a current gradient in the low-β regime. The resistive form of the kink instability is known as tearing instability. An instability with mode number $m = 1$ driven by a pressure gradient in the plasma core is the internal kink instability. There is also a resistive $m = 1$ instability.
- the interchange instabilities driven by pressure gradients in curved magnetic fields. In tokamaks, the modes are known as ballooning modes. The Mercier instability is a limiting case of the ballooning instability.

In the following, we shall discuss simple versions of these instabilities.

7.3.2
Kink and Sausage Instability

Let us start with a very simple situation when the plasma is directly bounded by a conducting wall. Then, the vacuum term will not be present. We further set $\gamma = 0$. We can conclude from the variational principle that if the plasma is stable for $\gamma = 0$, it will also be stable for $\gamma \neq 0$ since $p_0 > 0$. Furthermore, we investigate a force-free equilibrium which satisfies $\nabla \times \boldsymbol{B}_0 = \alpha \boldsymbol{B}_0$ with $\nabla p_0 = 0$. Under these assumptions, we have

$$\delta W = \delta W_{plasma} = \frac{1}{2} \int \left[\frac{Q^2}{\mu_0} + \frac{1}{\mu_0} (\nabla \times \boldsymbol{B}_0) \cdot (\boldsymbol{\xi} \times \boldsymbol{Q}) \right] d^3 r . \quad (7.272)$$

Introducing $\boldsymbol{R} := \boldsymbol{\xi} \times \boldsymbol{B}_0$, we may write

$$\delta W_{plasma} = \frac{1}{2\mu_0} \int \left[(\nabla \times \boldsymbol{R})^2 - \alpha \boldsymbol{R} \cdot \nabla \times \boldsymbol{R} \right] d^3 r . \quad (7.273)$$

We can discuss the sign by varying $\boldsymbol{R}$, keeping in mind that the variations should be perpendicular to $\boldsymbol{B}_0$. The displacements can be normalized. The obvious choice

$$\int |\boldsymbol{\xi}|^2 d^3 r = 1 \quad (7.274)$$

is perhaps not the most adequate one. It turns out that the possibility

$$\int \boldsymbol{R} \cdot \nabla \times \boldsymbol{R} \, d^3 r = \text{const} \quad (7.275)$$

is more appropriate, where const could be positive, negative, or zero. For the last case, $\delta W_{plasma} > 0$ would follow immediately. Handling the normalization as a constraint, we introduce the Lagrange parameter $\alpha \lambda$ and postulate for the minimum

$$\delta \int L d^3 r \equiv \delta \int \left[(\nabla \times \boldsymbol{R})^2 - (\lambda + 1) \alpha \boldsymbol{R} \cdot \nabla \times \boldsymbol{R} \right] d^3 r = 0 . \quad (7.276)$$

We determine the minimum from the Euler–Lagrange equations. This is possible because the functional L is bounded from below. We write the Euler–Lagrange equations in the form

$$\frac{\partial L}{\partial R_i} = \sum_k \frac{\partial}{\partial x_k} \frac{\partial L}{\partial R_{i,k}} , \tag{7.277}$$

with the definition $R_{i,k} := \partial R_i / \partial x_k$. For the minimum of L at $\boldsymbol{R} = \boldsymbol{R}_{min}$, we determine the latter from

$$\nabla \times \nabla \times \boldsymbol{R}_{min} = (\lambda + 1)\alpha \nabla \times \boldsymbol{R}_{min} . \tag{7.278}$$

For simplicity, in the following, the index "*min*" will be omitted. Now, λ is an eigenvalue. Scalar multiplication by $\boldsymbol{R}$ leads to

$$\alpha \boldsymbol{R} \cdot \nabla \times \boldsymbol{R} = \frac{1}{\lambda + 1} \left\{ (\nabla \times \boldsymbol{R})^2 + \nabla \cdot [(\nabla \times \boldsymbol{R}) \times \boldsymbol{R}] \right\} . \tag{7.279}$$

Inserting into δW, we find

$$min\, \delta W_{plasma} = \frac{1}{2\mu_0} \int \left[\frac{\lambda}{\lambda + 1} (\nabla \times \boldsymbol{R})^2 - \frac{1}{\lambda + 1} \nabla \cdot ((\nabla \times \boldsymbol{R}) \times \boldsymbol{R}) \right] d^3 r . \tag{7.280}$$

The last term on the right-hand side can be converted into a surface integral. Within the integrand, the term $\boldsymbol{n} \cdot [(\nabla \times \boldsymbol{R}) \times \boldsymbol{R}]$ will appear which has to be evaluated at the fixed surface. Since $\boldsymbol{R} \| \boldsymbol{n}$, the surface integral vanishes and we are left with

$$min\, \delta W_{plasma} = \frac{\lambda}{2\mu_0(\lambda + 1)} \int (\nabla \times R)^2 d^3 r . \tag{7.281}$$

The right-hand side will be positive for $\lambda > 0$ and $\lambda < -1$, respectively. So we should go back to the differential equation for $\boldsymbol{R}$ and discuss its eigenvalue spectrum. Introducing $\boldsymbol{Q} = \nabla \times \boldsymbol{R}$, we have in in the most simple (but perhaps too academic) Cartesian configuration

$$\frac{i}{\alpha}(k_z Q_y - k_y Q_z) = (\lambda + 1) Q_x , \tag{7.282}$$

$$\frac{i}{\alpha}(k_x Q_z - k_z Q_x) = (\lambda + 1) Q_y , \tag{7.283}$$

$$\frac{i}{\alpha}(k_y Q_x - k_x Q_y) = (\lambda + 1) Q_z . \tag{7.284}$$

Nontrivial eigensolutions exist for

$$(\lambda + 1)^2 = \frac{k^2}{\alpha^2} . \tag{7.285}$$

From here, it becomes obvious that for small k with $\boldsymbol{k} = (k_x, k_y, k_z)$, solutions with

$$-1 < \lambda < 0 \tag{7.286}$$

do exist. In other words, in large systems with extension $L \sim 2\pi/k$, unstable modes exist.

Before investigating the second (more realistic) example, the stabilization of a simple pinch discharge by an axial magnetic field, we have to analyze the simple pinch discharge first. We consider a cylindrical plasma confined by an azimuthal magnetic field component B_θ. The cylindrical coordinates are r, θ, z. The field component B_θ is being generated by surface currents flowing along $r = r_0$ in z-direction. No electric currents are assumed within the cylindrical column of the equilibrium plasma. Obviously, under these assumptions,

$$B_\theta = B_0 \frac{r_0}{r} , \tag{7.287}$$

and the equation

$$\rho_0 \ddot{\xi} = \boldsymbol{F}(\xi) \tag{7.288}$$

can be directly solved in the normal mode approximation since after Fourier transformation in time, it takes the very simple form

$$-\rho_0 \omega^2 \xi = \gamma p_0 \nabla\nabla \cdot \xi . \tag{7.289}$$

Fourier transforming also in θ and z, or in other words, looking for modes $\xi \sim \xi(r) \exp[i m\theta + i k z]$, we obtain for the z-component of ξ

$$\frac{d^2 \xi_z}{d r^2} + \frac{1}{r}\frac{d\xi_z}{dr} - \left(\alpha^2 + \frac{m^2}{r^2}\right) \xi_z = 0 , \tag{7.290}$$

with

$$\alpha^2 = \frac{-\omega^2 \rho_0}{\gamma p_0} + k^2 . \tag{7.291}$$

We shall confirm $\alpha^2 > 0$ with $\omega^2 < 0$ *a posteriori*. The eigenfunctions can be written in terms of modified Bessel functions,

$$\xi_z = A_{mk} I_m(\alpha r) e^{i m\theta + i k z} . \tag{7.292}$$

To correctly determine the eigenvalue (ω^2), we have to discuss the boundary conditions at the plasma-vacuum interface, where the total vacuum field is written as $\boldsymbol{B}_\theta + \delta \boldsymbol{B}$. Since there is no current in the vacuum, we may postulate $\nabla \times \delta \boldsymbol{B} = 0$ for the vacuum perturbation. As long as we do not pass through the current-carrying surface, we can introduce a scalar magnetic potential ψ such that $\delta \boldsymbol{B} = \nabla\psi$ with $\nabla^2\psi = 0$. The ansatz $\psi \sim \psi(r) \exp[i(kz + m\theta)]$ leads to exponentially decreasing solutions

$$\psi = C_{mk} K_m(k r) e^{i m\theta + i k z} \tag{7.293}$$

which determine $\delta \boldsymbol{B}$. The position of the plasma surface is given by

$$r = r_0 + \xi_r = r_0 + \frac{1}{ik}\frac{d}{dr}\xi_z = r_0 + \frac{\alpha}{ik} A_{mk} I'_m(\alpha r_0) e^{ikz + im\theta} . \tag{7.294}$$

The equilibrium surface is given by $r - r_0 = 0$. The condition changes into $\varphi(\boldsymbol{r}) = 0$ with

$$\varphi := r_0 + r_1 e^{im\theta + ikz} - r \,, \quad r_1 = \left(\frac{\alpha}{i} k\right) A_{mk} I'_m(\alpha r_0) \,. \tag{7.295}$$

The vector

$$\nabla\varphi = \left(-1, \frac{imr_1}{r_0} e^{im\theta + ikz}, ikr_1 e^{im\theta + ikz}\right) \tag{7.296}$$

is parallel to the normal vector on the surface. The total magnetic field in vacuum is to lowest order

$$\boldsymbol{B}(\boldsymbol{r}, t) = \boldsymbol{B}(r_0) + \delta\boldsymbol{B} + (\xi \cdot \nabla)\boldsymbol{B}(\boldsymbol{r}_0) \,, \tag{7.297}$$

and can be written as

$$\boldsymbol{B}(\boldsymbol{r}, t) = B_0\hat{\theta} + \delta\boldsymbol{B} + \xi_r(-1)\frac{B_0}{r_0}\hat{\theta} \,. \tag{7.298}$$

The boundary conditions requires that $\boldsymbol{B}(\boldsymbol{r}, t)$ is perpendicular to the surface, that is,

$$\nabla\varphi \cdot \boldsymbol{B} = 0 \approx -\delta B_r + \frac{im}{r_0} B_0 r_1 e^{im\theta + ikz} \,. \tag{7.299}$$

Inserting the known solutions, this is equivalent to

$$\frac{C_{mk}}{A_{mk}} = \frac{B_0}{r_0}\frac{\alpha m}{k^2}\frac{I'_m(\alpha r_0)}{K'_m(k r_0)} \,. \tag{7.300}$$

Finally, pressure continuity can be assumed. This leads to

$$-\gamma p_0 \nabla\cdot\xi + \frac{\boldsymbol{B}_{plasma}}{\mu_0}\cdot\left[\boldsymbol{Q} + (\xi\cdot\nabla)\boldsymbol{B}_{plasma}\right] = \frac{\boldsymbol{B}}{\mu_0}\cdot\left[\delta\boldsymbol{B} + (\xi\cdot\nabla)\boldsymbol{B}\right] , \tag{7.301}$$

where we can set $\boldsymbol{B}_{plasma} = 0$. Thus, the pressure balance at the surface becomes

$$\frac{\rho_0\omega^2}{ik} I_m(\alpha r_0) = \frac{B_0}{\mu_0 r_0}\left[\frac{C_{mk}}{A_{mk}} im K_m(k r_0) + i B_0 \frac{\alpha}{k} I'_m(\alpha r_0)\right] . \tag{7.302}$$

Together with Eq. (7.300), it can be combined to the dispersion relation

$$-\frac{\omega^2 r_0 \rho_0}{2\alpha p_0}\frac{I_m(\alpha r_0)}{I'_m(\alpha r_0)} = 1 + \frac{m^2}{k r_0}\frac{K_m(k r_0)}{K'_m(k r_0)} \,. \tag{7.303}$$

We always get unstable solutions with $\omega^2 < 0$. Take, for example, $m = 0$, that is, consider θ-independent modes. Then,

$$-\frac{\omega^2 r_0 \rho_0}{2\alpha p_0}\frac{I_0(\alpha r_0)}{I_1(\alpha r_0)} = 1 \tag{7.304}$$

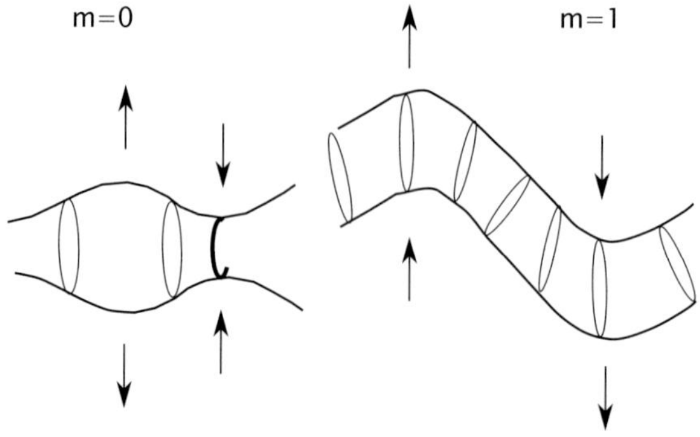

Figure 7.3 Demonstration of the $m = 0$ ("sausage") and $m = 1$ ("kink") instability.

can easily be discussed. Note $I_0 > 0$ and $I_0' = I_1 > 0$ for $-\omega^2$ between 0 and $+\infty$. Plotting the left-hand side for $0 \le -\omega^2 < +\infty$, for each k, we find a solution at some $-\omega^2 > 0$. The instability is known as the sausage or pinch instability. A similar analysis for $m = 1$ leads to the kink instability. In Figure 7.3, the principal forms of both instabilities are depicted. Having in mind that the simple pinch configuration is always unstable, we now apply the variational principle to demonstrate that an axial field B_z and a conducting wall may stabilize the cylindrical discharge. The B_z field in the interior of the plasma will prevent the $m = 0$ instability, and the conducting wall should preclude the "kinking." Let us introduce the notation

$$\boldsymbol{B}_{vacuum} = B_\theta \hat{\theta} + B_{z,vacuum}\hat{z} = B_0 \frac{r_0}{r}\hat{\theta} + b_{vacuum}\hat{z}\,, \quad \boldsymbol{B}_{plasma} = b_{plasma}\hat{z}\,. \tag{7.305}$$

The external cylindrical wall should be at position $r = R_0 = \Upsilon r_0$, with $\Upsilon > 1$. To keep the calculation as simple as possible, we assume $p_0 = \text{const} > 0$ and $\nabla \times b_{plasma}\hat{z} = 0$ (the interior magnetic field is generated by external sources). The plasma part of the potential energy is now

$$\delta W_{plasma} = \frac{1}{2\mu_0} \int \left[Q^2 + \mu_0 \gamma p_0 (\nabla \cdot \boldsymbol{\xi})^2 \right] d^3 r\,. \tag{7.306}$$

This expression is always positive. However, in general, for $\boldsymbol{n} \cdot \boldsymbol{\xi}$, we cannot assume $\boldsymbol{n} \cdot \boldsymbol{\xi} = 0$, but should allow for $\boldsymbol{n} \cdot \boldsymbol{\xi} \neq 0$ at the plasma-vacuum surface. Only if the sum of contributions to δW remains positive for any distribution of $\boldsymbol{n} \cdot \boldsymbol{\xi}$ at the surface, stability will follow. First, we rewrite the plasma contribution as

$$\delta W_{plasma} = \frac{1}{2\mu_0} \int \sum_{i,k} \left\{ b_k \xi_{i,k} Q_i - \sum_l b_l \xi_{i,k} Q_l \delta_{ik} - \xi_i b_{k,i} Q_k + \mu_0 \gamma p_0 \delta_{ik} \xi_{i,k} \nabla \cdot \boldsymbol{\xi} \right\}. \tag{7.307}$$

To determine its minimum, we use the Euler–Lagrange equations, this time without any constraint from normalization of the displacement. They can be written in the compact form

$$\frac{1}{\mu_0}(\nabla \times \boldsymbol{Q}) \times \boldsymbol{B}_{plasma} + \gamma\, p_0 \nabla\nabla \cdot \boldsymbol{\xi} = 0\,. \tag{7.308}$$

It is equivalent to $\boldsymbol{F}(\boldsymbol{\xi}) = 0$. The normalization is not required since we consider all possible boundary conditions $\boldsymbol{n} \cdot \boldsymbol{\xi}$ at the plasma-vacuum surface; actually, then we do not have the normalization of $\boldsymbol{\xi}$ at our disposal. For the minimum of δW_{plasma}, with the help of the Euler–Lagrange equations, we find

$$min\, \delta W_{plasma} = \frac{1}{2} \int \left(\gamma\, p_0 \nabla \cdot \boldsymbol{\xi} - \frac{\boldsymbol{B}_{plasma} \cdot \boldsymbol{Q}}{\mu_0} \right) \boldsymbol{\xi} \cdot d\boldsymbol{F} \tag{7.309}$$

since $\boldsymbol{B}_{plasma}$ is orthogonal to the normal of the surface. Let us have a look at the solution of Eq. (7.308) by making an ansatz in the form $\boldsymbol{\xi} = \boldsymbol{\xi}(r) \exp(i m\theta + i k z)$. Because of $\boldsymbol{B}_{plasma} \parallel \hat{z}$, we have for $k \neq 0$ the condition $\nabla \cdot \boldsymbol{\xi} = 0$. The case $k = 0$ has to be treated separately. Taking the divergence of Eq. (7.308), we then obtain

$$\nabla^2 (b_{plasma} Q_z) = 0\,. \tag{7.310}$$

Its solution is

$$Q_z = \frac{A}{b_{plasma}} I_m(kr) e^{im\theta + ikz}\,, \tag{7.311}$$

with a constant A which must still to be determined. Applying that solution, we get

$$min\, \delta W_{plasma} = -\frac{A}{2\mu_0} I_m(k r_0) \int \xi_r(r = r_0) e^{im\theta + ikz} dz\, r_0\, d\theta\,. \tag{7.312}$$

Next, we determine $\xi_r(r_0)$. Note that $\boldsymbol{Q}$ is related to $\boldsymbol{\xi}$ in the form $\boldsymbol{Q} = i k b_{plasma} \boldsymbol{\xi}$. We further have

$$\nabla \left(b_{plasma} Q_z \right) = i k b_{plasma} \boldsymbol{Q}\,. \tag{7.313}$$

Combining the last relations, we find for the displacement

$$\boldsymbol{\xi} = -\frac{1}{k^2 b_{plasma}^2} \nabla (b_{plasma} Q_z) = -\frac{A}{k^2 b_{plasma}^2} \nabla \left[I_m(kr) e^{im\theta + ikz} \right]. \tag{7.314}$$

From here, we determine $\xi_r(r = r_0) \equiv r_1 \exp[-i(m\theta + ikz)]$, which will be inserted into Eq. (7.312),

$$min\, \delta W_{plasma} = \frac{r_0 L k b_{plasma}^2 r_1^2}{8} \frac{I_m(k r_0)}{I'_m(k r_0)}\,, \quad r_1 = -\frac{A I'_m(k r_0)}{k b_{plasma}^2}\,. \tag{7.315}$$

Obviously, in the integrand, only, $\xi_r(r_0) \sim \exp(-im\theta - ikz)$ did contribute. The integration over θ leads to the factor 2π. However, when using the complex notation for ξ, only the real part should be taken into account, giving an additional factor $1/2$. Finally, the integration over z introduces the length L.

An additional contribution originates from δW_{vacuum}. By introducing the scalar potential ϕ via $\delta \boldsymbol{B}_v = \nabla\phi$, we use $(\nabla\phi)^2 = \nabla\cdot(\phi\nabla\phi) - \phi\nabla^2\phi = \nabla\cdot(\phi\nabla\phi)$ in vacuum, and we get the simplified version of the vacuum term

$$\delta W_{vacuum} = \frac{1}{2\mu_0}\int \phi\nabla\phi\cdot d\boldsymbol{F} \tag{7.316}$$

for further analysis. The vacuum contribution will be minimized for a function ϕ which will again be determined from the Euler–Lagrange equations

$$\frac{\partial}{\partial x_k}\frac{\partial}{\partial\phi_{,k}}\frac{1}{2}\phi_{,k}\phi_{,k} = \frac{\partial}{\partial x_k}\phi_{,k} = \nabla^2\phi = 0 \tag{7.317}$$

in the vacuum region $r_0 < r < R_0$. Here, $\phi_{,k} \equiv \partial\phi/\partial x_k$, and summation over repeated indices is assumed. Solutions can be written in the form

$$\phi = C I_m(kr)e^{im\theta+ikz} + D K_m(kr)e^{im\theta+ikz} \tag{7.318}$$

with coefficients C and D still to be determined by the boundary conditions. At the conducting wall, the normal component of the $\boldsymbol{B}$ field should vanish,

$$\left.\frac{\partial\phi}{\partial r}\right|_{r=R_0} = 0 = Ck I'_m(kR_0)e^{im\theta+ikz} + Dk K'_m(kR_0)e^{im\theta+ikz}\,. \tag{7.319}$$

On the other hand, at the plasma-vacuum surface, the total magnetic field $\boldsymbol{B}$ should also be tangential, meaning

$$\nabla\varphi\cdot\boldsymbol{B} = 0 = -\left.\frac{\partial\phi}{\partial r}\right|_{r_0} + \frac{imr_1}{r_0}B_0 + ikr_1 b_{vacuum}\,. \tag{7.320}$$

Inserting the general solution, we obtain a second equation for the coefficients C and D, namely,

$$Ck I'_m(kr_0) + Dk K'_m(kr_0) = ikr_1 b_{vacuum} + \frac{imr_1}{r_0}B_0\,. \tag{7.321}$$

Determining the coefficients from the linear, inhomogeneous set of equations, and inserting the result into the vacuum contribution, we finally get

$$\begin{aligned} \min\delta W_{vacuum} = {} & \frac{r_0 Lk}{8}\left(b_{vacuum} + \frac{mB_0}{kr_0}\right)^2 \\ & \times \frac{K'_m(kR_0)I_m(kr_0) - I'_m(kR_0)K_m(kr_0)}{K'_m(kR_0)I'_m(kr_0) - I'_m(kR_0)K'_m(kr_0)}r_1^2\,. \end{aligned}$$

The final remark concerns the interface term $\delta W_{interface}$. The assumptions on the zeroth order terms lead to

$$\nabla\left(p_0 + \frac{b^2_{plasma}}{2\mu_0}\right) = 0\,, \quad -\nabla\left(\frac{B_0^2}{2\mu_0}\frac{r_0^2}{r^2} + \frac{b^2_{vacuum}}{2\mu_0}\right) = \frac{B_0^2}{\mu_0 r_0}\hat{r} \tag{7.322}$$

at $r = r_0$, and the obvious result

$$min\,\delta W_{interface} = -\frac{B_0^2 \pi L}{2\mu_0} r_1^2 \,. \tag{7.323}$$

It is always negative. However, the combination of the minima of δW_{plasma}, δW_{vacuum}, and $\delta W_{interface}$ can lead to a positive value, and thus stability. In the limit $R_0 \to \infty$, the stability condition is known as the Kruskal–Shafranov criterion [71, 209]. MHD stability investigations are crucial for tokamak and stellarator performances [210–212].

7.3.2.1 Kink Instability in Toroidal Systems

In toroidal geometry, we have toroidal (B_ϕ) and poloidal (B_θ) magnetic fields. The tension of the toroidal field lines resists the kinking. For sufficiently strong toroidal fields, the kink instability can be suppressed. Kink instabilities thus place an upper limit on the plasma current for a given toroidal field [14, 68]. For a tokamak, this limit can be expressed in terms of the safety factor

$$q(r) = \frac{r B_\phi(r)}{R_0 B_\theta(r)} \,. \tag{7.324}$$

Here, R_0 denotes the center of the confinement region (in cylindrical coordinates R, Φ, Z). The coordinate r is the radial coordinate within the poloidal plane. The nonlocal helical perturbations in the plasma region ($r < a$) with vacuum ($a < r < b$) and a perfectly conducting wall at $r = b$ are written in the form

$$\xi(r, \theta, \phi) = \xi(r) e^{i(m\theta - n\phi)} \,. \tag{7.325}$$

The complete form of δW becomes [68]

$$\delta W = \frac{\pi^2 B_\phi^2}{\mu_0 R_0} \left\{ \int_0^a \left[\left(r \frac{d\xi}{dr} \right)^2 + (m^2 - 1)\,\xi^2 \right] \left(\frac{n}{m} - \frac{1}{q} \right)^2 r\,dr + \left[\frac{2}{q(a)} \left(\frac{n}{m} - \frac{1}{q(a)} \right) + (1 + m\lambda) \left(\frac{n}{m} - \frac{1}{q(a)} \right)^2 \right] a^2 \xi_a^2 \right\} \tag{7.326}$$

with

$$\lambda = \frac{1 + \left(\frac{a}{b}\right)^{2m}}{1 - \left(\frac{a}{b}\right)^{2m}} \,. \tag{7.327}$$

Kink modes or surface kink modes can only occur if there is a vacuum region between the plasma and the wall. A kink mode is stable if

$$q(a) > \frac{m}{n} \,. \tag{7.328}$$

Since $n \geq 1$ for kink modes [14], $q(a) > m$ is a sufficient condition for stability against mode m. The safety factor depends on the current distribution in the plasma (which determines the poloidal magnetic field component). The stability problem has been solved for different current distributions, for example,

$$j_\varphi(r) = j_{\varphi 0}\left(1 - \frac{r^2}{a^2}\right)^\nu , \quad \nu = 0, 1, 2, 3, \ldots \tag{7.329}$$

Let us summarize some of the obtained results [213, 214]. Growth rates have been obtained by solving the eigenvalue problem directly. When plotting the growth rates versus $nq(a)$ for $\nu = 1$, one obtains unstable bands being increasingly localized and decreasing in amplitude with increasing m; see Figure 7.4. The next Figure 7.5 shows for $b \to \infty$ and different current distributions, characterized by ν, the stability regions for surface kink modes and the internal kink mode ($m = n = 1$) [68]. Increased peaking of currents means increasing ν starting from $\nu = 0$. Increasing current peaking also means increasing $q(a)/q(0) \equiv q_a/q_0$ since

$$\frac{q(a)}{q(0)} = \nu + 1 . \tag{7.330}$$

The values on the horizontal axis are proportional (from left to right) to $1/q_a$ and therefore to the total current

$$I = \frac{1}{\nu + 1} A j_{\varphi 0} , \tag{7.331}$$

where A is the plasma cross section area. A current profile at least peaked as parabolic ($\nu \geq 1$) is necessary to insure stability against surface kink modes. The

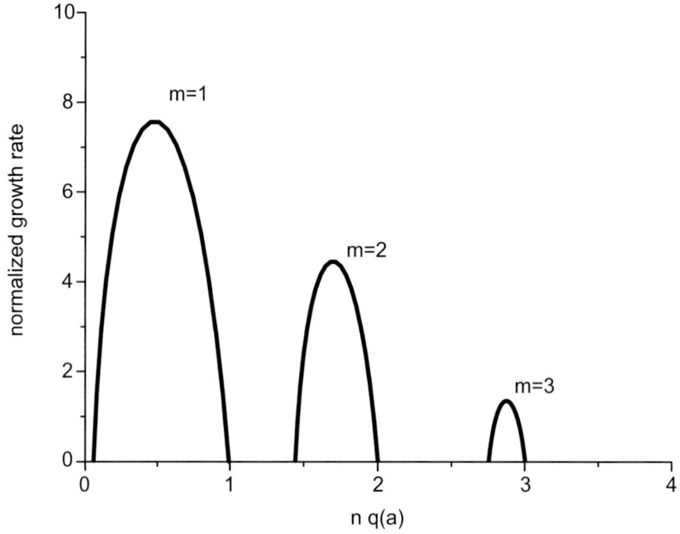

Figure 7.4 Growth rates of kink modes in toroidal configuration with a parabolic current distribution ($\nu = 1$); see [68].

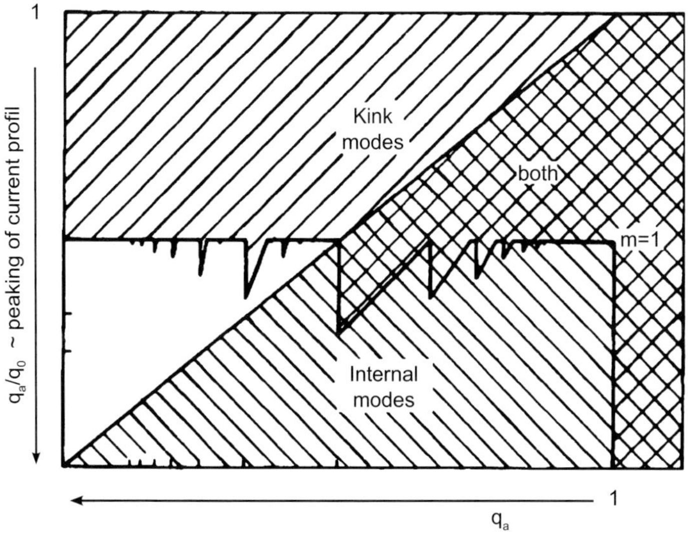

Figure 7.5 MHD stability diagram for kink instabilities; see [68].

condition $q(a) > 1$ is a minimal condition for stability (Kruskal–Shafranov stability limit). $q(0) > 1$ is sufficient to ensure stability of interchange modes (internal modes) for $dp/dr < 0$.

7.3.3
Interchange Instability

Let us now investigate the stability of a plasma confined by curved magnetic field lines. As we shall see, the relative direction of the pressure gradient with respect to the curvature vector will be important. We shall talk about bad and good curvature. The exact treatment of the stability problems is known as the ballooning mode problem, at least in tokamak physics. Here, we treat a simplified version of it, the interchange instability. Before presenting more sophisticated calculations, let us start with a very simple situation when within the plasma $\nabla \times \boldsymbol{B}_0 = 0$ and $\nabla p_0 = 0$. Then,

$$\begin{aligned} \delta W &\equiv \delta W_{plasma} + \delta W_{interface} + \delta W_{vacuum} \\ &> \delta W_{interface} = -\frac{1}{2} \int_S d^2 r (\boldsymbol{n} \cdot \boldsymbol{\xi})^2 \boldsymbol{n} \cdot \nabla \left(p + \frac{B_0^2}{2\mu_0} - \frac{B_{vacuum}^2}{2\mu_0} \right) . \end{aligned} \tag{7.332}$$

Here, $\boldsymbol{n}$ is the normal vector on the interface (pointing from plasma to vacuum). We shall further introduce $\boldsymbol{R}$ as the curvature vector of a magnetic field $\boldsymbol{B}$,

$$\frac{\boldsymbol{R}}{R^2} \equiv -\frac{(\boldsymbol{B} \cdot \nabla) \boldsymbol{B}}{B^2} . \tag{7.333}$$

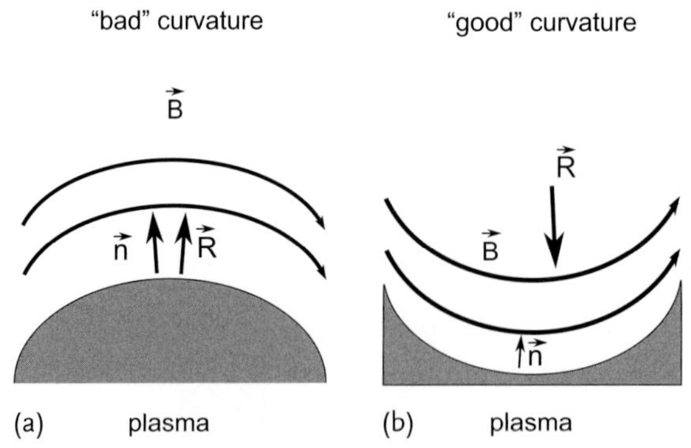

Figure 7.6 Plasma confinement by magnetic fields with bad (a) and good (b) curvature of the magnetic fields.

When the total pressure is assumed constant within the plasma volume, we estimate the term in the integrand by

$$\boldsymbol{n} \cdot \nabla \left(-\frac{B^2_{vacuum}}{2\mu_0} \right) = \frac{B^2_{vacuum}}{\mu_0} \frac{(\boldsymbol{R} \cdot \boldsymbol{n})}{R^2} . \tag{7.334}$$

Thus, for $\boldsymbol{n} \cdot \boldsymbol{R} < 0$, the system should be stable ("good confinement"). In Figure 7.6, we have sketched two typical situations. In the stable configuration, the magnetic field is curved away from the fluid. This is also called "good curvature." In the case of good curvature, the latter is dished (concave into the plasma), whereas for bad curvature, the latter is pulvinate (convex away from the plasma).

Thus far, we did not calculate the contributions from the volume terms (plasma plus vacuum). Instead of a very detailed analysis of the variational principle, we compare the *total* potential energy of two (interchanged) configurations; see Figure 7.7. The *total* potential energy consists of the magnetic energy

$$W_{magnetic} \equiv \int\limits_{system} \frac{\boldsymbol{B}^2}{2\mu_0} d^3 r \tag{7.335}$$

and the internal energy

$$W_{internal} = \int\limits_{plasma} \frac{p}{\gamma - 1} dV . \tag{7.336}$$

Let us start with the magnetic energy. We consider a volume of plasma with no field lines crossing the lateral surface area, that is, a flux tube, in area I of Figure 7.7 and "shift" it to area II. With l as the length along the flux tube and $A(l)$ as the cross sectional area, the flux $\phi = B(l)A(l)$ is constant. Within ideal MHD, also $d\phi/dt = 0$. When the plasma in area I is shifted to II, the magnetic flux of area I is also displaced to II, and vice versa. We characterize the difference in magnetic energy by

$$\delta W_{magnetic} = \delta \int \frac{B^2}{2\mu_0} A dl = \delta \left[\frac{\phi^2}{2\mu_0} \int \frac{dl}{A} \right] . \tag{7.337}$$

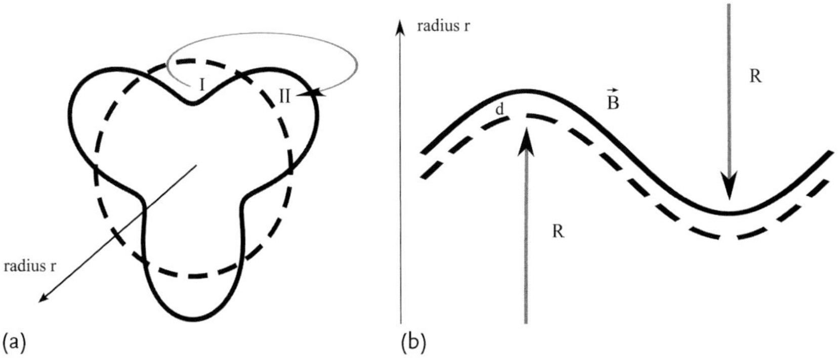

Figure 7.7 (a) Schematic plot of the cross section of a discharge with flute perturbations and corresponding interchanges of flux tubes between regions I and II. (b) The axial propagation of two neighboring (distance *d*) curved magnetic field lines is shown, together with the local curvature radii *R*.

Documented in more detail,

$$\begin{aligned}\delta W_{magnetic} &= \frac{1}{2\mu_0}\left[\phi_{\mathrm{II}}^2\left(\int\frac{dl_{\mathrm{I}}}{A_{\mathrm{I}}}-\int\frac{dl_{\mathrm{II}}}{A_{\mathrm{II}}}\right)+\phi_{\mathrm{I}}^2\left(\int\frac{dl_{\mathrm{II}}}{A_{\mathrm{II}}}-\int\frac{dl_{\mathrm{I}}}{A_{\mathrm{I}}}\right)\right]\\ &= \frac{1}{2\mu_0}\left(\phi_{\mathrm{II}}^2-\phi_{\mathrm{I}}^2\right)\left(\int\frac{dl_{\mathrm{I}}}{A_{\mathrm{I}}}-\int\frac{dl_{\mathrm{II}}}{A_{\mathrm{II}}}\right)\equiv-\frac{\delta\phi^2}{2\mu_0}\delta\int\frac{dl}{A}\,.\end{aligned} \tag{7.338}$$

In case of incompressibility, $\nabla\cdot\boldsymbol{u}=0$ or $\nabla\cdot\boldsymbol{\xi}=0$, that is, for $\gamma\to\infty$, with $A_{\mathrm{I}}dl_{\mathrm{I}}=A_{\mathrm{II}}dl_{\mathrm{II}}$, this is identical to

$$\delta W_{magnetic}\equiv\delta W=\frac{1}{2\mu_0}\left(\phi_{\mathrm{II}}^2-\phi_{\mathrm{I}}^2\right)\int\left[1-\left(\frac{dl_{\mathrm{II}}}{dl_{\mathrm{I}}}\right)^2\right]\frac{dl_{\mathrm{I}}}{A_{\mathrm{I}}}\,, \tag{7.339}$$

and we could discuss the stability properties for incompressible perturbations directly from here. However, let us be a little bit more general by including finite γ. We calculate the changes of internal energy for the flux tubes. To simplify the arguments, let us take perturbations with $\phi_{\mathrm{II}}=\phi_{\mathrm{I}}\equiv\phi$ for which $\delta W_{magnetic}=0$. Now, for adiabatic perturbations $[d(p\,dV^\gamma)/dt=0]$, we calculate ($dV\to V$ for a flux tube)

$$\begin{aligned}\delta W_{internal} &= \frac{1}{\gamma-1}\delta(pV)=\frac{1}{\gamma-1}\delta\left(\frac{pV^\gamma}{V^{\gamma-1}}\right)\\ &= \frac{1}{\gamma-1}\left\{p_{\mathrm{II}}V_{\mathrm{II}}^\gamma\left[\frac{1}{V_{\mathrm{I}}^{\gamma-1}}-\frac{1}{V_{\mathrm{II}}^{\gamma-1}}\right]+p_{\mathrm{I}}V_{\mathrm{I}}^\gamma\left[\frac{1}{V_{\mathrm{II}}^{\gamma-1}}-\frac{1}{V_{\mathrm{I}}^{\gamma-1}}\right]\right\}\\ &= \frac{1}{\gamma-1}\underbrace{\left[p_{\mathrm{II}}V_{\mathrm{II}}^\gamma-p_{\mathrm{I}}V_{\mathrm{I}}^\gamma\right]}_{\delta(pV^\gamma)}\left(\frac{1}{V_{\mathrm{I}}^{\gamma-1}}-\frac{1}{V_{\mathrm{II}}^{\gamma-1}}\right)\end{aligned} \tag{7.340}$$

or

$$\delta W_p \approx \frac{\delta (p V^\gamma)}{V^\gamma} \delta V = \delta p \delta V + \gamma p \frac{(\delta V)^2}{V} . \tag{7.341}$$

Since the second term is always positive, $\delta W_p \geq \delta p \delta V$, and a stability criterion is

$$\delta p \delta V = \delta p \delta \int A dl = \delta p \phi \delta \int \frac{dl}{B} > 0 . \tag{7.342}$$

Assuming for the confined plasma that the pressure decreases from the center outwards, we postulate $\delta p = p_{\mathrm{II}} - p_{\mathrm{I}} < 0$ such that the sufficient criterion for stability becomes

$$\delta \int \frac{dl}{B} < 0 . \tag{7.343}$$

Let us evaluate the condition for a magnetic mirror configuration as discussed in selected applications of single particle motion. We consider two magnetic field lines separated in radial direction by a distance d as depicted in Figure 7.7b. The flux in an annular tube, that is, $\phi = 2\pi r d B$, is constant. Applying the definition of the curvature radius, we estimate the change of the magnetic field strength according to

$$\left| \frac{\boldsymbol{R}}{R^2} \right| \equiv \left| -\frac{(\boldsymbol{B} \cdot \nabla)\boldsymbol{B}}{B^2} \right| \sim \left| \frac{\delta B}{d B} \right| \sim \left| \frac{1}{R} \right| , \quad \rightarrow \quad \frac{\delta B}{B} = -\frac{d}{R} . \tag{7.344}$$

The last conversion of the equation contains R including a sign, which is positive if the field lines are convex away from the plasma (left R in Figure 7.7) and negative if the field lines are concave into the plasma (right R in Figure 7.7). Finally, we assume that the magnetic field can be constructed from a scalar magnetic potential such that between equipotential surfaces separated by dl, the exchange of field lines leads to $\delta(B dl) = 0$. Using this, we write the condition Eq. (7.343) as

$$\begin{aligned} \delta \int \frac{dl}{B} &\approx \int \delta \left(\frac{1}{B^2} \right) B dl \approx -\int \frac{2\delta B}{B^2} dl \\ &\approx \int \frac{2d}{RB} dl \approx \frac{\phi}{\pi} \int \frac{1}{R r B^2} dl < 0 . \end{aligned} \tag{7.345}$$

Therefore, $R < 0$ defines the stable region, as previously obtained.

The interchange or the ballooning instabilities are strongly related to the classical Rayleigh–Taylor instability. When a heavy fluid is above a lighter fluid, instability occurs. In curved magnetic fields, drifting particles experience an centrifugal force which can be interpreted as the analogue to a gravitational force. In this picture, for $\boldsymbol{n} \cdot \boldsymbol{R} > 0$ ("bad curvature instability"), the situation is similar to the unstable Rayleigh–Taylor instability.

This analogue is mentioned for another reason. Variational principles allow one to identify regions of stability or instability, but they give no information on the dynamics of an unstable situation. Whether an instability is fast or weak has to be determined by other means. The standard way is to go back to the eigenvalue problem in case instability is predicted. We now present the solution of the eigenvalue problem for the Rayleigh–Taylor instability.

Let us consider a plasma as depicted in Figure 7.8. An external straight magnetic field $\boldsymbol{B}_0 = \hat{z} B_0(y)$ is applied. In y-direction, the effective gravitational acceleration $\boldsymbol{g}_0 = -\hat{y} g_0$ mimics the centrifugal acceleration. The stationary configuration is described by

$$\frac{\partial}{\partial y}\left(p_0 + \frac{B_0^2}{2\mu_0}\right) = -\rho_0 g_0 \, . \tag{7.346}$$

Compared to the previous MHD description, we have to add the gravitational term, $\rho_0 \dot{\boldsymbol{u}} = \cdots + \rho \boldsymbol{g}_0$. When we differentiate with respect to time, we use $\dot{\rho} = -\nabla(\rho_0 \boldsymbol{u})$. Thus, the additional term $-\boldsymbol{g}_0 \nabla \cdot (\rho_0 \boldsymbol{\xi}) = \hat{y} g_0 \nabla \cdot (\rho_0 \boldsymbol{\xi}) = \hat{y} g_0 [\boldsymbol{\xi} \cdot \nabla \rho_0 + \rho_0 \nabla \cdot \boldsymbol{\xi}]$ will be generated, leading to

$$\begin{aligned} -\rho_0 \omega^2 \boldsymbol{\xi} &= \nabla\left(\boldsymbol{\xi} \cdot \nabla p_0 + \gamma p_0 \nabla \cdot \boldsymbol{\xi}\right) \\ &\quad + \frac{1}{\mu_0}\left[(\nabla \times \boldsymbol{Q}) \times \boldsymbol{B}_0 + (\nabla \times \boldsymbol{B}_0) \times \boldsymbol{Q}\right] + \hat{y} g_0 \nabla \cdot (\rho_0 \boldsymbol{\xi}) \, . \end{aligned} \tag{7.347}$$

To simplify the following calculations, we shall only consider perturbations which do not depend on the z-coordinate. In addition, the perturbations should be incompressible ($\nabla \cdot \boldsymbol{\xi} = 0$). Introducing $\boldsymbol{Q}$ and making use of vector relations in the form

$$\boldsymbol{Q} = -\xi_y \frac{d B_0}{d y} \hat{z} \, , \quad (\nabla \times \boldsymbol{Q}) \times \boldsymbol{B}_0 + (\nabla \times \boldsymbol{B}_0) \times \boldsymbol{Q} = -\nabla(\boldsymbol{B}_0 \cdot \boldsymbol{Q}) \, , \tag{7.348}$$

we postulate $\boldsymbol{\xi} = \boldsymbol{\xi}(y) \exp(i k x)$. For the components, we get

$$\rho_0 \omega^2 \xi_x = i k \left(\boldsymbol{\xi} \cdot \nabla p_0 + \frac{1}{\mu_0} \boldsymbol{B}_0 \cdot \boldsymbol{Q}\right) , \tag{7.349}$$

$$\rho_0 \omega^2 \xi_y = \frac{d}{d y}\left(\boldsymbol{\xi} \cdot \nabla p_0 + \frac{1}{\mu_0} \boldsymbol{B}_0 \cdot \boldsymbol{Q}\right) - g_0 \xi_y \frac{d \rho_0}{d y} \, . \tag{7.350}$$

A simple combination leads to

$$\rho_0 \omega^2 \xi_y = \frac{\omega^2}{i k} \frac{d}{d y}(\rho_0 \xi_x) - g_0 \xi_y \frac{d \rho_0}{d y} \, . \tag{7.351}$$

Differentiating with respect to x and using the incompressibility condition, we can eliminate ξ_x such that for the y-component, we obtain the closed equation

$$\frac{d^2 \xi_y}{d y^2} + \left(\frac{1}{\rho_0} \frac{d \rho_0}{d y}\right) \frac{d \xi_y}{d y} - k^2 \left(1 + \frac{g_0}{\omega^2} \frac{1}{\rho_0} \frac{d \rho_0}{d y}\right) \xi_y = 0 \, . \tag{7.352}$$

An explicit solution of the eigenvalue equation requires the knowledge of $\rho_0 = \rho_0(y)$. For an exponential variation with $d \ln \rho_0 / d y = \kappa = \text{const}$, we can assume

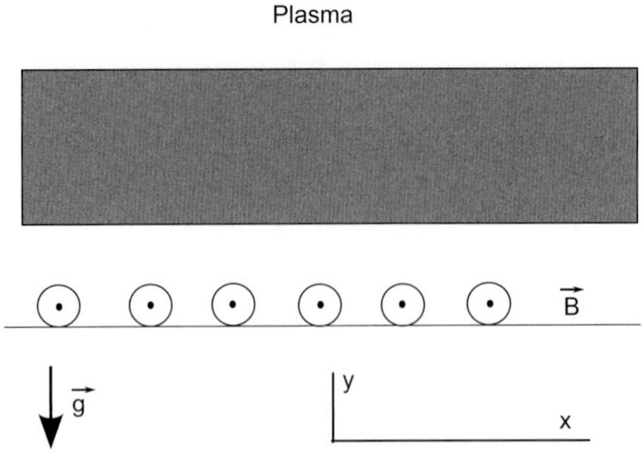

Figure 7.8 Geometry of the hydrodynamic Rayleigh–Taylor instability.

y-independent perturbations with

$$\frac{d^2 \xi_y}{dy^2} = \frac{d\xi_y}{dy} = 0 \,. \tag{7.353}$$

In that simple case, $\omega^2 = -g_0 \kappa$. That means instability for $\kappa > 0$ when the density decreases in the direction of the gravitational force. The result is known as the classical Rayleigh–Taylor instability.

As another example, being closer to the situation depicted in Figure 7.8, we may assume

$$\rho_0 = \begin{cases} \rho_{00} = \text{const} & \text{für} \quad y \geq 0 \,, \\ 0 & \text{für} \quad y < 0 \,. \end{cases} \tag{7.354}$$

In the region $y > 0$, the differential simplifies to

$$\frac{d^2 \xi_y}{dy^2} = k^2 \xi_y \,, \tag{7.355}$$

with the solution $\xi_y = \xi_y^{(0)} e^{-ky}$. At $y = 0$, we expect a significant contribution from ρ_0 since

$$\frac{d\rho_0}{dy} = \rho_0 \delta(y) \,. \tag{7.356}$$

We then integrate both sides of Eq. (7.352) over y from $-\varepsilon$ to $+\varepsilon$, subsequently with $\varepsilon \to 0$, we find from

$$\int_{-\varepsilon}^{+\varepsilon} \frac{d}{dy}\left(\rho_0 \frac{d\xi_y}{dy}\right) dy = \int_{-\varepsilon}^{+\varepsilon} k^2 \left(\rho_0 + \frac{g_0}{\omega^2}\frac{d\rho_0}{dy}\right) \xi_y \, dy \tag{7.357}$$

the relation

$$\left.\frac{k^2 g_0}{\omega^2} \rho_0 \xi_y\right|_{y=0} = \lim_{\varepsilon \to +0} \left.\rho_0 \frac{d\xi_y}{dy}\right|_{y=\varepsilon} \,. \tag{7.358}$$

Continuity of ξ_y at $y = 0$ leads to $\omega^2 = -k g_0$, which determines the Rayleigh–Taylor growth of perturbations in a plasma with a sharp boundary.

We have already mentioned that Rayleigh–Taylor-like instabilities have several applications in plasma physics. In magnetic fusion devices, ballooning modes appear on the bad curvature areas of a tokamak. In laser–plasma-interaction, expanding fronts cause growing modes. The corresponding instability with time-dependent driving is known as Richtmyer–Meshkov instability [215–217].

8
General Theory of Nonlinear Waves and Solitons

In this chapter, we investigate finite amplitude waves. We present an overview of the nonlinear versions of typical plasma waves and comment on characteristic methods in nonlinear wave theory with special emphasis on solitary waves and solitons.

8.1
Historical Remarks

Solitons are nonlinear waves. As a preliminary definition, a soliton is considered as a solitary, traveling wave pulse solution of a nonlinear partial differential equation (PED). It should have remarkable stability properties. Stability plays a central role in soliton physics.

The beginning of soliton physics in often dated back to the month of August 1834 when John Scott Russel observed the "great wave if translation" [218]. Due to the work of Stokes, Boussinesq, Rayleigh, Korteweg, de Vries, and many others we know that the "great wave of translation" is a special form of a surface water wave [219, 220].

Modern soliton physics has branches in practically all disciplines of physics [17, 219, 221–228]. The physically important point is that many complicated models of nonlinear processes can be reduced, for a basic understanding, to generic equations which allow for soliton solutions; see, for example, [229–237]. We distinguish between conservative solitons (in integrable systems) [227] and dissipative solitons [238, 239]. In the present book we concentrate on conservative solitons, assuming that the amplitudes are large enough that dissipation is a higher-order effect. In nonlinear optics with weak laser intensities dissipative solitons are of increasing interest, and it seems that three-dimensional objects can be created much easier [240].

In the present introduction, let us first concentrate on one of the most obvious principles underlying soliton propagation. A linear wave equation

$$\frac{\partial^2 \psi}{\partial t^2} - c^2 \frac{\partial^2 \psi}{\partial x^2} = 0 \tag{8.1}$$

High Temperature Plasmas, Theory and Mathematical Tools for Laser and Fusion Plasmas, First Edition.
Karl-Heinz Spatschek.

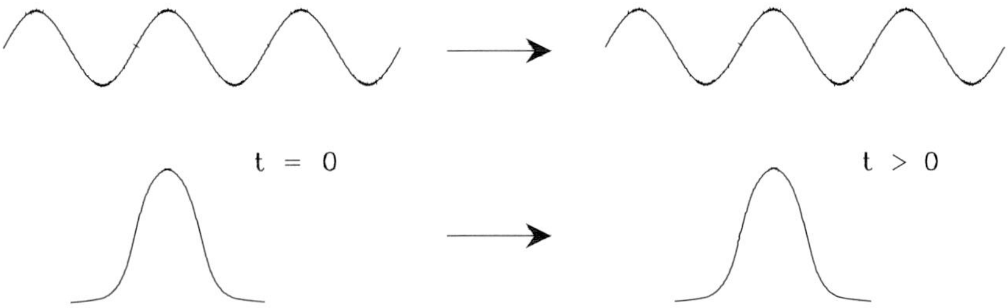

Figure 8.1 Solutions of an idealized linear wave equation.

represents the idealized model for wave motion in one space-dimension. The wave field ψ is a function of time t and the space-coordinate x. The (phase) velocity of the wave is c (not necessary the velocity of light in the present example). We can write the general solution as a combination of

$$\psi = \psi_1(x \pm ct) , \tag{8.2}$$

where the functional forms of ψ_1 are determined by the initial values. The $\pm$ signs designate the propagation directions. In Figure 8.1, it is demonstrated that any initial value can propagate without changes in form. It should be emphasized that we do not need a (harmonic) wave profile. Also, pulses will be transported without changes. However, the wave Eq. (8.1) is generally much too simple. The corresponding dispersion relation

$$\omega = \pm ck \tag{8.3}$$

can be understood as an idealization. Damping, dispersion, and wave-breaking are known phenomena which are not included in that idealized situation.

To make mathematics even simpler, in the following, we "reduce" the ideal wave equation to a simpler mathematical model

$$\frac{\partial \psi}{\partial t} + c\frac{\partial \psi}{\partial x} = 0 \tag{8.4}$$

which singles out one propagation direction of

$$\left(\frac{\partial}{\partial t} + c\frac{\partial}{\partial x}\right)\left(\frac{\partial}{\partial t} - c\frac{\partial}{\partial x}\right)\psi = 0 . \tag{8.5}$$

Formally, we can come to an equation of the form Eq. (8.4) by introducing

$$\xi = x + ct , \quad \tau = t , \quad c' = 2c , \quad \varphi = \frac{\partial \psi}{\partial \tau} \tag{8.6}$$

into Eq. (8.1), leading to

$$\frac{\partial \varphi}{\partial \tau} + c'\frac{\partial \varphi}{\partial \xi} = 0 . \tag{8.7}$$

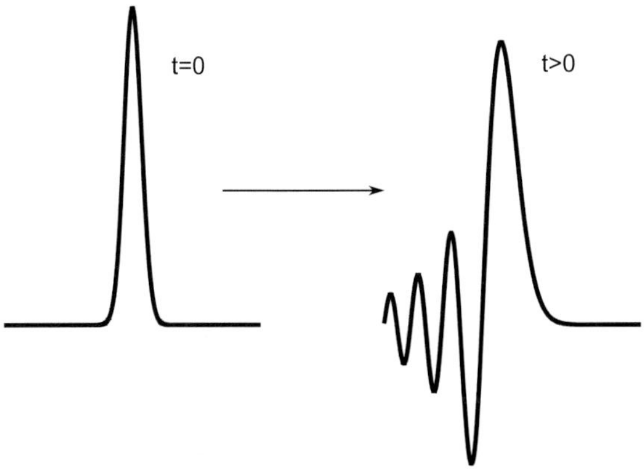

Figure 8.2 Effect of dispersion on linear wave propagation.

Solutions are now written in the form

$$\psi = \psi_1(x - ct) . \tag{8.8}$$

Dispersion modifies Eq. (8.4), for example, to

$$\frac{\partial \psi}{\partial t} + c\frac{\partial \psi}{\partial x} + \delta \frac{\partial^3 \psi}{\partial x^3} = 0 . \tag{8.9}$$

Here, we have introduced the parameter δ to characterize the deviations from the linear $\omega \sim k$ relation. By introducing the new coordinates

$$\xi = x - ct , \quad \tau = t \tag{8.10}$$

the dispersive wave equation simplifies to

$$\frac{\partial \psi}{\partial \tau} + \delta \frac{\partial^3 \psi}{\partial \xi^3} = 0 . \tag{8.11}$$

This is the Airy differential equation with the solution

$$\psi(\xi, \tau) = \frac{1}{\sqrt{\pi}(3\delta\tau)^{1/3}} \int_{-\infty}^{\infty} Ai\left[\frac{\xi - \xi'}{(3\delta\tau)^{1/3}}\right] \psi\left(\xi', \tau = 0\right) d\xi' . \tag{8.12}$$

Figure 8.2 sketches the properties of the solutions. We get dispersive broadening.

Nonlinearities also change the wave equation. Compared with Eq. (8.11) we consider

$$\frac{\partial \psi}{\partial \tau} + \kappa\psi \frac{\partial \psi}{\partial \xi} = 0 \tag{8.13}$$

with a quadratic nonlinearity. This equation is equivalent to

$$\psi(\xi, \tau) = \psi(\xi - \kappa\psi\tau) . \tag{8.14}$$

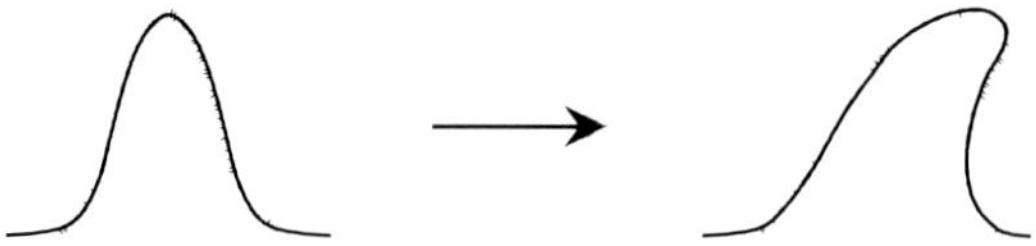

Figure 8.3 Wave-breaking due to nonlinearities.

The last relation clearly shows that wave-breaking can occur: Larger amplitudes propagate faster. We elucidate that behavior in Figure 8.3. In the following chapters, we shall discuss new effects which can occur when dispersion and nonlinearity are simultaneously present.

8.1.1 The Water-Wave Paradigm

As a preliminary definition, a soliton is considered as a solitary, traveling wave pulse solution of a nonlinear partial differential equation. It should have remarkable stability properties. That means that the stability (we have later to define what we mean by "remarkable stability") plays a central role in soliton physics. Before going into more detail, let us review some of the historical milestones regarding the discovery of the soliton.

We start with the "the great wave of translation" [218, 220, 241]. Due to the work of Stokes, Boussinesq, Rayleigh, Korteweg, de Vries [242], and many others, we know now that this "great wave of translation" is a special nonlinear form of a surface water wave. Therefore, it is appropriate to repeat some of the basic physical principles of water waves. The methods being exemplified here also apply to the plasma solitons discussed a little bit later.

For a curl-free fluid motion, the fluid velocity $\boldsymbol{v}$ can be deduced from a scalar potential ϕ via $\boldsymbol{v} = \nabla\phi$. In addition, we assume incompressibility,

$$\frac{d\rho}{dt} = 0\,, \tag{8.15}$$

which, because of the continuity equation for the mass density

$$\frac{\partial\rho}{\partial t} + \nabla\cdot(\rho\boldsymbol{v}) = 0\,, \tag{8.16}$$

means $\nabla\cdot\boldsymbol{v} = 0$. Therefore, we obtain for ϕ the Laplace equation

$$\nabla^2\phi \equiv \partial_x^2\phi + \partial_z^2\phi = 0\,. \tag{8.17}$$

Here, z is the coordinate measuring the height of the fluid element over the ground (at $z = -h$) and, for simplicity, we only assume a variation in a single perpendicular direction, let us say the x-direction. For an illustration, see Figure 8.4. Since the normal component of the velocity v_z vanishes on the ground, we have

$$\partial_z\phi = 0 \quad \text{at} \quad z = -h\,. \tag{8.18}$$

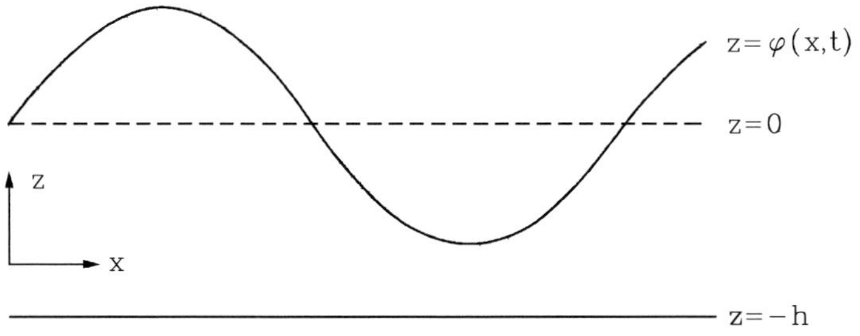

Figure 8.4 Schematic plot of the two-dimensional model of a water surface $z = \varphi(x, t)$ in a system of finite depth h.

These equations have to be supplemented by the so called kinematic free boundary condition at the surface $z = \varphi(x, t)$. We have

$$\partial_z \phi = \partial_t \varphi + \partial_x \varphi \partial_x \phi \quad \text{at} \quad z = \varphi \tag{8.19}$$

because of $v_z = dz/dt$. Finally, the force balance at the free boundary (neglecting surface tension) has to be considered. We start with

$$\partial_t \boldsymbol{v} + \boldsymbol{v} \cdot \nabla \boldsymbol{v} = -g\hat{z} - \nabla \left(\frac{p}{\rho} \right), \tag{8.20}$$

where ρ is the (incompressible) mass density, $\hat{z} \equiv \boldsymbol{e}_z$, and $\boldsymbol{v} \cdot \nabla \boldsymbol{v} = 1/2 \nabla \boldsymbol{v}^2$ because of $\nabla \cdot \boldsymbol{v} = 0$ and $\nabla \times \boldsymbol{v} = 0$. The balance in z-direction leads, to lowest order, on the free surface φ to

$$\partial_t \phi + \frac{1}{2} (\nabla \phi)^2 = -g\varphi - \frac{p - p_0}{\rho} \quad \text{at} \quad z = \varphi \,, \tag{8.21}$$

where p_0 is the atmospheric pressure. For $p \approx p_0$ at the surface, we have the dynamic free boundary condition

$$\partial_t \phi + \frac{1}{2} (\nabla \phi)^2 + g\varphi = 0 \quad \text{at} \quad z = \varphi \,. \tag{8.22}$$

Equation (8.17), together with the boundary conditions (8.18), (8.19), and (8.22), constitute the basic set of equations.

Let us now discuss the *linear* limit

$$\nabla^2 \phi = 0 \,, \tag{8.23}$$

$$\partial_z \phi = 0 \quad \text{at} \quad z = -h \,, \tag{8.24}$$

$$\partial_z \phi = \partial_t \varphi \quad \text{at} \quad z = \varphi \,, \tag{8.25}$$

$$\partial_t \phi + g\varphi = 0 \quad \text{at} \quad z = \varphi \,. \tag{8.26}$$

We can combine Eqs. (8.25) and (8.26) to obtain, at $z = \varphi$,

$$\partial_z \phi + g^{-1} \partial_t^2 \phi = 0 \,. \tag{8.27}$$

For a traveling wave in the x-direction

$$\phi = f(z)\cos(kx - \omega t), \tag{8.28}$$

with wave-number k and frequency ω, we obtain from Eq. (8.23)

$$\phi = \left(Ae^{kz} + Be^{-kz}\right)\cos(kx - \omega t)\,. \tag{8.29}$$

When using the boundary conditions (8.24) as well as Eq. (8.27) (at $z = 0$ in the linear limit), we finally obtain the linear system of two equations

$$\begin{bmatrix} k - \frac{\omega^2}{g} & -\left(k + \frac{\omega^2}{g}\right) \\ k\exp(-kh) & -k\exp(kh) \end{bmatrix} \begin{bmatrix} A \\ B \end{bmatrix} = 0\,. \tag{8.30}$$

For a nontrivial solution, the dispersion relation

$$\omega^2 = gk\tanh(kh) \tag{8.31}$$

follows. We clearly see that for long waves in shallow water, that is, $kh \ll 1$, Eq. (8.31) simplifies to

$$\omega \approx k(gh)^{1/2}\left[1 - \frac{1}{6}(kh)^2\right], \tag{8.32}$$

whereas in deep water, that is, for $kh \gg 1$,

$$\omega \approx (gk)^{1/2}\,. \tag{8.33}$$

Let us now aim for a simple *nonlinear* description in the shallow water case. Shallow means that the characteristic horizontal length l is large compared to the depth h, that is, we can scale the horizontal variation by a small parameter

$$\varepsilon = \frac{h}{l} \ll 1\,. \tag{8.34}$$

Furthermore, we shall consider the weakly nonlinear regime, assuming

$$\mu = \frac{v_0}{c_{ph}} \ll 1\,, \tag{8.35}$$

that is, the typical fluid velocity v_0 is much smaller than the phase velocity $c_{ph} \equiv (gh)^{1/2}$ of the wave.

Introducing the new variables

$$x' = \frac{x}{l}\,, \quad z' = \frac{z}{h}\,, \quad t' = \frac{c_{ph}t}{l}\,, \quad \phi' = \frac{\phi}{v_0 l}\,, \quad \text{and} \quad \varphi' = \frac{c_{ph}\varphi}{v_0 h}\,, \tag{8.36}$$

we obtain from Eqs. (8.17)–(8.19), and (8.22) a new set of scaled equations. Note that the scaling of φ is caused by the requirement $\phi'_{t'} \sim \varphi'$ on the surface. Here and in the following, we indicate the (partial) derivatives by indices, as along as no confusion is expected. Omitting the dashes, for the sake of simplicity, we write the basic nonlinear equations in the form

$$\phi_{zz} + \varepsilon^2 \phi_{xx} = 0 \,, \tag{8.37}$$

$$\phi_z = 0 \quad \text{at} \quad z = -1 \,, \tag{8.38}$$

$$\varphi_t + \mu \varphi_x \phi_x - \frac{1}{\varepsilon^2} \phi_z = 0 \quad \text{at} \quad z = \mu\varphi \,, \tag{8.39}$$

$$\phi_t + \varphi + \frac{1}{2} \mu \phi_x^2 + \frac{1}{2} \frac{\mu}{\varepsilon^2} \phi_z^2 = 0 \quad \text{at} \quad z = \mu\varphi \,. \tag{8.40}$$

Equation 8.37 admits a power series solution in z. We write

$$\phi(x, z, t) = F(x, t) + z\phi_{z0} + \frac{z^2}{2} \phi_{zz0} + \cdots \tag{8.41}$$

Let us now determine the orders of magnitudes (in ε) of the various terms on the right-hand side of Eq. (8.41). If we differentiate both sides of Eq. (8.41) with respect to z, we obtain

$$\phi_z = \phi_{z0} + z\phi_{zz0} + \cdots \,, \tag{8.42}$$

$$\phi_{zz} = \phi_{zz0} + z\phi_{zzz0} + \cdots \,, \tag{8.43}$$

and so on. Using Eq. (8.43) in Eq. (8.37), we have to order ε^2

$$\phi_{zz0} = -\varepsilon^2 F_{xx} \,. \tag{8.44}$$

In a similar way, we get (equating terms with equal powers in z)

$$\phi_{zzz0} = -\varepsilon^2 \partial_x^2 \phi_{z0} \,, \tag{8.45}$$

$$\phi_{zzzz0} = -\varepsilon^2 \partial_x^2 \phi_{zz0} = \varepsilon^4 F_{xxxx} \,, \tag{8.46}$$

and so on. Inserting these results into Eq. (8.38), we obtain, up to order ε^4,

$$\phi_{z0} + \varepsilon^2 F_{xx} - \frac{\varepsilon^2}{2} \partial_x^2 \phi_{z0} - \frac{\varepsilon^4}{6} F_{xxxx} = 0 \,. \tag{8.47}$$

From here, we find in various orders

$$\phi_{z0} = -\varepsilon^2 F_{xx} + \mathcal{O}(\varepsilon^4) \,, \tag{8.48}$$

$$\phi_{z0} = -\varepsilon^2 F_{xx} - \frac{1}{2} \varepsilon^4 F_{xxxx} + \frac{1}{6} \varepsilon^4 F_{xxxx} = -\varepsilon^2 F_{xx} - \frac{\varepsilon^4}{3} F_{xxxx} + \mathcal{O}(\varepsilon^6) \,, \tag{8.49}$$

and so on. Combining these results, we can rewrite Eq. (8.41) in a form where, on the right-hand side, only F and its spatial derivatives occur. Now, we are in a position to obtain Eq. (8.40) in the form

$$F_t + \varphi + \frac{\mu}{2} F_x^2 + \mathcal{O}\left(\varepsilon^4, \mu^2, \varepsilon^2 \mu\right) = 0 \,. \tag{8.50}$$

On the other hand, Eq. (8.39) reads

$$\varphi_t + \mu F_x \varphi_x + F_{xx} + \frac{\varepsilon^2}{3} F_{xxxx} + \mu \varphi F_{xx} + \mathcal{O}\left(\varepsilon^4, \mu^2, \varepsilon^2 \mu\right) = 0 \,. \tag{8.51}$$

At the present, we are interested in the case when nonlinearity (measured by μ) and dispersion (measured by ε^2) are both small and in balance. Setting $\mu = \varepsilon^2$, we get from Eqs. (8.50) and (8.51) two relatively simple equations for $\varphi(x,t)$ and $F(x,t)$,

$$F_t + \varphi + \frac{1}{2} \varepsilon^2 F_x^2 = 0 \,, \tag{8.52}$$

$$\varphi_t + \varepsilon^2 \left[F_x \varphi\right]_x + F_{xx} + \frac{1}{3} \varepsilon^2 F_{xxxx} = 0 \,. \tag{8.53}$$

Eliminating φ, we have

$$F_{tt} + 2\varepsilon^2 F_x F_{xt} + \varepsilon^2 F_t F_{xx} - F_{xx} - \frac{1}{3} \varepsilon^2 F_{xxxx} + \mathcal{O}(\varepsilon^4) = 0 \,. \tag{8.54}$$

To solve Eq. (8.54), we can use an iteration procedure. Let $F = f + \varepsilon^2 F_1 + \cdots$, then $f = f(x - t, \varepsilon^2 t)$ satisfies Eq. (8.54) to leading order. (Note that we have chosen a certain direction of propagation which causes an asymmetry in the flow.) In F_1, we have a fast variation with respect to x and t. We replace the latter variables by $\theta_+ = x - t$ and $\theta_- = x + t$ such that in order ε^2,

$$F_{1tt} - F_{1xx} = -4 \frac{\partial^2 F_1}{\partial \theta_+ \partial \theta_-} = 2 f_{\theta_+ T} + 3 f_{\theta_+} f_{\theta_+ \theta_+} + \frac{1}{3} f_{\theta_+ \theta_+ \theta_+ \theta_+} \,, \tag{8.55}$$

where $T = \varepsilon^2 t$, $\theta_\pm = x \mp t$. Since the right-hand side of Eq. (8.55) is independent of θ_-, we would obtain secular growth unless

$$2u_T + 3u u_X + \frac{1}{3} u_{XXX} = 0 \,, \tag{8.56}$$

where we have defined

$$X = \theta_+ \tag{8.57}$$

and

$$u = f_{\theta_+} = f_X \,. \tag{8.58}$$

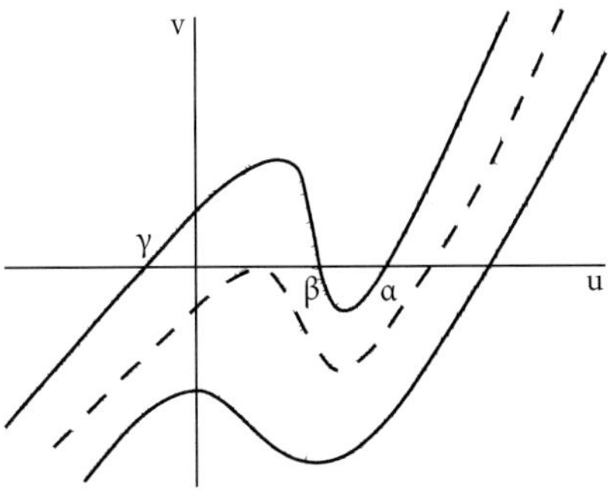

Figure 8.5 Effective potential V for the stationary soliton of the KdV equation. Three different possibilities, with one, two, and three intersections at the real u axis, are shown.

Thus, we have derived the famous KdV (Korteweg–deVries) equation [242]. Later, we shall investigate this equation in more detail. At the moment, we are interested in its stationary, solitary solutions. Let us introduce the coordinate

$$\xi = X - c'T \tag{8.59}$$

and look for solutions $u(X, T) = u(\xi)$. For convenience, we write the stationary equation in the form

$$-cu_\xi + uu_\xi + Ku_{\xi\xi\xi} = 0 , \tag{8.60}$$

with $c = 2c'/3$ and $K = 1/9$. Subsequent integration, multiplication by u_ξ, and integration yield

$$\frac{1}{2}u_\xi^2 + \frac{1}{6K}[-f(u)] = 0 , \tag{8.61}$$

where

$$f(u) := -u^3 + 3cu^2 + 6Au + 6B , \tag{8.62}$$

and A and B are constants of integration. Typical forms of the "potential" $V := -f(u)/6K$ are shown in Figure 8.5.

From the physical point of view, it is immediately clear that we can have bounded solutions only if three real roots of $V(u) = 0$ exist. Let us call them α, β, and γ. A sufficient criterion for that case can be derived in the following way. Letting

$$f(u) := (u - \gamma)(u - \beta)(\alpha - u) , \tag{8.63}$$

we have

$$c = \frac{1}{3}(\alpha + \beta + \gamma) , \quad A = -\frac{1}{6}(\alpha\beta + \beta\gamma + \gamma\alpha) , \quad B = \frac{1}{6}\alpha\beta\gamma . \tag{8.64}$$

From the identity

$$c^2 + 2A = \frac{1}{18}\left[(\alpha - \beta)^2 + (\beta - \gamma)^2 + (\gamma - \alpha)^2\right] , \tag{8.65}$$

we get the necessary condition

$$c^2 + 2A \geq 0 \,. \tag{8.66}$$

In Figure 8.5, we further recognize that $f(u)$ should have a positive maximum between β and α, and a negative minimum between γ and β. (If two solutions coincide, two extrema will become zero.) From $f_u = 0$, we find the positions

$$u = c \pm (c^2 + 2A)^{1/2} \tag{8.67}$$

of the extrema and get the sufficient criteria for three real roots

$$f\left[c + (c^2 + 2A)^{1/2}\right] \geq 0 \,, \quad f\left[c - (c^2 + 2A)^{1/2}\right] \leq 0 \,. \tag{8.68}$$

First, we treat the case $\alpha > \beta > \gamma$. Then, from Eq. (8.61), we obtain

$$\frac{du}{[(u-\gamma)(u-\beta)(\alpha-u)]^{1/2}} = \frac{d\xi}{(3K)^{1/2}} \,. \tag{8.69}$$

We can introduce

$$p^2 := \alpha - u \geq 0 \tag{8.70}$$

for bounded solutions. For

$$q := \frac{p}{(\alpha-\beta)^{1/2}} = \frac{(\alpha-u)^{1/2}}{(\alpha-\beta)^{1/2}} \tag{8.71}$$

and

$$s^2 := \frac{\alpha-\beta}{\alpha-\gamma} \,, \quad 0 < s^2 < 1 \,, \tag{8.72}$$

we finally obtain from Eq. (8.69)

$$\frac{d\xi}{(3K)^{1/2}} = -\frac{2dq}{\left[(1-s^2q^2)(1-q^2)\right]^{1/2}} \frac{1}{(\alpha-\gamma)^{1/2}} \,. \tag{8.73}$$

When integrating, it is obvious that the upper limit $u = \alpha$ corresponds to $q = 0$, and the lower limit $u = \beta$ corresponds to $q = 1$. Assuming $\xi = 0$ for $u = \alpha$, we have

$$\xi = \left(\frac{12K}{\alpha-\gamma}\right)^{1/2} \int_0^q \frac{dq}{[(1-s^2q^2)(1-q^2)]^{1/2}} \equiv \left(\frac{12K}{\alpha-\gamma}\right)^{1/2} F\left(\varphi|s^2\right) . \tag{8.74}$$

The integral on the right-hand side of Eq. (8.74) has standard form. Using $F(\varphi|s^2) = u$ and $\sin(\varphi) = q = sn(u)$ [243], we can write

$$\xi = \left[\frac{12K}{\alpha-\gamma}\right]^{1/2} sn^{-1}(q|s^2) \,, \tag{8.75}$$

$$u(\xi) = \beta + (\alpha-\beta)cn^2\left[\left(\frac{\alpha-\gamma}{12K}\right)^{1/2} \xi, s^2\right] , \tag{8.76}$$

where sn and cn are Jacobi elliptic functions. Thus, we have found the so called cnoidal waves. Their period P (in ξ) is given by

$$P = 2\left(\frac{12K}{\alpha - \gamma}\right)^{1/2} \int_0^1 \frac{dq}{[(1 - s^2q^2)(1 - q^2)]^{1/2}} \equiv 4\left(\frac{3K}{\alpha - \gamma}\right)^{1/2} K(s^2) , \tag{8.77}$$

where $K(s^2)$ is the complete elliptic integral of the first kind.

Obviously, this form of a stationary solution does not describe the "great wave of translation" observed by Scott Russell. However, there is still a case missing, $\gamma = \beta \neq \alpha$. (The other case, $\alpha = \beta \neq \gamma$, is of less interest to us.) We have real bounded solutions for $\gamma \leq u \leq \alpha$. Instead of Eq. (8.69), we now have

$$d\xi = (3K)^{1/2} \frac{du}{(u - \gamma)(\alpha - u)^{1/2}} . \tag{8.78}$$

The solution of Eq. (8.78) is

$$u(\xi) = \gamma + (\alpha - \gamma)\text{sech}^2\left[\left(\frac{\alpha - \gamma}{12K}\right)^{1/2} \xi\right] . \tag{8.79}$$

Note, that in this case, ($s = 0, \alpha = \beta$),

$$\xi = x - t - \frac{2\alpha + \gamma}{2} t . \tag{8.80}$$

The "period" of this solitary wave is

$$P = 4\left(\frac{3K}{\alpha - \gamma}\right)^{1/2} \int_0^1 \frac{dq}{1 - q^2} = \infty . \tag{8.81}$$

For $\xi \to \pm\infty$, we have $u \to \gamma \equiv u_\infty$; defining an amplitude

$$a := \alpha - \gamma , \tag{8.82}$$

we rewrite the solution Eq. (8.79) in the form

$$u(\xi) = u_\infty + a\text{sech}^2\left\{\left[\frac{a}{12K}\right]^{1/2}\left[x - \left(1 + \frac{3}{2}u_\infty - \frac{a}{2}\right)t\right]\right\} . \tag{8.83}$$

We clearly see that

a) the velocity, relative to the undisturbed case, is proportional to the amplitude;
b) the width of the solitary wave is inversely proportional to the amplitude;
c) the amplitude is independent of the asymptotic value u_∞,
d) an increase in the dispersive term in Eq. (8.60) increases the width ($\sim K^{1/2}$), but does not lead to a (dispersive) smearing out.

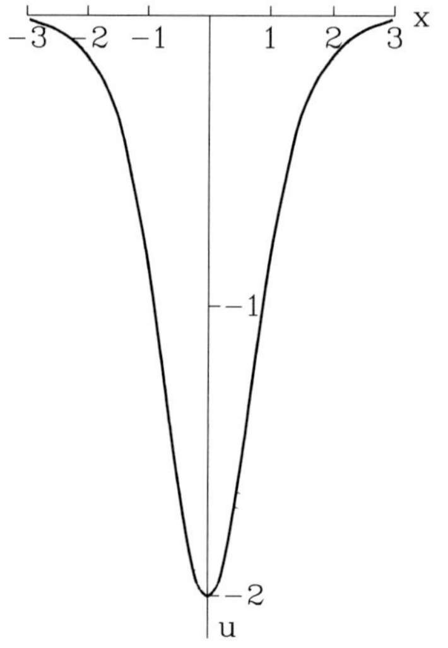

Figure 8.6 Plot of the KdV soliton $u(x,0) = -2\text{sech}^2 x$ for $a = -2$, $K = -1/6$, and $u_\infty = 0$.

The form of the solitary wave is sketched in Figure 8.6. We should mention that the realistic theory of nonlinear water waves is even more complicated than the model presented above. Nevertheless, one is already tempted to compare this with the observations by Scott Russell [218] in more detail. He observed a water soliton of approximately 30 feet width and 1–1.5 feet height with a speed of 8–9 mi/h.

From Eq. (8.52), we get

$$\varphi \approx -F_t \,, \tag{8.84}$$

and from Eq. (8.58), together with $F \approx f$,

$$u = f_{\theta_+} \approx F_x \approx -F_t \,. \tag{8.85}$$

Then,

$$u \approx \varphi \tag{8.86}$$

follows. Introducing the dimensional units (see Eq. (8.36)), and setting $\mu = \varepsilon^2$,

$$\alpha - \gamma = \frac{\varphi_0}{h}\frac{l^2}{h^2} \,, \tag{8.87}$$

we can rewrite the solution Eq. (8.83) in the form

$$\frac{\varphi}{h} = \frac{\varphi_0}{h}\text{sech}^2\left\{\frac{3^{1/2}\varphi_0^{1/2}}{2h^{3/2}}\left[\frac{x}{h} - \left(1 + \frac{l^2}{2h^2}\frac{\varphi_0}{h}\right)\frac{g^{1/2}}{h^{1/2}}t\right]\right\} \quad \text{at} \quad \varphi_\infty = 0 \,. \tag{8.88}$$

If we assume

$$g^{1/2}h^{1/2} \approx 8\text{–}9\,\frac{\text{mi}}{\text{h}}\,, \tag{8.89}$$

we obtain $h \approx 2\,\text{m}$. That means that φ_0/h is approximately 0.2. These values will produce a total width of

$$\frac{4h^{3/2}}{3^{1/2}\varphi_0^{1/2}} \approx 11\,\text{m}\,, \tag{8.90}$$

in a quite remarkable agreement with the observation of Scott Russell [218].

8.2 The Generalized KdV Equation for Ion-Acoustic Solitons

In the following sections of the present chapter, we will take a break from the mathematical techniques discussed in the previous section and present some more physical applications. We shall first develop the nonlinear model for ion-acoustic waves in plasmas.

Let us assume a one-dimensional macroscopic model of a system where the ion temperature T_i is much smaller than the electron temperature T_e. Since the phenomenon under consideration will turn out to be a low-frequency one (compared to the electron plasma frequency ω_{pe}), we can assume Boltzmann distributed electrons. (This approximation is sometimes referred to as the $m_e \to 0$ model, for obvious reasons). For simplicity, we also assume an isothermal equation of state: $p_e = n_e k_B T_e \sim n_e$, that is, $T_e = \text{const}$. The low-frequency dynamics are then determined by the ion motion. The latter follows from the momentum balance (with ion velocity v_i) and the continuity equation for the ion particle number density n_i. The ambipolar potential ϕ follows from the Poisson equation. In summary, we start with the following equations:

$$\partial_t n_i + \partial_x(n_i v_i) = 0\,, \tag{8.91}$$

$$\partial_t v_i + v_i \partial_x v_i = -\frac{e}{m_i}\partial_x \phi\,, \tag{8.92}$$

$$\partial_x^2 \phi = \frac{1}{\varepsilon_0} e\left[n_0 \exp\left(\frac{e\phi}{k_B T_e}\right) - n_i\right]. \tag{8.93}$$

At this stage, it is advisable to introduce nondimensional variables, that is,

$$\frac{x}{\lambda_{De}} \to x\,, \quad t\omega_{pi} \to t\,, \quad \frac{e\phi}{k_B T_e} \to \phi\,, \quad \frac{n_i}{n_0} \to n\,, \quad \frac{v_i}{c_s} \to v\,. \tag{8.94}$$

Here, we have introduced the electron Debye length

$$\lambda_{De} = \left(\frac{\varepsilon_0 k_B T_e}{n_0 e^2}\right)^{1/2}, \tag{8.95}$$

the ion plasma frequency

$$\omega_{pi} = \left(\frac{n_0 e^2}{\varepsilon_0 m_i} \right)^{1/2} , \tag{8.96}$$

and the ion-sound velocity

$$c_s = \frac{\lambda_{De}}{\omega_{pi}} = \left(\frac{k_B T_e}{m_i} \right)^{1/2} . \tag{8.97}$$

In the following, n_0 is the homogeneous average density (we assume at infinity $n_e = n_i = n_0$). Eqs. (8.91)–(8.93) can easily be transformed into nondimensional form,

$$\partial_t n + \partial_x (n v) = 0 , \tag{8.98}$$

$$\partial_t v + v \partial_x v = -\partial_x \phi , \tag{8.99}$$

$$\partial_x^2 \phi = e^{\phi} - n . \tag{8.100}$$

First, it is trivial to find the linear dispersion relation. A simple calculation yields $\omega = k/(1+k^2)^{1/2} \approx k$, which in dimensional quantities, reads $\omega \approx k c_s$. Since the sound velocity c_s is much smaller than the electron thermal velocity, the Boltzmann distribution for the electrons is appropriate. Within a weak-amplitude theory, we assume that the deviations from the homogeneous values $n = 1, \phi = 0, v = 0$ are small. We take care of this fact by introducing an amplitude factor ε, that is,

$$n = 1 + \varepsilon n^{(1)} + \varepsilon^{(2)} n + \cdots , \tag{8.101}$$

$$\phi = \varepsilon \phi^{(1)} + \varepsilon^2 \phi^{(2)} + \cdots , \tag{8.102}$$

$$v = \varepsilon v^{(1)} + \varepsilon^2 v^{(2)} + \cdots . \tag{8.103}$$

In addition, we have to scale the space- and time-variations. We do this in the form

$$\xi = \varepsilon^{1/2} (x - t) \tag{8.104}$$

and

$$\tau = \varepsilon^{3/2} t . \tag{8.105}$$

In the following, we present an argument for the latter scaling which might also guide us to treat other cases easily. In principle, we should introduce a second smallness parameter μ and find out the ε–μ-relation from the appropriate balancing of nonlinearity and dispersion. In practice, one often chooses a different way. Let us look for the stationary small-amplitude solutions of the system Eqs. (8.98)–(8.100). Assuming stationarity in a moving frame,

$$\partial_t \longrightarrow -M \partial_x , \tag{8.106}$$

we immediately find from Eq. (8.98),

$$n = \frac{M}{M - v} , \tag{8.107}$$

which satisfies soliton boundary conditions, $n \to 1, v \to 0, \phi \to 0$ for $x \to \pm\infty$. From Eq. (8.99), we obtain

$$(M - v)^2 = M^2 - 2\phi , \tag{8.108}$$

where again we have used the soliton boundary conditions. Multiplying both sides of Eq. (8.100) with $\partial_x \phi$, after expressing n via Eqs. (8.107) and (8.108) through ϕ, we obtain in integrated form

$$\frac{1}{2}(\partial_x \phi)^2 = \exp[\phi] + M^2 \left(1 - \frac{2\phi}{M^2}\right)^{1/2} - (M^2 + 1) . \tag{8.109}$$

We can interpret this equation as the energy equation for the motion of a "particle" with "position coordinate ϕ" and "time x" in a "potential $V(\phi)$," that is,

$$\frac{d^2\phi}{dx^2} = -\frac{dV(\phi)}{d\phi} \tag{8.110}$$

with potential V given by

$$V(\phi) = -\exp[\phi] - M^2 \left(1 - \frac{2\phi}{M^2}\right)^{1/2} + (M^2 + 1) . \tag{8.111}$$

In Figure 8.7, we have plotted this potential for two values of M. From an analysis of the potential V, it becomes clear that finite-amplitude localized solutions for ϕ exist in the regime

$$\exp\left[\frac{M^2}{2}\right] - 1 < M^2 \quad \text{or} \quad M < 1.6 . \tag{8.112}$$

Now, for the weakly nonlinear case, we set

$$\exp[\phi] \approx 1 + \phi + \frac{1}{2}\phi^2 + \frac{1}{6}\phi^3 + \cdots , \tag{8.113}$$

and only expect a small deviation from the linear propagation velocity, that is,

$$0 < \delta M = M - 1 \ll 1 . \tag{8.114}$$

Then, we expand

$$\begin{aligned}(M^2 - 2\phi)^{1/2} &= \left[1 + 2\delta M + (\delta M)^2 - 2\phi\right]^{1/2} \\ &\approx 1 + \delta M - \phi + \frac{1}{2}(\delta M)^2 - \frac{1}{2}(\delta M - \phi)^2 \\ &\quad - \frac{1}{2}(\delta M - \phi)(\delta M)^2 + \frac{1}{2}(\delta M - \phi)^3 - \cdots \end{aligned} \tag{8.115}$$

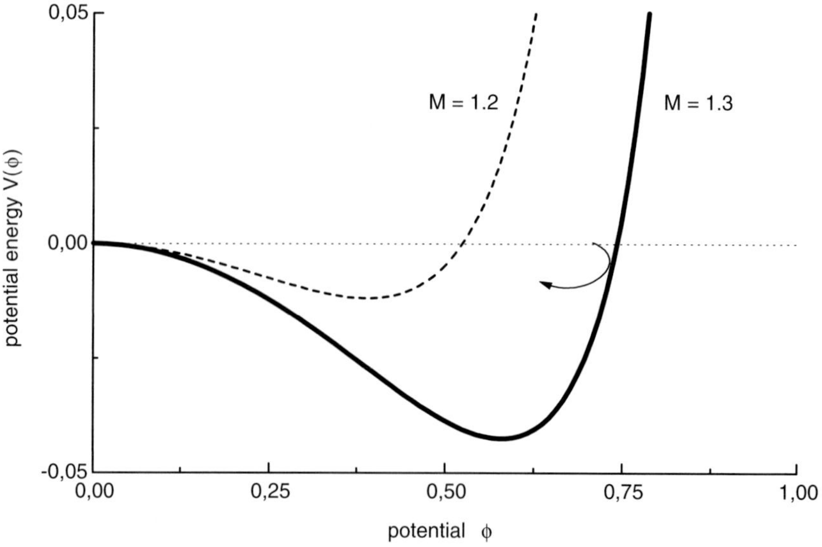

Figure 8.7 Graphs of $V(x)$ for $M = 1.3$ (solid line) and $M = 1.2$ (broken line).

Taking into account only the first order in δM and the lowest orders in ϕ, we obtain from Eqs. (8.115) and (8.113) in Eq. (8.109)

$$(\partial_x \phi)^2 = \frac{2}{3}\phi^2(3\delta M - \phi) . \tag{8.116}$$

It is not difficult to obtain from here the solution

$$\phi = 3\delta M \operatorname{sech}^2\left[\left(\frac{1}{2}\delta M\right)^{1/2}(x - Mt)\right] , \tag{8.117}$$

which clearly supports the scalings Eqs. (8.104) and (8.105).

Now, we introduce Eqs. (8.101)–(8.105) into Eqs. (8.98)–(8.100) and collect the various orders in ε. Using

$$\partial_x = \varepsilon^{1/2}\partial_\xi \tag{8.118}$$

and

$$\partial_t = \varepsilon^{3/2}\partial_\tau - \varepsilon^{1/2}\partial_\xi , \tag{8.119}$$

we obtain in lowest order (ε and $\varepsilon^{3/2}$)

$$\phi^{(1)} = n^{(1)} \tag{8.120}$$

$$\partial_\xi n^{(1)} = \partial_\xi v^{(1)} . \tag{8.121}$$

Assuming, again, vanishing boundary values at infinity, we have the solution

$$n^{(1)} = \phi^{(1)} = v^{(1)} . \tag{8.122}$$

The next orders in ε yield

$$\partial_\xi^2 \phi^{(1)} = \phi^{(2)} - \frac{1}{2}\left(\partial_x \phi^{(1)}\right)^2 - n^{(2)} , \tag{8.123}$$

$$-\partial_\xi n^{(2)} + \partial_\tau n^{(1)} + \partial_\xi \left[n^{(1)} v^{(1)}\right] + \partial_\xi v^{(2)} = 0 , \tag{8.124}$$

$$-\partial_\xi v^{(2)} + \partial_\tau v^{(1)} + v^{(1)} \partial_\xi v^{(1)} = -\partial_\xi \phi^{(2)} . \tag{8.125}$$

First, we introduce the finding Eq. (8.122). Secondly, we pick out Eq. (8.124) and substitute $n^{(2)}$ from Eq. (8.123) and $\partial_\xi v^{(2)}$ from Eq. (8.125). Then, luckily $\partial_\xi \phi^{(2)}$ cancels, and we are left with a closed equation for $n^{(1)}$, that is,

$$\partial_\tau n^{(1)} + n^{(1)} \partial_\xi n^{(1)} + \frac{1}{2} \partial_\xi^3 n^{(1)} = 0 . \tag{8.126}$$

Again, we arrived at the Korteweg–deVries (KdV) equation which can be easily brought to standard form. The KdV equation belongs to the class of exactly solvable models. It possesses an inverse scattering transform (IST). We present that method in the Appendix C. A stationary solution is

$$n^{(1)}(\xi - c\tau) = 3c\,\text{sech}^2\left[\left(\frac{1}{2}c\right)^{1/2} (\xi - c\tau)\right] . \tag{8.127}$$

The solution is of the form Eq. (8.117), that is, we get what we expected. However, we may raise the question as to whether the result is unique. A lot of tries and errors will show that there is no other pulse-like solution. Nevertheless, the development of nonlinear models always looks a little bit arbitrary.

However, in the case of the ion-acoustic waves in plasmas, we can be quite confident. The reason is that one of the first laboratory experiments on solitons was for ion-acoustic solitons. The predictions by the KdV equation have been verified. In Figure 8.8, we reproduce the experimental results which have been published in a famous paper by Ikezi, Taylor, and Baker [244].

Let us now generalize the model considerably. We discuss the two general aspects: (i) nonlinear equations in more than one space-dimension and (ii) appearance of a small parameter in the model. A small parameter appears, for example, for ion-acoustic waves in a plasma when an external magnetic field $B\hat{z}$ is present. The Lorentz force term has to be added to the right-hand side of Eq. (8.92). When we change to nondimensional quantities, Eq. (8.99) reads

$$\partial_t \boldsymbol{v} + \boldsymbol{v} \cdot \nabla \boldsymbol{v} + \nabla \phi + \Omega\, \hat{z} \times \boldsymbol{v} = 0 , \tag{8.128}$$

where

$$\Omega = \frac{\Omega_i}{\omega_{pi}} = \frac{eB}{m_i \omega_{pi}} . \tag{8.129}$$

Equations (8.98) and (8.100) are unchanged except for the different notation due to higher space-dimension. Instead of Eq. (8.98), we now have

$$\partial_t n + \nabla \cdot (n\boldsymbol{v}) = 0 , \tag{8.130}$$

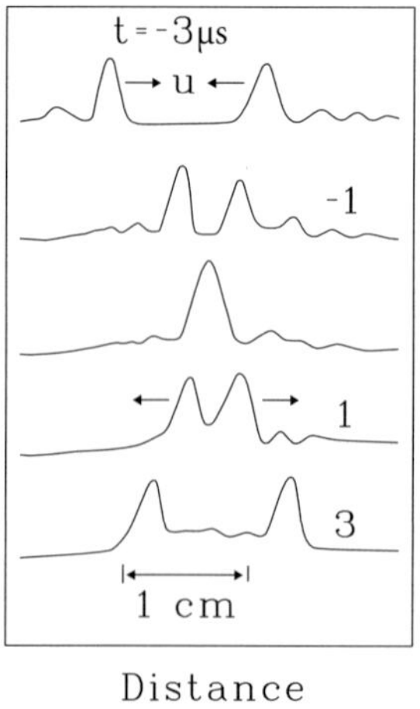

Figure 8.8 Interaction of ion-acoustic solitons at times $t = -3, -1, 0, 1, 3$ [μ s], as has been found by Ikezi, Taylor, and Baker [244]. Note that in this experiment, the Debye length was of the order of 0.2 mm.

and Eq. (8.100) reads

$$\nabla^2 \phi = e^{\phi} - n . \tag{8.131}$$

Equations (8.128), (8.130), and (8.131) constitute our basic set of equations.

In the linear limit, we obtain the dispersion relation

$$\omega^2 = \frac{1}{2}\left(\Omega^2 + \frac{k^2}{1+k^2}\right) \pm \left[\frac{1}{4}\left(\Omega^2 + \frac{k^2}{1+k^2}\right)^2 - \frac{k_z^2}{1+k^2}\Omega^2\right]^{1/2} . \tag{8.132}$$

First, for $\Omega = 0$, we recover the old result

$$\omega = \frac{k}{(1+k^2)^{1/2}} , \tag{8.133}$$

though now we have $\boldsymbol{k} = (\boldsymbol{k}_\perp, k_z) = (k_x, k_y, k_z)$.

For small k and weak transverse effects, for example, when propagation in z-direction is considered with $k_\perp \ll k_z$, we have

$$\begin{aligned}\omega &\approx k\left(1 - \frac{1}{2}k^2\right) \approx k_z\left(1 + \frac{1}{2}\frac{k_\perp^2}{k_z^2}\right)\left[1 - \frac{1}{2}k_z^2\left(1 + \frac{k_\perp^2}{k_z^2}\right)\right] \\ &\approx k_z - \frac{1}{2}k_z^3 + \frac{1}{2}\frac{k_\perp^2}{k_z} .\end{aligned} \tag{8.134}$$

This linear dispersion will appear in a nonlinear three-dimensional model which was first derived by Kadomtsev and Petviashvili [245]. (Remember that we are still treating the case $\Omega = 0$.) Let us now derive this model equation. We may assume the scaling as shown in Eqs. (8.101)–(8.105) for the quantities n, ϕ, v_z, z, and t. (Note that the direction of main propagation is z and not x as previously.) In addition, we have to scale the "transverse" variables $\boldsymbol{v}_\perp$, as well as x and y. Since we will concentrate here on weak transverse dependencies, we weight the transverse variables with an additional $\varepsilon^{1/2}$. (We do not choose ε as the additional weight factor since the transversal effects would not appear at all within leading order.) In summary, we choose

$$\xi = \varepsilon^{1/2}(z - t) \, , \zeta = \varepsilon x \, , \eta = \varepsilon y \, , \tau = \varepsilon^{3/2} t \tag{8.135}$$

as well as

$$n = 1 + \varepsilon n^{(1)} + \varepsilon^2 n^{(2)} + \cdots \, , \tag{8.136}$$

$$v_z = \varepsilon v_z^{(1)} + \varepsilon^2 v_z^{(2)} + \cdots \, , \tag{8.137}$$

$$\boldsymbol{v}_\perp = \varepsilon^{3/2} \boldsymbol{v}_\perp^{(1)} + \varepsilon^{5/2} \boldsymbol{v}_\perp^{(2)} + \cdots \, , \tag{8.138}$$

$$\phi = \varepsilon \phi^{(1)} + \varepsilon^{(2)} \phi^{(2)} + \cdots \tag{8.139}$$

If we introduce this into the basic equations ($\Omega = 0$), we obtain to order ε

$$\phi^{(1)} = n^{(1)} \, , \tag{8.140}$$

and in $\varepsilon^{3/2}$,

$$\partial_\xi n^{(1)} = \partial_\xi v_z^{(1)} = \partial_\xi \phi^{(1)} \, , \tag{8.141}$$

that is,

$$n^{(1)} = \phi^{(1)} = v_z^{(1)} \, . \tag{8.142}$$

In order ε^2, one gets

$$\partial_\xi v_y^{(1)} = \partial_\eta \phi^{(1)} \, , \tag{8.143}$$

$$\partial_\xi v_x^{(1)} = \partial_\zeta \phi^{(1)} \, , \tag{8.144}$$

$$n^{(2)} = \phi^{(2)} - \partial_\xi^2 \phi^{(1)} + \frac{1}{2} \left(\phi^{(1)}\right)^2 \, . \tag{8.145}$$

The final order we have to consider is $\varepsilon^{5/2}$. Then,

$$-\partial_\xi n^{(2)} + \partial_\tau n^{(1)} + \partial_\xi \left(n^{(1)} v_z^{(1)}\right) \, , \tag{8.146}$$

$$+\partial_\xi v_z^{(2)} + \partial_\zeta v_x^{(1)} + \partial_\eta v_y^{(1)} = 0 \, , \tag{8.147}$$

$$-\partial_\xi v_z^{(2)} + \partial_\tau v_z^{(1)} + v_z^{(1)} \partial_\xi v_z^{(1)} = -\partial_\xi \phi^{(2)} \, . \tag{8.148}$$

If we differentiate both sides of Eq. (8.145) with respect to ξ,

$$\partial_\xi n^{(2)} = -\partial_\xi^3 \phi^{(1)} + \phi^{(1)} \partial_\xi \phi^{(1)} + \partial_\xi \phi^{(2)} \, , \tag{8.149}$$

we can insert this, as well as $\partial_\xi v_z^{(2)}$ from Eq. (8.148), into Eq. (8.147) to obtain, after some simple manipulations for

$$n \equiv n^{(1)}, \tag{8.150}$$

a closed equation. The latter is known as the Kadomtsev–Petviashvili equation (KP equation) [245],

$$2\partial_\tau n + \partial_\xi^3 n + 2n\partial_\xi n + \int^\xi d\xi' \left(\partial_\zeta^2 + \partial_\eta^2\right) n = 0\,. \tag{8.151}$$

Note that the linear terms reflect the dispersion relation Eq. (8.134); Eq. (8.151) is a three-dimensional generalization of the KdV equation. One should discuss the possibility to solve it by an inverse scattering transform (IST) (see Appendix C). Here, we only mention that in the context of shallow water waves in two space-dimensions, we get the explicit form analogous to Eq. (8.151), that is, [246]

$$\partial_\tau n + \frac{3}{2} n\partial_\xi n + \frac{1}{6}\partial_\xi^3 n + \frac{1}{2}\int_{-\infty}^{\xi} d\xi' \partial_\eta^2 n(\xi';\eta) = 0\,. \tag{8.152}$$

Let us now turn to the case $\Omega \neq 0$. To be more concrete, we demand $\Omega \sim \mathcal{O}(1)$ and treat the case $\omega \ll \Omega$. Then,

$$\omega \approx k_z \left[1 - \frac{1}{2}k_z^2 - \frac{1}{2}\left(1 + \Omega^{-2}\right) k_\perp^2\right] \tag{8.153}$$

is an appropriate solution of the dispersion relation (8.132). Note that this relation only follows under the assumptions $\Omega \gg \omega, k \ll 1$, but no relation between k_z and $k_\perp$ is required.

To find the nonlinear wave model which corresponds to the branch Eq. (8.153), we therefore scale as follows

$$\xi = \varepsilon^{1/2}(z - t)\,, \quad \zeta = \varepsilon^{1/2} x\,, \quad \eta = \varepsilon^{\frac{1}{2}} y\,, \quad \tau = \varepsilon^{3/2} t \tag{8.154}$$

and expand

$$n = 1 + \varepsilon n^{(1)} + \varepsilon^2 n^{(2)} + \cdots\,, \tag{8.155}$$

$$v_z = \varepsilon v_z^{(1)} + \varepsilon^2 v_z^{(2)} + \cdots\,, \tag{8.156}$$

$$\boldsymbol{v}_\perp = \varepsilon^{3/2} \boldsymbol{v}_\perp^{(1)} + \varepsilon^2 \boldsymbol{v}_\perp^{(2)} + \cdots\,, \tag{8.157}$$

$$\phi = \varepsilon \phi^{(1)} + \varepsilon^2 \phi^{(2)} + \cdots \tag{8.158}$$

Note that in comparison to the previous scalings, x and y are scaled with $\varepsilon^{1/2}$ (to take into account the $\boldsymbol{E} \times \boldsymbol{B}$ drift), and in $\boldsymbol{v}_\perp$, the steps are of order $\varepsilon^{1/2}$ in order to include the polarization drift.

Putting the pieces together, in order ε, we get

$$n^{(1)} = \phi^{(1)} , \tag{8.159}$$

and from the $\varepsilon^{3/2}$ contributions

$$n^{(1)} = v_z^{(1)} = \phi^{(1)} , \tag{8.160}$$

together with the $\boldsymbol{E} \times \boldsymbol{B}$ drift

$$\Omega\, v_y^{(1)} = \partial_\zeta \phi(1) , \tag{8.161}$$

$$\Omega\, v_x^{(1)} = -\partial_\eta \phi^{(1)} . \tag{8.162}$$

The next order (ε^2) yields

$$\partial_\xi v_y^{(1)} = \Omega\, v_x^{(2)} , \tag{8.163}$$

$$-\partial_\xi v_x^{(1)} = \Omega\, v_y^{(2)} , \tag{8.164}$$

$$n^{(2)} = \phi^{(2)} - \left(\partial_\xi^2 + \partial_\zeta^2 + \partial_\eta^2\right) \phi^{(1)} + \frac{1}{2} \left(\phi^{(1)}\right)^2 . \tag{8.165}$$

Finally, the order $\varepsilon^{5/2}$ leads to

$$-\partial_\xi n^{(2)} + \partial_\tau n^{(1)} + \partial_\xi \left(n^{(1)} v_z^{(1)}\right) + \partial_\xi v_z^{(2)} , \tag{8.166}$$

$$+\partial_\zeta v_x^{(2)} + \partial_\eta v_y^{(2)} = 0 , \tag{8.167}$$

$$-\partial_\xi v_z^{(2)} + \partial_\tau v_z^{(1)} + v_z^{(1)} \partial_\xi v_z^{(1)} = -\partial_\xi \phi^{(2)} . \tag{8.168}$$

Combining in the usual manner, we obtain the Zakharov–Kuznetsov equation (ZK equation) [247]

$$2\partial_\tau n + \partial_\xi^3 n + 2n\partial_\xi n + \left(1 + \Omega^{-2}\right) \partial_\xi \left(\partial_{\zeta^2} + \partial_{\eta^2}\right) n = 0 . \tag{8.169}$$

Note that in the linear limit, we recover the dispersion relation (8.153). This teaches us that although the derivation of a nonlinear equation looks quite formal, a lot of physical background is needed to find the correct scaling in the various regimes.

We now ask the question as to whether there is a continuous transition from Eq. (8.169) to Eq. (8.151). Obviously, we cannot perform it in a trivial manner by setting $\Omega = 0$ in Eq. (8.169). We should not hesitate to scale Ω itself and, of course, one choice is

$$\Omega \sim \varepsilon^{1/2} , \tag{8.170}$$

and keeping the other scalings. At this stage, it is no longer necessary to repeat every step of the straightforward calculation; we just present the result

$$\partial_\tau n + \partial_\xi^3 n + 2n\partial_\xi n + \left(\partial_\xi^2 + \Omega^2\right)^{-1} \partial_\xi \left(\partial_\zeta^2 + \partial_\eta^2\right) n = 0 \,, \tag{8.171}$$

which can also be written in the equivalent form

$$2\partial_\tau n + \partial_\xi^3 n + 2n\partial_\xi n + \int^{\xi} d\xi' \cos\left[\Omega\left(\xi - \xi'\right)\right] \left(\partial_\zeta^2 + \partial_\eta^2\right) n\left(\xi'\right) = 0 \,. \tag{8.172}$$

We clearly see that for $\Omega = 0$, the KP equation is obtained. However, for $\Omega^2 \gg \partial_\xi^2$, the form is similar to the ZK equation.

The general conclusion is that if we consider physical relevant situations, model equations often arise which do not fit into the standard scheme, for example, the one-dimensional KdV equation.

Before closing this section, we want to discuss another point which is related to the physics of the basic hydrodynamic model. Without any further arguments, we have assumed that the electrostatic approximation is correct for the description of *nonlinear* ion-acoustic modes. Is this really true? This question is very relevant since we assume that weak nonlinearities should be taken into account, but electromagnetic contributions from the vector potential are still neglected. Such questions regarding the importance of the various effects always arise in nonlinear physics, and we should confess that they are generally very difficult to answer. In the present case, we shall find the answer that the plasma β (equals the ratio of the hydrodynamic pressure $p = nk_BT$ to the magnetic pressure $B^2/2\mu_0$) should be small, effectively $\beta \ll \varepsilon$. Here, $\varepsilon \ll 1$ denotes the relative amplitude of the nonlinear wave under consideration.

To support this argument, we perform a *linear* calculation in the limit $\omega/\Omega_i \ll 1$. We prefer to use dimensional variables to find the β explicitly. For the vector potential, we make the ansatz

$$\boldsymbol{A} \approx A\hat{z} \,. \tag{8.173}$$

Within the drift approximation ($\omega/\Omega_i \ll 1$), we use the drift velocity

$$v_{i\perp} \approx \frac{m_i}{eB_0^2}\frac{\partial \boldsymbol{E}_\perp}{\partial t} + \frac{1}{B_0^2}\boldsymbol{E} \times \boldsymbol{B}_0 \,, \tag{8.174}$$

where the index zero characterizes the external field. From

$$\partial_t v_{iz} \approx -\frac{e}{m_i}\left(\partial_z \phi + \partial_t A\right) , \tag{8.175}$$

we obtain for the ion density

$$\partial_t^2 n_i - \frac{1}{B_0 \Omega_i}\partial_t^2 \nabla_\perp^2 \phi + \partial_z \left[-\frac{e}{m_i}\left(\partial_z \phi + \partial_t A\right)\right] = 0 \,. \tag{8.176}$$

For $\omega / k_z \ll v_{the}$, we have the electron momentum balance

$$0 \approx \frac{e}{m_e} \partial_z \phi - \frac{k_B T_e}{m_e} \partial_z n_e + \frac{e}{m_e} \partial_t A \,. \tag{8.177}$$

These equations have to be supplemented by Poisson's equation

$$\nabla^2 \phi = \frac{e}{\varepsilon_0}(n_e - n_i) \tag{8.178}$$

and

$$\nabla \times \nabla \times A\hat{z} \approx \mu_0 \boldsymbol{j} \,, \tag{8.179}$$

when for low-frequency waves the displacement current is neglected.

First, in the *linear* limit, Eqs. (8.173)–(8.179) lead to the linear dispersion relation

$$\left(\omega^2 - \frac{k_z^2 c_s^2}{1 + k_\perp^2 \rho_s^2}\right) \left[\omega^2 - k_z^2 v_A^2 \left(1 + k_\perp^2 \rho_s^2\right)\right] = \frac{\omega^2}{\omega_{pi}^2} \frac{k_z^2 k_\perp^2 c_s^2 c^2 \beta}{1 + k_\perp^2 \rho_s^2} \,, \tag{8.180}$$

where $\rho_s = c_s / \Omega_i$. We clearly see that in the limit $\beta \to 0$, the electrostatic branch (first bracket on the left-hand side of Eq. (8.180)) decouples from the electromagnetic Alfvén branch (second bracket on the left-hand side of Eq. (8.180)).

Second, during the calculations of the nonlinear electrostatic branch, we neglected contributions from the vector potential. To justify that, we use quasineutrality $n_e \approx n_i \approx n$. Then, Eq. (8.179) leads approximately to

$$\begin{aligned} -\partial_z \nabla_\perp^2 A &\approx \mu_0 \frac{\partial j_z}{\partial z} \approx -\mu_0 \nabla_\perp \cdot \boldsymbol{j}_\perp \approx -\mu_0 \nabla_\perp \cdot (en\boldsymbol{v}_{i\perp}) \\ &\approx \frac{\mu_0 n m_i}{B_0^2} \partial_t \nabla_\perp^2 \phi \approx \frac{1}{v_A^2} \partial_t \nabla_\perp^2 \phi \,, \end{aligned} \tag{8.181}$$

where the Alfvén velocity is $v_A = B_0 / (\mu_0 n m_i)^{1/2}$. We use this for an estimate of the ion-acoustic branch, for which we approximately (and symbolically) use

$$\partial_t \sim \omega \sim k_z c_s \sim c_s \partial_z \,. \tag{8.182}$$

Then, we may deduce from Eq. (8.181)

$$\left|\frac{\partial A}{\partial t}\right| \approx \frac{m_i}{m_e} \beta \left|\nabla \phi\right| , \tag{8.183}$$

and therefore for $\beta \ll m_e / m_i$, the electromagnetic contributions can be neglected compared to the electrostatic ones, as has been done for the derivation of nonlinear ion-acoustic wave equations.

8.3
Envelope Solitons

Thus far, we have only studied soliton equations and soliton solutions of the KdV-type. The latter are solitary, *pulse-like* solutions. KdV-type equations generally occur if the linear waves are of low frequency ("acoustic branch"). High-frequency waves ("optical branch") can show a localization of the *envelope* of rapid oscillations. Let us first demonstrate the procedure at a simple (academic) model; later, we shall turn to realistic plasma systems.

In order to demonstrate, we choose a model equation for high-frequency waves whose linear dispersion relation is

$$\omega^2 = \omega_p^2 + c^2 k^2 \,. \tag{8.184}$$

The detailed nonlinear system of equations will now be replaced by a single one-field model equation

$$u_{tt} - c^2 u_{xx} + \omega_p^2 u = \frac{1}{6}\omega_p^2 u^3 \tag{8.185}$$

to demonstrate the main procedure. Note that the nonlinear term on the right-hand side of Eq. (8.184) is quite arbitrarily chosen. (In general, the detailed model is much more complicated so that a reduction, e.g., by making use of the multiple-scale analysis, is necessary; here, for the moment, we use the artificial equation as it stands.) We make the ansatz

$$u = \sum_{\nu=-\infty}^{\infty} \sum_{\mu=|\nu|}^{\infty} \varepsilon^{\mu} u_{\nu\mu}(\xi, \tau) e^{i\nu(kx-\omega t)} \,, \tag{8.186}$$

where $\tau = \varepsilon^2 t$ and $\xi = \varepsilon(x - v_g t)$ with $v_g = d\omega/dk \equiv \omega'$. For real u, we have $u_{-\nu\mu} = u^*_{\nu\mu}$. In the *order* ε^1, one immediately finds $u_{01} = 0$, and we are left with an equation where only u_{11} is involved. This equation is satisfied because of the linear dispersion relation. In *order* ε^2, we can collect the various harmonic contributions $\nu = 0, 1, 2$. Let us start with the zero harmonics. We can immediately conclude

$$u_{02} = 0 \,. \tag{8.187}$$

Next, we investigate the first harmonic. From Eq. (8.185), we obtain

$$\left(-\omega^2 + c^2 k^2 + \omega_p^2\right) u_{12} = 2i\left(c^2 k - \omega v_g\right) \partial_\xi u_{11} \tag{8.188}$$

because of

$$\partial_t = \varepsilon^2 \partial_\tau - \varepsilon v_g \partial_\xi - i\nu\omega \,, \tag{8.189}$$

and

$$\partial_x = \varepsilon \partial_\xi + i\nu k \,. \tag{8.190}$$

We have chosen

$$v_g = \frac{d\omega}{dk} = \frac{c^2 k}{\omega} ; \tag{8.191}$$

therefore, Eq. (8.188) is identically satisfied. Finally, for the second harmonic $\nu = 2$, an equation similar to Eq. (8.188) follows where now, however, the bracket on the left-hand side does not vanish. This leads to

$$u_{22} = 0 . \tag{8.192}$$

In the *order* ε^3, we have to investigate $\nu = 0, 1, 2, 3$. The zeroth harmonic yields

$$\left(v_g^2 - c^2\right) \partial_\xi^2 u_{01} + \omega_p^2 u_{03} = 0 , \tag{8.193}$$

from where

$$u_{03} = 0 \tag{8.194}$$

follows. The first harmonic contributions lead to

$$\left(-\omega^2 + c^2 k^2 + \omega_p^2\right) u_{13} - 2i(c^2 k - \omega v_g)\partial_\xi u_{12} - 2i\omega\partial_\tau u_{11} - \left(c^2 - v_g^2\right) \partial_\xi^2 u_{11} = \frac{1}{6}\omega_p^2 3 u_{11} u_{11} u_{11}^* . \tag{8.195}$$

For $U = u_{11}$, we thus have the closed equation

$$\partial_\tau U - \frac{i\omega''}{2}\partial_\xi^2 U - \frac{i\omega_p^2}{4\omega}|U|^2 U = 0 , \tag{8.196}$$

that is, the celebrated cubic nonlinear Schrödinger (NLS) equation. Note that

$$\frac{c^2 - v_g^2}{\omega} = \frac{c^2\omega_p^2}{\omega^3} = \frac{d^2\omega}{dk^2} = \omega'' . \tag{8.197}$$

Equation (8.196) has a simple physical meaning. In higher orders, the first harmonics react back on the slow variations of the envelopes. Since we have chosen an extremely simple starting model, we cannot see the whole richness which is usually hidden in the cubic nonlinear Schrödinger equation. In general, higher harmonics are generated (in a quite sophisticated manner) which then, in higher orders, can react back on the first harmonic.

Before discussing the cubic nonlinear Schrödinger equation as a model equation in more detail, we show that a formally different ansatz (compared with the ansatz Eq. (8.186)) is actually equivalent and leads to the same result Eq. (8.196). Let us expand

$$u = \varepsilon u_1 (\xi_1, \xi_2, \ldots ; \tau_1, \tau_2, \ldots) \exp\left[i(kx - \omega t)\right] + \varepsilon^2 u_2(x, \xi_1, \xi_2, \ldots ; t, \tau_1, \tau_2, \ldots) + \cdots + c.c. , \tag{8.198}$$

where we have the different space- and time-scales $\xi_1, \xi_2, \ldots, \tau_1, \tau_2, \ldots$ with

$$\partial_x \to \partial_x + \varepsilon\partial_{\xi_1} + \varepsilon^2\partial_{\xi_2} + \cdots , \quad \partial_t \to \partial_t + \varepsilon\partial_{\tau_1} + \varepsilon^2\partial_{\tau_2} + \cdots \tag{8.199}$$

Note that we introduced a mixed notation $\xi_0 \to x$, $\tau_0 \to t$. In *order* ε^1, we recover the linear dispersion relation. In *order* ε^2, we get

$$\left[\partial_t^2 - c^2\partial_x^2 + \omega_p^2\right] u_2 = \left[2i\omega\partial_{\tau_1} u_1 + 2ikc^2\partial_{\xi_1} u_1\right] e^{i(kx-\omega t)} . \tag{8.200}$$

If the bracket on the right-hand side of Eq. (8.200) does not vanish, secular behavior will occur. Secular behavior means that the iterates $u_2, u_3, \ldots$ grow algebraically (e.g., $\sim t \exp i(kx - \omega t)$) in the fast time- or space-variables τ_0 or ξ_0, respectively, and then the asymptotic series Eq. (8.198) is not uniformly valid over long times and distances. However, the vanishing of the right-hand side of Eq. (8.200) means

$$\partial_{\tau_1} u_1 = -\frac{c^2 k}{\omega}\partial_{\xi_1} u_1 = -v_g\partial_{\xi_1} u_1 ; \tag{8.201}$$

the formulation may be written in the variable ξ which has previously been introduced. Then, we can set $u_2 = 0$ since otherwise we could have absorbed that contribution in u_1. In *order* ε^3, we obtain

$$\begin{aligned} \left[\partial_t^2 - c^2\partial_x^2 + \omega_p^2\right] u_3 &= \frac{\omega_p^2}{6} u_1^3 e^{3i(kx-\omega t)} \\ &+ \left[2i\omega\partial_{\tau_2} u_1 - \partial_{\tau_1}^2 u_1 + 2ikc^2\partial_{\xi_2} u_1 + c^2\partial_{\xi_1}^2 u_1 + \frac{1}{2}\omega_p^2 |u_1|^2 u_1\right] \\ &\times e^{i(kx-\omega t)} + c.c. \end{aligned} \tag{8.202}$$

We have two different harmonic terms on the right-hand side. If we consider $\sim \exp[i(kx - \omega t)]$, we again get secular growth for u_3 if the coefficient in the bracket does not vanish. (To prove this, make the ansatz $u_3 = f(x,t) \exp[i(kx - \omega t)]$ to find for f the simple relation $\partial_\xi f(\xi = x - v_g t) = \text{const.}$) Note that this argument does not apply to the third harmonic. Therefore, we can choose u_3 to be simply dependent on $\exp[3i(kx - \omega t)]$, provided

$$2i\omega\partial_{\tau_2} u_1 - \partial_{\tau_1}^2 u_1 + 2ikc^2\partial_{\xi_2} u_1 + c^2\partial_{\xi_1}^2 u_1 + \frac{1}{2}\omega_p^2 |u_1|^2 u_1 = 0 . \tag{8.203}$$

If we remember Eq. (8.201), we immediately recover the cubic nonlinear Schrödinger equation (8.196). The argument is as follows. From Eq. (8.201), we have

$$\partial_{T_1} u_1 \equiv \left(\partial_{\tau_1} + v_g\partial_{\xi_1}\right) u_1 = 0 . \tag{8.204}$$

When we transform into a moving frame

$$X_1 = \xi_1 - v_g\tau_1 , \quad T_1 = \tau_1 , \tag{8.205}$$

we should use $\partial_{\xi_1} \to \partial_{X_1}$, $\partial_{\tau_1} \to \partial_{T_1} - v_g \partial_{X_1}$, that is, in the moving frame, the first order time-derivative with respect to T_1 vanishes. From Eq. (8.203), we obtain

$$2i\left(\partial_{T_2} - v_g \partial_{\xi_2}\right) u_1 - \frac{1}{\omega}\left(\partial_{T_1} - 2v_g \partial_{\xi_1}\right)\partial_{T_1} u_1 + \frac{c^2 - v_g^2}{\omega}\partial_{\xi_1}^2 u_1$$
$$+ 2i\frac{kc^2}{\omega}\partial_{\xi_2} u_1 + \frac{\omega_p^2}{2\omega}|u_1|^2 u_1 = 0\,, \tag{8.206}$$

or

$$\partial_t u_1 - \frac{i\omega''}{2}\partial_x^2 u_1 - \frac{i\omega_p^2}{4\omega}|u_1|^2 u_1 = 0\,. \tag{8.207}$$

Thus, we clearly see that this method is equivalent to the first one.

Let us now investigate the stationary solutions of the cubic nonlinear Schrödinger equation. The latter can always be written in the standard form

$$i\phi_\tau + \frac{1}{2}\phi_{\xi\xi} + \kappa|\phi|^2\phi = 0\,, \tag{8.208}$$

with a constant κ, which for obvious reasons is often displayed as

$$\kappa = -\frac{\dfrac{\partial\omega}{\partial|\phi_0|^2}}{\dfrac{\partial^2\omega}{\partial k^2}}\,. \tag{8.209}$$

The complex envelope ϕ is now separated into an amplitude and a phase factor,

$$\phi = \rho^{1/2} e^{i\sigma}\,, \tag{8.210}$$

and we find from Eq. (8.208) the real and imaginary parts

$$\rho_\tau + (\rho\sigma_\xi)_\xi = 0\,, \tag{8.211}$$

$$\kappa\rho - \frac{1}{8}\frac{1}{\rho^2}\left(\frac{\partial\rho}{\partial\xi}\right)^2 + \frac{1}{4}\frac{1}{\rho}\frac{\partial^2\rho}{\partial\xi^2} = \frac{\partial\sigma}{\partial\tau} + \frac{1}{2}\left(\frac{\partial\sigma}{\partial\xi}\right)^2. \tag{8.212}$$

A stationary solution is a function with $\rho_\tau = 0$, and therefore from Eq. (8.211), $\sigma\rho_\xi$ is not a function of ξ, that is,

$$\rho\sigma_\xi = c(\tau)\,. \tag{8.213}$$

Furthermore, we define

$$\sigma_\tau + \frac{1}{2}\sigma_\xi^2 = f(\xi)\,, \tag{8.214}$$

where c and f are yet undetermined functions of τ and ξ, respectively. We shall now prove that $c =$ const is the only permitted choice. From Eq. (8.214), we have

$$\sigma_{\tau\tau\xi} = -\frac{1}{2}\frac{\partial^2}{\partial\xi\partial\tau}\left(\frac{\partial\sigma}{\partial\xi}\right)^2 = -\frac{1}{2}\frac{\partial^2}{\partial\xi\partial\tau}\frac{c^2}{\rho^2} = \frac{1}{\rho^3}\frac{d\rho}{d\xi}\frac{dc^2}{d\tau}\,, \tag{8.215}$$

whereas from Eq. (8.213),

$$\sigma_{\tau\tau\xi} = \frac{1}{\rho}\frac{d^2 c}{d\tau^2} \,. \tag{8.216}$$

Equating the right-hand side of Eqs. (8.215) and (8.216), we obtain

$$\frac{\frac{d^2 c}{d\tau^2}}{\frac{d c^2}{d\tau}} = \frac{1}{\rho^2}\frac{d\rho}{d\xi} \,, \tag{8.217}$$

where the variables are separated. Since $\rho^{-2}\rho_\xi$ = const is not suitable ($\rho \sim 1/[\rho_0^{-1} + \text{const}(\xi - \xi_0)]$ will diverge!), we can conclude that $c(\tau) = \text{const} = c_1$ is the only possible choice. Of course, in the latter situation, we have to go back to Eqs. (8.215) and (8.216) separately. Then, we can integrate Eq. (8.213) to obtain

$$\sigma = \int^{\xi} \frac{c_1}{\rho} d\xi' + A(\tau) \,. \tag{8.218}$$

On the other hand, since σ_ξ is a function of ξ only, then by Eq. (8.214), σ_τ is only a function of ξ, leading to

$$\frac{dA}{d\tau} = \Omega = \text{const} \,. \tag{8.219}$$

We write Eq. (8.210) in the form

$$\sigma = \int^{\xi} \frac{c_1}{\rho} d\xi' + \Omega\tau \,. \tag{8.220}$$

Inserting this into Eq. (8.212), we have

$$\kappa\rho - \frac{1}{8}\frac{1}{\rho^2}\left(\frac{d\rho}{d\xi}\right)^2 + \frac{1}{4}\frac{1}{\rho}\frac{d^2\rho}{d\xi^2} = \Omega + \frac{1}{2}\left(\frac{c_1}{\rho}\right)^2 . \tag{8.221}$$

After multiplication by $8\rho_\xi$, we can rewrite the last equation in the form

$$\frac{d}{d\xi}\left[\frac{1}{\rho}\left(\frac{d\rho}{d\xi}\right)^2\right] = -4\kappa\frac{d\rho^2}{d\xi} + 8\Omega\frac{d\rho}{d\xi} - 4c_1^2\frac{d}{d\xi}\frac{1}{\rho} \,. \tag{8.222}$$

Integrating with the boundary condition $\rho_\xi = 0$ at $\xi = \pm\infty$, we find, after multiplication by ρ,

$$\left(\frac{d\rho}{d\xi}\right)^2 = -4\kappa\rho^3 + 8\Omega\rho^2 + c_2\rho - 4c_1^2 = -4\kappa(\rho - \rho_\infty)^2(\rho - \rho_s) \,, \tag{8.223}$$

where c_2 is the integration constant. We have furthermore required that $d\rho/d\xi$ only vanishes for two ρ values: ρ_s (extremum) and ρ_∞ (at $x \to \pm\infty$).

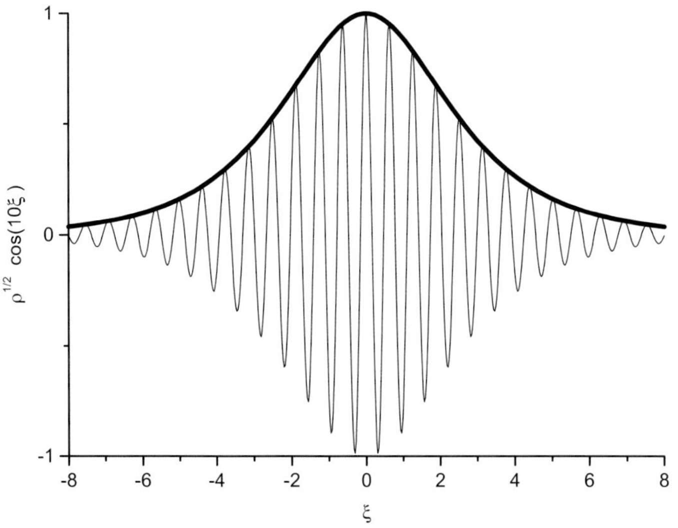

Figure 8.9 Plot of a typical envelope soliton solution of the cubic nonlinear Schrödinger equation. The heavy line represents the sech-profile. The thin line depicts the actual field variation as a function of the space coordinate ξ.

Let us first consider the case $\kappa > 0$. Then, because of $\rho \geq 0$ and $4\kappa\rho_\infty^2 \rho_s = -4c_1^2$, from Eq. (8.223), $c_1 = 0$ follows. Then, $\rho_\infty = 0$, and immediately $c_2 = 0$, is the consequence. We write Eq. (8.223) in the form

$$\left(\frac{d\rho}{d\xi}\right)^2 = -4\kappa\rho^2(\rho - \rho_s) , \tag{8.224}$$

where

$$\rho_s = 2\frac{\Omega}{\kappa} . \tag{8.225}$$

The solution of Eq. (8.224) is

$$\rho = \rho_s \text{sech}^2\left[(\kappa\rho_s)^{1/2}\,\xi\right] . \tag{8.226}$$

The solution is sketched in Figure 8.9. From Eq. (8.220), we have

$$\sigma = \Omega\tau . \tag{8.227}$$

Note that the velocity of this solitary wave solution is independent of the amplitude. We should also state that if $\phi(\xi, \tau)$ is a solution of the (time-dependent) cubic nonlinear Schrödinger equation, then by Galilei transformation,

$$\tilde{\phi}(\xi, \tau) = e^{i\left(u\xi - \frac{1}{2}u^2\tau\right)}\phi(\xi - u\tau, \tau) \tag{8.228}$$

is also a solution.

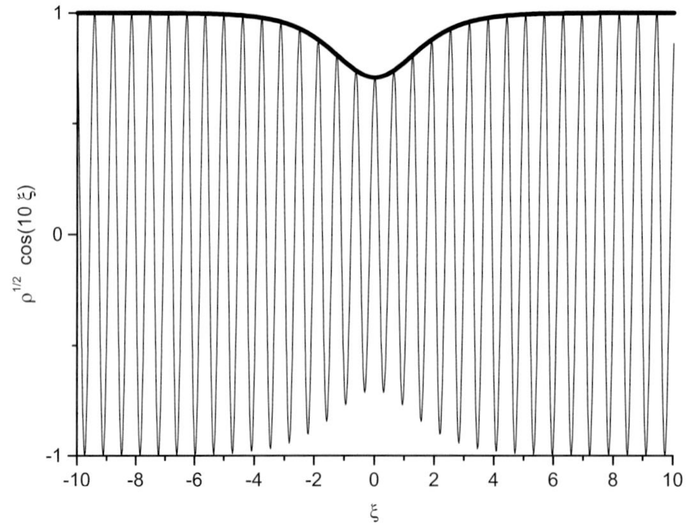

Figure 8.10 Same as Figure 8.9, but for an envelope hole with a $\sqrt{1-a^2\text{sech}^2}$-profile. We have chosen $a^2 = 0.5$.

Next, the *case* $\kappa < 0$. The choice $c_1 \neq 0$ is acceptable now, and we rewrite Eq. (8.223) in the form

$$\left(\frac{d\rho}{d\xi}\right)^2 = 4|\kappa|(\rho-\rho_\infty)^2(\rho-\rho_s) . \tag{8.229}$$

The solution of this equation can be put into the form

$$\rho = \rho_\infty \left\{1 - a^2\text{sech}^2\left[(|\kappa|\rho_\infty)^{1/2}\, a\xi\right]\right\} , \tag{8.230}$$

where

$$a^2 = \frac{\rho_\infty - \rho_s}{\rho_\infty} \leq 1 , \quad c_1^2 = |\kappa|\rho_\infty^3\left(1-a^2\right) , \quad |\kappa|\rho_\infty\left(3-a^2\right) = -2\Omega , \tag{8.231}$$

and

$$\sigma = \sin^{-1}\left\{a\frac{\tanh[(|\kappa|\rho_\infty)^{1/2}\, a\xi]}{1-a^2\text{sech}^2\left[(|\kappa|\rho_\infty)^{1/2}\, a\xi\right]}\right\} + \Omega\tau . \tag{8.232}$$

This solution represents an envelope hole (gray soliton); see Figure 8.10. Compared with the envelope soliton Eq. (8.226), the envelope hole has an additional parameter a. When $a = 1$,

$$\rho = \rho_\infty \tanh^2\left[(|\kappa|\rho_\infty)^{1/2}\, \xi\right] . \tag{8.233}$$

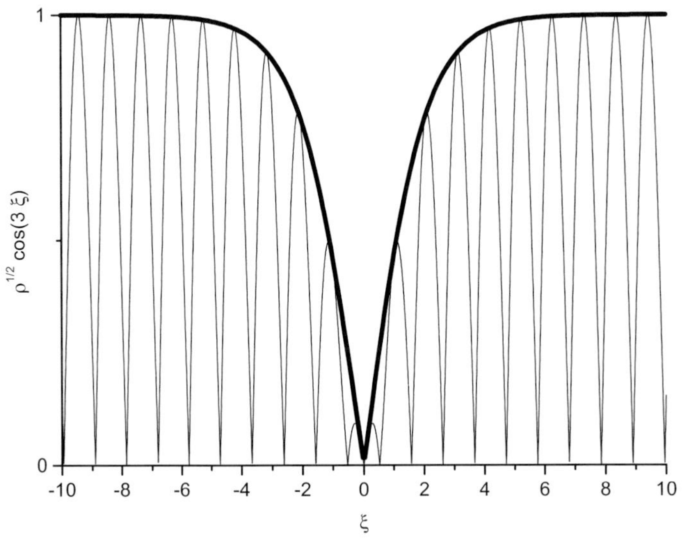

Figure 8.11 Same as Figure 8.10, but for $a = 1$. In addition, here the absolute value of the field is shown.

This is often called a black soliton; see Figure 8.11. A generalization by a Galilean transformation is also possible. One should note again the fundamental difference to a KdV soliton where amplitude, width, and velocity are related.

We continue with the following important remarks. The cubic nonlinear Schrödinger equation belongs to the class of exactly solvable models. It possesses a (generalized) inverse scattering transform (IST). An introduction into that method can be found in Appendix C. Appendices F and G contain additional information on the appearance of nonlinear Schrödinger-type equations in physical systems.

8.3.1 Modulational Instability

The next point is related to the significance of envelope solitons. Let us study the linear dynamics of a continuous train of modulated waves (see Figure 8.12). Then, we assume a finite value for ϕ at $\xi \to \pm\infty$, and the nonlinear Schrödinger equation (8.208) should be changed to

$$i\phi_\tau + \frac{1}{2}\phi_{\xi\xi} + \kappa\left(|\phi|^2 - |\phi_0|^2\right)\phi = 0\,. \tag{8.234}$$

Introducing the same notation for the modulus and nonlinear phase as before, we make the ansatz

$$\begin{pmatrix}\rho\\ \sigma\end{pmatrix} = \begin{pmatrix}\rho_0\\ 0\end{pmatrix} + \mathrm{Re}\begin{pmatrix}\rho_1\\ \sigma_1\end{pmatrix}\exp\left[i\left(K\xi - \Omega\tau\right)\right] \tag{8.235}$$

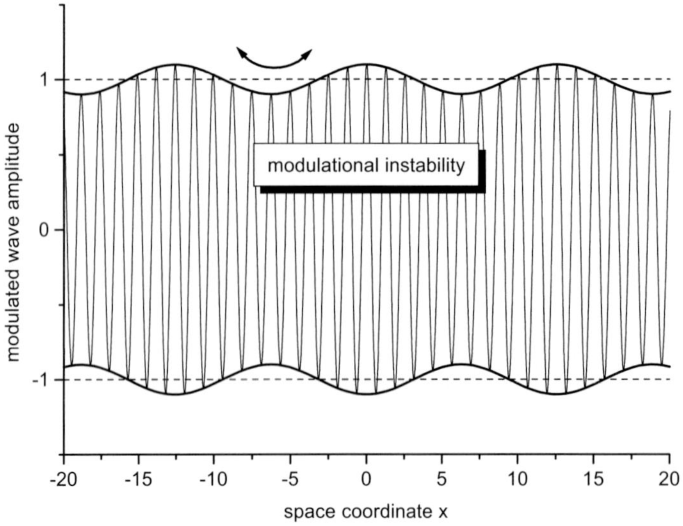

Figure 8.12 Sketch of a modulated wave-train. For modulational instability, the (at first) constant amplitude is slightly modulated by an harmonic perturbation.

and linearize. That is to say, we have to linearize the equations

$$\rho_\tau + (\rho\sigma_\xi)_\xi = 0 \,, \tag{8.236}$$

$$\kappa(\rho - \rho_0) + \frac{\rho_{\xi\xi}}{4\rho} - \sigma_\tau - \frac{1}{2}(\sigma_\xi)^2 - \frac{1}{8}\frac{\rho_\xi^2}{\rho^2} = 0 \,, \tag{8.237}$$

and then after some algebra we obtain the dispersion relation for the modulation

$$\Omega^2 = \frac{1}{4}K^4 - \kappa\rho_0 K^2 \,. \tag{8.238}$$

From here, we clearly see that for $\kappa > 0$ and $K \to 0$, we can have $\Omega^2 < 0$. The corresponding instability is called modulational instability. The maximum growth occurs for

$$K_{max} = (2\kappa\rho_0)^{1/2} \,; \tag{8.239}$$

the corresponding growth rate is

$$\gamma_{max} = \kappa\rho_0 \equiv \kappa|\phi_0|^2 \,. \tag{8.240}$$

Thus, modulational instability occurs in a nonlinear medium in which the nonlinear frequency shift and the group dispersion have different signs, that is, $\kappa > 0$. For $\kappa > 0$ in Eq. (8.208), the effective potential is attractive. If we increase the quasiparticle density $|\phi|^2$, the potential depth increases, becomes more attractive, attracts more quasiparticles, and so on: instability develops.

8.3.2
Historical Remark on Envelope Water Solitons

We conclude this section by a historical example for envelope solitons. We choose the water wave model of Section 8.1,

$$\nabla^2 \phi \equiv \left(\partial_x^2 + \partial_y^2 + \partial_z^2\right) \phi = 0 \,, \tag{8.241}$$

$$\partial_z \phi = 0 \quad \text{at} \quad z = -h \,, \tag{8.242}$$

$$\partial_z \phi = \partial_t \psi + \partial_x \psi \partial_x \phi + \partial_y \psi \partial_y \phi \quad \text{at} \quad z = \psi(x, y, t) \,, \tag{8.243}$$

$$\partial_t \phi + \frac{1}{2}(\nabla \phi)^2 + g\psi = 0 \quad \text{at} \quad z = \psi(x, y, t) \,. \tag{8.244}$$

Note that we generalize the previous case to two spatial surface dimensions and finite kh.

We start with an initial distribution at the surface

$$g\psi(x, y, t = 0) = i\varepsilon\omega a(\varepsilon x, \varepsilon y) \exp(ikx) + c.c. \tag{8.245}$$

and apply the slowly varying envelope approximation. Let us introduce

$$\xi = \varepsilon(x - c_g t) \,, \quad \eta = \varepsilon y \,, \quad \tau = \varepsilon^2 t \,, \tag{8.246}$$

where

$$c_g = \left(\frac{g}{2\omega}\right) \left[\sigma + kh\left(1 - \sigma^2\right)\right], \tag{8.247}$$

with

$$\sigma = \tanh(kh) \,. \tag{8.248}$$

We make the ansatz

$$\psi = \sum_{\nu=-\infty}^{+\infty} \sum_{\mu=|\nu|}^{\infty} \varepsilon^\mu \psi_{\nu\mu} e^{i\nu(kx-\omega t)} \,, \quad \phi = \sum_{\nu=-\infty}^{+\infty} \sum_{\mu=|\nu|}^{\infty} \varepsilon^\mu \phi_{\nu\mu} e^{i\nu(kx-\omega t)} \,. \tag{8.249}$$

The coefficients are functions of ξ, η, and τ. From Eq. (8.245), we have $\psi_{00} = 0$, and we can also set $\phi_{00} = 0$. After some algebra, we find, when collecting the orders in ε and the contributions to the harmonics,

$$\phi_{11} = A\frac{\cosh k(z+h)}{\cosh kh}\,, \tag{8.250}$$

$$\phi_{12} = D\frac{\cosh k(z+h)}{\cosh kh} - i\frac{\partial A}{\partial \xi}\left\{\frac{(z+h)\sinh k(z+h) - h\sigma\cosh k(z+h)}{\cosh kh}\right\}, \tag{8.251}$$

$$\phi_{22} = F\frac{\cosh 2k(z+h)}{\cosh 2kh}\,, \tag{8.252}$$

where A, D and F are functions of ξ, η, and τ. It is further assumed that

$$\varepsilon kh \leq 1\,. \tag{8.253}$$

Next, ϕ_{01} and ϕ_{02} are found to be independent of z, and

$$\frac{\partial \phi_{03}}{\partial z} = -(z+h)\left[\frac{\partial^2 \phi_{01}}{\partial \xi^2} + \frac{\partial^2 \phi_{01}}{\partial \eta^2}\right]. \tag{8.254}$$

The steps become more and more complicated; a detailed evaluation has been performed by Davey and Stewartson [246]. They get

$$\left(gh - c_g^2\right)\frac{\partial^2 \phi_{01}}{\partial \xi^2} + gh\frac{\partial^2 \phi_{01}}{\partial \eta^2} = -k^2\left[2c_p + c_g(1-\sigma^2)\right]\frac{\partial |A|^2}{\partial \xi}\,, \tag{8.255}$$

and

$$\begin{aligned} 2i\omega\frac{\partial A}{\partial \tau} &- \left[c_g^2 - gh\left(1-\sigma^2\right)\left(1-kh\sigma\right)\right]\frac{\partial^2 A}{\partial \xi^2} + c_p c_g\frac{\partial^2 A}{\partial \eta^2} \\ &= \frac{1}{2}k^4\left(9\sigma^{-2} - 12 + 13\sigma^2 - 2\sigma^4\right)|A|^2 A \\ &+ k^2\left[2c_p + c_g\left(1-\sigma^2\right)\right]A\frac{\partial \phi_{01}}{\partial \xi}\,. \end{aligned} \tag{8.256}$$

In addition, we have

$$\psi_{01} = 0 \quad \text{and} \quad g\psi_{11} = i\omega A\,. \tag{8.257}$$

We now see that the initial distribution Eq. (8.245) corresponds to

$$A(\xi,\eta,\tau=0) = a(\xi,\eta)\,. \tag{8.258}$$

The usual form of the Davey–Stewartson equations [246] is obtained by the transformation

$$c_g\frac{\partial \phi_{01}}{\partial \xi} = k^2\left(1-\sigma^2\right)|A|^2 + g\left\{\frac{k^2}{g}Q - \frac{k\left[\sigma + 2kh\left(1-\sigma^2\right)\right]}{gh - c_g^2}|A|^2\right\}, \tag{8.259}$$

where ϕ_{01} is replaced by Q. The result is

$$i\frac{\partial A}{\partial \tau} + \lambda\frac{\partial^2 A}{\partial \xi^2} + \mu\frac{\partial^2 A}{\partial \eta^2} = \nu|A|^2 A + \nu_1 A Q\,, \tag{8.260}$$

$$\lambda_1\frac{\partial^2 Q}{\partial \xi^2} + \mu_1\frac{\partial^2 Q}{\partial \eta^2} = k_1\frac{\partial^2|A|^2}{\partial \eta^2}\,, \tag{8.261}$$

where

$$\lambda = \frac{1}{2}\omega'' \le 0\,, \quad \mu = \frac{1}{2k}\omega' \ge 0\,, \tag{8.262}$$

$$\nu = \frac{k^4}{4\omega\sigma^2}\left\{9 - 10\sigma^2 + 9\sigma^4 - \frac{2\sigma^2}{gh - c_g^2}\right.$$
$$\left.\times\left[4c_p^2 + 4c_p c_g\left(1-\sigma^2\right) + gh\left(1-\sigma^2\right)^2\right]\right\}, \tag{8.263}$$

$$\nu_1 = \frac{k^4}{\omega c_g}\left[2c_p + c_g\left(1-\sigma^2\right)\right], \quad \lambda_1 = gh - c_g^2 \ge 0\,, \tag{8.264}$$

$$\mu_1 = gh\,, \quad k_1 = ghc_g\frac{2c_p + c_g\left(1-\sigma^2\right)}{gh - c_g^2} \ge 0\,. \tag{8.265}$$

With

$$\Delta^{-1} := \left(\lambda_1\partial_\xi^2 + \mu_1\partial_\eta^2\right)^{-1}, \tag{8.266}$$

we can combine Eqs. (8.260) and (8.261) to obtain

$$i\frac{\partial A}{\partial \tau} + \lambda\frac{\partial^2 A}{\partial \xi^2} + \mu\frac{\partial^2 A}{\partial \eta^2} - \nu|A|^2 A = k_1\nu_1 A\Delta^{-1}\frac{\partial^2|A|^2}{\partial \eta^2}\,. \tag{8.267}$$

Finally, when rescaling

$$\tilde{\xi} = (-\lambda)^{-1/2}\xi\,, \quad \tilde{\eta} = \mu^{-1/2}\eta\,, \quad \tilde{A} = \nu^{1/2}A\,, \tag{8.268}$$

Eq. (8.267) reads

$$i\frac{\partial \tilde{A}}{\partial \tau} - \frac{\partial^2 \tilde{A}}{\partial \tilde{\xi}^2} + \frac{\partial^2 \tilde{A}}{\partial \tilde{\eta}^2} - |\tilde{A}|^2\tilde{A} = \tilde{A}\frac{k_1\nu_1}{\nu\mu}\left(\frac{\lambda_1}{\lambda}\frac{\partial^2}{\partial \tilde{\xi}^2} + \frac{\mu_1}{\mu}\frac{\partial^2}{\partial \tilde{\eta}^2}\right)^{-1}\frac{\partial^2|\tilde{A}|^2}{\partial \tilde{\eta}^2}\,. \tag{8.269}$$

This expression is appropriate for performing the limit $kh \to \infty$. We note

$$\frac{k_1\nu_1}{\nu\mu} \sim k^{1/2}\,, \tag{8.270}$$

but

$$\frac{\lambda_1}{\lambda} \sim \frac{\mu_1}{\mu} \sim k^{3/2}\,. \tag{8.271}$$

Therefore, the contribution on the right-hand side of Eq. (8.269) vanishes in the deep-water limit. If we go back to the quantities without wriggles, we obtain

$$i\partial_\tau A - \frac{1}{8}g^{1/2}k^{-3/2}\partial_\xi^2 A + \frac{1}{4}g^{1/2}k^{-3/2}\partial_\eta^2 A - 2k^{7/2}g^{-1/2}|A|^2 A = 0\,, \tag{8.272}$$

which has been first derived by Zakharov [248]. The soliton solution of Eq. (8.272) is

$$A = A_0 \mathrm{sech}\left[2\sqrt{2}kA_0\xi\right]\exp\left(-ik^{7/2}g^{-\frac{1}{2}A_0^2\tau}\right). \tag{8.273}$$

We will stop our general discussion of envelope solitons at this point. It is hoped that the simple model calculation as well as the historical review of water wave investigations have shown us the typical procedures and difficulties. It should be emphasized that nonlinear oscillations in plasmas show a similar richness in model areas just as the presented fluid example.

8.3.3
Nonlinear Dispersion Relation and Schrödinger Equation

Formally, one may "deduce" the cubic nonlinear Schrödinger equation for the wave envelope from a nonlinear dispersion relation. Let us abbreviate the linear dispersion by

$$\varepsilon_L\left(\omega, k; \alpha_{j0}\right) E(k,\omega) = 0\,, \tag{8.274}$$

where the linear dispersion function ε_L depends on plasma parameters α_j like n, T, and so on. Finite amplitudes of the waves may alter the plasma parameters α_j. We denote with the index zero the lowest order values α_{j0} being independent of the wave amplitude. Then, we may formally write the nonlinear dispersion as

$$e^{i(kx-\omega t)}\varepsilon_{NL}\left\{\left(\omega + i\frac{\partial}{\partial t}\right), \left(k - i\frac{\partial}{\partial x}\right); (\alpha_{j0} + \Delta\alpha_j)\right\} \tilde{E}(x,t) = 0\,. \tag{8.275}$$

The substitutions $\omega \to \omega + i\partial/\partial t$ and $k \to k - i\partial/\partial x$ take care of the change in the wave amplitude due to the changed plasma parameters. Expanding ε_{NL} in the weakly nonlinear limit, we obtain

$$\begin{aligned}\varepsilon_{NL} \approx \varepsilon_L &+ i\frac{\partial\varepsilon_L}{\partial\omega}\frac{\partial}{\partial t} - i\frac{\partial\varepsilon_L}{\partial k}\frac{\partial}{\partial x} - \frac{1}{2}\frac{\partial^2\varepsilon_L}{\partial k^2}\frac{\partial^2}{\partial x^2} - \frac{1}{2}\frac{\partial^2\varepsilon_L}{\partial\omega^2}\frac{\partial^2}{\partial t^2}\\ &+ \frac{\partial^2\varepsilon_L}{\partial\omega\partial k}\frac{\partial}{\partial x}\frac{\partial}{\partial t} + \sum_j \varDelta\alpha_j \frac{\partial\varepsilon_L}{\partial\alpha_{j0}} + \cdots\end{aligned} \tag{8.276}$$

Note $\varepsilon_L \equiv \varepsilon_L(\omega, k; \alpha_{j0}) = 0$. We may transform into a frame with the group velocity v_g by using $\xi = x - v_g t$, $\tau = t$. The group velocity is

$$v_g = \left.\frac{d\omega}{dk}\right|_{\varepsilon_L=0} = -\frac{\dfrac{\partial\varepsilon_L}{\partial k}}{\dfrac{\partial\varepsilon_L}{\partial\omega}}\,. \tag{8.277}$$

For the envelope $\tilde{E}$ of the phase $\exp[i(kx - \omega t)]$, one obtains in first order of ∂_τ and second order in ∂_ξ

$$i\frac{\partial \tilde{E}}{\partial \tau} + \frac{1}{2}\frac{\partial^2 \omega}{\partial k^2}\frac{\partial^2 \tilde{E}}{\partial \xi^2} - \frac{\partial \omega}{\partial |\tilde{E}|^2}|\tilde{E}|^2 \tilde{E} = 0\,. \tag{8.278}$$

We have used

$$\frac{\partial^2 \omega}{\partial k^2} = \frac{\partial}{\partial k}\frac{\partial \omega}{\partial k}\bigg|_{\varepsilon_L=0} = -\frac{\partial}{\partial k}\left(\frac{\frac{\partial \varepsilon_L}{\partial k}}{\frac{\partial \varepsilon_L}{\partial \omega}}\right)_{\varepsilon_L=0}$$
$$= -\frac{\frac{\partial^2 \varepsilon_L}{\partial k^2}}{\frac{\partial \varepsilon_L}{\partial \omega}} - \frac{\frac{\partial^2 \varepsilon_L}{\partial \omega^2}}{\frac{\partial \varepsilon_L}{\partial \omega}}\frac{\left(\frac{\partial \varepsilon_L}{\partial k}\right)^2}{\left(\frac{\partial \varepsilon_L}{\partial \omega}\right)^2} + 2\frac{\frac{\partial^2 \varepsilon_L}{\partial \omega \partial k}}{\frac{\partial \varepsilon_L}{\partial \omega}}\frac{\frac{\partial \varepsilon_L}{\partial k}}{\frac{\partial \varepsilon_L}{\partial \omega}}\,. \tag{8.279}$$

Furthermore,

$$\sum_j \frac{\partial \varepsilon_L}{\partial \alpha_{j0}}\left(\frac{\partial \varepsilon_L}{\partial \omega}\right)^{-1} \Delta \alpha_j = -\frac{\partial \omega}{\partial \left|\tilde{E}\right|^2}\left|\tilde{E}\right|^2 \tag{8.280}$$

for $\Delta \alpha_j \sim |\tilde{E}|^2$. The latter becomes evident, for example, for the simple example $n = n_0 - \delta|\tilde{E}|^2$ and $\varepsilon_L(k = 0) = \varepsilon_L = 1 - ne^2/(\varepsilon_0 m_e \omega^2)$. Renormalizing the time via

$$\frac{\partial^2 \omega}{\partial k^2}\tau \to \tau \tag{8.281}$$

leads to the standard form of the cubic nonlinear Schrödinger equation

$$i\frac{\partial \tilde{E}}{\partial \tau} + \frac{1}{2}\frac{\partial^2 \tilde{E}}{\partial \xi^2} + \kappa\left|\tilde{E}\right|^2 \tilde{E} = 0 \tag{8.282}$$

with

$$\kappa = -\frac{\frac{\partial \omega}{\partial |\tilde{E}|^2}}{\frac{\partial^2 \omega}{\partial k^2}}\,. \tag{8.283}$$

8.4 Nonlinear Langmuir Waves

The "simplest" optical branch in a plasma consists of Langmuir waves. The latter are high-frequency oscillations in an unmagnetized plasma. The characteristic frequency (at $k = 0$) is the electron plasma frequency ω_{pe}. In the nonlinear regime, via the ponderomotive pressure, they also influence the low-frequency dynamics of the system. Here, we describe nonlinear Langmuir waves in a Maxwell fluid model using particle density and momentum equations for electrons and ions. From the equations of states, the scalar (electron or ion) pressure is $p_{i,e} \sim n_{i,e}^{\gamma_{i,e}}$ where for adiabatic changes $\gamma_{i,e} = (2+N)/N$, where N is the degree of freedom. For isothermal changes, $\gamma_{i,e} = 1$.

We subdivide the relevant quantities into high-frequency (h) and low-frequency, slow (s) contributions,

$$n_{i,e} = n_0 + n_{i,e}^s + n_{i,e}^h\,, \qquad \boldsymbol{v}_{i,e} = \boldsymbol{v}_{i,e}^s + \boldsymbol{v}_{i,e}^h\,, \qquad \boldsymbol{E} = \boldsymbol{E}^s + \boldsymbol{E}^h\,. \tag{8.284}$$

In unmagnetized plasmas, the magnetic field $\boldsymbol{B}$ is assumed to contribute only via $\boldsymbol{B}^h$. Furthermore, we separate the fast time-dependence in all high-frequency parts, for example, for the particle density

$$n_{i,e}^h = \frac{1}{2}\left[\tilde{n}_{i,e}(\boldsymbol{r},t)e^{-i\omega t} + c.c.\right], \tag{8.285}$$

where the amplitudes $\tilde{n}_{i,e}$, $\tilde{\boldsymbol{v}}_{i,e}$, $\tilde{\boldsymbol{E}}$, $\tilde{\boldsymbol{B}}$ should only weakly depend on time. The averaging over the fast time-variations is indicated by $\langle\ldots\rangle$. The slow components thus obey

$$\nabla \cdot \boldsymbol{E}^s = \frac{1}{\varepsilon_0} e(n_i^s - n_e^s)\,, \tag{8.286}$$

$$\nabla \times \boldsymbol{E}^s = 0\,, \tag{8.287}$$

$$\frac{1}{\varepsilon_0}\boldsymbol{j}^s + \frac{\partial \boldsymbol{E}^s}{\partial t} = 0\,, \tag{8.288}$$

$$\frac{\partial n_e^s}{\partial t} + \nabla \cdot \left[\left(n_0 + n_e^s\right)\boldsymbol{v}_e^s + \left\langle n_e^h \boldsymbol{v}_e^h\right\rangle\right] = 0\,, \tag{8.289}$$

$$\frac{\partial n_i^s}{\partial t} + \nabla \cdot \left[\left(n_0 + n_i^s\right)\boldsymbol{v}_i^s + \left\langle n_i^h \boldsymbol{v}_i^h\right\rangle\right] = 0\,, \tag{8.290}$$

$$\frac{\partial \boldsymbol{v}_e^s}{\partial t} + \boldsymbol{v}_e^s \cdot \nabla \boldsymbol{v}_e^s = -\frac{e}{m_e}\left\{\boldsymbol{E}^s + \left\langle \boldsymbol{v}_e^h \times \boldsymbol{B}^h\right\rangle\right\} - \left\langle \boldsymbol{v}_e^h \cdot \nabla \boldsymbol{v}_e^h\right\rangle - \frac{\gamma_e k_B}{m_e}\left\langle T_e \nabla \ln\left(n_0 + n_e^s\right)\right\rangle, \tag{8.291}$$

$$\frac{\partial \boldsymbol{v}_i^s}{\partial t} + \boldsymbol{v}_i^s \cdot \nabla \boldsymbol{v}_i^s = \frac{e}{m_i}\left\{\boldsymbol{E}^s + \left\langle \boldsymbol{v}_i^h \times \boldsymbol{B}^h\right\rangle\right\} - \left\langle \boldsymbol{v}_i^h \cdot \nabla \boldsymbol{v}_i^h\right\rangle - \frac{\gamma_i k_B}{m_i}\left\langle T_i \nabla \ln(n_0 + n_i^s)\right\rangle. \tag{8.292}$$

Since the low-frequency electric field is curl-free,

$$\boldsymbol{E}^s = -\nabla \phi^s\,. \tag{8.293}$$

Next, we write up the high-frequency equations

$$\nabla \cdot \boldsymbol{E}^h = \frac{1}{\varepsilon_0} e(n_i^h - n_e^h) , \tag{8.294}$$

$$\nabla \times \boldsymbol{E}^h = -\frac{\partial \boldsymbol{B}^h}{\partial t} , \tag{8.295}$$

$$\nabla \times \boldsymbol{B}^h = \mu_0 \boldsymbol{j}^h + \frac{1}{c^2}\frac{\partial \boldsymbol{E}^h}{\partial t} , \tag{8.296}$$

$$\frac{\partial n_e^h}{\partial t} + \nabla \cdot \left[(n_0 + n_e^s)\boldsymbol{v}_e^h + n_e^h \boldsymbol{v}_e^s \right] = 0 , \tag{8.297}$$

$$\frac{\partial n_i^h}{\partial t} + \nabla \cdot \left[(n_0 + n_i^s)\boldsymbol{v}_i^h + n_i^h \boldsymbol{v}_i^s \right] = 0 , \tag{8.298}$$

$$\frac{\partial \boldsymbol{v}_e^h}{\partial t} + \boldsymbol{v}_e^h \cdot \nabla \boldsymbol{v}_e^s + \boldsymbol{v}_e^s \cdot \nabla \boldsymbol{v}_e^h + \frac{e}{m_e}\boldsymbol{E}^h + \frac{e}{m_e}\boldsymbol{v}_e^s \times \boldsymbol{B}^h + \frac{\gamma_e k_B}{m_e}\left(\frac{T_e \nabla n_e}{n_e}\right)^h = 0 , \tag{8.299}$$

$$\frac{\partial \boldsymbol{v}_i^h}{\partial t} + \boldsymbol{v}_i^h \cdot \nabla \boldsymbol{v}_i^s + \boldsymbol{v}_i^s \cdot \nabla \boldsymbol{v}_i^h - \frac{e}{m_i}\boldsymbol{E}^h - \frac{e}{m_i}\boldsymbol{v}_i^s \times \boldsymbol{B}^h + \frac{\gamma_i k_B}{m_i}\left(\frac{T_i \nabla n_i}{n_i}\right)^h = 0 . \tag{8.300}$$

Provided the amplitudes are not too large, we may assume $v_{i,e}^s \ll v_{the}$. On the other hand, for the phase velocity, the estimate $v_{ph} \gg v_{the}$ is appropriate. Thus, $v_{i,e}^s/v_{ph} \ll 1$ holds. Since

$$\left| \boldsymbol{v}^s \cdot \nabla \boldsymbol{v}^h \right| \sim \mathcal{O}\left(\frac{v^s}{v_{ph}}\right) \left| \frac{\partial \boldsymbol{v}^h}{\partial t} \right| , \tag{8.301}$$

$$\left| \boldsymbol{v}^h \cdot \nabla \boldsymbol{v}^s \right| \sim \mathcal{O}\left(\frac{v^s}{v_{ph}}\right) \left| \frac{\partial \boldsymbol{v}^h}{\partial t} \right| , \tag{8.302}$$

$$\left| \frac{e}{m_{i,e}} \boldsymbol{v}^s \times \boldsymbol{B}^h \right| \sim \mathcal{O}\left(\frac{v^s}{v_{ph}}\right) \left| \frac{e}{m_{i,e}} \boldsymbol{E}^h \right| , \tag{8.303}$$

we may neglect the convective terms and the Lorentz force in the high-frequency equations. A similar situation occurs in the particle density continuity equations since

$$\left| \nabla \cdot \left(n_{e,i}^h \boldsymbol{v}_{e,i}^s \right) \right| \sim \mathcal{O}\left(\frac{v_{e,i}^s}{v_{ph}}\right) \left| \frac{\partial n_{e,i}^h}{\partial t} \right| . \tag{8.304}$$

Putting everything together, the high-frequency equations can be simplified to

$$\nabla \cdot \boldsymbol{E}^h = \frac{1}{\varepsilon_0} e \left(n_i^h - n_e^h \right) , \tag{8.305}$$

$$\nabla \times \boldsymbol{E}^h = -\frac{\partial \boldsymbol{B}^h}{\partial t} , \tag{8.306}$$

$$\nabla \times \boldsymbol{B}^h = \mu_0 \boldsymbol{j}^h + \frac{1}{c^2}\frac{\partial \boldsymbol{E}^h}{\partial t}, \tag{8.307}$$

$$\nabla \cdot \boldsymbol{B}^h = 0, \tag{8.308}$$

$$\frac{\partial n_e^h}{\partial t} + \nabla \cdot \left[\left(n_0 + n_e^s \right) \boldsymbol{v}_e^h \right] = 0, \tag{8.309}$$

$$\frac{\partial n_i^h}{\partial t} + \nabla \cdot \left[\left(n_0 + n_i^s \right) \boldsymbol{v}_i^h \right] = 0, \tag{8.310}$$

$$\frac{\partial \boldsymbol{v}_e^h}{\partial t} + \frac{e}{m_e}\boldsymbol{E}^h + \frac{\gamma_e k_B}{m_e}\left(\frac{T_e \nabla n_e}{n_e} \right)^h = 0, \tag{8.311}$$

$$\frac{\partial \boldsymbol{v}_i^h}{\partial t} - \frac{e}{m_i}\boldsymbol{E}^h + \frac{\gamma_i k_B}{m_i}\left(\frac{T_i \nabla n_i}{n_i} \right)^h = 0. \tag{8.312}$$

Using, from the continuity equations, the estimates

$$n_e^h \sim \mathcal{O}\left(\frac{n_0 + n_e^s}{v_{ph}} v_e^h \right), \tag{8.313}$$

$$n_i^h \sim \mathcal{O}\left(\frac{n_0 + n_i^s}{v_{ph}} v_i^h \right), \tag{8.314}$$

we estimate the pressure terms as

$$\frac{\gamma_e k_B}{m_e}\left(T_e \frac{\nabla n_e}{n_e} \right)^h \sim \mathcal{O}\left(\frac{v_{the}^2}{v_{ph}^2} \right) \frac{\partial v_e^h}{\partial t}, \tag{8.315}$$

$$\frac{\gamma_i k_B}{m_i}\left(T_i \frac{\nabla n_i}{n_i} \right)^h \sim \mathcal{O}\left(\frac{v_{thi}^2}{v_{ph}^2} \right) \frac{\partial v_i^h}{\partial t}. \tag{8.316}$$

We neglect temperature fluctuations and skip all terms of orders m_e/m_i and v_{the}^2/v_{ph}^2. This leads to a further simplification of the high-frequency equations. Since the contributions from the ion high-frequency ion velocity and ion density perturbation are small, we obtain

$$\nabla \cdot \boldsymbol{E}^h = -\frac{1}{\varepsilon_0} e n_e^h, \quad \nabla \times \boldsymbol{E}^h = -\frac{\partial \boldsymbol{B}^h}{\partial t}, \tag{8.317}$$

$$\frac{1}{c^2}\frac{\partial^2 \boldsymbol{E}^h}{\partial t^2} + \nabla \times \nabla \times \boldsymbol{E}^h = -\mu_0 \frac{\partial \boldsymbol{j}^h}{\partial t}, \tag{8.318}$$

$$\frac{\partial \boldsymbol{v}_e^h}{\partial t} + \frac{e}{m_e}\boldsymbol{E}^h + \frac{\gamma_e k_B T_e}{m_e}\frac{\nabla n_e^h}{\left(n_0 + n_e^s \right)} = 0. \tag{8.319}$$

The time-derivative of the high-frequency electric current density

$$\frac{\partial \boldsymbol{j}^h}{\partial t} = \frac{\partial}{\partial t}\left[e\left(n_i \boldsymbol{v}_i - n_e \boldsymbol{v}_e \right) \right]^h \tag{8.320}$$

can be written as

$$\frac{\partial \boldsymbol{j}^h}{\partial t} = e\frac{\partial}{\partial t}\left[\left(n_0 + n_i^s\right)\boldsymbol{v}_i^h + n_i^h \boldsymbol{v}_i^s - \left(n_0 + n_e^s\right)\boldsymbol{v}_e^h - n_e^h \boldsymbol{v}_e^s\right]. \tag{8.321}$$

The terms $(n_0 + n_i^s)\boldsymbol{v}_i^h, n_i^h \boldsymbol{v}_i^s$ can be neglected since they are smaller by a factor m_e/m_i than the corresponding electron terms. Furthermore, we note

$$n_e^h v_e^s \leq \mathcal{O}\left(\frac{v_e^s}{v_{ph}}\right)(n_0 + n_e^s)v_e^h \tag{8.322}$$

and remember $\omega^h \gg \omega^s$ for the time-variations. Then,

$$\frac{\partial \boldsymbol{j}^h}{\partial t} \approx -e\left(n_0 + n_e^s\right)\frac{\partial \boldsymbol{v}_e^h}{\partial t}. \tag{8.323}$$

Inserting into the equation for the high-frequency electric field, we obtain

$$\frac{\partial^2 \boldsymbol{E}^h}{\partial t^2} + c^2\nabla\times\nabla\times\boldsymbol{E}^h + \omega_{pe}^2\left(1 + \frac{n_e^s}{n_0}\right)\boldsymbol{E}^h - 3v_{the}^2\nabla\nabla\cdot\boldsymbol{E}^h = 0\,. \tag{8.324}$$

For the high-frequency phenomena, we use the adiabatic equation of state. Assuming one degree of freedom ($N = 1$), we set $\gamma_e = 3$.

Next, we have to consider the low-frequency contributions. After straightforward simplifications, the basic equations are

$$\nabla\cdot\boldsymbol{E}^s = \frac{1}{\varepsilon_0}e\left(n_i^s - n_e^s\right)\,, \tag{8.325}$$

$$\frac{\partial \boldsymbol{v}_i^s}{\partial t} + \boldsymbol{v}_i^s\cdot\nabla\boldsymbol{v}_i^s = -\frac{e}{m_i}\nabla\phi^s - \frac{\gamma_i k_B T_i}{m_i}\nabla\ln\left(n_0 + n_i^s\right)\,, \tag{8.326}$$

$$\frac{\partial n_i^s}{\partial t} + \nabla\cdot\left[\left(n_0 + n_i^s\right)\boldsymbol{v}_i^s\right] = 0\,, \tag{8.327}$$

$$\frac{\gamma_e k_B T_e}{m_e}\nabla\ln\left(n_0 + n_e^s\right) = \frac{e}{m_e}\nabla\phi^s - \frac{1}{m_e}\boldsymbol{F}_e\,. \tag{8.328}$$

Because of m_e/m_i, the terms $\langle n_i^h \boldsymbol{v}_i^h\rangle$ and $\partial\boldsymbol{v}_e^s/\partial t + \boldsymbol{v}_e^s\cdot\nabla\boldsymbol{v}_e^s$ have been neglected. We have introduced the ponderomotive force

$$\boldsymbol{F}_e = \left\langle \boldsymbol{v}_e^h\cdot\nabla\boldsymbol{v}_e^h\right\rangle + e\left\langle \boldsymbol{v}_e^h\times\boldsymbol{B}^h\right\rangle. \tag{8.329}$$

The ponderomotive force mainly acts on the electrons. To lowest order $(v_{the}^2/v_{ph}^2) \ll 1$, we use

$$\frac{\partial \boldsymbol{v}_e^h}{\partial t} \approx -\frac{e}{m_e}\boldsymbol{E}^h\,. \tag{8.330}$$

With

$$\boldsymbol{B}^h \sim -\frac{1}{\omega}\nabla\times\boldsymbol{E}^h\,,\quad \boldsymbol{v}_e^h \sim -\frac{e}{\omega m_e}\boldsymbol{E}^h\,, \tag{8.331}$$

we obtain

$$F_e \approx \frac{\varepsilon_0 \omega_{pe}^2}{n_0 \omega^2} \left\langle E^h \cdot \nabla E^h + E^h \times \nabla \times E^h \right\rangle . \tag{8.332}$$

At this stage, we have to be a little bit more careful with the factors. In the high-frequency components, we have terms being proportional to $\exp(-i\omega t)$ and $\exp(i\omega t)$ (with $\partial_t \sim -i\omega$ and $\partial_t \sim +i\omega$, respectively). Low-frequency components "collect" terms being proportional to $\exp(-i\omega t + i\omega t)$. Now, using the vector identity $\nabla(a \cdot a) = 2[a \cdot \nabla a + a \times (\nabla \times a)]$ and $\langle E^h \cdot E^h \rangle = 1/2|\tilde{E} \cdot \tilde{E}^*|$, we finally find

$$F_e = \frac{\varepsilon_0 \omega_{pe}^2}{4\omega^2} \frac{\nabla |\tilde{E} \cdot \tilde{E}^*|}{n_0} . \tag{8.333}$$

Letting $\omega \approx \omega_{pe}$, we rewrite the equation for the high-frequency electric field in the slowly varying envelope approximation as

$$-2i\omega_{pe} \frac{\partial \tilde{E}}{\partial t} + c^2 \nabla \times \nabla \times \tilde{E} + \omega_{pe}^2 \frac{n_e^s}{n_0} \tilde{E} - 3 v_{the}^2 \nabla \nabla \cdot \tilde{E} = 0 . \tag{8.334}$$

Note that in the slowly varying envelope approximation,

$$\frac{\partial^2 \tilde{E}}{\partial t^2} \ll 2\omega_{pe} \frac{\partial \tilde{E}}{\partial t} . \tag{8.335}$$

For normalization, let us introduce the length $\sqrt{3}\lambda_{De}$, time $\sqrt{3}/\omega_{pi}$, potential $k_B T_e/e$, particle density n_0, electric field strength $(4 n_0 k_B T_e/\varepsilon_0)^{1/2}$, and velocity $c_s = (k_B T_e/m_i)^{1/2}$ as units. Then, the basic set of equations can be summarized as

$$i\varepsilon \frac{\partial E}{\partial t} + \nabla^2 E - Q \nabla \times \nabla \times E - (n_e - 1) E = 0 , \tag{8.336}$$

$$\frac{\partial n_i}{\partial t} + \nabla(n_i v_i) = 0 , \tag{8.337}$$

$$\frac{\partial v_i}{\partial t} + v_i \cdot \nabla v_i = -\nabla \phi - \frac{T_i}{T_e} \nabla \ln n_i , \tag{8.338}$$

$$\frac{\nabla n_e}{n_e} = \nabla \phi - \nabla |E|^2 , \tag{8.339}$$

$$\frac{1}{3} \nabla^2 \phi = n_e - n_i . \tag{8.340}$$

For simplicity, we have omitted the index s. We have further introduced the notation $n_{i,e} \hat{=} 1 + n_{i,e}^s/n_0$ and $E \hat{=} \tilde{E}$. Furthermore, $\varepsilon \equiv 2(m_e/3m_i)^{1/2}$ and $Q = (m_e c^2/3k_B T_e) - 1$. Note that often, definitions other than $\sqrt{3}\lambda_{De}$ (for the length unit) and $\sqrt{3}/\omega_{pi}$ (for the time unit) are in use (especially without the facto $\sqrt{3}$). Also, in the definition of ε, the factor two is often omitted. We shall come back to that later.

Thus far, we did not comment the equations of states in the low-frequency dynamics. For electrons,

$$\frac{\omega^s}{k^s v_{the}} \approx M \frac{c_s}{v_{the}} \ll 1 \tag{8.341}$$

holds, where M is the Mach number (soliton velocity measured in ion-sound speed c_s). For the electrons, all variations are slow. Therefore, the isothermal equation of state is appropriate. On the other hand, for ions,

$$\frac{\omega^s}{k^s v_{thi}} = M \frac{c_s}{v_{thi}} = M \left(\frac{T_e}{T_i}\right)^{1/2} . \tag{8.342}$$

We have to distinguish two cases. For

$$M < \left(\frac{T_i}{T_e}\right)^{1/2} , \tag{8.343}$$

an isothermal description of ions is also possible. If, on the other hand, the adiabatic equation would apply, that is,

$$M > \left(\frac{T_i}{T_e}\right)^{1/2} , \tag{8.344}$$

the ion pressure term is small anyhow and the approximation "$T_i = 0$" is adequate.

Next, we investigate the one-dimensional situation when

$$i\varepsilon \frac{\partial E}{\partial t} + \frac{\partial^2 E}{\partial x^2} - (n_e - 1) E = 0 , \tag{8.345}$$

$$\frac{\partial n_i}{\partial t} + \frac{\partial}{\partial x}(n_i v_i) = 0 , \tag{8.346}$$

$$\frac{\partial v_i}{\partial t} + v_i \frac{\partial v_i}{\partial x} = -\frac{\partial \phi}{\partial x} , \tag{8.347}$$

$$\frac{\partial n_e}{\partial x} = n_e \left(\frac{\partial \phi}{\partial x} - \frac{\partial |E|^2}{\partial x}\right) , \tag{8.348}$$

$$\frac{1}{3}\frac{\partial^2 \phi}{\partial x^2} = n_e - n_i , \tag{8.349}$$

and the ions are adiabatic with $T_i = 0$.

When we change the normalization to λ_{De} for the length unit and ω_{pi}^{-1} for the time unit and use $\varepsilon \equiv (m_e/m_i)^{1/2}$, the often used form

$$2i\varepsilon \frac{\partial E}{\partial t} + 3\frac{\partial^2 E}{\partial x^2} - (n_e - 1) E = 0 , \tag{8.350}$$

$$\frac{\partial n_i}{\partial t} + \frac{\partial}{\partial x}(n_i v_i) = 0 , \tag{8.351}$$

$$\frac{\partial v_i}{\partial t} + v_i \frac{\partial v_i}{\partial x} = -\frac{\partial \phi}{\partial x} , \tag{8.352}$$

$$\frac{\partial n_e}{\partial x} = n_e \left(\frac{\partial \phi}{\partial x} - \frac{\partial |E|^2}{\partial x} \right), \tag{8.353}$$

$$\frac{\partial^2 \phi}{\partial x^2} = n_e - n_i \, , \tag{8.354}$$

appears. Since the solitons will be slowly moving, the quasineutrality assumption $n_e \approx n_i \mathrel{\hat{=}} n$ will be used. Identifying $v_i \equiv v$, the basic set of 1D equations is

$$2i\varepsilon \frac{\partial E}{\partial t} + 3\frac{\partial^2 E}{\partial x^2} - (n-1)E = 0 \, , \tag{8.355}$$

$$\frac{\partial n}{\partial t} + \frac{\partial}{\partial x}(nv) = 0 \, , \tag{8.356}$$

$$\frac{\partial v}{\partial t} + v\frac{\partial v}{\partial x} = -\frac{\partial}{\partial x}\left(\ln n + |E|^2\right), \tag{8.357}$$

where we used boundary conditions for localized solutions. Localized solutions of Eqs. (8.355)–(8.357) exist [249]. Their stability will be discussed in the next section.

In the small amplitude limit, we introduce a smallness parameter ε and expand all terms in Eqs. (8.345)–(8.349) according to

$$n_e = 1 + \varepsilon^2 n_e^{(1)} + \varepsilon^3 n_e^{(2)} + \cdots \, , \tag{8.358}$$

$$n_i = 1 + \varepsilon^2 n_i^{(1)} + \varepsilon^3 n_i^{(2)} + \cdots \, , \tag{8.359}$$

$$v_i = \varepsilon^2 v_i^{(1)} + \varepsilon^3 v_i^{(2)} + \cdots \, , \tag{8.360}$$

$$\phi = \varepsilon^2 \phi^{(1)} + \varepsilon^3 \phi^{(2)} + \cdots \, , \tag{8.361}$$

$$E = \varepsilon E^{(1)} + \varepsilon^2 E^{(2)} + \cdots \, , \tag{8.362}$$

$$\xi = \varepsilon(x - Mt) \, , \tag{8.363}$$

$$\tau = \varepsilon^2 t \, . \tag{8.364}$$

The systematic collection of terms of the same order leads to the cubic nonlinear Schrödinger equation

$$i\varepsilon\partial_\tau E^{(1)} - iM\partial_\xi E^{(1)} + \partial_\xi^2 E^{(1)} - n_e^{(1)} E^{(1)} = 0 \, , \tag{8.365}$$

$$n_e^{(1)} = -\frac{|E^{(1)}|^2}{1-M^2} \, . \tag{8.366}$$

The appearance of the smallness parameter ε is a "historical relict." It can be easily cured by a redefinition of the slow time variable τ.

Stationary soliton solutions are of the form

$$E^{(1)} = E_s \exp\left\{ i \left[\left(\frac{\eta^2}{\varepsilon} + \frac{M^2 \varepsilon}{4} \right) \tau + \frac{M\varepsilon}{2} \xi \right] \right\} \tag{8.367}$$

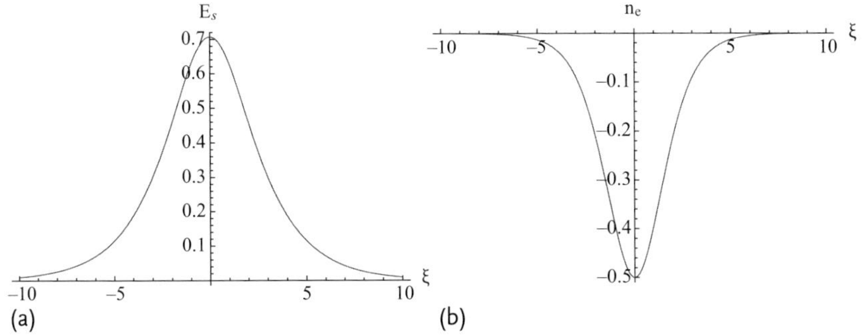

Figure 8.13 (a) Envelope of the electric field E_s for $\eta = 0.5$ and $M = 0$. (b) Density depression n_e for $\eta = 0.5$ and $M = 0$.

and

$$n_e^{(1)} = -\frac{E_s^2}{1 - M^2} . \tag{8.368}$$

Here,

$$E_s = \sqrt{2}\sqrt{1 - M^2}\eta \operatorname{sech}(\eta\xi) . \tag{8.369}$$

The corresponding ambipolar potential is

$$\phi^{(1)} = -\frac{2\eta^2 M^2}{1 - M^2}\operatorname{sech}^2(\eta\xi) . \tag{8.370}$$

The solutions E_s and $n_e^{(1)}$ are shown in Figure 8.13. The form of E_s is called a bell soliton, whereas $n_e^{(1)}$ is an inverted bell density depression. Changing, compared to Eqs. (8.358)–(8.364), the scaling of x and t into

$$\xi = \varepsilon x , \quad \tau = \varepsilon t , \tag{8.371}$$

a short calculation leads to the (one-dimensional) Zakharov equations

$$i\partial_t E^{(1)} + \partial_\xi^2 E^{(1)} - n_e^{(1)} E^{(1)} = 0 , \tag{8.372}$$

$$\partial_\tau^2 n_e^{(1)} - \partial_\xi^2 n_e^{(1)} = \partial_\xi^2 |E^{(1)}|^2 . \tag{8.373}$$

Their stationary solutions have the same form as the solutions of the cubic Schrödinger equation.

For $M \to 1$, another scaling has to be used. For

$$n_e = 1 + \varepsilon n_e^{(1)} + \varepsilon^2 n_e^{(2)} + \cdots , \tag{8.374}$$

$$n_i = 1 + \varepsilon n_i^{(1)} + \varepsilon^2 n_i^{(2)} + \cdots , \tag{8.375}$$

$$v_i = \varepsilon v_i^{(1)} + \varepsilon^2 v_i^{(2)} + \cdots , \tag{8.376}$$

$$\phi = \varepsilon\phi^{(1)} + \varepsilon^2\phi^{(2)} + \cdots , \tag{8.377}$$

$$E = \varepsilon E^{(1)} + \varepsilon^2 E^{(2)} + \cdots , \tag{8.378}$$

$$\xi = \varepsilon^{1/2}(x - t) , \tag{8.379}$$

$$\tau = \varepsilon^{3/2} t , \tag{8.380}$$

we get to lowest order (ε) the identity $n_e^{(1)} = n_i^{(1)}$. The next order ($\varepsilon^{3/2}$) leads to

$$\partial_\xi n_i^{(1)} = \partial_\xi v_i^{(1)} , \tag{8.381}$$

$$\partial_\xi v_i^{(1)} = \partial_\xi \phi^{(1)} , \tag{8.382}$$

$$\partial_\xi n_e^{(1)} = \partial_\xi \phi^{(1)} . \tag{8.383}$$

For localized solutions,

$$n_e^{(1)} = n_i^{(1)} = v_i^{(1)} = \phi^{(1)} \tag{8.384}$$

should hold. The next order ε^2 results in

$$i\varepsilon^{3/2}\partial_\tau E^{(1)} - \varepsilon^{1/2}\partial_\xi E^{(1)} + \partial_\xi^2 E^{(1)} - n_e^{(1)} E^{(1)} = 0 , \tag{8.385}$$

$$\frac{1}{3}\partial_\xi^2 \phi^{(1)} = n_e^{(2)} - n_i^{(2)} . \tag{8.386}$$

Finally, the order $\varepsilon^{5/2}$ leads to a closed system of equations

$$\partial_\tau n_i^{(1)} - \partial_\xi n_i^{(2)} + \partial_\xi (n_i^{(1)} v_i^{(1)}) + \partial_\xi v_i^{(2)} = 0 , \tag{8.387}$$

$$\partial_\tau v_i^{(1)} - \partial_\xi v_i^{(2)} + v_i^{(1)}\partial_\xi v_i^{(1)} + \partial_\xi \phi^{(2)} = 0 , \tag{8.388}$$

$$\partial_\xi n_e^{(2)} = n_e^{(1)}\partial_\xi \phi^{(1)} + \partial_\xi \phi^{(2)} - \partial_\xi |E^{(1)}|^2 . \tag{8.389}$$

By combination, we may formulate

$$2\partial_\tau n_e^{(1)} + \partial_\xi \left[n_e^{(1)}\right]^2 + \frac{1}{3}\partial_\xi^3 n_e^{(1)} + \partial_\xi \left|E^{(1)}\right|^2 = 0 , \tag{8.390}$$

which is of KdV type. Equations (8.390) and (8.385) form a closed system for nearly sonic Langmuir solitons. Stationary solutions result from the ansatz

$$E^{(1)} = E_s(\xi) \exp\left[i\left(\frac{\eta^2}{\varepsilon^{3/2}} + \frac{M\varepsilon^{1/2}}{4}\right)\tau + \frac{M\varepsilon^{1/2}}{2}\xi\right]. \tag{8.391}$$

The functional forms of the solutions have to be determined from

$$\partial_\xi^2 E_s - \eta^2 E_s - n_e^{(1)} E_s = 0 , \tag{8.392}$$

$$\frac{1}{3}\partial_\xi^2 n_e^{(1)} + \left[n_e^{(1)}\right]^2 - 2(M-1)n_e^{(1)} + |E^{(1)}|^2 = 0 . \tag{8.393}$$

One easily finds the solutions

$$E_s = \sqrt{48}\eta^2 \mathrm{sech}(\eta\xi)\tanh(\eta\xi)\,, \quad n_e^{(1)} = -6\eta^2 \mathrm{sech}^2(\eta\xi)\,, \tag{8.394}$$

which are correct, provided

$$\eta^2 = \frac{3}{10}(1-M)\,. \tag{8.395}$$

This is the typical outcome of a discrete eigenvalue problem. We shall find similar behaviors in laser-plasma interaction scenarios. The electric field envelope is a odd solution while the density still shows the inverse bell behavior. Some typical forms are depicted in Figure 8.14.

We conclude this section with a remark on the cubic nonlinear Schrödinger model for Langmuir solitons. In principle, we start from Eqs. (8.336)–(8.340), but we make an approximation similar to Eq. (8.373) for the density response. Then, the 3D Zakharov equations are

$$i\partial_\tau \boldsymbol{E} + \nabla^2 \boldsymbol{E} - Q\nabla\times\nabla\times\boldsymbol{E} - \delta n \boldsymbol{E} = 0\,, \tag{8.396}$$

$$\partial_t^2 \delta n - \nabla^2 \delta n = \nabla^2 |\boldsymbol{E}|^2\,. \tag{8.397}$$

For nonrelativistic plasmas, Q is a large parameter, and Q^{-1} may be used as an expansion parameter. Rearranging the first equation into the form

$$\nabla\times\nabla\times\boldsymbol{E} = \frac{1}{Q}\left(i\partial_\tau \boldsymbol{E} + \nabla^2\boldsymbol{E} - \delta n\boldsymbol{E}\right)\,, \tag{8.398}$$

we find

$$\boldsymbol{E} = -\nabla\varphi + \frac{1}{Q}\boldsymbol{E}_1 + \cdots\,, \tag{8.399}$$

where $\boldsymbol{E}_1$ is related to the scalar electric potential φ via

$$-\nabla\times\nabla\times\boldsymbol{E}_1 = i\partial_\tau\nabla\varphi + \nabla^2\nabla\varphi - \delta n\nabla\varphi\,. \tag{8.400}$$

The solvability condition for $\boldsymbol{E}_1$ is

$$\nabla\cdot\left(i\partial_\tau\nabla\varphi + \nabla^2\nabla\varphi - \delta n\nabla\varphi\right) = 0\,. \tag{8.401}$$

It has to be supplemented with

$$\partial_t^2\delta n - \nabla^2\delta n = \nabla^2|\nabla\varphi|^2\,. \tag{8.402}$$

Only when we neglect the ion inertia, a cubic nonlinear Schrödinger equation

$$\nabla\cdot\left(i\partial_t\nabla\varphi + \nabla^2\nabla\varphi + |\nabla\varphi|^2\nabla\varphi\right) = 0 \tag{8.403}$$

appears. The form

$$i\partial_\tau\varphi + \nabla^2\varphi + |\varphi|^2\varphi = 0 \tag{8.404}$$

is by no means a trivial simplification of it.

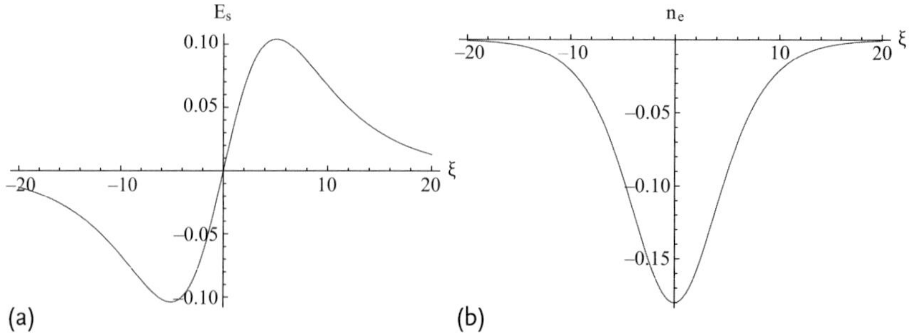

Figure 8.14 (a) Envelope of the electric field E_s for $\eta = 0.03$ and $M = 0.9$. (b) Density depression n_e for $\eta = 0.03$ and $M = 0.9$.

8.5
Longitudinal Stability of Generalized Langmuir Solitons

One-dimensional, localized, and stationary Langmuir waves are solutions of the cubic nonlinear Schrödinger equation in the small-amplitude limit. They haven been proven to be solitons since an inverse scattering transform (IST) existed; see Appendix C. Mostly based on the cubic nonlinear Schrödinger equation, the stability of Langmuir solitons is concluded. Actually, in experiments [250], density depressions of more than 10% have been observed. Then, we can no longer talk about small-amplitude solitons, and the stability of finite amplitude stationary localized solutions has to be investigated. We will address this here on the basis of a quasineutral model.

We start from

$$2i\varepsilon\frac{\partial E}{\partial t} + 3\frac{\partial^2 E}{\partial x^2} - (n-1)E = 0\,, \tag{8.405}$$

$$\frac{\partial n}{\partial t} + \frac{\partial}{\partial x}(nv) = 0\,, \tag{8.406}$$

$$\frac{\partial v}{\partial t} + v\frac{\partial v}{\partial x} = -\frac{\partial}{\partial x}\left(\ln n + |E|^2\right), \tag{8.407}$$

which has been derived in the previous section.

Stationary localized solitary solutions of Eqs. (8.405)–(8.407) may be written in the form

$$E_0 = G\exp\left[i\frac{\eta_0^2}{2\varepsilon}t + i\frac{\varepsilon M}{3}\left(x - \frac{M}{2}t\right)\right], \tag{8.408}$$

where $G = G(x - Mt)$ is an even, nodeless solution of the Schrödinger equation

$$-3\frac{d^2 G}{dx^2} + (n_0 - 1)G + \eta_0^2 G = 0\,. \tag{8.409}$$

Here, $u_0 = v_0 - M$ and n_0 follow from the algebraic equations

$$n_0 u_0 = -M\,, \quad \frac{1}{2}u_0^2 + \ln n_0 + G^2 = \frac{1}{2}M^2\,. \tag{8.410}$$

In order to prove stability of the one-dimensional solution (8.408), we use the Liapunov[1)] method [251]. From the physical point of view, the Liapunov theory for stability is nothing else but a generalization of the physical principle that a stable state should be characterized by minimum free energy. However, the mathematically rigorous procedure [252–260] is often quite laborious. More details of the Liapunov stability concept are presented in Appendix H.

We should mention that it would be unphysical to call a solitary wave solution unstable just when the distance between the perturbed and unperturbed states grows in time. The reason is that for solitary waves, amplitude, Mach number, and nonlinear frequency shift are correlated. An amplitude perturbation will lead to a different propagation velocity and phasor compared to the unperturbed soliton. To exclude effects originating from this behavior, we consider stability with respect to form [261] by introducing an invariant set consisting of all functions which differ from the original unperturbed state by arbitrary translations, $x - x'$ and rotations in phase.

Guided by the analogy with the free energy, for a stable soliton one has to find a functional L with the following characteristic properties:

a) For perturbed states f out of the neighborhood of the unperturbed state (more generally, the invariant set when considering stability with respect to form), upper and lower bounds for $L(f)$ can be given in terms of the norm of the perturbations. L should be positive definite.
b) L is nonincreasing with time, that is, $dL/dt \leq 0$ should hold.

When such a Liapunov functional exists, the invariant set (soliton) is stable. Physically, this conclusion is plausible. A perturbed state being close to the unperturbed state will have a small free energy difference due to condition (a). Because of condition (b), the free energy will not grow, implying via condition one that the perturbed state will stay in a certain neighborhood of the unperturbed state, meaning stability.

For a reversible system, the Liapunov functional should be constructed from the constants of motion for number density, momentum, and energy. Then, condition (b) is automatically satisfied. In the present case, we use

$$I_1 = \int dx\, E E^* \,, \tag{8.411}$$

$$I_2 = \int dx \left[nv + i\varepsilon \left(E \frac{\partial E^*}{\partial x} - E^* \frac{\partial E}{\partial x} \right) \right], \tag{8.412}$$

$$I_3 = \int dx \left[\frac{1}{2} n v^2 + \frac{1}{2} \left(\frac{\partial \varphi}{\partial x} \right)^2 + n \ln n - n + 1 + 3 \frac{\partial E}{\partial x} \frac{\partial E^*}{\partial x} + (n-1) E E^* \right], \tag{8.413}$$

1) Different forms, e.g., Liapunov, Ljapunov, and Lyapunov, are in use for the transcript of the name of the Russian mathematician and physicist.

where $\varphi = \ln n + EE^*$. From here we construct the functional $L = L_1 - L_0$, with

$$L_1 = \left(\eta_0^2 + \frac{\varepsilon^2 M^2}{3}\right) I_1 - M I_2 + I_3 \,. \tag{8.414}$$

L_0 denotes the corresponding zeroth order contribution. We write the perturbed quantities in the form

$$E = (G + a + ib)\exp\left\{i\left[\frac{\eta_0^2 t}{2\varepsilon} + \frac{\varepsilon M}{3}\left(x - \frac{1}{2}Mt\right)\right]\right\}, \tag{8.415}$$

$$v = v_0 + v_1 \,, \qquad n = n_0 + n_1 \,, \qquad \varphi = \varphi_0 + \varphi_1 \,. \tag{8.416}$$

Since L is constructed from constants of motion, $dL/dt \le 0$ is trivially satisfied. It remains to be seen as to whether L is positive definite, that is, upper and lower bounds in terms of the norm of the perturbations exist. To show this, we reformulate L as

$$L = \int dx \left[\frac{1}{2}n_0 v_1^2 + n_1 v_1 u_0 + \frac{1}{2}\frac{n_1^2}{n_0} + 2an_1 G + \left(\frac{\partial \varphi_1}{\partial x}\right)^2 + aH_+ a + bH_+ b\right] + L_3 \,. \tag{8.417}$$

Here, L_3 combines all higher order contributions. The operators

$$H_+ = -3\frac{\partial^2}{\partial x^2} - (1 - n_0) + \eta_0^2 \,, \tag{8.418}$$

$$H_- = H_+ - \frac{2n_0}{1 - u_0^2}G^2 \tag{8.419}$$

have been introduced. Note that

$$H_+ G = 0 \,, \quad H_- \frac{\partial G}{\partial x} = 0 \,. \tag{8.420}$$

Thus, since G is a nodeless function, H_+ is positive semidefinite. H_- is positive semidefinite for odd functions since for odd functions F, we may write

$$H_- F = -\frac{3}{\frac{\partial G}{\partial x}}\frac{\partial}{\partial x}\left[\left(\frac{\partial G}{\partial x}\right)^2 \frac{\partial}{\partial x}\frac{F}{\frac{\partial G}{\partial x}}\right]. \tag{8.421}$$

Next, we rewrite L as

$$L = \int dx \left[\frac{1}{2n_0}(n_0 v_1 + n_1 u_0)^2 + \frac{1}{2n_0}(1 - u_0^2)\left(n_1 + \frac{2an_0 G}{1 - u_0^2}\right)^2 + \left(\frac{\partial \varphi_1}{\partial x}\right)^2 + aH_- a + bH_+ b\right] + L_3 \,, \tag{8.422}$$

with higher-order terms being summarized in L_3. For odd functions a, L is obviously positive semidefinite. We must now discuss the definiteness of

$$\Delta L = \int dx\, a H_- a \tag{8.423}$$

for even functions a. We define m as the infimum of ΔL, that is,

$$m = \inf_a \Delta L\,. \tag{8.424}$$

When measuring the deviations of the perturbations from the invariant set corresponding to the initial soliton, we use the norm

$$N^2 = \int dx \left[\eta_0^2 |E|^2 + 3 \left| \frac{\partial E}{\partial x} \right|^2 + (n-1)^2 + \left(\frac{\partial \varphi}{\partial x} \right)^2 + n_0 v^2 \right]. \tag{8.425}$$

First, for perturbations which satisfy $I_1 = I_{10}$, we have to lowest order

$$\int dx\, a G = 0\,. \tag{8.426}$$

If for these perturbations $m > 0$, stability follows immediately. Thus, let us assume the opposite case $m \leq 0$. Since now $m < \eta_0^2$, a function a_m exists which realizes the minimum. The function a_m can be determined from the Euler equations belonging to the corresponding variational principle,

$$H_- a_m = \alpha a_m + \beta G\,. \tag{8.427}$$

The Lagrange parameter β takes care of the constraint Eq. (8.426); normalized perturbations imply the Lagrange parameter α. Obviously, $\alpha = m$, and therefore

$$a_m = \beta (H_- - m)^{-1} G\,. \tag{8.428}$$

Note that $H_- - m$ indeed can be inverted since H_- only has one negative eigenvalue λ_-, and the corresponding eigenfunction ψ_- is even and has no nodes. Therefore,

$$\int dx\, \psi_- G \neq 0\,, \tag{8.429}$$

and together with Eq. (8.426), the condition $m > \lambda_-$ follows. Introducing a_m into Eq. (8.426) leads to

$$\int dx\, G (H_- + |m|)^{-1} G = 0\,. \tag{8.430}$$

The case $\beta = 0$ can be excluded since it would lead to the contradiction that a_m is an additional eigenfunction of H_- with eigenvalue m.

Decomposing G into a component $G_{\|}$ being parallel to ψ_-, and a perpendicular component $G_{\perp}$, from Eq. (8.430), we obtain

$$\int dx G H_-^{-1} G \geq 0 , \quad \text{for} \quad m < 0 \tag{8.431}$$

since

$$\begin{aligned} 0 &= (\lambda_- + |m|)^{-1} \int dx G_{\|}^2 + \int dx G_{\perp} (H_- + |m|)^{-1} G_{\perp} \\ &\leq \lambda_-^{-1} \int dx G_{\|}^2 + \int dx G_{\perp} H_-^{-1} G_{\perp} = \int dx G H_-^{-1} G . \end{aligned} \tag{8.432}$$

That means that

$$\int dx G H_-^{-1} G < 0 \tag{8.433}$$

is a sufficient criterion for $m > 0$. Making use of

$$H_- \frac{\partial G}{\partial \eta_0} + 2\eta_0 G = 0 , \tag{8.434}$$

the stability criterion Eq. (8.433) can be formulated as

$$\frac{\partial}{\partial \eta_0^2} \int dx G^2 > 0 \quad \text{for stability.} \tag{8.435}$$

That criterion Eq. (8.435) has been evaluated [262] for finite amplitude solitons with the result of stability for all amplitudes.

The finite amplitude model will reduce to the cubic nonlinear Schrödinger equation (NLS) for small amplitudes and low Mach numbers if we approximate the density response by $\delta n \approx -|E|^2$. Let us exemplify the stability criterion on the example of standing soliton solutions of the cubic Schrödinger equation

$$2i \frac{\partial E}{\partial \tau} + 3 \frac{\partial^2 E}{\partial \xi^2} + |E|^2 E = 0 . \tag{8.436}$$

Though, in principle, no new stability consideration is necessary because of the results from the inverse scattering transform (IST); see Appendix C. For standing ($\xi = x - Mt \to x$) solutions

$$G = \sqrt{2(1 - M^2)} \eta_0 \text{sech} \left[\eta_0 \frac{(x - Mt)}{\sqrt{3}} \right] \to G = \sqrt{2} \eta_0 \text{sech} \left[\eta_0 \frac{x}{\sqrt{3}} \right] , \tag{8.437}$$

we construct the Liapunov functional ($L = L_1 - L_0$ for $M = 0$) as

$$L_1 = \int dx \left[\eta_0^2 E E^* + 3 \frac{\partial E}{\partial x} \frac{\partial E^*}{\partial x} - \frac{1}{2} (E E^*)^2 \right] . \tag{8.438}$$

A short calculation leads to

$$L = \int dx(aH_-a + bH_+b) + L_3 \,. \tag{8.439}$$

The operators are

$$H_+ = -3\frac{d^2}{dx^2} - 2\eta_0^2\mathrm{sech}^2\left(\frac{\eta_0 x}{\sqrt{3}}\right) + \eta_0^2 \,, \tag{8.440}$$

$$H_- = -3\frac{d^2}{dx^2} - 6\eta_0^2\mathrm{sech}^2\left(\frac{\eta_0 x}{\sqrt{3}}\right) + \eta_0^2 \,. \tag{8.441}$$

Since

$$H_-\mathrm{sech}^2\left(\frac{\eta_0 x}{\sqrt{3}}\right) = -3\eta_0^2\mathrm{sech}^2\left(\frac{\eta_0 x}{\sqrt{3}}\right), \tag{8.442}$$

the criterion Eq. (8.435) takes the form

$$\frac{\partial}{\partial\eta_0^2}\int d\xi\, \eta_0\mathrm{sech}^2\xi > 0 \tag{8.443}$$

which is trivially satisfied.

8.6 Transverse Instabilities

After having analyzed one-dimensional soliton solutions, the obvious question arises as to whether the one-dimensional (plane) soliton solutions are stable in two and/or three space dimensions. We call that problem the transverse stability problem. It will finally lead us to the question of the existence of stable two and/or three dimensional soliton solutions.

8.6.1 Transverse Instabilities of KdV Solitons

Let us start here with KdV solitons for which we know that (one-dimensional) plane solutions are one-dimensionally stable. We pose the question regarding what happens if such solutions (plus small perturbations) are used as initial distributions in 2D generalized KdV equations, for example, the KP and ZK equation, respectively.

We start with the Kadomtsev–Petviashvili (KP) equation. The instability of solitary waves in the KP equation was considered in [263, 264]. One-dimensional solitary wave solutions of the KP equation, that is, plane solutions of the one-dimensional KdV-equation, were shown to be unstable to long-wavelength transverse disturbances for the positive dispersion case

$$(u_t + uu_x + u_{xxx})_x = u_{yy} \,. \tag{8.444}$$

The existence of a cut-off, the form of the exponentially unstable modes of the linearized equation of motion for the perturbations of the solitary wave, and the analytical values of the growth rates were discussed.

Let us briefly summarize the results. We consider the (plane) one-dimensional solitary wave solution

$$u_s = 12\eta^2 \mathrm{sech}^2\left[\eta\left(x - x_0 - 4\eta^2 t\right)\right] \tag{8.445}$$

and introduce the variables

$$\begin{aligned} &\xi = \eta\left(x - x_0 - 4\eta^2 t\right), \quad Y = \sqrt{3}\eta^2 y\,, \\ &\tau = 4\eta^3 t\,, \quad \psi = 4\eta^2 u\,. \end{aligned} \tag{8.446}$$

Perturbing in the form

$$\psi(\xi, Y, \tau) = \psi_s(\xi) + f(\xi)e^{\Gamma\tau + iKY}, \tag{8.447}$$

for f, we obtain

$$\frac{d^4 f}{d\xi^4} - 4\frac{d^2}{d\xi^2}\left[\left(1 - 3\mathrm{sech}^2\xi\right) f\right] + 4\Gamma\frac{df}{d\xi} - 3K^2 f = 0\,. \tag{8.448}$$

The corresponding equation for

$$f(\xi) := \frac{d^2 g}{d\xi^2} \equiv g'' \tag{8.449}$$

reads

$$g'''' - 4\left(1 - 3\mathrm{sech}^2\xi\right) g'' + 4\Gamma g' + 3K^2 g = 0\,. \tag{8.450}$$

One can check that it has the following bounded solutions

$$g(\xi) = e^{\mu\xi}\left(\mu^3 + 2\mu + \Gamma - 3\mu^2 \tanh\xi\right) \tag{8.451}$$

for

$$\Gamma = K\sqrt{1 - K} \tag{8.452}$$

in the region $0 \le K \le 1$, $\mu = 1 + \sqrt{1 - K}$.

There are many methods for deriving these results. The first one is outlined in the paper by Kuznetsov *et al.* [263]. Another one, which does not make use of integrability, has been used in the later paper by Alexander *et al.* [264]. The main idea of the latter procedure is to separate, in the solution for g, the asymptotic exponential behavior,

$$g(\xi) = e^{\mu\xi} h(\xi)\,, \tag{8.453}$$

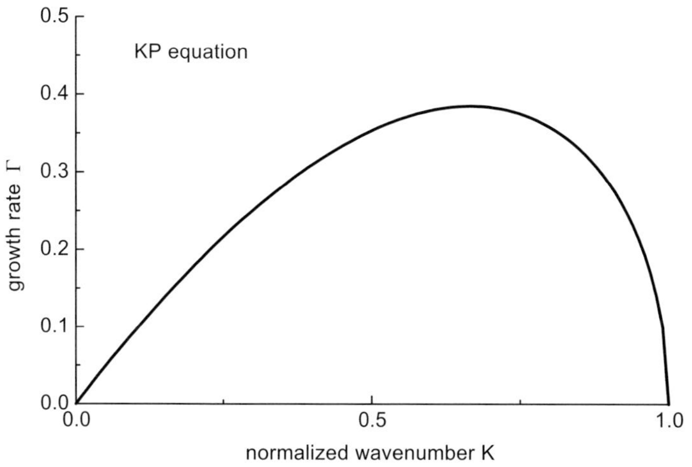

Figure 8.15 Analytically known normalized growth rate Γ versus transverse wavenumber K in the KP case.

and then to introduce the new variable

$$z = \frac{1}{2}(1 + \tanh \xi) . \tag{8.454}$$

The equation for $h(z)$ can be solved by a power series solution

$$h(z) = a + bz . \tag{8.455}$$

One gets the solutions Eq. (8.451) for appropriate constants a and b. Thus, the unstable domain is $0 \leq K \leq 1$. We call $K_c = 1$ the cut-off. For long-wavelength perturbations with wave-numbers below the cut-off wavenumber K_c, the one-dimensional soliton solution of the KdV equation breaks. It is interesting to see that by changing K, the eigensolutions are continuously deformed; however, they always have two nodes.

A plot of the growth rates Γ versus the normalized wavenumbers K is shown in Figure 8.15.

Let us now examine the Zakharov–Kuznetsov (ZK) equation

$$\frac{\partial \varphi}{\partial t} + \varphi \frac{\partial \varphi}{\partial x} + \frac{\partial^3 \varphi}{\partial x^3} + \frac{\partial}{\partial x}\frac{\partial^2 \varphi}{\partial y^2} = 0 , \tag{8.456}$$

which also has a (plane) one-dimensional solitary wave solution

$$\varphi_s = 12\eta^2 \text{sech}^2 \left[\eta \left(x - x_0 - 4\eta^2 t\right)\right] . \tag{8.457}$$

In order to make contact with the corresponding calculations for the Kadomtsev–Petviashvili (KP) equation, we shall introduce

$$\xi = \eta(x - x_0 - 4\eta^2 t) , \quad Y = \eta y , \quad \tau = 4\eta^3 t , \quad \psi = 4\eta^2 \varphi . \tag{8.458}$$

Then, the starting equations is

$$4\left(\frac{\partial\psi}{\partial\tau}-\frac{\partial\psi}{\partial\xi}+\psi\frac{\partial\psi}{\partial\xi}\right)+\frac{\partial^3\psi}{\partial\xi^3}+\frac{\partial}{\partial\xi}\frac{\partial^2\psi}{\partial Y^2}=0\,. \tag{8.459}$$

Perturbing in the form

$$\psi\,(\xi,Y,\tau)=\psi_s(\xi)+f(\xi)e^{\Gamma\tau+iKY}\,, \tag{8.460}$$

for f, we obtain

$$\frac{d^3 f}{d\xi^3}-4\frac{d}{d\xi}\left\{\left[1-3\text{sech}^2(\xi)\right]f+\frac{K^2}{4}f\right\}+4\Gamma f=0\,. \tag{8.461}$$

Note that this eigenvalue problem was considered in several papers, for example, in [265] and [266]. We have instability in the region

$$0\le K^2\le 5\,. \tag{8.462}$$

However, we do not know the analytical solutions for the eigenvalues and the eigenfunctions. Nevertheless, we can determine the dynamical behavior approximately. In the following, we shall prove: At $K=0$, we have $\Gamma=0$ with the eigenmode

$$f_{K^2=0}=\text{sech}^2(\xi)\tanh(\xi)\,. \tag{8.463}$$

For small $K^2>0$, growth occurs and the growth rate is

$$\Gamma^2\approx\frac{4}{15}K^2\,. \tag{8.464}$$

The dynamics near the cut-off $K_c^2=5$ are dominated by

$$f_{K^2=5}=\text{sech}^3(\xi)\quad\text{with}\quad\Gamma=0\,. \tag{8.465}$$

For K^2 close to $K_c^2=5$,

$$\Gamma\approx\frac{64}{15\pi^2}K_c(K_c-K)\,. \tag{8.466}$$

Infeld and Rowlands [267] first published the suspicion of transverse stability. They corrected this prediction in their book [224]. In the following, we summarize the instability results.

First, we present the eigensolutions for $\Gamma=0$. Again, it may be useful to introduce the variable

$$z=\frac{1}{2}\left[1+\tanh(\xi)\right]. \tag{8.467}$$

After some lengthy algebra, Eq. (8.461) can be written as

$$8\frac{d}{dz}\left[z^3(1-z)^3\frac{d^2 f}{dz^2}\right]+48\frac{d}{dz}\left[z^2(1-z)^2 f\right]$$
$$-2K^2 z(1-z)\frac{df}{dz}+4\Gamma f=0\,. \tag{8.468}$$

It is an interesting equation since it shows that for $\Gamma\neq 0$, new singularities appear.

Next, we introduce the function $h(z)$ via

$$f(z) \equiv z^p (1-z)^p h(z)\,. \tag{8.469}$$

Note that this ansatz takes care of an exponential asymptotic behavior for $\xi \to \pm\infty$ since

$$e^{2\xi} = \frac{z}{1-z}\,. \tag{8.470}$$

After some lengthy algebra, one obtains the differential equation for h

$$\begin{aligned} &z^3(1-z)^3 h''' + 3z^2(1-z)^2(p+1)(1-2z)h'' \\ &\quad - \frac{1}{4} z(1-z)\left[K^2 - 12p - 12p^2 + z(-24 + 72p + 48p^2)(1-z)\right] h' \\ &\quad + \frac{1}{4}\Big[2\Gamma - 4p - K^2 p + 4p^3 + z(48 + 48p + 2K^2 p - 24p^2 - 24p^3) \\ &\quad + z^2(-144 - 120p + 72p^2 + 48p^3)\left(1 - \frac{2}{3}z\right)\Big] h = 0\,. \end{aligned} \tag{8.471}$$

From here, it is easy to determine the mentioned eigenfunctions by power series expansion. Let us start with $h \equiv 1$. Then, $p^3 + 3/2p^2 - 5/2p - 3 = 0$ has the solutions $p = 3/2$, -1, and -2, and only the positive value leads to localized solutions. We determine K^2 for $p = 3/2$ from $K^2 p = 12p^3 + 12p^2 - 24p - 24 = 15/2$, leading to $K^2 = 5$. The growth rate follows from

$$\Gamma = 2p + \frac{K^2}{2} p - 2p^3 = 0\,. \tag{8.472}$$

Note that the solution

$$f = z^{3/2}(1-z)^{3/2} = \frac{1}{8}\operatorname{sech}^3(\xi) \tag{8.473}$$

agrees with the previously mentioned eigensolution at the cut-off. (Of course, the amplitude factor is not important in the linear eigenvalue problem!)

Next, we make the more complicated ansatz $h = 1 + bz$. We obtain from $p^3 + 3p^2 - 1/4p - 15/4 = 0$ the values $p = 1$, $-3/2$, and $-5/2$. The bounded solution $p = 1$ leads to $b = -2$, and the coefficient of z^2 gives $K^2 = 0$. The other equations are satisfied for $\Gamma = 0$. Thus, we have found the solution

$$f(z) = z(1-z)[1-2z] = -\frac{1}{4}\operatorname{sech}^2(\xi)\tanh(\xi)\,, \tag{8.474}$$

which agrees with the previously mentioned eigenfunction.

It is straightforward to further generalize the procedure by introducing into Eq. (8.471) the form $h = 1 + bz + cz^2$, which leads the solution

$$p = \frac{1}{2}\,, \quad b = -5\,, \quad c = 5\,, \quad K^2 = -3\,, \quad \Gamma = 0 \tag{8.475}$$

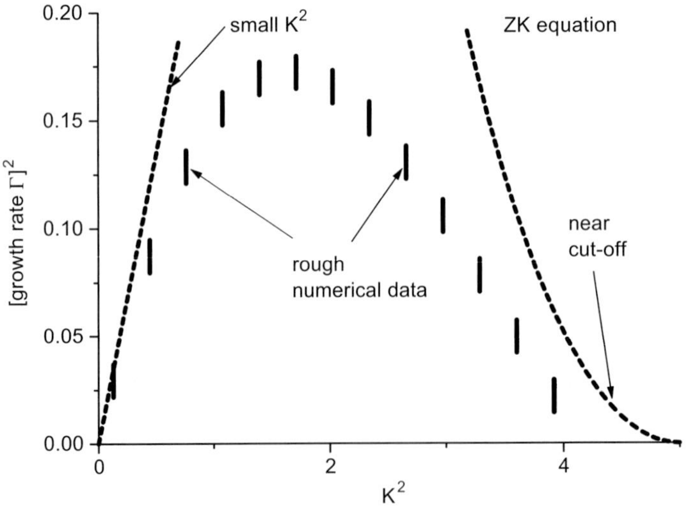

Figure 8.16 The growth rate curve as a function of the transverse wavenumber K is not known analytically in the ZK case. In this figure, we show some rough numerical data (error bars) and the functions obtained by perturbation theory (dashed lines).

that corresponds to

$$f = z^{1/2}(1-z)^{1/2}\left[1-5z+5z^2\right] \equiv \frac{1}{8}\left[4\text{sech}(\xi) - 5\text{sech}^3(\xi)\right] . \tag{8.476}$$

However, because of $K^2 < 0$, this solution is not relevant. It turns out that also further generalizations do not give acceptable eigenfunctions with $K^2 \geq 0$. Thus, by the standard method, we have only found the three discrete solutions. The question arises regarding what can be done in situations when no exact solution is known analytically. Several strategies are possible. In the following, we shall concentrate on perturbative treatments.

First of all, one is tempted to just solve the eigenvalue equation numerically. This, of course, is possible. However, we should remark that numerical solutions are very often quite inaccurate and/or time consuming. Nevertheless in Figure 8.16, we have, just for illustration, shown some error bars for typical numerical values obtained with some shooting method. For the data shown there, of course one could try to optimize the accuracy. The fact we want to demonstrate with the data is that for the ZK case, a similar behavior as in the KP case should be expected.

Analytically, we can determine the growth rate $\Gamma = \Gamma(K)$ for long-wavelength perturbations as well as close to the cut-off K_c. Let us start with the small K^2 case. We go back to the linear eigenvalue problem Eq. (8.461), that is,

$$\Gamma f = \partial_\xi \hat{L} f , \tag{8.477}$$

with

$$\hat{L} = -\frac{1}{4}\frac{\partial^2}{\partial \xi^2} - 3\text{sech}^2\xi + 1 + \frac{K^2}{4} , \tag{8.478}$$

and we split f into its odd (a) and even (b) perturbations. We obtain

$$\Gamma a = \partial_\xi \hat{L} b, \tag{8.479}$$

$$\Gamma b = \partial_\xi \hat{L} a. \tag{8.480}$$

At $K = 0$, we know the odd solution $M \operatorname{sech}^2 \xi \tanh \xi$ which belongs to $\Gamma = 0$. Here, M is an arbitrary amplitude factor. Thus, for small K, we try a solution by perturbation theory, starting with

$$a_0 = 0\,, \quad b_0 = M \operatorname{sech}^2 \xi \,\tanh \xi \tag{8.481}$$

for $\Gamma = 0$. We use K as an expansion parameter (around zero), and write

$$a = a_0 + K a_1 + K^2 a_2 + \cdots\,, \tag{8.482}$$

$$b = b_0 + K b_1 + K^2 b_2 + \cdots\,, \tag{8.483}$$

$$\Gamma = \Gamma_0 + K \Gamma_1 + K^2 \Gamma_2 + \cdots \tag{8.484}$$

By construction, the zeroth order in K is satisfied. Next, we proceed to the first order in K,

$$0 = \partial_\xi \hat{L}(K = 0) b_1\,, \tag{8.485}$$

$$\Gamma_1 b_0 = \partial_\xi \hat{L}(K = 0) a_1\,. \tag{8.486}$$

This is a set of inhomogeneous equations for a_1, b_1, which, of course, posses a solvability condition. Note that the kernel function of $[\partial_\xi \hat{L}(K = 0)]^+$ is $\operatorname{sech}^2 \xi$. Therefore, we multiply Eqs. (8.485) and (8.486) from the left by $\operatorname{sech}^2 \xi$ and integrate. It turns out that at this stage, the solvability condition is trivially satisfied since the kernel function $\operatorname{sech}^2 \xi$ is even and b_0 is odd. The solutions to Eqs. (8.485) and (8.486) can be obtained analytically; they read

$$a_1 = -\frac{\Gamma_1}{2} M \left(\xi \operatorname{sech}^2 \xi \,\tanh \xi - \operatorname{sech}^2 \xi \right), \quad b_1 = 0\,. \tag{8.487}$$

In the next order (K^2), we obtain

$$\Gamma_1 a_1 = \partial_\xi \hat{L}(K = 0) b_2 - \frac{1}{4} \partial_\xi b_0\,, \tag{8.488}$$

$$0 = \partial_\xi \hat{L}(K = 0) a_2\,. \tag{8.489}$$

This set of inhomogeneous differential equations for a_2 and b_2 now has some nontrivial solvability condition. The inhomogeneous term should be perpendicular to the kernel of $[\partial_\xi \hat{L}(K=0)]^+$. This leads to

$$-\frac{\Gamma_1^2}{2}\int_{-\infty}^{+\infty} d\xi \left(\xi \operatorname{sech}^4\xi \tanh\xi - \operatorname{sech}^4\xi\right)$$

$$= -\frac{1}{4}\int_{-\infty}^{+\infty} d\xi \left[\operatorname{sech}^2\xi \frac{d}{d\xi}\left(\operatorname{sech}^2\xi \tanh\xi\right)\right]. \tag{8.490}$$

It gives $\Gamma_1^2 = 4/15$. The curve $\Gamma^2 \approx 4/15\,K^2$ is shown in the left part of Figure 8.16.

Let us now turn to the region near the cut-off K_c. We again use the formulation given in Eqs. (8.479) and (8.480). However, close to the cut-off, we now expand in the form

$$a = a_0 + \varepsilon a_1 + \varepsilon^2 a_2 + \cdots , \tag{8.491}$$

$$b = b_0 + \varepsilon b_1 + \varepsilon^2 b_2 + \cdots , \tag{8.492}$$

$$\Gamma = \Gamma_0 + \varepsilon \Gamma_1 + \varepsilon^2 \Gamma_2 + \cdots \tag{8.493}$$

Here, we define $\varepsilon = K - K_c$. In lowest order in ε, we obtain (note that we take the first order in ε; starting with the zeroth order would not change the conclusions)

$$0 = \partial_\xi \hat{L}(K = K_c) b_1 , \tag{8.494}$$

$$0 = \partial_\xi \hat{L}(K = K_c) a_1 , \tag{8.495}$$

where

$$\hat{L}(K = K_c) = -\frac{1}{4}\frac{\partial^2}{\partial \xi^2} - 3\operatorname{sech}^2\xi + \frac{9}{4} .$$

We have already replaced $\partial_Y \to iK$, and therefore only the solutions

$$a_1 = \operatorname{sech}^3(\xi)\left(N e^{i\sqrt{5}Y} + c.c.\right), \quad b_1 = 0 \tag{8.496}$$

appear. We ignore the solution $b_1 = \operatorname{sech}^2(\xi)\tanh(\xi)$ which does not have any Y-dependence. A nonlinear calculation (see Appendices F and G) will justify the simplified procedure which we accept for the following.

At this stage, it is advisable to introduce the notation

$$a \equiv \partial_\xi \varphi , \tag{8.497}$$

$$b \equiv \partial_\xi \psi . \tag{8.498}$$

Obviously we have the nonlocalized representations

$$\varphi_1 = \left(N e^{i\sqrt{5}Y} + c.c.\right)\int_{-\infty}^{\xi} d\xi' \operatorname{sech}^3(\xi') , \tag{8.499}$$

$$\psi_1 = 0 . \tag{8.500}$$

In second order in ε, we write

$$\Gamma_1(\varphi_1 + \psi_1) = \hat{L}(K = K_c)\partial_\xi[\varphi_2 + \psi_2] - \frac{1}{2}K_c\partial_\xi[\varphi_1 + \psi_1]. \tag{8.501}$$

The relevant kernel function of $\hat{L}(K = K_c)$ is $\text{sech}^3(\xi)[Ne^{i\sqrt{5}Y} + c.c.]$; note that we also integrate with respect to Y. Making use of the formulas

$$\int_{-\infty}^{+\infty} d\xi \,\text{sech}^3(\xi) = \frac{\pi}{2}, \tag{8.502}$$

$$\int_{-\infty}^{+\infty} d\xi \,\text{sech}^6(\xi) = \frac{16}{15}, \tag{8.503}$$

we finally obtain

$$\Gamma_1 = -\frac{64\sqrt{5}}{15\pi^2}. \tag{8.504}$$

Therefore, we have derived in lowest order the result

$$\Gamma \approx \frac{64}{15\pi^2}K_c(K_c - K). \tag{8.505}$$

This curve is shown in Figure 8.16. We should mention that the present perturbative treatment is rather incomplete. We have not presented the most general case. For example, even up to the second order in ε, we did not make the most general ansatz. In addition, we did not care about localization of φ and ψ, which can be accomplished by a multiple scale analysis. Most important, however, is the convergence problem which is not treated explicitly in most applications. For a very detailed investigation of convergence, we refer the reader to the previously mentioned book by Infeld and Rowlands [224].

8.6.2 Transverse Instability of Envelope Solitons (NLS)

The simplest generalization of the integrable 1D cubic nonlinear Schrödinger (NLS) equation to two and three space dimensions is

$$i\partial_t\psi + \nabla^2\psi + |\psi|^2\psi = 0. \tag{8.506}$$

Here, the nabla operator is, for example, in Cartesian coordinates, $\nabla^2 = \partial_x^2 + \partial_y^2 + \partial_z^2$. For our present purpose, we write it in the form $\nabla^2 = \partial_x^2 + \nabla_\perp^2$, and for simplicity, we assume here $\nabla_\perp^2 = \partial_y^2$. We should keep in mind that the present form of a multidimensional NLS equation should only be understood as the starting point of a model calculation. Actual envelope soliton models, arising in different areas of physics, are usually much more complicated.

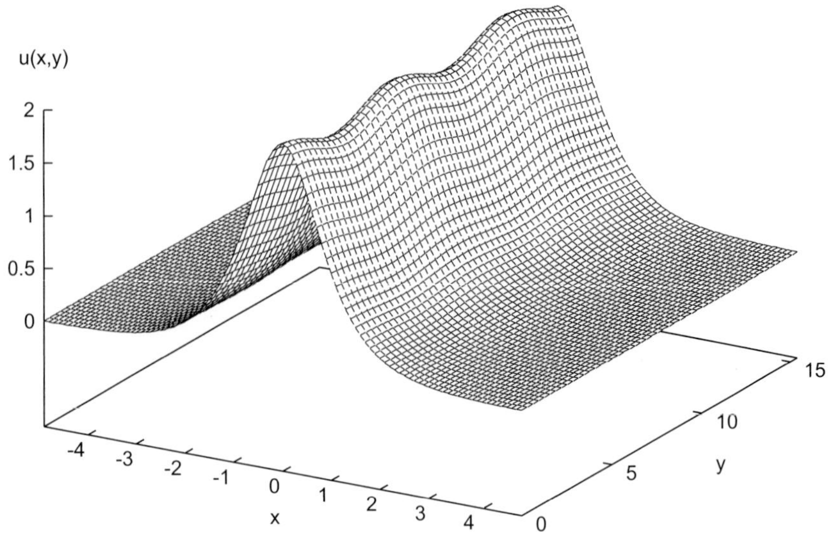

Figure 8.17 Transverse modulation of a one-dimensional plane soliton.

As we have shown in previous sections, the one-dimensional NLS equation ($\nabla_\perp^2 \equiv 0$) is well-understood. Here, we study the transverse ($\partial_y \neq 0$) instability of a one-dimensional standing soliton solution

$$\psi_s \equiv G e^{i\eta_0^2 t} = \sqrt{2}\eta_0 \operatorname{sech}(\eta_0 x) e^{i\eta_0^2 t} \; ; \tag{8.507}$$

η_0 is a free parameter. In Figure 8.17, we visualize what is meant by a transverse instability.

Before applying variational principles, let us again start with a perturbative treatment [247] which is quite illustrative. Introducing perturbations in the form

$$\psi = (G + a + ib)\, e^{i\eta_0^2 t} \,, \tag{8.508}$$

we assume a and b to be real. They will depend on the transverse coordinate y (besides the obvious x- and t-dependencies). For small perturbations, we linearize the system and consider modes

$$a, b \sim e^{iky} \,. \tag{8.509}$$

For a linearized system with only x-dependent coefficients, the Fourier-mode description is appropriate. Some straightforward manipulations lead to

$$\partial_t a \equiv \dot{a} = H_+^{(k)} b \,, \tag{8.510}$$

$$\partial_t b \equiv \dot{b} = -H_-^{(k)} a \,, \tag{8.511}$$

where we have introduced the operators

$$H_+^{(k)} = -\partial_x^2 - G^2 + \eta_0^2 + k^2 \equiv H_+ + k^2 \,, \tag{8.512}$$

$$H_-^{(k)} = -\partial_x^2 - 3G^2 + \eta_0^2 + k^2 \equiv H_- + k^2 \,. \tag{8.513}$$

For the following discussion, the spectral properties of the operators H_+ and H_- at $k = 0$ are essential. H_+ is positive semidefinite with

$$H_+ G = 0 \,. \tag{8.514}$$

The operator H_- has one negative eigenvalue,

$$H_- \mathrm{sech}^2(\eta_0 x) = -3\eta_0^2 \mathrm{sech}^2(\eta_0 x) \,, \tag{8.515}$$

and the translation mode dG/dx as kernel function,

$$H_- \frac{dG}{dx} = 0 \,. \tag{8.516}$$

The latter can, for example, be verified by calculating

$$\frac{d}{dx} H_+ G = 0 \,. \tag{8.517}$$

Equation (8.515) can be verified easily, and thereby – together with the node theorem – the definiteness properties of H_+ and H_- become obvious.

At this stage, some more relations can be straightforwardly derived which will be of use at some later stages. For example, again starting from $H_+ G = 0$, differentiation with respect to η_0^2 leads to

$$H_- \frac{\partial G}{\partial \eta_0^2} = -G \,. \tag{8.518}$$

In similar ways, we deduce

$$H_+(-xG) = 2\frac{dG}{dx} \,, \quad H_- G = -2G^3 \,, \tag{8.519}$$

$$H_- \left(x\frac{dG}{dx} \right) = -2\frac{d^2 G}{dx^2} \,, \quad H_- \left(G + x\frac{dG}{dx} \right) = -2\eta_0^2 G \,. \tag{8.520}$$

The latter relation immediately leads to

$$H_-^{-1} G = -\frac{1}{2\eta_0^2} \left(G + x\frac{dG}{dx} \right) . \tag{8.521}$$

Let us now come back to the system Eqs. (8.510) and (8.511), which we rewrite as

$$\partial_t^2 a = -H_+^{(k)} H_-^{(k)} a \tag{8.522}$$

or

$$\partial_t^2 b = -H_-^{(k)} H_+^{(k)} b \,. \tag{8.523}$$

We now investigate the time-dependencies of the perturbations by choosing

$$a, b \sim e^{\gamma t} \tag{8.524}$$

and performing a small-k expansion within a perturbative treatment:

$$\gamma^2 = \gamma_{(2)}^2 k^2 + \gamma_{(4)}^2 k^4 + \cdots \,, \tag{8.525}$$

$$a = a_{(0)} + a_{(2)} k^2 + \cdots \,, \tag{8.526}$$

$$b = b_{(0)} + b_{(2)} k^2 + \cdots \,. \tag{8.527}$$

Treating k as a small parameter actually means $k^2 \ll \eta_0^2$.

In zeroth order, we obtain

$$H_+ H_- a_{(0)} = 0 \tag{8.528}$$

or

$$H_- H_+ b_{(0)} = 0 \,. \tag{8.529}$$

These equations can be solved by the following (nontrivial) functions

$$a_{(0)} = C_1 \left(-\frac{\partial G}{\partial \eta_0^2} \right) + C_2 \frac{dG}{dx} \equiv C_1 a_{(0)}^{even} + C_2 a_{(0)}^{odd} \,, \tag{8.530}$$

$$b_{(0)} = D_1 G + D_2(-xG) \equiv D_1 b_{(0)}^{even} + D_2 b_{(0)}^{odd} \,. \tag{8.531}$$

Simple inspection shows that we can consider either even (in x) or odd (in x) functions, and it suffices to investigate only one of the two Eqs. (8.522) and (8.523). Let us continue with the next order perturbation terms,

$$\gamma_{(2)}^2 a_{(0)}^{even} = -(H_+ + H_-)\, a_{(0)}^{even} - H_+ H_- a_{(2)}^{even} \,. \tag{8.532}$$

This equation for $a_{(0)}^{even}$ has a nontrivial solvability condition following

$$H_- H_+ b_{(0)}^{even} = 0 \,. \tag{8.533}$$

Multiplying from the left with $b_{(0)}^{even}$ and integrating over x leads to

$$\gamma_{(2)}^2 \left\langle b_{(0)}^{even} | a_{(0)}^{even} \right\rangle = -\left\langle b_{(0)}^{even} | H_- | a_{(0)}^{even} \right\rangle \,. \tag{8.534}$$

This condition can be rewritten as

$$\gamma_{(2)}^2 = \frac{\langle G|G \rangle}{\frac{1}{2} \frac{d}{d\eta_0^2} \langle G|G \rangle} = 4\eta_0^2 \,. \tag{8.535}$$

Thus, the even mode leads to an instability in the long-wavelength limit with a growth rate

$$\gamma \approx 2\eta_0 k \,. \tag{8.536}$$

We must now repeat the calculation for the odd mode. Instead of Eq. (8.534), we get

$$\gamma_{(2)}^2 \left\langle b_{(0)}^{odd} | a_{(0)}^{odd} \right\rangle = -\left\langle b_{(0)}^{odd} | H_+ | a_{(0)}^{odd} \right\rangle \,, \tag{8.537}$$

which can be rewritten as

$$\gamma_{(2)}^2 = -\frac{2\left\langle \frac{dG}{dx} | \frac{dG}{dx} \right\rangle}{\langle G|G\rangle} \,. \tag{8.538}$$

Clearly, this does not lead to an unstable behavior, at least for the model equation we started with.

Next, we discuss the instability predictions by variational principles. Using the standard form as discussed in the Appendix I, that is,

$$N\partial_t^2 a = Fa \,, \tag{8.539}$$

we may apply that with the interpretations

$$N \hat{=} \left[H_+^{(k)}\right]^{-1} \,, \quad F = -H_-^{(k)} \,, \tag{8.540}$$

provided $\langle a|G\rangle = 0$ (because of $H_+ G = 0$). Then,

$$\gamma_k^2 = \sup_{a_k} \frac{-\langle a_k|\, H_-^{(k)} \,|a_k\rangle}{-\langle a_k| \left[H_+^{(k)}\right]^{-1} |a_k\rangle} \,, \tag{8.541}$$

$$\gamma_k^2 = \inf_{a_k} \frac{\langle a_k|\, H_-^{(k)} [H_+^{(k)}]^{-1} H_-^{(k)} \,|a_k\rangle}{-\langle a_k|\, H_-^{(k)} \,|a_k\rangle} \,, \tag{8.542}$$

with $\langle a_k|H_-^{(k)}|a_k\rangle < 0$ follows.

These relations, together with the spectral properties of H_+ and H_-, prove instability in the region

$$0 < k^2 < 3\eta_0^2 \equiv k_c^2 \,. \tag{8.543}$$

Considering the supremum principle, we evaluate it in the limit $k^2 \to 0$ in the following way. First, we note

$$\left[H_+^{(k)}\right]^{-1} a_k = \frac{1}{k^2} \frac{\langle a_k|G\rangle}{\langle G|G\rangle} G + g_\perp \,. \tag{8.544}$$

Here, by $g_\perp$, we mean a function being orthogonal to G and of zeroth order in k. For the proof, we multiply from the left with $H_+^{(k)}$:

$$a_k = \frac{\langle a_k | G \rangle}{\langle G | G \rangle} G + a_{k\perp} \,, \tag{8.545}$$

$$a_{k\perp} = H_+^{(k)} g_\perp \,. \tag{8.546}$$

Thus, we obtain

$$\gamma_k^2 \approx k^2 \sup_{\langle \varphi | G \rangle \neq 0} \frac{-\langle \varphi | H_- | \varphi \rangle}{\langle \varphi | G \rangle^2} \langle G | G \rangle \quad \text{for} \quad k \ll \eta_0 \,. \tag{8.547}$$

To evaluate the latter supremum, let

$$\varphi = H_-^{-1} G + \tilde{\varphi} \,. \tag{8.548}$$

We are free to normalize the test functions according to

$$\langle \varphi | G \rangle^2 = \langle G | H_-^{-1} | G \rangle^2 \,. \tag{8.549}$$

Then,

$$\sup_{\langle \varphi | G \rangle \neq 0} \frac{-\langle \varphi | H_- | \varphi \rangle}{\langle \varphi | G \rangle^2} = \sup_{\langle \tilde{\varphi} | G \rangle = 0} \frac{-\langle G | H_-^{-1} | G \rangle - \langle \tilde{\varphi} | H_- | \tilde{\varphi} \rangle}{\langle G | H_-^{-1} | G \rangle^2} \,. \tag{8.550}$$

Note that

$$\langle \tilde{\varphi} | H_- | \tilde{\varphi} \rangle \geq 0 \quad \text{for} \quad \langle \tilde{\varphi} | G \rangle = 0 \tag{8.551}$$

(the only negative eigenvalue of H_- corresponds to an even eigenfunction in x), and

$$\tilde{\varphi} = \frac{dG}{dx} \,, \qquad H_- \tilde{\varphi} = 0 \tag{8.552}$$

will determine the maximum value,

$$\gamma_k^2 \approx k^2 \frac{\langle G | G \rangle}{-\langle G | H_-^{-1} | G \rangle} \,. \tag{8.553}$$

With the result (8.521), we find

$$\gamma_k^2 \approx 4 k^2 \eta_0^2 , \tag{8.554}$$

which agrees with the prediction of the perturbation theory.

In a similar way, we can evaluate the infimum principle Eq. (8.542), and without further details, we cite that again Eq. (8.554) turns out to be the correct prediction. Since for small k, upper and lower bounds are identical, and Eq. (8.554) is the correct small-k dependence of the growth rate for transverse instability.

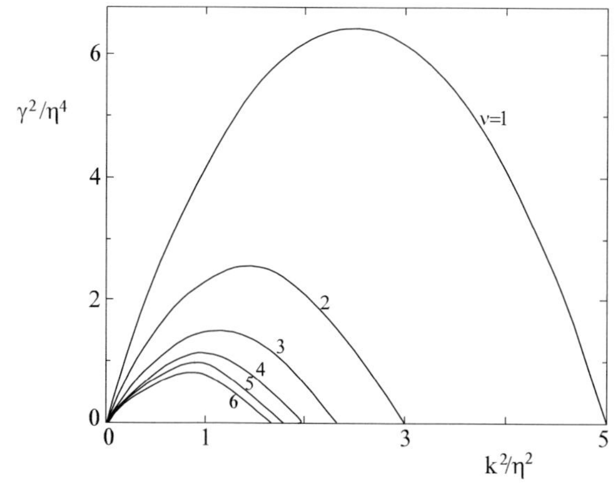

Figure 8.18 Growth rate curves for various parameters ν versus transverse wavenumber k for plane soliton solutions.

Similarly to the KdV discussion of analytical methods for determining the growth rates of transverse instabilities, an expansion in the neighborhood of $k_c^2 = 3\eta_0^2$ is also possible. We do not present these details here. Figure 8.18 presents the correct growth rate curve for the NLS soliton with $\nu = 2$ in dependence on the transverse wavenumber k.

Finally, let us proceed by presenting an approximate method which is widely used in this area. The (approximate) variation of action method (VAM) has been developed by Makhankov and collaborators [268]. For reasons which will become clear later, we elucidate the procedure for the generalized NLS equation

$$i\partial_t \psi + \nabla^2 \psi + |\psi|^\nu \psi = 0 , \tag{8.555}$$

where the exponent $\nu \geq 1$. The results for $\nu = 2$ can now be generalized by first introducing the stationary solution

$$G_s = \left[\frac{2(\nu+2)}{\nu^2}\right]^{\frac{1}{\nu}} \eta_s^{\frac{2}{\nu}} \left[\mathrm{sech}(\eta_s x)\right]^{\frac{2}{\nu}} , \quad \psi_s = G_s e^{i4\eta_s^2 \frac{t}{\nu^2}} . \tag{8.556}$$

The variation of action method starts from the Lagrangian density

$$\mathcal{L} = -\frac{i}{2}\left(\psi \partial_t \psi^* - \psi^* \partial_t \psi\right) + \nabla\psi \cdot \nabla\psi^* + \frac{2}{\nu+2}|\psi|^{\nu+2} \tag{8.557}$$

and the action

$$S = \int dy\, dt\, \mathcal{S} = \int dx\, dy\, dt\, \mathcal{L} . \tag{8.558}$$

Next, we make an ansatz for the time-dependent solution,

$$\psi = A\,\mathrm{sech}^{\frac{2}{\nu}}(Dx) e^{-i\phi} , \tag{8.559}$$

with unknown coefficients $A = A(y,t)$, $D = D(y,t)$, and $\phi = \phi(y,t)$. Inserting that ansatz into Lagrangian density $\mathcal{L}$ and subsequent integration over x yields

$$\mathcal{S} = \frac{A^2\partial_t\phi}{D} - \frac{4I(\nu)}{\nu^2 B\left(\frac{2}{\nu},\frac{1}{2}\right)}\frac{A^2(\partial_y D)^2}{D^3} + \frac{A(\partial_y A)(\partial_y D)}{D^2} - \left[\frac{(\partial_y A)^2}{A^2} + (\partial_y\phi)^2\right]\frac{A^2}{D} - \frac{4DA^2}{\nu(\nu+4)} + \frac{8A^{\nu+2}}{(\nu+2)(\nu+4)D}\,, \tag{8.560}$$

where $B(2/\nu, 1/2)$ is the Euler beta function and

$$I(\nu) = \int_{-\infty}^{+\infty} \frac{\xi^2\tanh^2\xi}{\cosh^{\frac{4}{\nu}}\xi}d\xi\,. \tag{8.561}$$

We consider $\mathcal{S} = \mathcal{S}(A, \partial_y A, \phi, \partial_y\phi, \partial_t\phi, D, \partial_y D)$ and require $\delta\mathcal{S} = 0$, which leads to Euler equations, for example,

$$\frac{\partial\mathcal{S}}{\partial\phi} - \frac{\partial}{\partial t}\frac{\partial\mathcal{S}}{\partial(\partial_t\phi)} - \frac{\partial}{\partial y}\frac{\partial\mathcal{S}}{\partial(\partial_y\phi)} = 0\,, \tag{8.562}$$

and similar ones for A and D.

Linearizing the Euler equations near the soliton solution

$$A_0 = \left[\frac{2(\nu+2)}{\nu^2}\right]^{\frac{1}{\nu}}\eta_s^{\frac{2}{\nu}}\quad D_0 = \eta_s\,,\quad \phi_0 = 4\frac{\eta_s^2}{\nu^2} \tag{8.563}$$

in the form

$$A = A_0 + \delta A e^{-i\omega t+iky}\,,\quad D = D_0 + \delta D e^{-i\omega t+iky}\,,$$
$$\phi = \phi_0 + \delta\phi e^{-i\omega t+iky}\,, \tag{8.564}$$

one gets a dispersion relation $\omega^2 = \omega^2(k^2)$ describing the transverse stability ($\omega^2 > 0$) or instability ($\omega^2 < 0$) of the soliton solution. For example, for $k \to 0$ and $\nu \neq 4$, we obtain

$$\omega^2 \approx \frac{8\nu A_0^\nu}{(\nu+2)(\nu-4)}k^2\,, \tag{8.565}$$

implying that for $\nu < 4$, the solitons are unstable with respect to perturbations in the transverse direction. For $\nu = 2$, the result agrees with the previously derived one.

For $\nu = 4$,

$$\omega^2 \approx -\frac{64}{45}A_0^8 - 2A_0^4k^2 + k^4\,, \tag{8.566}$$

meaning that solitons for $\nu = 4$ are exponentially unstable in both the longitudinal and transverse direction. However, the more exact calculations by variational

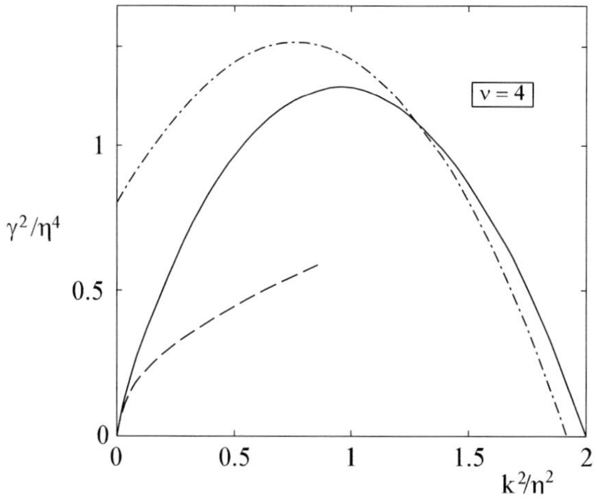

Figure 8.19 Growth rate curve for $\nu = 4$ in the case of a plane soliton solution. The line ––––– depicts the small-k expansion whereas the curve – · – · – shows the approximate result by the VAM.

principles show that the growth rates are only approximately valid and the stability results by the VAM are erroneous for $\nu \geq 4$. Figure 8.19 shows the k-dependencies for $\nu = 4$ obtained by the different methods.

In general, stability cannot be concluded by the VAM since it restricts the perturbed states to a certain subclass. On the other hand, for instability, the variational principle is for the action and not for the growth rate. Therefore, trial functions approximating the action do not necessarily imply a proper dispersion relation. In other words, it is by no means clear that those trial functions are approximate solutions of the dynamic equations when the whole x-dependence is retained.

Thus far, we have only analyzed the transverse stability behavior of plane solitons. Radial solutions in higher dimensions are also of interest. Their existence has been shown by Strauss [260]. Let us make the following remarks. For cylindrical solitons, transverse perturbations can be characterized by two different modes: the azimuthal ϕ-dependent mode and the z-mode. Instability always occurs with respect to the latter one (by similar arguments, as presented for the plane case), whereas the ϕ-mode does not introduce any new instability (note the discrete mode numbers m). The 3D system is longitudinally stable for $0 < \nu < 4/3$. A transverse instability (for φ- and θ-dependent perturbations) will not occur [269].

8.7
The Collapse Phenomenon and the Existence of Stable 3D Solitons

It turns out that one-dimensional Langmuir envelope solitons are always multidimensionally (transversely) unstable. The unstable behavior of solitary solutions of the cubic nonlinear Schrödinger in higher space-dimensions is a generic phenomenon, as will be shown in the following subsection on the collapse phenomenon. On the other hand, the collapse is restricted to the algebraic form of the nonlinearity. Physically, nonlinearities may saturate. This will allow one to predict stable three-dimensional solitary envelope solutions.

8.7.1
The Collapse Phenomenon

Let us demonstrate the multidimensional instability (collapse) on a very simple version of a scalar nonlinear Schrödinger (NLS)-equation, namely,

$$i\partial_t \psi + \nabla^2 \psi + |\psi|^p \psi = 0 . \tag{8.567}$$

Here, ∇^2 denotes the d–dimensional ($d \geq 1$) Laplace operator, and the exponent p should be positive, $p > 0$. It has been discussed [270, 271] under what conditions the complex-valued wave-function ψ can develop a singularity within a finite time (depending on values of the parameters d and p). This singular behavior is commonly called "collapse" or "blow up" and has important physical applications, for example, in plasma physics the energy transfer to particles via the "Langmuir-collapse" and in nonlinear optics, the "self-focusing" of light pulses in a nonlinear, dispersive medium.

More detailed discussions of the collapse phenomenon can be found in the literature, for example, [270, 271] and references therein. Here, we only present a qualitative argument, making use of the conserved integrals

$$H = \int d\Gamma \left(|\nabla \psi|^2 - \frac{2}{p+2} |\psi|^{p+2} \right) , \tag{8.568}$$

$$P = \int d\Gamma \, \psi \nabla \psi^* , \tag{8.569}$$

$$N = \int d\Gamma |\psi|^2 , \tag{8.570}$$

where $d\Gamma = d^d r$, depending on dimension d.

Considering Eq. (8.567), it is evident that two competing terms are essential. One is the dispersion term ($\nabla^2 \psi$) which leads to a spreading of the wave packet, the other is the nonlinearity ($|\psi|^p \psi$) which causes a wave-steepening. Heuristically, one can study the balance between dispersion and nonlinearity by a simple scaling argument. Introducing the scaled variable $\tilde{x} \equiv x/L$ with a characteristic length L, the conservation law (8.570) gives $|\psi|^2 \sim L^{-d}$. Then, it follows that $|\psi|^p \sim L^{-pd/2}$. On the other hand, we have $\nabla^2 \sim L^{-2}$. This clearly shows that for $pd < 4$ (the so-called subcritical case), dispersion dominates, while for $pd > 4$ (the supercritical case), the "focusing" nonlinearity dominates. A balance between both effects is achieved in the critical case $pd = 4$. If we apply this to the cubic nonlinear Schrödinger equation ($p = 2$) in three space dimensions ($d = 3$), we find that all localized solutions will collapse. It was shown by Zakharov [270] that in the critical and supercritical regime $p \geq 4/d$, a focusing singularity may develop in finite time.

8.7.2 Stable Three-Dimensional Envelope Solitons

The nonlinear Langmuir wave models considered thus far were mostly investigated in the weak-amplitude limit. In algebraic nonlinearity, for example, the cubic nonlinearity, appears. When a collapse sets in, a tendency to a singularity with locally large amplitudes is observed. Then, of course, the small-amplitude limit does not apply anymore. Nonlinearities have a tendency to saturate. With saturating nonlinearities, one might expect a different scenario which, for example, could be investigated with the finite amplitude model. In order not to complicate the algebra, we ease the model. In order to mimic the 3D nature of the phenomena, we write the electric field as a vector and approximate the saturating nonlinearity by an exponential ansatz.

We base our investigation on the exponential nonlinear Schrödinger equation [272]

$$i\partial_t \boldsymbol{E} + \nabla\nabla\cdot\boldsymbol{E} - q\nabla\times\nabla\times\boldsymbol{E} + \left[1 - \exp(-\boldsymbol{E}\cdot\boldsymbol{E}^*)\right]\boldsymbol{E}\,, \tag{8.571}$$

where $\boldsymbol{E}$ is the electric field envelope vector and $q = c^2/(3v_{the}^2) - 1$ is a large parameter, at least for nonrelativistic situations. Explicit forms of 3D solutions have been first given by Kaw *et al.* [273]. Solutions of similar scalar field equations have also been discussed [274]. With the ansatz

$$\boldsymbol{E}_s = G(r)\exp(i\eta^2 t)\hat{r}\,, \tag{8.572}$$

the function G has to satisfy the radial equation

$$H_+^r G(r) = 0\,, \quad H_+^r = -\frac{\partial}{\partial r}r^{-2}\frac{\partial}{\partial r}r^2 + \eta^2 - 1 + \exp(-G^2)\,. \tag{8.573}$$

The natural boundary conditions for G are $G(r)/r \to$ const for $r \to 0$ and $G(r) \to 0$ for $r \to \infty$. G is called the ground state when it has no nodes for $r \in (0,\infty)$. The operator H_+^r is positive semidefinite since it can be written in the form

$$H_+^r g \equiv -\frac{1}{r^2 G}\frac{\partial}{\partial r}\left[r^2 G^2 \frac{\partial}{\partial r}\frac{g}{G}\right]\,, \tag{8.574}$$

which holds for any function $g(r)$ in $(0,\infty)$.

The stability of $G(r)$ will now be investigated with the result that (i) for large amplitudes (corresponding to large η^2) three-dimensional solitons are completely stable, (ii) a stability threshold in η^2 exists, and (iii) below the threshold instability occurs.

The following functional will act as a Liapunov functional for stability

$$L = \int d^3r \left\{ \sum_j \frac{\partial \boldsymbol{E}}{\partial x_j}\cdot\frac{\partial \boldsymbol{E}^*}{\partial x_j} + q(\nabla\times\boldsymbol{E})\cdot(\nabla\times\boldsymbol{E}^*) - \left[\exp\left(-\boldsymbol{E}\cdot\boldsymbol{E}^*\right) - 1\right] \right.$$
$$\left. - \left(1-\eta^2\right)\boldsymbol{E}\cdot\boldsymbol{E}^* \right\} - L_0\,, \tag{8.575}$$

where L_0 designates the value of the integral for the stationary solution under consideration. This choice of L_0 guarantees that L is zero at the stationary point. Furthermore, $\partial_t L = 0$ since L is constructed from constants of motion. Defining real and imaginary parts of the perturbations as

$$E = E_s + (a + ib)\exp\left(i\eta^2 t\right), \tag{8.576}$$

we may rewrite L as $L = L_2 + R$, where L_2 contains the second-order terms. R is the rest which can be estimated in terms of the second-order contributions. By means of Sobolev inequalities [257], one can show that the higher-order terms R are much smaller than L_2. We have

$$L_2 = \langle \boldsymbol{a}|H_+|\boldsymbol{a}\rangle + \langle \boldsymbol{b}|H_-|\boldsymbol{b}\rangle\,, \tag{8.577}$$

with

$$\begin{aligned} H_+ &= -\nabla^2 + q\nabla\times\nabla\times + \eta^2 - \left[1 - \exp(-G^2)\right], \\ H_- &= H_+ - 2G^2\exp(-G^2)\hat{r}\hat{r}\,. \end{aligned} \tag{8.578}$$

Furthermore, $\langle\ldots\rangle \equiv \int d^3r\ldots$, and the only r-dependent parts of H_+ and H_- will be denoted by H_+^r and H_-^r, respectively.

Similar to the calculations in previous sections, it is not difficult to show

$$\langle \boldsymbol{a}|H_+|\boldsymbol{a}\rangle \geq \quad \text{for} \quad \langle \boldsymbol{a}|G\hat{r}\rangle = 0\,. \tag{8.579}$$

The constraint results from stability with respect to form.

We must also discuss the sign of $\langle \boldsymbol{b}|H_-|\boldsymbol{b}\rangle$. We decompose $\boldsymbol{b}$ into electrostatic and electromagnetic components,

$$\boldsymbol{b} = \boldsymbol{b}_{es} + \boldsymbol{b}_{em} = -\nabla\chi - \frac{\partial \boldsymbol{A}}{\partial t} \quad \text{with} \quad \nabla\cdot\boldsymbol{A} = 0\,, \tag{8.580}$$

and only assume r-dependent perturbations. Because of $\boldsymbol{b}_{em}\cdot\hat{r} = 0$, we find

$$\langle \boldsymbol{b}|H_-|\boldsymbol{b}\rangle = \langle \boldsymbol{b}_{em}|H_+|\boldsymbol{b}_{em} + \langle \boldsymbol{b}_{es}|H_-|\boldsymbol{b}_{es}\rangle\,. \tag{8.581}$$

Thus, for stability, only the sign of the term containing the electrostatic components es must be examined. For perturbations only depending on the radial coordinate r, we have

$$\langle \boldsymbol{b}_{es}|H_-|\boldsymbol{b}_{es}\rangle = \left\langle \frac{\partial\chi}{\partial r}\right| H_-^r \left|\frac{\partial\chi}{\partial r}\right\rangle\,. \tag{8.582}$$

By arguments similar to the ones presented for one-dimensional stability, one finds

$$\left\langle \frac{\partial\chi}{\partial r}\right| H_-^r \left|\frac{\partial\chi}{\partial r}\right\rangle > 0 \quad \text{for} \quad \langle G\left|(H_-^r)^{-1}\right|G\rangle < 0\,. \tag{8.583}$$

Since

$$(H_-^r)^{-1}G = -\frac{\partial G}{\partial\eta^2}\,, \tag{8.584}$$

we get the stability criterion

$$\frac{\partial N}{\partial \eta^2} > 0 \quad \text{where} \quad N = \int_0^\infty dr r^2 G^2 \,. \tag{8.585}$$

This criterion is also necessary for longitudinal stability (note that thus far we only considered r-dependent perturbations for radially symmetric ground states). When a variational principle (see Appendix I) is formulated [258, 259, 262, 275, 276] for the instability growth rate, the test function $\langle G|G\rangle(H_-^r)^{-1}G - \langle G|(H_-^r)^{-1}G\rangle G$, which satisfies $\langle a|G\rangle = 0$, demonstrates instability in the region $\langle G|(H_-^r)^{-1}|G\rangle > 0$.

For arbitrary perturbations, for example, for the electrostatic part with an expansion in terms of spherical harmonics in the form $\chi = \sum_{lm} \chi_{lm}(r) Y_{lm}(\theta, \varphi)$, it was shown in [272] that the stability criterion Eq. (8.585) remains valid. Thus, it is sufficient to evaluate Eq. (8.585) for the existence of stable three-dimensional solitons. One can analytically show that the derivative of N with respect to η^2 changes sign when η^2 is increased from small values (where $\partial N/\partial \eta^2 < 0$) to larger values. For example, when making use of integral relations of the virial type, one finds

$$\frac{\partial N}{\partial \eta^2} = -\frac{1}{\eta^2} \frac{\partial}{\partial \eta^2} \int_0^\infty dr \left[2G^2 - 3\eta^2 G^2 + 2\exp(-G^2) - 2\right]. \tag{8.586}$$

Obviously, for $\eta^2 > 2/3$, the integral on the right-hand side is negative. On the other hand, for small η^2, the ground state is a solution of the form $G(r) \sim \eta f(\eta r)$, and the integral is positive and proportional to η. Thus, for $0 < \eta^2 < 2/3$, a transition from an unstable to the stable behavior occurs. Numerically, the critical value

$$\eta_c^2 \approx 0.10125: \quad \frac{\partial N}{\partial \eta^2} > 0 \quad \text{for} \quad \eta > \eta_c \tag{8.587}$$

has been found [272]. The existence of such a threshold in amplitude (energy) can be interpreted similarly to particle physics [277]. Below η_c^2, a lower energy state than the soliton exists such that a decay of the soliton can occur.

Finally, we emphasize that the collapse argument [270] does not apply for $\eta > \eta_c$. It can be shown that the collapse criterion requires values for the invariants which do not contradict the existence of stable three-dimensional solitons. In addition, it is interesting to note that already for the scalar cubic nonlinear Schrödinger equation the invariant

$$I_3 = \int d^3 r \left[(\nabla E)^2 - \frac{1}{2} E^4\right] \tag{8.588}$$

is not negative for a spherical soliton solution.

9
Nonlinear Wave Aspects in Laser–Matter Interaction

9.1
History and Perspectives of Laser–Plasma Interaction

High-intensity laser interaction with matter is a broad field which has experienced an explosive scientific development during the last three decades. Wavelength, energy, pulse duration, and focal spot size are the four "magic numbers" [278] characterizing laser development with respect to nonlinear laser–plasma physics. High-intensity, long pulse laser–plasma interaction is mainly relevant to inertial confinement fusion. On the other hand, since the invention of the chirped pulse amplification (CPA) technique [279], short pulse laser-matter interaction has led to an entirely new area of research, namely, what is called now relativistic optics [280]. Femtosecond ($1\,\text{fs} = 10^{-15}\,\text{s}$) and even attosecond ($1\,\text{as} = 10^{-18}\,\text{s}$) time-scales start to become accessible at relatively high powers. Figure 9.1 gives an overview of the highest intensity laser pulses currently available, and the corresponding new areas of physics which could be investigated.

A new area of laser–plasma interaction started approximately 40 years ago when laser fields became strong enough to directly ionize matter. With lasers operating at wavelengths 0.25–13.4 μm, one cannot expect the photoelectric effect to be effective on normal material since the photon energy $\hbar\omega$ is much smaller than the atomic potential barrier which the electron experiences in the vicinity of the nucleus (e.g., 13.6 eV for the ground state in hydrogen). With more powerful laser systems in the 1960s and 1970s, multiphoton ionization became available. The rich physics encountered in the weakly nonlinear low-intensity regime is, for example, summarized in [20]. Before the CPA-area, the possibility of amplifying pulses directly was limited since intensities of some GW/cm^2 would damage the gain medium. Stretching the pulse in a controlled (reversible) manner by adding a chirp (temporal variation of the frequency) is the first stage in CPA. The chirped pulse is then amplified. For that, usually one or more conventional laser amplifier stages can be used to increase the pulse energy by a factor 10^7–10^9. Finally, a compressor performs the exact optical inverse of the stretcher to deliver an amplified pulse of the same duration as the oscillator at the beginning. Optical parametric amplification (OPCPA) in a nonlinear crystal promises even higher intensities than conventional CPA.

High Temperature Plasmas, Theory and Mathematical Tools for Laser and Fusion Plasmas, First Edition.
Karl-Heinz Spatschek.

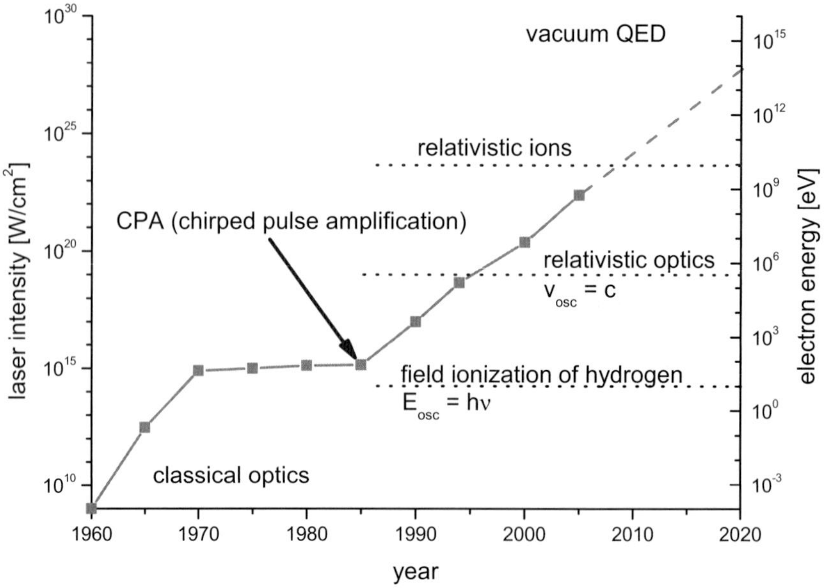

Figure 9.1 Schematic plot of the history of large intensity laser pulses, including extrapolation into the coming years with new applications in physics.

To present an example, the "magic numbers" of the Jena TW laser are [278]: wavelength 800 nm, energy 0.8 J, pulse duration 80 fs, and focal spot size 5 μm^2, leading to an approximate intensity of 10^{20} W/cm^2. In this regime, the generated laser–plasma becomes relativistic. It is convenient to characterize the laser field by the dimensionless parameter $a_0 = eA_0/m_e c$, where A_0 is the peak vector potential of the laser pulse. Since a_0 corresponds to the normalized (by $m_e c$) maximum transverse momentum of an electron in the laser field, $a_0 \geq 1$ is often referred to as the entry into the relativistic optics (see below). University facilities like the ARCTURUS at the University of Düsseldorf are able to produce $a_0 \sim 10$. A next big step will be reached at $a_0 \sim 1000$ when ions will also be accelerated to relativistic velocities. For higher intensities, it will be possible to even study electron–positron pair creation and other quantum-electrodynamic (QED) effects.

Though definitely, intensities of 10^{19} W/cm^2 are now available in teraWatt (tera $\hat{=}\, 10^{12}$) tabletop laser systems based on chirped-pulse amplification. Let us consider, for example, a joule of light compressed into a picosecond (pico $\hat{=}\, 10^{-12}$) to produce a terawatt of peak power. The laser may be focused to a few wavelengths squared, for example, 10 μm^2. That leads to an intensity of

$$I_0 \sim \frac{1\,\text{J}}{1\,\text{ps}\ 10\,\mu\text{m}^2} \sim \frac{1}{10^{-12} 10\ 10^{-8}} \sim 10^{19} \left[\frac{\text{W}}{\text{cm}^2}\right]. \tag{9.1}$$

The photon number density $n_{photon} = N/V$ follows from the photon energy

$$\hbar\omega \equiv \frac{hc}{\lambda} \approx \frac{6.6 \times 10^{-34}\,\mathrm{J\,s}\; 3 \times 10^{8}\,\mathrm{m\,s^{-1}}}{10^{-6}\,\mathrm{m}} \approx 2 \times 10^{-19}\,\mathrm{J} \quad \text{at} \quad \lambda = 1\,\mu\mathrm{m} \tag{9.2}$$

as $N = 5 \times 10^{18}$. Thus, the number density is

$$n_{photon} = \frac{5 \times 10^{18}}{10 \times 10^{-8}\,\mathrm{cm}^2\; 3 \times 10^{10}\,\mathrm{cm\,s^{-1}}\; 10^{-12}\,\mathrm{s}} \approx 1.6 \times 10^{27}\,\frac{\text{photons}}{\mathrm{cm}^3}\,. \tag{9.3}$$

9.1.1
Areas of Relativistic Optics

When investigating the motion of a single particle (electron) in a plane electromagnetic wave, we already started with the discussion of the importance of relativistic effects during laser–particle interaction. Equation (2.213) is central in this respect. Therefore, we repeat it here for convenience, that is,

$$I_0\lambda^2 = \zeta\left[1.37 \times 10^{18}\,\frac{\mathrm{W}}{\mathrm{cm}^2}\,\mu\mathrm{m}^2\right]a_0^2\,. \tag{9.4}$$

When we specify circular polarization $\zeta = 2$ and wavelength $\lambda = 1\,\mu$m, we have

$$I_0 \approx 2.7 \times 10^{18} a_0^2 \left[\frac{\mathrm{W}}{\mathrm{cm}^2}\right] \quad \text{for} \quad \lambda = 1\,\mu\mathrm{m}\,, \quad \zeta = 2\,. \tag{9.5}$$

For $\lambda = 1\,\mu$m, we relate the electric field strength E_{field} to a_0 via

$$E_{field} \approx 3.14 \times 10^{12} a_0 \left[\frac{\mathrm{V}}{\mathrm{m}}\right] \quad \text{for} \quad \lambda = 1\,\mu\mathrm{m}\,. \tag{9.6}$$

Then, the intensity can also expressed by

$$I_0 \approx \frac{\left(E_{field}\left[\frac{\mathrm{V}}{\mathrm{cm}}\right]\right)^2}{377\,[\Omega]}\left[\frac{\mathrm{W}}{\mathrm{cm}^2}\right] \quad \text{for} \quad \lambda = 1\,\mu\mathrm{m}\,, \quad \zeta = 2\,. \tag{9.7}$$

Since a_0^2 is proportional to $I_0\lambda^2$, relativistic effects on particles favor large wavelengths λ (or small frequencies ω). Thus, higher frequencies which can be focused down to smaller spots are not so suitable for observation of ion relativistic effects, but may well be used to reach the QED-area. For the latter, I_0 itself is important. Higher-harmonics generation at solid surfaces may allow one to reach the QED-area much easier.

9.1.1.1 The Weakly Relativistic Regime

From the approximate solution of the motion of a single electron in a laser field, for $|v| \ll c$, we find the condition

$$a_0 = \frac{eA_0}{m_e c} \ll 1\,. \tag{9.8}$$

In that limit, the γ-factor can be expanded

$$\gamma = \sqrt{1 + \frac{|\boldsymbol{p}_e|^2}{m_e^2 c^2}} \approx \sqrt{1 + a_0^2} \approx 1 + \frac{a_0^2}{2} . \tag{9.9}$$

The nonrelativistic case has $\gamma = 1$.

9.1.1.2 The High-Intensity Relativistic Regime

The following terminology has been used by Mourou *et al.* [280]. The high-intensity regime (for electrons) is defined as the region

$$\hbar\omega \ll m_e c^2 \left[\sqrt{1 + a_0^2} - 1 \right] < m_e c^2 \tag{9.10}$$

when we use $\gamma \approx \sqrt{1 + a_0^2}$. Here, relativistic effects are very important, but the physics is not yet ultrarelativistic. Obviously, the upper limit is $a_0 \approx 1$. For $\lambda = 1\ \mu\text{m}$, from the condition $a_0 = 1$, the estimate

$$I_{0e} \approx 2.7 \times 10^{18} \frac{\text{W}}{\text{cm}^2} \tag{9.11}$$

follows. We note that higher intensities of order

$$I_{0i} \approx 10^{25} \frac{\text{W}}{\text{cm}^2} \tag{9.12}$$

are required for ion quiver motion becoming strongly relativistic. An evaluation for relativistic electrons leads to the estimate

$$5 \times 10^{-6} \ll a_0^2 \leq 1 , \tag{9.13}$$

or with a factor 100 for the lower limit

$$2.2 \times 10^{-2} \leq a_0 \leq 1 . \tag{9.14}$$

In terms of intensity, we have

$$5 \times 10^{14} \left[\frac{\text{W}}{\text{cm}^2} \right] \leq I_0 \leq 2.7 \times 10^{18} \left[\frac{\text{W}}{\text{cm}^2} \right] \quad \text{for} \quad \lambda = 1\ \mu\text{m} . \tag{9.15}$$

Note that the numerical values depend on the laser wavelength. We used $\lambda = 1\ \mu\text{m}$, having in mind that for an excimer laser $\lambda \approx 248$ nm leading to an upper limit of order $10^{19}\ \text{W/cm}^2$, whereas for a CO_2 laser $\lambda \approx 10.6\ \mu\text{m}$ leading to an upper limit of order $10^{16}\ \text{W/cm}^2$ for entering the so called ultrahigh intensity regime.

9.1.1.3 The Ultrarelativistic Regime for Ultrahigh Intensities

The ultrahigh intensity regime – above the high-intensity regime (Eq. 9.13) – covers the region $a_0 > 1$ or

$$e E_{field} \lambda > 2\pi m_e c^2 \tag{9.16}$$

for the laser electric field E_{field}. Depending on the dominating physical processes, we distinguish between different areas.

The Relatively Low-Amplitude Regime

An accelerated relativistic electron propagating in the laser field loses energy through synchotron radiation. Anticipating the standard relativistic formula [281, 282] for the radiation intensity

$$P = \frac{2}{3}\frac{e^2}{4\pi\varepsilon_0 m_e^2 c^3}\gamma^2\omega^2|\boldsymbol{p}|^2 , \tag{9.17}$$

one may approximate $|\boldsymbol{p}| \approx m_e c a_0$ for relatively low laser amplitudes (see below). The radius of the electron orbit is approximately $R = c/\omega_0 = \lambda/2\pi$, where a_0 is the normalized vector potential amplitude, $m \equiv m_e$ is the electron rest mass, ω_0 the laser frequency. In addition, for $a_0 \gg 1$ and as long as $|\boldsymbol{p}| \approx m_e c a_0$, we may use $\gamma \approx a_0$. Putting everything together, we arrive at

$$P \approx \underbrace{\frac{4\pi r_e}{3\lambda}}_{\varepsilon_{rad}} \omega_0 m_e c^2 a_0^4 \tag{9.18}$$

with

$$\varepsilon_{rad} = \frac{4\pi r_e}{3\lambda} \approx 1.2 \times 10^{-8} \quad \text{at} \quad \lambda = 1\,\mu\text{m} , \tag{9.19}$$

and the classical electron radius

$$r_e = \frac{1}{4\pi\varepsilon_0}\frac{e^2}{m_e c^2} \approx 2.82 \times 10^{-13}\,\text{cm} . \tag{9.20}$$

In the laser field, the electron energy

$$E = \gamma m_e c^2 \approx m_e c^2 a_0 \tag{9.21}$$

changes. Based on the energy formula, we approximate the time change by

$$\frac{\partial E}{\partial t} \sim \omega_0 m_e c^2 a_0 . \tag{9.22}$$

Comparing Eq. (9.18) with Eq. (9.22), we find that radiation becomes dominant for

$$\varepsilon_{rad} a_0^3 \gg 1 . \tag{9.23}$$

The region below this limit, that is,

$$1 \le a_0 \le \frac{1}{\varepsilon_{rad}^{1/3}} \equiv a_{radiation} \approx 440 \quad \text{for} \quad \lambda = 1\,\mu\text{m} , \tag{9.24}$$

is called the relatively low-amplitude regime. Above that regime, radiation effects become important. At the upper limit, the electric field has the order of magnitude

$$E_{field}^{radiation} \approx 1.4 \times 10^{15} \left[\frac{\text{V}}{\text{m}}\right] ; \tag{9.25}$$

the corresponding intensity is

$$I_0^{radiation} \approx 5.2 \times 10^{23} \left[\frac{\mathrm{W}}{\mathrm{cm}^2}\right] \quad \text{at} \quad \lambda = 1\,\mu\mathrm{m}\,. \tag{9.26}$$

In conclusion,

$$2.7 \times 10^{18} \left[\frac{\mathrm{W}}{\mathrm{cm}^2}\right] \leq I_0 \leq 5.2 \times 10^{23} \left[\frac{\mathrm{W}}{\mathrm{cm}^2}\right] \quad \text{at} \quad \lambda = 1\,\mu\mathrm{m} \tag{9.27}$$

characterizes the relatively low-intensity region of the ultrarelativistic regime.

The Intermediate Regime

Before discussing the new physical effects appearing at even larger amplitudes, let us come back to the relation between $|\boldsymbol{p}|$ and a_0. Without proof, we state (see [280])

$$|\boldsymbol{p}| \approx \begin{cases} m_e c a_0 & \text{for } 1 < a_0 < a_{radiation}\,, \\ m_e c \left(\frac{a_0}{\varepsilon_{rad}}\right)^{1/4} & \text{for } a_{radiation} < a_0 < a_{quantum}\,. \end{cases} \tag{9.28}$$

Quantum effects become important when the photons being generated due to Compton scattering have energies of the order (or larger) of the electron energy $E_e = \gamma m_e c^2$. The energy of a photon is

$$E_{photon} = \hbar \omega_m\,. \tag{9.29}$$

The frequency of the photon being generated by an electron rotating with frequency ω is [283]

$$\omega_m = \gamma^3 \omega\,. \tag{9.30}$$

Equating the electron energy with the photon energy, $E = E_{photon}$, we find the critical γ-value

$$\gamma_{quantum} \equiv \sqrt{\frac{m_e c^2}{\hbar \omega}} \tag{9.31}$$

such that for

$$\gamma > \gamma_{quantum} \approx 600 \quad \text{for} \quad \lambda = 1\,\mu\mathrm{m}, \tag{9.32}$$

quantum effects come into play. Using Eq. (9.28) and

$$\gamma \approx \left(\frac{a_0}{\varepsilon_{rad}}\right)^{1/4} \approx \left(\frac{3 a_0 \lambda}{4\pi r_e}\right)^{1/4}\,, \tag{9.33}$$

we may translate that condition into

$$a_0 > a_{quantum} = \frac{1}{4\pi\varepsilon_0} \frac{2e^2 m_e c}{3\hbar^2 \omega} = \frac{1}{3\pi} \frac{r_e \lambda}{\lambda_c^2} \approx 2000 \quad \text{for} \quad \lambda = 1\,\mu\mathrm{m}\,, \tag{9.34}$$

where

$$\lambda\!\!\!^{-}_c = \frac{\hbar}{m_e c} \approx 3.86 \times 10^{-11}\,\text{cm} \tag{9.35}$$

is the Compton length. The region

$$a_{radiation} \leq a_0 \leq a_{quantum} \tag{9.36}$$

is called the intermediate ultrarelativistic region. At the upper limit, the electric field has the order of magnitude

$$E_{field}^{\text{quantum}} \approx \frac{1}{4\pi\varepsilon_0}\frac{2em_e^2c^2}{3\hbar^2} \approx 6.5 \times 10^{15}\left[\frac{\text{V}}{\text{m}}\right]; \tag{9.37}$$

the corresponding intensity is

$$I_0^{\text{quantum}} \approx 10^{25}\left[\frac{\text{W}}{\text{cm}^2}\right] \quad \text{at} \quad \lambda = 1\,\mu\text{m}\,. \tag{9.38}$$

To summarize,

$$5.2 \times 10^{23}\left[\frac{\text{W}}{\text{cm}^2}\right] \leq I_0 \leq 10^{25}\left[\frac{\text{W}}{\text{cm}^2}\right] \quad \text{at} \quad \lambda = 1\,\mu\text{m} \tag{9.39}$$

characterizes the intermediate region of the ultrarelativistic regime.

The Nonlinear QED Regime

For even larger intensities, we reach the region where breakdown of the vacuum with spontaneous pair production becomes possible. Noting the condition that the field can do sufficient work on a virtual electron–positron pair to produce breakdown of the vacuum, that is,

$$e\,E_{field}\lambda\!\!\!^{-}_c > 2m_e c^2\,, \tag{9.40}$$

we may define the Schwinger field as

$$E_{field}^{Schwinger} = \frac{m_e^2c^3}{e\hbar} \approx 1.3 \times 10^{16}\left[\frac{\text{V}}{\text{cm}}\right]. \tag{9.41}$$

The corresponding normalized vector potential is

$$a_{Schwinger} = \frac{m_e c^2}{\hbar\omega} \approx 4 \times 10^5\,. \tag{9.42}$$

That field will be reached for the laser intensity

$$I_0^{Schwinger} \approx 4.5 \times 10^{29}\left[\frac{\text{W}}{\text{cm}^2}\right] \quad \text{at} \quad \lambda = 1\,\mu\text{m}\,. \tag{9.43}$$

At this intensity, we shall reach the nonlinear QED area.

Hawking–Unruh Radiation

The original papers which initiated the discussion of Unruh radiation are the following [284–287]. Subsequently, many papers, for example, [288], addressed that interesting topic. The existence of Unruh radiation is not universally accepted. Some claim that it has already been observed, while others claim that it is not emitted at all. While the skeptics accept that an accelerating object thermalizes at the Unruh temperature, they do not believe that this leads to the emission of photons, arguing that the emission and absorption rates of the accelerating particle are balanced.

Here, we follow the interpretation by McDonald [289]. According to Hawking, an observer outside a strong gravitational field, as it appears in a black hole, experiences a bath of thermal radiation with temperature T given by

$$k_B T_{Hawking} = \frac{\hbar g}{2\pi c}, \tag{9.44}$$

where g is the local acceleration due to gravity. The gravitational field interacts with the quantum fluctuations of the electromagnetic field with the result that energy can be transferred to an observer. If the temperature is equivalent to 1 MeV or more, virtual electron–positron pairs emerge from the vacuum into real particles. The idea of Unruh was that, similar to the Hawking radiation, a uniformly accelerated detector also experiences a thermal bath of radiation. An accelerated observer in a gravity-free environment experiences the same physics (locally) as an observer at rest in a gravitational field. The temperature should be given by

$$k_B T_{Unruh} = \frac{\hbar a^*}{2\pi c}, \tag{9.45}$$

where a^* is the acceleration as measured in the observer's instantaneous rest frame. The observer could be an electron accelerated by an electromagnetic field E_{field}. Suppose that the characteristic energy $k_B T_{Unruh}$ is much less than 1 MeV. Then, Thomson scattering of the electron off photons in the apparent thermal bath would be interpreted by a laboratory observer as an extra contribution to the radiation rate of the accelerated charge. The power of the extra (Unruh) radiation can be calculated by

$$\frac{dU_{Unruh}}{dt} = F_{energy\,flux} \times \sigma_{Thomson}, \tag{9.46}$$

where

$$F_{energy\,flux} = U_{Unruh}\, c, \quad \sigma_{Thomson} = \frac{8\pi r_e^2}{3}. \tag{9.47}$$

Let us assume that the radiation is in thermal equilibrium such that the energy density of thermal radiation is given by the Planck expression

$$\frac{dU_{Unruh}}{d\nu} = \frac{8\pi}{c^3} \frac{h\nu^3}{e^{\frac{h\nu}{k_B T_{Unruh}}} - 1}, \tag{9.48}$$

where ν is the frequency. Note that these relations hold in the instantaneous rest frame of the electron. Combining Eq. (9.46) with Eq. (9.48) leads to

$$\frac{dU_{Unruh}}{dt d\nu} = \frac{8\pi}{c^2} \frac{h\nu^3}{e^{\frac{h\nu}{k_B T_{Unruh}}} - 1} \frac{8\pi r_e^2}{3} . \tag{9.49}$$

After integrating over ν, we find the Stefan–Boltzmann type relation

$$\frac{dU_{Unruh}}{dt} = \frac{8\pi^3 \hbar r_e^2}{45 c^2} \left(\frac{k_B T_{Unruh}}{\hbar} \right)^4 . \tag{9.50}$$

Making use of Eq. (9.45), one gets

$$\frac{dU_{Unruh}}{dt} = \frac{\hbar r_e^2 a^{*4}}{90\pi c^6} \approx \frac{e^4 E_{field}^4 \hbar r_e^2}{90\pi m^4 c^6} . \tag{9.51}$$

The proportionality to the fourth power of $a^* \approx e E_{field}/m$ is very important. When compared to other radiation, for example, Larmor radiation with

$$\frac{dU_{Larmor}}{dt} = \frac{1}{4\pi\varepsilon_0} \frac{2e^2 a^{*2}}{3c^3} , \tag{9.52}$$

one finds

$$\frac{dU_{Unruh}}{dt} \geq \frac{dU_{Larmor}}{dt} \tag{9.53}$$

for

$$E_{field} \geq E_{field}^{Unruh} \equiv \sqrt{\frac{60\pi}{\alpha} \frac{m^2 c^3}{e\hbar}} \approx 3 \times 10^{18} \left[\frac{\mathrm{V}}{\mathrm{cm}} \right] . \tag{9.54}$$

Here, $\alpha = e^2/4\pi\varepsilon_0 \hbar c$ is the fine-structure constant. The field being necessary for significant Unruh radiation is much larger than the Schwinger field. The required acceleration is also very large, $a^{*Unruh} \approx \mathcal{O}(10^{31})g$.

Such fields are not common in nature, but are thought to occur at the surface of neutron stars and may play a role in pulsar physics. Critical fields can be temporarily created in the laboratory by the superposition of Coulomb fields during the collision of two heavy nuclei. It is known that plasma wake fields excited by either a laser pulse or an intense electron beam can in principle provide an acceleration gradient as high as 100 GeV/cm, or 10^{23} g. Such acceleration relies on the collective perturbations of the plasma density excited by the driving pulse and restored by the immobile ions, and therefore is an effect over a plasma period. There is in fact another aspect of laser-driven electron acceleration. Namely, when a laser is ultrarelativistic, an electron under the direct influence of the laser can be instantly accelerated (and decelerated) in every laser cycle, resulting in an intermittent acceleration that is much more violent than that provided by the plasma wake fields. For the petawatt class (peta $\hat{=} 10^{15}$), lasers focused, for example, on 10 μm^2 leading to $I_0 \approx 10^{22}$–10^{23} W/cm^2 for $\lambda \sim \mathcal{O}(1\ \mu\mathrm{m})$ fields of order $E_{field} \approx 6 \times 10^{12}$ V/cm will be available. The corresponding acceleration is of order 2×10^{25} g.

9.2
Time- and Space-Dependent Maxwell Fluid Models

9.2.1
Fully Relativistic Maxwell Electron Fluid Model

When investigating high-frequency plasma phenomena, we take the electron motion into account, though we ignore the ion motion (immobile ion approximation). From Maxwell's equations in SI units, for the vector potential $\boldsymbol{A}$ and the scalar potential ϕ in Coulomb gauge $\nabla \cdot \boldsymbol{A} = 0$, we get

$$-\nabla^2 \boldsymbol{A} + \frac{1}{c^2}\frac{\partial^2 \boldsymbol{A}}{\partial t^2} = -\frac{1}{c^2}\frac{\partial \nabla \phi}{\partial t} + \mu_0 \boldsymbol{j} \, , \tag{9.55}$$

where we have used $\boldsymbol{B} = \nabla \times \boldsymbol{A}$, $\boldsymbol{E} = -\nabla\phi - \partial \boldsymbol{A}/\partial t$, $\nabla \times \nabla \times \boldsymbol{a} = \nabla\nabla \cdot \boldsymbol{a} - \nabla^2 \boldsymbol{a}$, and $\varepsilon_0 \mu_0 = c^{-2}$. For fixed ions, the electric current density is

$$\boldsymbol{j} \approx -e n_e \boldsymbol{v}_e \, ; \tag{9.56}$$

it follows from the electron velocity $\boldsymbol{v}_e$ and particle density n_e. The wave equation is coupled to the continuity equation for the electron density

$$\frac{\partial n_e}{\partial t} + \nabla \cdot (n_e \boldsymbol{v}_e) = 0 \tag{9.57}$$

and the electron momentum balance

$$\left(\frac{\partial}{\partial t} + \boldsymbol{v}_e \cdot \nabla\right) \boldsymbol{p}_e = -e\left[-\nabla\phi - \frac{\partial \boldsymbol{A}}{\partial t} + \boldsymbol{v}_e \times (\nabla \times \boldsymbol{A})\right], \tag{9.58}$$

which describe the nonlinear response of the medium. The pressure term has been neglected. Because of the relativistic mass factor, with

$$\boldsymbol{v}_e = \frac{\boldsymbol{p}_e}{m_e \gamma_e} \, , \tag{9.59}$$

where m_e is the electron rest mass and γ_e is the relativistic factor

$$\gamma_e = \frac{1}{\sqrt{1 - \left(\frac{v_e}{c}\right)^2}} = \sqrt{1 + \left(\frac{p_e}{m_e c}\right)^2} \, , \tag{9.60}$$

the momentum balance contains several nonlinear terms.

By some straightforward manipulations, we can write the momentum balance in the form

$$\frac{\partial}{\partial t}(\boldsymbol{p}_e - e\boldsymbol{A}) = e\nabla\phi - m_e c^2 \nabla \gamma_e + \frac{1}{m_e \gamma_e} \boldsymbol{p}_e \times \left[\nabla \times (\boldsymbol{p}_e - e\boldsymbol{A})\right], \tag{9.61}$$

when we make use of the vector identity

$$\boldsymbol{a} \times (\nabla \times \boldsymbol{a}) = \frac{1}{2}\nabla a^2 - \boldsymbol{a} \cdot \nabla \boldsymbol{a} \, . \tag{9.62}$$

To normalize the equations, we use a constant density n_0 (for example, the constant ion density which should be identical to the zeroth order electron density), the length $L = c\omega_{pe}^{-1}$, and the time $T = \omega_{pe}^{-1}$, where

$$\omega_{pe} \equiv \sqrt{\frac{n_0 e^2}{\varepsilon_0 m_e}} \tag{9.63}$$

is the electron plasma frequency.

The scalar potential is measured in $m_e c^2/e$, whereas the vector potential is measured in units of $m_e c/e$. Velocities are measured in units of c, and momenta are normalized with $m_e c$. From now on, we omit the index e for electrons. Making use of the normalization, the Maxwell electron fluid model for a plasma with immobile ions may be written in the form

$$\frac{\partial^2 \boldsymbol{A}}{\partial t^2} - \nabla^2 \boldsymbol{A} + \frac{\partial \nabla \phi}{\partial t} = -n\frac{\boldsymbol{p}}{\gamma} , \tag{9.64}$$

$$\nabla^2 \phi = n - 1 , \tag{9.65}$$

$$\frac{\partial n}{\partial t} + \nabla \cdot \left(n \frac{\boldsymbol{p}}{\gamma} \right) = 0 , \tag{9.66}$$

$$\frac{\partial}{\partial t}(\boldsymbol{p} - \boldsymbol{A}) - \frac{\boldsymbol{p}}{\gamma} \times \nabla \times (\boldsymbol{p} - \boldsymbol{A}) = \nabla(\phi - \gamma) , \tag{9.67}$$

where $\gamma = \sqrt{1 + \boldsymbol{p}^2}$ is the relativistic factor. We use the Coulomb gauge condition $\nabla \cdot \boldsymbol{A} = 0$.

9.2.2 Fully Relativistic Maxwell Two-Fluid Model

It is rather straightforward to generalize the Maxwell electron fluid model to a two-component plasma with mobile ions. For many applications, we may assume the plasma to be cold (with "zero" ion and electron temperatures). In the following, however, we include simple scalar pressure terms (finite temperatures) for later use in stability considerations. The used forms of the temperature (and pressure) terms only hold in the weakly relativistic ($A \ll 1$) and "low temperature" ($T_e \ll 1$, meaning $k_B T_e \ll m_e c^2$) case. Also, the isothermal equation of state (constant temperature approximation) is only appropriate if the phase velocities of the phenomena are smaller then the thermal velocities. Otherwise, an adiabatic approximation may be appropriate. Dimensionless quantities are used in the same form as before, now denoting species by α, and introducing $\varepsilon = m_e/m_i$ as a smallness parameter. In the Coulomb gauge, the Maxwell equations for the vector and scalar potentials $\boldsymbol{A}$ and ϕ, and the hydrodynamic equations (continuity and momentum balance) for the densities n_α and the canonical momentum $\boldsymbol{P}_\alpha$ of electrons and ions, respectively, can be written in dimensionless form as

$$\nabla^2 \boldsymbol{A} - \frac{\partial^2 \boldsymbol{A}}{\partial t^2} - \frac{\partial \nabla \phi}{\partial t} = n_e \boldsymbol{v}_e - n_i \boldsymbol{v}_i \,, \tag{9.68}$$

$$\nabla^2 \phi = n_e - n_i \,, \tag{9.69}$$

$$\frac{\partial n_\alpha}{\partial t} + \nabla \cdot (n_\alpha \boldsymbol{v}_\alpha) = 0 \,, \tag{9.70}$$

$$\frac{\partial \boldsymbol{P}_e}{\partial t} - \boldsymbol{v}_e \times \nabla \times \boldsymbol{P}_e = \nabla(\phi - \gamma_e) - T_e \nabla \ln n_e \,, \tag{9.71}$$

$$\frac{\partial \boldsymbol{P}_i}{\partial t} - \boldsymbol{v}_i \times \nabla \times \boldsymbol{P}_i = \nabla\left(-\phi - \frac{\gamma_i}{\varepsilon}\right) - T_i \nabla \ln n_i \,, \tag{9.72}$$

where $\boldsymbol{P}_\alpha$ and γ_α are related to the kinetic momentum $\boldsymbol{p}_\alpha$ by $\boldsymbol{P}_e = \boldsymbol{p}_e - \boldsymbol{A}$, $\boldsymbol{P}_i = \boldsymbol{p}_i + \boldsymbol{A}$, $\gamma_e = \sqrt{1 + \boldsymbol{p}_e^2}$, $\gamma_i = \sqrt{1 + \varepsilon^2 \boldsymbol{p}_i^2}$, with $\boldsymbol{v}_e = \boldsymbol{p}_e / \gamma_e$ and $\boldsymbol{v}_i = \varepsilon \boldsymbol{p}_i / \gamma_i$ being the (dimensionless) fluid velocities. The factor $\varepsilon = m_e / m_i$ appears because of the normalization $\boldsymbol{p}_e / m_e c \to \boldsymbol{p}_e$ and $\boldsymbol{p}_i / m_e c \to \boldsymbol{p}_i$. As mentioned, we shall neglect the pressure terms (by setting $T_e = T_i = 0$) in most applications.

9.2.3
1D Propagation in Space-Direction *x*

Let us now write the basic equations for immobile ions under the assumption that the wave propagates in x-direction, and thus all variables only depend on one space-coordinate, that is,

$$\boldsymbol{A} = \boldsymbol{A}(x, t) \,, \quad n = n(x, t) \,, \quad \phi = \phi(x, t) \,. \tag{9.73}$$

Furthermore, we designate $p \equiv p_x$. Within the Coulomb gauge, the purely transverse nature of the wave ($A_x = 0$, $\boldsymbol{A} = \boldsymbol{A}_\perp$) follows. The nondimensional form of the wave equation (9.55) for the transverse component $\boldsymbol{A}_\perp$ now reads

$$\frac{\partial^2}{\partial x^2} \boldsymbol{A}_\perp - \frac{\partial^2}{\partial t^2} \boldsymbol{A}_\perp = n \frac{\boldsymbol{p}_\perp}{\gamma} \,. \tag{9.74}$$

The longitudinal part of the wave equation simplifies to

$$\frac{\partial^2 \phi}{\partial t \partial x} + n \frac{p}{\gamma} = 0 \,. \tag{9.75}$$

The perpendicular component of the electron momentum balance

$$\frac{\partial}{\partial t} (\boldsymbol{p}_\perp - \boldsymbol{A}_\perp) + \left(\frac{p}{\gamma}\right) \frac{\partial (\boldsymbol{p}_\perp - \boldsymbol{A}_\perp)}{\partial x} = 0 \tag{9.76}$$

has the special solution

$$\boldsymbol{p}_\perp = \boldsymbol{A}_\perp \,. \tag{9.77}$$

Provided the initial values are chosen accordingly, the last relation simplifies the longitudinal electron momentum balance

$$\frac{\partial p}{\partial t} = \boldsymbol{p}_\perp \cdot \frac{\partial (\boldsymbol{p}_\perp - \boldsymbol{A}_\perp)}{\partial x} + \frac{\partial (\phi - \gamma)}{\partial x} = \frac{\partial (\phi - \gamma)}{\partial x} . \tag{9.78}$$

This leads to the basic set of equations for 1D fully relativistic wave propagation in a plasma with immobile ions, that is,

$$\frac{\partial^2}{\partial x^2}\boldsymbol{A}_\perp - \frac{\partial^2}{\partial t^2}\boldsymbol{A}_\perp = n\frac{\boldsymbol{A}_\perp}{\gamma} , \tag{9.79}$$

$$\frac{\partial^2 \phi}{\partial t \partial x} + n\frac{p}{\gamma} = 0 , \tag{9.80}$$

$$\frac{\partial^2 \phi}{\partial x^2} = n - 1 , \tag{9.81}$$

$$\frac{\partial n}{\partial t} + \frac{\partial}{\partial x}\left(\frac{np}{\gamma}\right) = 0 , \tag{9.82}$$

$$\frac{\partial p}{\partial t} = \frac{\partial (\phi - \gamma)}{\partial x} . \tag{9.83}$$

Note that Eqs. (9.81) and (9.82) lead to

$$\frac{\partial}{\partial x}\left[\frac{\partial^2 \phi}{\partial t \partial x} + n\frac{p}{\gamma}\right] = 0 . \tag{9.84}$$

In principle, by integration, we get Eq. (9.80) with an arbitrary constant on the right-hand side. However, in view of Eq. (9.80), the constant should be zero.

Assuming *linearly polarized* waves ($\boldsymbol{A}_\perp = A\hat{y}$), letting $E = -\partial\phi/\partial x$, ignoring the ion response, and setting $T_e = T_i = 0$, these equations agree with the model [290]

$$\frac{\partial E}{\partial t} = \frac{n_e p_e}{\gamma_e} , \quad \frac{\partial E_y}{\partial t} + \frac{\partial B_z}{\partial x} = \frac{n_e A}{\gamma_e} , \tag{9.85}$$

$$\frac{\partial B_z}{\partial t} + \frac{\partial E_y}{\partial x} = 0 , \quad \frac{\partial A}{\partial t} = -E_y , \tag{9.86}$$

$$\frac{\partial n_e}{\partial t} = -\frac{\partial}{\partial x}\left(\frac{n_e p_e}{\gamma_e}\right), \quad \frac{\partial p_e}{\partial t} = -E - \frac{\partial \gamma_e}{\partial x} , \tag{9.87}$$

$$\gamma = \sqrt{1 + p_e^2 + A^2} . \tag{9.88}$$

Besides, to discuss the interplay between backward and forward Raman scattering, modulation of broad light pulses, down-cascade in the frequency spectrum, photon condensation, and break-up of the original laser beam, this model is also suitable for investigating 1D slow solitons on the electron time-scale.

The 1D relativistic Maxwell two-fluid model can be handled in a similar way as the previously discussed Maxwell electron-fluid model. We abbreviate $\hat{x} \cdot \boldsymbol{p}_\alpha \equiv p_\alpha$.

For a laser pulse starting in vacuum, we get $\boldsymbol{P}_{\perp\alpha} = 0$. The set of equations

$$\frac{\partial^2 \boldsymbol{A}_\perp}{\partial x^2} - \frac{\partial^2 \boldsymbol{A}_\perp}{\partial t^2} = n_e \frac{\boldsymbol{A}_\perp}{\gamma_e} + \varepsilon n_i \frac{\boldsymbol{A}_\perp}{\gamma_i} , \tag{9.89}$$

$$\frac{\partial^2 \phi}{\partial t \partial x} + \frac{n_e p_e}{\gamma_e} - \varepsilon \frac{n_i p_i}{\gamma_i} = 0 , \tag{9.90}$$

$$\frac{\partial^2 \phi}{\partial x^2} = n_e - n_i , \tag{9.91}$$

$$\frac{\partial n_e}{\partial t} + \frac{\partial}{\partial x}\left(\frac{n_e p_e}{\gamma_e}\right) = 0 , \tag{9.92}$$

$$\frac{\partial n_i}{\partial t} + \varepsilon \frac{\partial}{\partial x}\left(\frac{n_i p_i}{\gamma_i}\right) = 0 , \tag{9.93}$$

$$\frac{\partial p_e}{\partial t} = \frac{\partial(\phi - \gamma_e)}{\partial x} - T_e \frac{\partial \ln n_e}{\partial x} , \tag{9.94}$$

$$\frac{\partial p_i}{\partial t} = \frac{\partial\left(-\phi - \frac{\gamma_i}{\varepsilon}\right)}{\partial x} - T_i \frac{\partial \ln n_i}{\partial x} \tag{9.95}$$

form a closed system generalizing the previous one-fluid model. Whether we can use an isothermal equation of state (as assumed here) depends on the phase velocity in relation to the thermal velocity.

9.2.4
The Weakly Relativistic Limit

It is not always appropriate (of course, in vacuum it is even not possible) to use the inverse plasma frequency as a time unit. It makes sense to use the inverse laser frequency ω_0^{-1} as a new time unit. Formally, we may relate it to a density, the so called critical density n_c via

$$\omega_0 \equiv \sqrt{\frac{n_c e^2}{\varepsilon_0 m_e}} . \tag{9.96}$$

In the following, we shall use the inverse laser frequency for time normalization and $c\omega_0^{-1}$ for length normalization. With the new units, the Maxwell fluid Eqs. (9.64)–(9.67) in full 3D space dimension change into

$$\frac{\partial^2 \boldsymbol{A}}{\partial t^2} - \nabla^2 \boldsymbol{A} + \frac{\partial \nabla \phi}{\partial t} = -\frac{n_0}{n_c} n \frac{\boldsymbol{p}}{\gamma} , \tag{9.97}$$

$$\nabla^2 \phi = -\frac{n_0}{n_c}(1 - n) , \tag{9.98}$$

$$\frac{\partial n}{\partial t} + \nabla \cdot \left(n \frac{\boldsymbol{p}}{\gamma}\right) = 0 , \tag{9.99}$$

$$\frac{\partial}{\partial t}(\boldsymbol{p} - \boldsymbol{A}) - \frac{\boldsymbol{p}}{\gamma} \times \nabla \times (\boldsymbol{p} - \boldsymbol{A}) = \nabla(\phi - \gamma) , \tag{9.100}$$

where we do not use the index e at the electron density and used $\gamma = \sqrt{1+p^2}$ as the relativistic factor. As before, we applied the Coulomb gauge condition $\nabla \cdot A = 0$. The vector potential A is measured in the unit $e/m_e c$, whereas for the nondimensional electrostatic potential ϕ, we use $e/m_e c^2$. The unit for the density is n_0. In principle, it may be arbitrary, but in the presence of plasma, we shall choose the ion background density, being generally different from the critical density n_c. For laser propagation in an underdense plasma, $n_0/n_c < 1$ holds. The momentum $p \equiv p_e$ is measured in $m_e c$, where m_e, e, and c are electron mass, elementary charge, and speed of light, respectively. Sometimes, for the Laplace operator ∇^2, we shall also use the notation Δ.

Let us emphasize once more that, compared to the previous section, the time unit has been changed from ω_{pe}^{-1} to ω_0^{-1}. This has the advantage of being easily comparable with the vacuum limit $n_0 = 0$.

The momentum balance Eq. (9.100) can be simplified further by choosing an initial gauge condition (which, of course, should not restrict generality). We define the projection operators Π_c and Π_g such that any vector field u is decomposed into $u = v + w$ with the following properties

$$\Pi_c u = v \equiv u_c \,, \quad \nabla \times v = 0 \,, \quad \text{but generally} \quad \nabla \cdot v \neq 0 \,, \tag{9.101}$$

$$\Pi_g u = w \equiv u_g \,, \quad \nabla \cdot w = 0 \,, \quad \text{but generally} \quad \nabla \times w \neq 0 \,, \tag{9.102}$$

with

$$\Pi_c + \Pi_g = \mathbf{1} \,. \tag{9.103}$$

Clearly, v is a gradient field, and w is a rotation field. The operators can be represented as

$$\Pi_c = \nabla \Delta^{-1} \nabla \cdot \tag{9.104}$$

$$\Pi_g = 1 - \nabla \Delta^{-1} \nabla \cdot \,. \tag{9.105}$$

Applying the projection operators to the momentum balance Eq. (9.100) allows one to split the equation in a divergence-free and a curl-free part. The equation

$$\frac{\partial}{\partial t}(p_g - A) - \Pi_g \left[\frac{p}{\gamma} \times \left\{ \nabla \times (p_g - A) \right\} \right] = 0 \tag{9.106}$$

describes the convective transport of the divergence-free part of the canonical momentum $P_{can} = p - A$. This implies that for the initial condition $p_g = A$, the canonical momentum stays curl-free for all times, that is,

$$P_{can} = p_g + p_c - A = p_c \,. \tag{9.107}$$

This initial condition simplifies the curl-free part to

$$\frac{\partial}{\partial t} p_c = \nabla(\phi - \gamma) \,. \tag{9.108}$$

Since $\nabla \times \boldsymbol{p}_c = 0$, $\boldsymbol{p}_c$ can be written in terms of a Clebsch potential, $\boldsymbol{p}_c = \nabla \psi$. The integration of Eq. (9.108) then leads to

$$\frac{\partial \psi}{\partial t} = \phi - \gamma + 1 \,. \tag{9.109}$$

Applying the splitting via Π_g and Π_c to the wave equation (9.97) for $\boldsymbol{A}$, we obtain under the condition $\boldsymbol{p}_g = \boldsymbol{A}$ for the divergence-free part

$$\frac{\partial^2 \boldsymbol{A}}{\partial t^2} - \nabla^2 \boldsymbol{A} = -\frac{n_0}{n_c}(1 - \nabla \Delta^{-1} \nabla \cdot) \left\{ \frac{n}{\gamma}(\boldsymbol{A} + \nabla \psi) \right\}, \tag{9.110}$$

and for the curl-free part,

$$\frac{\partial}{\partial t} \nabla \phi = -\frac{n_0}{n_c} \nabla \Delta^{-1} \nabla \cdot \left\{ \frac{n}{\gamma}(\boldsymbol{A} + \nabla \psi) \right\}. \tag{9.111}$$

Straightforward simplifications of the vector operations of the right-hand sides lead to

$$\frac{\partial^2 \boldsymbol{A}}{\partial t^2} - \nabla^2 \boldsymbol{A} = -\frac{n_0}{n_c} \left[\frac{n}{\gamma} \boldsymbol{A} - \Delta^{-1} \left\{ \nabla \left(\boldsymbol{A} \cdot \nabla \frac{n}{\gamma} \right) + \nabla \times \left[\left(\nabla \frac{n}{\gamma} \right) \times (\nabla \psi) \right] \right\} \right], \tag{9.112}$$

$$\frac{\partial \nabla \phi}{\partial t} = -\frac{n_0}{n_c} \left[\frac{n}{\gamma} \nabla \psi + \Delta^{-1} \left\{ \nabla \left(\boldsymbol{A} \cdot \nabla \frac{n}{\gamma} \right) + \nabla \times \left[\left(\nabla \frac{n}{\gamma} \right) \times (\nabla \psi) \right] \right\} \right]. \tag{9.113}$$

The equations for $\boldsymbol{A}$ and ϕ are coupled to n and ψ. The evolution of the latter is given by Eq. (9.109) in the form

$$\frac{\partial \psi}{\partial t} = \phi - \gamma + 1 = \phi - \sqrt{1 + (\boldsymbol{A} + \nabla \psi)^2} + 1 \,. \tag{9.114}$$

The density n can be obtained from the Poisson equation (9.98) in the form

$$n = 1 + \left(\frac{n_0}{n_c} \right)^{-1} \nabla^2 \phi \,. \tag{9.115}$$

In the following, we shall use the Maxwell fluid equations in the just presented reformulated form to derive simpler models. Let us consider the weakly relativistic limit. For small amplitudes, we scale the variables.

Let us first introduce different smallness parameters. We scale the perpendicular variation with respect to the parallel changes in the propagation direction x (scaling parameter α) and introduce smallness parameters ε, μ, β, ρ, and δ for the amplitudes of the physical variables (note that ε is a general smallness parameter and not anymore fixed to m_e/m_i):

$$A(\mathbf{r}, t) = \varepsilon\left\{A_\perp(x, \alpha \mathbf{r}_\perp, t) + \mu \mathbf{e}_x A_\parallel(x, \alpha \mathbf{r}_\perp, t)\right\}, \tag{9.116}$$

$$n(\mathbf{r}, t) = 1 + \beta n_e^1(x, \alpha \mathbf{r}_\perp, t), \tag{9.117}$$

$$\phi(\mathbf{r}, t) = \rho\phi^1(x, \alpha \mathbf{r}_\perp, t), \tag{9.118}$$

$$\psi(\mathbf{r}, t) = \delta\psi^1(x, \alpha \mathbf{r}_\perp, t), \tag{9.119}$$

$$\gamma(\mathbf{r}, t) = \sqrt{1 + \left(\varepsilon A_\perp + \varepsilon\mu \mathbf{e}_x A_\parallel + \delta\nabla\psi^1\right)^2}. \tag{9.120}$$

The different smallness parameters are of course interrelated. The relations

$$\mu \sim \alpha \sim \varepsilon \ll 1 \quad \text{and} \quad \delta \sim \rho \sim \beta \sim \varepsilon^2 \ll 1 \tag{9.121}$$

can be justified as follows. From the Coulomb gauge, we get

$$\nabla \cdot A = \varepsilon\left\{\alpha\nabla_\perp \cdot A_\perp + \mu\partial_x A_\parallel\right\} = 0 \Rightarrow \mu \sim \alpha, \tag{9.122}$$

unless $A_\parallel \equiv 0$. The Laplace equation (9.98) for ϕ yields

$$\rho\nabla^2\phi^1 = \frac{n_0}{n_c}(n-1) = \frac{n_0}{n_c}\beta n_e^1 \Rightarrow \rho \sim \beta. \tag{9.123}$$

The reduced momentum balance Eq. (9.109)

$$\delta\partial_t\psi^1 = \rho\phi^1 - (\gamma - 1) = \rho\phi^1 + \mathcal{O}(\varepsilon^2) + \cdots \Rightarrow \delta \sim \rho \sim \varepsilon^2 \tag{9.124}$$

and the combination of Poisson's equation (9.123) with the continuity equation (9.99)

$$\rho\nabla \cdot \frac{\partial}{\partial t}\nabla\phi^1 = \frac{n_0}{n_c}\beta\frac{\partial}{\partial t}n_e^1 = -\frac{n_0}{n_c}\delta\nabla^2\psi^1 + \cdots \Rightarrow \rho \sim \delta, \tag{9.125}$$

are consistent with $\delta \sim \rho \sim \beta \sim \varepsilon^2$. This scaling is also compatible with the wave equation for A, that is,

$$\varepsilon\left\{\frac{\partial^2}{\partial t^2}A - \frac{\partial^2}{\partial x^2}A - \alpha^2\Delta_\perp A\right\} = -\varepsilon\left\{\frac{n_0}{n_c}(1 + \beta n_e^1)\left[1 - \frac{\varepsilon^2}{2}A^2\right]A\right\} + \cdots \tag{9.126}$$

Based on the just presented arguments for a consistent scaling, we now introduce one smallness parameter ε and scale the physical quantities as follows:

$$A(\mathbf{r}, t) = \varepsilon\left\{A_\perp(x, \varepsilon \mathbf{r}_\perp, t) + \varepsilon \mathbf{e}_x A_\parallel(x, \varepsilon \mathbf{r}_\perp, t)\right\}, \tag{9.127}$$

$$n(\mathbf{r}, t) = 1 + \varepsilon^2\delta n(x, \varepsilon \mathbf{r}_\perp, t), \tag{9.128}$$

$$\phi(\boldsymbol{r}, t) = \varepsilon^2 \delta\phi(x, \varepsilon \boldsymbol{r}_\perp, t) , \tag{9.129}$$

$$\psi(\boldsymbol{r}, t) = \varepsilon^2 \delta\psi(x, \varepsilon \boldsymbol{r}_\perp, t) , \tag{9.130}$$

$$\gamma(\boldsymbol{r}, t) = \sqrt{1 + \left(\varepsilon \boldsymbol{A}_\perp + \varepsilon^2 \boldsymbol{e}_x A_\parallel + \varepsilon^2 \nabla\delta\psi\right)^2} \approx 1 + \frac{\varepsilon^2}{2} \boldsymbol{A}_\perp^2 + \mathcal{O}\left(\varepsilon^3\right) . \tag{9.131}$$

Then, because of

$$\nabla \frac{n}{\gamma} = \varepsilon^2 \left(\nabla\delta n - n_e^0 \nabla \frac{\boldsymbol{A}_\perp^2}{2}\right) + \mathcal{O}\left(\varepsilon^3\right) \tag{9.132}$$

and

$$\boldsymbol{A} \cdot \nabla = \varepsilon^2 \boldsymbol{A}_\perp \cdot \nabla_\perp + \varepsilon^2 A_\parallel \partial_x , \tag{9.133}$$

we have

$$\nabla \left(\boldsymbol{A} \cdot \nabla \frac{n}{\gamma}\right) = \mathcal{O}\left(\varepsilon^4\right) , \tag{9.134}$$

and

$$\nabla \times \left[\left(\nabla \frac{n}{\gamma}\right) \times (\nabla\psi)\right] = \mathcal{O}\left(\varepsilon^5\right) . \tag{9.135}$$

The inverse Laplace operator does not change the order of the dominating terms since

$$\Delta^{-1} \mathrel{\hat{=}} \mathcal{F}^{-1} \frac{1}{k_\parallel^2 + \varepsilon^2 k_\perp^2} \approx \mathcal{F}^{-1} \frac{1}{k_\parallel^2} \left(1 - \varepsilon^2 \frac{k_\perp^2}{k_\parallel^2}\right) \mathrel{\hat{=}} \left(\frac{\partial^2}{\partial x^2}\right)^{-1} + \mathcal{O}\left(\varepsilon^2\right) , \tag{9.136}$$

where $\mathcal{F}^{-1}$ is the inverse Fourier transform.

To get consistent equations, we include all terms up to order ε^3 and neglect terms of order ε^4 and higher.

From the continuity Eq. (9.99), we obtain with $\boldsymbol{p} = \boldsymbol{A} + \nabla\psi$ after a short calculation

$$\varepsilon^2 \frac{\partial^2 \delta n}{\partial t^2} \approx -\varepsilon^2 \frac{\partial}{\partial t} \nabla^2 \delta\psi + \mathcal{O}\left(\varepsilon^4\right) . \tag{9.137}$$

On the other hand, taking the Laplacian of Eq. (9.109), with the Poisson equation (9.115), we find

$$\varepsilon^2 \frac{\partial}{\partial t} \nabla^2 \delta\psi = \frac{n_0}{n_c} \varepsilon^2 \delta n - \frac{\varepsilon^2}{2} \nabla^2 \boldsymbol{A}_\perp^2 + \mathcal{O}\left(\varepsilon^4\right) . \tag{9.138}$$

Combining these two equations, we get to lowest order

$$\frac{\partial^2 \delta n}{\partial t^2} + \frac{n_0}{n_c}\delta n \approx \frac{1}{2}\nabla^2 \boldsymbol{A}_\perp^2 \,. \tag{9.139}$$

Together with the lowest order wave equation in perpendicular direction

$$\frac{\partial^2 \boldsymbol{A}_\perp}{\partial t^2} - \nabla^2 \boldsymbol{A}_\perp \approx -\frac{n_0}{n_c}\left\{1 - \frac{1}{2}\boldsymbol{A}_\perp^2 + \delta n\right\}\boldsymbol{A}_\perp , \tag{9.140}$$

we obtain a closed system of two equations. Instead of a series expansion, we may retain the full γ-factor, keeping in mind the valid accuracy of the results at the end. For the initial condition $A_\parallel = 0$, the parallel component will approximately stay at zero during time evolution.

Some explanation is needed to fully understand Eq. (9.140). Because of the zeroth order dispersion relation, we find to lowest order

$$\frac{\partial^2 \boldsymbol{A}_\perp}{\partial t^2} - \frac{\partial^2 \boldsymbol{A}_\perp}{\partial x^2} + \frac{n_0}{n_c}\boldsymbol{A}_\perp = \varepsilon 0 + \mathcal{O}\left(\varepsilon^3\right) , \tag{9.141}$$

showing that the system is really correctly closed in the intended order.

The weakly relativistic model for a *linearly* polarized electromagnetic wave with $\boldsymbol{A} \approx \boldsymbol{A}_\perp = a\boldsymbol{e}_z$ is

$$\frac{\partial^2 a}{\partial t^2} = \nabla^2 a - \frac{n_0}{n_c}\frac{(1 + \delta n)}{\gamma} a \,, \tag{9.142}$$

$$\frac{\partial^2 \delta n}{\partial t^2} = -\frac{n_0}{n_c}\delta n + \nabla^2 \gamma \,, \tag{9.143}$$

where γ is the relativistic factor,

$$\gamma \approx 1 + \frac{1}{2}a^2 \quad \text{and} \quad \gamma^{-1} \approx 1 - \frac{1}{2}a^2 \,. \tag{9.144}$$

The consistency of the model requires that one use the latter approximate expressions and truncates consistently at orders ε^3 and ε^2, respectively.

For circular polarization, it is possible to use the complex-valued scalar field a to describe the vector potential via

$$a(\boldsymbol{r}, t) = A_y(\boldsymbol{r}, t) + iA_z(\boldsymbol{r}, t) \quad \text{with} \quad \boldsymbol{A}^2 = |a|^2 \,. \tag{9.145}$$

Then,

$$\frac{\partial^2 a}{\partial t^2} = \nabla^2 a - \frac{n_0}{n_c}\frac{(1 + \delta n)}{\gamma} a \,, \tag{9.146}$$

$$\frac{\partial^2 \delta n}{\partial t^2} = -\frac{n_0}{n_c}\delta n + \nabla^2 \gamma \tag{9.147}$$

where now the relativistic factor is

$$\gamma \approx 1 + \frac{1}{2}|a|^2 \quad \text{and} \quad \gamma^{-1} \approx 1 - \frac{1}{2}|a|^2 \,. \tag{9.148}$$

Note the difference in the relativistic factor to the model for linear polarization.

To summarize, the space-variation of the γ-factor causes density modulations and drives plasma oscillations. Ion motion is neglected for fast propagation. The parameter n_0/n_c (ion background density over critical density) should be less than one for wave propagation in an underdense medium, and greater than 1/4 to avoid Raman instability. Note, that in vacuum, $n_0 \equiv 0$ and therefore $\delta n \equiv n_0 \delta n \equiv 0$.

9.2.5
The Weakly Relativistic 1D Maxwell Two-Fluid Model

Let us briefly reconsider Eq. (9.141). For its physical understanding, we recapitulate the main steps in its derivation. One starts from the wave equation in Coulomb gauge

$$\nabla^2 \boldsymbol{A}_\perp - \mu_0 \varepsilon_0 \frac{\partial^2 \boldsymbol{A}_\perp}{\partial t^2} = -\mu_0 \boldsymbol{j}_\perp \,, \tag{9.149}$$

and approximates

$$\boldsymbol{j}_\perp \approx -e n_e \boldsymbol{v}_{e\perp} = -\frac{e n_e}{m_e \gamma} \boldsymbol{p}_{e\perp} \approx -\frac{e^2 n_e}{m_e \gamma} \boldsymbol{A}_\perp \,, \tag{9.150}$$

where the last relation originates from

$$\frac{d \boldsymbol{p}_{e\perp}}{dt} \approx e \frac{\partial \boldsymbol{A}_\perp}{\partial t} \leftrightarrow \boldsymbol{p}_{e\perp} \approx e \boldsymbol{A}_\perp \,. \tag{9.151}$$

Since $\gamma \equiv \gamma_e \approx \sqrt{1 + e^2 A_\perp^2}$, the expansion of the γ-factor leads to the form Eq. (9.141) derived before. Now let us introduce

$$\boldsymbol{A}_\perp \approx \frac{1}{e} \boldsymbol{p}_{e\perp} \approx \frac{m_e}{e} \boldsymbol{v}_{e\perp} \left(1 + \frac{1}{2c^2} v_{e\perp}^2\right) \tag{9.152}$$

and $\boldsymbol{v}_\perp \equiv \boldsymbol{v}_{e\perp}$ to obtain in 1D

$$c^2 \frac{\partial^2 \boldsymbol{v}_\perp}{\partial x^2} - \frac{\partial^2 \boldsymbol{v}_\perp}{\partial t^2} - \omega_{pe}^2 \boldsymbol{v}_\perp \approx -\frac{1}{2} \frac{\partial^2 \boldsymbol{v}_\perp^3}{\partial x^2} + \frac{1}{2c^2} \frac{\partial^2 \boldsymbol{v}_\perp^3}{\partial t^2} + \frac{\omega_{pe}^2}{n_0} \delta n_e \boldsymbol{v}_\perp \,, \tag{9.153}$$

where $\boldsymbol{v}_\perp^3 = (\boldsymbol{v}_\perp \cdot \boldsymbol{v}_\perp) \boldsymbol{v}_\perp$. This form was first derived by Gorbunov and Kirsanov [291].

The other Eq. (9.139) describes the *high-frequency* electron response. Remember that we have derived it only from the electron dynamics with immobile ions. A simplified derivation for 1D starts from the electron continuity equation

$$\frac{\partial n_e}{\partial t} + \frac{\partial}{\partial x} \left(\frac{n_e p_{ex}}{m_e \gamma} \right) = 0 \tag{9.154}$$

and the momentum balance in x-direction

$$\frac{\partial p_{ex}}{\partial t} + v_{ex} \frac{\partial}{\partial x} p_{ex} = -e E_x - e [\boldsymbol{v}_e \times \boldsymbol{B}]_x \,. \tag{9.155}$$

With the approximate solution $\boldsymbol{p}_{e\perp} \approx e\boldsymbol{A}_{e\perp}$, the last term on the right-hand side can be written as

$$-e\,[\boldsymbol{v}_e \times \boldsymbol{B}]_x \approx -\boldsymbol{v}_{e\perp} \cdot \frac{\partial}{\partial x}\boldsymbol{p}_{e\perp} \,. \tag{9.156}$$

All together, we obtain

$$\frac{\partial p_{ex}}{\partial t} \approx -e\,E_x - m_e c^2 \frac{\partial \gamma}{\partial x} \,. \tag{9.157}$$

Combining with the time-derivative of Eq. (9.154), we obtain in the weakly relativistic case

$$\frac{\partial^2 \delta n_e}{\partial t^2} + \frac{\omega_{pe}^2}{\gamma}\delta n_e \approx \frac{n_0 c^2}{\gamma}\frac{\partial^2 \gamma}{\partial x^2} \,. \tag{9.158}$$

When ordering in the same way as discussed previously, immediately Eq. (9.139) follows.

The question remains as to when a low-frequency ion density response should be included. Clearly, when we transform from the lab frame with x, t to a moving frame with $\tau = t, \xi = x - Vt$, the partial time-derivative at constant ξ becomes $\partial/\partial\tau$, and the time-derivative in the lab frame $\partial/\partial t$ (at constant x) transforms as

$$\frac{\partial}{\partial t} \to \frac{\partial}{\partial \tau} - V\frac{\partial}{\partial \xi} \,. \tag{9.159}$$

For example, when we consider stationary waves or pulses in a moving frame for

$$\left| V\frac{\partial}{\partial \xi} \right| \sim \frac{V}{L} \gg \omega_{pi} \,, \tag{9.160}$$

ion reactions can be neglected. That condition is satisfied for fast moving and sufficiently narrow (characteristic length L) pulses. If the condition is not satisfied for $\tau > \omega_{pi}^{-1}$, the ion dynamics come into play. The low-frequency ion response (e.g., nonrelativistic) follows from

$$\frac{\partial n_i}{\partial t} + \nabla \cdot (n_i \boldsymbol{v}_i) = 0 \,, \tag{9.161}$$

$$\frac{\partial \boldsymbol{v}_i}{\partial t} + \boldsymbol{v}_i \cdot \nabla \boldsymbol{v}_i = -\frac{e}{m_i}\nabla\phi - \frac{T_i}{m_i}\nabla \ln n_i \tag{9.162}$$

if we include the ion pressure term with an isothermal equation of state (see below). For small ion temperatures (writing $T_i \approx 0$) and small ion velocities (because of the large ion mass), we obtain

$$\begin{aligned}
&\frac{\partial^2 n_i}{\partial t^2} + \nabla \cdot \frac{\partial}{\partial t}(n_i \boldsymbol{v}_i) \\
&\approx \frac{\partial^2 n_i}{\partial t^2} + \nabla \cdot \left\{ \boldsymbol{v}_i \nabla \cdot (n_i \boldsymbol{v}_i) - n_i \boldsymbol{v}_i \cdot \nabla \boldsymbol{v}_i - \frac{e}{m_i} n_i \nabla\phi \right\} \\
&\approx \frac{\partial^2 n_i}{\partial t^2} - \frac{e}{m_i} n_0 \nabla^2 \phi \approx 0 \,.
\end{aligned} \tag{9.163}$$

For low-frequency phenomena, the electron inertia term may be neglected in the electron momentum balance

$$\frac{\partial}{\partial t}\boldsymbol{p}_e + \underbrace{m_e c^2 \nabla \tilde{\gamma} - \boldsymbol{v}_e \times \nabla \times \boldsymbol{p}_e}_{\boldsymbol{v}_e \cdot \nabla \boldsymbol{p}_e} \approx e\nabla\phi + e\frac{\partial \boldsymbol{A}}{\partial t} - e\boldsymbol{v}_e \times \nabla \times \boldsymbol{A} - T_e \nabla \ln n_e \, . \tag{9.164}$$

Using $\boldsymbol{p}_e \approx e\boldsymbol{A}_\perp$, we find

$$m_e c^2 \nabla \tilde{\gamma} \approx e\nabla\phi - T_e \nabla \ln n_e \, , \tag{9.165}$$

where $\tilde{\gamma} = \sqrt{1+\boldsymbol{a}^2}$ and $\boldsymbol{a} = e\boldsymbol{A}_\perp / m_e c$. Quasineutrality $n_e \approx n_i$ should hold for low-frequency reactions such that a combination of Eq. (9.165) with Eq. (9.163) leads to

$$\frac{\partial^2 \delta n}{\partial t^2} - c_s^2 \nabla^2 \delta n \approx \frac{m_e}{m_i} c^2 \frac{n_0}{2} \left. \nabla^2 \boldsymbol{a}^2 \right|_{lf} \tag{9.166}$$

with the ion-sound velocity $c_s = \sqrt{T_e/m_i}$. We have indicated by the index lf that the ions only react on the low-frequency (lf) part of the forces. When we investigate phase velocities larger than the thermal electron velocity, the adiabatic law $P_e \sim n_e^\gamma$ should be used instead for the kinetic electron pressure P_e. Note that here γ is the adiabatic index which should not be confused with the relativistic γ-factor. That means, we should replace in Eq. (9.164)

$$\underbrace{-T_e \nabla \ln n_e}_{\text{isothermal}} \longrightarrow \underbrace{-\gamma n_e^{\gamma-2} \frac{P_{e0}}{n_0^\gamma} \nabla n_e}_{\text{adiabatic}} \, . \tag{9.167}$$

Then, instead of Eq. (9.165), we will get

$$m_e c^2 \nabla \tilde{\gamma} \approx e\nabla\phi - \gamma T_0 \frac{1}{n_0} \nabla n_e \, , \tag{9.168}$$

and we end up with

$$\frac{\partial^2 \delta n}{\partial t^2} - c_s^2 \gamma \nabla^2 \delta n \approx \frac{m_e}{m_i} c^2 \frac{n_0}{2} \left. \nabla^2 \boldsymbol{a}^2 \right|_{lf} \tag{9.169}$$

for $|V/L| \leq \omega_{pi}$. Both Eq. (9.166) as well as Eq. (9.169) have the consequence that on the slow time-scale weakly relativistic laser amplitudes

$$|\boldsymbol{a}| \sim \frac{v_{the}}{c} \tag{9.170}$$

can already produce density cavities emptying some region in space. Approximating $c \approx 3 \times 10^{10}$ cm/s and $v_{the} \approx 4.19 \times 10^7 \sqrt{T_e\,[\text{eV}]}$ cm/s, we recognize that temperatures of order 50 keV are necessary to avoid complete evacuation at $|\boldsymbol{a}| \approx 0.3$.

When we replace $-\omega_{pe} A \to E$ and use the isothermal equation of state, Eq. (9.166) will transform into

$$\frac{\partial^2 \delta n}{\partial t^2} - c_s^2 \nabla^2 \delta n \approx \varepsilon_0 \frac{\nabla^2 E^2}{2 m_i} , \tag{9.171}$$

an equation which has first been introduced by Zakharov in a model which is known as the Zakharov equations [247, 270, 292–294].

Concluding this part, we note that depending on the physics under consideration, the equation describing the nonlinear density reaction within a two-field description has different forms.

9.2.6
The Nonrelativistic Limit

The formulation chosen in the previous subsection could make us believe that there is no effect on the density when relativistic effects are absent ($\gamma \equiv 1$). However, we should be careful in performing the limit $\gamma \to 1$. The correct procedure is, for example, for linear polarization,

$$\frac{\partial^2 a}{\partial t^2} = \nabla^2 a - \frac{n_0}{n_c} \frac{(1 + \delta n)}{\gamma} a \to \nabla^2 a - \frac{n_0}{n_c} (1 + \delta n) a , \tag{9.172}$$

$$\frac{\partial^2 \delta n}{\partial t^2} = -\frac{n_0}{n_c} \delta n + \nabla^2 \gamma \mathrel{\hat{=}} -\frac{n_0}{n_c} \delta n + \frac{1}{2} \nabla \cdot \frac{\nabla a^2}{\gamma} \to -\frac{n_0}{n_c} \delta n + \frac{1}{2} \nabla^2 a^2 . \tag{9.173}$$

In the last equation, we showed that within the present scaling, the effect of the electromagnetic wave on the density perturbation originates from the *relativistic ponderomotive force* ($\sim -1/(2\gamma) \nabla a^2$) which correctly reduces to its known nonrelativistic form. In that sense, the mathematically correct term $\nabla^2 \gamma$ (up to order a^2) suggests a somewhat misleading origin.

9.2.7
One-Field Models

The coupled weakly relativistic equations sometimes allow a further reduction to one-field models. In the case of circular polarization (Eqs. 9.146 and 9.147), the driving term in the density equation is of low frequency, and one might reduce the system to a standard cubic nonlinear Schrödinger equation. The situation is different for linear polarization (Eqs. (9.142) and (9.143)), when both low-frequency reactions and higher harmonics appear.

For low-frequency reactions, the coupled weakly relativistic equations (9.146) and (9.147) for circularly polarized waves have to be supplemented by the ion density dynamics, provided the pulse length is longer than ω_{pi}^{-1}. On the other hand, for short pulses, the purely electron response can be used.

For a nearly stationary electron response, we get (using ω_{pe}^{-1} for time normalization)

$$\frac{\partial^2 a}{\partial t^2} = \frac{\partial^2 a}{\partial x^2} - \left(1 + \delta n - \frac{1}{2}|a|^2\right) a \,, \qquad \delta n \approx \frac{1}{2}\frac{\partial^2 |a|^2}{\partial x^2} \,. \tag{9.174}$$

For 1D propagation, we obtain the one-field model

$$\frac{\partial^2 a}{\partial t^2} - \frac{\partial^2 a}{\partial x^2} + a = \frac{1}{2}\left(\frac{\partial^2 |a|^2}{\partial x^2} - |a|^2\right) a \,, \tag{9.175}$$

which applies for $V/L \gg \omega_{pi}$.

For linear polarization, we start with Eq. (9.142)

$$\frac{\partial^2 a}{\partial t^2} - \frac{\partial^2 a}{\partial x^2} = -\left(1 + \delta n + \frac{1}{2}a^2\right) a \tag{9.176}$$

and Eq. (9.143) in the form

$$\frac{\partial^2 \delta n}{\partial t^2} = -\delta n + \frac{1}{2}\frac{\partial^2 a^2}{\partial x^2} \tag{9.177}$$

with ω_{pe}^{-1} as the time normal. That means that we assume short pulses with $|V/L| \gg \omega_{pi}$ such that we can neglect the ion dynamics. For the density, we have to solve a driven linear oscillator equation. The general procedure will be discussed in the section on wake field generation. Here, we present a less general ansatz.

Let us consider the modification of the first harmonic of the linear transverse mode by writing

$$a \equiv a^{(0)} = \frac{1}{2}\left[A(x,t)e^{-i\omega_\perp^{(0)} t} + A^*(x,t)e^{+i\omega_\perp^{(0)} t}\right]. \tag{9.178}$$

We make for the driven density response the ansatz

$$\delta n = N_0 + \frac{1}{2}\left(N_2 e^{-2i\omega_\perp^{(0)} t} + N_2^* e^{2i\omega_\perp^{(0)} t}\right). \tag{9.179}$$

The frequency of the transverse mode is designated by $\omega_\perp^{(0)}$; we shall come back to this definition later. Inserting Eqs. (9.179) and (9.178) into Eq. (9.177), a short calculation leads to

$$N_0 = \frac{1}{4}\frac{\partial^2 |A|^2}{\partial x^2} \,, \qquad N_2 = -\frac{1}{16\omega_\perp^{(0)2} - 4}\frac{\partial^2 A^2}{\partial x^2} \,. \tag{9.180}$$

We insert Eqs. (9.178) and (9.179) with Eq. (9.180) into Eq. (9.176) to obtain, after some calculations for the coefficient proportional to $\exp[-i\omega_\perp^{(0)} t]$,

$$\begin{aligned} &\frac{1}{2}\frac{\partial^2 A}{\partial t^2} - i\omega_\perp^{(0)}\frac{\partial A}{\partial t} + \frac{1}{2}\left(1 - \omega_\perp^{(0)2}\right) A - \frac{1}{2}\frac{\partial^2 A}{\partial x^2} + \frac{1}{8}\frac{\partial^2 |A|^2}{\partial x^2} A \\ &- \frac{1}{64\omega_\perp^{(0)2} - 16}\frac{\partial^2 A^2}{\partial x^2} A^* - \frac{3}{16}|A|^2 A = 0 \,. \end{aligned} \tag{9.181}$$

In principle, since a low-frequency density response occurs, the ion dynamics should be included too. However, for short pulses with pulse-length shorter than the inverse of the ion plasma frequencies, ions can be treated as immobile.

9.3 Stationary Wave Solutions and Their Stability

9.3.1 Fully Relativistic Maxwell Fluid Systems

Nonlinear wave solutions within the Maxwell electron-fluid model were first presented by Akhiezer and Polovin [295]. For a discussion of these solutions, we transform Eqs. (9.79)–(9.83) to a frame of reference that is moving with the normalized phase velocity $\beta = V/c$ of an electromagnetic wave by introducing $\xi = x - \beta t$ (replacing the variable x) and $\tau \equiv t$. Within this moving frame, the partial derivatives change to

$$\frac{\partial}{\partial x} \to \frac{\partial}{\partial \xi}\,, \quad \frac{\partial}{\partial t} \to \frac{\partial}{\partial t} - \beta \frac{\partial}{\partial \xi} \equiv \frac{\partial}{\partial \tau} - \beta \frac{\partial}{\partial \xi}\,. \tag{9.182}$$

Straightforward calculations lead to

$$\left(1-\beta^2\right) \frac{\partial^2 \boldsymbol{A}_\perp}{\partial \xi^2} - \frac{\partial^2 \boldsymbol{A}_\perp}{\partial t^2} + 2\beta \frac{\partial^2 \boldsymbol{A}_\perp}{\partial \xi \partial t} = \frac{n \boldsymbol{A}_\perp}{\gamma}\,, \tag{9.183}$$

$$\frac{\partial^2 \phi}{\partial \xi^2} = n - 1\,, \tag{9.184}$$

$$\frac{\partial n}{\partial t} + \frac{\partial}{\partial \xi}\left(\frac{np}{\gamma} - \beta n\right) = 0\,, \tag{9.185}$$

$$\frac{\partial p}{\partial t} - \beta \frac{\partial p}{\partial \xi} = \frac{\partial}{\partial \xi}(\phi - \gamma)\,. \tag{9.186}$$

We are looking for solutions that do not depend explicitly on t, only on ξ. In case of such stationary ($\partial/\partial t \equiv 0$ in the moving frame) solutions, n, γ, and p can be expressed as functions of $\boldsymbol{A}_\perp$ and ϕ. Integration of the stationary forms of Eqs. (9.185) and (9.186) gives

$$\frac{np}{\gamma} - \beta n = C_1\,, \tag{9.187}$$

and

$$-\beta p = \phi - \gamma + C_2\,. \tag{9.188}$$

In its stationary form, Eq. (9.80) leads together with Eq. (9.184) to

$$\beta \frac{\partial^2 \phi}{\partial \xi^2} = \frac{np}{\gamma} = \beta n - \beta\,. \tag{9.189}$$

From here, we immediately conclude

$$C_1 = -\beta\,. \tag{9.190}$$

Next, using

$$\gamma = \sqrt{1 + p^2 + A_\perp^2} \tag{9.191}$$

in Eq. (9.188), we obtain a quadratic equation for p,

$$1 + p^2 + A_\perp^2 = (C_2 + \phi + \beta p)^2 \ . \tag{9.192}$$

Still, the electrostatic potential ϕ may have a free additive constant. When introducing

$$\psi \equiv \phi + C_2 \ , \tag{9.193}$$

we determine the constant in such a way that $\phi = \phi_0 = 0$ (without loss of generality we set $\phi_0 = 0$) when $p = 0$. At that position, $\boldsymbol{A}_\perp = \boldsymbol{a}_0$ is the amplitude of the vector potential. Within this calibration of the electrostatic potential, we obtain the constant

$$C_2 = \sqrt{1 + \boldsymbol{a}_0^2} \ . \tag{9.194}$$

To summarize, by using Eqs. (9.187) and (9.188) appropriately, we can express p and n in terms of ϕ and $\boldsymbol{A}_\perp$. A short calculation leads to

$$p = \frac{\beta\psi - R}{1 - \beta^2} \ , \tag{9.195}$$

$$n = \beta \frac{\psi - \beta R}{R\,(1 - \beta^2)} \ , \tag{9.196}$$

where

$$\psi = \sqrt{1 + \boldsymbol{a}_0^2} + \phi \ , \tag{9.197}$$

$$R = \left[\psi^2 - (1 - \beta^2)\left(1 + A_\perp^2\right)\right]^{1/2} \ . \tag{9.198}$$

Next, the 1D relativistic Maxwell two-fluid model is the starting point for considering high-frequency wave motion of an electron–ion plasma in the hydrodynamical approximation. The one-dimensional (1D) propagation geometry for the waves means that all quantities only vary along the direction of propagation, let us say x. Vector quantities, of course, may have components pointing in other directions than x. The plasma is assumed to be cold, kinetic effects are neglected. For later interpretations, we formulate the system with a slight generalization to a three-species plasma consisting of electrons (e), ions (i), and positrons (p) using the index $\alpha = e, i, p$. Dimensionless quantities are introduced in the usual way, that is, lengths x, times t, velocities $\boldsymbol{v}_\alpha$, momenta $\boldsymbol{p}_\alpha$, vector $\boldsymbol{A}$ and scalar potential ϕ, and particle densities n_α are normalized by c/ω_{pe}, ω_{pe}^{-1}, c, $m_e c$, $m_e c/e$, $m_e c^2/e$, and n_0 respectively. Here, $\omega_{pe} = (n_{e0} e^2/\varepsilon_0 m_e)^{1/2}$ is the electron plasma frequency, m_e the electron rest mass, e the (absolute value of the) electron charge, and

n_{e0} the unperturbed electron density. Charges q_α are normalized to $-e$, so $q_e = 1$, $q_i = q_p = -1$; masses are normalized by m_e. In order to assure global charge neutrality, we introduce the positron to ion density ratio $\chi = n_{0p}/n_{0i}$ such that $n_{0e} = (1-\chi)n_{0i} + \chi n_{0p}$. Maxwell's equations (in SI units) will be expressed in the Coulomb gauge which leads to $\boldsymbol{A} \equiv \boldsymbol{A}_\perp$ as a result of the 1D propagation model. A further consequence of the 1D geometry is that $\boldsymbol{p}_{\alpha\perp} = q_\alpha \boldsymbol{A}_\perp$. The hydrodynamic equations for the particle densities n_α and the (parallel) momentum p_α as well as the Maxwell equations for the vector and scalar potentials $\boldsymbol{A}_\perp$ and ϕ, can be written in dimensionless form as

$$\frac{\partial^2 \boldsymbol{A}_\perp}{\partial x^2} - \frac{\partial^2 \boldsymbol{A}_\perp}{\partial t^2} = \sum_\alpha q_\alpha \varepsilon_\alpha \frac{n_\alpha}{\gamma_\alpha} \boldsymbol{p}_{\alpha\perp}, \tag{9.199}$$

$$\frac{\partial^2 \phi}{\partial x^2} = \sum_\alpha q_\alpha n_\alpha, \tag{9.200}$$

$$\frac{\partial^2 \phi}{\partial t \partial x} + \sum_\alpha q_\alpha \varepsilon_\alpha \frac{n_\alpha p_\alpha}{\gamma_\alpha} = 0, \tag{9.201}$$

$$\frac{\partial n_\alpha}{\partial t} + \varepsilon_\alpha \frac{\partial}{\partial x}\left(\frac{n_\alpha p_\alpha}{\gamma_\alpha}\right) = 0, \tag{9.202}$$

$$\frac{\partial p_\alpha}{\partial t} = \frac{\partial}{\partial x}\left(q_\alpha \phi - \frac{\gamma_\alpha}{\varepsilon_\alpha}\right), \tag{9.203}$$

where $\varepsilon_\alpha = m_e/m_\alpha$. The normalized (rest) masses are $m_e = m_p = 1$ and $m_i = 1836$ (when the ions are protons). The relativistic γ-factors are $\gamma_\alpha = \sqrt{1 + \varepsilon_\alpha^2(\boldsymbol{A}_\perp^2 + p_\alpha^2)}$ for the different species α.

Transforming to a frame of reference moving with the normalized phase-velocity $\beta = V/c \gg 1$ of an electromagnetic wave, by introducing $\xi = x - \beta t$, one gets

$$(1-\beta^2)\frac{\partial^2 \boldsymbol{A}_\perp}{\partial \xi^2} - \frac{\partial^2 \boldsymbol{A}_\perp}{\partial t^2} + 2\beta \frac{\partial^2 \boldsymbol{A}_\perp}{\partial \xi \partial t} = \sum_\alpha q_\alpha \varepsilon_\alpha \frac{n_\alpha}{\gamma_\alpha} \boldsymbol{p}_{\alpha\perp}, \tag{9.204}$$

$$\frac{\partial^2 \phi}{\partial \xi^2} = \sum_\alpha q_\alpha n_\alpha, \tag{9.205}$$

$$\frac{\partial}{\partial \xi}\left(\frac{\partial}{\partial t} - \beta \frac{\partial}{\partial \xi}\right)\phi + \sum_\alpha q_\alpha \varepsilon_\alpha \frac{n_\alpha p_\alpha}{\gamma_\alpha} = 0, \tag{9.206}$$

$$\frac{\partial n_\alpha}{\partial t} + \frac{\partial}{\partial \xi}\left(\varepsilon_\alpha \frac{n_\alpha p_\alpha}{\gamma_\alpha} - \beta n_\alpha\right) = 0, \tag{9.207}$$

$$\frac{\partial p_\alpha}{\partial t} - \beta \frac{\partial p_\alpha}{\partial \xi} = \frac{\partial}{\partial \xi}\left(q_\alpha \phi - \frac{\gamma_\alpha}{\varepsilon_\alpha}\right). \tag{9.208}$$

The linear dispersion relation suggests $\beta \to 1$ when the plasma density approaches zero, whereas $\beta \to \infty$ occurs when one approaches the critical density. In the new frame of reference, we will discuss stationary solutions. Then, all quantities only depend on ξ and not explicitly on t (effectively meaning $\partial/\partial t = 0$). We therefore start from

$$\left(1-\beta^2\right)\frac{\partial^2 \boldsymbol{A}_\perp}{\partial\xi^2} = \sum_\alpha q_\alpha \varepsilon_\alpha \frac{n_\alpha}{\gamma_\alpha}\boldsymbol{p}_{\alpha\perp}\,, \tag{9.209}$$

$$\frac{\partial^2\phi}{\partial\xi^2} = \sum_\alpha q_\alpha n_\alpha\,, \tag{9.210}$$

$$-\beta\frac{\partial^2\phi}{\partial\xi^2} + \sum_\alpha q_\alpha\varepsilon_\alpha\frac{n_\alpha p_\alpha}{\gamma_\alpha} = 0\,, \tag{9.211}$$

$$\frac{\partial}{\partial\xi}\left(\varepsilon_\alpha\frac{n_\alpha p_\alpha}{\gamma_\alpha} - \beta n_\alpha\right) = 0\,, \tag{9.212}$$

$$\frac{\partial}{\partial\xi}\left(\beta p_\alpha + q_\alpha\phi - \frac{\gamma_\alpha}{\varepsilon_\alpha}\right) = 0\,. \tag{9.213}$$

It is possible to express n_α, p_α, and γ_α as functions of $\boldsymbol{A}_\perp$ and ϕ. Integrating Eqs. (9.212) and (9.213), one gets

$$\varepsilon_\alpha\frac{n_\alpha p_\alpha}{\gamma_\alpha} = C_{\alpha 1} + \beta n_\alpha \tag{9.214}$$

and

$$q_\alpha\phi - \frac{\gamma_\alpha}{\varepsilon_\alpha} + \beta p_\alpha = -C_{\alpha 2}\,. \tag{9.215}$$

We determine $C_{\alpha 1}$ by postulating in Eq. (9.214) for a uniform plasma $p_\alpha = 0$ at $n_e = 1$, $n_i = (1-\chi)$, $n_p = \chi$. This results into $C_{e1} = -\beta$, $C_{i1} = -\beta(1-\chi)$, and $C_{p1} = -\beta\chi$.

Next, inserting

$$\gamma_\alpha = \sqrt{1 + \varepsilon_\alpha^2(A_\perp^2 + p_\alpha^2)} \tag{9.216}$$

into Eq. (9.215), we obtain a quadratic equation for p_α,

$$1 + \varepsilon_\alpha^2(A_\perp^2 + p_\alpha^2) = \varepsilon_\alpha^2(q_\alpha\phi + \beta p_\alpha + C_{\alpha 2})^2\,. \tag{9.217}$$

We determine $C_{\alpha 2}$ in such a way that $\phi = 0$ where $p_\alpha = 0$ and $A_\perp^2 = a_0^2$. With this calibration, we get

$$C_{\alpha 2} = \sqrt{\frac{1}{\varepsilon_\alpha^2} + a_0^2}\,. \tag{9.218}$$

When focusing on linearly polarized waves, we introduce $\boldsymbol{A}_\perp = a\hat{\boldsymbol{e}}_z$. Defining

$$\psi_\alpha = \varepsilon_\alpha(q_\alpha\phi + C_{\alpha 2})\,, \tag{9.219}$$

$$R_\alpha = \left[\psi_\alpha^2 - \left(1-\beta^2\right)\left(1 + \varepsilon_\alpha^2 a^2\right)\right]^{1/2}, \tag{9.220}$$

we find

$$p_\alpha = \frac{\beta\psi_\alpha - R_\alpha}{\varepsilon_\alpha(1-\beta^2)}\,, \tag{9.221}$$

$$n_\alpha = \frac{C_{\alpha 1}(\psi_\alpha - \beta R_\alpha)}{R_\alpha(\beta^2-1)}\,, \tag{9.222}$$

$$\gamma_\alpha = \frac{\psi_\alpha - \beta R_\alpha}{1-\beta^2}\,. \tag{9.223}$$

We end up with the following two coupled differential equations for the (normalized) potentials a and ϕ:

$$\frac{d^2 a}{d\xi^2} = -a\frac{\beta}{\beta^2-1}\left[\frac{1}{R_e} + \frac{\chi}{R_p} + \varepsilon_i\frac{1-\chi}{R_i}\right], \tag{9.224}$$

$$\frac{d^2\phi}{d\xi^2} = \frac{-\beta}{\beta^2-1}\left[\frac{\psi_e}{R_e} - \chi\frac{\psi_p}{R_p} - (1-\chi)\frac{\psi_i}{R_i}\right]. \tag{9.225}$$

Equations (9.224) and (9.225) describe the nonlinear relativistic coupling between transverse and longitudinal oscillations. The system Eqs. (9.224) and (9.225) constitute the starting point for the following investigation.

For immobile ions and linear polarization, the equations of motion reduce to

$$\frac{d^2 a}{d\xi^2} = -a\frac{\beta}{\beta^2-1}\frac{1}{R_e}\,, \tag{9.226}$$

$$\frac{d^2\phi}{d\xi^2} = \frac{-\beta}{\beta^2-1}\left[\frac{\psi_e}{R_e} - \frac{1}{\beta}\right]. \tag{9.227}$$

Equations (9.226) and (9.227) follow from Eqs. (9.224) and (9.225) for $\chi = 0$ in the limit $\varepsilon_i \to 0$.

Within a Lagrangian formulation, Eqs. (9.226) and (9.227) follow from the Lagrange density

$$L = (\beta^2-1)\frac{\dot{a}^2}{2} + \frac{\dot{\phi}^2}{2} - \frac{\beta}{\beta^2-1}(R_e - \beta) + \frac{\phi}{\beta^2-1} \equiv L_0\,. \tag{9.228}$$

Introducing the canonical momenta

$$p_a \equiv \frac{\partial L}{\partial \dot{a}} = (\beta^2-1)\dot{a}\,, \quad p_\phi \equiv \frac{\partial L}{\partial \dot{\phi}} = \dot{\phi}\,, \tag{9.229}$$

we obtain the Hamiltonian

$$\begin{aligned} H &\equiv p_a\dot{a} + p_\phi\dot{\phi} - L = \frac{p_a^2}{2(\beta^2-1)} + \frac{p_\phi^2}{2} + \frac{\beta}{\beta^2-1}(R_e-\beta) - \frac{\phi}{\beta^2-1} \\ &\equiv H_0 \equiv T + V_0 \end{aligned} \tag{9.230}$$

for laser intensities and phase-velocities β where the ion motion is negligible [296, 297]. The dot indicates the derivative with respect to ξ. The canonical equations

$$\dot{a} = \frac{\partial H}{\partial p_a}\,, \quad \dot{\phi} = \frac{\partial H}{\partial p_\phi}\,, \quad \dot{p}_a = -\frac{\partial H}{\partial a}\,, \quad \dot{p}_\phi = -\frac{\partial H}{\partial \phi} \tag{9.231}$$

are equivalent to Eqs. (9.226) and (9.227). The Hamiltonian formulation is very useful to study coupled electromagnetic wave solutions. From the canonical equations, we conclude $H = const$ because of

$$\frac{dH}{d\xi} = \frac{\partial H}{\partial a}\dot{a} + \frac{\partial H}{\partial \phi}\dot{\phi} + \frac{\partial H}{\partial p_a}\dot{p}_a + \frac{\partial H}{\partial p_\phi}\dot{p}_\phi = 0\,. \tag{9.232}$$

The system of Eqs. (9.226) and (9.227) is equivalent to the description that Akhiezer and Polovin [295] obtained when studying relativistic plasma motion in coupled longitudinal and transversal electromagnetic waves. Many articles on relativistic plasma wave phenomena used a set of equations equivalent to Eqs. (9.226) and (9.227). However, the range of applicability is limited to a parameter regime where ion motion can be neglected. Certainly for laser amplitudes $a \sim 1/\varepsilon_i$, ions can gain relativistic momenta and should be included as mobile species into the model. Since much effort is devoted to push laser intensities into and beyond such regimes, it seems important to develop an understanding of wave phenomena in plasmas where the ion response may be relativistic. However, one should note that the electrons already become strongly relativistic for $a \sim 1$. Therefore, for $1 < a \ll 1/\varepsilon_i$, the electron inertia already increases and the lower mobility of ions becomes less important.

It is possible to calculate the motion of a plasma fluid element by expressing the momenta $\boldsymbol{p}_\alpha$ in terms of the potentials. In general, the motion of a fluid element consists of an average drift in propagation direction of the wave and a superposed motion in the frame where the average momenta are zero. A systematic discussion of plasma motion in linearly polarized periodic waves assuming an electron plasma with immobile ions can be found in [295, 297, 298], and will partially be outlined in the following.

9.3.2 Hamiltonian Formulation for Linearly Polarized Waves

In the following, let us assume a *linearly* polarized[1)] electromagnetic wave with $\boldsymbol{A}_\perp = a\boldsymbol{e}_z$. We calculate the γ-factor

$$\gamma = \frac{\psi - \beta R}{1 - \beta^2} \tag{9.233}$$

1) The following discussions are for linear polarization. Similar analysis can be performed for circular polarization.

such that the equations for the potentials are

$$\left(1-\beta^2\right)\frac{d^2 a}{d\xi^2} = a\frac{\beta}{R}\,, \tag{9.234}$$

$$\frac{d^2\phi}{d\xi^2} = n - 1\,. \tag{9.235}$$

The equations show the known coupling between transverse and longitudinal oscillations.

We will follow [296] and rewrite Eqs. (9.234) and (9.235) in terms of new variables

$$X = a\sqrt{\beta^2-1}\,,\quad Z = -\sqrt{1+a_0^2}-\phi\,,\quad \zeta = \frac{\xi}{\sqrt{\beta^2-1}}\,. \tag{9.236}$$

This leads to two nonlinear coupled oscillators X and Z of the form

$$\ddot{X} + \frac{\beta X}{\sqrt{\beta^2-1+X^2+Z^2}} = 0\,, \tag{9.237}$$

$$\ddot{Z} + \frac{\beta Z}{\sqrt{\beta^2-1+X^2+Z^2}} + 1 = 0\,, \tag{9.238}$$

where the dots denote derivatives with respect to ζ. These oscillators can be derived from the Hamiltonian

$$H = \frac{1}{2}\left(\dot{X}^2+\dot{Z}^2\right) + \beta\sqrt{\beta^2-1+X^2+Z^2} + Z\,. \tag{9.239}$$

Equations (9.237) and (9.238) are the corresponding canonical equations for $P_X = \dot{X}$ and $P_Z = \dot{Z}$. They are equivalent to the equations derived by Akhiezer and Polovin [295, Eq. 16]. Instead of expressing the stationary solutions in terms of the fluid momenta p_x, p_y, p_z, everything is formulated in equations for the potentials.

The integrability of the Hamiltonian system Eq. (9.239) has been studied analytically and numerically [299, 300]. The numerical studies suggest that the system is not integrable since chaotic trajectories have been found in the high-energy limit. For magnetized systems, a similar set of equations was derived and the aperiodic modulation of the coupled transverse-longitudinal plasma oscillations was found [301].

For the solutions ϕ and a, we may resort to the solutions X and Z provided we determine the parameter a_0 which corresponds to a solution X and Z at fixed energy $E = H$ from

$$Z(\zeta) = -\sqrt{1+\frac{X^2(\zeta)}{\beta^2-1}} \quad \text{at} \quad \zeta = \zeta_0\,, \tag{9.240}$$

such that

$$a_0 = \frac{X(\zeta_0)}{\sqrt{\beta^2-1}}\,. \tag{9.241}$$

It is clear that for fixed energy, a specific solution of X and Z corresponds to solutions of ϕ and a with fixed maximum amplitude $\tilde{a}_0$. The latter follows from the obtained value of a_0.

9.3.3
Plasma Motion in Linearly Polarized Waves

When characterizing the plasma motion associated with the coupled transverse-longitudinal plasma motion, we follow [297]. One starts from the mean electron momentum p and defines a mean position $\boldsymbol{r}$ with components x, y, z via the expression for the velocity,

$$\frac{dx(t)}{dt} \equiv \frac{p}{\gamma}, \quad \frac{dz(t)}{dt} \equiv \frac{a}{\gamma}. \tag{9.242}$$

Making use of the results for $p_z = a$ and $p_x = p$, we find

$$\begin{aligned} \gamma \frac{dx(t)}{dt} &= \gamma \frac{dx}{d\zeta}\frac{d\zeta}{dt} = \frac{1}{\sqrt{\beta^2 - 1}}\gamma(\dot{x} - \beta)\frac{dx}{d\zeta} \\ &= \frac{1}{\sqrt{\beta^2 - 1}}(p - \beta\gamma)\frac{dx}{d\zeta} = -\frac{R}{\sqrt{\beta^2 - 1}}\frac{dx}{d\zeta}, \end{aligned} \tag{9.243}$$

and similarly

$$\gamma \frac{dz(t)}{dt} = \gamma \frac{dz}{d\zeta}\frac{d\zeta}{dt} = -\frac{R}{\sqrt{\beta^2 - 1}}\frac{dz}{d\zeta}. \tag{9.244}$$

Thus, the equations of motion can be written as

$$\frac{dx(\zeta)}{d\zeta} = -\sqrt{\beta^2 - 1}\frac{p}{R} = -\frac{\beta Z + \sqrt{Z^2 + X^2\beta^2 - 1}}{\sqrt{\beta^2 - 1}}, \tag{9.245}$$

$$\frac{dz(\zeta)}{d\zeta} = -\sqrt{\beta^2 - 1}\frac{a}{R} = -\frac{X}{\sqrt{Z^2 + X^2\beta^2 - 1}}. \tag{9.246}$$

Interpreting ζ as the parameter for the curve $x = x(z)$, one can draw the path $x = x(z)$.

Solutions to Eqs. (9.237) and (9.238) can be periodic, quasiperiodic or chaotic. All solutions can be expressed in terms of momenta p_x, p_y, p_z of fluid elements. In the laboratory frame, the motion can be decomposed into an average drift in the propagation direction of the electromagnetic wave and a motion within this average drift frame. It is common to describe periodic solutions to Eqs. (9.237) and (9.238) as trajectories within this average drift frame. The most prominent periodic solutions are probably the well-known figure-eight and the circular motion of the fluid elements.

Akhiezer and Polovin [295] presented exact solutions in the form of elliptic functions when the electromagnetic wave is absent ($X = \dot{X} = 0$). More general cases, for $\beta \geq 1$, were reported by Kaw and Dawson [302], Max and Perkins [303], and Chian and Clemmow [304]. The linearly polarized solutions have a small longitudinal component with twice the frequency of the transverse momentum. In the average drift frame, the fluid elements perform a motion along an eight-like trajectory in the plane spanned by propagation- and polarization-direction.

The possibility of circular trajectories was also noted by Akhiezer and Polovin [295], and was worked out by Clemmow [305, 306] and Pesch [298]. For these solutions, the parallel and the longitudinal momenta are of the same order. The solutions exist below and above the critical plasma density. Above the critical density, they are purely electrostatic, below the critical plasma density an additional electromagnetic component is present. For small plasma densities, the circular solution disappears. Additional periodic solutions exist. Equations (9.237) and (9.238) describe trajectories in a four-dimensional phase space. One may render projections in order to identify new structures. The trajectories are calculated by standard Runge–Kutta methods from initial conditions. If not stated otherwise, one must always consider $X = 0$, $\dot{Z} = 0$ for $\zeta = 0$. The projections are carried out by using one of the oscillators as a clock, while plotting the coordinates of the other one.

Figure 9.2 shows a Poincaré surface plot for $\beta = 2.55$ and $H = 53.4$. X versus $\dot{X}$ is plotted every time that the Z oscillator has a minimum. Five sets of points within the section plot are marked. They correspond to periodic solutions. The two elliptic fixed-points labeled (i) originate from the figure-eight motion, Z has twice

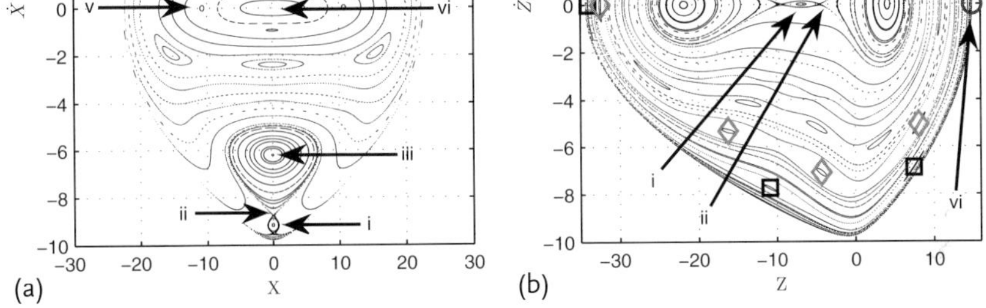

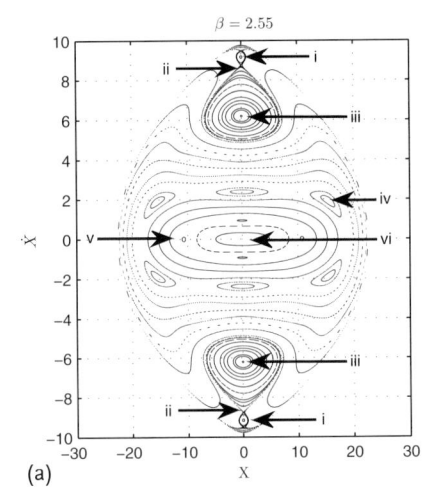

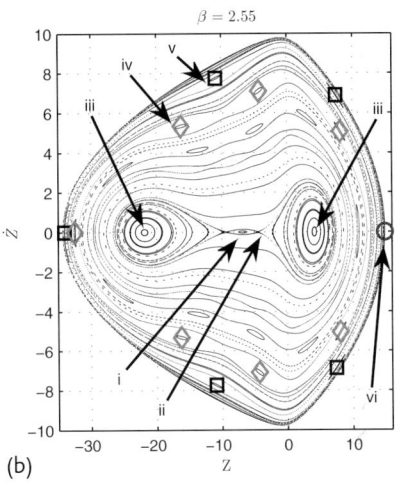

Figure 9.2 Poincaré surface plot of X versus $\dot{X}$ (for $Z = Z_{min}$), for $H = 53.4$ and $\beta = 2.55$ [297]. The fixed-points which correspond to the periodic wave-motions are labeled: (i) figure-eight, (ii) deformed figure-eight, (iii) circular. The points (iv) and (v) correspond to periodic higher-order amplitude modulated solutions. The island chains for the solutions (iv) and (v) are marked by diamonds and squares in the Z vs. $\dot{Z}$ plot, respectively. A purely longitudinal solution is represented by the point (vi) (additionally marked by a circle in the Z vs. $\dot{Z}$ plot). For these periodic solutions, the oscillators X and Z, and the wave-motion trajectories in the average drift frame are shown in Figures 9.3 and 9.4.

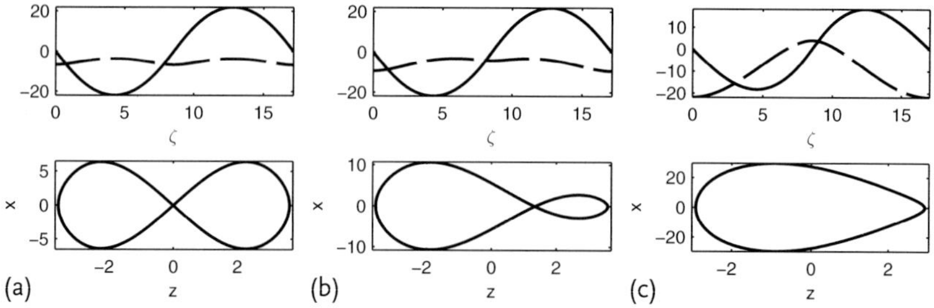

Figure 9.3 Oscillators X (solid line) and Z (dashed line) and the corresponding wave-motion trajectory (lower row) in the average drift frame for basic periodic solutions (Figure 9.2 i–iii) at $\beta = 2.55$ with $H = 53.4$ [297]. (a) Figure-eight; (b) deformed figure-eight; (c) circular solution.

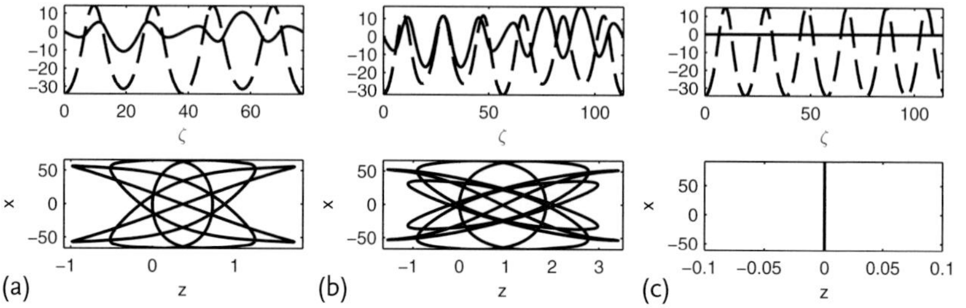

Figure 9.4 Oscillators X (solid line) and Z (dashed line) and the corresponding wave-motion trajectory (lower row) in the average drift frame for higher-order periodic solutions (Figure 9.2 iv–vi) at $\beta = 2.55$ with $H = 53.4$ [297]. Higher order solution iv (a); higher order solution v (b); purely longitudinal solution vi (c).

the frequency of X. For every β and H, the two points can be identified within the corresponding surface plot. Below the critical plasma density, there exists a figure-eight solution with any amplitude $\tilde{a}_0$ at any density. The hyperbolic fixed-points labeled (ii) correspond to a motion which resembles a deformed figure-eight motion, where one loop of the eight is smaller than the other one. This type of solution has already been observed by the authors of [298]. Just like for the point labeled (iii), which belongs to the circular motion, these solutions do not exist below certain critical velocities. The island chains with the labels (iv) and (v) describe higher-order amplitude modulated solutions for which neither X or Z resemble harmonic oscillations. Those much more complex trajectories exist only within a certain velocity window as well. The fixed-point labeled (vi) defines a purely longitudinal plasma wave, in which the fluid elements perform oscillations only in the propagation direction of the wave.

Oscillator solutions are shown in Figures 9.3 and 9.4 together with the corresponding trajectories of fluid elements in the average drift frame. The trajectories have been calculated by Eqs. (9.245) and (9.246). The shown solutions correspond to the points of Figure 9.2 which carry labels. All the different solutions exist simultaneously for fixed H and β. In particular, there is no transition from figure-eight to deformed figure-eight, to circular trajectories with increasing plasma density (starting from the vacuum limit). For any plasma density, we can find figure-eight trajectories which are generalizations of the vacuum figure-eight solution. The solutions exist up to the critical plasma density. Solutions describing circular and deformed figure-eight trajectories can not be found below certain plasma densities. The deformed figure-eight motion, like the higher-order solutions, will vanish at higher plasma densities.

9.3.4
Influence of Mobile Ions on Stationary Wave Solutions

We will now generalize to a plasma consisting of electrons and *mobile* ions [307]. To demonstrate, we use the largest value of ε_i, that is, $\varepsilon_i = 1/1836$ and linear polarization. The equations of motion follow for $\chi = 0$ in the form

$$\frac{d^2 a}{d\xi^2} = -a\frac{\beta}{\beta^2-1}\left(\frac{1}{R_e} + \varepsilon_i\frac{1}{R_i}\right), \tag{9.247}$$

$$\frac{d^2\phi}{d\xi^2} = \frac{-\beta}{\beta^2-1}\left(\frac{\psi_e}{R_e} - \frac{\psi_i}{R_i}\right). \tag{9.248}$$

They can be derived from the Lagrangian

$$L = \frac{\beta^2-1}{2}\dot{a}^2 + \frac{\dot{\phi}^2}{2} + \frac{\beta}{1-\beta^2}\left(R_e - \beta + \frac{R_i-\beta}{\varepsilon_i}\right). \tag{9.249}$$

Introducing canonical momenta in the same way as above, we find the Hamiltonian

$$H = \frac{p_a^2}{2(\beta^2-1)} + \frac{p_\phi^2}{2} + \frac{\beta}{\beta^2-1}\left(R_e - \beta + \frac{R_i-\beta}{\varepsilon_i}\right). \tag{9.250}$$

Compared to the immobile ion approximation, we have a perturbed Hamiltonian

$$H = H_0 + \Delta H\,, \quad H_0 = T + V_0\,, \quad \Delta H = \frac{\beta}{\beta^2-1}\left(\frac{R_i-\beta}{\varepsilon_i} + \frac{\phi}{\beta}\right). \tag{9.251}$$

In the limit $\varepsilon_i \to 0$, the following limits can be calculated for finite a:

$$\psi_i \to 1\,, \quad R_i \to \beta\,, \quad C_{i2} \to \frac{1}{\varepsilon_i}\,, \tag{9.252}$$

$$\frac{\psi_i}{R_i} \to \frac{1}{\beta}\,, \quad \frac{R_i-\beta}{\varepsilon_i} \to -\frac{\phi}{\beta}\,. \tag{9.253}$$

Therefore,

$$\Delta H \to 0 \quad \text{for} \quad \varepsilon_i \to 0\,. \tag{9.254}$$

However, for fixed $\varepsilon_i \ll 1$,

$$\Delta H \ll V_0 \quad \text{only for} \quad max\,(a) \to 0\,. \tag{9.255}$$

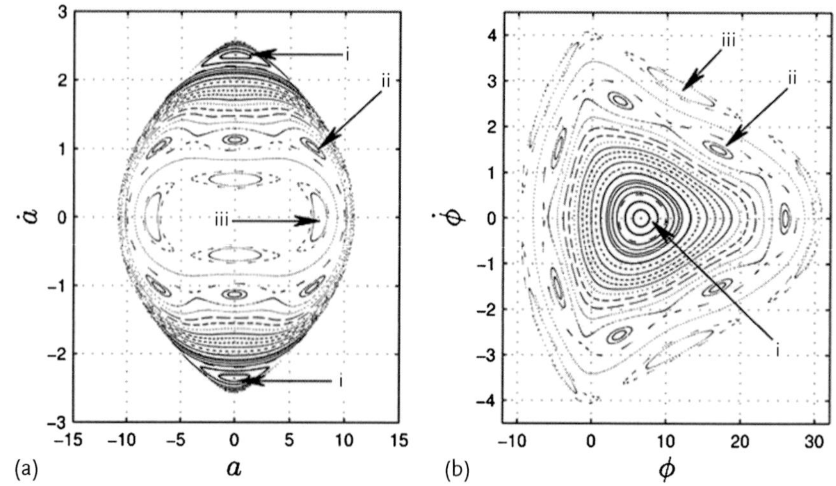

Figure 9.5 Poincaré surface plot of $\dot{a}$ versus a (for $\phi = \phi_{min}$) and $\dot{\phi}$ versus ϕ (for $a = 0, \dot{a} > 0$), for $H = 10$ and $\beta = 2$ with mobile ions [307]. The fixed-point (i) and the island-chains ((ii), (iii)), corresponding to periodic wave motions are labeled.

On the other hand, when the laser amplitude a becomes larger, even for small ε_i, the "perturbation" ΔH cannot be neglected anymore compared to the zeroth order Hamiltonian.

In order to study the variety of possible solutions of Eqs. (9.247) and (9.248) at fixed values of H and β, again, one may render Poincaré section plots of the four-dimensional phase space. Equations (9.247) and (9.248) are nonlinearly coupled oscillators which can be solved by standard Runge–Kutta algorithms. In phase-space projections, we use one oscillator as a clock while plotting the other one every time the clock ticks. Using a as a clock, ticking every time $a = 0$ and $\dot{a} > 0$, we plot ϕ vs. $\dot{\phi}$. When using ϕ as a clock, we plot a vs. $\dot{a}$ every time $\dot{\phi} = 0$ and $\ddot{\phi} > 0$. With the help of these projections, it is easy to identify periodic solutions as they correspond to fixed-points or island-chains. The initial conditions for all trajectories are such that we require $a = 0, \dot{\phi} = 0$ at $\xi = 0$.

Figure 9.5 shows Poincaré section plots for $\beta = 2$ and $H = 10$. Three periodic solutions (i)–(iii) are labeled in the plots. Solution (i) is the famous figure-eight solution, having an electromagnetic field oscillating with twice the frequency of the electrostatic field. The island-chains (ii) and (iii) correspond to higher-order amplitude modulated solutions. It is interesting to note that one obtains qualitatively the same types of solutions as the ones appearing in the Akhiezer–Polovin model [297]. Of course, a detailed quantitative comparison of the phase-space structures reveals small changes. However, the conclusion is that for the chosen parameter regime, the ion motion is still negligible.

Next, we further increase the energy H. With increasing energies, more and more formerly closed KAM (Kolmogorov–Arnold–Moser) surfaces break-up and form island-chains. For $H = 600$, the phase-space topology has changed signifi-

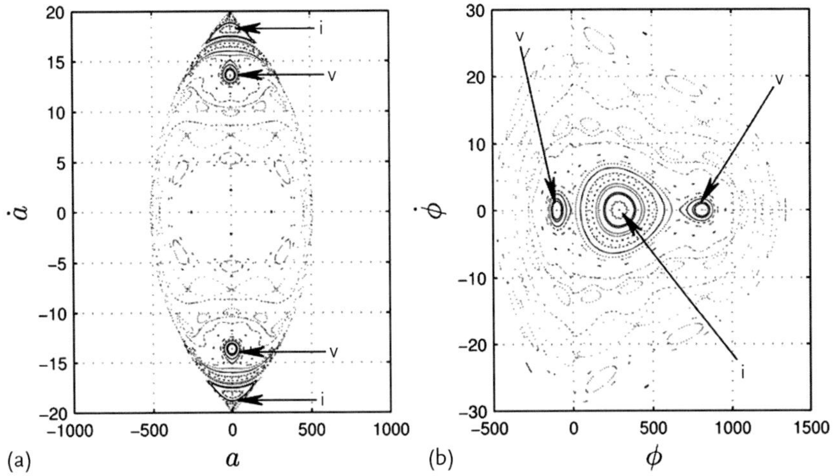

Figure 9.6 Poincaré surface plot of $\dot{a}$ versus a (for $\phi = \phi_{min}$) and $\dot{\phi}$ versus ϕ (for $a = 0, \dot{a} > 0$), for $H = 600$ and $\beta = 2$ [307]. The fixed-points (i), (v), corresponding to periodic wave motions are labeled.

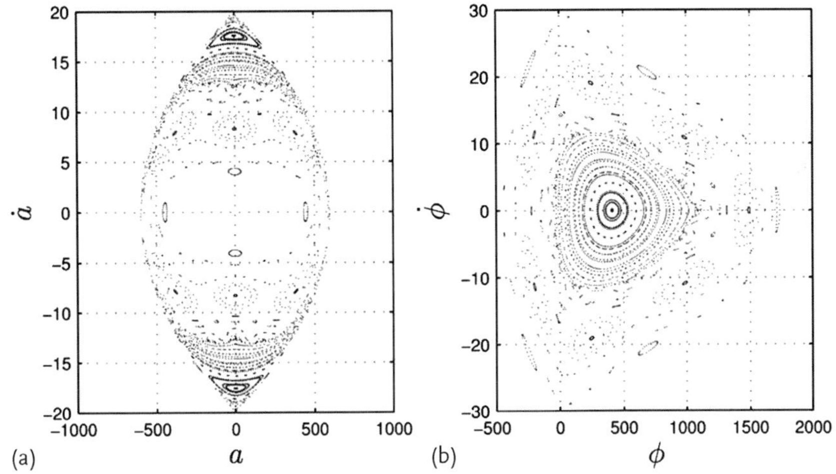

Figure 9.7 Poincaré surface plot of $\dot{a}$ versus a (for $\phi = \phi_{min}$) (a) and $\dot{\phi}$ versus ϕ (for $a = 0, \dot{a} > 0$) (b), for $H = 600$ and $\beta = 2$ [307]. The ion motion has been neglected here. Notice the different structure of the phase-space compared to Figure 9.6.

cantly; very pronounced new fixed-points appear. Figure 9.6 demonstrates the beginning of a new fixed-point labeled (v). The solutions a and ϕ are of the same magnitude and share a common frequency. Although electrons are strongly relativistic, ions are still weakly relativistic. The plasma motion is a circular trajectory in the average drift frame. Besides the beginning of the new fixed-points, we observe an increase in the amount of island-chains.

Let us discuss whether the change in phase-space topology, and thereby the appearance of new solutions (v) should be attributed to ion motion. For that, we carry out Poincaré plots for the same values $\beta = 2$ and $H = 600$, however using the Hamiltonian equation (9.230) for fixed ions. Figure 9.7 shows that the solution (v) does not appear in the case of immobile ions.

A sense of caution is, however, appropriate here. As long as the energy is not too large, and depending on the choice of the other control parameter β, solutions of type (v) can also be observed in systems consisting of electrons with a fixed ion background. Regimes for solutions (v) were detected in [297, 298, 308]. For example, a phase-space structure similar to the one shown in Figure 9.6 can be found at $\beta = 2.5$ and $H = 100$ for a system with immobile ions. In the latter case, the separatrix that is formed around the islands containing (i) and (v) is a closed KAM surface [309]. Then, the ion-motion leads to a stochastization of the separatrix and the vanishing of hyperbolic fixed-points along the separatrix.

Hyperbolic fixed-points represent (deformed) figure-eight motions [297]. With increasing H, the volume contained by the central island containing the fixed-point (i) becomes smaller, while the volumes of the islands around the fixed-points (v) grow. At higher intensities, the ion-motion becomes strongly relativistic ($\varepsilon_i \phi > 1$). One observes that with increasing energy, the islands around the fixed-point (i) shrink, while the islands around solution (v) grow. The process continues with increasing values of H, until only the islands of (v) are left, separated by a stochastic separatrix. Figure 9.8 shows Poincaré section plots for $\beta = 2$ and $H = 2000$. Compared to Figure 9.6, the fixed-point corresponding to solution (i) no longer exists. Solutions of type (v) and higher-order amplitude modulated solutions still existent. This is a generic result. Strongly ultrarelativistic plasmas do not show the figure-eight solution anymore. The manifold of solutions becomes relatively simple. Besides the higher-order amplitude modulated solutions which occupy little

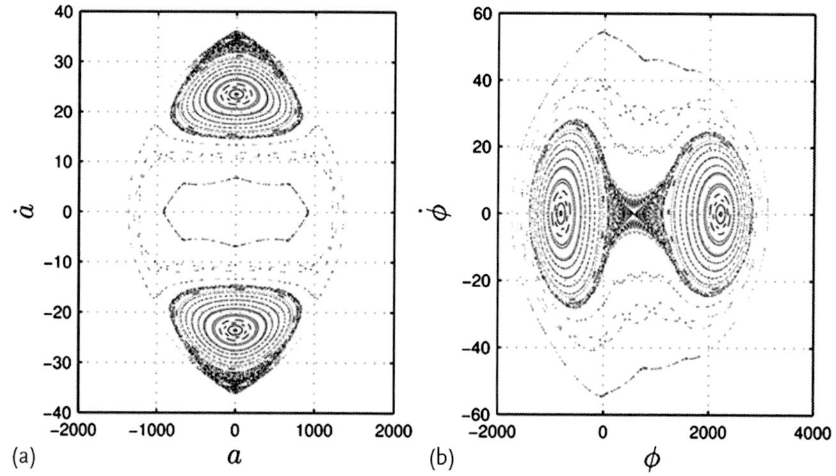

Figure 9.8 Poincaré surface plot of $\dot{a}$ versus a (a) (for $\phi = \phi_{min}$) and $\dot{\phi}$ versus ϕ (b) (for $a = 0, \dot{a} > 0$), for $H = 2000$ and $\beta = 2$ with mobile ions [307].

phase space, the circular motion and the surrounding island are the dominating surviving structures.

9.3.5
Electron–Positron Plasmas

When H becomes large and the typical amplitude a approaches $1/\varepsilon_i$, electrons quivering in the strongly ultrarelativistic waves effectively become heavy. The quantitative differences in effective masses between electrons and ions are expected to diminish. In the present section, we quantify that scenario for relativistic plasma waves.

Electron and ion motion within the relativistic waves follows from Eqs. (9.214) and (9.215) for the parallel component together with $\boldsymbol{p}_{\alpha\perp} = q_\alpha \boldsymbol{A}_\perp$ for the perpendicular component. The velocities are varying, depending on the position in the wave. Then, the relativistic γ-factor also changes with ξ. We obtain the relativistic γ-factor via expression (9.223). Typical examples for γ_e and γ_i at large wave energies are shown in Figure 9.9 [307].

Because of the variations with ξ, we should average γ over one oscillation period for obtaining an averaged γ-factor. After multiplication with the corresponding rest mass, the averaged effective masses can be calculated. Evaluating the averaged effective masses for a series of waves, depending on the Hamiltonian H, the graphs shown in Figure 9.10 appear [307]. The effective electron to ion mass ratio is altered by relativistic effects in such a way that the ratio increases with higher intensities. For $H \to \infty$, the effective mass ratio $m_e\langle\gamma_e\rangle/m_i\langle\gamma_i\rangle \to 1$. However, for large H quantum effects, being ignored in the present formulation, will enter into the picture. Quantum effects become important when the photons being generated due to Compton scattering have energies of the order of the electron energy $\gamma_e m_e c^2$.

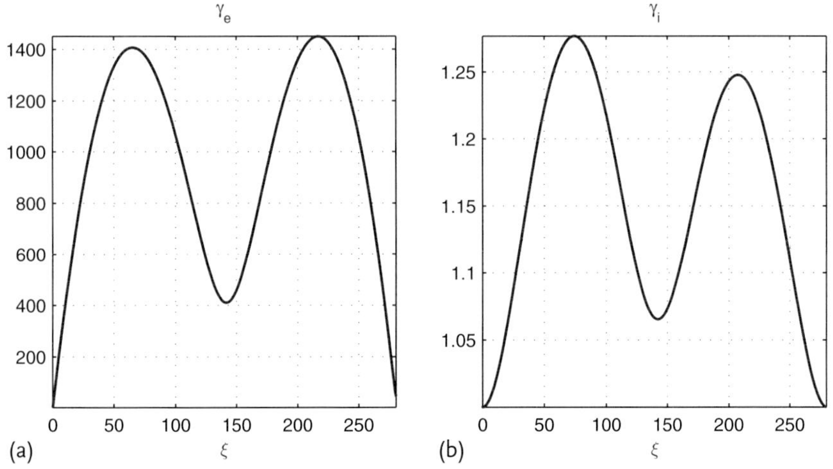

Figure 9.9 The relativistic γ-factors γ_e (a) and γ_i (b) calculated from Eq. (9.223) for a specific wave solution with $\beta = 2$, $H = 2000$.

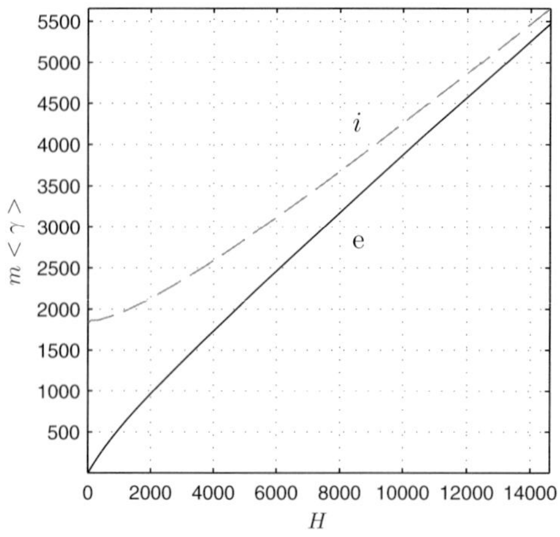

Figure 9.10 Averaged effective masses $m\langle\gamma\rangle \hat{=} m_\alpha\langle\gamma_\alpha\rangle$ of electrons ($\alpha = e$) and ions (protons) ($\alpha = i$) as functions of H for $\beta = 2$.

The frequency of the photon being generated by an electron rotating with frequency ω is $\omega_n = \gamma_e^3\omega$. From here, we may estimate that for $\gamma_e \geq \mathcal{O}(10^3)$ (when using wavelength $\lambda = 1\,\mu\text{m}$), quantum effects come into play and the purely classical treatment breaks down. At $\lambda = 1\mu m$, the condition for quantum effects may be translated into the condition $a \geq \mathcal{O}(10^3)$ for the amplitude a. These estimates show that at least up to $H \leq \mathcal{O}(2 \times 10^3)$, the present classical treatment will apply. Of course, radiation damping which already might set in at lower values is not taken into account within the Hamiltonian formulation.

The results for ultrarelativistic electron–ion plasmas suggest that for high intensities, the electron–ion system behaves qualitatively like a electron–positron plasma. The analogy to an electron–positron plasma with respect to the topology of the phase-space will be worked out in the following to support this expectation. Pair plasmas have attracted special attention mainly because of the enormous astrophysical applications [310].

Setting $\chi = 1$ in the basic formulas (9.224) and (9.225), we obtain the equations of motion for an electron–positron plasma

$$\frac{d^2 a}{d\xi^2} = -a\frac{\beta}{\beta^2 - 1}\left(\frac{1}{R_e} + \frac{1}{R_p}\right), \tag{9.256}$$

$$\frac{d^2\phi}{d\xi^2} = \frac{-\beta}{\beta^2 - 1}\left(\frac{\psi_e}{R_e} - \frac{\psi_p}{R_p}\right). \tag{9.257}$$

They follow from the Lagrangian

$$L = \frac{\beta^2 - 1}{2}\dot{a}^2 - \frac{\dot{\phi}^2}{2} + \frac{\beta}{\beta^2 - 1}\left(R_e - \beta + R_p - \beta\right). \tag{9.258}$$

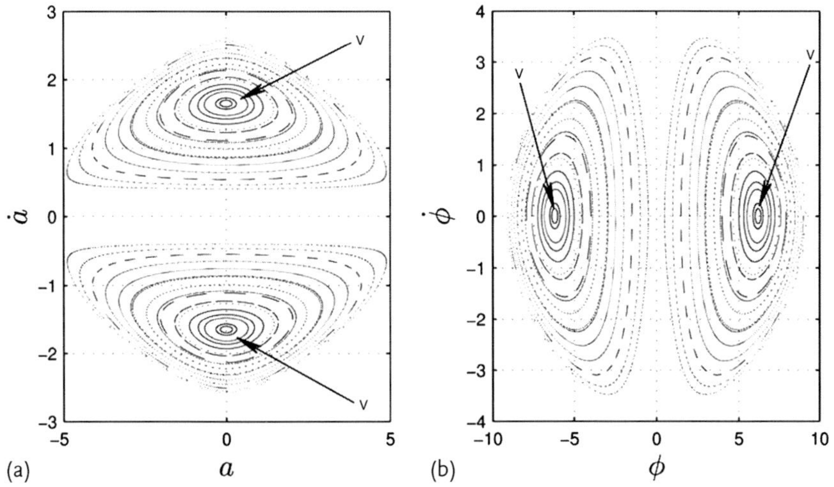

Figure 9.11 Poincaré section plots of $\dot{a}$ versus a (a) and $\dot{\phi}$ versus ϕ (b) for electron–positron plasmas with $H = 10$ and $\beta = 2$.

The corresponding Hamiltonian is

$$H = \frac{p_a^2}{2(\beta^2 - 1)} + \frac{p_\phi^2}{2} - \frac{\beta}{\beta^2 - 1}\left(R_e - \beta + R_p - \beta\right). \tag{9.259}$$

Typical Poincaré section plots of $\dot{a}$ versus a and $\dot{\phi}$ versus ϕ for electron–positron plasma at low energies are shown in Figure 9.11 [307]. We recognize the strong similarity with Figure 9.8. That means that the expected similarity of an ultrarelativistic electron–ion plasma with a relativistic electron–positron plasma is endorsed.

One can mimic the reported bifurcation scenario for ultrarelativistic electron–ion plasmas up to the agreement in phase-space for large energies (when H is the control parameter) by considering an electron–ion–positron plasma [311]. Let us discuss the influence of positrons (with control parameter $0 < \chi \leq 1$) on a relativistic wave solution. Positrons carry the same amount of positive charge as we considered for the ions, but have the same mass as the electrons. This implies that nonlinearities due to positrons appear on the same scale as they do for electrons. The general Hamiltonian corresponding to Eqs. (9.224) and (9.225) is

$$H = \frac{p_a^2}{2(\beta^2 - 1)} + \frac{p_\phi^2}{2} + \frac{\beta}{\beta^2 - 1}\left(R_e - \beta + \chi(R_p - \beta) + (1 - \chi)\frac{R_i - \beta}{\varepsilon_i}\right). \tag{9.260}$$

The presence of positrons leads to a reduction of the maximum value of ϕ at a fixed maximum value of a since the positrons can react stronger to the radiation pressure than the heavier ions. This for example is known to influence the possibility of having localized solutions in a system consisting only of electrons and positrons [312].

The positrons already have a pronounced impact for low positron density. When comparing electron–ion plasmas (with control parameter H) with electron–ion–positron plasmas (with control parameter χ), one observes similar bifurcation scenarios. The phase-space topologies are very much alike when one alters the control parameters.

9.3.6
Wave Solutions in Weakly Relativistic Two-Field Models

In the weakly relativistic regime, we start from the system Eqs. (9.142) and (9.143). We are exclusively interested in linear polarized plasma wave motion in the presence of plasma, and change to ω_{pe}^{-1} as the time unit and c/ω_{pe}^{-1} as the corresponding space unit. Therefore, we get rid of the factor n_0/n_c. We introduce the coordinate $\xi = x - \beta t$ (instead of x), as in the previous calculations. Now, we obtain

$$(1-\beta^2)\frac{\partial^2 a}{\partial \xi^2} - \frac{\partial^2 a}{\partial t^2} + 2\beta\frac{\partial^2 a}{\partial \xi \partial t} = \left(1 + \delta n - \frac{a^2}{2}\right) a \,, \tag{9.261}$$

$$\beta^2 \frac{\partial^2 \delta n}{\partial \xi^2} + \frac{\partial^2 \delta n}{\partial t^2} - 2\beta \frac{\partial^2 \delta n}{\partial \xi \partial t} + \delta n = \frac{1}{2}\frac{\partial^2 a^2}{\partial \xi^2} \,. \tag{9.262}$$

Next, we look for stationary wave solutions in the form

$$a(x,t) = a(\xi) \,, \quad \delta n(x,t) = \delta n(\xi) \,, \tag{9.263}$$

that is, we set the explicit time-dependence to zero in the moving frame. This leads to

$$(1-\beta^2)\frac{\partial^2 a}{\partial \xi^2} = \left(1 + \delta n - \frac{a^2}{2}\right) a \,, \tag{9.264}$$

$$\beta^2 \frac{\partial^2 \delta n}{\partial \xi^2} + \delta n = \frac{1}{2}\frac{\partial^2 a^2}{\partial \xi^2} \,. \tag{9.265}$$

Replacing δn by

$$\delta n = \frac{\partial^2 \phi}{\partial \xi^2} , \tag{9.266}$$

we can integrate Eq. (9.265) twice to obtain

$$\beta^2 \frac{\partial^2 \phi}{\partial \xi^2} + \phi + c_2 \xi + c_1 = \frac{1}{2} a^2 \,. \tag{9.267}$$

Two constants appear. Looking for *periodic* solutions in ξ, we set

$$c_2 = 0 \,. \tag{9.268}$$

The constant c_1 will be determined from the condition that $\phi = \phi_0 = 0$ when $p = 0$ with $a = a_0$. Making use of Eq. (9.196), we find

$$\left.\frac{\partial^2 \phi}{\partial \xi^2}\right|_{a=a_0,\phi=0} = n|_{a=a_0,\phi=0} - 1 = 0 \tag{9.269}$$

such that

$$c_1 = \frac{1}{2} a_0^2 . \tag{9.270}$$

The two coupled equations for stationary wave solutions in a moving frame for a and ϕ then read

$$(1-\beta^2)\frac{\partial^2 a}{\partial \xi^2} = \left(1 + \frac{1-\beta^2}{2\beta^2} a^2 - \frac{1}{\beta^2}\phi - \frac{a_0^2}{2\beta^2}\right) a , \tag{9.271}$$

$$\beta^2 \frac{\partial^2 \phi}{\partial \xi^2} = -\phi + \frac{1}{2}\left(a^2 - a_0^2\right) . \tag{9.272}$$

With the new variables

$$\tilde{Z} = -\phi - \frac{1}{2} a_0^2 - 1 , \quad \tilde{X} \equiv X = \sqrt{\beta^2 - 1}\, a , \quad \zeta = \frac{\xi}{\sqrt{\beta^2 - 1}} , \tag{9.273}$$

we finally obtain

$$\ddot{X} = -\frac{1}{\beta^2}\left(\tilde{Z} - \frac{1}{2} X^2 + \beta^2 + 1\right) X , \tag{9.274}$$

$$\ddot{\tilde{Z}} = -\frac{1}{\beta^2}\left((\beta^2 - 1)\tilde{Z} + \frac{1}{2} X^2 + \beta^2 - 1\right) . \tag{9.275}$$

When X, $\tilde{Z}$ are interpreted as coordinates, and $\dot{X}$, $\dot{\tilde{Z}}$ as the corresponding momenta, Eqs. (9.274) and (9.275) are the canonical equations with the Hamiltonian

$$\begin{aligned} \tilde{H} &= \frac{1}{2}\left(\dot{X}^2 + \dot{\tilde{Z}}^2\right) + \frac{1}{2}\frac{\beta^2 + 1}{\beta^2} X^2 - \frac{1}{8}\frac{1}{\beta^2} X^4 \\ &\quad + \frac{1}{2}\frac{1}{\beta^2}\tilde{Z} X^2 + \frac{\beta^2 - 1}{\beta^2}\left(\frac{1}{2}\tilde{Z}^2 + \tilde{Z}\right) + C . \end{aligned} \tag{9.276}$$

The free constant C will be determined when comparing with the fully relativistic Hamiltonian equation (9.239),

$$H(X = 0, Z = -1) = -1 + \beta^2 = \tilde{H}\left(X = 0, \tilde{Z} = -1\right) \tag{9.277}$$

with the result

$$C = -\frac{1}{2} + \beta^2 - \frac{1}{2\beta^2} . \tag{9.278}$$

It should be noted that both the Hamiltonian equation (9.276) as well as the canonical Eqs. (9.274) and (9.275) follow from Eqs. (9.239) and (9.237) together with Eq. (9.238), respectively, by appropriate expansion in the weakly relativistic limit.

Equations (9.271) and (9.272) describe the coupling of linearly polarized transverse modes with longitudinal oscillations. In the linear limit, that is, without coupling, we have

$$(1-\beta^2)\frac{d^2a}{d\xi^2} = a\ , \tag{9.279}$$

$$\beta^2\frac{d^2\phi}{d\xi^2} = -\phi\ . \tag{9.280}$$

Obviously, for $\beta > 1$, we obtain harmonic oscillations

$$a(\xi) \sim a_1 \cos\left(\frac{\xi}{\sqrt{\beta^2-1}}\right) \Leftrightarrow a(x,t) \sim a_1 \cos(\omega_\perp \tau)\ , \tag{9.281}$$

$$\phi(\xi) \sim \phi_1 \cos\left(\frac{\xi}{\beta} + \varphi_1\right) \Leftrightarrow \phi(x,t) \sim \phi_1 \cos(\omega_\parallel \tau)\ , \tag{9.282}$$

with

$$\xi = x - Vt\ , \quad \tau = t - \frac{x}{V}\ , \tag{9.283}$$

and the dimensional frequencies

$$\omega_\perp^2 \equiv \omega_\perp^{(0)2}\omega_{pe}^2 = \frac{\omega_{pe}^2\beta^2}{\beta^2-1}\ , \quad \omega_\parallel^2 \equiv \omega_\parallel^{(0)2}\omega_{pe}^2 = \omega_{pe}^2\ . \tag{9.284}$$

We recognize within the present scaling that any transverse wave will drive electrostatic oscillations, whereas the electrostatic oscillations will need a seed electromagnetic wave to cause a nonlinear frequency shift of the latter.

Let us have a closer look on the driven electrostatic oscillations coupled to the vector potential. In the following, we start from Eqs. (9.176)–(9.181). When we are looking for periodic waves (which only depend on $\tau = t - x/\beta$ or equivalently on $\xi = x - \beta t$), we should use

$$A \sim e^{-i\delta\omega t} e^{+i(\omega_\perp^{(0)}+\delta\omega)\frac{x}{\beta}} \tag{9.285}$$

to determine the nonlinear frequency shift $\delta\omega$. From Eq. (9.181), we obtain the quadratic equation

$$-\frac{1}{2}(\delta\omega)^2 - \omega_\perp^{(0)}\delta\omega + \frac{1}{2}(1-\omega_\perp^{(0)2}) + \frac{(\omega_\perp^{(0)}+\delta\omega)^2}{2\beta^2} + \frac{(\omega_\perp^{(0)}+\delta\omega)^2}{\beta^2(64\omega_\perp^{(0)2}-16)} - \frac{3}{16} = 0 \tag{9.286}$$

which can be solved easily.

9.3.7
Instability of Stationary Wave Solutions

The effects of localized perturbations on the dynamics of the periodic solutions can be interpreted in terms of scattering instabilities (see next section). Here, we

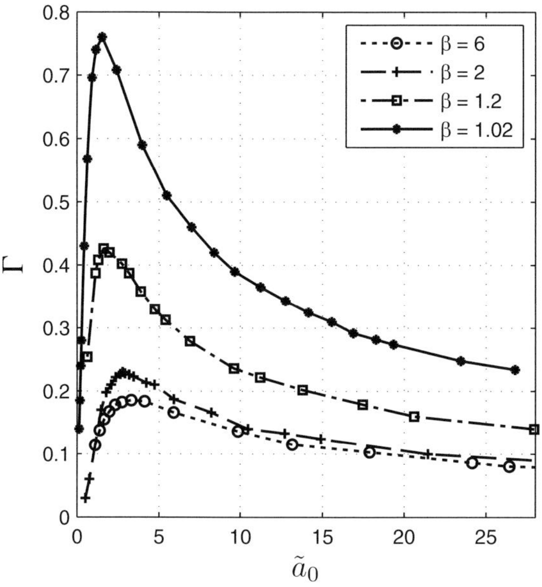

Figure 9.12 Exponential growth rate Γ depending on the maximum vector potential amplitude $\tilde{a}_0$ of the unperturbed solution for $\beta = 1.02$ (solid line), $\beta = 1.2$ (dash-dotted line), $\beta = 2$ (dashed line), and $\beta = 6$ (dotted line), respectively [297].

summarize the results from numerical simulations. The unperturbed solutions, in the frame of reference moving with the phase-velocity β, are time-independent. Within this frame, one may carry out a linear stability analysis. The initial perturbations being introduced into the system have nonzero energy, are spatially localized, time-dependent, and will perturb all quantities, that is, $\boldsymbol{A}$, ϕ (respectively n) and p. Due to the explicit time-dependence of all perturbed quantities, the stability analysis differs from the problem of structural stability. Structural stability deals with the evolution of neighboring trajectories (i.e., stationary solutions with slightly different initial conditions) within the Akhiezer–Polovin model itself [313].

As, in general, the periodic stationary solutions are only given numerically, one can perform only a a numerical linear stability analysis. The equations governing the nonlinear evolution of the perturbed system are Eqs. (9.204)–(9.208). One linearizes these equations about the unperturbed solution and solves the resulting system of equations numerically. For integrating these equations, one may use a spectral method for the spatial derivatives and a leap-frog scheme to advance in time. When one offers the system an initial perturbation which has a component along the direction of the most unstable mode, it is possible to grow the most unstable mode by integration in time. This is due to the fact that the fastest growing linear mode will dominate all others when time progresses. In regular intervals, one rescales the solution back to unity, where the norm of a perturbation is measured by a uniform norm. By rescaling the distribution, one avoids the growth of numerical errors. The results from the linear model are then validated by nonlinear simulations. One starts with an unperturbed state with a small amount of the fastest growing mode added. As long as the perturbation is small, its growth should resemble the predictions from the linear model. For more details about this method see [314].

For every perturbed periodic solution, one observes a growth of the perturbation proportional to $\exp(\lambda t)$, where $\lambda = \Gamma + i\omega$. Here, we will focus on the real part, that is, on the exponential growth factor due to Γ.

The growth rate Γ depends on the velocity β and on the amplitude of the unperturbed state. Let us exemplify this for the figure-eight motion. Figure 9.12 shows the relation between maximum amplitude $\tilde{a}_0$ of the unperturbed solution and the growth rate Γ of the fastest growing perturbation for states at four different velocities $\beta = 1.02$, $\beta = 1.2$, $\beta = 2$, and $\beta = 6$, respectively. From these curves, one can draw the conclusion that the figure-eight motion becomes more and more unstable the closer the plasma density approaches vacuum. Furthermore, the amplitude $\tilde{a}_0$ for which the maximum growth rate is obtained at fixed β becomes smaller with decreasing plasma density. The maximum growth rates range from $\Gamma = 0.1$ to $\Gamma = 0.8$. A perturbation with 1% of the fastest growing mode would grow to the order of one within a time of about 10 to 50 ω_{pe}^{-1}. In the low density limit ($\beta \to 1$), it is useful to normalize the growth rates Γ by the laser frequency ω_0, rather than the plasma frequency ω_{pe}. The frequency ω_0 of the stationary state in the laboratory frame is defined as $\omega_0 = k\beta$, where k is the fundamental wavenumber of the unperturbed solution. Figure 9.13 shows Γ vs. $\tilde{a}_0$ normalized by ω_0 for $\beta = 1.02$ and

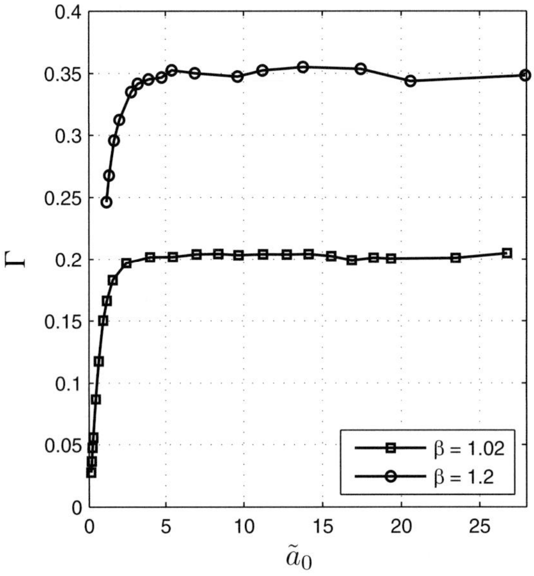

Figure 9.13 Exponential growth rate Γ (normalized by ω_0) depending on the maximum vector potential amplitude $\tilde{a}_0$ of the unperturbed solution for $\beta = 1.02$ (line marked by squares) and $\beta = 1.2$ (line marked by circles) [297].

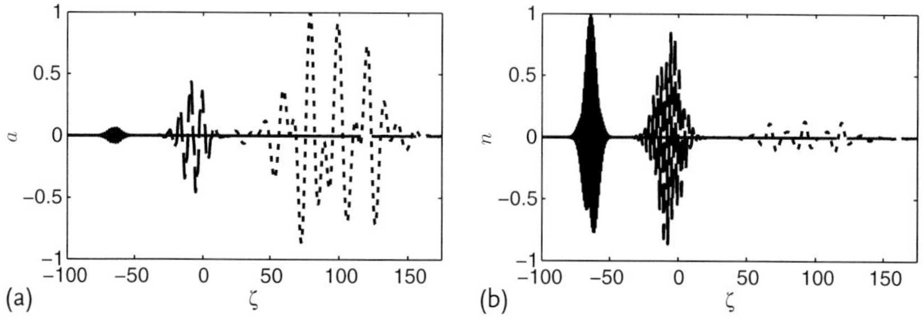

Figure 9.14 Vector potential $\boldsymbol{a}$ (a) and electron density n (b) of the most unstable linear mode for $\beta = 1.02$ (solid line), $\beta = 1.2$ (dashed line) and $\beta = 6$ (dotted line) [297].

$\beta = 1.2$. At higher densities, the higher harmonics of the fundamental wavenumber k contribute stronger to the unperturbed state, and no clear time-scale can be defined by the laser. Then, Γ is normalized by the plasma frequency ω_{pe}.

Since the figure-eight motion in vacuum is stable, the plasma effect should be emphasized. A look at the individual structures of the fastest growing modes shows that the energy of the modes is distributed differently over electromagnetic and electrostatic contributions for different β parameters; see Figure 9.14. In order to compare the composition of the modes, one may normalize them to unity by using a uniform norm. The higher the velocity, the more dominant is the electromagnetic part of the fastest growing perturbation, whereas for smaller velocities, that is, smaller plasma densities, the electrostatic part of the mode becomes increasingly important. In the limit of $\beta \to 1$ (transition to vacuum), the whole perturbation leading to instability would be of pure electrostatic nature. Exactly this electrostatic component is absent in pure vacuum calculations, hence there is no contradiction between the stability of the vacuum figure-eight motion and the instability of figure-eight motion inside plasma.

One can perform stability analysis for the other periodic solutions and different polarizations as well. Within their range of existence, they show a similar unstable behavior as the figure-eight motion. At a fixed value of β, Γ reaches a maximum at a certain amplitude value. With decreasing β, the values of the maxima increase and the positions of the maxima gets shifted to smaller $\tilde{a}_0$. The growth rates are comparable in magnitude to those for the figure-eight motion, however they are a little bit smaller.

The linear results may be validated by nonlinear simulations. For small times, one observes a very good agreement with the predictions from the linear model.

9.4 Parametric Instabilities in the Relativistic Regime

Nonlinear effects preclude, in general, analytical forms for wave solutions of the Maxwell fluid systems. Therefore, an exact stability analysis of stationary relativistic wave solutions has to also be performed numerically, as has been discussed in the previous section. One may consider the found instabilities as generalizations of known nonrelativistic instabilities for plane wave solutions. Of course, relativistic effects also generate new types of instabilities, like the relativistic modulational instability. Nevertheless, it seems that parametric scattering instabilities dominate, and therefore in the following sections, we present the relativistic versions of Raman instability for plane waves.

Waves in plasmas are the sources of many types of instabilities [20, 315, 316]. Electrostatic decay and modulational instabilities occur. When electromagnetic modes are primarily involved, the two-plasmon decay instability as well as Raman and Brillouin scattering instabilities evolve. Kinetic limits, when quasimodes are involved, are known as Compton scattering. From the rich variety of instabilities, we will discuss the Raman and Brillouin instabilities in the following.

9.4.1 Stimulated Raman and Brillouin Scattering in the Classical Regime

The physics of the Raman instability is straightforward to understand [20]. When a light wave is propagating through a plasma, it may generate electron plasma wave density ripples in the direction of propagation. The oscillating electrons cause a transverse current which may generate a scattered light wave when wavenumbers and wave frequencies are properly matched. The scattered light wave interferes with the incident light wave to produce a ponderomotive pressure which mainly acts on the electrons, influencing the incident density ripples. Due to this feedback loop, an instability is possible.

We start with the wave equation for the vector potential $\boldsymbol{A}$ in Coulomb gauge $\nabla \cdot \boldsymbol{A} = 0$,

$$\left(\frac{1}{c^2}\frac{\partial^2}{\partial t^2} - \nabla^2\right) \boldsymbol{A} = \mu_0 \boldsymbol{j} - \mu_0 \varepsilon_0 \frac{\partial}{\partial t} \nabla \phi \, . \tag{9.287}$$

We split the current density in the wave equation into a transversal part $\boldsymbol{j}_\perp$ (associated with the light wave) and a longitudinal part $\boldsymbol{j}_\parallel$. The longitudinal part originates from the space charge (electron density) oscillations. Space charge continuity and the Poisson equation lead to

$$\nabla \cdot \left(\frac{\partial}{\partial t}\nabla\phi - \frac{1}{\varepsilon_0}\boldsymbol{j}\right) \approx \nabla \cdot \left(\frac{\partial}{\partial t}\nabla\phi - \frac{1}{\varepsilon_0}\boldsymbol{j}_\parallel\right) = 0 \tag{9.288}$$

since $\nabla \cdot \boldsymbol{j}_\perp = 0$ to zeroth order. Equation (9.287) then reduces to

$$\left(\frac{1}{c^2}\frac{\partial^2}{\partial t^2} - \nabla^2\right) \boldsymbol{A} = \mu_0 \boldsymbol{j}_\perp \approx -\mu_0 e n_e \boldsymbol{v}_{e\perp} \approx -\frac{e^2}{\varepsilon_0 m_e} n_e \boldsymbol{A} \, , \tag{9.289}$$

where, as in previous models, we have substituted for the nonrelativistic electron momentum $\boldsymbol{p}_{e\perp} = m_e \boldsymbol{v}_{e\perp} \approx e\boldsymbol{A} \hat{=} e\boldsymbol{A}_\perp$. We can use this equation for the scattered electromagnetic wave after substituting $n_e = n_0 + \delta n$ and $\boldsymbol{A} = \boldsymbol{A}_0 + \delta \boldsymbol{A}$ to find in lowest order for δA

$$\left(\frac{\partial^2}{\partial t^2} - c^2\nabla^2 + \omega_{pe}^2\right)\delta A = -\frac{e^2}{\varepsilon_0 m_e}\delta n A_0 \,. \tag{9.290}$$

The density response follows from the conservation laws for electron number density and momentum,

$$\frac{\partial n_e}{\partial t} + \nabla\cdot(n_e \boldsymbol{v}_e) = 0\,, \tag{9.291}$$

$$\frac{\partial \boldsymbol{v}_e}{\partial t} + \boldsymbol{v}_e\cdot\nabla\boldsymbol{v}_e = -\frac{e}{m_e}(\boldsymbol{E} + \boldsymbol{v}_e\times\boldsymbol{B}) - \frac{1}{n_e m_e}\nabla p_e\,, \tag{9.292}$$

together with the equation of state $p_e \sim n_e^3$. Letting $\boldsymbol{v}_e = \boldsymbol{v}_{e\|} + e\boldsymbol{A}/m_e$, the parallel component of Eq. (9.291) leads to

$$\frac{\partial \boldsymbol{v}_{e\|}}{\partial t} = \frac{e}{m_e}\nabla\phi - \frac{1}{2}\left(\boldsymbol{v}_{e\|} + \frac{e\boldsymbol{A}}{m_e}\right)^2 - \frac{1}{n_e m_e}\nabla p_e\,. \tag{9.293}$$

Identifying $\boldsymbol{v}_{e\|} \equiv \delta\boldsymbol{v}$ and $\phi = \delta\phi$, we obtain for the fluctuations in lowest order

$$\frac{\partial \delta n}{\partial t} + n_0\nabla\cdot\delta\boldsymbol{v} = 0\,, \tag{9.294}$$

$$\frac{\partial \delta\boldsymbol{v}}{\partial t} = \frac{e}{m_e}\nabla\delta\phi - \frac{e^2}{m_e^2}\nabla\left(\boldsymbol{A}_0\cdot\delta\boldsymbol{A}\right) - \frac{3v_{the}^2}{n_0}\nabla\delta n\,. \tag{9.295}$$

These equations can be combined to

$$\left(\frac{\partial^2}{\partial t^2} + \omega_{pe}^2 - 3v_{the}^2\nabla^2\right)\delta n = \frac{n_0 e^2}{m_e^2}\nabla^2\left(\boldsymbol{A}_0\cdot\delta\boldsymbol{A}\right). \tag{9.296}$$

Equations (9.290) and (9.296) are the coupled equations for the electrostatic (Langmuir) and electromagnetic (scattered) waves. For the dispersion relation of the Raman instability, we take $\boldsymbol{A}_0 = A_0\cos(\boldsymbol{k}_0\cdot\boldsymbol{r} - \omega_0 t)$ and Fourier-transform to

$$\begin{aligned}&\left(\omega^2 - c^2k^2 - \omega_{pe}^2\right)\delta A(\boldsymbol{k},\omega)\\&= \frac{e^2}{2\varepsilon_0 m_e}A_0\left[\delta n(\boldsymbol{k}-\boldsymbol{k}_0,\omega-\omega_0) + \delta n(\boldsymbol{k}+\boldsymbol{k}_0,\omega+\omega_0)\right],\end{aligned} \tag{9.297}$$

$$\begin{aligned}&\left(\omega^2 - \omega_k^2\right)\delta n(\boldsymbol{k},\omega)\\&= \frac{k^2 e^2 n_0}{2m_e^2}A_0\left[\delta A(\boldsymbol{k}-\boldsymbol{k}_0,\omega-\omega_0) + \delta A(\boldsymbol{k}+\boldsymbol{k}_0,\omega+\omega_0)\right],\end{aligned} \tag{9.298}$$

with $\omega_k = \sqrt{\omega_{pe}^2 + 3k^2v_{the}^2}$. Considering the modes $\boldsymbol{k},\omega$ with Re $\omega \approx \omega_{pe}$ (plasma wave), the sidebands $\boldsymbol{k}-\boldsymbol{k}_0, \omega-\omega_0$ and $\boldsymbol{k}+\boldsymbol{k}_0, \omega+\omega_0$, and assuming that the harmonics $\boldsymbol{k}\pm 2\boldsymbol{k}_0, \omega\pm 2\omega_0$ are strongly nonresonant, we can easily eliminate δA

in Eq. (9.298) by using Eq. (9.297) to obtain for nontrivial solutions a sixth-order polynomial in ω,

$$\omega^2 - \omega_k^2 = \frac{\omega_{pe}^2 k^2 v_0^2}{4} \left[\frac{1}{D_-} + \frac{1}{D_+} \right] . \tag{9.299}$$

Here, $v_0 = eE_0/m_e\omega_0 = eA_0/m_e$, $D_\pm = \omega_\pm^2 - k_\pm^2 c^2 - \omega_{pe}^2$, $\omega_\pm^2 = (\omega \pm \omega_0)^2$, and $k_\pm^2 = (\boldsymbol{k} \pm \boldsymbol{k}_0)^2$. Scattering may occur in all directions. For back- or side-scatter, we neglect the nonresonant upshifted light wave as nonresonant, and make the ansatz $\omega = \omega_k + \delta\omega$ with $\delta\omega \ll \omega_k$. Maximum growth occurs when the scattered light wave is also resonant ($D_- \approx 0$). Then, the growth rate $\gamma = -i\delta\omega$ maximizes for back-scattering, which has maximum growth

$$\gamma = \frac{k v_0}{4} \sqrt{\frac{\omega_{pe}^2}{\omega_k(\omega_0 - \omega_k)}} \quad \text{at} \quad k = k_0 + \frac{\omega_0}{c} \sqrt{1 - \frac{2\omega_{pe}}{\omega_0}} . \tag{9.300}$$

One clearly recognizes that $n \leq n_{crit}/4$ (quarter-critical density) is necessary for Raman backward scattering to occur. More details, especially on forward Raman and stimulated Compton scattering, can be found in the more specialized literature [20, 315, 317, 318].

Stimulated Raman scattering describes scattering off a high-frequency electron plasma oscillation (Langmuir wave). For the Brillouin instability, the density fluctuation δn is the low-frequency oscillation associated with an ion-acoustic wave. For the (more or less immediate) electron response, we may use Eq. (9.295), however with an isothermal instead of the adiabatic equation of state. In addition, electron inertia is negligible such that

$$0 \approx \frac{e}{m_e} \nabla \delta\phi - \frac{e^2}{m_e^2} \nabla (\boldsymbol{A}_0 \cdot \delta \boldsymbol{A}) - \frac{v_{the}^2}{n_0} \nabla \delta n . \tag{9.301}$$

The scalar potential transmits the ponderomotive force to the ions. In the ion equations, we neglect the ion temperature for simplicity,

$$\frac{\partial n_i}{\partial t} + \nabla \cdot (n_i \boldsymbol{v}_i) = 0 , \tag{9.302}$$

$$\frac{\partial \boldsymbol{v}_i}{\partial t} + \boldsymbol{v}_i \cdot \nabla \boldsymbol{v}_i \approx -\frac{Ze}{m_i} \nabla \phi . \tag{9.303}$$

To lowest order, the equations for the ion perturbations become

$$\frac{\partial \delta n_i}{\partial t} + n_{i0} \nabla \cdot \delta \boldsymbol{v}_i = 0 , \tag{9.304}$$

$$\frac{\partial \delta \boldsymbol{v}_i}{\partial t} = -\frac{Ze}{m_i} \nabla \delta\phi . \tag{9.305}$$

They can be combined to (note $Z n_{i0} \approx n_{e0} \equiv n_0$)

$$\frac{\partial^2 \delta n_i}{\partial t^2} - \frac{n_0 Z e}{m_i} \nabla^2 \delta\phi = 0 . \tag{9.306}$$

For quasineutral perturbations $Z\delta n_i \approx \delta n_e \equiv \delta n$, we may eliminate $\delta\phi$ from (9.301) to finally obtain

$$\left(\frac{\partial^2}{\partial t^2} - c_s^2\nabla^2\right)\delta n = \frac{n_0 Z e^2}{m_e m_i}\nabla^2\left(\boldsymbol{A}_0 \cdot \delta \boldsymbol{A}\right), \tag{9.307}$$

where $c_s = \sqrt{Z T_e/m_i}$ is the ion-sound velocity. The equation for δn is coupled to the wave equation (9.290)

$$\left(\frac{\partial^2}{\partial t^2} - c^2\nabla^2 + \omega_{pe}^2\right)\delta \boldsymbol{A} = -\frac{e^2}{\varepsilon_0 m_e}\delta n \boldsymbol{A}_0 . \tag{9.308}$$

The further analysis is similar to that for Raman scattering. We now obtain the dispersion relation

$$\omega^2 - k^2 c_s^2 = \frac{\omega_{pi}^2 k^2 v_0^2}{4}\left[\frac{1}{D_-} + \frac{1}{D_+}\right], \tag{9.309}$$

where $\omega_{pi} = \sqrt{Z n_0 e^2/\varepsilon_0 m_i}$. Proceeding as before, one arrives at the instability result for back-scattering

$$\gamma = \frac{1}{2\sqrt{2}}\frac{k_0 v_0 \omega_{pi}}{\sqrt{\omega_0 k_0 c_s}} \quad \text{at} \quad k = 2k_0 - \frac{2\omega_0}{c}\frac{c_s}{c} . \tag{9.310}$$

Generalizing, the general form of the dispersion relation, taking into account the full spectrum of electromagnetic excitations (electrostatic as well as electromagnetic) can be cast into the form [317]

$$\begin{aligned} &1 + \chi_e(\boldsymbol{k},\omega) + \chi_i(\boldsymbol{k},\omega) \\ &= -\chi_e(\boldsymbol{k},\omega)\left[1 + \chi_i(\boldsymbol{k},\omega)\right] \\ &\times \frac{k^2}{4}\left\{\frac{|\boldsymbol{k}_+ \times \boldsymbol{v}_0|^2}{k_+^2 D_+} + \frac{|\boldsymbol{k}_- \times \boldsymbol{v}_0|^2}{k_-^2 D_-} - \frac{|\boldsymbol{k}_+ \cdot \boldsymbol{v}_0|^2}{k_+^2 \omega_+^2 \varepsilon_+} - \frac{|\boldsymbol{k}_- \cdot \boldsymbol{v}_0|^2}{k_-^2 \omega_-^2 \varepsilon_-}\right\} . \end{aligned} \tag{9.311}$$

Here, $\boldsymbol{v}_0 = e\boldsymbol{A}_0/m_e = e\boldsymbol{E}_0/m_e\omega_0$, χ_j designates the electric susceptibilities in the respective frequency domains, and $\varepsilon_\pm = 1 + \chi_\pm$ is the high-frequency relative permittivity. For example, in the case of Raman scattering, we take

$$\chi_i(\boldsymbol{k},\omega) \approx 0 , \quad \chi_e(\boldsymbol{k},\omega) \approx -\frac{\omega_{pe}^2}{\omega^2} + \cdots (\text{thermal effects}) , \quad \varepsilon_\pm = 1 - \frac{\omega_{pe}^2}{\omega^2} . \tag{9.312}$$

On the other hand, for Brillouin scattering,

$$\begin{aligned} &\chi_i(\boldsymbol{k},\omega) \approx -\frac{\omega_{pi}^2}{\omega^2} , \quad \chi_e(\boldsymbol{k},\omega) \approx \frac{1}{k^2\lambda_{De}^2} , \\ &1 + \chi_e + \chi_i \approx \chi_e + \chi_i , \quad 1 + \chi_i \approx \chi_i , \end{aligned} \tag{9.313}$$

and $\varepsilon_\pm$ is the same as before. The more general formulation Eq. (9.311) has the advantage that damping effects can easily be incorporated to determine threshold values for the scattering instabilities [315].

9.4.2
Stimulated Scattering Instabilities in the Relativistic Regime

In the weakly relativistic regime, we should replace the rest mass by the relativistic mass containing the relativistic factor γ. Besides this "rather trivial" substitution, in the wave equation an additional term originating from a derivative of the γ-factor appears. Then, in case of circular polarization in the weakly relativistic regime (as long as the ions behave nonrelativistically), it is straightforward to generalize the nonrelativistic results. We should replace the high-frequency wave equation (9.290) by

$$\left(\frac{\partial^2}{\partial t^2} - c^2\nabla^2 + \omega_{pe}^2\right)\delta A = -\frac{e^2}{\varepsilon_0 m_e}\delta n A_0 + \frac{e^2 n_0}{\varepsilon_0 m_e} A_0 \frac{e^2}{m_e^2 c^2}(A_0 \cdot \delta A) . \tag{9.314}$$

The second term on the right-hand side originates from the relativistic mass variation. The plasma reactions can be followed by the relativistic fluid equations (using the momentum $\boldsymbol{p}_e$ instead of the velocity $\boldsymbol{v}_e$) to obtain similar results as in the nonrelativistic cases (of course replacing at the end rest masses by relativistic masses). In case of Raman scattering, one can consult [319–321] where the following dispersion relation was obtained:

$$\omega^2 - \frac{\omega_{pe}^2}{\gamma_e} = \frac{\omega_{pe}^2 k^2 v_0^2}{4\gamma_e^3} \underbrace{\left(\frac{k^2c^2 - \omega^2 + \frac{\omega_{pe}^2}{\gamma_e}}{k^2c^2}\right)}_{\approx 1 \text{ for } c\to\infty} \left[\frac{1}{\tilde{D}_-} + \frac{1}{\tilde{D}_+}\right], \tag{9.315}$$

with m_e as the electron rest mass, $v_0 = eE_0/(m_e\omega_0) = eA_0/m_e$, $\tilde{D}_\pm = \omega_\pm^2 - k_\pm^2 c^2 - \omega_{pe}^2/\gamma_e$, $\omega_\pm^2 = (\omega \pm \omega_0)^2$, $k_\pm^2 = (\boldsymbol{k} \pm \boldsymbol{k}_0)^2$, and $\gamma_e = \sqrt{1 + v_0^2/(2c^2)}$. Thermal effects were neglected. Note that this dispersion relation reduces to Eq. (9.299), that is,

$$\omega^2 - \omega_k^2 = \frac{\omega_{pe}^2 k^2 v_0^2}{4}\left[\frac{1}{D_-} + \frac{1}{D_+}\right], \tag{9.316}$$

when $\gamma_e \to 1$ and the nonrelativistic limit is taken for $c \to \infty$. The time-dependence of the relativistic factor is only "simple" for circular polarization. Thus, the shown dispersion relations are only valid for circular polarization. Otherwise, one should, for example, expand the relativistic factor (for small amplitudes) and put into the Fourier matching appropriately. The amplitude dependence of the dispersion relation in the relativistic regime leads to interesting conclusions about the possibility of realistic wave propagation [321]. The conclusions help to interpret the numerically obtained growth rates presented in the section on the stability of nonlinear waves.

The scheme for relativistic effects on Raman scattering can be transferred (for circular polarization) to Brillouin scattering. We have to substitute the relativistic electron mass into the dispersion relation (as long as the ions can be treated nonrelativistically). Thus, for Brillouin scattering in the (weakly) relativistic regime, one finds

$$\omega^2 - k^2 c_s^2 = \frac{\omega_{pi}^2 k^2 v_0^2}{4\gamma_e^2} \underbrace{\left(\frac{k^2 c^2 - \frac{m_i}{\gamma_e m_e}(\omega^2 - k^2 c_s^2)}{k^2 c^2} \right)}_{\approx 1 \text{ for } c \to \infty} \left[\frac{1}{\tilde{D}_-} + \frac{1}{\tilde{D}_+} \right] , \quad (9.317)$$

where $\omega_{pi} = \sqrt{Z n_0 e^2 / \varepsilon_0 m_i}$ and $c_s = \sqrt{Z T_e / m_i}$. The other notations are the same as in the Raman case, especially m_e and m_i are the rest masses. In the nonrelativistic limit $\gamma_e \to 1$ and $c \to \infty$, this dispersion relation reduces to Eq. (9.309), that is,

$$\omega^2 - k^2 c_s^2 = \frac{\omega_{pi}^2 k^2 v_0^2}{4} \left[\frac{1}{D_-} + \frac{1}{D_+} \right] . \quad (9.318)$$

9.5 Solitary Envelope Solutions and Their Stability

Relativistic self-modulation and self-focusing, though known for a long time [295, 302, 322–326], are still important processes in actual laser–plasma configurations. One severe consequence of the instabilities is trapping of radiation through soliton formation. When a laser pulse enters a slightly overdense plasma ($\omega_0 < \omega_{pe}$, where ω_0 is the laser frequency and ω_{pe} is the electron plasma frequency), penetration of relativistically intense laser radiation may occur in the form of soliton-like structures [327]. A high-power laser pulse propagating in an underdense plasma ($\omega_0 > \omega_{pe}$) will be also influenced by nonlinearity, for example, compressed. Stimulated Raman scattering and a Raman cascade may lead to large amplitude relativistic electromagnetic solitons in an underdense plasma [328, 329]. In addition to the nonlinearly shaped leading pulse, a laser beam propagating in underdense plasmas also creates slow, nearly standing narrow structures behind the leading edge. The new processes are especially present in the ultrashort pulse regime [330]. In underdense plasmas, for high intensities and short pulses, the energy of an ultrarelativistic laser pulse may be partially transferred into fields behind it [291]. Bubble-like structures have been observed in numerical simulations [290, 330–334]. Within the latter, ponderomotive pressure leads to a strong depression of the electron density. It is predicted that up to 40% of the laser energy can be trapped. The structures consisting of electron depressions and intense electromagnetic field concentrations (with an occasionally much bigger amplitude and a lower frequency than those of the original laser pulse) are called (slow) solitons.

Macroscopic evidence of soliton formation in multiterawatt laser–plasma interaction has been reported from experiments [335, 336]. Typical sizes of the spatial structures are of the order of the collisionless electron skin depth c/ω_{pe} of the surrounding plasma. The fields inside the solitons include oscillating electric fields [336] which arise from the charge separation (due to ponderomotive forces). On the longer time-scale, besides the electrons, also the ions are pushed out of the density holes, and the solitons evolve into postsolitons [333]. The expansion of the postsolitons under the push of the electromagnetic radiation (being trapped inside) has been analyzed within the snowplow approximation [333, 337]. Plasma-ion evolution in the wake of a high-intensity ultrashort laser pulse was also observed experimentally [338]. The ion dynamics are responsible for a slowly expanding plasma cavity [314]. Low-frequency relativistic electromagnetic stationary states in collisionless plasma were predicted theoretically [339, 340], and the influences of plasma parameters were discussed [341–345]. The special forms of very narrow solitons could be analyzed [330, 346–348].

Based on all these works, one gets quite a general idea of soliton solutions in 1D, especially for circular polarization [341, 349, 350]. Within the cold relativistic hydrodynamic approximation, the properties of solitons have been catalogued with respect to the number p of zeros of the vector potential, the velocity V, and the frequency ratio $\omega_0 \sqrt{1 - V^2/c^2}/\omega_{pe}$ ($\equiv \omega \sqrt{1 - V^2}$ in nondimensional form). Note that $\omega \sqrt{1 - V^2}$ is the frequency in the comoving frame. Solitons do exist for $\omega^2(1 - V^2) < 1$; the density depression allows trapping of radiation. Single-humped ($p = 0$) solitons have been found [350], for example, for $V = 0$, when the ion response is neglected. Subcycle solitons ($p = 1, 2, \ldots$) do exist for finite velocities V with a discrete ω-spectrum. On the ion time-scale, slow solitons do exist only above a certain threshold velocity V. The *1D stability* of solitons was investigated by several authors [314, 330, 341, 343–345, 351–353]. Most of the results have been obtained for circular polarization of the laser field. Models for the linear polarization [290, 351] were mainly restricted to the electron time-scale and to a weak plasma response [351]. In general, it is difficult to find exact analytical stability criteria for finite amplitude solitons [258, 275, 354, 355]. Therefore, many stability investigations are carried out numerically.

9.5.1 The Farina–Bulanov Model for Circularly Polarized Solitons

The starting point for soliton solutions in the fluid Maxwell description are Eqs. (9.68)–(9.72). For localized solutions propagating in x-direction, we have $A_x \sim A_\parallel = 0$. For obtaining solitary envelope solutions, we follow Farina and Bulanov [341, 350, 356] and introduce for 1D propagation in x-direction, instead of x and t, the new variables ξ and τ. Normalizing velocities by c and the space coordinates by c/ω_{pe}, we define

$$\xi = x - Vt\,, \quad \tau = t - Vx\,. \tag{9.319}$$

Here, V is the group velocity of the laser pulse, and in dimensionless form, $1/V$ is the phase velocity. That inverse behavior becomes obvious when we consider the linear dispersion relation

$$\omega^2 = \omega_{pe}^2 + c^2 k^2 \tag{9.320}$$

in dimensional form such that

$$v_{group}v_{phase} = \frac{d\omega(k)}{dk}\frac{\omega(k)}{k} = \frac{d\omega^2}{dk^2} = c^2 \,. \tag{9.321}$$

For circular polarization, we look for solutions in the form

$$A_y + iA_z = a(\xi)e^{-i\omega\tau} \,, \tag{9.322}$$

splitting the vector potential into an envelope (which for solitary solutions will be localized in ξ) and a phase factor. The other quantities, for example, particle densities, momenta, and so on should only depend on variable ξ. Note that the pure ξ-dependence of all other quantities, including the relativistic γ-factors, is only true for circularly polarized situations. For linear polarization, the situation is much more complicated because of the higher harmonics.

Boundary conditions at $\xi = -\infty$ are: $a = \pm a_0$, $\phi = 0$, $n_\alpha = 1$ and $p_{\|\alpha} = 0$. We do not exclude the value $a_0 = 0$. Compared to the previous normalization, we now normalize the momenta $\boldsymbol{p}_\alpha$ by $m_\alpha c$ (and not $m_e c$ as previously). Other normalizations are unchanged. Then, we obtain from Eqs. (9.68)–(9.72) the following set of equations for 1D solitons [341]

$$(1 - V^2)\left(a'' + \omega^2 a\right) = a\left(\frac{n_e}{\gamma_e} + \varepsilon\frac{n_i}{\gamma_i}\right), \tag{9.323}$$

$$V\phi'' = n_e v_{\|e} - n_i v_{\|i} \,, \tag{9.324}$$

$$\phi'' = n_e - n_i \,, \tag{9.325}$$

$$\left[n_e(V - v_{\|e})\right]' = 0 \,, \tag{9.326}$$

$$\left[n_i(V - v_{\|i})\right]' = 0 \,, \tag{9.327}$$

$$\left(\gamma_e - Vp_{\|e} - \phi\right)' = 0 \,, \tag{9.328}$$

$$\left(\gamma_i - Vp_{\|i} + \varepsilon\phi\right)' = 0 \,, \tag{9.329}$$

where $\varepsilon = m_e/m_i$ is the ratio of the rest masses, and the prime designates a derivative with respect to ξ. In the present normalization,

$$\gamma_e = \sqrt{1 + p_{\|e}^2 + a^2} \,, \quad \gamma_i = \sqrt{1 + p_{\|i}^2 + \varepsilon^2 a^2} \,. \tag{9.330}$$

By similar algebra, as discussed in the section on stationary wave solutions of the 1D Maxwell fluid model, we may eliminate $p_{\|\alpha}, \gamma_\alpha, n_\alpha$ to obtain a closed set of equations for the potentials,

$$\phi'' = \frac{V}{1 - V^2}\left(\frac{\psi_e}{R_e} - \frac{\psi_i}{R_i}\right), \tag{9.331}$$

$$a'' + \omega^2 a = a\frac{V}{1 - V^2}\left(\frac{1}{R_e} + \frac{\varepsilon}{R_i}\right). \tag{9.332}$$

Here, one has defined

$$\psi_e = \sqrt{1 + a_0^2} + \phi\,, \quad \psi_i = \sqrt{1 + \varepsilon^2 a_0^2} - \varepsilon\phi\,, \tag{9.333}$$

$$R_e = \sqrt{\psi_e^2 - (1 - V^2)(1 + a^2)}\,, \quad R_i = \sqrt{\psi_i^2 - (1 - V^2)(1 + \varepsilon^2 a^2)}\,, \tag{9.334}$$

when at $x \to \pm\infty$ the boundary values are $\phi = 0$ and $a = \pm a_0$. In terms of these quantities,

$$p_{\|\alpha} = \frac{V\psi_\alpha - R_\alpha}{1 - V^2}\,, \tag{9.335}$$

$$\gamma_\alpha = \frac{\psi_\alpha - V R_\alpha}{1 - V^2}\,, \tag{9.336}$$

$$n_\alpha = V\frac{\psi_\alpha - V R_\alpha}{R_\alpha (1 - V^2)}\,. \tag{9.337}$$

Before proceeding, let us come back to the ansatz Eq. (9.322). Effectively, we have treated a as real, although it should be complex, in general. Thus, in general, we should replace $a(\xi) \to a(\xi)e^{i\theta(\xi)}$ with real a in the new formulation. In that case, we obtain, instead of Eqs. (9.331) and (9.332), the following system

$$\phi'' = \frac{V}{1 - V^2}\left(\frac{\psi_e}{R_e} - \frac{\psi_i}{R_i}\right), \tag{9.338}$$

$$a'' + a\left(\bar{\omega}^2 - \bar{k}^2\frac{a_0^4}{a^4}\right) = a\frac{V}{1 - V^2}\left(\frac{1}{R_e} + \frac{\varepsilon}{R_i}\right), \tag{9.339}$$

$$\theta' = -\bar{k}\left(1 - \frac{a_0^2}{a^2}\right)\,, \tag{9.340}$$

where $\bar{\omega} = (\omega - kV)/(1 - V^2)$ and $\bar{k} = (k - \omega V)/(1 - V^2)$. The phase Eq. (9.340) becomes linear for $a_0 = 0$, that is, the case we consider in the following. It applies for localized bright solitons. Since then,

$$\theta = \theta_0 - \bar{k}\xi\,, \tag{9.341}$$

and the ansatz follows in the form

$$A_y + iA_z = a(\xi)e^{-i\bar{\omega}\tau + i\theta_0}\,. \tag{9.342}$$

The ansatz Eq. (9.322) becomes valid when, for the sake of simplicity, we consider the case $\bar{k} = 0$ so that $k = \omega V$ and $\bar{\omega} = \omega$. Localized solutions of the system Eqs. (9.331) and (9.332) will be obtained for $a' = 0$, $a'' = 0$, $\phi' = 0$, $\phi'' = 0$ at $\xi = \pm\infty$.

In the case of large amplitude bright solitons ($a_0 = 0$), a wide class of solutions has been found [341, 343, 344, 350, 356, 357]. When the analysis is restricted to

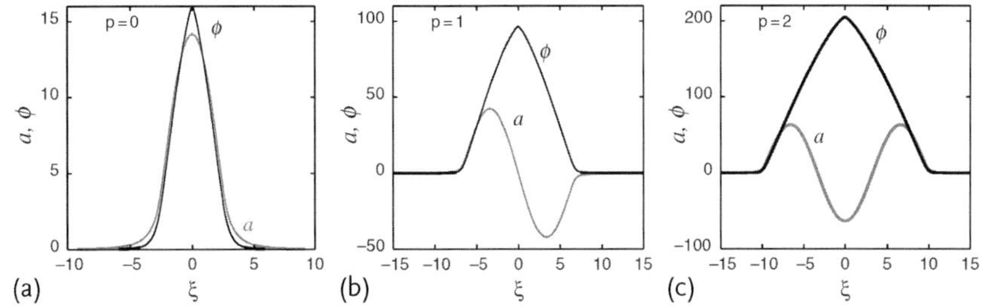

Figure 9.15 Typical solutions for electrostatic potential ϕ and vector potential a are shown for bright solitons with $p = 0$ (a), 1 (b), and 2 (c) for velocities V close to breaking [350]. Courtesy of IOP.

single-humped profiles in the electrostatic potential ϕ, the solutions are characterized by the number p of nodes in the vector potential a. Typical solutions are shown in Figure 9.15 for bright solitons with $p = 0, 1, 2$ for velocities V close to breaking [350].

For fixed V, the frequency ω is the eigenvalue when solving Eqs. (9.331) and (9.332) with suitable boundary conditions [341, 349]. The eigenvalue spectrum for low p-values is reproduced in Figure 9.16 from [350]. The cases with fixed ions are compared to the situations with mobile ions. When ion dynamics are neglected, for the bright soliton eigenspectrum, the following features were reported [350]. Single-peaked solitons in Figure 9.15a admit a continuous spectrum for any velocity value. This was already shown in [343]. In particular, standing ($V = 0$) solitons are found with frequency in the range $2/3 \leq \omega^2 \leq 1$. In addition, $max\{a\} \leq \sqrt{3}$ has been shown analytically in [340]. Solitons with node numbers $p = 1, 2, \ldots$ have a discrete spectrum for velocities larger than a small critical velocity value. Figure 9.16a shows this behavior. Quite different results are found [350] when the ion dynamics is taken into account. At very high velocities V, the ion dynamics should not play a significant role, and the behavior of Figure 9.16a should remain valid. However, single-humped solitons are found only for velocities larger $V \approx \sqrt{\varepsilon}$, both in the continuous and the discrete spectrum [341, 344]. Again, for $p = 1, 2, \ldots$, only a discrete spectrum is found. At low p-values, the eigenvalue is a nonmonotonic function of V. A lowest velocity V_{br} exists where each branch ends and the soliton breaks. Typical forms of the soliton near the breaking velocity V_{br} are shown in Figure 9.15. The corresponding eigenvalue spectrum is shown in Figure 9.16b. It is found [350] that the ions pile at the soliton center and the electrons at the edges, giving rise to very peaked density distributions. When the ions diverge at the soliton center, breaking occurs. Then, both electrons and ions almost move with V_{br} when $V \approx V_{br}$. Solitons at large p-numbers have been investigated numerically and analytically in various approximations [339, 340, 344, 358, 359].

For $a_0 \neq 0$, the situation is more complicated. Dark solitons are only found when the ion dynamics are taken into account. In addition, the propagation velocities are very low, $V \leq \sqrt{\varepsilon}$. For a discussion of the numerically obtained solutions, we refer to [350, 357] and the references therein.

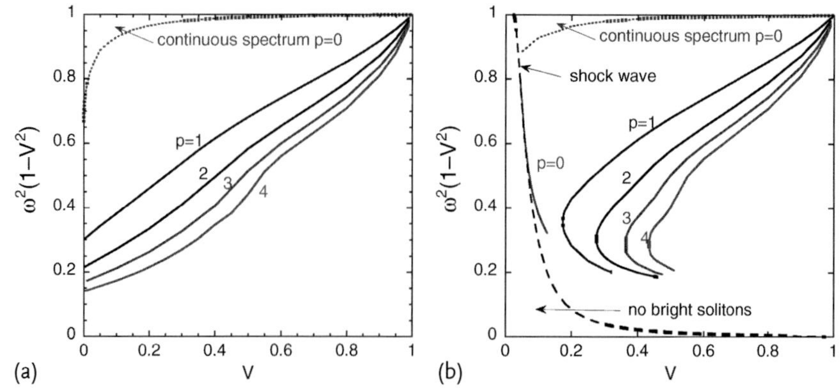

Figure 9.16 Eigenvalue spectrum of bright solitons in dependence of the group velocity V. (a) corresponds to immobile ions ($\varepsilon \equiv 0$), whereas in (b), mobile ions are included with $\varepsilon = 1/1836$ [350]. Courtesy of IOP.

9.5.1.1 Remark on the Small-Amplitude Limit

In Figure 9.16, the upper parts of the figures correspond to small amplitudes. Consistent with the considerations presented in the general chapter on solitons, for small amplitudes and low velocities, the eigenvalue spectrum is continuous. In the case of Langmuir solitons, we already reported on discrete $p = 1$ solutions for large velocities $M \to 1$. As can be seen, a similar situation occurs for relativistic solitons.

In principle, we may use Eqs. (9.174). However, their validity is restricted to pure electron response with $V/L \gg \omega_{pi}$. For standing solitons with immobile ions, one should go back directly to Eqs. (9.331) and (9.332). For immobile ions ($\varepsilon = 0$), in the limit $V \to 0$ when one simultaneously takes $p_{\parallel e} \to 0$, one gets

$$\phi \approx \sqrt{1 + a^2} - 1 \,. \tag{9.343}$$

Note that in the small-amplitude limit, this is equivalent to Eq. (9.174). The coupled Eqs. (9.331) and (9.332) can then be combined, forming

$$\frac{a''}{1 + a^2} + a\left[\omega^2 - \frac{a\,(a')^2}{(1 + a^2)^2} - \frac{1}{\sqrt{1 + a^2}}\right] = 0 \,. \tag{9.344}$$

Esirkepov *et al.* [340] have presented the exact analytical standing solution

$$a = 2\sqrt{1 - \omega^2}\frac{\cosh\left(\sqrt{1 - \omega^2}\xi\right)}{\cosh^2\left(\sqrt{1 - \omega^2}\xi\right) + \omega^2 - 1} \,. \tag{9.345}$$

It is a single-humped $p = 0$ solution with a continuous spectrum in ω. A typical solution is shown in Figure 9.17. The amplitude decreases with growing $\omega \to 1$. There is an upper limit on amplitude, that is, a lower limit on ω from the requirement that the electron density remains nonnegative. Standing solutions exist for $1 < \omega \leq \sqrt{2/3}$.

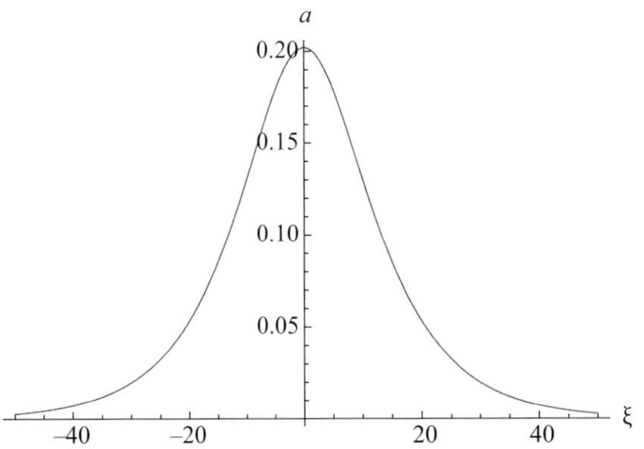

Figure 9.17 Plot of the solution Eq. (9.345) for $\omega^2 = 0.99$ and $V = 0$.

For moving solitons, we may return to Eqs. (9.174) in the small-amplitude $a^2 \ll 1$ limit. The soliton solution

$$a = \sqrt{\frac{4\left[1 - V^2 - \omega^2(1 - V^2)^2\right]}{4\omega^2(1 - V^2) - (3 + V^2)}} \operatorname{sech}\left[\sqrt{\frac{1 - \omega^2(1 - V^2)}{1 - V^2}}\,\xi\right] \tag{9.346}$$

is known [339, 349]. This solution also belongs to a continuous eigenvalue spectrum; in the limit $\omega(1 - V^2) \to 1$ for $V \to 0$, it agrees with Eq. (9.345).

9.5.2
Linearly Polarized Solitons of the Maxwell Fluid System

As has been mentioned, for linear polarization, the situation is more complicated because of the harmonic generation. Nevertheless, numerical solutions have been obtained for the fully relativistic case (see below), but as far as we know, the detailed quantitative structure of the eigenvalue spectrum was not discussed in literature. We therefore only exemplify the weakly relativistic case.

For $\omega_0 \approx \omega_{pe}$, weakly relativistic amplitudes, and immobile ions, we may rewrite Eqs. (9.142) and (9.143) for 1D propagation in the form

$$\left(\frac{\partial^2}{\partial t^2} - \frac{\partial^2}{\partial x^2}\right) a = -\left(1 + \delta n_e - \frac{a^2}{2}\right) a\,, \tag{9.347}$$

$$\frac{\partial^2 \delta n_e}{\partial t^2} + \delta n_e = \frac{1}{2}\frac{\partial^2}{\partial x^2} a^2\,. \tag{9.348}$$

First, we approximately combine these two equations to a one-field model [351]. Introducing a slowly-varying complex envelope in the form

$$a = \frac{1}{2}\left(A e^{-it} + A^* e^{it}\right), \tag{9.349}$$

$$\delta n_e = N_0 + \frac{1}{2}\left(N_2 e^{-2it} + N_2^* e^{2it}\right), \tag{9.350}$$

and substituting Eqs. (9.349) and (9.350) into Eq. (9.348), leads to (see below)

$$N_0 = \frac{1}{4}(|A|^2)_{xx}\,, \quad N_2 = -\frac{1}{12}(A^2)_{xx}\,. \tag{9.351}$$

Slowly varying means

$$\frac{\partial A}{\partial t} \ll A\,, \quad \frac{\partial N_0}{\partial t} \ll N_0\,, \quad \frac{\partial N_s}{\partial t} \ll N_2\,, \tag{9.352}$$

within the nondimensional formulation. Evaluating all terms in this approximation leads to

$$iA_t + \frac{1}{2}A_{xx} - \frac{1}{8}A(|A|^2)_{xx} + \frac{1}{48}A^*(A^2)_{xx} + \frac{3}{16}|A|^2 A = 0\,. \tag{9.353}$$

However, it is not necessary to make this reduction. We may still use the two-field model, even introducing for better comparisons higher order terms into Eq. (9.347) in the form

$$\left(\frac{\partial^2}{\partial t^2} - \frac{\partial^2}{\partial x^2}\right) a = -\left(1 + \delta n_e - \frac{a^2}{2} + \frac{a^4}{8}\right) a\,, \tag{9.354}$$

and leaving Eq. (9.348) unchanged.

For stationary (nonmoving) solitons, we make the ansatz

$$a(x,t) = \frac{1}{2}\left(\alpha(x)e^{-iEt} + \bar{\alpha}(x)e^{iEt}\right). \tag{9.355}$$

Since higher harmonics will be generated, to lowest order it is appropriate to write the electron density perturbation in the form

$$\delta n_e(x,t) = N_0(x) + \frac{1}{2}\left(N_2(x)e^{-2iEt} + \bar{N}_2(x)e^{2iEt}\right). \tag{9.356}$$

Introducing both representations into Eq. (9.354), and collecting terms being proportional to e^{-iEt}, we find the contributions

$$\partial_{tt}a \to \frac{1}{2}\left(\alpha_{tt} - E^2\alpha - 2iE\alpha_t\right)e^{-iEt}\,, \quad \partial_{xx}a \to \frac{1}{2}\alpha_{xx}e^{-iEt}\,, \tag{9.357}$$

$$-a \to -\frac{1}{2}\alpha e^{-iEt}\,, \quad -\delta n a \to -\left(\frac{1}{2}\alpha N_0 + \frac{1}{4}\bar{\alpha}N_2\right)e^{-iEt}\,, \tag{9.358}$$

$$\frac{a^3}{2} \to \frac{3}{16}|\alpha|^2\alpha e^{-iEt}\,, \quad -\frac{a^5}{8} \to -\frac{5}{128}|\alpha|^4\bar{\alpha}e^{-iEt}\,, \tag{9.359}$$

to finally obtain

$$\eta^2\alpha - \alpha_{xx} = -N_0\alpha - \frac{1}{2}N_2\bar{\alpha} + \frac{3}{8}|\alpha|^2\alpha - \frac{5}{64}|\alpha|^4\alpha\,, \tag{9.360}$$

with $\eta^2 = 1 - E^2$. On the other hand, Eq. (9.348) leads to

$$-2E^2(N_2e^{-2iEt} + \bar{N}_2e^{2iEt}) + N_0 = \frac{1}{8}\left((\alpha^2)_{xx}e^{-2iEt} + 2(|\alpha|^2)_{xx} + \bar{\alpha}^2_{xx}e^{2iEt}\right). \tag{9.361}$$

Equating coefficients leads to

$$N_0 = \frac{1}{4}|\alpha|^2_{xx} , \quad N_2 = \frac{1}{8}(\alpha^2)_{xx}\frac{1}{\eta^2 - \frac{3}{2}} . \tag{9.362}$$

Expanding for small amplitudes N_2 with respect to η^2, we find

$$N_2 \approx -\frac{1}{12}(\alpha^2)_{xx}\left(1 + \frac{\eta^2}{3}\right) . \tag{9.363}$$

When $a \sim \varepsilon\alpha(\varepsilon x)$, the second term is of negligible order ε^6 such that

$$N_2 \approx -\frac{1}{12}\left(\alpha^2\right)_{xx} . \tag{9.364}$$

To determine the slope $\alpha(x)$, the ordinary differential equation

$$\eta^2\alpha - \alpha_{xx} = -\frac{1}{4}|\alpha|^2\alpha + \frac{1}{24}(\alpha^2)_{xx}\bar{\alpha} + \frac{3}{8}|\alpha|^2\alpha - \frac{5}{64}|\alpha|^4\bar{\alpha} \tag{9.365}$$

has to be solved. It can be integrated once, that is,

$$(\alpha_x)^2 = \frac{\eta^2\alpha^2\left(1 - \frac{3}{16}\frac{\alpha^2}{\eta^2} + \frac{5}{192}\frac{\alpha^4}{\eta^2}\right)}{1 - \frac{5}{12}\alpha^2} . \tag{9.366}$$

Just as a side remark, the one-field model would lead to

$$(\alpha_x)^2 = \frac{2\lambda^2\alpha^2\left(1 - \frac{3}{32}\frac{\alpha^2}{\lambda^2} + \frac{5}{384}\frac{\alpha^4}{\lambda^2}\right)}{1 - \frac{5}{12}\alpha^2} , \tag{9.367}$$

with $\eta \approx 2\lambda$.

For nodeless $p = 0$ solutions, the maximum amplitude α_0 is related to η, and can be calculated from $\alpha = \alpha_0$ for $\partial\alpha(x)/\partial x = 0$, leading to

$$\eta = \frac{\sqrt{36\alpha_0^2 - 5\alpha^4}}{8\sqrt{3}} . \tag{9.368}$$

In Figure 9.18, we have plotted the solution

$$a(x, t) = \alpha(x)\cos(Et) , \tag{9.369}$$

$$\delta n_e(x, t) = \frac{1}{4}\left(\alpha^2\right)_{xx} - \frac{1}{12}\left(\alpha^2\right)_{xx}\cos(2Et) , \tag{9.370}$$

for $t = 0$. We shall come back to this – and more general linearly polarized solutions – in the following sections on stability.

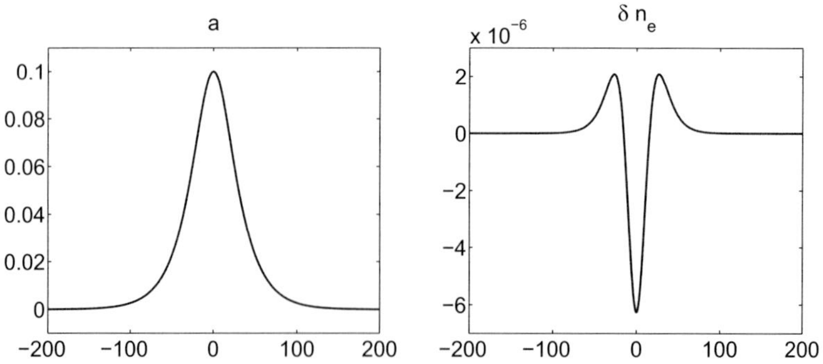

(a) (b)

Figure 9.18 Plot of the solution Eq. (9.369) (a) and Eq. (9.370) (b) at $t = 0$.

9.5.3 Longitudinal Stability of Solitary Envelope Solutions

In various investigations [341, 343–345, 351, 356, 361], the *stability* of solitons was investigated. Most of the results have been obtained for circular polarization of the laser field. In the present subsection, we follow [314, 360] and investigate the stability of existing planar soliton solutions. Ion motion is included. We use the (cold) fluid Maxwell model. The latter, of course, cannot take care of discrete particle effects like acceleration and heating of particles. We present the influence of the ion dynamics on the stability properties of slow solitons for both cases of linear and circular polarization. For circular polarization, it is known that on the electron time-scale solitons (in the following, solitons on the electron time-scale will be called presolitons) are stable in a broad parameter range [340]. For linear polarization, the stability region has been calculated in the literature [351] on the basis of the Vakhitov–Kolokolov [362] stability criterion. Presolitons become unstable on the ion time-scale. This is shown for both types of polarization. We discuss the evolution into 1D postsolitons and a possible stabilization mechanisms with finite temperatures.

We would like to emphasize that we numerically solve the basic equations without any slowly varying envelope approximation. Explicit schemes as well as implicit schemes cannot be used with step sizes which are much larger than the inverse of the highest frequency arising in the model. New methods for special oscillatory problems have been proposed, see, for example, [328], and we shall use them here whenever applicable.

Within a 1D relativistic fluid Maxwell model, which will be used for the so called longitudinal stability analysis, we assume that all quantities only depend, besides time t, on the single space variable x. The Coulomb gauge immediately leads to $\boldsymbol{A} \equiv \boldsymbol{A}_\perp$, where $\perp$ designates the directions perpendicular to x. We abbreviate $\hat{x} \cdot \boldsymbol{p}_\alpha \equiv p_\alpha$. For a laser pulse starting in vacuum, we get $\boldsymbol{P}_{\perp\alpha} = 0$. As has been discussed, the following set of dynamical equations will be used

$$\frac{\partial^2 \mathbf{A}_\perp}{\partial x^2} - \frac{\partial^2 \mathbf{A}_\perp}{\partial t^2} = n_e \frac{\mathbf{A}_\perp}{\gamma_e} + \varepsilon n_i \frac{\mathbf{A}_\perp}{\gamma_i} , \tag{9.371}$$

$$\frac{\partial^2 \phi}{\partial t \partial x} + \frac{n_e p_e}{\gamma_e} - \varepsilon \frac{n_i p_i}{\gamma_i} = 0 , \tag{9.372}$$

$$\frac{\partial^2 \phi}{\partial x^2} = n_e - n_i , \tag{9.373}$$

$$\frac{\partial n_e}{\partial t} + \frac{\partial}{\partial x}\left(\frac{n_e p_e}{\gamma_e}\right) = 0 , \tag{9.374}$$

$$\frac{\partial n_i}{\partial t} + \varepsilon \frac{\partial}{\partial x}\left(\frac{n_i p_i}{\gamma_i}\right) = 0 , \tag{9.375}$$

$$\frac{\partial p_e}{\partial t} = \frac{\partial(\phi - \gamma_e)}{\partial x} - T_e \frac{\partial \ln n_e}{\partial x} , \tag{9.376}$$

$$\frac{\partial p_i}{\partial t} = \frac{\partial\left(-\phi - \frac{\gamma_i}{\varepsilon}\right)}{\partial x} - T_i \frac{\partial \ln n_i}{\partial x} . \tag{9.377}$$

They form a closed system; temperature effects have been included in the most simple form. For linearly polarized waves ($\mathbf{A}_\perp = A\hat{y}$), letting $E = -\partial\phi/\partial x$, ignoring the ion response, and setting $T_e = T_i = 0$, these equations agree with the previously proposed model [290].

9.5.3.1 General Remarks on Numerical Stability Analysis

In general, soliton solutions of the two-fluid system are nonlinear, time-dependent states which are known either analytically or numerically. The relevant concept of stability for such states is the stability of invariant sets [251].

To examine stability by only using the nonlinear equations can lead to the wrong results. The reasons are numerical noise and the limited numerical stability of the schemes when integrating over long times. If the initial perturbation only contains a very small part of the unstable mode, very long integration times are necessary to recognize the unstable behavior. Numerical instabilities would usually prevail much earlier and may lead to the wrong conclusions.

The standard (analytic) approach to examine stability with respect to small perturbations is the following: Linearize with respect to an unperturbed state, transform-out a possible time-dependence, and find the eigenvalues of the linear time evolution operator related to the problem. A positive real eigenvalue corresponds to an exponentially growing mode, an imaginary one to an oscillatory mode. For the most complex systems, it is very complicated, or even impossible, to determine the eigenvalues analytically. In any numerical approach, one has to choose a proper basis, numerically calculate the matrix representation of the time evolution operator within this basis, to find the eigenvalues. For systems consisting of many equations, and for a basis that would consist of a sufficiently large number of functions to represent the solutions accurately, this approach is usually very inefficient.

To show exponential instability of a solution, it is sufficient to know the fastest growing perturbation and its exponential growth rate γ. The von-Mises iterative method [363] calculates the mode corresponding to the largest (in general complex) eigenvalue. Since we are not interested in the mode with the largest absolute eigenvalue but in the mode with the largest real part of the eigenvalue, one should use the following method. By numerical integration of the linearized equation, one "grows" the most unstable mode. This is possible since it dominates all other modes after some time. The initial conditions are almost arbitrary, but must have a component in the direction of the most unstable mode. After some timesteps of the integration, a normalization becomes necessary to avoid overflows. One may use the renormalized distribution again as an initial condition for the linear integrator. This has the additional advantage that the numerical noise will be damped out in case of instability. Thus, the numerical long time stability is much better for the linear integrator than for the nonlinear one. Finally, the distribution will converge towards the mode corresponding to the eigenvalue with the largest real part. By measuring the time-dependence of the maximum of the distribution during its time-development, it is possible to identify the rate γ of instability. If dealing with time-dependent periodic coefficients in the linearized equations (as happens in the linear polarized case), it is important to measure the maximum amplitude only at multiples of the period [253].

By this method, as for any other numerical method, it is not possible to prove stability in the strict mathematical sense. However, one can estimate regions of confidence for the possible rates of instability in terms of a bound for detectable rates of instability. The typical time for the instability to develop is of the order $T_{inst} \approx 1/\gamma$. If the scheme is numerically stable up to times T_{num}, we can estimate the minimum rate that can be resolved by $\gamma \approx -\ln(\varepsilon)/T_{num}$. Here, ε is the relative amount of the fastest growing mode within the initial perturbation. If ε would be just due to numerical errors, very long times would be necessary in the nonlinear integration to see small instabilities.

The linearized equations may, however, have growing solutions (e.g., translation modes) following a power-law growth in time. They do not imply instability of the nonlinear problem. For this reason, it is important to confirm instability by the nonlinear system. One should always solve the nonlinear system using a significant part of the linearly detected growing mode as initial perturbation. If instability prevails, an amount of 1% of the unstable mode in the (total) initial contribution is definitely sufficient to show the unstable behavior.

Besides the stability of solutions with respect to form, it is interesting to investigate the structural stability. Structural instability means that the system exhibits new behavior if additional terms, for example, originating from the ion component or temperature, are added to the basic model equations. The instability of presolitons (see below) due to ion space charge effects is an example of a structural instability.

9.5.3.2 Solitons on the Electron Time-Scale

Here, we first consider soliton solutions on the electron time-scale (presolitons). As shown in the previous section, stationary soliton solutions have been found numerically in [341] for circular polarization. The properties of solitons have been catalogued with respect to the number p of zeros of the vector potential, the velocity V, and the frequency ratio ω. In the following, we summarize results for single-humped ($p = 0$), standing ($V = 0$) solitons without any ion response in overdense ($\omega < 1$) plasmas.

It is known [341] that in addition to single-humped solitons, also solitons with nodes (in the vector potential) exist. We shall focus on these solutions within a more general context in the next subsection.

Standing or nearly standing solitons are especially important since they have been observed in experiments [335] and PIC simulations [333]. For times smaller than the characteristic ion time $t_i = 1/\omega_{pi} = \sqrt{m_i/m_e}t_e$, one may consider the ions to be at rest. Depending on the polarization, different solutions exist.

For *circular* polarization, we may use the analytically known solution [340] which contains the frequency ω as a free parameter. The value of ω determines the maximum soliton amplitude a_0 and the minimum electron density in the center of the soliton. The minimum density leads to a lowest value $\omega_{min} = \sqrt{2/3}$ which is related to $a_0 \approx 1.73$. Below this minimal value ω_{min}, negative densities would appear in the analytical solution [340]. Hydrodynamic numerical simulations of the solitons show that the solutions are stable (i.e., no significant growth rate can be detected by the method outlined in the previous subsection) over the whole parameter regime $\omega_{min} \le \omega \le 1$. This is in agreement with the PIC simulations presented in [340].

For *linear* polarization, it is much more difficult to solve the fully relativistic 1D model due to the generation of higher harmonics of the incident wave. A simplified approach is to solve, instead of the fully relativistic system, the previously discussed weakly nonlinear system implying the scaling $A = \varepsilon a(\varepsilon x)$, where ε is a smallness parameter.

We emphasize that, depending on the η-value, the solutions are stable or not (on the electron time-scale). Previous investigations [351] predicted stability over a broad range of η-values. The predictions were made by using the Vakitov–Kolokolov criterion [362]. The derivation of the latter, however, is based on assumptions which are not met here; see the remarks in the Appendix H on the stability theorems. The Vakitov–Kolokolov criterion states that solutions are stable with respect to small perturbations as long as $dP/d\eta^2 > 0$, where P is given by

$$P = \int |a(x)|^2 dx \,. \tag{9.378}$$

Numerical simulations [314] show that the range of η in which stability prevails is much smaller than the previously predicted one [351]. The Vakitov–Kolokolov criterion predicts instability only for high intensities $a_0 \ge 1.44$. Actually, one finds that the instability sets in already for $a_0 \approx 0.2$, that is, well below the ultrarelativistic regime, as shown in Figure 9.19.

The "numerical proof of instability" shows instability for amplitudes of the order of, or larger than, $a_0 \approx 0.2$. Figure 9.20 shows results from the nonlinear integration of an unperturbed and a perturbed linear polarized initial condition, respectively, for $a_0 = 0.2$ after the time $t = 190$. The unperturbed initial condition (solid line) at $t = 0$ is also shown. During the time of integration, the unperturbed

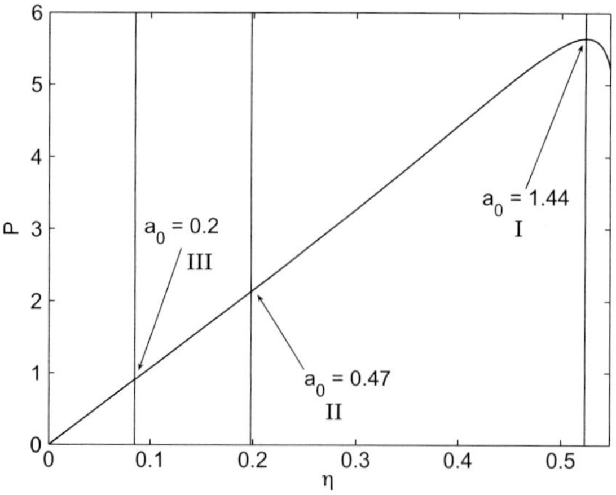

Figure 9.19 Evaluation of P as a function of amplitude. The prediction of stability up to $a_0 \approx 1.44$ is by far too optimistic [360]. Unstable solutions have already been found, for example, $a_0 = 0.47$. The stability limit is approximately $a_0 \approx 0.2$.

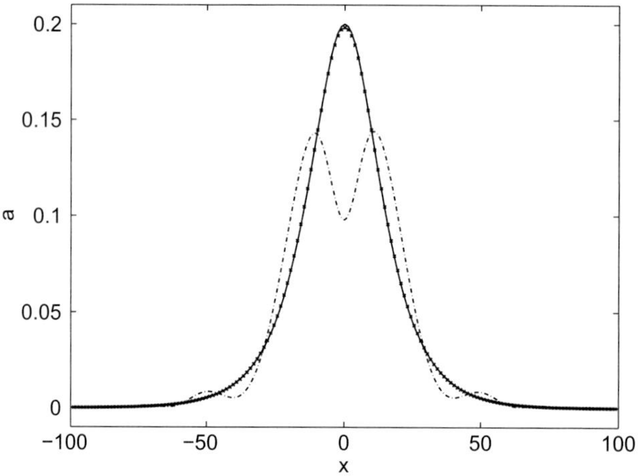

Figure 9.20 Results from nonlinear integration of an unperturbed and a perturbed linear polarized initial condition with $a_0 = 0.2$ after the time $t = 190$, and the unperturbed initial condition (solid line) at $t = 0$. The result for the unperturbed distribution is shown as dotted line, for the perturbed as dash–dotted line. The amount of perturbation was 1%.

distribution is practically unchanged while a perturbation of 1% leads to significant unstable behavior.

9.5.3.3 **Stability of Farina–Bulanov Solitons**

In this subsection, we comment on stationary solutions of the full set of Eqs. (9.371)–(9.377) for $T_e = T_i = 0$. That means that we include the ion response into the soliton solutions. Stationary soliton solutions of Eqs. (9.371)–(9.377) have been found numerically in [341] for circular polarization on the basis of the formulation in [349]. Multiple-hump solitons ($p = 0, 1, 2, \ldots$) do exist on the electron time-scale for all velocities $V < 1$. Including the ion motion, solitons do exist only above a certain threshold velocity V.

A detailed numerical investigation [314] leads to the result that all solutions $p = 1, 2, \ldots$ including ions are *unstable* with respect to small perturbations fulfilling $\boldsymbol{P}_{\perp a} = 0$. For $p = 0$, the solutions are *stable*. This conclusion is based on the "negative result" in the search for linearly unstable modes. We can exclude significant growth rates which would alter the behavior of the Farina–Bulanov solitons for $p = 0$ over physically relevant times.

Summarizing, solitons with $p = 0$ are longitudinally stable [353]. Solitons with node numbers $p = 1, 2, \ldots$ are 1D unstable [353] and show two different types of instabilities. In general, the growth rate increases as the soliton velocity decreases. For $p = 1$, there is a local minimum of the growth rate for $V \approx 0.7$. This point separates two kinds of instabilities. Perturbations for solitons above this velocity are purely growing $\sim e^{\Gamma t}$, while those below exhibit an additional oscillatory evolution and grow $\sim e^{\Gamma t} \cos(\tilde{\omega} t)$. The separation between the two types of instabilities can be seen for solitons with higher node numbers p too, but takes place at higher velocities.

Let us discuss the meaning of $\tilde{\omega}$ for the evolution of the perturbation in the laboratory frame. If $\tilde{\omega}$ is zero, the evolution of the perturbation in the laboratory frame at a fixed position is only due to the spatial distribution of the mode in the comoving frame. For the vector potential, the additional phase factor has to be included. To grow this mode, one should transform out the explicit time dependence due to this factor. The oscillations of the vector potential of the perturbation at a fixed position in the laboratory frame are therefore determined by the soliton frequency ω_0 and the spatial distribution in the comoving frame.

In the case $\tilde{\omega} \neq 0$, the unstable mode oscillates in the comoving frame. That oscillation introduces additional oscillations in the laboratory frame. It leads to the generation of frequency sidebands in the vector potential. Let us again consider the vector potential in more detail. First, the spatial distribution in the comoving frame leads to a temporal evolution in the laboratory frame at a fixed position. The oscillations are then shifted by ω_0 due to the phase factor. Second, the frequency $\tilde{\omega}$ in the comoving frame leads to sidebands $\tilde{\omega} \pm \omega_0$. These sidebands can be interpreted as an indication of Raman scattering [353].

9.5.3.4 Soliton Generation on the Ion Time-Scale

Ion movement can be neglected for times smaller than the ion time t_i. Presolitons may be formed. For larger times, one has to use the full set of model Eqs. (9.371)–(9.377), which include ion density and momentum together with the ion current. One could still make the cold plasma approximation. We choose circularly polarized solutions for demonstration. It is by no means evident that the presolitons will evolve into the Farina–Bulanov solitons discussed in Section 9.5.3.3. Whether they do or not is the main topic of this section.

When looking at the time-development of solitons, we have the following picture in mind. Ions are not contributing to the initial phase of soliton creation. It is expected that presolitons, like those described in Section 9.5.3.2, are generated in the beginning. For small times, the ion distribution would be unperturbed and equal to the background density. After a certain time of the order of t_i, we can expect that the ions react to the space charge field associated with the soliton. Ion motion is important for sufficiently slow structures. In the following, we shall use $V = 0$ for demonstration.

9.5.3.5 From Presolitons to Postsolitons

Simulations of the system Eq. (9.371)–(9.377) for $T_e = T_i = 0$ with a presoliton and an uniform ion density as initial conditions are shown in Figure 9.21. Both, a and δn_e, are depicted for an initial, *circularly* polarized $p = 0$ soliton with amplitude $a_0 = 1.73$. One finds an unstable evolution of the presoliton on the ion time-scale. After a short time (order t_i), the ions react to the field and move towards the electrons. The system intents to evolve into a quasineutral state. As the ions become pushed out of the center of the soliton, the electrons become even more evacuated. This leads to an almost complete evacuation of density in the middle. At the same time, the trapped vector potential gets narrower and steeper. With larger initial amplitudes a_0, the whole process becomes faster.

Figure 9.22 shows the evolution of the density perturbations δn_e and δn_i, respectively, for an initial soliton with $a_0 = 0.2$. The electrons, being pushed outwards, pile up at the border of the depression. Very large density gradients are achieved. Then, of course, finite pressure (temperature) effects may come into play. Finite electron temperatures will ultimately stop the density steepening [314]. With finite electron temperature, a complete density evacuation no longer takes

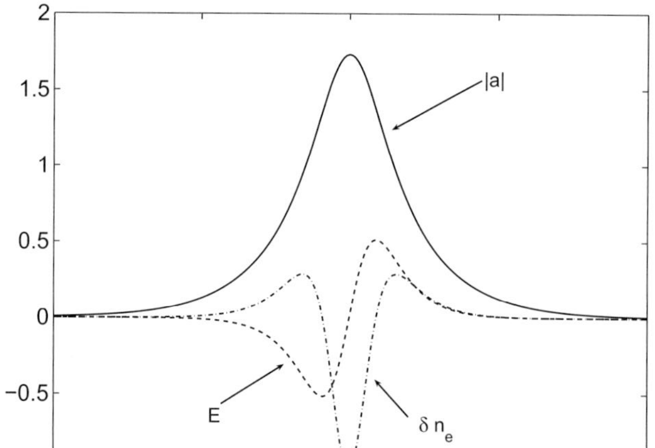

Figure 9.21 Vector potential a, electron density perturbation δn_e, and electrostatic field E for a standing, circular polarized presoliton with $a_0 = 1.73$ as initial distributions.

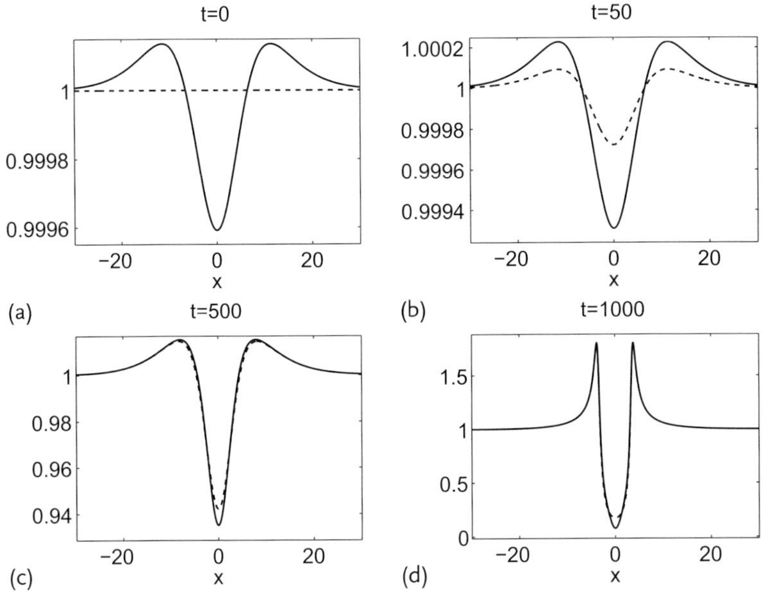

Figure 9.22 Density perturbations for ions (dashed line) and electrons (solid line) at different times $t = 0$ (a), $t = 50$ (b), $t = 500$ (c), and $t = 1000$ (d), respectively, when starting with an initial presoliton with $a_0 = 0.2$. Circular polarization is assumed.

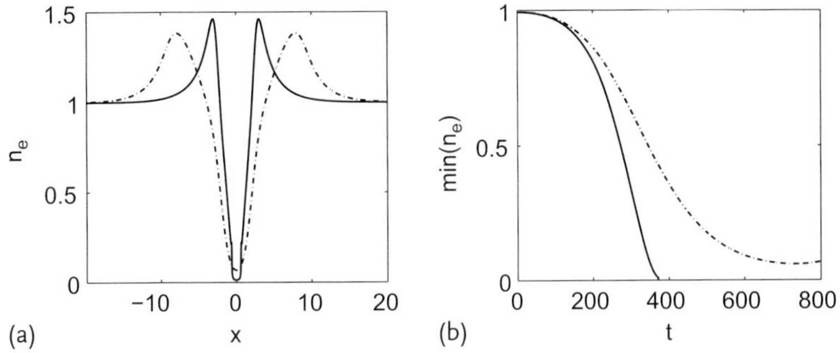

Figure 9.23 (a) Electron density distribution without (solid line) and with (dash-dotted line) electron temperature $T_e = 50$ keV at times $t = 370$ and $t = 800$, respectively. (b) Evolution of the minimum density with time.

place. The required temperatures are quite large, for example, of order 50 keV. Figure 9.23 shows a typical example.

The transition from a presoliton to a postsoliton on the ion time-scale was observed experimentally and has already been shown by multidimensional PIC simulations [333].

For *linear* polarization one should start with initial distributions for a and δn_e which follow from Eq. (9.366). The amplitudes are chosen within the range of stable presolitons. In the case of linear polarization, a comparison of the results from

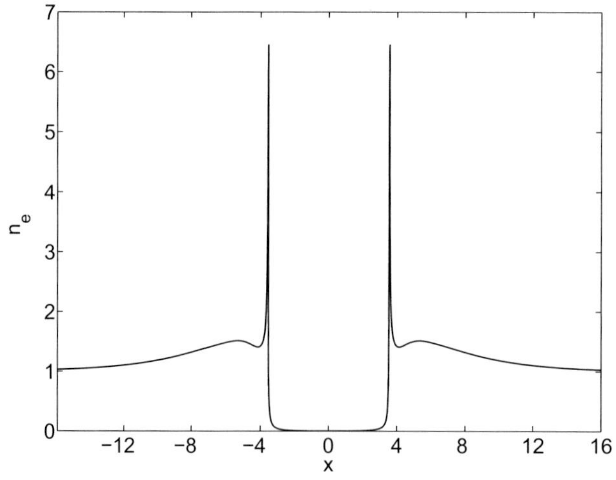

Figure 9.24 Density cavity (postsoliton), as produced by an initial presoliton, at time $t = 1100$ for *linear* polarization.

two different models leads to the following picture. First, one may extend the linearized system with an equation for the ion density perturbation δn_i and include the ion current into the wave equation. Secondly, one can simulate the fully relativistic system Eqs. (9.371)–(9.377) for $T_e = T_i = 0$. Both models predict similar behaviors of ions and electrons as were already observed in the circular polarized case. As a matter of fact, linearly polarized presolitons are unstable on the ion time-scale.

For similar initial amplitudes, the two polarization cases lead to comparable time-scales in which the evacuation of density (see Figure 9.24) takes place in the center. Since the results look similar to those presented in Figure 9.22, we do not show more details here.

9.5.3.6 **Generation of Postsolitons in Inhomogeneous Plasmas**

The generation of postsolitons is a generic process in laser–plasma interaction. Postsolitons may appear from the original laser pulse after frequency-downshift and slowing-down of the group velocity, accompanied by relativistic modulational instability and forward Raman scattering [364–373]. Here, we present *simulations* of postsoliton generation by laser pulse propagation in inhomogeneous plasmas. Linearly, a laser pulse propagating in inhomogeneous underdense plasmas towards the critical density would be reflected. Nonlinearly, the laser pulse will dig a hole into the plasma density. First, electrons are pushed out. At later times, the ions will follow.

Simulations have been performed as follows. A density profile of the form $n(x) = 1 + 0.21\exp(-\tilde{x}^6/\sigma^2)$ with $\tilde{x} = x - 280$ and $\sigma = 300$ is used. Equations (9.371)–(9.377) are taken in the cold plasma approximation (plus the background density variation). An initial propagating circular polarized Gaussian laser

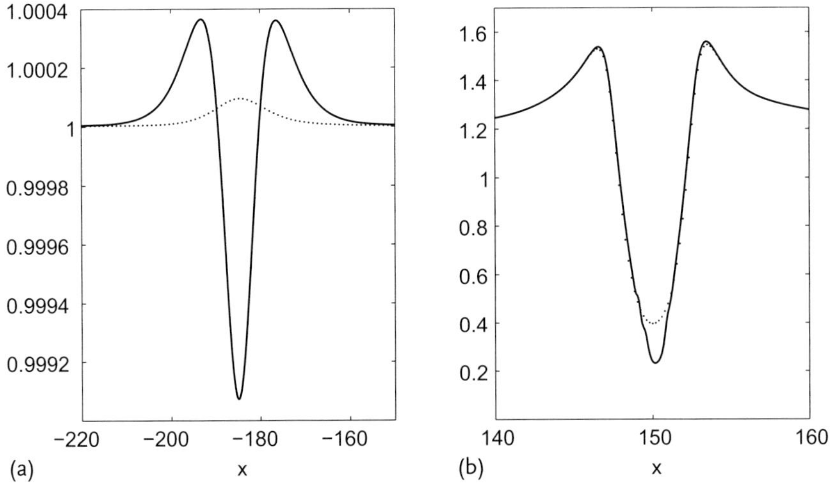

Figure 9.25 Density perturbations of electrons (solid line) and ions (dotted line) due to an initial Gaussian laser pulse with $a_0 = 0.2$ and FWHM $= 40$. In (a), we see density perturbations moving in underdense plasma, whereas (b) illustrates density depressions standing at critical density when radiation is being trapped.

pulse with $a_0 = 0.2$, FWHM = 40 and $\omega = 1.1$ is started at $x = -525$, propagating the density gradient upwards. At $x = 280$, the critical density is reached.

During propagation, electrons are pushed out. For slow propagation, ions pile up in the middle. Note that this behavior is similar to the soliton motion in overdense plasmas [341]. Figure 9.25 shows two snapshots of the transition from underdense to overdense plasma. When the pulse reaches the critical density, it almost stops and an interesting dynamical evolution takes place. While bouncing back and force, the pulse becomes narrower and the electron density depression increases dramatically. In the Figure 9.25b, we are at a time when the ions are also beginning to be evacuated from the central part. The further development becomes very similar to the transition from pre- to postsolitons. The conjecture that the appearance of postsolitons is induced by trapping of radiation on the electron time-scale was confirmed by many simulations.

9.5.3.7 **Stability of Solitons in Warm Plasmas**

Relativistic electromagnetic solitons in a warm quasineutral electron–ion plasma were investigated in [345] within a kinetic model. Circularly polarized, nondrifting localized solutions have been obtained. Within the quasineutral approximation, the solutions are calculated from Eq. (16) of [345], that is,

$$a_{xx} - \frac{\partial^2}{\partial t^2} a = G(a) , \tag{9.379}$$

where $G(a)$ is given by Eq. (17) of [345]. Note that the nondrifting soliton solutions are written in the form $a(x,t) = a_s(x)e^{i\omega t}$. The temperatures appear via the parameters $\lambda_e = T_e/m_e$ and $\lambda_i = T_i/m_e$ in the expansion of $G(a)$.

For various values of the parameters ω, λ_e, and λ_i, one first determines the soliton solutions [345]. Then, one linearizes Eq. (9.379) with respect to a soliton solution. One recognizes a *linear* growth of the perturbation. It corresponds to a solution $i\delta\omega t a_s(x)\exp(i\omega t)$ (the so called phasor mode) of the linearized version of Eq. (9.379). We may obtain the mode by a replacement $\omega \to \omega + \delta\omega$ in the soliton solution, with subsequent linearization. The interpretation as a phasor mode, and not as an exponential instability, becomes obvious. The stability of the solitons in warm plasmas also follows from the Q theorem [265, 268, 272, 374–376] which applies here. The solution $a(x,t) = a_s(x)e^{i\omega t}$ is stable, provided

$$Q = i\int dx\left(a\frac{\partial a^*}{\partial t} - a^*\frac{\partial a}{\partial t}\right) \tag{9.380}$$

is a monotonically decreasing function of ω. This behavior can be easily checked.

9.5.4
Solitary Envelope Solutions in Higher Dimensions

One-dimensional solitons have a tendency to *transverse (2D)* filamentation [377–380]. In the present subsection we investigate the transverse instability for the known 1d soliton solutions in laser–plasma interaction. Ion motion is included. We shall use the (cold) fluid Maxwell model. The "numerical proof of stability" requires codes which follow all possible initial perturbations with high resolution. We use here the strategy which has been applied already at longitudinal stability [314] to identify the (linearly) most unstable modes and to demonstrate exponential instabilities (if they exist) by fast algorithms. The calculations of transverse filamentation are import for two reasons. First, they show us which solutions may be long-living (i.e., if we can prove stability). Second, in the case of instability, the unstable modes give us a hint for the dynamics of structures in higher dimension. Following the so called slaving principle [381], the most unstable mode will dominate the dynamics, slave the others, and show up in the topology of the possible end states. Stable soliton solutions seem to exist in higher dimensions [269, 272, 376], also in laser plasmas [334].

If we consider a laser pulse propagating along x and allow for transversal dependency of all quantities along y we arrive at a two-dimensional (2D) relativistic fluid Maxwell description. We consider a comoving frame by introducing $\xi = x - Vt$, $\tau = t$. Then, the set of equations

$$\frac{\partial^2}{\partial\tau^2}A - 2V\frac{\partial^2}{\partial\xi\partial\tau}A - (1-V^2)\frac{\partial^2}{\partial\xi^2}A - \frac{\partial^2}{\partial y^2}A = \mathsf{P}_{df}\boldsymbol{j}\ , \tag{9.381}$$

$$\left(\frac{\partial^2}{\partial\xi^2} + \frac{\partial^2}{\partial y^2}\right)\phi = n_e - n_i\ , \tag{9.382}$$

$$\frac{\partial}{\partial\tau}n_\alpha - V\frac{\partial}{\partial\xi}n_\alpha + \frac{\partial}{\partial\xi}j_\xi + \frac{\partial}{\partial y}j_y = 0\ , \tag{9.383}$$

$$\frac{\partial}{\partial \tau}\begin{pmatrix} P_{a\xi} \\ P_{ay} \\ P_{az} \end{pmatrix} - V\frac{\partial}{\partial \xi}\begin{pmatrix} P_{a\xi} \\ P_{ay} \\ P_{az} \end{pmatrix} - \frac{\varepsilon_a}{\gamma_a}\begin{pmatrix} p_{ay}\left(\partial_\xi P_{ay} - \partial_y P_{a\xi}\right) + p_{az}\partial_\xi P_{az} \\ p_{az}\partial_y P_{az} - p_{a\xi}\left(\partial_\xi P_{ay} - \partial_y P_{a\xi}\right) \\ -p_{a\xi}\partial_\xi P_{az} - p_{ay}\partial_y P_{az} \end{pmatrix}$$

$$= \begin{pmatrix} \partial_\xi \\ \partial_y \\ 0 \end{pmatrix}\left(q_a \phi - \frac{\gamma_a}{\varepsilon_a}\right), \tag{9.384}$$

$$\boldsymbol{j} = \varepsilon_i \frac{n_i \boldsymbol{p}_i}{\gamma_i} - \frac{n_e \boldsymbol{p}_e}{\gamma_e}, \tag{9.385}$$

forms a closed system. $\mathsf{P}_{df} = \mathrm{Id} - \nabla(\nabla^2)^{-1}\nabla\cdot$ is a projector which gives the divergence-free part of a given vector field. In the present notation, $\nabla = (\partial_\xi, \partial_y, 0)^T$.

By neglecting derivatives along the y-direction and assuming $P_y = P_z = 0$, the standard 1D relativistic description emerges. The latter reads

$$\frac{\partial^2}{\partial \tau^2}\boldsymbol{A} - 2V\frac{\partial^2}{\partial \xi \partial \tau}\boldsymbol{A} - \left(1 - V^2\right)\frac{\partial^2}{\partial \xi^2}\boldsymbol{A} = \left(\varepsilon_i \frac{n_i}{\gamma_i} - \frac{n_e}{\gamma_e}\right)\boldsymbol{A}, \tag{9.386}$$

$$\frac{\partial^2}{\partial \xi^2}\phi = n_e - n_i, \tag{9.387}$$

$$\frac{\partial}{\partial \tau}n_a - V\frac{\partial}{\partial \xi}n_a + \frac{\partial}{\partial \xi}j_\xi = 0, \tag{9.388}$$

$$\frac{\partial}{\partial \tau}p_{a\xi} - V\frac{\partial}{\partial \xi}p_{a\xi} = \partial_\xi\left(q_a \phi - \frac{\gamma_a}{\varepsilon_a}\right). \tag{9.389}$$

As has been discussed already, for the 1D set of Eqs. (9.386)–(9.389), *circular* polarized soliton solutions are known [339, 341]. The envelope $\hat{a}(\xi)$ of the vector potential $\boldsymbol{a}(\xi, t) = A_y + iA_z$ of the solitons may have different numbers $p = 0, 1, 2, \ldots$ of nodes. In the comoving frame of reference, these solitons have a vector potential of the form

$$a(\xi, t) = \hat{a}(\xi)\exp\left[i\omega_0(1 - V^2)t + ik_0\xi\right], \tag{9.390}$$

where $k_0 = \omega_0 V$. When ion motion is included, there is a minimum soliton velocity above which these solutions exist. The faster the soliton, the smaller and narrower is the associated depletion in the electron density.

In the following, we use circularly polarized solitons for demonstration. The results for linearly polarized solitons are similar. From *nonlinear* simulations, linear growth rates can only be determined with relatively large errors since linear, nonlinear, and numerical effects will appear together. Thus, a (numerical) stability analysis using a *linear* 2D integrator is often appropriate. We linearize the set of Eqs. (9.381)–(9.385) with respect to the unperturbed 1D soliton solution (the 1D solution is constant in transversal y-direction). We investigate the linear growth of modes with wavenumbers $k_\perp$ in transversal direction. Since the unperturbed state does not depend on the y-coordinate, it is appropriate to Fourier transform in transversal direction. For any fixed $k_\perp$, we end up with a stability problem which is effectively 1D. For details on how to determine the linear growth rate Γ from the simulations, we refer the reader to [314].

Stability codes are based on second-order finite-difference schemes in time and spectral methods to calculate derivatives in space. Because of the spectral methods, we anticipate aliasing effects [382]. To keep aliasing errors as small as possible, one can apply a filter function in Fourier space that cuts 10% of the highest modes. This cutting is only done for the densities n_e and n_i. It turns out to be sufficient enough to avoid numerical instabilities.

Simulating soliton solutions by solving the nonlinear system Eqs. (9.386)–(9.389) with such types of codes allows one to follow the unperturbed evolution for times up to several thousands of $1/\omega_{pe}$. Finite difference schemes in space are often much more limited in their numerical stability. Then, the unperturbed soliton may only be followed for several hundreds of $1/\omega_{pe}$ until perturbations are observed. The numerical errors due to finite differences in space may serve as initial conditions for a numerical or physical instability.

9.5.4.1 Transversal Instability of $p = 0$ Solitons

The simulations carried out in [353] showed that solitons with no nodes in the envelope of the vector potential, that is, $p = 0$ solutions, are longitudinal stable for every velocity V. Raman scattering is prohibited due to the fact that the depleted density of these solutions is too large. Plane wave calculations predict that for $n_{min} > \gamma\omega_0^2/4$, no Raman scattering appears. In [314], it was shown that this also holds true for $p = 0$ solitons when ion response is included into the model.

Summarizing, the $p = 0$ solitons show no longitudinal instability since their density distribution allows no Raman-like processes. From the simulation of the linearized system, however, one finds transversely unstable modes for all possible velocities V. The $p = 0$ solitons have a continuous spectrum of existence in ω_0 for a given velocity V, which differentiates them from solitons with $p = 1, 2, \ldots$ The growth rate of the transversal instability varies with $k_\perp$ in a way that is typ-

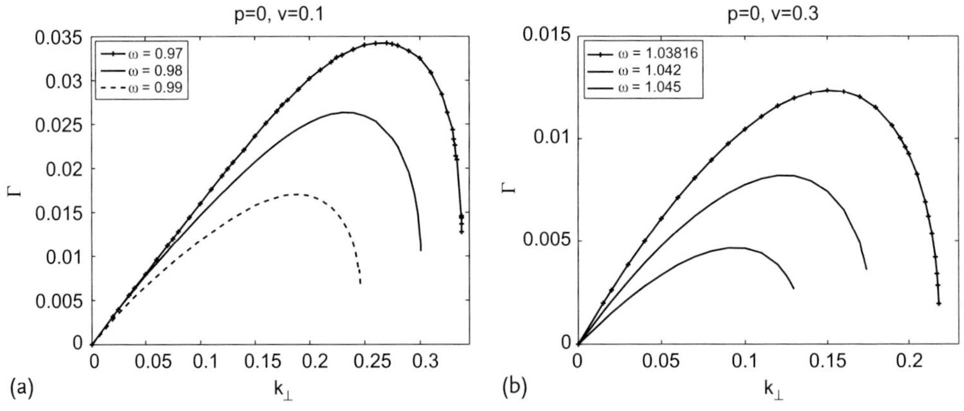

Figure 9.26 Linear growth rates Γ for two $p = 0$ solitons with velocities $V = 0.1$ (a) and $V = 0.3$ (b). For each velocity, the growth rates for the transversal instability (in dependence on the wavenumbers $k_\perp$) are shown for three solitons which differ in their frequencies ω_0.

ical for transversal instabilities. It increases with increasing $k_\perp$ until it reaches a maximum and shows a clear cut-off for larger $k_\perp$. The quantitative behaviors are depicted in Figure 9.26. In the linear regime, the growth of perturbations with small wavelengths is prohibited. Nevertheless, small wavelengths may grow in the nonlinear regime due to mode conversion. Since for all velocities V there is a range in the frequency ω_0 for which the solitons exist, one should investigate different frequencies. It has been found [360] that the maximum growth rate increases as the frequency decreases, which is also displayed in Figure 9.26. The slower the solitons are, the larger their maximum growth rate is and the larger the value of $k_\perp$ for which this maximum value is reached.

The maximum growth rates are quite significant since a growth rate of 0.035 would allow a 1% perturbation of this wavelength to grow to the order of one in a time of only $t \approx 135$.

9.5.4.2 Transversal Instability of $p = 1, 2, \ldots$ Solitons

Most of the solitons with at least one node in the vector potential exhibit a forward Raman instability in the linear 1D stability analysis [353]. 2D linear simulations show [360] that these solitons also suffer from a transversal instability in the same way as the $p = 0$ solitons do. The maximum transversal growth rate is always larger than the longitudinal growth rate by a factor of at least two (up to a factor of 14). Figure 9.27 shows the ratio of the (transversal over longitudinal) growth rates in dependency on the soliton velocity V. The peak in Figure 9.27 is due to a

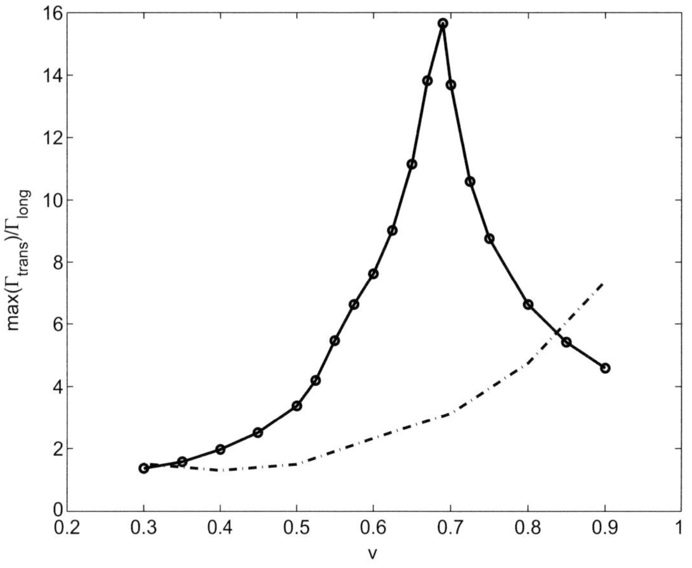

Figure 9.27 Ratio of the maximum transversal growth rate and the longitudinal growth rate for $p = 1$ (solid line) and $p = 2$ (dashed line) solitons with different velocities V. The peak for $p = 1$ is associated with a minimum in the longitudinal growth rate which separates solitons suffering from different types of longitudinal instability.

minimum in the longitudinal growth rate which separates two regions of different unstable behaviors in the 1D case. For velocities smaller than $V \approx 0.7$, the 1D unstable mode shows frequency sidebands. On the other hand, for solitons faster than $V \approx 0.7$, the instability shows no frequency sidebands. It is therefore not of oscillatory type.

9.5.4.3 Nonlinear Simulations

To verify the findings from the linear simulation, one should carry out nonlinear 2D simulations. As an initial condition, a 1D soliton is used which is constant along y. A small amount of a few percent of the most unstable mode (for a given $k_\perp$) is then added to this distribution. In the following, we will discuss this for cases when the wavenumbers $k_\perp$ have been chosen such that the growth rate has a maximum. The comparison with the linear integrator leads to an independent verification of the linear result. As long as the perturbation is small, linear and nonlinear evolution have to be the same. All linear results have been verified by this method and gave excellent agreement with the nonlinear results.

Eventually the perturbation will become large and the nonlinear evolution will set in. Because of their different stability properties in the purely longitudinal case, we distinguish in this section between nodeless solitons and solitons with nodes.

When the nonlinear development prevails for a nodeless solution, a complete transversal filamentation of the perturbed soliton appears. The electrons start being expelled from the regions where the intensity $|a|^2$ is large and begin to bunch beside the pulse. The ion density forms peaks at the front and is depleted inside the filaments.

Since solitons with $p = 1, 2, \ldots$ are both, transversely as well as longitudinal, unstable, it is important to check the nonlinear evolution of the two instabilities. The evolution of the two instabilities could be different from the linear ones. The latter only gives the ratio of the linear growth rates presented in Section 9.5.4.2.

The simulations show that the transversal perturbation always prevails over the longitudinal one in the nonlinear regime. We demonstrate this for a $p = 1$ soliton with $V = 0.8$. For example, the initial perturbations were constructed in such a way that they contain each of the most unstable modes for every direction, as determined by linear simulations. The norm of the perturbations in each direction was 1% of the norm of the soliton. The longitudinal growth rate is $\Gamma_l = 0.010\,393$, and the transversal $\Gamma_t = 0.068\,90$. After a time of 56, the pulse already shows clear transversal filamentation but almost no longitudinal deformation.

9.5.4.4 Transversal Instability from Noise

Thus far, we always assumed knowledge of the most unstable modes from the linearized systems. To demonstrate the development of the transversal instability for systems without exact initial excitation of those modes, a soliton with random noise added to the initial vector potential $\boldsymbol{a}$ was simulated [360]. The noise $\boldsymbol{a}_{noise}$ is Gaussian distributed around a mean value. Besides fulfilling the Coulomb gauge $\nabla \cdot \boldsymbol{a}_{noise} = 0$, the amplitudes of the Fourier modes are such that the average magnitude of the corresponding magnetic field $\boldsymbol{B}_{noise}$ is the same for every wavenumber $\boldsymbol{k} = (k_\xi, k_y)$. Thus, there is no preferred direction within the noise, and all Fourier modes of $\boldsymbol{B}_{noise}$ are equally strong. The energy of the added noise is about one per mil of that of the total energy of the soliton. The highest frequency contained in the initial noise is in the order of a few percent of the highest frequency the computational grid can resolve. Initial excitations of larger frequencies would lead to strong aliasing errors from the numerical integration.

The simulations show that within a time of the order of a few tens of $1/\omega_{pe}$, a transversal perturbation with a distinct $k_\perp$ dominates. The value of the transversal wave number $k_\perp$ corresponds to the maximum transversal growth rate that has been determined from the linear analysis. After the transversal instability has evolved for some time, eventually wave-breaking sets in. This breaking is not due to a longitudinal instability, but is part of the transversal dynamics. Longitudinal instabilities can be excluded as a source of this early breaking because of the different time-scale on which the breaking sets in. The longitudinal growth rates are too small to allow noteworthy growing of the longitudinal modes within the time considered here.

9.5.4.5 Three-Dimensional Solitons

Simulations indicate that unstable planar solitons do not evolve into (stable) three-dimensional localized structures. Stable soliton solutions do exist in nonlinear systems of higher dimensions [269, 272, 376]. There is some indication that stable 3D solitons may also exist in laser–plasma systems [334]. However, this field is not yet settled.

9.6 Wake Field Excitation

The discussion of the Maxwell fluid models has shown that a strong coupling between electromagnetic and electrostatic components exist. An originally purely transverse laser light (in vacuum) may excite electrostatic oscillations in a plasma. When we consider a laser pulse, it is therefore expected that at the location of the main electromagnetic field electrostatic oscillations will be excited. Besides the expected local behavior, also nonlocal phenomena, such as wake field excitation, may occur.

9.6.1 Excitation of Quasistationary Wake Fields

The density reactions are driven by the laser field. In the weakly relativistic limit, the driven density follows, for example, from Eq. (9.139). In the case of linear polarization, the right-hand side contains higher harmonics and has to be treated separately. Let us first look for stationary solutions in the moving frame $\xi = x - Vt$ for circular polarization. Using the dimensional form Eq. (9.158) in the new variables $\tau = t$ and ξ, we have

$$\left(\frac{\partial}{\partial\tau} - V\frac{\partial}{\partial\xi}\right)^2 \delta n_e + \omega_{pe}^2 \delta n_e \approx \frac{n_0 e^2}{2m_e^2}\frac{\partial^2 \mathbf{A}_\perp^2}{\partial\xi^2} . \tag{9.391}$$

For $\partial/\partial\tau \equiv 0$, relation Eq. (9.391) can be abbreviated as a driven oscillator equation

$$V^2\ddot{y} + \tilde{\omega}^2 y = \ddot{f} , \tag{9.392}$$

where the dot represents a derivative with respect to ξ, $\tilde{\omega} \equiv \omega_{pe}$, and $\ddot{f}$ denotes the driving term on the right-hand side of Eq. (9.391). The homogeneous part of the differential equation describes plasma oscillations $y \sim \exp[i(\tilde{\omega}/V)\xi]$. To determine the driven (forced) oscillations, we make use of the variation of parameter method. Let us introduce

$$y(\xi) = Y(\xi)e^{i\frac{\tilde{\omega}}{V}\xi} \tag{9.393}$$

into Eq. (9.392). Integrating once, we find

$$V^2\dot{Y} + 2i\tilde{\omega}VY = \int_\infty^\xi d\xi'\,\ddot{f}e^{-i\frac{\tilde{\omega}}{V}\xi'} . \tag{9.394}$$

Integrating the last term on the right-hand side twice, we obtain

$$\int_\infty^\xi d\xi'\,\ddot{f}e^{-i\frac{\tilde{\omega}}{V}\xi'} = \dot{f}e^{-i\frac{\tilde{\omega}}{V}\xi} + i\frac{\tilde{\omega}}{V}fe^{-i\frac{\tilde{\omega}}{V}\xi} - \frac{\tilde{\omega}^2}{V^2}\int_\infty^\xi d\xi'\,fe^{-i\frac{\tilde{\omega}}{V}\xi'} . \tag{9.395}$$

We have assumed vanishing constants of integration at $\xi = \infty$. If the laser front propagates from left to right along the x-axis, for finite times we require that nothing happened far ahead. Next, the change of notation

$$Y(\xi) = W(\xi)e^{-2i\frac{\tilde{\omega}}{V}\xi} \tag{9.396}$$

leads to

$$V^2\dot{W} = \frac{d}{d\xi}\left[fe^{i\frac{\tilde{\omega}}{V}\xi}\right] - \frac{\tilde{\omega}^2}{V^2}e^{2i\frac{\tilde{\omega}}{V}\xi}\int_\infty^\xi d\xi'\,fe^{-i\frac{\tilde{\omega}}{V}\xi'} . \tag{9.397}$$

Integration and reformulating the last term (integration by parts) results in

$$W(\xi) = \frac{1}{V^2}fe^{i\frac{\tilde{\omega}}{V}\xi} + \frac{\tilde{\omega}}{V^3}fe^{i\frac{\tilde{\omega}}{V}\xi}\int_\infty^\xi d\xi'\,f\sin\left[k_p\left(\xi' - \xi\right)\right], \tag{9.398}$$

where

$$k_p = \frac{\tilde{\omega}}{V} \equiv \frac{\omega_{pe}}{V} . \tag{9.399}$$

For circular polarization, we use

$$a = \frac{e}{m_e c}(A_y + i A_z) , \quad \frac{e^2}{m_e^2 c^2} A_\perp^2 = |a|^2 , \quad f = \frac{n_0 e^2}{2 m_e^2} A_\perp^2 = \frac{n_0 c^2}{2} |a|^2 , \tag{9.400}$$

leading to

$$\frac{\delta n_e}{n_0} = \frac{c^2}{2V^2} \left\{ |a(\xi)|^2 - k_p \int_\infty^\xi d\xi' |a(\xi')|^2 \sin[k_p(\xi - \xi')] \right\} . \tag{9.401}$$

In this case, we only have a low-frequency driving force for which the assumption $a = a(\xi)$ is meaningful. The above solution can be used for the excitation of wake fields. Different limits through light on the physical content of the solution. First, for narrow laser pulses, for example, in the simple approximation (width $l \to 0$)

$$\left|a(\xi')\right|^2 = |\tilde{a}|^2 \, \delta(\xi') \tilde{L} , \tag{9.402}$$

with $V \approx v_{gr}$ as the group velocity of the laser, we obtain for finite wavenumbers k_p the constant envelope wake field

$$\frac{\delta n_e}{n_0} = -\frac{c^2 k_p}{2V^2} |\tilde{a}|^2 \tilde{L} \sin(k_p x - \omega_{pe} t) \approx -\frac{c^2 k_p \tilde{L}}{2 v_{gr}^2} |\tilde{a}|^2 \sin(k_p x - \omega_{pe} t) . \tag{9.403}$$

In general, the wake field generation vanishes when $k_p l \gg 1$, where l is the width of the driver. Significant contributions occur for

$$k_p l \sim \mathcal{O}(1) . \tag{9.404}$$

On the other hand, in the limit $k_p \to 0$, V finite, we obtain

$$\frac{\delta n_e}{n_0} = \frac{c^2}{2V^2} |a|^2 , \tag{9.405}$$

that is, a density hump instead of the ponderomotive driven density depression.

Finally, for $V \approx (c^2 k)/\omega_{pe} \to 0$, ω_{pe} finite, $k_p \to \infty$, we can recover from Eq. (9.401) the ponderomotive density depression of a moving laser pulse [351, 352]

$$\frac{\delta n_e}{n_0} = \frac{c^2}{2V^2 k_p^2} \frac{d^2 |a|^2}{d\xi^2} \tag{9.406}$$

by the following arguments. A first integration by parts in Eq. (9.401) leads to

$$\frac{\delta n_e}{n_0} = \frac{c^2}{2V^2} \int_\infty^\xi d\xi' \frac{d \, |a \, (\xi')|^2}{d\xi'} \cos\left[k_p \left(\xi - \xi'\right)\right] . \tag{9.407}$$

Continuing with an additional integration by parts, we obtain

$$\frac{\delta n_e}{n_0} = \frac{c^2}{2V^2}\frac{1}{k_p}\int_\infty^\xi d\xi' \frac{d^2\,|a\,(\xi')|^2}{d\xi'^2} \sin\left[k_p\,(\xi-\xi')\right]. \tag{9.408}$$

The next integration by parts gives

$$\begin{aligned}\frac{\delta n_e}{n_0} &= \frac{c^2}{2V^2k_p^2}\left\{\frac{d^2|a|^2}{d\xi^2} - \int_\infty^\xi d\xi' \frac{d^3|a(\xi')|^2}{d\xi'^3}\cos\left[k_p\,(\xi-\xi')\right]\right\}\\ &\approx \frac{c^2}{2V^2k_p^2}\frac{d^2|a|^2}{d\xi^2} + \mathcal{O}\left(\frac{1}{k_p}\right).\end{aligned} \tag{9.409}$$

This is the result which has been used for soliton investigations.

For linear polarization, we use

$$\boldsymbol{A}_\perp(x,t) = A_\perp(x,t)\hat{\boldsymbol{e}}_z\,, \qquad a = \frac{eA_\perp}{m_e c}\,. \tag{9.410}$$

Next, we make the ansatz

$$a(x,t) = \frac{1}{2}\left[\tilde{a}(\xi)e^{-i\omega t+ikx} + \tilde{a}^*(\xi)e^{i\omega t-ikx}\right], \tag{9.411}$$

leading to

$$a^2 = \underbrace{\frac{1}{2}|\tilde{a}|^2}_{\text{low-frequency}} + \underbrace{\frac{1}{4}\tilde{a}^2e^{-2i\omega t+2ikx} + \frac{1}{4}\tilde{a}^{*2}e^{2i\omega t-2ikx}}_{\text{high-frequency}}. \tag{9.412}$$

Since we have different frequencies in the driving term, we decompose the density perturbation δn_e in the two parts $\delta n_e = \delta n_e^{(0)} + \delta n_e^{(2)}$ where the index zero denotes the low-frequency part and the index two indicates that also a second harmonic perturbation will occur. From the low-frequency density part $\delta n_e^{(0)}$, we obtain a similar expression as in the case of circular polarization, that is, we may use

$$f \sim \frac{n_0 e^2}{2m_e^2}\boldsymbol{A}_\perp^2 \sim \frac{n_0 c^2}{4}|\tilde{a}|^2\,, \tag{9.413}$$

leading to

$$\frac{\delta n_e^{(0)}}{n_0} = \frac{c^2}{4V^2}\left\{|\tilde{a}(\xi)|^2 - k_p\int_\infty^\xi d\xi'|\tilde{a}(\xi')|^2\sin[k_p(\xi-\xi')]\right\}. \tag{9.414}$$

For the high-frequency density part $\delta n_e^{(2)}$, we go back to Eq. (9.158) which then reads

$$\frac{\partial^2\delta n_e^{(2)}}{\partial t^2} + \omega_{pe}^2\delta n_e^{(2)} \approx \frac{n_0 c^2}{8}\frac{\partial^2}{\partial x^2}\left\{\tilde{a}^2e^{-2i\omega t+2ikx} + c.c.\right\}. \tag{9.415}$$

We abbreviate that equation (including the $c.c.$ part afterwards) by the form

$$\ddot{y} + \tilde{\omega}^2 y = \left[h e^{-2i\omega t + 2ikx} \right]'' \tag{9.416}$$

where the dot stands for a partial time-derivative, the prime represents the space derivative, $y = \delta n_e^{(2)}$, $\tilde{\omega} = \omega_{pe}$, and $h = n_0 c^2 \tilde{a}^2/8$. Equation (9.416) will again be solved by the method of variation of parameters. Let us introduce

$$y(x,t) = Y(x,t) e^{i\tilde{\omega}t}. \tag{9.417}$$

Integrating once, we find

$$\dot{Y} + 2i\tilde{\omega} Y = \int_{-\infty}^{t} dt' \left[h e^{-2i\omega t' + 2ikx} \right]'' e^{-i\tilde{\omega}t'} . \tag{9.418}$$

Next,

$$Y(x,t) = W(x,t) e^{-2i\tilde{\omega}t} \tag{9.419}$$

leads to

$$W = \int_{-\infty}^{t} dt'' e^{2i\tilde{\omega}t''} \int_{-\infty}^{t''} dt' \left[h(x,t') e^{-2i\omega t' + 2ikx} \right]'' e^{-i\tilde{\omega}t'} . \tag{9.420}$$

When evaluating the integrals, we begin with a systematic expansion for weak space-dependence of h. To lowest order, we approximate

$$\begin{aligned} \int_{-\infty}^{t''} dt' \left[h(x,t') e^{-2i\omega t' + 2ikx} \right]'' e^{-i\tilde{\omega}t'} &\approx -4k^2 h e^{2ikx} \int_{-\infty}^{t''} dt' e^{-i(2\omega + \tilde{\omega})t'} \\ &= -4ik^2 h e^{2ikx} \frac{e^{-i(2\omega + \tilde{\omega})t''}}{2\omega + \tilde{\omega}} . \end{aligned} \tag{9.421}$$

Inserting that into the formula for W, we find

$$W \approx -\frac{4ik^2 h e^{2ikx}}{2\omega + \tilde{\omega}} \int_{-\infty}^{t} dt'' e^{-i(2\omega - \tilde{\omega})t''} = \frac{4k^2 h}{4\omega^2 - \tilde{\omega}^2} e^{2ikx} e^{-i(2\omega - \tilde{\omega})t} . \tag{9.422}$$

To lowest order, we therefore obtain

$$\frac{\delta n_e^{(2)}}{n_0} \approx \frac{c^2 k^2}{2} \frac{1}{4\omega^2 - \tilde{\omega}^2} \tilde{a}^2 e^{-2i\omega t + 2ikx} . \tag{9.423}$$

For $\omega \gg \tilde{\omega} \equiv \omega_{pe}$, we may approximate

$$\frac{1}{4\omega^2 - \tilde{\omega}^2} \approx \frac{1}{4\omega^2}\left(1 + \frac{\omega_{pe}^2}{4\omega^2}\right), \tag{9.424}$$

leading to

$$\delta n_e^{(2)} \approx \frac{n_0 c^2 k^2}{8\omega^2} e^{-2i\omega t + 2ikx} \tilde{a}^2 \left(1 + \frac{\omega_{pe}^2}{4\omega^2}\right) + c.c. \tag{9.425}$$

when we include the $c.c.$ which has been suppressed so far. The main approximation in the present evaluation was in Eq. (9.421). In principle, the bracket term in the integrand

$$\left[h\left(x, t'\right) e^{-2i\omega t' + 2ikx}\right]'' = \left[h'' + 4ikh' - 4k^2 h\right] e^{-2i\omega t' + 2ikx} \tag{9.426}$$

leads to more contributions than considered up to now. The last term in the bracket on the right-hand side of Eq. (9.426) was considered so far by approximating the integral (after integration by parts)

$$\int_{-\infty}^{t} dt' h e^{-i(2\omega - \tilde{\omega})t'} = h \int_{-\infty}^{t} dt' e^{-i(2\omega - \tilde{\omega})t'} - \int_{-\infty}^{t} dt' \frac{\partial h}{\partial t'} \frac{e^{-i(2\omega - \tilde{\omega})t'}}{-i(2\omega - \tilde{\omega})} \tag{9.427}$$

only by the first part. However, within a systematic expansion in terms of derivatives of h, in the next order, the second term may be included as

$$\int_{-\infty}^{t} dt' \frac{\partial h}{\partial t'} \frac{e^{-i(2\omega - \tilde{\omega})t'}}{-i(2\omega - \tilde{\omega})} \approx V \frac{\partial h}{\partial \xi} \frac{1}{i(2\omega - \tilde{\omega})} \int_{-\infty}^{t} dt' e^{-i(2\omega - \tilde{\omega})t'} . \tag{9.428}$$

In addition, the second term in the bracket on the right-hand side of Eq. (9.426) should also be considered. When we do this, for $\omega \gg \tilde{\omega} \equiv \omega_{pe}$, we get

$$\delta n_e^{(2)} \approx \frac{n_0 c^2 k}{8\omega^2} e^{-2i\omega t + 2ikx} \left[k\tilde{a}^2 \left(1 + \frac{\omega_{pe}^2}{4\omega^2}\right) - i\left(1 - \frac{kV}{\omega}\right) \frac{d\tilde{a}^2}{d\xi}\right] + c.c. + \mathcal{O}\left(\frac{d^2\tilde{a}^2}{d\xi^2}\right). \tag{9.429}$$

For $V = v_{gr} \equiv c^2 k/\omega$, we might approximate

$$\left(1 - \frac{kV}{\omega}\right) \approx \frac{\omega_{pe}^2}{\omega^2} . \tag{9.430}$$

The next corrections of order $d^2\tilde{a}^2/d\xi^2$ are not shown. However, they might become important in the limit $k \to 0$. In that case, the evaluation leads to

$$\left.\frac{\delta n_e^{(2)}}{n_0}\right|_{k\to 0} \approx -\frac{c^2}{32\omega^2}\left(1+\frac{\omega_{pe}^2}{4\omega^2}\right)\frac{d^2\tilde{a}^2}{d\xi^2}e^{-2i\omega t+2ikx} + c.c. \tag{9.431}$$

We remind the reader of the definition $\xi = x - Vt$.

When comparing the zeroth-harmonic contributions with the second-harmonic ones, we roughly recognize

$$\frac{\delta n_e^{(0)}}{n_0} \sim \frac{c^2}{V^2}\tilde{a}^2 , \quad \frac{\delta n_e^{(2)}}{n_0} \sim \frac{c^2k^2}{\omega^2}\tilde{a}^2 \tag{9.432}$$

such that for $V \approx v_{gr}$ (group velocity) and $v_{ph} = \omega/k$, the ratio

$$\frac{\delta n_e^{(2)}}{\delta n_e^{(0)}} \sim \frac{V^2k^2}{\omega^2} \sim \frac{v_{gr}^2}{v_{ph}^2} \sim \frac{v_{gr}^4}{c^4} \ll 1 \tag{9.433}$$

in many applications of wake field excitation.

Summarizing for later applications, we first state that the phase-velocity of the excited wake field agrees with the group velocity v_{gr} of the laser pulse. Determining the laser frequency ω_L from $\omega_L^2 = \omega_p^2 + k^2c^2$, we find

$$v_{gr} = \frac{kc^2}{\omega_L} \mathrel{\hat{=}} v_{phase}^{wake} \equiv v_{ph} . \tag{9.434}$$

Inserting the expression for the laser frequency and afterwards replacing the wavenumber via the laser-frequency formula, we obtain

$$v_{gr} = c\sqrt{1-\frac{\omega_p^2}{\omega_L^2}} . \tag{9.435}$$

For particle acceleration, we may consider particle velocities close to the phase-velocity of the wave (particles surfing the wave) and introduce

$$\beta_p = \frac{v_{ph}}{c} = \frac{v_{gr}}{c} = \sqrt{1-\frac{\omega_p^2}{\omega_L^2}} . \tag{9.436}$$

The gamma-factor of these particles is

$$\gamma_p = \frac{1}{\sqrt{1-\beta_p^2}} = \frac{\omega_L}{\omega_p} = \sqrt{\frac{n_c}{n_0}} ; \tag{9.437}$$

it is completely determined by the electron density n_0 and the critical density n_c at which the laser frequency equals the plasma frequency. The plasma should be strongly undercritical ($n_0 < n_c$) for reaching large velocities.

9.6.2
Strongly Relativistic Nonlinear Electrostatic Wake Fields

We have calculated the amplitude of the exited density oscillations in the weakly relativistic regime. Let us now go to the strongly relativistic case. Since the exited wake field is mainly of high-frequency and is electrostatic, we start from an *electrostatic* Maxwell electron-fluid model

$$\frac{\partial n}{\partial t} + \frac{\partial}{\partial x}(nv) = 0 , \tag{9.438}$$

$$\left(\frac{\partial}{\partial t} + v\frac{\partial}{\partial x}\right)(\gamma m v) = -eE , \tag{9.439}$$

$$\frac{\partial E}{\partial x} = \frac{1}{\varepsilon_0} e(n_0 - n) , \tag{9.440}$$

where we omitted the index e (for electrons). Furthermore, we ignored temperature effects (cold plasma approximation) in the model for 1D propagation. We shall use

$$\gamma = \frac{1}{\sqrt{1-\beta^2}} , \quad \beta = \frac{v}{c} , \tag{9.441}$$

and make the assumptions

$$n(x,t) = n(\tau) , \quad v(x,t) = v(\tau) , \quad E(x,t) = E(\tau) , \tag{9.442}$$

with

$$\tau = \omega_p \left(t - \frac{x}{V}\right) , \quad V \approx v_{gr} . \tag{9.443}$$

Furthermore, normalizing

$$\frac{n}{n_0} \to n , \quad \frac{v}{c} \to v \hat{=} \beta , \quad \frac{eE}{m\omega_p c} \to E , \tag{9.444}$$

from Eqs. (9.438)–(9.440), we obtain

$$n\left(1 - \frac{\beta}{\beta_p}\right) = \text{const} \equiv 1 , \tag{9.445}$$

$$\left(1 - \frac{\beta}{\beta_p}\right)\frac{d(\beta\gamma)}{d\tau} = -E , \tag{9.446}$$

$$\frac{dE}{d\tau} = \frac{\beta}{1 - \frac{\beta}{\beta_p}} . \tag{9.447}$$

Let us combine the last two equations to

$$E\frac{dE}{d\tau} \equiv \frac{d}{d\tau}\left(\frac{E^2}{2}\right) = -\beta\frac{d\beta\gamma}{d\tau} . \tag{9.448}$$

Now, calculating

$$-\frac{d\gamma}{d\tau} = -\frac{\beta\gamma}{1-\beta^2}\frac{d\beta}{d\tau} \tag{9.449}$$

and

$$-\beta\frac{d(\beta\gamma)}{d\tau} = -\frac{\beta}{1-\beta^2}\gamma\frac{d\beta}{d\tau}\,, \tag{9.450}$$

we find

$$\frac{d}{d\tau}\left(\frac{E^2}{2}\right) = -\frac{d\gamma}{d\tau}\,. \tag{9.451}$$

That is to say, we have found an integral of motion which we express in the form

$$E = E(\gamma) = \sqrt{2(\gamma_m - \gamma)}\,. \tag{9.452}$$

Here, γ_m is the integration constant. We have $E = 0$ at $\gamma = \gamma_m$. The constant γ_m corresponds to the maximum fluid velocity v_m with $\beta_m = v_m/c$ and

$$\gamma_m = \frac{1}{\sqrt{1-\beta_m^2}}\,. \tag{9.453}$$

The maximum electric field

$$E_m = \sqrt{2(\gamma_m - 1)} \tag{9.454}$$

is obtained at locations where $v = 0$ and $\gamma = 1$. From Eqs. (9.446) and (9.452), we obtain

$$\pm d\tau = \frac{\left(1-\frac{\beta}{\beta_p}\right)d(\beta\gamma)}{\sqrt{2(\gamma_m-\gamma)}}\,. \tag{9.455}$$

Comparing with the weakly relativistic case

$$\beta \ll 1\,, \quad \gamma \approx 1\,, \quad \sqrt{2(\gamma_m-\gamma)} \approx \sqrt{\beta_m^2-\beta^2}, \tag{9.456}$$

we approximate

$$\frac{\left(1-\frac{\beta}{\beta_p}\right)d(\beta\gamma)}{\sqrt{2(\gamma_m-\gamma)}} \approx \frac{\left(1-\frac{\beta}{\beta_p}\right)d\left(\frac{\beta}{\beta_m}\right)}{\sqrt{1-\left(\frac{\beta}{\beta_m}\right)^2}}\,. \tag{9.457}$$

Using this and introducing the abbreviation $x = \beta/\beta_m$, we can reformulate Eq. (9.455) as

$$\pm d\tau \approx \frac{dx}{\sqrt{1-x^2}} - \frac{\beta_m}{\beta_p}\frac{x\,dx}{\sqrt{1-x^2}}\,, \tag{9.458}$$

leading to

$$\pm(\tau - \tau_0) \approx \arcsin\left(\frac{\beta}{\beta_m}\right) + \frac{\beta_m}{\beta_p}\sqrt{1 - \left(\frac{\beta}{\beta_m}\right)^2} . \tag{9.459}$$

For small wave amplitudes $\beta_m \ll \beta_p$, Eq. (9.459) can be inverted, and one recovers the linear wave result

$$\beta \approx \beta_m \sin(\tau - \tau_0) , \tag{9.460}$$

$$n \approx \frac{1}{1 - \frac{\beta}{\beta_p}} + \frac{\beta_m}{\beta_p} \sin(\tau - \tau_0) , \tag{9.461}$$

$$E \approx \sqrt{\beta_m^2 - \beta^2} \approx \beta_m \cos(\tau - \tau_0) . \tag{9.462}$$

The extrema of $\beta(\tau)$, $n(\tau)$, and $E(\tau)$ do not shift in τ when increasing β_m, while the zeros of $\beta(\tau)$ and $n(\tau)$ and the extrema of $E(\tau)$ are shifted in such a way that in the nonlinear high-amplitude regime, velocity and density develop sharp wave crests with wide troughs in between. The electric field acquires a sawtooth shape. In the half-waves with negative $E(\tau)$, electrons can be accelerated. That effect is being used for plasma wave accelerators.

9.7
Breaking of Wake Fields

Wave-breaking is an important and fundamental process in nonlinear wave dynamics. The latter becomes especially important when electron acceleration in electrostatic plasma oscillations, whose phase velocities are close to the speed of light, is investigated. For fast electron acceleration, stability criteria for localized (inhomogeneous) and relativistic nonlinear electrostatic fields have to be known. Wave-breaking criteria define the stability of excited wake fields. Wave-breaking can also be an efficient injection mechanism. Thus, general criteria for wave-breaking are highly needed for both a better understanding of fundamental plasma processes and applications in accelerator physics.

Particles being properly injected into the wake field [383] may be trapped in the wake. They can be accelerated to high energies since the phase velocity of the wake field oscillations is close to the speed of light. Wake field generation [291] became a very interesting phenomenon of short pulse laser physics, with a huge potential for practical applications. Laser-plasma accelerators have been proposed as a next generation of compact accelerators because of the large electric fields they can sustain [368, 384, 385]. The accelerating fields in conventional accelerators are limited to energy gains of a few MeV m^{-1}, owing to material breakdown at the walls. Laser-plasma accelerators may sustain 100 GeV m^{-1} and more [368]. However, the quality of the beam, including large energy spreads, is the main problem [386–392].

Besides breakdown phenomena, the maximum acceleration rate is determined by the maximum amplitude of oscillation that can be supported by the plasma over sufficient times. Wave-breaking is a well-known limitation. It generally occurs where the amplitude of the wave reaches the point that the crest of the wave actually overturns. This picture originates from water surface waves. For other waves, for example, electromagnetic waves, also breaking definitions exist [393]. They have been applied to waves with slowly varying envelopes. A new, highly nonlinear regime occurs for laser pulses shorter than $\lambda_p = 2\pi c/\omega_{pe}$ (where ω_{pe} is the electron plasma frequency) for relativistic intensities high enough to break the plasma wave at the first oscillation [394, 395].

Many very interesting papers exist in which wave-breaking and injection mechanisms into plasma waves have been discussed. The maximum amplitude of relativistic plasma oscillations in a thermal plasma was obtained in a combined one-dimensional waterbag and warm-fluid model [396]. Conditions for the onset of wave-breaking due to sublight-velocity pulses were given for both leading and trailing edge pulses [397]. Trapping and acceleration of a test electron in a nonlinear plasma wave was analyzed in one dimension using Hamiltonian dynamics [398]. Transverse-wake wave-breaking [399], electron injection into plasma wake fields by colliding laser pulses [400] as well as injection mechanisms due to nonlinear wake wave-breaking [401] were discussed. The structure and dynamics of the wake field in a plasma column [402–404] were studied analytically and numerically, explaining, for example, channel formation. Recently, the advantages and disadvantages of various injection mechanisms of particles into wake fields were investigated theoretically [369, 405–409]. The principal acceleration mechanisms have been generalized to positrons, photons, and ions [338, 369, 410, 411].

Wave-breaking analysis goes back to a classical papers by Dawson [412], and Davidson and Schram [413]. Plane oscillations in a uniform plasma were found to be stable below a critical amplitude. For larger amplitudes, it was shown that multistream flow or fine-stream-mixing sets in on the first oscillation. Later, a one-dimensional nonlinear relativistic second-order differential equation for the electron fluid velocity was derived in [414] in Lagrangian coordinates. This formulation allows one to analyze the dynamics until wave-breaking. The Lagrangian analysis, combined with an appropriate numerical integration of the second-order differential equation, gave further insight into the wave-breaking process.

9.7.1 Classical, Nonrelativistic Wave-Breaking Analysis

The following Lagrangian analysis is based on Dawson's picture [412] that longitudinal wave-breaking in a cold one-dimensional (1D) plasma occurs when elements of the plasma electron fluid that started out in different positions overtake each other. From the literature [415], we cite that for both nonrelativistic and relativistic plasmas, this overtaking happens when the peak fluid velocity equals the phase speed of the plasma wave [295, 412].

The classical, nonrelativistic wave-breaking analysis starts from the simple electron fluid equations

$$\frac{\partial n_e}{\partial t} + \frac{\partial}{\partial x} n_e v_e = 0 \, , \tag{9.463}$$

$$\frac{\partial v_e}{\partial t} + v_e \frac{\partial v_e}{\partial x} = -\frac{e}{m_e} E \, , \tag{9.464}$$

$$\frac{\partial E}{\partial x} = \frac{e}{\varepsilon_0}(n_o - n_e) \, . \tag{9.465}$$

In contrast to Euler coordinates x and t, Lagrange coordinates ξ and τ measure the position of each fluid element with respect to its initial position. We define

$$\tau = t \quad , \quad \xi = x - \int_0^\tau d\tau' v_e(\xi, \tau') \, . \tag{9.466}$$

Making use of the differentiation formulas

$$\frac{\partial}{\partial \xi} = \frac{\partial x}{\partial \xi}\frac{\partial}{\partial x} \, , \quad \frac{\partial}{\partial \tau} = \frac{\partial t}{\partial \tau}\frac{\partial}{\partial t} + \frac{\partial x}{\partial \tau}\frac{\partial}{\partial x} \, , \tag{9.467}$$

one immediately finds

$$\frac{\partial}{\partial x} = \left[1 + \int_0^\tau d\tau' \frac{\partial v_e(\xi, \tau')}{\partial \xi} \right]^{-1} \frac{\partial}{\partial \xi} \, , \tag{9.468}$$

$$\frac{\partial}{\partial t} = \frac{\partial}{\partial \tau} - v_e(\xi, \tau) \left[1 + \int_0^\tau d\tau' \frac{\partial v_e(\xi, \tau')}{\partial \xi} \right]^{-1} \frac{\partial}{\partial \xi} \, . \tag{9.469}$$

Transforming Eq. (9.463) into Lagrange coordinates leads to

$$\frac{\partial}{\partial \tau} \left\{ n_e \left[1 + \int_0^\tau d\tau' \frac{\partial v_e(\xi, \tau')}{\partial \xi} \right] \right\} = 0 \, . \tag{9.470}$$

In a similar way, Eq. (9.464) is transformed into

$$\frac{\partial v_e(\xi, \tau)}{\partial \tau} = -\frac{e}{m_e} E(\xi, \tau) \, . \tag{9.471}$$

Finally, taking the τ-derivative of Eq. (9.465),

$$\frac{\partial}{\partial \tau}\frac{\partial E}{\partial \xi} = \frac{1}{\varepsilon_0} e \frac{\partial}{\partial \tau} \left\{ \left[1 + \int_0^\tau d\tau' \frac{\partial v_e(\xi, \tau')}{\partial \xi} \right] (n_0 - n_e) \right\} = \frac{e n_0}{\varepsilon_0} \frac{\partial v_e(\xi, \tau)}{\partial \xi} \, , \tag{9.472}$$

we obtain the simple oscillator equation

$$\frac{\partial^2 v_e(\xi,\tau)}{\partial \tau^2} + \omega_{pe}^2 v_e(\xi,\tau) = 0 . \tag{9.473}$$

At the first glance, this is rather surprising. Starting from a nonlinear system, we end up with a linear equation. Unfortunately, we have to apply the nonlinear transformation Eq. (9.467) to return to the Euler coordinate x. The solution of Eq. (9.473) may be written as

$$v_e(\xi,\tau) = V_e(\xi)\cos(\omega_{pe}\tau) + \omega_{pe} X_e(\xi)\sin(\omega_{pe}\tau) . \tag{9.474}$$

When

$$v_e(\xi,\tau=0) = V_e(\xi) \equiv 0 , \tag{9.475}$$

the momentum Eq. (9.471) leads to

$$E(\xi,\tau) = -\frac{m_e}{e}\omega_{pe}^2 X_e(\xi)\cos(\omega_{pe}\tau) . \tag{9.476}$$

Next, from the Poisson equation (9.472) at time $\tau = 0$ follows

$$\frac{\partial X_e(\xi)}{\partial \xi} = \frac{n_e(\xi,\tau=0)}{n_0} - 1 . \tag{9.477}$$

Now, the density continuity Eq. (9.470) can be rewritten as

$$n_e(\xi,\tau) = \frac{n_e(\xi,0)}{\left\{1 + \frac{\partial X_e(\xi)}{\partial \xi}\left[1 - \cos(\omega_{pe}\tau)\right]\right\}} . \tag{9.478}$$

When the free function $X_e(\xi)$ is assumed in the form

$$X_e(\xi) = \frac{\Delta}{k}\sin(k\xi) , \tag{9.479}$$

it contains the parameters Δ and k. Their meaning becomes obvious from the corresponding initial density distribution

$$n_e(\xi,0) = n_0\left[1 + \Delta\cos(k\xi)\right] . \tag{9.480}$$

In Lagrange coordinates v_e, E and n_e are simple harmonic functions of τ. The back-transformation to the laboratory (Euler) coordinates is

$$t = \tau, \quad x = \xi + \frac{2\Delta}{k}\sin^2\left(\frac{\omega_{pe}\tau}{2}\right)\sin(k\xi) . \tag{9.481}$$

It becomes linear for $\Delta \to 0$; otherwise, it is nonlinear. For $|\Delta| > 1/2$, the transformation is not unique, corresponding to wave-breaking.

Let us have an explicit look at the solution $E(x,t)$. We expand

$$\sin\left[k\xi(x,t)\right] = \sum_{n=1}^{\infty} a_n(t)\sin(nkx)\,, \tag{9.482}$$

where the coefficients are given by

$$a_n(t) = \frac{k}{\pi}\int_0^{2\frac{\pi}{k}} dx\,\sin(nkx)\sin[k\xi(x,t)]\,. \tag{9.483}$$

When evaluating the integral on the right-hand side, we use $dx = d\xi[1 + \alpha\cos(k\xi)]$ with $\alpha = 2\Delta\sin^2(\omega_{pe}\tau/2) \equiv \alpha(t)$. Besides the addition theorems for sine and cosine functions, the representation of the Bessel functions J_n of order n

$$J_n(z) = \frac{1}{\pi}\int_0^{\pi}\cos[z\sin\theta - n\theta]d\theta \tag{9.484}$$

will be utilized. We note

$$J_{-n}(z) = (-1)^n J_n(z) \tag{9.485}$$

and

$$J_{\nu-1}(z) + J_{\nu+1}(z) = \frac{2\nu}{z}J_\nu(z)\,. \tag{9.486}$$

A short calculation of the coefficients leads to

$$a_n(t) = (-1)^{n+1}\frac{2}{n\alpha(t)}J_n[n\alpha(t)]\,. \tag{9.487}$$

Then, we may represent the electric field as

$$E(x,t) = -\frac{m_e}{e}\frac{\omega_{pe}^2}{k}\sum_{n=1}^{\infty}(-1)^{n+1}\frac{1}{n\sin^2\left(\frac{\omega_{pe}t}{2}\right)}J_n\left[2n\Delta\sin^2\left(\frac{\omega_{pe}t}{2}\right)\right] \times\sin(nkx)\cos(\omega_{pe}t)\,. \tag{9.488}$$

This is not a pure harmonic function. Higher harmonics in k appear, and also the time-dependence is quite complicated. Of course, the description only makes sense as long as $|\Delta| < 1/2$.

9.7.2
Relativistic Wave-Breaking Analysis

Dawson [412] succeeded in showing that wave-breaking (within the first period) may occur provided that the distance traveled by the fluid with peak fluid velocity V

over the time T of the order of the inverse plasma frequency exceeds the inhomogeneity length L (being of the order of the inverse wavenumber k). In this scenario, the breaking time T is restricted to the first period. For given k, a threshold in amplitude A of the wave ($V \sim A$) exists. If we are below that threshold, no wave-breaking should occur [412]. However, relativistic nonlinearity creates a nonlinear frequency shift which may become space-dependent. Thus, a nonlinear inhomogeneity length L, depending on the amplitude of the oscillation and thereby on A, occurs. The condition $VT > L$ can now be satisfied without threshold. For small amplitudes A, the breaking time T may become large, scaling as $T \sim 1/A^3$ for $A \to 0$.

We start with a cold fluid description and Lagrange coordinates in one space dimension. The fluid element was at x_0 at $t = 0$ and is at x_L at t, that is, the Lagrange coordinate is $x_L = x_L(x_0, t)$ with $x_L(x_0, t = 0) = x_0$. Defining also Lagrangian electron momentum $p_L = p_L(x_0, t) = p(x_L, t)$, the Lagrange electrostatic field $E_L = E_L(x_0, t) = E(x_L, t)$, and electron density $n_L = n_L(x_0, t) = n(x_L, t)$, the ion density N is assumed to be fixed.

Maxwell's equations (we consider electrostatic wake fields and neglect the magnetic fields) lead to

$$\frac{\partial E_L}{\partial t} = \frac{1}{\varepsilon_0} e n_L v_L \,. \tag{9.489}$$

The total time-derivative of E_L is

$$\frac{d E_L}{dt} = \frac{\partial E}{\partial x_L}\frac{d x_L}{dt} + \frac{\partial E}{\partial t} \,. \tag{9.490}$$

The Poisson equation

$$\frac{\partial E}{\partial x_L} = \frac{1}{\varepsilon_0}(N - n_L)e \tag{9.491}$$

and the definition

$$\frac{d x_L}{dt} = v_L = \frac{p_L}{m_e \gamma} \tag{9.492}$$

lead together with Eq. (9.489) to

$$\frac{d E_L}{dt} = \frac{1}{\varepsilon_0} N e v_L \,. \tag{9.493}$$

We treat the case of homogeneous ion density N first. After integration, we have

$$E_L = \frac{1}{\varepsilon_0} N e x_L + E_0(x_0) - \frac{1}{\varepsilon_0} N e x_0 \,, \tag{9.494}$$

where we have defined $E_0(x_0) = E_L(x_0, t = 0)$. Equation (9.494) is a basic relation which we shall use below. The (longitudinal) momentum balance reads

$$\frac{d p_L}{dt} = -e E_L = -\frac{1}{\varepsilon_0} N e^2 x_L - e E_0(x_0) + \frac{1}{\varepsilon_0} N e^2 x_0 \,. \tag{9.495}$$

Together with Eq. (9.492), this forms the basic dynamical equations for constant ion density. Here,

$$\gamma = \sqrt{1 + \frac{p_L^2}{m_e^2 c^2}} . \tag{9.496}$$

When the ion density distribution is not homogeneous (but the ion dynamics is still neglected), we have $N_L = N_L(x_0, t) = N(x_L)$. The generalization of Eq. (9.493) is

$$\frac{d E_L}{dt} = \frac{1}{\varepsilon_0} N_L e \frac{d x_L}{dt} . \tag{9.497}$$

Introducing the function $Y_i(\xi)$ such that

$$N(x_L) = \left. \frac{d Y_i}{d\xi} \right|_{\xi = x_L} , \tag{9.498}$$

we find

$$E_L = \frac{1}{\varepsilon_0} e \left[Y_i(x_L) - Y_i(x_0) \right] + E_0(x_0) , \tag{9.499}$$

which replaces the previous Eq. (9.494). Now, Eq. (9.495) can be generalized to

$$\frac{d p_L}{dt} = -\frac{1}{\varepsilon_0} e^2 \left[Y_i(x_L) - Y_i(x_0) \right] - e E_0(x_0) , \tag{9.500}$$

which, together with Eq. (9.492), forms the basic system of equations for inhomogeneous plasmas.

We now exemplify wave-breaking for $N = \text{const}$; the conclusions and calculations are the same for inhomogeneous situations $N(x)$. Using Poisson's equation, we have

$$\frac{\partial E_L}{\partial x_0} = \frac{\partial E}{\partial x_L} \frac{\partial x_L}{\partial x_0} = \frac{1}{\varepsilon_0} (N - n_L) e \frac{\partial x_L}{\partial x_0} . \tag{9.501}$$

On the other hand,

$$\frac{\partial E_L}{\partial x_0} = \frac{\partial}{\partial x_0} \left(\frac{1}{\varepsilon_0} N e x_L + E_0(x_0) - \frac{1}{\varepsilon_0} N e x_0 \right) . \tag{9.502}$$

A short calculation leads to

$$\frac{\partial E_L}{\partial x_0} = \frac{1}{\varepsilon_0} e \left(\frac{\partial x_L}{\partial x_0} N - n_0 \right) , \tag{9.503}$$

where we have used

$$\frac{\partial E_0}{\partial x_0} = \frac{1}{\varepsilon_0} e (N - n_0) \tag{9.504}$$

with $n_0 = n_L(x_0, t = 0)$. Comparing the right-hand sides of Eqs. (9.501) and (9.503), we find

$$n_L = \frac{n_0}{\frac{\partial x_L}{\partial x_0}} \,. \tag{9.505}$$

The condition

$$\frac{\partial x_L}{\partial x_0} = 0 \tag{9.506}$$

defines wave-breaking. Relation (9.505) shows that infinite density corresponds to wave-breaking.

For the case $N = \text{const}$, we insert into Eq. (9.495) the definition

$$\gamma_L := \frac{1}{\varepsilon_0} N e^2 x_L + e E_0 - \frac{1}{\varepsilon_0} e^2 x_0 N \,. \tag{9.507}$$

We then can write the basic dynamical equations as

$$\frac{d\gamma_L}{dt} = \frac{N e^2}{\varepsilon_0 m_e} \frac{p_L}{\gamma} \,, \tag{9.508}$$

$$\frac{dp_L}{dt} = -\gamma_L \,. \tag{9.509}$$

Because of the appearance of γ, this is a nonlinear oscillator equation. Instead of ω_{pe0}, an amplitude-dependent frequency generally appears.

From the nonlinear oscillator equation

$$\frac{d^2 p}{dt^2} = -\frac{p}{\sqrt{1+p^2}} \,, \tag{9.510}$$

we calculate the nonlinear frequency. Now, momentum and time have been normalized by mc and ω_{pe}^{-1}, respectively. The amplitude-dependent frequency can be determined by perturbation analysis. First, we expand the square-root for small amplitudes to obtain for the oscillator (with linear frequency $\omega = 1$)

$$\frac{d^2 p}{dt^2} + \omega^2 p = -(1-\omega^2)p - \frac{1}{2} p^3 \,. \tag{9.511}$$

Multiplication of both sides with $\cos(\omega t)$ and integration over t from $-\pi/\omega$ to $+\pi/\omega$, and approximating $p(t) \approx \tilde{A}\cos(\omega t)$ within the integrals, leads to the approximate result for the frequency at small amplitudes

$$\omega^2 \approx 1 - \frac{3}{8}\tilde{A}^2 \,. \tag{9.512}$$

This is the result shown in Eq. (9.533). It is depicted in Figure 9.28 by the crosses. We can even obtain a better approximation, being valid for larger amplitudes, when

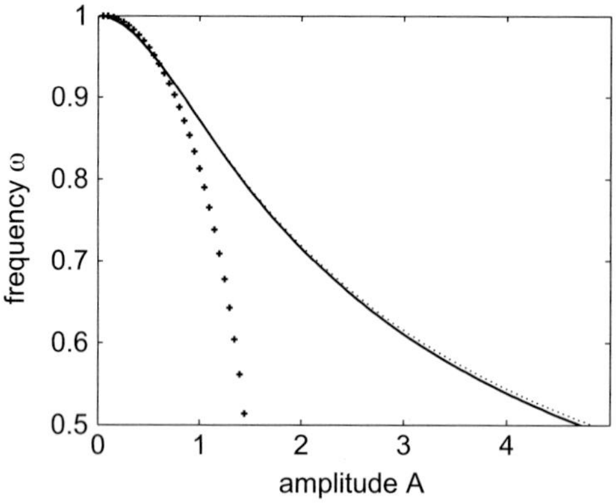

Figure 9.28 Amplitude dependence of the frequency of the nonlinear oscillator Eq. (9.510). The exact result (solid line) is compared with the small amplitude approximation Eq. (9.512) (crosses) and Eq. (9.513) (dotted curve) [416].

not expanding the square-root. By performing the same steps as before, we arrive at

$$\omega^2 \approx \frac{1}{\pi} \int_{-\pi}^{\pi} d\tau \frac{\cos^2(\tau)}{\sqrt{1 + \tilde{A}^2 \cos^2(\tau)}} . \tag{9.513}$$

This result is also shown in Figure 9.28 by the lightly dotted curve. When compared to the exact numerical result, we recognize excellent agreement up to quite large amplitudes.

For inhomogeneous $N = N(x)$, an approximate calculation is possible, for example, when $|x_L - x_0| \ll 1$. Then, we start from the equations

$$\frac{dp_L}{dt} = -\frac{1}{\varepsilon_0} e^2 N(x_0)\,[x_L - x_0] - e\,E_0(x_0) \equiv -\tilde{y}_L , \tag{9.514}$$

$$\frac{d\tilde{y}_L}{dt} = \frac{e^2 N(x_0)}{\varepsilon_0 m_e \gamma} p_L . \tag{9.515}$$

Now, the frequency is not only amplitude-, but also space-dependent. The solutions for $\tilde{y}_L$ and $x_L - x_0$ are straightforward to obtain.

For the nonrelativistic limit, we start with an harmonic field at $t = 0$

$$E_0 = A(x_0) \sin(k x_0 + \varphi) \tag{9.516}$$

and momentum

$$p_0 = -\sqrt{\frac{\varepsilon_0 m_e}{N}} A(x_0) \cos(k x_0 + \varphi) , \tag{9.517}$$

where φ is an arbitrary phase, we can write the wave-solution as

$$E_L = A(x_0) \sin(k x_0 - \omega_{pe0} t + \varphi) . \tag{9.518}$$

The reason is that in the present case, Eqs. (9.508) and (9.509) simplify to

$$\frac{d^2 E_L}{dt^2} = -\omega_{pe0}^2 E_L \,. \tag{9.519}$$

Then, the basic relation (9.494) leads to

$$x_L = x_0 + \frac{\varepsilon_0}{Ne} A(x_0) \left[\sin(kx_0 - \omega_{pe0}t + \varphi) - \sin(kx_0 + \varphi)\right], \tag{9.520}$$

and we have for constant amplitudes $A(x_0) = A$

$$\frac{\partial x_L}{\partial x_0} = 1 + \frac{\varepsilon_0 k}{Ne} A \left[\cos(kx_0 - \omega_{pe0}t + \varphi) - \cos(kx_0 + \varphi)\right] \,. \tag{9.521}$$

A necessary condition for wave-breaking is

$$\frac{\varepsilon_0 k}{Ne} A \geq \frac{1}{2} \,, \tag{9.522}$$

which corresponds to the previous finding [412] (note the factor 2). Wave-breaking occurs when the peak fluid velocity change is larger than the phase velocity, that is,

$$\Delta v_L|_{max} = \frac{2\varepsilon_0 A \omega_{pe0}}{Ne} > \frac{\omega_{pe0}}{k} \,. \tag{9.523}$$

Dawson [412] has obtained the result $\varepsilon_0 A/Ne \geq 1/k$ for specific initial conditions.

Let us make the following remarks on the famous result Eq. (9.522). First, we may use (9.522) for determining the wave-breaking onset in dependence of, for example, the amplitude A. Introducing $T = \omega_{pe0}^{-1}$, $V = \Delta v_L|_{max}$, and $L = k^{-1}$, we find from

$$VT \approx L \tag{9.524}$$

the threshold amplitude for fixed k. When approximating *for wake fields* $\omega_{pe0}/k \approx c$ and using the normalized (by mc) maximum momentum electric field amplitude $a = (\varepsilon_0 A \omega_{pe0})/(Nec)$, the criterion Eq. (9.522) reads

$$a \geq \frac{1}{2} \,. \tag{9.525}$$

Thus, according to Eq. (9.522), small amplitudes do not lead to wake field breaking. On the other hand, Eq. (9.525) for wake fields requires a relativistic treatment.

In the homogeneous nonrelativistic limit (not necessarily wake fields), one may compare the threshold prediction Eq. (9.522) for wave-breaking in finite time with numerical simulations and find excellent agreement [416].

In the fully relativistic description, instead of Eq. (9.520), we obtain a nonlinear oscillator solution which we abbreviate in the form

$$x_L = x_0 + \frac{\varepsilon_0}{Ne} A \sin(kx_0) \left[F(\omega t, x_0) - 1\right], \tag{9.526}$$

where ω is the nonlinear frequency which in the weakly relativistic limit is $\omega \approx \omega_{pe0}$. The function F is still finite and 2π-periodic, that is, $F(y \equiv \omega t + 2\pi, x_0) = F(y \equiv \omega t, x_0)$. When we start with $E_0 = A\sin(k x_0)$, differentiation now leads to

$$\begin{aligned}\frac{\partial x_L}{\partial x_0} = 1 + \frac{\varepsilon_0}{Ne}\Bigg\{&\frac{\partial E_0}{\partial x_0}\left[F(y \equiv \omega t, x_0) - 1\right] \\ &+ E_0 \frac{\partial F(y \equiv \omega t, x_0)}{\partial y}\frac{\partial \omega}{\partial x_0} t \\ &+ E_0 \frac{\partial F(y \equiv \omega t, x_0)}{\partial x_0}\Bigg\}\,. \end{aligned} \tag{9.527}$$

The right-hand side contains, in addition to the Dawson criterion Eq. (9.522), a second source for wave-breaking. In any case, for $t \to \infty$, the term being linear in t will dominate. For a given

$$\frac{\partial \omega}{\partial x_0} \neq 0, \tag{9.528}$$

we can always find a series of intervals such that

$$\frac{\varepsilon_0}{Ne} E_0 \frac{\partial F(y \equiv \omega t, x_0)}{\partial y}\frac{\partial \omega}{\partial x_0} < 0 \tag{9.529}$$

and wave-breaking can occur without threshold. This modifies the statement [415] that "for both nonrelativistic and relativistic plasmas, this overtaking happens when the peak fluid velocity equals the phase speed." The breaking considered here may not occur on the first oscillation, but quite later, that is, after many electron plasma periods. In general, Eq. (9.529) allows one to estimate the time for breaking. We would like to emphasize that now the time for wave-breaking follows from

$$VT \approx L\,, \tag{9.530}$$

(see Eq. (9.524) which was used there for the amplitude threshold) where, however, now the inhomogeneity length

$$L \approx \left|\left[\frac{\partial \ln \omega}{\partial x_0}\right]^{-1}\right| \tag{9.531}$$

should be used. Quantitative predictions require the detailed knowledge of the nonlinear oscillation.

Obviously, the frequency becomes x_0-dependent when the plasma is inhomogeneous. Similarly, a nonlinearity can also introduce a space-dependent frequency. We demonstrate this analytically when, for example, solving the (expanded) equation

$$\frac{d^2 p_L}{dt^2} = -\frac{\omega_{pe0}^2}{\gamma} p_L \approx -\omega_{pe0}^2\left(1 - \frac{1}{2}\frac{p_L^2}{m_e^2 c^2}\right) p_l \tag{9.532}$$

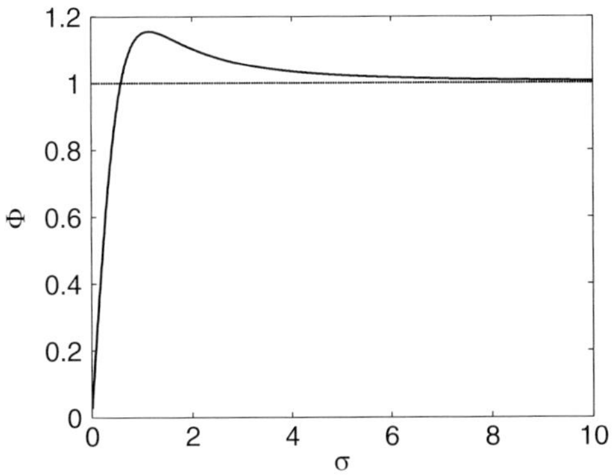

Figure 9.29 Graph of Φ as a function of σ for $(kc)/\omega_{pe} = 2$. Fields with $\varepsilon_0 A_\infty k/(Ne) \geq \Phi$ are supposed to break [416].

for small amplitudes. For an initial momentum $p_0 \sim \tilde{A}$, we have approximately

$$\omega^2 \approx \omega_{pe0}^2 \left(1 - \frac{3}{8}\frac{\tilde{A}^2}{m_e^2 c^2}\right) . \tag{9.533}$$

For example, with $\tilde{A} = \tilde{A}(x_0) = A(x_0)\cos(kx_0)$, we can determine the explicit space-dependence of the frequency.

Applying this to the wake field being excited by a pulsating laser beam, we recognize that the initially excited wake field has a space-dependent frequency. Thus, the wake field will break (sooner or later, depending on the strength of the exited field) even if Eq. (9.522) is not satisfied. For $\gamma \to 1$, the time for breaking tends to infinity. Using wake field relations, we can find the scaling

$$T \sim a^{-3} \tag{9.534}$$

in the small amplitude limit. Here, $a = (e\tilde{A})/(m_e c) \hat{=} (E_0 e)/(\omega_{pe} m_e c)$. The result Eq. (9.522) can be understood as the (nonrelativistic) prediction of breaking of (general) electrostatic oscillations in finite time. When we choose relativistic wake fields, numerical simulations always show wave-breaking in accordance with the analytical predictions.

Dawson [412] considered a nonrelativistic field of infinite length. To study the effect of finite field length, one may start with an electric field at $t = 0$ in the form

$$E_L = A_\infty e^{-\frac{x_0^2}{\sigma^2}} \cos(kx_0)\sin(\omega_{pe0}t) . \tag{9.535}$$

The basic relation Eq. (9.494) leads to

$$x_L = x_0 + \frac{\varepsilon_0 A_\infty}{Ne}\cos(kx_0)\sin(\omega_{pe0}t)e^{-\frac{x_0^2}{\sigma^2}} , \tag{9.536}$$

which after differentiation gives

$$\frac{\partial x_L}{\partial x_0} = 1 - \frac{\varepsilon_0 A_\infty}{Ne}\left[k\sin(kx_0) + 2\frac{x_0}{\sigma^2}\cos(kx_0)\right]\sin(\omega_{pe0}t)e^{-\frac{x_0^2}{\sigma^2}} . \tag{9.537}$$

Therefore, we expect breaking if

$$\frac{\varepsilon_0 A_\infty}{Ne}\left[k\sin(kx_0)+2\frac{x_0}{\sigma^2}\cos(kx_0)\right]e^{-\frac{x_0^2}{\sigma^2}}\geq 1\,. \tag{9.538}$$

For $\sigma \to \infty$, we recover Dawson's criterion [412] for breaking, namely, $\varepsilon_0 A_\infty k/(Ne) \geq 1$. In the limit $\sigma \to 0$, the field is expected to break even for arbitrarily small field amplitudes. Between these two limits, it is most convenient to evaluate the inequality Eq. (9.538) numerically. We define

$$\Phi = \frac{1}{\max\limits_{x_0} \tilde{f}}\,,$$

$$\text{with}\quad \tilde{f} \equiv \tilde{f}(x_0; k, \sigma) = \left(\sin(kx_0)+2\frac{x_0}{k\sigma^2}\cos(kx_0)\right)e^{-\frac{x_0^2}{\sigma^2}}\,. \tag{9.539}$$

In Figure 9.29, Φ is shown as a function of σ for $kc/\omega_{pe} = 2$. The breaking criterion reads

$$\frac{\varepsilon_0 A_\infty k}{Ne} \geq \Phi\,, \tag{9.540}$$

where, for simplicity, we treat k as a parameter. For very broad fields, that is, large σ, the criterion resembles Dawson's result [412]. If σ gets smaller, some fields with amplitudes A_∞ fulfilling the breaking condition for infinite length will break no more in the nonrelativistic limit. For very small σ, even in the nonrelativistic case, fields with very small amplitudes may suffer from breaking.

9.7.3
Numerical Results for Wave-Breaking

In the following, we present some typical examples for wave-breaking during laser–plasma interaction. In [416], a self-consistently determined wake field behind a short laser pulse in one space dimension was investigated. A laser pulse propagating in an underdense plasma oscillates in amplitude and width. Within a certain parameter regime, the variations may occur near the threshold for wake field generation. Then, during the times when the width becomes small enough, localized electrostatic wavepackets are generated. The latter may have amplitudes in the relativistic regime. After some time, they are decoupled from the propagating laser pulse. The breaking of these decoupled, nearly standing, localized wake field wavepackets was further analyzed. It was shown that relativistic, inhomogeneous packets break without threshold. The time for wave-breaking was compared with analytical estimates. First, ion dynamics were neglected. A typical numerical result is depicted in Figure 9.30. The numerical results show the following behavior. The electromagnetic pulse changes its shape periodically. When its width decreases, a significant wake field can be generated. On the other hand, when the pulse is broad, no wake field appears. The wake field has a finite wavenumber $k = k_p$,

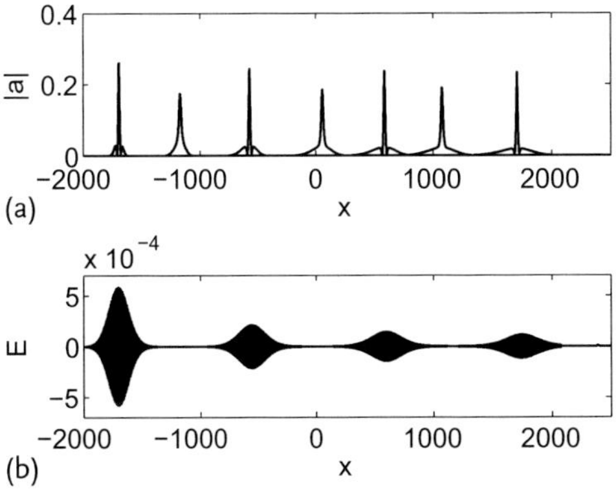

Figure 9.30 Wake field excitation with immobile ions [416]. The magnitude of the normalized vector potential of an initially Gaussian shaped laser pulse with circular polarization is shown in (a) at different times when propagating in an underdense plasma. The maximum initial amplitude is $a_0 = 0.15$, the laser frequency is $\omega_0 = 1.4$, and FWHM $= 40$ for the initial Gaussian. From left to right, profiles are plotted at times $t = 750$, $t = 1500$, $t = 2350$, $t = 3250$, $t = 4000$, $t = 4700$, and $t = 5600$. (b) Wake field for $t = 5500$ created by the pulse, corresponding to the different shapes of the laser pulse shown at (a). The spots are created at times where the compression of the laser pulse is at maximum.

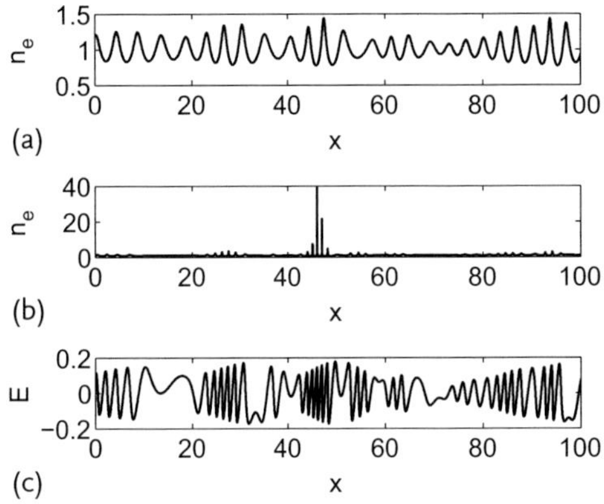

Figure 9.31 Electron density n_e (at $t = 1050$ (a) and $t = 6300$ (b)) and electric field E at $t = 6300$ (c), respectively, from numerical simulation of a relativistic fluid model with immobile ions [416]. The initial values correspond to an initially Gaussian shaped laser pulse with circular polarization. The maximum initial amplitude is $a_0 = 0.32$, the laser frequency is $\omega_0 = 1.9$, the wavenumber is $k = 1.62$, and FWHM $= 20$ for the initial Gaussian.

a frequency $\omega = \omega_{pe}$, and a phase velocity

$$\frac{\omega}{k} \approx \frac{\omega_{pe}}{k_p} \approx v_g \,. \tag{9.541}$$

The group velocity v_g of the laser pulse is close to the speed of light. The numerical results are in accordance with the analytical predictions. Especially, the estimate of the wake field localization via the wake field excitation threshold Eq. (9.404), determining the width of the wake field envelope in terms of the time-dependent pulse width, is quite good. Due to the cold plasma approximation, the decoupled, localized wake field is not propagating in space.

The dynamics of the wake field being left behind the laser pulse was analyzed in detail [416]. Figure 9.31 presents a typical example. The numerical simulations show the breaking of the wake field.

A small amplitude localized wake field shows breaking after many plasma periods. The breaking of the wave is easily diagnosed by the rapid growth of the local electron density.

Wave-breaking of wake fields was thus far discussed only within a purely electron fluid model. It has been checked [416] as to whether the ion dynamics will change the main conclusions. Breaking criterion based on electron dynamics gives a proper upper limit for breaking.

Appendix A
Units

When we consider a classical plasma, we could start with the Newtonian equations of motion. Since the particles are charged, they are influenced by electric and magnetic fields. The latter, however, are generated by the particles themselves, in addition to the so called external fields (produced by external coils, electrodes, etc.).

Various systems of units are in use. Theoreticians and plasma physicists often prefer the Gaussian system, whereas general recommendations of international societies and publishers are given for the SI or Rationalized MKSA system. The Gaussian system uses the centimeter, gram, and second as its fundamental units for length, mass, and time, respectively. With cgs-units, length is measured in cm, mass in g, and time in s $\hat{=}$ sec. For the charge, no new dimension is introduced. One may understand this when recognizing that the Coulomb force law between charges can be used to define the charge unit in cgs, that is, the dimension of charge is $\mathrm{g}^{1/2}\,\mathrm{cm}^{3/2}\,\mathrm{s}^{-1}$. The MKSA system uses the meter, kilogram, and second, plus current (I) as the fourth dimensional quantity, with the ampere (A) as the unit. The charge has the dimension A s. Another derived quantity is, for example, the voltage $1\,\mathrm{V} = 1\,\mathrm{m}^2\,\mathrm{kg}\,\mathrm{A}^{-1}\,\mathrm{s}^{-3}$.

In Gaussian units, the Lorentz force is

$$\boldsymbol{F} = q\left(\boldsymbol{E} + \frac{1}{c}\boldsymbol{v} \times \boldsymbol{B}\right) \tag{A.1}$$

and therefore the equation of motion of particle i of species s with position vector $\boldsymbol{r}_i^s$ reads in Gaussian units as

$$m^s \frac{d^2\boldsymbol{r}_i^s}{dt^2} = q^s \boldsymbol{E}\left(\boldsymbol{r}_i^s, t\right) + \frac{q^s}{c}\frac{d\boldsymbol{r}_i^s}{dt} \times \boldsymbol{B}\left(\boldsymbol{r}_i^s, t\right). \tag{A.2}$$

In SI units, the Lorentz force is

$$\boldsymbol{F} = q\left(\boldsymbol{E} + \boldsymbol{v} \times \boldsymbol{B}\right), \tag{A.3}$$

and we can easily write the corresponding equations of motion.

The electromagnetic fields follow from the Maxwell equations, sometimes also called *macroscopic* Maxwell equations [417]. Maxwell equations are formulated for the electric field $\boldsymbol{E}$, the electric displacement $\boldsymbol{D}$, the magnetic field $\boldsymbol{H}$, and the

High Temperature Plasmas, Theory and Mathematical Tools for Laser and Fusion Plasmas, First Edition.
Karl-Heinz Spatschek.

magnetic induction $\boldsymbol{B}$. They are interrelated via the polarization $\boldsymbol{P}$ and the magnetization $\boldsymbol{M}$, that is,

$$\boldsymbol{D} = \epsilon_0 \boldsymbol{E} + \lambda \boldsymbol{P} \,, \tag{A.4}$$

$$\boldsymbol{H} = \frac{1}{\mu_0} \boldsymbol{B} - \lambda' \boldsymbol{M} \,. \tag{A.5}$$

In rationalized systems (e.g., SI), we have

$$\lambda = \lambda' = 1 \,, \tag{A.6}$$

whereas in unrationalized systems (e.g., Gaussian),

$$\lambda = \lambda' = 4\pi \,. \tag{A.7}$$

There are two more parameters which can be fixed. For rationalized MKSA (SI),

$$\frac{1}{4\pi\epsilon_0} = 10^{-7} c^2 \,, \quad \frac{\mu_0}{4\pi} = 10^{-7} \quad \text{for MKSA system} \,, \tag{A.8}$$

where c is the speed of light in meters per second. The dielectric constant (electric permittivity of free space) ϵ_0 is now explicitly

$$\epsilon_0 = 8.854 \times 10^{-12} \frac{\text{A s}}{\text{V m}} \quad \text{for MKSA system} \,, \tag{A.9}$$

and the permeability μ_0 is

$$\mu_0 = 4\pi \times 10^{-7} \frac{\text{kg m}}{\text{A}^2\,\text{s}^2} \quad \text{for MKSA system} \,. \tag{A.10}$$

On the other hand,

$$\mu_0 = 1 \,, \quad \epsilon_0 = 1 \quad \text{for Gaussian system} \,. \tag{A.11}$$

Summarizing, we have in *Gaussian units*

$$\nabla \cdot \boldsymbol{D} = 4\pi\rho \,, \quad \nabla \cdot \boldsymbol{B} = 0 \,, \tag{A.12}$$

$$\nabla \times \boldsymbol{E} = -\frac{1}{c}\frac{\partial \boldsymbol{B}}{\partial t} \,, \quad \nabla \times \boldsymbol{H} = \frac{4\pi}{c}\boldsymbol{j} + \frac{1}{c}\frac{\partial \boldsymbol{D}}{\partial t} \,, \tag{A.13}$$

$$\boldsymbol{D} = \boldsymbol{E} + 4\pi \boldsymbol{P} \,, \quad \boldsymbol{H} = \boldsymbol{B} - 4\pi \boldsymbol{M} \,, \tag{A.14}$$

$$\boldsymbol{B} = \nabla \times \boldsymbol{A} \,, \quad \boldsymbol{E} = -\nabla \Phi - \frac{1}{c}\frac{\partial \boldsymbol{A}}{\partial t} \,. \tag{A.15}$$

In the *MKSA system*, the Maxwell equations are

$$\nabla \cdot \boldsymbol{D} = \rho \, , \quad \nabla \cdot \boldsymbol{B} = 0 \, , \tag{A.16}$$

$$\nabla \times \boldsymbol{E} = -\frac{\partial \boldsymbol{B}}{\partial t} \, , \quad \nabla \times \boldsymbol{H} = \boldsymbol{j} + \frac{\partial \boldsymbol{D}}{\partial t} \, , \tag{A.17}$$

$$\boldsymbol{D} = \epsilon_0 \boldsymbol{E} + \boldsymbol{P} \, , \quad \boldsymbol{H} = \frac{1}{\mu_0} \boldsymbol{B} - \boldsymbol{M} \, , \tag{A.18}$$

$$\boldsymbol{B} = \nabla \times \boldsymbol{A} \, , \quad \boldsymbol{E} = -\nabla \Phi - \frac{\partial \boldsymbol{A}}{\partial t} \, . \tag{A.19}$$

The (free) charge density ρ and the current density $\boldsymbol{j}$, respectively, are

$$\rho \equiv \rho(\boldsymbol{r}, t) = \sum_s q^s \sum_{i=1}^{N_s} \delta\left[\boldsymbol{r} - \boldsymbol{r}_i^s(t)\right] \, , \tag{A.20}$$

$$\boldsymbol{j} \equiv \boldsymbol{j}(\boldsymbol{r}, t) = \sum_s q^s \sum_{i=1}^{N_s} \frac{d\boldsymbol{r}_i^s}{dt} \delta\left[\boldsymbol{r} - \boldsymbol{r}_i^s(t)\right] \, . \tag{A.21}$$

For linear, isotropic media, so-called constitutive relations

$$\boldsymbol{D} = \epsilon \boldsymbol{E} \, , \quad \boldsymbol{B} = \mu \boldsymbol{H} \tag{A.22}$$

are in use. The dimensionless ratios ϵ/ϵ_0 (relative permittivity) and μ/μ_0 (relative permeability) characterize the deviations in media from the vacuum values.

In the present book, we use SI units, although especially in plasma physics, the Gaussian units are widely used. The use of SI units is based on international recommendations. However, in addition, there is also a special practical reason for using SI. Every formula in the SI system is easily converted into Gaussian units by using the following translation table (quantities on the left-hand side are in SI and quantities on the right-hand side are in the Gaussian system):

$$\varepsilon_0 \Rightarrow \frac{1}{4\pi} \, , \quad \mu_0 \Rightarrow \frac{4\pi}{c^2} \, , \tag{A.23}$$

$$\boldsymbol{B} \Rightarrow \frac{1}{c} \boldsymbol{B} \, , \quad \boldsymbol{H} \Rightarrow \frac{c}{4\pi} \boldsymbol{H} \, , \tag{A.24}$$

$$\boldsymbol{D} \Rightarrow \frac{1}{4\pi} \boldsymbol{D} \, , \quad \boldsymbol{M} \Rightarrow c\boldsymbol{M} \, , \tag{A.25}$$

$$\Phi \Rightarrow \Phi \, , \quad \boldsymbol{A} \Rightarrow \frac{1}{c} \boldsymbol{A} \, . \tag{A.26}$$

On the other hand, the symbols ρ, $\boldsymbol{j}$, $\boldsymbol{E}$, and $\boldsymbol{P}$ are invariant during the transformation.

After presenting the exact formulation, the question may be raised as to why the dynamics cannot be easily solved. The reason is the huge number of particles, namely, more than $N = 10^{21}$ particles for a typical tokamak plasma. Even if computers would allow one to solve 10^{21} equations of motion (there is no chance that they can do it!), we do not know the exact initial conditions. Thus, we should not waste the time with futile expectations. An additional point worth emphasizing, is that the equations of motion are coupled to the Maxwell equations. Because of this coupling, the system becomes highly nonlinear.

In summary, we are not able to evaluate the dynamics of all the (e.g., 10^{21}) particles, and actually we are not interested in all the microscopic details of the complex system. The statistical description is the only real outcome out of this dilemma. Though again, in general, we are confronted with a nonequilibrium situation, and good statistical approaches must be developed for a promising course of action.

Appendix B
Fourier and Laplace Transforms for Pedestrians

In this section, we present a very short and cursory summary of integral transforms which have been used in the previous chapters. More information can be found in standard books on mathematics and mathematical physics, for example, [54].

A general integral transform of a function $f(t)$ takes the form

$$F(s) = \int_a^b K(s,t) f(t) dt \; . \tag{B.1}$$

$F(s)$ is called the transform of $f(t)$ with respect to the kernel $K(s,t)$ and the transform variable s. Known integral transforms vary in K and the limits a, b. For example, we may define the

- Fourier transform with

$$a = -\infty \, , \quad b = \infty \, , \quad K(s \equiv i\omega, t) = e^{-i\omega t} \, , \quad F(s) \equiv \tilde{f}(\omega) \; ; \tag{B.2}$$

- Laplace transform with

$$a = 0 \, , \quad b = \infty \, , \quad K(s \equiv p, t) = e^{-pt} \, , \quad F(s) \equiv F(p) \; ; \tag{B.3}$$

- Hankel transform with

$$a = 0 \, , \quad b = \infty \, , \quad K(s \equiv k, t) = t J_n(kt) \, , \quad F(s) \equiv F(k) \; ; \tag{B.4}$$

- Mellin transform with

$$a = 0 \, , \quad b = \infty \, , \quad K(s \equiv z, t) = t^{z-1} \, , \quad F(s) \equiv F(z) \; . \tag{B.5}$$

In the following, we shall discuss the Fourier and Laplace transforms in more detail, though without any mathematical rigor, for example, the existence proofs and so on.

The Fourier transform is perhaps the most popular integral transform. It is extremely useful in many areas, from frequency analysis up to numerical integration. The function $f(t)$ should exist over the infinite interval $-\infty \leq t \leq \infty$. No particu-

High Temperature Plasmas, Theory and Mathematical Tools for Laser and Fusion Plasmas, First Edition.
Karl-Heinz Spatschek.

lar periodicity is required. The only requirement on $f(t)$ will be that $\int_{-\infty}^{\infty} |f(t)| dt$ is finite. We define

$$\tilde{f}(\omega) = \int_{-\infty}^{+\infty} f(t) e^{-i\omega t} dt \tag{B.6}$$

for real ω, assuming that the integral on the right-hand side exists. Note that t is not necessarily a time variable. Furthermore, we present here the unsymmetric form of the Fourier transform by not evenly distributing the factors $1/\sqrt{2\pi}$ between forth- and back-transformation. Having obtained the Fourier transform Eq. (B.6), we must now address the back-transformation. Let us "prove" that

$$f(t) = \frac{1}{2\pi} \int_{-\infty}^{+\infty} \tilde{f}(\omega) e^{i\omega t} d\omega \tag{B.7}$$

is correct. We remark that in the main part of the present book, we often assumed a mode in the form $\sim \exp(-i\omega t + ikx)$ such that the sign of ω is treated different in some applications.

The line of argument for Eq. (B.7) is as follows. Insert Eq. (B.7) into Eq. (B.6) and show that both sides are identical, that is,

$$\tilde{f}(\omega) \stackrel{?}{=} \frac{1}{2\pi} \int_{-\infty}^{+\infty} \int_{-\infty}^{+\infty} \tilde{f}(\omega') e^{i(\omega' - \omega)t} d\omega' dt \,. \tag{B.8}$$

Making use of

$$\delta(\omega' - \omega) = \frac{1}{2\pi} \int_{-\infty}^{+\infty} e^{i(\omega' - \omega)t} dt \tag{B.9}$$

immediately leads to

$$\tilde{f}(\omega) \stackrel{?}{=} \int_{-\infty}^{+\infty} \int_{-\infty}^{+\infty} \tilde{f}(\omega') \delta(\omega' - \omega) d\omega' \stackrel{!}{=} \tilde{f}(\omega) \,. \tag{B.10}$$

Thus, we may use the back-transform as stated above.

For obvious reasons, we cannot use the Fourier transform when $f(t)$ is known only on a semi-infinite interval $a \le t \le \infty$, where a is finite. Without loss of generality, we choose $a = 0$. Such cases occur for initial value problems in time. Another reason may be that the Fourier transform does not exist because of $f \nrightarrow 0$ as $t \to \infty$. For example, the function $f(t) = t$ does not possess a Fourier transform. In such cases, we have to "modify" Eq. (B.6) into the Laplace transform

$$F(p) = \int_0^{\infty} f(t) e^{-pt} dt \,, \quad \operatorname{Re} p > \gamma \,, \tag{B.11}$$

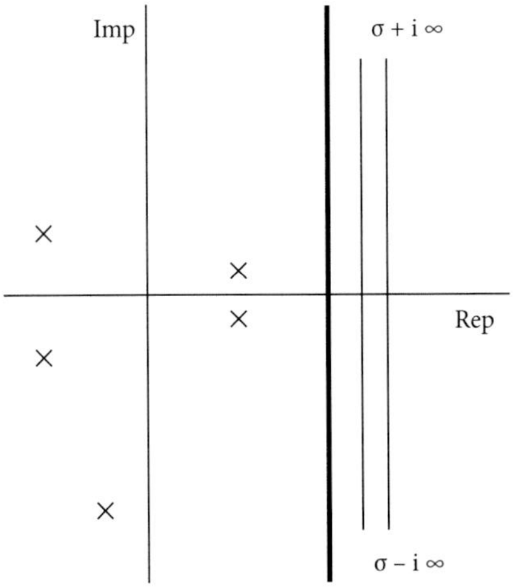

Figure B.1 The back-transformation Eq. (B.12) along Re $p = \sigma$ is shown in the complex p-plane. Crosses designate the singularities of $F(p)$.

provided that the integral on the right-hand side exists. The latter implies Re $p > \gamma$, where γ is the fastest growing exponential term contained in $f(t)$. Often in the demonstrations, one assumes $\gamma = 0$ without explicitly stating it. In some applications, we took into consideration that time runs for $t \geq 0$ and introduced $p \equiv s \equiv i\omega$ for convenience (having in mind a different choice of sign of ω in some applications). The Laplace variable p is generally a complex (number) variable. Let us now argue that

$$f(t) = \frac{1}{2\pi i} \int\limits_{\sigma - i\infty}^{\sigma + i\infty} F(p) e^{pt} dp \,, \quad \sigma > 0 \tag{B.12}$$

is correct when the back-transformation in the p-plane is performed along a line parallel to the imaginary axis, being situated right to all singularities of $F(p)$ (Bromwich integral). That situation is depicted in Figure B.1.

We first proceed similar to the proven method for the Fourier transform. Again, we insert Eq. (B.12) into Eq. (B.11),

$$F(p) \stackrel{?}{=} \frac{1}{2\pi i} \int\limits_{0}^{\infty} \int\limits_{\sigma - i\infty}^{\sigma + i\infty} F(p') e^{(p'-p)t} dp' dt \,. \tag{B.13}$$

The time integration leads to

$$F(p) \stackrel{?}{=} \frac{1}{2\pi i} \int\limits_{\sigma - i\infty}^{\sigma + i\infty} F(p') \left. \frac{e^{(p'-p)t}}{p' - p} \right|_{0}^{\infty} dp' \,. \tag{B.14}$$

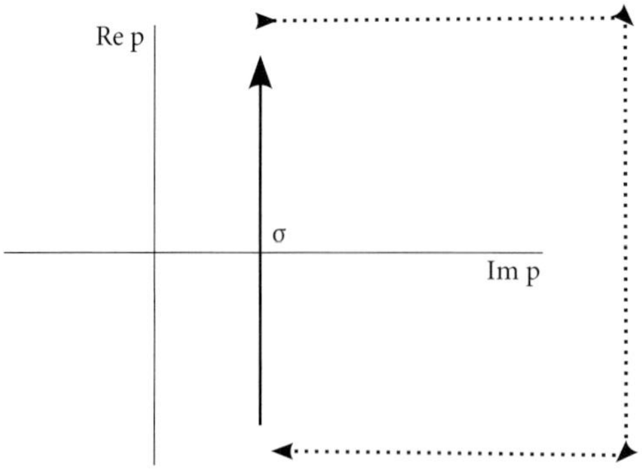

Figure B.2 Evaluation of Eq. (B.15) after closing the path at infinity in the complex p-plane via contour C. Note that the direction of the path is in the negative sense (interior to the right).

When evaluating the right-hand side, we choose $\text{Re}\, p > \text{Re}\, p' \equiv \sigma$. The value at the upper limit vanishes such that

$$F(p) \stackrel{?}{=} -\frac{1}{2\pi i} \int_{\sigma-i\infty}^{\sigma+i\infty} \frac{F(p')}{p'-p} dp' . \tag{B.15}$$

We close the contour (along the straight line from $\sigma - i\infty$ to $\sigma + i\infty$) at infinity as indicated in Figure B.2. Within the integrand, by definition, the only singularity occurs at $p' = p$. Moving along the closed contour C, the interior lies to the right. We may evaluate the integral according to the residuum theorem, leading to

$$F(p) \stackrel{?}{=} -\frac{1}{2\pi i} \oint_C \frac{F(p')}{p'-p} dp' = \frac{1}{2\pi i} \oint_{-C} \frac{F(p')}{p'-p} dp' \stackrel{!}{=} F(p)$$

$$\text{for} \quad \text{Re}\, p > \sigma . \tag{B.16}$$

Thus, we suggested the back-transformation in the form

$$f(t) \stackrel{!}{=} \frac{1}{2\pi i} \int_{\sigma-i\infty}^{\sigma+i\infty} F(p) e^{pt} dp , \tag{B.17}$$

and found some hint as to why the Bromwich integral should be chosen.

We may also proceed the other way around by inserting Eq. (B.11) into Eq. (B.12), leading to

$$f(t) \stackrel{?}{=} \frac{1}{2\pi i} \int_{\sigma-i\infty}^{\sigma+i\infty} dp\, e^{pt} \int_0^{\infty} f(t') e^{-pt'} dt' . \tag{B.18}$$

A simple rearrangement leads to

$$f(t) \stackrel{?}{=} \frac{1}{2\pi i} \int_0^{\infty} dt' f(t') \int_{\sigma-i\infty}^{\sigma+i\infty} e^{p(t-t')} dp . \tag{B.19}$$

By introducing new the integration variable $\omega \equiv \operatorname{Im} p$, we may write

$$f(t) \stackrel{?}{=} \frac{1}{2\pi} \int_0^\infty dt' \, f(t') e^{\sigma(t-t')} \int_{-i\infty}^{i\infty} e^{i\omega(t-t')} d\omega$$
$$\stackrel{?}{=} \int_0^\infty dt' \, f(t') e^{\sigma(t-t')} \delta(t-t') \,. \tag{B.20}$$

Since

$$\int_0^\infty dt' \, f(t') e^{\sigma(t-t')} \delta(t-t') = \begin{cases} f(t) & \text{for} \quad t \geq 0 \,, \\ 0 & \text{for} \quad t < 0 \,, \end{cases} \tag{B.21}$$

the assertion follows.

Let us conclude with two simple examples. The Fourier transform of a normalized Gaussian distribution

$$f(t) = \frac{1}{\tau\sqrt{2\pi}} \exp\left(-\frac{t^2}{2\tau^2}\right), \quad -\infty < t < \infty \tag{B.22}$$

is $\tilde{f}(\omega) = \exp(-\tau^2\omega^2/2)$ for $-\infty < \omega < \infty$.

The Laplace transform of $f(t) = e^{at}$ is

$$F(p) = \frac{1}{p-a} \quad \text{if} \quad p > a \,. \tag{B.23}$$

Appendix C
The Inverse Scattering Transform (IST) for Nonlinear Waves

The Fourier and Laplace transform techniques are applicable to linear systems. In this section, we discuss generalizations to nonlinear, integrable systems. We present an introduction into the inverse scattering transform (IST) theory following the historical developments [241, 248, 418–422].

It was shown [220] that the IST, in the linear limit, reduces to the Fourier transform. The formalism, originally developed for the KdV equation, became very powerful in its later generalizations [423] as pointed out in [220].

Here, we introduce the systematic procedure to obtain analytical solutions to the nonlinear wave equations derived in the main text. We concentrate on the one-component scattering theory being applicable to low-frequency modes described by the Korteweg–deVries equation. Generalizations to two-component scattering, enabling solutions to, for example, the cubic nonlinear Schrödinger equation, are beyond the scope of the present introduction into the IST.

Let us try to solve the initial value problem of the KdV equation

$$u_t - 6uu_x + u_{xxx} = 0 \,. \tag{C.1}$$

We assume $u(x,0)$ to be given and want to determine the solution $u(x,t)$ for all $t > 0$ at $-\infty < x < +\infty$. Of course, $u(x,0) = g(x)$ should obey some conditions [223, 424], for example,

$$\int_{-\infty}^{+\infty} \left| \frac{d^n g}{dx^n} \right|^2 dx < \infty \quad \text{for} \quad n = 0, 1, 2, 3, 4 \,, \tag{C.2}$$

or

$$\int_{-\infty}^{+\infty} (1 + |x|)|g| dx < \infty \,. \tag{C.3}$$

The solution of the KdV equation is related to a Schrödinger eigenvalue problem. An outline is presented in Figure C.1. One of the basic steps is due to Lax [421] who showed that if $u(x,t)$ is a solution of the KdV Eq. (C.1), then the eigenvalue

High Temperature Plasmas, Theory and Mathematical Tools for Laser and Fusion Plasmas, First Edition.
Karl-Heinz Spatschek.

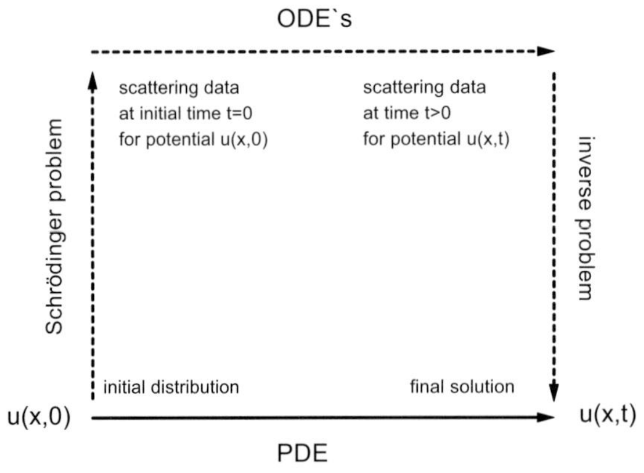

Figure C.1 Diagram of the solution of the KdV equation by the inverse scattering transform. ODE means ordinary differential equation while PDE abbreviates the partial differential equation.

spectrum of the Schrödinger operator, with potential $u(x,t)$,

$$L\psi \equiv \left[-\partial_x^2 + u(x,t)\right]\psi(x,t) = \lambda\psi(x,t)\,, \tag{C.4}$$

is independent of t. Note that here t is regarded as a parameter.

For a deeper understanding, let us first make a few remarks about Gardner's modification [223, 225] of the Miura transformation [425]. If we introduce u instead of the new variable w into Eq. (C.1), via the Riccati equation

$$u = -w - \epsilon w_x + \epsilon^2 w^2\,, \tag{C.5}$$

we obtain

$$\begin{aligned} &-(u_t - 6uu_x + u_{xxx}) = w_t + \epsilon w_{xt} - 2\epsilon^2 w w_t \\ &+ 6(w + \epsilon w_x - \epsilon^2 w^2)(w_x + \epsilon w_{xx} - 2\epsilon^2 w w_x) \\ &+ w_{xxx} + \epsilon w_{xxxx} - 2\epsilon^2 (w w_x)_{xx} \\ &= (1 + \epsilon\partial_x - 2\epsilon^2 w)\left[w_t + 6(w - \epsilon^2 w^2)w_x + w_{xxx}\right]. \end{aligned} \tag{C.6}$$

Here, ϵ is any real parameter, but note that w is not uniquely defined by Eq. (C.5). We can conclude that the KdV equation is satisfied by u if w satisfies the Gardner equation

$$w_t + 6(w - \epsilon^2 w^2)w_x + w_{xxx} = 0\,. \tag{C.7}$$

Now, linearize Eq. (C.5) with

$$w - \frac{1}{2\epsilon^2} = \frac{-1}{\epsilon}\frac{\phi_x}{\phi}\,. \tag{C.8}$$

Then, Eq. (C.5) becomes

$$\phi_{xx} + \left(u + \frac{1}{4\epsilon^2}\right)\phi = 0\,. \tag{C.9}$$

By a Galilean transformation, $u \to u - \lambda - 1/4\epsilon^2$, leaving Eq. (C.1) invariant, we immediately get a scattering problem analogous in form to (C.4). Before analyzing the latter, let us mention another very interesting aspect.

Suppose we can expand

$$w = \sum_{n=0}^{\infty} \epsilon^n w_n(u) \tag{C.10}$$

for small ϵ. Inserting this into Eq. (C.5) and equating coefficients with equal powers in ϵ, we obtain

$$w_0 = -u\,, \tag{C.11}$$

$$w_1 = -w_{0x} = +u_x\,, \tag{C.12}$$

$$w_2 = -w_{1x} + w_0^2 = -u_{xx} + u^2\,, \tag{C.13}$$

and so on. At this stage, it is important to note that Gardner's equation (C.7) has the form of a conservation law, and introducing the ansatz (C.10) into Eq. (C.7) leads to

$$u_t + (-3u^2 + u_{xx})_x = 0\,, \tag{C.14}$$

$$(-u_x)_t + (6uu_x + u_{xxx})_x = 0\,, \tag{C.15}$$

$$(u_{xx} - u^2)_t + (4u^3 - 8uu_{xx} - 5u_x^2 + u_{xxxx})_x = 0\,, \tag{C.16}$$

and so on. Assuming that u and its derivatives approach zero as $x \to \pm\infty$, one can capture the infinite set of constants of motion, for example, from the even powers in ϵ,

$$\int_{-\infty}^{+\infty} u\, dx = \text{const}\,, \tag{C.17}$$

$$\int_{-\infty}^{+\infty} u^2 dx = \text{const}\,, \tag{C.18}$$

$$\int_{-\infty}^{+\infty} \left(u^3 + \frac{1}{2}u_x^2\right) dx = \text{const}\,, \tag{C.19}$$

and so on. The existence of this infinite set of conservation laws is a strong indication for integrability and the existence of the IST.

Now, let us come back to the scattering problem Eq. (C.4). We assume that $u \to 0$ sufficiently rapidly as $x \to \pm\infty$. Then, we asymptotically have $\psi_{xx} \sim -\lambda\psi$, and ψ is of the form $\exp(\pm i\sqrt{\lambda}x)$ at infinity. The eigensolutions are of two kinds: bound states if $\lambda < 0$ and free states if $\lambda > 0$. Let us treat the bound states first. We define

$$\kappa = \sqrt{-\lambda} > 0 \tag{C.20}$$

and

$$\psi(x) \sim \alpha e^{\kappa x} \quad \text{at} \quad x \to -\infty \,. \tag{C.21}$$

In general, the corresponding solution of Eq. (C.4) is such that

$$\psi(x) \sim \beta e^{\kappa x} + \gamma e^{-\kappa x} \quad \text{at} \quad x \to +\infty \,; \tag{C.22}$$

the constants β and γ are proportional to α, and depend on κ and the form of the potential u. Only for special values of κ (eigenvalues λ) will ψ be bounded at $x \to +\infty$. If κ^2 is larger than $-u$ for all x, then $\psi_{xx}/\psi > 0$ everywhere and ψ cannot vanish at both sides. Thus,

$$0 < \kappa^2 < -u_{min} \tag{C.23}$$

is necessary for the boundary conditions $\psi < \infty$ at $x \to \pm\infty$. This situation is briefly summarized in Figure C.2. We further know that there is a finite number $p > 0$ (for $u_{min} < 0$) of discrete eigenvalues λ_n belonging to square-integrable eigenfunctions ψ_n,

$$u_{min} < \lambda_1 < \lambda_2 < \cdots < \lambda_p < 0 \,, \tag{C.24}$$

and ψ_n has $n-1$ finite zeros. The eigenfunctions can be normalized in different ways. One possibility is, of course, to take the integral of the squared function; another one is

$$\psi_n \approx \exp(+\kappa_n x) \quad \text{at} \quad x \to -\infty \,, \tag{C.25}$$

with the implication

$$\psi_n \approx b_n(t) \exp(-\kappa_n x) \quad \text{at} \quad x \to +\infty \,, \tag{C.26}$$

which is chosen here.

Now, we move to the unbound (free) states. If $\lambda > 0$, then the asymptotic behavior is oscillatory and the condition of boundedness is satisfied (but we do not have an exponential decay). Therefore, eigenfunctions exist for all $\lambda > 0$ which are bounded, but not square-integrable. Any solution can be expressed as a superposition, let us say of an incident wave from the right ($x = \infty$) together with transmitted and reflected waves. This scenario is shown in Figure C.3. Letting $k = \sqrt{\lambda} > 0$, we write

$$\psi \sim \begin{cases} e^{-ikx} + R(k,t)e^{ikx} & \text{at} \quad x \to +\infty \,, \\ T(k,t)e^{-ikx} & \text{at} \quad x \to -\infty \,, \end{cases} \tag{C.27}$$

with complex transmission (T) and reflection (R) coefficients. The data κ_n, $b_n(n = 1, \ldots, p)$, $R(k,t)$ and $T(k,t)$ are called *scattering data*.

Now, we prove the Lax result [421] that λ is independent of t if u obeys the KdV equation. The straightforward way is to substitute u from Eq. (C.4), that is,

$$u = \frac{\psi_{xx}}{\psi} + \lambda \,, \tag{C.28}$$

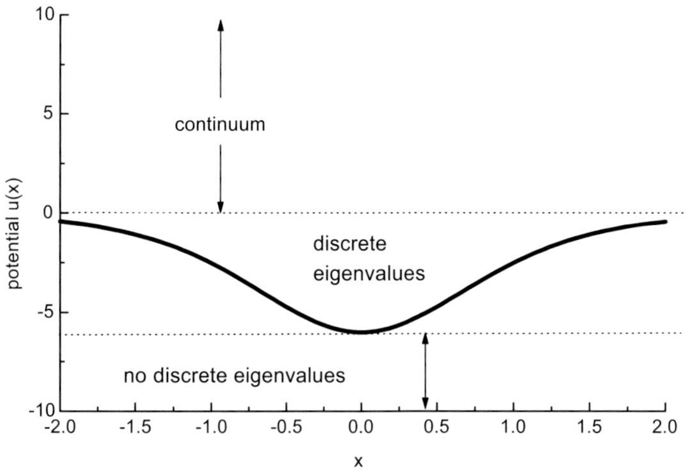

Figure C.2 For a potential $u(x) = -6\text{sech}^2(x)$, the eigenvalue spectrum is characterized.

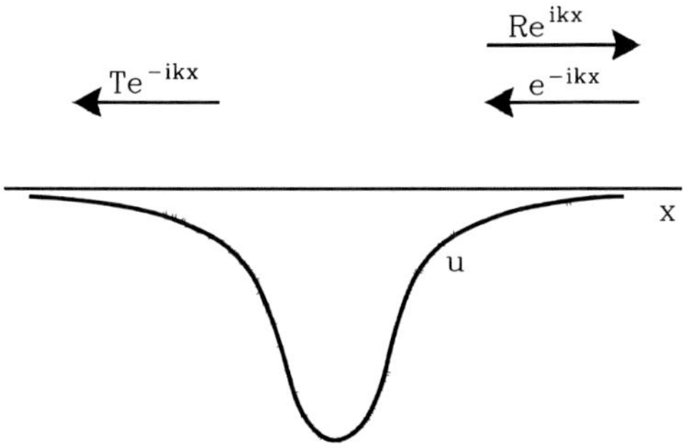

Figure C.3 The Schrödinger scattering problem with a plane wave incident from the right, the reflected wave (R), and the transmitted wave (T).

into the KdV Eq. (C.1). We have

$$u_t = \frac{\psi_{xxt}}{\psi} - \frac{\psi_{xx}\psi_t}{\psi^2} + \lambda_t\,, \tag{C.29}$$

$$u_x = \frac{\psi_{xxx}}{\psi} - \frac{\psi_{xx}\psi_x}{\psi^2}\,, \tag{C.30}$$

and so on. The final result is

$$\lambda_t \psi^2 + \left[\psi Q_x - \psi_x Q\right]_x = 0\,, \tag{C.31}$$

where

$$Q = \psi_t + \psi_{xxx} - 3(u + \lambda)\psi_x\,, \tag{C.32}$$

and u is again given by Eq. (C.28). If ψ vanishes at $x \to \pm\infty$ and if it is square integrable, then

$$\lambda_t = 0\,. \tag{C.33}$$

Then, from Eq. (C.31), we can conclude that $\psi Q_x - \psi_x Q$ is only a function of time. We have

$$\psi Q_{xx} - Q\psi_{xx} = 0 \tag{C.34}$$

or

$$Q_{xx} - \frac{\psi_{xx}}{\psi} Q = Q_{xx} + (\lambda - u) Q = 0 . \tag{C.35}$$

Thus, Q itself fulfills the Schrödinger equation. We therefore immediately know one solution $Q = c\psi$. However, the Schrödinger equation is of second order, and we can additionally have an unbounded solution. By the method of variation of parameters with the ansatz $Q = X\psi$, we find from Eqs. (C.35) and (C.4)

$$\frac{X_{xx}}{X_x} + \frac{2\psi_x}{\psi} = 0 . \tag{C.36}$$

This gives

$$X_x = \frac{D}{\psi^2} \tag{C.37}$$

or

$$X = D \int \frac{dx}{\psi^2} + C . \tag{C.38}$$

Therefore, we have the other (independent of ψ) solution

$$\psi \int^x \frac{dx}{\psi^2} . \tag{C.39}$$

As the lower border of the integral (C.39), we can choose a x-value where the asymptotic expansion of ψ at $x \to -\infty$ is already valid. Going back to Eq. (C.32), we find

$$\psi_t + \psi_{xxx} - 3(u + \lambda)\psi_x = C\psi + D\psi \int \frac{dx}{\psi^2} . \tag{C.40}$$

For the discrete eigenvalues, $\lambda = \lambda_n$, ψ_n vanishes at $x = \pm\infty$, and therefore

$$D_n = 0 . \tag{C.41}$$

The normalization of the bounded states (see Eq. (C.25)) leads via Eq. (C.40) to

$$C_n = 4(-\lambda_n)^{3/2} . \tag{C.42}$$

(Note that we have *no* contribution from ψ_{nt} at $x \to -\infty$.) Then, Eq. (C.40) has the simpler form (when we drop the subscript n at some intermediate stages)

$$\psi_t + \psi_{xxx} - 3\frac{\psi_{xx}\psi_x}{\psi} - 6\lambda\psi_x = 4(-\lambda)^{3/2}\psi . \tag{C.43}$$

Using the boundary condition Eq. (C.26), we find the time development of the coefficient $b_n(t)$ from

$$b_{nt} = \left[\kappa_n^3 - 3\kappa_n^3 - 6\lambda\kappa + 4(-\lambda)^{3/2}\right] b_n = 8(-\lambda)^{3/2} b_n \,, \qquad (C.44)$$

with the solution

$$b_n(t) = b_n(0) \exp\left[8\,(-\lambda_n)^{3/2}\, t\right] . \qquad (C.45)$$

For the continuous eigenvalues $\lambda = k^2$, we can again start with Eq. (C.40), but use the boundary conditions Eq. (C.27). At $x \to -\infty$, we find

$$T_t + 4ik^3 T = CT + \frac{D}{T} \int^x dx e^{2ikx} . \qquad (C.46)$$

Since all the terms, except the last one on the right-hand side of Eq. (C.46), do not depend on x,

$$D = 0 \qquad (C.47)$$

follows, and we have

$$T_t - (C - 4ik^3)T = 0 . \qquad (C.48)$$

At $x \to \infty$, we use Eq. (C.27) and insert it into Eq. (C.40) at $D = 0$ to obtain

$$\left(R_t - 4ik^3 R - CR\right) e^{ikx} + \left(4ik^3 - C\right) e^{-ikx} = 0 . \qquad (C.49)$$

From Eq. (C.49), we get

$$C = 4ik^3 \,, \qquad (C.50)$$

$$R_t = 8ik^3 R \,, \qquad (C.51)$$

and then from Eq. (C.48)

$$T_t = 0 . \qquad (C.52)$$

The solutions of Eqs. (C.51) and (C.52) are

$$T(k,t) = T(k,0) \qquad (C.53)$$

and

$$R(k,t) = R(k,0) \exp(8ik^3 t) . \qquad (C.54)$$

Thus,

$$\lambda_n(t) = \lambda_n(0) \qquad (C.55)$$

(see Eq. (C.33)) as well as the solutions Eqs. (C.45), (C.53), and Eq. (C.54) represent the basic results for the scattering data.

Before proceeding, we must make a useful remark concerning the normalization of bound states. Of course, we can now determine the time dependence of $\int_{-\infty}^{+\infty} \psi_n^2 dx$. Defining

$$c_n(t) = \left[\int_{-\infty}^{+\infty} \psi_n^2 dx \right]^{-1} , \tag{C.56}$$

we find

$$\begin{aligned} \frac{d}{dt}\left(c_n^{-1}\right) &= 2 \int_{-\infty}^{+\infty} \psi \psi_t dx \\ &= -2 \int_{-\infty}^{+\infty} \psi \psi_{xxx} dx + 6 \int_{-\infty}^{+\infty} \psi_{xx} \psi_x + 12\lambda_n \\ &\quad \times \int_{-\infty}^{+\infty} \psi \psi_x dx + 8(-\lambda_n)^{3/2} \int_{-\infty}^{+\infty} \psi \psi dx \\ &= 8(-\lambda_n)^{3/2} c_n^{-1} , \end{aligned} \tag{C.57}$$

with the solution

$$c_n(t) = c_n(0) \exp\left[-8(-\lambda_n)^{3/2} t\right] . \tag{C.58}$$

From $\lambda_n, b_n, n = 1, \ldots, p$, and $R(k)$ with k real, the transmission coefficient can be deduced. One, although not sufficient, relation we already know from quantum mechanics is

$$1 = |R|^2 + |T|^2 . \tag{C.59}$$

We conclude this first part with a short summary of the essence of the IST for the KdV equation. The KdV Eq. (C.1) is written as the integrability condition of two linear equations. First,

$$(-L + \lambda)\psi \equiv \psi_{xx} + [\lambda - u(x,t)]\psi = 0 , \tag{C.60}$$

which is identical with Eq. (C.4), and secondly,

$$\psi_t = -4\psi_{xxx} + 3u\psi_x + 3(u\psi)_x + C\psi , \tag{C.61}$$

which follows from Eq. (C.40) (with $D = 0$) and

$$\lambda \psi_x = -\psi_{xxx} + (u\psi)_x . \tag{C.62}$$

Now, we can make contact with the Lax equation. Lax [421] noted that the KdV Eq. (C.1) can be written in the form

$$L_t = AL - LA \equiv [A, L] , \tag{C.63}$$

where the operators L and A (the so called Lax pairs) are defined through

$$L = -\partial_x^2 + u \tag{C.64}$$

and

$$A = -4\partial_x^3 + 3(u\partial_x + \partial_x u) + C . \tag{C.65}$$

Note the equivalence with the operators appearing in Eqs. (C.60) and (C.61). The nice consequence of the more general formulation Eq. (C.63) is that we can decide for a whole class of flows $u(x, t)$ that the spectrum of L is preserved. This follows from the following sequence of calculations:

$$L_t\psi = A\lambda\psi - LA\psi , \tag{C.66}$$

$$\partial_t(L\psi) - L\psi_t = \lambda A\psi - LA\psi , \tag{C.67}$$

$$\lambda_t\psi + \lambda\psi_t - L\psi_t = (\lambda - L)A\psi , \tag{C.68}$$

$$\lambda_t\psi = (\lambda - L)(A\psi - \psi_t) . \tag{C.69}$$

Thus, for Eq. (C.61), that is,

$$\psi_t = A\psi , \tag{C.70}$$

$\lambda_t = 0$ follows. More generally, from Eq. (C.69), it follows that

$$\lambda_t \int \psi^* \psi dx = 0 , \tag{C.71}$$

if L is self-adjoint.

Now, we turn to the second important part of the IST, namely, that from the scattering data at time t, $u(x, t)$ can be constructed at any arbitrary time t. The basic ideas originate back to the mathematicians Gel'fand, Levitan, Marchenko, Kay, and Moses; refer to [220, 223, 225, 241] for more details. The *linear* integral equation for K,

$$K(x, y; t) + B(x + y; t) + \int_x^{\infty} K(x, s; t) B(s + y; t) ds = 0 \tag{C.72}$$

for $y > x$, is referred to as the Gel'fand–Levitan–Marchenko equation [220]. Here, B is defined by

$$B(z; t) = -i \sum_{n=1}^{p} \Gamma_n(t) \exp[i\zeta_n z] + \frac{1}{2\pi} \int_{-\infty}^{+\infty} R(k; t) e^{ikz} dk . \tag{C.73}$$

The discrete eigenvalues are denoted by $\zeta_n = i\kappa_n = i\sqrt{-\lambda_n}$, and

$$-i\Gamma_n(t) = c_n(0)\exp\left[8(-\lambda_n)^{3/2}t\right] \tag{C.74}$$

follows from the time-dependence of c_n. The solution of the KdV equation is obtained from K in the form

$$u(x,t) = -2\,\frac{d}{dx}K(x,y;t)\bigg|_{y=x}\,, \tag{C.75}$$

where we put $y = x$ into $K(x,y;t)$, and then differentiate with respect to x. The solitons are the physical manifestations of the discrete spectrum. To each eigenvalue corresponds a soliton which, in isolation, has a height, width and speed proportional to $-\lambda_n$, $\sqrt{-\lambda_n}$, and $-\lambda_n$, respectively. The continuous part of the spectrum corresponds to the so called radiation.

Let us now solve the integral equation for a very simple initial condition,

$$u(x,0) = -2\mathrm{sech}^2 x\,. \tag{C.76}$$

This is a special form of a reflectionless potential. In general, reflectionless potentials and their eigenvalues $\lambda = -(n-p)^2$ are associated with the Schrödinger equation

$$\psi_{xx} + \left[-(n-p)^2 + n(n+1)\mathrm{sech}^2 x\right]\psi = 0 \tag{C.77}$$

for $n = 1, 2, \ldots$ and $p = 0, 1, \ldots, n-1$. We shall not prove this general result [243], but rather perform an explicit calculation for Eq. (C.76), that is,

$$\psi_{xx} + \left(2\mathrm{sech}^2 x + \lambda\right)\psi = 0\,. \tag{C.78}$$

Let us substitute

$$\psi(x) = w(x)\mathrm{sech}^s x \tag{C.79}$$

into Eq. (C.78) in order to obtain

$$w_{xx} - 2s(\tanh x)w_x + w\left[(2-s-s^2)\mathrm{sech}^2 x + \lambda + s^2\right] = 0\,. \tag{C.80}$$

For simplicity, we set

$$2 - s - s^2 = 0\,, \tag{C.81}$$

with the two solution $s = 1$ and $s = -2$. To ensure bounded eigenfunctions at infinity, we choose $s = 1$ in the following, and Eq. (C.80) simplifies to

$$w_{xx} - 2(\tanh x)w_x + (\lambda+1)w = 0\,. \tag{C.82}$$

This can be transformed to the standard form of a hypergeometric equation,

$$x(1-x)y_{xx} + \left[\gamma - (\alpha+\beta+1)x\right]y_x - \alpha\beta y = 0\,. \tag{C.83}$$

By the transformation

$$\xi = \sinh^2 x\,, \tag{C.84}$$

we obtain from Eq. (C.82)

$$\xi(1+\xi)w_{\xi\xi} + \frac{1}{2}w_\xi + \frac{1}{4}\left(1-\kappa^2\right)w = 0\,, \tag{C.85}$$

where

$$\lambda = -\kappa^2 \tag{C.86}$$

has been defined (without any restriction). Note further that we have

$$w = (1+\xi)^{1/2}\psi\,. \tag{C.87}$$

The even and odd solutions of Eq. (C.85) can be written in the form

$$w_1 = F\left(-\frac{1}{2}+\frac{\kappa}{2}, -\frac{1}{2}-\frac{\kappa}{2}, \frac{1}{2}; -\xi\right), \tag{C.88}$$

$$w_2 = \sqrt{\xi}\,F\left(\frac{\kappa}{2}, -\frac{\kappa}{2}, \frac{3}{2}; -\xi\right). \tag{C.89}$$

We shall not discuss the details of that type of function. One can show that $\psi_2 = w_2/(1+\xi)^{1/2}$ will not be bounded at $x \to \pm\infty$ ($\xi \to \infty$) and ψ_1 only gives a well-behaved solution for

$$\kappa = 1\,. \tag{C.90}$$

As is known from quantum mechanics [426], we have $R = 0$ and $\lambda = -1$, that is, only one discrete (negative) eigenvalue and no reflection. The eigenfunction is obtained in the form

$$\psi = \alpha\,\mathrm{sech}x\,, \tag{C.91}$$

which, of course can be checked by inspection. The constant α follows from the normalization condition

$$\psi \sim e^{-x} \quad \text{at} \quad x \to +\infty\,. \tag{C.92}$$

Since $\mathrm{sech}x = 1/\cosh x = 2/(e^x + e^{-x})$, we have

$$\alpha = \frac{1}{2}\,. \tag{C.93}$$

Furthermore,

$$\int_{-\infty}^{+\infty} \psi^2 dx = \frac{1}{4}\int_{-\infty}^{+\infty} \mathrm{sech}^2 x\, dx = \frac{1}{2}\,, \tag{C.94}$$

such that the normalization

$$-i\Gamma(0) = 2 \tag{C.95}$$

follows. Therefore, we have

$$B(z) = 2\exp(8t - z) , \tag{C.96}$$

and the Gel'fand–Levitan–Marchenko equation explicitly reads as

$$K(x, y) + 2\exp(8t - x - y) + 2\exp(8t - y) \int_x^\infty e^{-s} K(x, s) ds = 0 \tag{C.97}$$

for $y > x$. To simplify the notation, we have abbreviated $K(x, y; t) \to K(x, y)$ and $B(z; t) \to B(z)$. From Eq. (C.97), we can conclude by differentiation

$$\frac{\partial K(x, y)}{\partial y} = -K(x, y) , \tag{C.98}$$

and therefore we make the ansatz

$$K(x, y) = L(x)e^{-y} . \tag{C.99}$$

Inserting this into Eq. (C.97), we obtain

$$L + 2e^{8t-x} + Le^{8t-2x} = 0 , \tag{C.100}$$

or

$$L(x) = \frac{-2e^x}{1 + \exp(2x - 8t)} . \tag{C.101}$$

By combining, according to the ansatz Eq. (C.99), we get the result

$$K(x, y) = \frac{-2\exp(x - y)}{1 + \exp(2x - 8t)} . \tag{C.102}$$

The solution $u(x, t)$ follows from Eq. (C.102) by differentiation,

$$u(x, t) = -2\frac{d}{dx}\left[\frac{-2}{1 + \exp(2x - 8t)}\right] = -2\text{sech}^2(x - 4t) . \tag{C.103}$$

This is a very interesting result. If we start with a reflectionless initial distribution with only one discrete eigenvalue, we obtain the moving one-soliton solution. Note that

$$u_t - 6uu_x + u_{xxx} = 0 \tag{C.104}$$

has the steady state solution

$$u = -\frac{1}{2}c\,\text{sech}^2\left[\frac{1}{2}\sqrt{c}(x - ct)\right], \tag{C.105}$$

which agrees with Eq. (C.103) for $c = 4$. Thus, the first example Eq. (C.76) was somehow trivial since Eq. (C.76) satisfies at $t = 0$ the amplitude-width relation for solitons. We could have written the solution Eq. (C.103) directly. Let us therefore consider a reflectionless potential which at $t = 0$ does not satisfy the amplitude-width relation as shown in Eq. (C.105), for example,

$$u(x, t) = -6\mathrm{sech}^2 x \,. \tag{C.106}$$

From the general result Eq. (C.77), we see that this potential corresponds to $n = 2$ and $p = 0, 1$. Thus, without further calculations, we found the two negative eigenvalues $\lambda_1 = -4$ and $\lambda_2 = -1$. The not yet properly normalized eigenfunctions are

$$\psi_1 = \alpha_1 \mathrm{sech}^2 x \quad \text{at} \quad \kappa_1 = 2 \,, \tag{C.107}$$

$$\psi_2 = \alpha_2 \mathrm{sech}^2 x \sinh x \quad \text{at} \quad \kappa_2 = 1 \,. \tag{C.108}$$

Now, we apply the boundary conditions which lead to

$$\alpha_1 = \frac{1}{4} \,, \quad \alpha_2 = \frac{1}{2} \,. \tag{C.109}$$

Noting the integral values

$$\int_{-\infty}^{+\infty} \mathrm{sech}^4 x \, dx = \frac{4}{3} \,, \quad \int_{-\infty}^{+\infty} \mathrm{sech}^4 x \sinh^2 x \, dx = \frac{2}{3} \,, \tag{C.110}$$

we find

$$-i\Gamma_1(t = 0) = 12 \,, \quad -i\Gamma_2(t = 0) = 6 \,. \tag{C.111}$$

Since the potential is reflectionless, the kernel of the Gel'fand–Levitan–Marchenko equation is

$$B(z) = 12 \exp(64t - 2z) + 6 \exp(8t - z) \,, \tag{C.112}$$

and the Gel'fand–Levitan–Marchenko equation explicitly reads

$$\begin{aligned} &K(x, y) + 12 \exp(64t - 2x - 2y) + 6 \exp(8t - x - y) \\ &\quad + \int_x^{\infty} [12 \exp(64t - 2y - 2s) + 6 \exp(8t - y - s)] K(x, s) ds = 0 \end{aligned} \tag{C.113}$$

for $y > x$. By similar arguments, as in the previous case, we make the ansatz

$$K(x, y) = c_1 L_1(x) e^{-2y} + c_2 L_2(x) e^{-y} \,, \tag{C.114}$$

where we abbreviate $c_1 = 12\exp(64t)$ and $c_2 = 6\exp(8t)$. If we collect the contributions to $\exp(-2y)$ and $\exp(-y)$, respectively, we obtain

$$\left(1 + \frac{c_1}{4} e^{-4x}\right) L_1 + \frac{c_2}{3} e^{-3x} L_2 + e^{-2x} = 0\,, \tag{C.115}$$

$$\frac{c_1}{3} e^{-3x} L_1 + \left(1 + \frac{c_2}{2} e^{-2x}\right) L_2 + e^{-x} = 0\,. \tag{C.116}$$

The solutions are

$$L_1 = -\frac{1}{D}\left(e^{-2x} + \frac{1}{6} c_2 e^{-4x}\right)\,, \tag{C.117}$$

$$L_2 = \frac{1}{D}\left(-e^{-x} + \frac{1}{12} c_1 e^{-5x}\right)\,, \tag{C.118}$$

where

$$D = 1 + \frac{c_1}{4} e^{-4x} + \frac{c_2}{2} e^{-2x} + \frac{c_1 c_2}{72} e^{-6x}\,. \tag{C.119}$$

By inserting Eqs. (C.117) and (C.118) into Eq. (C.114), we obtain in the usual way

$$u(x,t) = -12 \frac{3 + 4\cosh(8t - 2x) + \cosh(64t - 4x)}{\left[\cosh(36t - 3x) + 3\cosh(28t - x)\right]^2}\,. \tag{C.120}$$

To interpret this result, we let t and x tend to infinity in a certain way. Defining first

$$\xi = x - 4\kappa_1^2 t = x - 16t, \tag{C.121}$$

we get (ξ fixed)

$$\begin{aligned}\lim_{t\to-\infty} u(x,t) &= -12 \frac{2e^{-24t} e^{-2\xi}}{\frac{1}{4}[e^{-3\xi} e^{-12t} + 3e^{-12t} e^{\xi}]^2} = -96 \frac{e^{4\xi}}{(1 + 3e^{4\xi})^2} \\ &= -32 \frac{e^{4(\xi-\xi_1)}}{[1 + e^{4(\xi-\xi_1)}]^2} = -8\mathrm{sech}^2[2(x - 16t - \xi_1)]\,,\end{aligned} \tag{C.122}$$

with $e^{-4\xi_1} = 3$. A similar calculation yields at $t \to +\infty$ and ξ fixed

$$\lim_{t\to\infty} u(x,t) = -8\,\mathrm{sech}^2\left[2(x - 16t - \xi_1')\right]\,, \quad \xi \text{ fixed}\,, \tag{C.123}$$

with $e^{4\xi_1'} = 3$. Thus, "a first soliton" moves with fixed velocity and asymptotically recovers its shape with a phase shift which is described by the difference in ξ_1' and ξ_1. However, there is a "second soliton" which belongs to the eigenvalue λ_2. If we define

$$\xi = x - 4\kappa_2^2 t = x - 4t\,, \tag{C.124}$$

we obtain (ξ fixed)

$$\lim_{t \to -\infty} u(x,t) = -2\text{sech}^2(x - 4t - \xi_2) \,, \tag{C.125}$$

with $e^{2\xi_2} = 3$ and

$$\lim_{t \to +\infty} u(x,t) = -2\text{sech}^2(x - 4t - \xi_2') \,, \tag{C.126}$$

with $e^{-2\xi_2'} = 3$. The correct interpretation is that, all together, we have a (complicated) two-soliton solution. At times long before "the two solitons" interact, $u(x,t)$ reduces to

$$u(x,t) = -2\kappa_1^2\text{sech}^2(\gamma_1 + \Delta) - 2\kappa_2^2\text{sech}^2(\gamma_2 - \Delta) \,, \tag{C.127}$$

with

$$\gamma_i = \kappa_i x - 4\kappa_i^3 t + \delta_i \,, \tag{C.128}$$

where

$$\delta_i = \frac{1}{2}\ln\left[\frac{-i\Gamma_i(0)}{2\kappa_i}\left(\frac{\kappa_1 - \kappa_2}{\kappa_2 + \kappa_1}\right)\right] \tag{C.129}$$

and

$$\Delta = \frac{1}{2}\ln\left(\frac{\kappa_2 + \kappa_1}{\kappa_1 - \kappa_2}\right) \,. \tag{C.130}$$

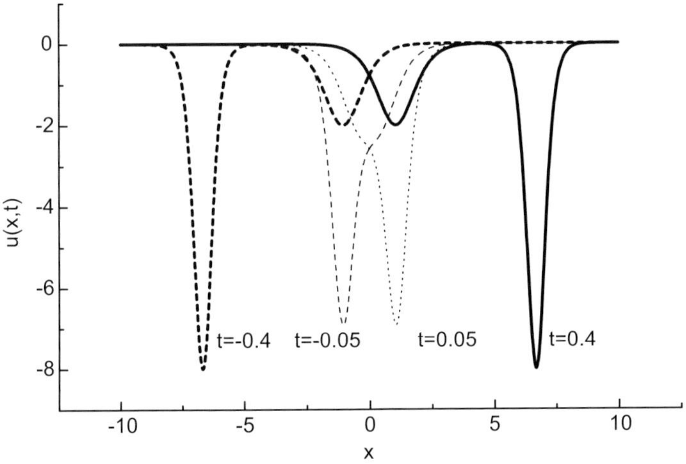

Figure C.4 Schematic plot of a two-soliton solution of the KdV equation at different times t, which can be interpreted as the interaction scenario of two solitons. The curves correspond to a reflectionless initial distribution $u(x,0) = -6\text{sech}^2(x)$.

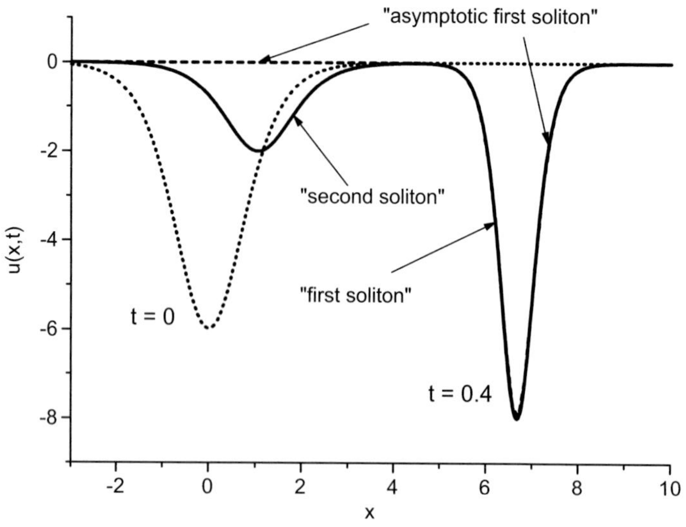

Figure C.5 Comparison of the exact solution (solid line) and the asymptotic one-soliton solution Eq. (C.123) (dashed line) at $t = 0.4$ for a reflectionless initial distribution $u(x,0) = -6\text{sech}^2(x)$ (dotted line). The curves at $t = 0.4$ show the same positions and forms of the "first soliton".

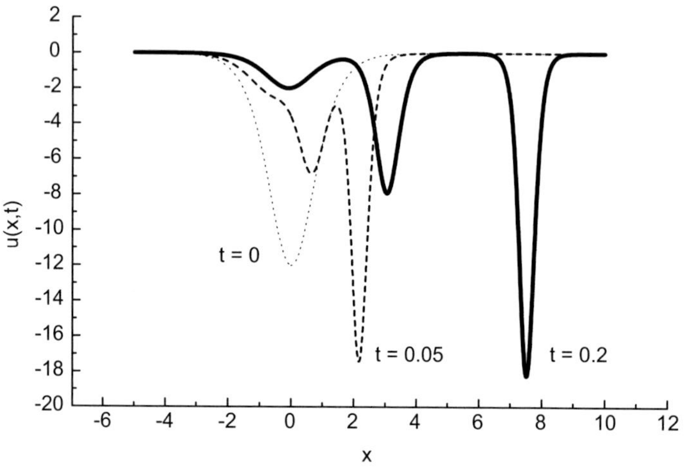

Figure C.6 Generation of three solitons for a reflectionless initial distribution $u(x,0) = -12\text{sech}^2(x)$ (dotted line). The broken curve is at $t = 0.05$, whereas the full development of three solitons can already be seen at $t = 0.2$ (solid line).

At times long after interaction, we obtain

$$u(x,t) = -2\kappa_1^2\text{sech}^2(\gamma_1 - \Delta) - 2\kappa_2^2\text{sech}^2(\gamma_2 + \Delta)\,. \tag{C.131}$$

The interaction is shown in Figure C.4 and is further analyzed in Figure C.5.

In summary, we have analytically demonstrated what we mean by the stable behavior of interacting solitons. Another aspect is worth mentioning. The results shown in Figures C.4 and C.5 can also be interpreted as the development of two solitons out of a reflectionless initial distribution $u(x,t) = -6\text{sech}^2(x)$. As we know, we can have different types of pure many-soliton generations from reflectionless potentials. Figure C.6 shows as another example, that is, the generation of three solitons from the initial distribution $u(x,t) = -12\text{sech}^2(x)$. This initial potential corresponds to $n = 3$ and $p = 0, 1, 2$ in Eq. (C.77). Thus, we expect asymptotically ($t \to +\infty$) three solitons with amplitudes $-2(n-p)^2$, that is, -18, -8, and -2. This behavior is clearly seen in Figure C.6.

Before concluding, we would like to make some additional remarks. First, one can generalize the foregoing results for reflectionless potentials with only discrete eigenvalues to include radiation. The generation of solitons in accordance with the discrete eigenvalues of the initial distribution also occurs for initial potentials with reflection. Secondly, for nonintegrable systems, very often the IST can be used as a starting point for a perturbation theory [427]. Finally, the IST can be generalized to many systems in mathematics and physics [220].

Appendix D
Lie Transform Techniques for Eliminating Fast Variations

When discussing drift motion in magnetized plasmas, we have applied averaging techniques for the motion of the gyrocenter. It was mentioned that Lie transforms [428] are very suitable for determining the mean motion in a mathematically sound way [41–44, 226, 429]. Thus, the purpose of the present section is to introduce the perturbation technique which is designed to perform an averaging via a coordinate transformation.

Suppose one has a physical process which consists of rapid variations overlaying a smooth long-time tendency. One might think – for example, outside physics – of the value of a stock which shows strong intraday fluctuations on the one hand and a generally clear average behavior. Within physics, the gyromotion of a charged particle in magnetic and electric fields is a prominent example. We have the fast gyration and the slow drift. Another important example is the motion of a soliton in the presence of damping and rapid amplification [226, 430, 431].

We first show the principle structure of a coordinate transform near identity, and then we discuss the simplification arising from the operator language. In mathematics, the Lie transform technique was primarily used to simplify differential equations, that is, to find their normal forms.

The basic idea of the Lie transform [429] is to make a transformation of the dependent variable, close to identity, that is, to go from $\boldsymbol{x}$ to $\boldsymbol{y}$ via

$$\boldsymbol{x} = \boldsymbol{y} + \epsilon \boldsymbol{X}_1(\boldsymbol{y}) + \epsilon^2 \boldsymbol{X}_2(\boldsymbol{y}) + \cdots \tag{D.1}$$

We could apply it, for example, to go from a nonlinear differential equation (where $\boldsymbol{x}$ is a vector, N and A are matrices)

$$\frac{d\boldsymbol{x}}{dt} = A\boldsymbol{x} + N \,, \tag{D.2}$$

to a linear one,

$$\frac{d\boldsymbol{y}}{dt} = A\boldsymbol{y} \,, \tag{D.3}$$

if possible. There are many more applications, and we shall discuss some of them now on the basis of the simple, one-dimensional differential equation

$$\frac{dx}{dt} = f(x; t) \,. \tag{D.4}$$

High Temperature Plasmas, Theory and Mathematical Tools for Laser and Fusion Plasmas, First Edition.
Karl-Heinz Spatschek.

We make the ansatz

$$x = y + \epsilon X_1(y) + \epsilon^2 X_2(y) + \cdots \tag{D.5}$$

Suppose we anticipate such a transformation with a smallness parameter ϵ. How do we invert it? This question can be answered by a systematic expansion. For example, in the first order in ϵ

$$x = y + \epsilon X_1(y) \approx y + \epsilon X_1(x) \tag{D.6}$$

leads to

$$y \approx x - \epsilon X_1(x)\,. \tag{D.7}$$

To generalize, one could introduce the generating function (see below). Before that, we allow for time-dependent transformations,

$$X_1(x) + \epsilon X_2(x) + \cdots \to W_1(x;t) + \epsilon W_2(x;t) + \cdots \tag{D.8}$$

For example, up to first order in ϵ, we again obtain

$$x - \epsilon W_1(x;t) = y\,. \tag{D.9}$$

Next, we want to determine the differential equation in the new coordinate. Differentiating with respect to t, making use of $dx/dt = f(x;t)$, and demanding

$$\frac{dy}{dt} = g(y;t)\,, \tag{D.10}$$

we can write in the present ordering

$$f(x;t)\left[1 - \epsilon\frac{\partial W_1(x;t)}{\partial x}\right] - \epsilon\frac{\partial W_1(x;t)}{\partial t} = g(x - \epsilon W_1, t)\,. \tag{D.11}$$

On the right-hand side, we Taylor-expand with respect to ϵ and arrive at

$$f(x;t) - \epsilon\left[f(x;t)\frac{\partial W_1(x;t)}{\partial x} - W_1(x;t)\frac{\partial g(x;t)}{\partial x}\right] - \epsilon\frac{\partial W_1(x;t)}{\partial t} = g(x;t)\,. \tag{D.12}$$

Let us further assume that the original differential equation also contains the parameter ϵ, and we may expand

$$f(x;t) = f_0(x;t) + \epsilon f_1(x;t)\,, \tag{D.13}$$

$$g(x;t) = g_0(x;t) + \epsilon g_1(x;t)\,. \tag{D.14}$$

We proceed with the systematic ordering,

$$f_0(x;t) = g_0(x;t)\,, \tag{D.15}$$

$$f_1(x;t) - \left[f_0(x;t)\frac{\partial W_1(x;t)}{\partial x} - W_1(x;t)\frac{\partial f_0(x;t)}{\partial x}\right] - \frac{\partial W_1(x;t)}{\partial t} = g_1(x;t)\,, \tag{D.16}$$

and define the Lie bracket

$$[u(x;t), v(x;t)] = u(x;t)\frac{\partial v(x;t)}{\partial x} - v(x;t)\frac{\partial u(x;t)}{\partial x} . \tag{D.17}$$

Now, we can write the first order relation in the form

$$f_1(x;t) - [f_0(x;t), W_1(x;t)] - \frac{\partial W_1(x;t)}{\partial t} = g_1(x;t) . \tag{D.18}$$

We can expect that higher orders could be calculated systematically. In doing so, we have some freedom in the meaning of the g_ν. For example, if elimination of nonlinearities is aimed for, we try solutions with $g_\nu = 0, \nu = 1, 2, \ldots$ Another example would be to require that g_1 is only slowly varying.

Let us pause here a little bit and consider the question as to whether a more elegant formulation of the procedure is possible. Let us make use of the operator formulation

$$x = e^{\phi \frac{\partial}{\partial y}} y = \left(1 + \phi\frac{\partial}{\partial y} + \frac{1}{2}\phi\frac{\partial}{\partial y}\phi\frac{\partial}{\partial y} + \cdots\right) y = y + \phi + \frac{1}{2}\phi\frac{\partial \phi}{\partial y} + \cdots , \tag{D.19}$$

assuming that x as well as ϕ are functions of y and t, that is,

$$x = x(y,t) , \quad \phi = \phi(y,t) . \tag{D.20}$$

In addition,

$$\mathcal{O}(\phi) = \epsilon \tag{D.21}$$

is required. The basic equation is then

$$\frac{d}{dt}\left(e^{\phi \frac{\partial}{\partial y}} y\right) = f_0\left(e^{\phi \frac{\partial}{\partial y}} y, t\right) + \epsilon f_1\left(e^{\phi \frac{\partial}{\partial y}} y, t\right), \tag{D.22}$$

when $g = g_0 + \epsilon g_1$ and $f = f_0 + \epsilon f_1$ is being used. Suppose we can make a series expansion of f_0 and f_1, then

$$f_0\left(e^{\phi \frac{\partial}{\partial y}} y, t\right) + \epsilon f_1\left(e^{\phi \frac{\partial}{\partial y}} y, t\right) = e^{\phi \frac{\partial}{\partial y}}\left[f_0(y,t) + \epsilon f_1(y,t)\right]. \tag{D.23}$$

In other words,

$$\frac{dy}{dt} + \frac{\partial \phi}{\partial y}\frac{dy}{dt} + \frac{\partial \phi}{\partial t} + \cdots = e^{\phi \frac{\partial}{\partial y}}\left(f_0(y,t) + \epsilon f_1(y,t)\right) . \tag{D.24}$$

Proceeding up to the order of interest, for example, first order, we obtain

$$\begin{aligned} &g_0(y,t) + \epsilon g_1(y,t) + g_0(y,t)\frac{\partial \phi(y,t)}{\partial y} + \frac{\partial \phi(y,t)}{\partial t} \\ &= f_0(y,t) + \epsilon f_1(y,t) + \phi(y,t)\frac{\partial f_0(y,t)}{\partial y} . \end{aligned} \tag{D.25}$$

Collecting terms of appropriate orders,

$$g_0(y, t) = f_0(y, t) , \tag{D.26}$$

$$g_1(y, t) = f_1(y, t) - f_0(y, t)\frac{\partial \phi(y, t)}{\partial y} + \phi(y, t)\frac{\partial f_0(y, t)}{\partial y} - \frac{\partial \phi(y, t)}{\partial t} \tag{D.27}$$

follow. That is, exactly Eq. (D.18) with $W_1 \equiv \phi$.

Thus far, we have assumed

$$f(x, t) = f_0(x, t) + \epsilon f_1(x, t) + \epsilon^2 f_2(x, t) + \cdots \tag{D.28}$$

Very often, another situation is of interest,

$$\frac{dx(t)}{dt} = f(x, t) = f_0(x, t) + \tilde{f}\left(x, \frac{t}{\epsilon}\right) , \tag{D.29}$$

that is, we have both a slowly varying and a rapidly varying force. A typical (simple) example is

$$\frac{dx}{dt} = x + x\cos\left(\frac{t}{\epsilon}\right) = x\left[1 + \cos\left(\frac{t}{\epsilon}\right)\right], \quad x(t = 0) = x_0 . \tag{D.30}$$

The exact solution of that equation is known as

$$x(t) = x_0 e^t e^{\epsilon \sin\left(\frac{t}{\epsilon}\right)} . \tag{D.31}$$

Looking at it, it becomes clear that a Taylor expansion around $\epsilon = 0$ does not exist (the function cannot be differentiated in ϵ at $\epsilon = 0$). However, we can write

$$x(t) = x_0 e^t \left[1 + \epsilon \sin\left(\frac{t}{\epsilon}\right) + \frac{1}{2}\epsilon^2 \sin^2\left(\frac{t}{\epsilon}\right) + \cdots\right]. \tag{D.32}$$

The structure of the solution becomes evident. We have a slowly varying part

$$y(t) = x_0 e^t \tag{D.33}$$

and small rapid oscillations. Figure D.1 shows $y(t)$ in addition to the exact solution $x(t)$. Having in mind

$$x(t) = y + \epsilon y \sin\left(\frac{t}{\epsilon}\right) + \epsilon^2 \frac{y}{2} \sin^2\left(\frac{t}{\epsilon}\right) + \cdots , \tag{D.34}$$

the following ansatz for a coordinate transformation seems to be appropriate in the general case Eq. (D.29):

$$x(t) = y(t) + \epsilon X_1\left(y, \frac{t}{\epsilon}\right) + \epsilon^2 X_2\left(y, \frac{t}{\epsilon}\right) + \cdots \tag{D.35}$$

From (D.29), we obtain up to first order

$$f(x, t) = \frac{dx}{dt} = \frac{dy}{dt} + \epsilon \frac{\partial X_1}{\partial y}\frac{dy}{dt} + \frac{\partial X_1}{\partial t} + \epsilon \frac{\partial X_2}{\partial t} . \tag{D.36}$$

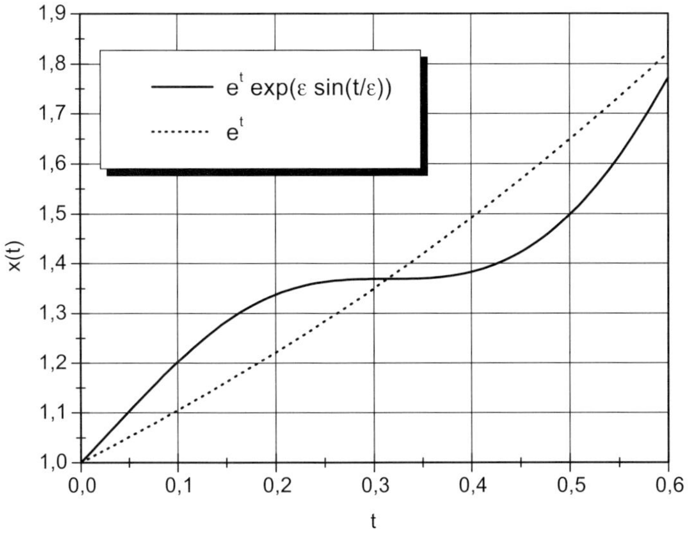

Figure D.1 Comparison of the approximate (average) solution $y(t)$ with the exact one $x(t)$ for the simple example. Parameters are $\epsilon = 1/10$ and $x_0 = 1$.

Expanding f around y yields

$$f(x,t) = f(y,t) + \epsilon X_1 \frac{\partial f}{\partial y} . \tag{D.37}$$

Next, we use $dy/dt = g_0(y,t) + \epsilon g_1(y,t)$ to obtain

$$f(y,t) + \epsilon X_1 \frac{\partial f}{\partial y} = g_0(y,t) + \epsilon g_1(y,t) + \epsilon \frac{\partial X_1}{\partial y} g_0(y,t) + \frac{\partial X_1}{\partial t} + \epsilon \frac{\partial X_2}{\partial t} . \tag{D.38}$$

Order by order, this leads to

$$f = g_0 + \frac{\partial X_1}{\partial t} , \tag{D.39}$$

$$X_1 \frac{\partial f}{\partial y} = g_1 + g_0 \frac{\partial X_1}{\partial y} + \frac{\partial X_2}{\partial t} , \tag{D.40}$$

and so on. Combining, and using the definition of the Lie bracket, gives

$$[X_1, f] = g_1 - \frac{\partial X_1}{\partial y} \frac{\partial X_1}{\partial t} + \frac{\partial X_2}{\partial t} . \tag{D.41}$$

To further rest a little bit, let us apply the above procedure to our simple example Eq. (D.30). To lowest order, we obtain

$$f(y,t) = y \left[1 + \cos\left(\frac{t}{\epsilon}\right) \right] = g_0(y) + \frac{\partial X_1}{\partial t} . \tag{D.42}$$

We identify a slowly varying and a rapidly varying part, respectively,

$$g_0(y) = y \, , \quad \frac{\partial X_1}{\partial t} = y \cos\left(\frac{t}{\epsilon}\right) . \tag{D.43}$$

The expression X_1 can be obtained by integration, provided y can be considered as constant during an oscillation period. This corresponds to averaging. Compared to the linear partial differential equation (D.18), we only have to solve a linear ordinary differential equation. We have

$$\epsilon X_1 = \epsilon y \sin\left(\frac{t}{\epsilon}\right) + \epsilon X_{10}(y) . \tag{D.44}$$

The integration constant $X_{10}(y)$ may depend on y. That expression will be obtained by the following procedure. Insert X_1 into the first order equation and integrate over the period $2\pi\epsilon$ (during that integration, y is again considered as constant). From (D.40), we obtain

$$\left[y \sin\left(\frac{t}{\epsilon}\right) + X_{10}\right]\left[1 + \cos\left(\frac{t}{\epsilon}\right)\right] = g_1(y) + \left[\sin\left(\frac{t}{\epsilon}\right) + \frac{\partial X_{10}}{\partial y}\right] y + \frac{\partial X_2}{\partial t} , \tag{D.45}$$

and the integration leads to

$$2\pi\epsilon X_{10} = 2\pi\epsilon g_1 + 2\pi\epsilon \frac{\partial X_{10}}{\partial y} y . \tag{D.46}$$

Here, g_1 or X_{10} can be chosen. One possibility would be to choose X_{10} in such a way that $g_1 = 0$. For the present simple example,

$$X_1\left(y, \frac{t}{\epsilon}\right) = y \sin\left(\frac{t}{\epsilon}\right) , \tag{D.47}$$

that is, the first order of Eq. (D.31). Figure D.2 makes a comparison of the exact with the approximate solution. In contrast to Figure D.1, we already recognize an impressive improvement. We should emphasize the following. The new "averaged" equation is

$$\frac{dy}{dt} = y + \mathcal{O}(\epsilon^2) \tag{D.48}$$

with the straightforward solution

$$y = e^t + \mathcal{O}(\epsilon^2) . \tag{D.49}$$

The original variable x is obtained via

$$x(t) = y + \epsilon X_1 + \mathcal{O}(\epsilon^2) = e^t + \epsilon e^t \sin\left(\frac{t}{\epsilon}\right) + \mathcal{O}(\epsilon^2) . \tag{D.50}$$

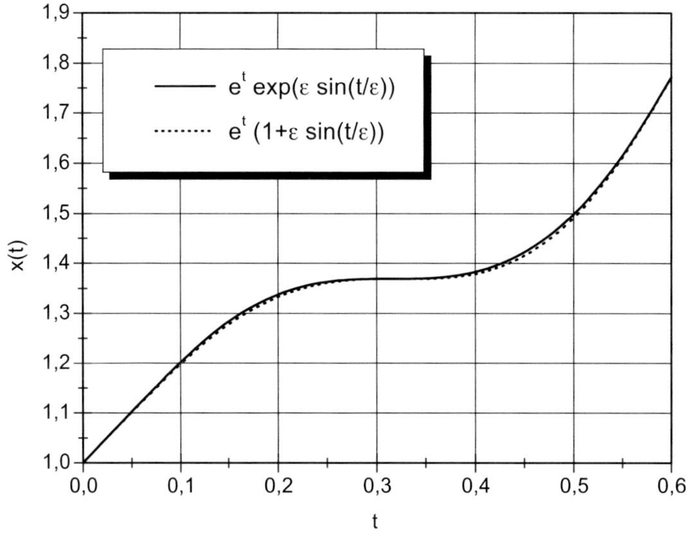

Figure D.2 Comparison of the first order perturbative result with the exact solution of the simple example when $\epsilon = 1/10$ and $x_0 = 1$.

The solution obtained in this manner already agrees very precisely with the exact solution; see Figure D.2

Let us return to the more general considerations and find out whether also here an elegant operator formalism is possible. We write

$$x = e^{\phi \frac{\partial}{\partial y}} y \,.$$

A little algebra, similar to that shown above, leads to

$$\frac{d}{dt}\left(y + \phi + \frac{1}{2}\phi\frac{\partial \phi}{\partial y} + \cdots\right) = \left(1 + \phi\frac{\partial}{\partial y} + \frac{1}{2}\phi\frac{\partial}{\partial y}\phi\frac{\partial}{\partial y} + \cdots\right) f(y,t)\,. \tag{D.51}$$

To first order, with

$$\mathcal{O}(\phi) = \epsilon\,, \quad \mathcal{O}\left(\frac{\partial \phi}{\partial t}\right) = \mathcal{O}(1)\,, \tag{D.52}$$

and the notation $dy/dt = g(y)$, we have

$$g + g\frac{\partial \phi}{\partial y} + \frac{\partial \phi}{\partial t} + \frac{1}{2}\frac{\partial \phi}{\partial t}\frac{\partial \phi}{\partial y} + \frac{1}{2}\phi\frac{\partial^2 \phi}{\partial y \partial t} + \cdots = f + \phi\frac{\partial f}{\partial y} + \cdots \tag{D.53}$$

This equation for g is treated in the following manner. First, we rewrite it as

$$g = f + \phi\frac{\partial f}{\partial y} - g\frac{\partial \phi}{\partial y} - \frac{\partial \phi}{\partial t} - \frac{1}{2}\frac{\partial \phi}{\partial t}\frac{\partial \phi}{\partial y} - \frac{1}{2}\phi\frac{\partial^2 \phi}{\partial y \partial t} + \cdots \approx f - \frac{\partial \phi}{\partial t} + \cdots \tag{D.54}$$

From here, we see that we can iterate Eq. (D.53) in order to obtain

$$g+f\frac{\partial\phi}{\partial y}-\frac{\partial\phi}{\partial t}\frac{\partial\phi}{\partial y}+\frac{\partial\phi}{\partial t}+\frac{1}{2}\frac{\partial\phi}{\partial t}\frac{\partial\phi}{\partial y}+\frac{1}{2}\phi\frac{\partial^2\phi}{\partial y\partial t}+\cdots = f+\phi\frac{\partial f}{\partial y}+\cdots \quad (D.55)$$

We can make use of the Lie bracket to get the formulation

$$g+\frac{\partial\phi}{\partial t}+\frac{1}{2}\left[\phi,\frac{\partial\phi}{\partial t}\right]\approx f+[\phi,f]\,. \quad (D.56)$$

Next, we begin with a perturbative treatment of that relation. Introducing

$$g = g_0+\epsilon g_1+\cdots\,,\qquad \phi=\phi_1+\phi_2+\cdots\,, \quad (D.57)$$

we make use of the ordering

$$\mathcal{O}(\phi_n)=\epsilon^n\,,\quad \mathcal{O}\left(\frac{\partial\phi}{\partial t}\right)=\mathcal{O}(\phi_{n-1})\,. \quad (D.58)$$

This leads immediately to

$$g_0+\frac{\partial\phi_1}{\partial t}=f\,, \quad (D.59)$$

$$g_1+\frac{\partial\phi_2}{\partial t}+\frac{1}{2}\left[\phi_1,\frac{\partial\phi_1}{\partial t}\right]=[\phi_1,f]\,, \quad (D.60)$$

and so on. We end up with a slight generalization of the previous formulas. First, $\phi_1\equiv X_1$ is still true. Next, two commutators appear, whereas in the previous presentation, the term $-\partial X_1/\partial y\partial X_1/\partial t$ was present. The reason is $\phi_2\not\equiv X_2$. Actually, we have

$$x=y+\epsilon X_1+\epsilon^2 X_2\,, \quad (D.61)$$

$$x=y+\phi+\frac{1}{2}\phi\frac{\partial\phi}{\partial y}=y+\phi_1+\phi_2+\frac{1}{2}\phi_1\frac{\partial\phi_1}{\partial y}\,, \quad (D.62)$$

leading to

$$X_2=\phi_2+\frac{1}{2}\phi_1\frac{\partial\phi_1}{\partial y}\,. \quad (D.63)$$

One can now easily prove the equivalence between the previous formulation and the operator formalism via

$$\frac{\partial X_2}{\partial t}=\frac{\partial\phi_2}{\partial t}+\frac{1}{2}\frac{\partial\phi_1}{\partial t}\frac{\partial\phi_1}{\partial y}+\frac{1}{2}\phi_1\frac{\partial^2\phi_1}{\partial y\partial t}\,. \quad (D.64)$$

That can be also written as

$$[X_1,f]=g_1-\frac{\partial X_1}{\partial y}\frac{\partial X_1}{\partial t}+\frac{\partial X_2}{\partial t}\,. \quad (D.65)$$

Straightforward calculations lead to

$$[\phi_1, f] = g_1 - \frac{\partial \phi_1}{\partial y}\frac{\partial \phi_1}{\partial t} + \frac{\partial \phi_2}{\partial t} + \frac{1}{2}\frac{\partial \phi_1}{\partial t}\frac{\partial \phi_1}{\partial y} + \frac{1}{2}\phi_1 \frac{\partial^2 \phi_1}{\partial y \partial t} \tag{D.66}$$

and

$$[\phi_1, f] = g_1 + \frac{\partial \phi_2}{\partial t} + \frac{1}{2}\left[\phi_1, \frac{\partial \phi_1}{\partial t}\right], \tag{D.67}$$

which demonstrates the equivalence.

For physical applications, a generalization to the multidimensional case is necessary. In the following, some steps of the generalization are shown [226].

Let us start with a system of ordinary differential equations in the form

$$\frac{dx_k}{dt} = f_k(x_1, x_2, \ldots; t) = f_{0k}(x_1, x_2, \ldots; t) + \tilde{f}_k\left(x_1, x_2, \ldots; \frac{t}{\epsilon}\right). \tag{D.68}$$

We write it in vector form

$$\frac{d}{dt}\boldsymbol{x} = \boldsymbol{f}(\boldsymbol{x}; t) = \boldsymbol{f}_0(\boldsymbol{x}; t) + \tilde{\boldsymbol{f}}\left(\boldsymbol{x}; \frac{t}{\epsilon}\right). \tag{D.69}$$

The Lie transform will be applied accordingly, that is,

$$\boldsymbol{x}(t) = \boldsymbol{y}(t) + \epsilon \boldsymbol{X}^{(1)}\left(\boldsymbol{y}; \frac{t}{\epsilon}\right) + \epsilon^2 \boldsymbol{X}^{(2)}\left(\boldsymbol{y}; \frac{t}{\epsilon}\right). \tag{D.70}$$

In order to avoid confusion with the index numbering of the equations, the ordering is now indicated by upper indices. Differentiating with respect to time yields

$$\frac{d}{dt}\boldsymbol{x}(t) = \frac{d}{dt}\boldsymbol{y}(t) + \epsilon \frac{\partial \boldsymbol{X}^{(1)}}{\partial \boldsymbol{y}}\frac{d\boldsymbol{y}}{dt} + \epsilon^2 \frac{\partial \boldsymbol{X}^{(2)}}{\partial \boldsymbol{y}}\frac{d\boldsymbol{y}}{dt} + \frac{\partial \boldsymbol{X}^{(1)}}{\partial t} + \epsilon \frac{\partial \boldsymbol{X}^{(2)}}{\partial t}. \tag{D.71}$$

Here, $\partial \boldsymbol{X}^{(1)}/\partial \boldsymbol{y}$ is the transformation matrix

$$\boldsymbol{X}^{(1)} = \begin{pmatrix} X_1^{(1)} \\ X_2^{(1)} \\ \vdots \\ X_n^{(1)} \end{pmatrix}, \quad \boldsymbol{y} = \begin{pmatrix} y_1 \\ y_2 \\ \vdots \\ y_n \end{pmatrix}, \quad \frac{\partial \boldsymbol{X}^{(1)}}{\partial \boldsymbol{y}} = \begin{pmatrix} \frac{\partial X_1^{(1)}}{\partial y_1} & \frac{\partial X_1^{(1)}}{\partial y_2} & \cdots & \frac{\partial X_1^{(1)}}{\partial y_n} \\ \frac{\partial X_2^{(1)}}{\partial y_1} & \frac{\partial X_2^{(1)}}{\partial y_2} & \cdots & \vdots \\ \vdots & \vdots & \ddots & \vdots \\ \frac{\partial X_n^{(1)}}{\partial y_1} & \cdots & \cdots & \frac{\partial X_n^{(1)}}{\partial y_n} \end{pmatrix}. \tag{D.72}$$

Expanding $\boldsymbol{f}$ and $\boldsymbol{g}$ in the form

$$\frac{d}{dt}\boldsymbol{y} = \boldsymbol{g}^{(0)}(\boldsymbol{y}) + \epsilon \boldsymbol{g}^{(1)}(\boldsymbol{y}), \quad \boldsymbol{f}(\boldsymbol{x}; t) = \boldsymbol{f}(\boldsymbol{y}; t) + \epsilon \frac{\partial \boldsymbol{f}}{\partial \boldsymbol{y}} \boldsymbol{X}^{(1)}, \tag{D.73}$$

and inserting into the differential equation $d/dt x = f(x; t)$, we obtain the intended relation between f and g:

$$f + \epsilon \frac{\partial f}{\partial y} X^{(1)} = g^{(0)}(y) + \epsilon g^{(1)}(y) + \epsilon \frac{\partial X^{(1)}}{\partial y} g^{(0)} + \frac{\partial X^{(1)}}{\partial t} + \epsilon \frac{\partial X^{(2)}}{\partial t} . \tag{D.74}$$

Collecting the various orders, we get

$$f = g^{(0)} + \frac{\partial X^{(1)}}{\partial t} , \tag{D.75}$$

$$\frac{\partial f}{\partial y} X^{(1)} = g^{(1)}(y) + \frac{\partial X^{(1)}}{\partial y} g^{(0)} + \frac{\partial X^{(2)}}{\partial t} . \tag{D.76}$$

Now, the generalization of the Lie bracket of u and v is

$$u = \begin{pmatrix} u_1 \\ u_2 \\ \vdots \\ u_n \end{pmatrix} , \quad v = \begin{pmatrix} v_1 \\ v_2 \\ \vdots \\ v_n \end{pmatrix} ,$$

$$[u, v] = \left(\sum_{i=1}^{n} u_i \frac{\partial}{\partial y_i} \right) \begin{pmatrix} v_1 \\ v_2 \\ \vdots \\ v_n \end{pmatrix} - \left(\sum_{i=1}^{n} v_i \frac{\partial}{\partial y_i} \right) \begin{pmatrix} u_1 \\ u_2 \\ \vdots \\ u_n \end{pmatrix} . \tag{D.77}$$

We can write

$$[u, v] = \frac{\partial v}{\partial y} u - \frac{\partial u}{\partial y} v \tag{D.78}$$

since

$$\frac{\partial v}{\partial y} u = \begin{pmatrix} \frac{\partial v_1}{\partial y_1} & \cdots & \frac{\partial v_1}{\partial y_n} \\ \vdots & \ddots & \vdots \\ \frac{\partial v_n}{\partial y_1} & \cdots & \frac{\partial v_n}{\partial y_n} \end{pmatrix} \begin{pmatrix} u_1 \\ \vdots \\ u_n \end{pmatrix}$$

$$= \begin{pmatrix} \frac{\partial v_1}{\partial y_1} u_1 + \frac{\partial v_1}{\partial y_2} u_2 + \ldots + \frac{\partial v_1}{\partial y_n} u_n \\ \vdots \\ \frac{\partial v_n}{\partial y_1} u_1 + \frac{\partial v_n}{\partial y_2} u_2 + \ldots + \frac{\partial v_n}{\partial y_n} u_n \end{pmatrix} = \left(\sum_{i=1}^{n} u_i \frac{\partial}{\partial y_i} \right) \begin{pmatrix} c v_1 \\ v_2 \\ \vdots \\ v_n \end{pmatrix} . \tag{D.79}$$

The first order now appears in the form

$$[X^{(1)}, f] = g^{(1)} + \frac{\partial X^{(2)}}{\partial t} - \left(\sum_{i=1}^{n} \frac{\partial X_i^{(1)}}{\partial t} \frac{\partial}{\partial y_i} \right) X^{(1)} . \tag{D.80}$$

From here, it becomes obvious why the operator formalism has some (practical) advantage. In the expansion used thus far, only half of a Lie bracket appears. Now, we show that the operator formalism leads to more symmetric results.

We introduce the notation

$$\boldsymbol{\phi} = \begin{pmatrix} \phi_1 \\ \phi_2 \\ \vdots \\ \phi_n \end{pmatrix}, \quad \nabla \equiv \begin{pmatrix} \frac{\partial}{\partial y_1} \\ \frac{\partial}{\partial y_2} \\ \vdots \\ \frac{\partial}{\partial y_n} \end{pmatrix}, \quad \boldsymbol{\phi} \cdot \nabla = \sum_{i=1}^{n} \phi_i \frac{\partial}{\partial y_i}, \tag{D.81}$$

and write the transformation as

$$\begin{aligned} x &= e^{\phi \cdot \nabla} y = \left(1 + \boldsymbol{\phi} \cdot \nabla + \frac{1}{2} (\boldsymbol{\phi} \cdot \nabla)(\boldsymbol{\phi} \cdot \nabla) + \cdots \right) y \\ &= y + \boldsymbol{\phi} + \frac{1}{2} (\boldsymbol{\phi} \cdot \nabla) \boldsymbol{\phi} + \cdots . \end{aligned} \tag{D.82}$$

Introducing that into the original differential equation (and defining $d/y dt = \mathbf{g}$), we easily obtain in first order (in $\boldsymbol{\phi}$)

$$g + \frac{\partial \boldsymbol{\phi}}{\partial t} + \frac{1}{2} \left[\boldsymbol{\phi}, \frac{\partial \boldsymbol{\phi}}{\partial t} \right] = f + [\boldsymbol{\phi}, f] . \tag{D.83}$$

Again,

$$g = \mathbf{g}^{(0)} + \epsilon \mathbf{g}^{(1)}, \quad \boldsymbol{\phi} = \boldsymbol{\phi}^{(1)} + \boldsymbol{\phi}^{(2)} \tag{D.84}$$

will be used, and we obtain order by order

$$\mathbf{g}^{(0)} + \frac{\partial \boldsymbol{\phi}^{(1)}}{\partial t} = f , \tag{D.85}$$

$$\mathbf{g}^{(1)} + \frac{\partial \boldsymbol{\phi}^{(2)}}{\partial t} + \frac{1}{2} \left[\boldsymbol{\phi}^{(1)}, \frac{\partial \boldsymbol{\phi}^{(1)}}{\partial t} \right] = \left[\boldsymbol{\phi}^{(1)}, f \right] , \tag{D.86}$$

and so on.

Comparing with the straightforward expansion, we find the relations

$$x(t) = y(t) + \epsilon \boldsymbol{X}^{(1)} + \epsilon^2 \boldsymbol{X}^{(2)} , \tag{D.87}$$

$$x(t) = y(t) + \boldsymbol{\phi}^{(1)} + \boldsymbol{\phi}^{(2)} + \frac{1}{2} \left(\boldsymbol{\phi}^{(1)} \cdot \nabla \right) \boldsymbol{\phi}^{(1)} . \tag{D.88}$$

From here, we conclude

$$\boldsymbol{X}^{(1)} = \boldsymbol{\phi}^{(1)}, \quad \boldsymbol{X}^{(2)} = \boldsymbol{\phi}^{(2)} + \frac{1}{2} (\boldsymbol{\phi}^{(1)} \cdot \nabla) \boldsymbol{\phi}^{(1)} . \tag{D.89}$$

The time derivative of $\boldsymbol{X}^{(2)}$ can now be written in the form

$$\frac{\partial \boldsymbol{X}^{(2)}}{\partial t} = \frac{\partial \boldsymbol{\phi}^{(2)}}{\partial t} + \frac{1}{2} \left(\frac{\partial \boldsymbol{\phi}^{(1)}}{\partial t} \cdot \nabla \right) \boldsymbol{\phi}^{(1)} + \frac{1}{2} (\boldsymbol{\phi}^{(1)} \cdot \nabla) \frac{\partial \boldsymbol{\phi}^{(1)}}{\partial t} . \tag{D.90}$$

Introducing that into

$$[\boldsymbol{X}^{(1)}, f] = g^{(1)} + \frac{\partial \boldsymbol{X}^{(2)}}{\partial t} - \left(\sum_{i=1}^{n} \frac{\partial X_i^{(1)}}{\partial t} \frac{\partial}{\partial y_i} \right) \boldsymbol{X}^{(1)} , \tag{D.91}$$

we immediately find the expected relation

$$\left[\boldsymbol{\phi}^{(1)}, f\right] = g^{(1)} + \frac{\partial \boldsymbol{\phi}^{(2)}}{\partial t} + \frac{1}{2} \left[\boldsymbol{\phi}^{(1)}, \frac{\partial \boldsymbol{\phi}^{(1)}}{\partial t} \right] . \tag{D.92}$$

At this point, we will stop discussing the Lie transforms and the (related) average of equations. As mentioned, the guiding center of motion as well as certain types of nonlinear wave motion (with rapid perturbations) are some of the more important applications. In nonlinear high-frequency wave motion in the presence of rapid perturbations, we start with a perturbed equation (in the present notation for $\boldsymbol{x}$) and transform to an averaged equation (in the present notation for $\boldsymbol{y}$), which – and that is the most interesting point – is to a certain order of magnitude, the integrable cubic nonlinear Schrödinger equation. That topic is known as the guiding center soliton [226].

We conclude with a remark. As has been discussed in detail, the present formalism allows one to treat an equation of the form Eq. (D.30), that is,

$$\frac{dx}{dt} = x + x \cos\left(\frac{t}{\epsilon}\right) = x \left[1 + \cos\left(\frac{t}{\epsilon}\right) \right] , \quad x(t = 0) = x_0 . \tag{D.93}$$

Note that the rapidly varying perturbation can be of the same order as the other terms. We cannot directly apply the method to

$$\frac{dx}{dt} = x + \frac{1}{\epsilon} x \cos\left(\frac{t}{\epsilon}\right), \quad x(t = 0) = x_0 \tag{D.94}$$

when very strong rapid perturbations are present. Here, the exact solution

$$x = x_0 e^t e^{\sin\left(\frac{t}{\epsilon}\right)} \tag{D.95}$$

does not allow a similar expansion as previously. In the case of very strong perturbations, some additional manipulations are necessary which are beyond the scope of the present introduction into Lie transform techniques.

Appendix E
Choices of Low-Dimensional Basis Systems

In this subsection, we address the problem of reduced descriptions of infinite-dimensional systems in conventional ways. *Conventional* means that we proceed with the most straightforward idea to expand any solution in an (appropriate) system of basis functions. The questions are: (i) what are appropriate basis functions, and (ii) how many do we need for a satisfactory description of the problem. We have already touched on this problem when investigating the eigenfunctions within the IST scheme. Here, we proceed in a less specific and less mathematical manner. This question is strongly related to numerical schemes, and we refer to the numerical literature, for example, [432] for further reading on those aspects. Now, we will only discuss some simple physical applications.

E.1
Galerkin Approximation

The "most naive" idea is the following. Let us assume that we have an orthonormal (complete) set of basis functions $\phi_i(x)$, $i = 1, 2, \ldots$ In general, we have an infinite number of basis functions such that we may expand a solution $X(x; t)$ in the form

$$X(x; t) = \sum_{i=1}^{\infty} a_i \phi_i(x) \,. \tag{E.1}$$

Inserting this expansion into the basic equation under investigation, for example, an equation which contains a linear part ($\mathcal{L}$) and a nonlinear part ($\mathcal{N}$) in the form

$$\partial_t X = \mathcal{L} X + \mathcal{N}(X) \,, \tag{E.2}$$

and subsequent multiplication with the orthogonal functions $\phi^{(i)}(x)$, leads to the infinite system of ordinary differential equations for the coefficients

$$\frac{d a_i}{d t} = \sum_{j=1}^{\infty} A_i^j a_j + \sum_{j,l=1}^{\infty} A_i^{jl} a_j a_l + \cdots \,, \quad i = 1, 2, \ldots \tag{E.3}$$

Here, the coefficients are, for example,

$$A_i^j = \langle \phi_i | \mathcal{L} \phi_j \rangle \,. \tag{E.4}$$

High Temperature Plasmas, Theory and Mathematical Tools for Laser and Fusion Plasmas, First Edition.
Karl-Heinz Spatschek.

For simplicity, we assume all variables as real-valued. The higher order terms appear after, for example, a series expansion of $\mathcal{N}(X)$.

Truncation, that is, working with a *finite* basis, means to approximate the solution by

$$X(x;t) \approx \sum_{i=1}^{N} a_i \phi_i(x) , \tag{E.5}$$

for a finite number N, and to work with a reduced system

$$\frac{da_i}{dt} = \sum_{j=1}^{N} A_i^j a_j + \sum_{j,l=1}^{N} A_i^{jl} a_j a_l + \cdots , \quad i = 1, 2, \ldots, N . \tag{E.6}$$

One could call this a (general) Galerkin approximation. The basic question, of course, is which basis functions should be chosen. In the next subsection, we shall present an answer to that question.

E.2 Karhunen–Loève Expansion

As discussed, we want to approximate the solution X by a finite series. The finite sum S_N,

$$X(x;t) \approx \sum_{n=1}^{N} a_n \phi_n \equiv S_N, \tag{E.7}$$

will produce different approximations depending on the chosen basis functions and the truncation depth N. We pose the question as to whether there exists a basis system with the following properties [433–435]:

1. independence of modes,
2. uniform convergence for $N \to \infty$,
3. optimal projection,
4. maximum information for finite N.

In order to quantify the quality of the approximation, we introduce

$$\langle \epsilon_N \rangle := \left\langle \int dx (S_N - X)^2 \right\rangle . \tag{E.8}$$

The averaging $\langle \cdots \rangle$ is introduced as time-averaging. With a time-step Δt and a fixed number M (which could go to infinity), we define

$$\langle f(t) \rangle := \frac{1}{M} \sum_{j=1}^{M} f(j \Delta t) . \tag{E.9}$$

As a first consequence, we can note

$$\langle \epsilon_N \rangle := \left\langle \int dx\,(S_N - X)^2 \right\rangle = \sum_{n=N+1}^{\infty} \langle a_n^2 \rangle \tag{E.10}$$

since we are using an orthonormal basis. For later purposes, we also introduce the correlation function

$$K(x, x') := \langle X(x,t) X(x',t) \rangle \,. \tag{E.11}$$

In general, we have a correlation matrix when the solution X is not a scalar function. When evaluating $\langle \epsilon_N \rangle$, we shall use

$$X(x;t) = \sum_{n=1}^{\infty} a_n(t)\phi_n(x) \approx \sum_{n=1}^{N} a_n(t)\phi_n(x) \equiv S_N(x;t) \tag{E.12}$$

and

$$a_n(t) = \int dx\, X(x;t)\phi_n(x) \,. \tag{E.13}$$

The evaluation

$$\begin{aligned}
\langle \epsilon_N \rangle &\equiv \left\langle \int (S_N - X)^2 dx \right\rangle = \sum_{n=N+1}^{\infty} \langle a_n^2 \rangle \\
&= \sum_{n=N+1}^{\infty} \left\langle \int dx\, X(x)\phi_n(x) \int dx'\, X(x')\phi_n(x') \right\rangle \\
&\equiv \sum_{n=N+1}^{\infty} \int dx\, dx'\, \phi_n(x) K(x,x') \phi_n(x')
\end{aligned} \tag{E.14}$$

shows that the correlation function appears in a natural way.

Next, we determine the minimum of $\langle \epsilon_N \rangle$ under the constraint $\int \phi_n^2 dx = 1$. For that, we must introduce Lagrange multipliers λ_j which take care of the normalization of each basis function. The minimizing basis functions obey the relation

$$\lambda_j \phi_j(x) = \int K(x,x')\phi_j(x')dx' \equiv \int \langle X(x)X(x')\rangle \phi_j(x')dx' \,. \tag{E.15}$$

As a very important conclusion, we obtain the knowledge that the minimizing basis functions are the eigenfunctions of the correlation (matrix) K. These functions are called "empiric eigenfunctions" or "Karhunen–Loève modes." The minimum of $\langle \epsilon_N \rangle$ will be

$$\langle \epsilon_N \rangle = \sum_{n=N+1}^{\infty} \left\langle \int X(x')\phi_n(x')dx' \int X(x)\phi_n(x)dx \right\rangle = \sum_{n=N+1}^{\infty} \lambda_n \,. \tag{E.16}$$

The error $\langle \epsilon_N \rangle$ will be minimal if $\lambda_j, j = N+1, \ldots$ are the smallest eigenvalues with ordering $\lambda_1 \geq \lambda_2 \geq \cdots \geq \lambda_N \geq \lambda_{N+1} \geq \cdots \geq 0$. It is quite obvious that all contributions to $\langle \epsilon_N \rangle$ in the sum over n are positive or zero, and thus the eigenvalues are ≥ 0.

The independence of modes (item (a)) follows from the following consideration:

$$\langle a_n a_{n'} \rangle = \left\langle \int dx\, X(x)\phi_n(x) \int dx'\, X(x')\phi_{n'}(x') \right\rangle$$
$$= \int dx \int dx' \phi_n(x) K(x, x') \phi_{n'}(x') = \lambda_n \delta_{n,n'} \,. \qquad \text{(E.17)}$$

The expansion parameters are statistically independent. The mean values are the eigenvalues λ_n. Now, we will start to prove item (b), that is, uniform convergence. For that, we have to show

$$D \equiv \left\langle (X(x,t) - S_N(x;t))^2 \right\rangle \longrightarrow 0 \quad \text{uniformly in x for} \quad N \to \infty \,. \qquad \text{(E.18)}$$

A short calculation leads to

$$D = \left\langle S_N^2 \right\rangle - 2 \left\langle X S_N \right\rangle + \left\langle X^2 \right\rangle = \sum_{k=1}^{N} \lambda_k \phi_k^2 - 2 \sum_{k=1}^{N} \langle X a_k \rangle \, \phi_k + K(x,x)$$
$$= K(x,x) - \sum_{k=1}^{N} \lambda_k \phi_k^2$$
$$\text{(E.19)}$$

when we make explicit use of the independence of modes. The uniform convergence can now be shown by using Mercer's theorem; see, for example, [436]. The theorem states:

"Let A be an integral operator on $L^2[a,b]$ associated with K, defined by

$$(A\varphi)(x) = \int_a^b K(x,x')\varphi(x')dx' \,, \quad a \leq x \leq b \,, \quad \varphi \in L^2[a,b] \,. \qquad \text{(E.20)}$$

Let $\{\phi_n, n = 1, 2, \ldots\}$ be an orthonormal basis for the space spanned by the eigenvectors corresponding to the nonzero eigenvalues. If the basis is taken so that ϕ_k is an eigenvector corresponding to the eigenvalues $\lambda_k (> 0)$, then

$$K(x,x') = \sum_{k=1}^{\infty} \lambda_k \phi_k(x) \phi_k(x') \,, \quad x, x' \in [a,b] \,, \qquad \text{(E.21)}$$

where the series converges absolutely and also converges uniformly in both variables."

Making use of that, we immediately find

$$D = \sum_{k=1}^{\infty} \lambda_k \phi_k^2(x) - \sum_{k=1}^{N} \lambda_k \phi_k^2(x) \longrightarrow 0 \tag{E.22}$$

uniformly for $x \in [a, b]$.

Now, let us derive a conclusion from the property that $\langle \epsilon_N \rangle$ is minimal when the (empiric) eigenfunctions of the correlation are used as a basis system. The second variation of ϵ_N follows as

$$\begin{aligned} \delta^2 \langle \epsilon_N \rangle = \Bigg\langle \int \Bigg[\sum_{n=1}^{N} \sum_{n'=1}^{N} (\delta a_n \delta a_{n'} \phi_n \phi_{n'} + a_n a_{n'} \delta \phi_n \delta \phi_{n'} \\ +2 a_n \phi_n \delta a_{n'} \delta \phi_{n'} + 2 a_n \phi_{n'} \delta a_{n'} \delta \phi_n) - 2X \sum_{n=1}^{N} \delta a_n \delta \phi_n \Bigg] dx \Bigg\rangle . \end{aligned} \tag{E.23}$$

When starting from the expansion of X, we can derive (to lowest order)

$$\delta a_l \approx - \sum_{k=1}^{N} a_k \int \phi_l \delta \phi_k dx . \tag{E.24}$$

Inserting that into Eq. (E.23), we can evaluate

$$\begin{aligned} &\left\langle \sum_{n=1}^{N} \sum_{n'=1}^{N} a_n \phi_n \delta a_{n'} \delta \phi_{n'} - X \sum_{n'=1}^{N} \delta a_{n'} \delta \phi_{n'} \right\rangle \\ &= \sum_{n=N+1}^{\infty} \phi_n \sum_{n'=1}^{N} \delta \phi_{n'} \sum_{k=1}^{N} \langle a_n a_k \rangle \int \phi_{n'}(x') \delta \phi_k(x') dx' = 0 . \end{aligned} \tag{E.25}$$

A short calculation leads to

$$\begin{aligned} \delta^2 \langle \epsilon_N \rangle &= \sum_{n=1}^{N} \left[\lambda_n \int (\delta \phi_n)^2 dx - \langle (\delta a_n)^2 \rangle \right] \\ &= \sum_{n=1}^{N} \lambda_n \left[\int (\delta \phi_n)^2 dx - \sum_{m=1}^{N} \left(\int \phi_m \delta \phi_n dx \right)^2 \right] \\ &= \sum_{n=1}^{N} \lambda_n \sum_{m=N+1}^{\infty} \left(\int \phi_m \delta \phi_n dx \right)^2 \geq 0 . \end{aligned} \tag{E.26}$$

As it should be, the second variation is positive at a minimum. From here, the optimal projection (item (c)) on the (empiric) eigenfunctions follows.

Reformulating the result, we can arrive at another conclusion for the information entropy. Let us compare $\langle \epsilon_N \rangle$ for the optimal set of basis functions with the corresponding expression for an arbitrary set of basis functions by calculating

$$
\begin{aligned}
&\left\langle \int [S_N - X]^2\, dx \right\rangle - \left\langle \int [T_N - X]^2\, dx \right\rangle \\
&= \sum_{k=1}^{N} \langle a_k^2 \rangle \int \phi_k^2 dx - 2\left\langle \sum_{k=1}^{N} \int X(x)\phi_k(x)dx\, a_k \right\rangle \\
&- \sum_{k=1}^{N} \langle b_k^2 \rangle \int \psi_k^2 dx + 2\left\langle \sum_{k=1}^{N} \int X(x)\psi_k(x)dx\, b_k \right\rangle \\
&= \left[-\sum_{k=1}^{N} p_k + \sum_{k=1}^{N} q_k \right] \int X^2 dx \equiv \left\langle \epsilon_N^{\phi} \right\rangle - \left\langle \epsilon_N^{\psi} \right\rangle \leq 0\,.
\end{aligned}
\tag{E.27}
$$

We have defined

$$
p_k \equiv \frac{\lambda_k}{\langle \int X^2 dx \rangle} \quad \text{and} \quad q_k = \frac{\langle \int (X\psi_k)^2 dx \rangle}{\langle \int X^2 dx \rangle}\,, \tag{E.28}
$$

such that the relation

$$
\sum_{k=1}^{M} p_k \geq \sum_{k=1}^{M} q_k \quad \text{for all} \quad M \tag{E.29}
$$

follows, that is, provided q_k is calculated with respect to another orthonormal basis $\{\psi_k\}$ with the abbreviation $T_N \equiv \sum_{k=1}^{N} b_k(t)\psi_k(x)$.

The last results allow one to draw some conclusion with respect to the information entropy (item (4)). We shall show that with the Karhunen–Loève basis, we receive the maximum information. In other words, when

$$
\sum_{k=1}^{M} p_k \geq \sum_{k=1}^{M} q_k \quad \text{for all} \quad M \left[\sum_{k=1}^{\infty} p_k = \sum_{k=1}^{\infty} q_k = 1 \right], \tag{E.30}
$$

then

$$
S_p \equiv -\sum_{k=1}^{\infty} p_k \ln p_k \leq -\sum_{k=1}^{\infty} q_k \ln q_k \equiv S_q \tag{E.31}
$$

follows, meaning that the representational entropy is minimal if the basis set consists of empirical eigenfunctions. (The empirical eigenfunctions ideally compress the information content of the flow.)

To prove the transition from Eqs. (E.30) to (E.31), let us introduce

$$
\Theta_m \equiv \sum_{k=1}^{M} q_k\,, \quad q_1 \geq q_2 \geq \cdots \geq q_M \geq q_{M+1} \geq \cdots \geq 0\,. \tag{E.32}
$$

Then, we can write

$$
S_q = -\sum_{k=1}^{\infty} (\Theta_k - \Theta_{k-1}) \ln (\Theta_k - \Theta_{k-1})\,. \tag{E.33}
$$

Taking the first derivatives,

$$\frac{\partial S_q}{\partial \Theta_j} = \ln \frac{\Theta_{j+1} - \Theta_j}{\Theta_j - \Theta_{j-1}} \equiv \ln \frac{q_{j+1}}{q_j} \leq 0\,, \tag{E.34}$$

follows. That means that the larger Θ_j is, the smaller S_q is. However, the Θ_j, for all j, will be largest for the Karhunen–Loève-expansion. Therefore,

$$S_q \geq S_p \tag{E.35}$$

follows.

E.3 Determination of the Basis Functions in Practice

The basis functions ϕ_j follow, in principle, from the eigenvalue equation for the correlation matrix. In practice, especially when a high resolution in space is required (or if we investigate multidimensional problems), a reformulation of the problem is advisable.

The alternative determination of ϕ_j and λ_j starts from the decomposition of the ϕ_j in terms of the different times t,

$$\phi_k(x) = \sum_{j=1}^{M} c_j^{(k)} X(x, j\Delta t)\,. \tag{E.36}$$

The coefficients $c_j^{(k)}$ and the eigenvalues λ_j can be determined from a different eigenvalue problem, namely,

$$\lambda_j c_k^{(j)} = \sum_l H_{kl} c_l^{(j)}\,, \tag{E.37}$$

where

$$H_{kl} = \frac{1}{M} \int dx\, X(x, k\Delta t) X(x, l\Delta t)\,. \tag{E.38}$$

To prove this assertion, let us multiply Eq. (E.37) by $X(x, k\Delta t)$ and sum over k,

$$\lambda_j \sum_k X(x, k\Delta t) c_k^{(j)} = \sum_k X(x, k\Delta t) \times \sum_l \frac{1}{M} \int dx' X(x', k\Delta t) X(x', l\Delta t) c_l^{(j)}\,. \tag{E.39}$$

On the left-hand side, we obtain $\lambda_j \phi_j$ when we use the definition Eq. (E.36). On the right-hand side, we rearrange and thus

$$\int dx' \langle X(x,t) X(x',t) \rangle \sum_l X(x', l\Delta t) c_l^{(j)}\,. \tag{E.40}$$

Therefore, we obtain from Eq. (E.39)

$$\lambda_j \phi_j(x) = \int dx' K(x, x') \phi_j(x') , \tag{E.41}$$

that is, exactly the eigenvalue problem Eq. (E.15).

Numerical advantages (less storage requirements) lead to a preference for Eq. (E.37) compared to Eq. (E.15).

We conclude this section by a remark which follows the introductory remark on the Galerkin procedure. Even if we determine an optimal basis in the sense of the Karhunen–Loève model, we truncate the system after a finite number N without taking into account (at least in an approximate manner) the influence of the rest of the modes. That aspect of optimal prediction has been discussed by Chorin *et al.* [437] (see also the references therein).

Appendix F
Center Manifold Theory

The procedures discussed in the following should mainly be applied when linearization will not lead to an unambiguous answer as to the dynamical behavior of a dynamical system near a marginal point. Then, only a few modes will determine the dynamics. In physics, this is called the slaving principle [381]. The mathematical foundations are given by the center manifold theory; see, for example, [438–440]. As a result of normal forms determined by center manifold theory, one can find out whether nonlinearities will lead to stabilization or unlimited growth in the neighborhood of a marginal point.

The complex dynamics of a nonlinear system in the neighborhood of an equilibrium will be projected into a low-dimensional phase space. One starts with the eigenvalue spectrum λ of the linearized system. In the linear evolution $\sim \exp(\lambda t)$, eigenvalues with $\mathrm{Re}\,\lambda < 0$ ($\mathrm{Re}\,\lambda > 0$) are called stable (unstable), whereas those with $\mathrm{Re}\,\lambda = 0$ are called central.

In the neighborhood of an equilibrium point P of a dynamical system, generally three different types of invariant manifolds exist. The trajectories belonging to the stable manifold M^s are being attracted by P, whereas those of the unstable manifold M^u are being repelled. The dynamics on the center manifold M^c depend on the nonlinearities. For the linearized problem, $E^s \equiv M^s$, $E^u \equiv M^u$, and $E^c \equiv M^c$ are uniquely determined linear subspaces which span the whole space. The transition to the nonlinear system causes (only) deformations of the linearly determined manifolds E^s, E^u, and E^c to M^s, M^u, and M^c. However, the form of the latter crucially depends on the nonlinear terms.

Let us first elucidate the behavior for a very simple system, namely, [441]

$$\dot{x} = -xy\,, \quad \dot{y} = -y + x^2\,. \tag{F.1}$$

Here, the dot means differentiation with respect to time t. The equilibrium point P is $(0, 0)$, and the linearized system turns out to be

$$\dot{x} = 0\,, \quad \dot{y} = -y\,. \tag{F.2}$$

The stable manifold E^s is identical with the y-axis and the center manifold E^c is identical with the x-axis. The linearized problem can be visualized by the graph shown in Figure F.1. Both manifolds will be deformed in the transition to the nonlinear system. The (nonlinear, perturbed) center manifold can be described in the

High Temperature Plasmas, Theory and Mathematical Tools for Laser and Fusion Plasmas, First Edition.
Karl-Heinz Spatschek.

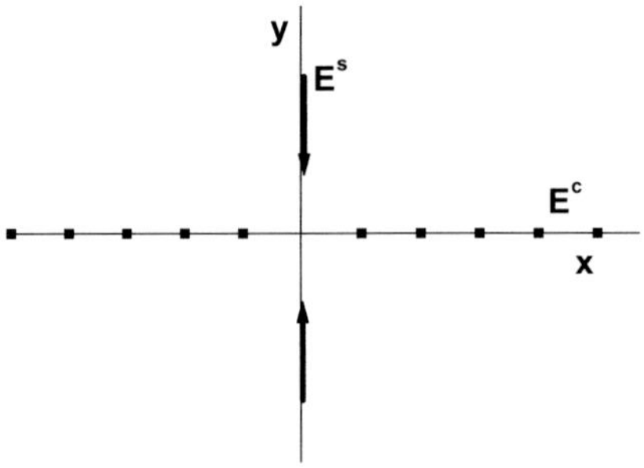

Figure F.1 The linear stable (E^s) and center manifold (E^c) for Eq. (F.1).

present example by

$$M^c : y = h(x) \tag{F.3}$$

with $h(0) = h'(0) = 0$. Using that ansatz for the center manifold in Eq. (F.1), we obtain

$$\dot{x} = -x\ h(x)\ . \tag{F.4}$$

Differentiating Eq. (F.3) with respect to t leads to

$$\dot{y} = h'(x)\dot{x} \tag{F.5}$$

or

$$-h(x) + x^2 = h'(x)\left[-x h(x)\right]\ , \tag{F.6}$$

that is, a differential equation for $h = h(x)$. Making a power series ansatz $h(x) = cx^2 + dx^3 + \cdots$, we find $c = 1$. The (nonlinear) center manifold is thus given by

$$y = x^2 + \cdots\ , \tag{F.7}$$

and the dynamics on it follow from

$$\dot{x} = -x^3 + \cdots\ , \tag{F.8}$$

that is, the trajectories are being attracted by P. For an illustration, see Figure F.2. For the present (simple) example, it is possible to directly compare the exact solution with the center manifold reduction. First, let us formally solve the equation for y by the variation of parameters method. The result is

$$y = y_0 e^{-t} + e^{-t} \int_0^t e^s \left[x(s)\right]^2 ds\ . \tag{F.9}$$

Here, y_0 is the initial value $y_0 = y(t = 0)$. Obviously, at larger times, the initial value will be "forgotten," and in the asymptotic regime, we may only use the second

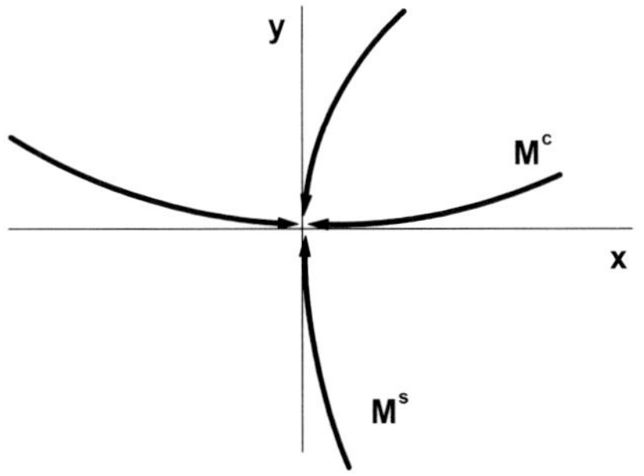

Figure F.2 The stable (M^s) and center manifold (M^c) for Eq. (F.1).

term on the right-hand side. Having in mind the smallness of x, we may expand (by integration by parts)

$$e^{-t}\int_0^t e^s x^2 ds = e^{-t}\left[e^s x^2\right]_{s=0}^{s=t} - e^{-t}\int_0^t ds e^s \frac{d}{ds}x^2$$
$$\approx \left[x(t)\right]^2 + \text{“higher order terms”} . \quad \text{(F.10)}$$

Thus, with

$$y \approx x^2,$$

we directly obtain the center manifold equation

$$\dot{x} = -x^3 + \text{“higher order terms”} . \quad \text{(F.11)}$$

From this discussion, two facts become obvious. First, the center manifold reduction works in the asymptotic limit of large t, and secondly, the influence of the initial value(s) of the unresolved variables(s) disappears with increasing time. In Figure F.3, we show a comparison of the exact solution with the asymptotic center manifold reduction. We clearly recognize the difference between the initial (strongly damped) phase and the typical dependence in the region of applicability of the center manifold reduction. The forms of the (nonlinear) stable manifold are depicted in Figure F.4. We clearly recognize the asymptotic form $y \approx x^2$.

Let us now generalize the idea and consider a system of ordinary differential equations (ODEs)

$$\dot{a} = Aa + N(a, b) , \quad \text{(F.12)}$$

$$\dot{b} = Bb + M(a, b) , \quad \text{(F.13)}$$

describing the dynamics of amplitudes $a_1, \ldots, a_n$ and $b_1, \ldots, b_m$ of n linear *marginal* stable modes and m linear *stable* modes, respectively $[(a_1, \ldots, a_n) \mathrel{\hat{=}} a, (b_1, \ldots, b_m) \mathrel{\hat{=}} b]$. This implies that the real parts of eigenvalues of the matrix A

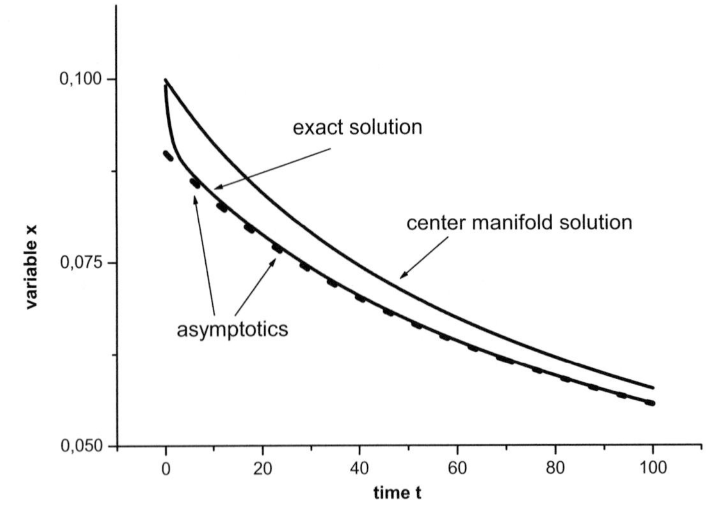

Figure F.3 The exact solution of Eq. (F.1) is compared with the center manifold solution of Eq. (F.8). The dotted line shows the asymptotics as follows from Eq. (F.8) with a modified initial value.

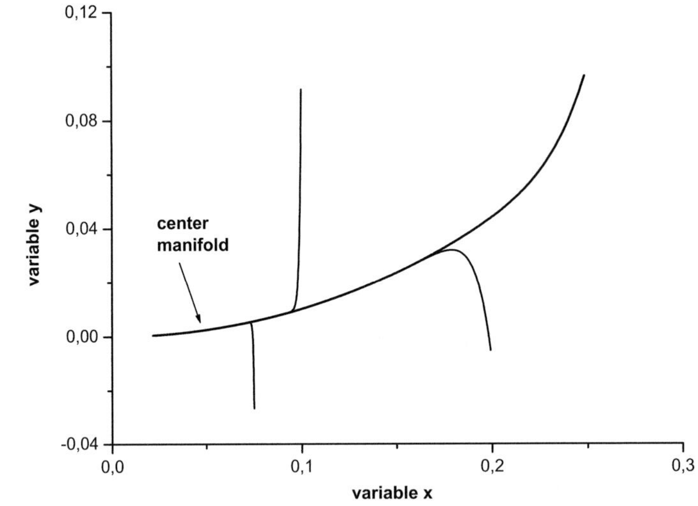

Figure F.4 Solution of the simple model Eq. (F.1) for four different initial conditions. For small x and y, $y \approx x^2$ always holds.

vanish and the real parts of the eigenvalues of the matrix B are negative. The functions $N(a, b), M(a, b) \in C^r$ on the right-hand sides of Eqs. (F.12) and (F.13), respectively, represent the nonlinear terms. Let E^c be the n-dimensional (generalized) eigenspace of A and E^s the m-dimensional (generalized) eigenspace of B. Under these assumptions, the center manifold theorem provides the following statement [439]

"There exists an invariant C^r manifold M^s and an invariant C^{r-1} manifold M^c which are tangent at $(a, b) = (0, 0)$ to the eigenspaces E^s and E^c, respectively. The stable manifold M^s is unique, but the center manifold M^c is not necessarily unique."

Locally, the center manifold M^c can be represented as a graph,

$$M^c = \{(a, b)|b = h(a)\}\,, \quad h(0) = 0\,, \quad Dh(0) = 0\,, \tag{F.14}$$

where the C^{r-1} function h is defined in a neighborhood of the origin; Dh denotes the Jacobi matrix. Introducing Eq. (F.14) into Eqs. (F.12) and (F.13), we obtain

$$\dot{a} = Aa + N(a, h(a))\,, \tag{F.15}$$

$$Dh(a)\left\{Aa + N\left[a, h(a)\right]\right\} = Bh(a) + M[a, h(a)]\,. \tag{F.16}$$

The solution h of Eq. (F.16) can be approximated by a power series. The ambiguity of the center manifold is manifested by the fact that h is determined only modulo a C^∞ nonanalytic function; so the power series approximation of the function h is unique. The importance of the center manifold theory is reflected by the following theorem [442, 443]

"If there exists a neighborhood U^c of $(a, b) = (0, 0)$ on M^c, so that every trajectory starting in U^c never leaves it, then there exists a neighborhood U of $(a, b) = (0, 0)$ in $\mathcal{R}^n \times \mathcal{R}^m$, and thus every trajectory starting in U converges to a trajectory on the center manifold."

Therefore, close to the onset of instability, it is sufficient to discuss the dynamics on the center manifold described by Eq. (F.15). If all solutions are bounded to some neighborhood of the origin, then we have described all features of the asymptotic behavior of Eqs. (F.12) and (F.13). In order to fulfill the condition, the function $N[a, h(a)]$ has to be expanded up to a sufficiently high order. We end up with normal forms, for example, the third order may be adequate.

Very often, the problems contain parameters and, in addition, the systems may be infinite-dimensional. In both cases, one can generalize the theory presented so far. Parameters can be taken into account by enlarging Eqs. (F.12) and (F.13) to

$$\dot{a} = A(\Lambda)a + N(a, b, \Lambda)\,, \tag{F.17}$$

$$\dot{b} = B(\Lambda)b + M(a, b, \Lambda)\,, \tag{F.18}$$

$$\dot{\Lambda} = 0\,, \tag{F.19}$$

where $\Lambda = (a_{n+1}, \ldots, a_{n+l})$ contains l parameters. The center manifold now has the dimension $n + l$.

The theory is also valid in the infinite-dimensional case, that is, if the spectrum of the linear operator can be split into two parts. The first part contains a finite number of eigenvalues whose real parts are zero, the second part contains (an infi-

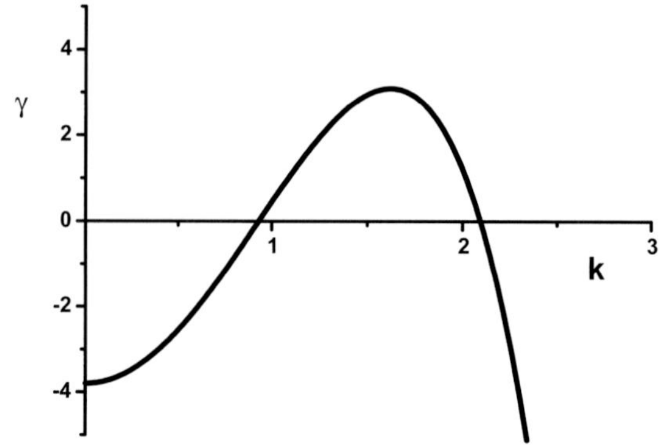

Figure F.5 Growth rate curve, with two unstable modes at $k = 1$ and $k = 2$ for the PDE Eq. (F.20).

nite number of) eigenvalues with negative real parts which are bounded away from zero.

To elucidate the power of center manifold reduction, let us consider a generalization of the famous KdV equation [444]. In 1D, we add second and fourth order derivative terms as well as a dissipative part (ν). The model has applications in wave physics. Here, we concentrate on the mathematical analysis when starting from the partial differential equation (PDE)

$$\frac{\partial \phi}{\partial t} + \phi \frac{\partial \phi}{\partial y} + \alpha \frac{\partial^2 \phi}{\partial y^2} + \beta \frac{\partial^3 \phi}{\partial y^3} + \frac{\partial^4 \phi}{\partial y^4} + \nu \phi = 0 . \tag{F.20}$$

All coefficients α, β, and ν are assumed to be nonnegative. In the following, we treat β as a fixed parameter and consider the dynamics in dependence on α and ν. The linearization with respect to $\phi \equiv 0$ as the equilibrium solution leads for perturbations $\sim \exp\{-i\omega t + iky\}$ to

$$\omega = -\beta k^3 + i\left(-k^4 + \alpha k^2 - \nu\right), \tag{F.21}$$

when we assume a unit cell of length 2π with periodic boundary conditions. A typical dependence of the linear growth (or damping) rate $\gamma := \text{Im}\,\omega$ is shown in Figure F.5 for $\alpha = 5.25$ and $\nu = 3.8$.

The case of two unstable modes ($k = 1, k = 2$) is already highly nontrivial. Let us choose $\alpha = \alpha_c = 5$ and $\nu = \nu_c = 4$. Then, the modes $\phi^{(1)} = \sin y, \phi^{(2)} = \cos y, \phi^{(3)} = \sin 2y$, and $\phi^{(4)} = \cos 2y$ belonging to $k = 1$ and $k = 2$, respectively, are marginally stable. We introduce the four (real) amplitudes a_1, a_2, a_3, and a_4 as well as $a_5 = \alpha - \alpha_c$ and $a_6 = \nu - \nu_c$. Center manifold theory will allow one to derive a closed set of nonlinear amplitude equations

$$\dot{a}_n = f_n(a_1, \ldots, a_6) , \quad n = 1, \ldots, 6 , \tag{F.22}$$

which are valid in the neighborhood of the critical point α_c, ν_c. One has $f_5 \equiv f_6 \equiv 0$. The other functions f_n are written as a power series in a_n,

$$f_n = \sum_{1 \leq m \leq 6} A_n^m a_m + \sum_{1 \leq m \leq p \leq 6} A_n^{mp} a_m a_p + \cdots \tag{F.23}$$

The dynamics on the center manifold are characterized by $a_1, \ldots, a_6$. Thus, we can make the ansatz

$$\phi(y,t) = \sum_{1 \leq n \leq 4} a_n(t)\phi^{(n)}(y) + \sum_{1 \leq n \leq m \leq 6} a_n(t) a_m(t) \phi^{(nm)}(y) , \tag{F.24}$$

where the $\binom{7}{2} = 21$ new functions $\phi^{(nm)}$ and, of course, the next $\binom{8}{3} = 56$ functions $\phi^{(nmp)}$, and so on, can be chosen orthogonal to $\phi^{(n)}$, $n = 1, \ldots, 4$.

The technical procedure is now the following. One inserts Eq. (F.24) into the basic Eq. (F.20) and compares equal orders in the amplitudes. For example, in the second order one collects equal powers $a_r a_s$; the "coefficients" (being equated to zero) will determine the unknown functions $\phi^{(nm)}$ via ODEs. Taking into account the (periodic) boundary conditions, we have to satisfy the solvability conditions. Collecting equal powers of the amplitudes a_n, we find the solutions for the coefficients A_r^{np}. With these values, we can solve for $\phi^{(mn)}$. This procedure should be continued to higher orders.

Actually, when written explicitly, one faces a huge amount of work (in second order, we have to solve for 84, in third order, for 224 coefficients $A_r^{\cdot\cdot}$, and so on). One can simplify the calculations by making use of symmetries.

Translational invariance implies the following. If $\phi(y)$ is a solution, then also

$$T_{y_0}\phi(y) := \phi(y + y_0) \tag{F.25}$$

will satisfy the dynamical Eq. (F.20), where y_0 is a real shift parameter. (In the case $\beta \equiv 0$, we also have the mirror symmetry $\phi(y) = \phi(-y)$.)

Remember the structure of the center manifold reduction: The modes $\phi^{(nm)}$, $\phi^{(nmp)}, \ldots$ have to be determined from inhomogeneous differential equations. The inhomogeneities contain (in nonlinear forms) the marginal modes $\phi^{(r)}$, $r = 1, \ldots, 4$. The so called slaved modes can be written in symbolic forms as

$$\phi^{(m\ldots)} = h^{(m\ldots)}\left[\{\phi^{(r)}\}\right]. \tag{F.26}$$

Thus, the following symmetry should hold, that is,

$$T_{y_0} h^{(m\ldots)}\left[\{\phi^{(r)}\}\right] = h^{(m\ldots)}\left[T_{y_0}\{\phi^{(r)}\}\right] . \tag{F.27}$$

The consequence of the translational symmetry is easiest seen when combing the marginal modes

$$\varphi := \sum_{r=1}^{4} a_r \phi^{(r)} \equiv \operatorname{Im}\left(c_1 e^{iy} + c_2 e^{2iy}\right) \tag{F.28}$$

with the complex amplitudes $c_1 := a_1 + i a_2$, $c_2 := a_3 + i a_4$.

The (complex) amplitude equations are

$$\dot{c}_n = g_n\,(c_1, c_2, a_5, a_6)\ , \qquad n = 1, 2\ , \tag{F.29}$$

$$\dot{a}_m = 0\ , \qquad m = 5, 6\ . \tag{F.30}$$

The translational symmetry Eq. (F.25) requires

$$e^{iny_0} g_n\,(c_1, c_2, a_5, a_6) = g_n\left(e^{iy_0} c_1, e^{i2y_0} c_2, a_5, a_6\right) \tag{F.31}$$

for $n = 1, 2$. The vector field $(\mathbf{g}_1, \mathbf{g}_2)$ is called equivariant with respect to the operation

$$(c_1, c_2) \to \left(e^{iy_0} c_1, e^{i2y_0} c_2\right). \tag{F.32}$$

The most general form of vector fields being equivariant under the operation Eq. (F.31) is

$$(g_1, g_2) = \left(c_1 P_1 + \bar{c}_1 c_2 Q_1, c_2 P_2 + c_1^2 Q_2\right), \tag{F.33}$$

where P_1, P_2, Q_1, and Q_2 are polynomials in $|c_1|^2, |c_2|^2$, and $\mathrm{Re}(c_1^2 \bar{c}_2)$; of course, they can also depend on a_5 and a_6.

Having in mind the symmetry properties, the general form of the amplitude equations reduces to

$$\dot{c}_1 = \lambda c_1 + \mathcal{A}\bar{c}_1 c_2 + \mathcal{C} c_1|c_1|^2 + \mathcal{E} c_1|c_2|^2 + \mathcal{O}(|c|^4)\ , \tag{F.34}$$

$$\dot{c}_2 = \mu c_2 + \mathcal{B} c_1^2 + \mathcal{D} c_2|c_1|^2 + \mathcal{F} c_2|c_2|^2 + \mathcal{O}(|c|^4)\ , \tag{F.35}$$

$$\dot{a}_5 = \dot{a}_6 = 0\ . \tag{F.36}$$

A straightforward analysis leads to

$$\begin{aligned} \lambda &= a_5 - a_6 - i\beta\ , & \mu &= 4a_5 - a_6 + i8\beta\ , \\ \mathcal{A} &= \frac{1}{2}\ , & \mathcal{B} &= -\frac{1}{2}\ , \\ \mathcal{C} &= 0\ , & \mathcal{D} &= -\frac{3}{4(20 - i9\beta)}\ , \\ \mathcal{E} &= \frac{1}{2}\mathcal{D}\ , & \mathcal{F} &= -\frac{1}{12(15 - i4\beta)}\ . \end{aligned} \tag{F.37}$$

This completes the center manifold reduction.

In the present case, from the systematic treatment with center manifold theory, very interesting conclusions result, for example, with respect to the number of modes and their interplay in time. For example, one interesting aspect is that the present codimension-two analysis can describe successive bifurcations of one unstable mode, which in some cases can lead to chaos in time.

The advantage of center manifold reduction – compared to direct numerical simulations – is the possibility of a clear and systematic classification of the complex dynamics in the neighborhood of (a) marginal point(s). Of course, once a clarification of the principal dynamical behavior has been reached, it is easier to compare with the flood of data created by numerical simulations. It is also clear that the center manifold predictions are more exact the closer we are to the marginal point. Applications of center manifold theory exist in many branches of plasma physics; see, for example, [445].

Appendix G
Newell–Whitehead Procedure

The center manifold theory is a rigorous mathematical procedure which perturbatively determines the dynamics close to the onset of an instability. Originally, it was developed for ordinary differential equations, but as mentioned, under certain conditions, it can be also applied to partial differential equations. One condition for the applicability to partial differential equations is that zero is not a cumulation point in the spectrum of the linear operator. If, on the other hand, the spectrum is continuous, that assumption is not fulfilled, and one has to look for other procedures. The Newell–Whitehead procedure [446], an older method which has not the same mathematical rigor as the center manifold theory, can be then applied. The Newell–Whitehead procedure has shown its power in several weakly dispersive, weakly nonlinear systems, with many applications in different physical situations [447].

In the following, we shall present a general outline [226]. We shall not relate to a specific form of the basic dynamical (differential) equation. The main idea is that, to the lowest order, the marginal modes dominate the dynamical behavior. Thus we present the solution of the equation as a linear combination of the marginal modes. Corrections are taken into account via (weak) space- and time-dependencies of the coefficients and, in the higher order, also via the generation of harmonics. Formally, one introduces a scaling parameter $\epsilon \ll 1$.

Let us start from the general form

$$Lq = N(q, q) \tag{G.1}$$

for a nonlinear differential equation. Here,

$$L = L\left(\partial_t, \partial_y, \partial_z, R; \partial_x\right) \tag{G.2}$$

is a linear differential operator which contains time- (t) and space-derivatives (y, z, x) as well as a bifurcation parameter R. The latter has the value R_c at the critical point. We do not treat all space coordinates on the same footing. The system may be infinite in y- and z-directions whereas in x, we assume finite boundary conditions. $N(q, q)$ designates the (quadratic) nonlinearity; for simplicity, we assume a

High Temperature Plasmas, Theory and Mathematical Tools for Laser and Fusion Plasmas, First Edition.
Karl-Heinz Spatschek.

quadratic nonlinearity in the solution (vector) $\boldsymbol{q}$. We apply the ordering

$$\boldsymbol{q} \sim \mathcal{O}(\epsilon)\,, \tag{G.3}$$

$$R = R_c(1 + \epsilon^2 \chi)\,, \tag{G.4}$$

and make the ansatz

$$\boldsymbol{q} = A\left(\epsilon t_1, \epsilon \boldsymbol{r}_1, \epsilon^2 t_2, \ldots\right) Q(x, \boldsymbol{r}, t)\,. \tag{G.5}$$

The assumption Eq. (G.3) reflects the fact that $\boldsymbol{q}$ should be a weakly nonlinear solution (vector). In $\boldsymbol{r}$, we combine the space coordinates y and z. Q represents the marginal solution at the critical point,

$$LQ = 0\,. \tag{G.6}$$

When the linear operator consists of time- and space-derivatives with constant coefficients, the marginal modes have, for example, the simple structure

$$Q(x, \boldsymbol{r}, t) = Q_0 \sin(k_x x) e^{i\boldsymbol{k}\cdot\boldsymbol{r} - i\omega t}\,, \tag{G.7}$$

plus the complex conjugate form. Characteristic for the Newell–Whitehead procedure is the multiple-scale analysis with the variables

$$t_0 = t\,, \quad t_1 = \epsilon t\,; \quad t_2 = \epsilon^2 t\,, \ldots\,, \tag{G.8}$$

$$\boldsymbol{r}_0 = \boldsymbol{r}\,, \quad \boldsymbol{r}_1 = \epsilon \boldsymbol{r}\,, \ldots \tag{G.9}$$

which are introduced to denote slow and fast variations with respect to time and space.

By $\partial_1 L$, we denote the variation of L with respect to $\partial_t \equiv \partial/\partial t$, that is, $\partial_1 L = \delta L/\delta \partial_t$, and use a similar notation with respect to the space variations when introducing by $\partial_2 L$ and $\partial_3 L$ the variations of L with respect to the space derivatives ∂_y and ∂_z, respectively. Then, we combine, that is,

$$\nabla_{23} L \equiv (\partial_2 L, \partial_3 L) \equiv \left(\frac{\delta L}{\delta \partial_y}, \frac{\delta L}{\delta \partial_z}\right). \tag{G.10}$$

Correspondingly, we introduce

$$\nabla \equiv (\partial_{y_1}, \partial_{z_1}) \tag{G.11}$$

and

$$\partial_4 L = \left.\frac{\partial L}{\partial R}\right|_{R_c}\,. \tag{G.12}$$

Just to present a (trivial) example, let us demonstrate the notations on

$$L = \partial_t + \frac{\partial^2}{\partial y^2}\,. \tag{G.13}$$

Then, we get

$$\partial_1 L = 1 \quad \text{and} \quad \partial_2 L = 2\partial_y \,. \tag{G.14}$$

In general, the operator L can be expanded up to the various orders in ϵ, for example,

$$L = L_0 + \epsilon \left[(\partial_1 L)\, \partial_{t_1} + (\nabla_{23} L) \cdot \nabla \right] + \epsilon^2 \left[\frac{1}{2} (\partial_1 \partial_1 L)\, \partial^2_{t_1} + (\partial_1 \nabla_{23} L) \cdot \partial_{t_1} \nabla \right. \\ \left. + \frac{1}{2} (\nabla_{23} \nabla_{23} L) : \nabla\nabla + (\partial_1 L)\, \partial_{t_2} + \chi R_c \partial_4 L \right] . \tag{G.15}$$

Obviously, the lowest order (everywhere in L the time-derivatives are written as time-derivatives with respect to t_0, and similar for the space derivatives with respect to y_0 and z_0) differential equation is satisfied by the ansatz Eq. (G.7),

$$L_0 A Q = 0 \,. \tag{G.16}$$

Now, we can exemplify in even more detail what we mean by marginal modes and the critical (bifurcation) point. The linear dispersion can be written in the form

$$\omega = \omega(\boldsymbol{k}, R) \,. \tag{G.17}$$

Note that $\boldsymbol{k} = (k_y, k_z)$. We have used the parameter k_x in Eq. (G.7); the notation is just for convenience. Marginal modes follow from $\text{Im}\,\omega \equiv \omega_i = 0$. The condition

$$\omega_i(\boldsymbol{k}, R) = 0 \Leftrightarrow R = R(\boldsymbol{k}) \tag{G.18}$$

defines the surface of marginal modes in the three-dimensional parameter space. For bifurcation analysis, we are interested in the behavior near R_c, the so called critical point. The latter is defined on the R-axis as the lowest R value where instability sets in; see Figure G.1. At the critical point,

$$\partial_{\boldsymbol{k}} R = 0 \quad \text{for} \quad R = R_c \,. \tag{G.19}$$

Here, ∂_{k_y}, ∂_{k_z} denote partial derivatives with respect to k_y and k_z, respectively. On the other hand, we can define "derivatives on the surface $R = R(\boldsymbol{k})$". When a function depends on k_y, k_z, and R, we define that derivative as

$$\nabla_{\boldsymbol{k}} = \partial_{\boldsymbol{k}} + (\partial_{\boldsymbol{k}} R)\, \partial_R \,. \tag{G.20}$$

Operating with $\nabla_{\boldsymbol{k}}$ on $LQ = 0$ yields

$$0 = \nabla_{\boldsymbol{k}}(LQ) = L\nabla_{\boldsymbol{k}} Q + [-i\, (\nabla_{\boldsymbol{k}} \omega)\, \partial_1 L + i\nabla_{23} L + (\partial_{\boldsymbol{k}} R)\, \partial_4 L]\, Q \,. \tag{G.21}$$

Using (G.20), we can write

$$-L\nabla_{\boldsymbol{k}} Q = -\left(-\nabla_{\boldsymbol{k}} \omega \partial_1 + \nabla_{23}\right) LQ \,. \tag{G.22}$$

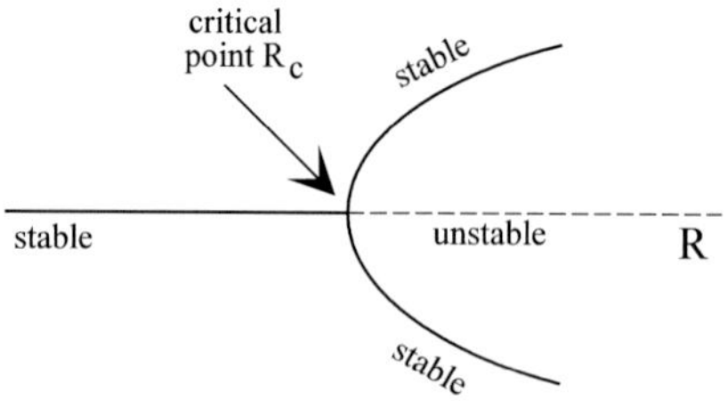

Figure G.1 A typical example of a bifurcation which can be described by the Newell–Whitehead procedure [446].

That relation will be used later. However, one remark is appropriately already here. The group velocity is

$$u = \nabla_k \omega \ . \tag{G.23}$$

That relation can be calculated directly from the dispersion relation; it also follows from the solvability condition of Eq. (G.22). Multiplying both sides of the latter equation by Q^+ from the left and integrating, we assume that Q^+ is the kernel function of the adjoint operator L^+. Then,

$$\langle Q|L\cdots\rangle \equiv \int d\Gamma Q^+ L\cdots = \int d\Gamma (L^+ Q^+)\cdots = 0\,, \tag{G.24}$$

and $\boldsymbol{u}$ may be also written as

$$\boldsymbol{u} = \frac{\langle Q|\nabla_{23} L Q\rangle}{\langle Q|\partial_1 L Q\rangle}\ . \tag{G.25}$$

Let us now consider the original differential equation (G.1) in the order ϵ^2 when the solution is written as

$$q = \epsilon q_1 + \epsilon^2 q_2 + \cdots\,, \quad q_1 = AQ\,. \tag{G.26}$$

We obtain an inhomogeneous differential equation for q_2:

$$L_0 q_2 + \left[(\partial_1 L)\,\partial_{t_1} + (\nabla_{23} L)\cdot\nabla\right] L q_1 = N\left(q_1 + q_1^*, q_1 + q_1^*\right)\ . \tag{G.27}$$

At this stage, an expansion in terms of harmonics is actually required. We abbreviate these steps a little bit when calculating the (dominating) zeroth-harmonic contribution to q_2 directly. Then,

$$N\left(q_1 + q_1^*, q_1 + q_1^*\right) = N\left(q_1, q_1^*\right) + c.c. \tag{G.28}$$

is the relevant inhomogeneity contribution for the zeroth harmonic. When solving the linear inhomogeneous Eq. (G.27) for q_2, we can make the ansatz

$$q_2 = q_{2,1} + q_{2,0}\,, \tag{G.29}$$

leading to

$$L_0 q_{2,1} + \left[(\partial_1 L)\,\partial_{t_1} + (\nabla_{23} L)\cdot\nabla\right] L q_1 = 0\,, \tag{G.30}$$

$$L_0 q_{2,0} = |A|^2 N(Q, Q^*) + c.c. \tag{G.31}$$

The first equation requires the solvability condition

$$\partial_{t_1} A = -\boldsymbol{u}\cdot\nabla A\,. \tag{G.32}$$

If that is satisfied, the explicit solution can be found with the help of Eq. (G.22) in the form

$$q_{2,1} = -i\nabla A\cdot\nabla_k Q\,. \tag{G.33}$$

Any contributions $q_{2,1}||Q$ can be ignored since they have been taken into account already via q_1.

Equation (G.31) does not impose a new solvability condition. The reason is that the kernel mode of L^+ is proportional to $\sin(k_x x)$ whereas the inhomogeneity is proportional to $\sin(2k_x x)$. For convenience, we use the notation

$$q_{2,0} = |A|^2 Q_2\,. \tag{G.34}$$

In order to find the dynamical equation for the amplitude A, we have to proceed to next order ϵ^3. When evaluating the relevant solvability conditions, some relations are helpful which we will determine first. Operating again with ∇_k, this time twice, on both sides of Eq. (G.6), we find

$$\begin{aligned} 0 &= \nabla_k\nabla_k(LQ) \\ &= L\nabla_k\nabla_k Q + i\,(-\boldsymbol{u}\omega\partial_1 + \nabla_{23})\,L\nabla_k Q + i\left[(-\boldsymbol{u}\omega\partial_1 + \nabla_{23})L\nabla_k Q\right]^t \\ &\quad + (-i\nabla_k\nabla_k\omega\partial_1 - \boldsymbol{u}\boldsymbol{u}\partial_1\partial_1 + \boldsymbol{u}\nabla_{23} + \nabla_{23}\boldsymbol{u} - \nabla_{23}\nabla_{23} + \partial_k\partial_k R\partial_4)\,LQ\,. \end{aligned} \tag{G.35}$$

The solution q_3 will again be split with respect to the various harmonic contributions,

$$q_3 = q_{3,0} + q_{3,1} + \cdots \tag{G.36}$$

Most relevant is the first harmonic contribution

$$q_{3,1} \sim e^{i(\boldsymbol{k}\cdot\boldsymbol{r}-\omega t)}\,. \tag{G.37}$$

Making use of Eq. (G.35), the equation for $q_{3,1}$ can be written in the form

$$\begin{aligned} &L_0 q_{3,1} + \partial_1 L Q\partial_{t_2} A + \chi R_c\partial_4 LQA + \frac{1}{2}L(\nabla\cdot\nabla_k)^2 QA \\ &\quad - \frac{i}{2}\,\partial_1 L(\nabla\cdot\nabla_k)^2\omega QA + \frac{1}{2}\partial_4 L(\nabla\cdot\nabla_k)^2 RQA \\ &= N(q_1, q_{2,0}) + N(q_{2,0}, q_1)\,. \end{aligned} \tag{G.38}$$

The solvability condition will provide us with the amplitude equation. A short calculation leads to

$$\partial_{t_2} A + a\chi R_c A - \frac{i}{2}(\nabla \cdot \nabla_k)^2 \omega A + \frac{1}{2} a(\nabla \cdot \nabla_k)^2 R A$$
$$= |A|^2 A \frac{\langle Q | N(Q, Q_2) + N(Q_2, Q)\rangle}{\langle Q | \partial_1 L Q\rangle}. \tag{G.39}$$

The coefficient a is given by

$$a = \frac{\langle Q | \partial_4 L Q\rangle}{\langle Q | \partial_1 L Q\rangle} \equiv i \partial_R \omega \,. \tag{G.40}$$

Remember that we consider Q_2 as a known solution such that all coefficients are precisely determined. The coefficients can be simplified. Because of (G.19), we may use

$$\nabla_k \nabla_k = \partial_k \partial_k + \partial_k \partial_k R \partial_R \,. \tag{G.41}$$

This immediately leads to

$$-\frac{i}{2}(\nabla \cdot \nabla_k)^2 \omega A + \frac{1}{2} a (\nabla \cdot \nabla_k)^2 R A = -\frac{i}{2} \partial_k \partial_k \omega : \nabla\nabla A \,. \tag{G.42}$$

Provided that at the critical point we have $k_z = 0$, leading to the vanishing of all odd derivatives with respect to k_z, we can rewrite (G.39) as the well-known complex Ginzburg–Landau equation

$$\left(\partial_t - \alpha - \gamma_y \partial_{yy} - \gamma_z \partial_{zz} + \beta |A|^2\right) A = 0 \,, \tag{G.43}$$

with the coefficients

$$\alpha = -i R_c \partial_R \omega \frac{R - R_c}{R_c}, \tag{G.44}$$

$$\gamma_y = \frac{i}{2} \partial_{k_y k_y} \omega \,, \tag{G.45}$$

$$\gamma_z = \frac{i}{2} \partial_{k_z k_z} \omega \,, \tag{G.46}$$

$$\beta = -\frac{\langle Q | N(Q, Q_2) + N(Q_2, Q)\rangle}{\langle Q | \partial_1 L Q\rangle} \,. \tag{G.47}$$

The advantage of that result is that we can look for new solutions after a bifurcation. Let us assume that the real real parts of $\alpha, \gamma_y, \gamma_z$, and β are positive. Then, the solution

$$A = A_0 e^{-i\omega t} \tag{G.48}$$

exists, with

$$|A_0|^2 = \frac{\alpha_r}{\beta_r} > 0 \,, \quad \omega = -\alpha_i + \beta_i \frac{\alpha_r}{\beta_r} \,. \tag{G.49}$$

We can get rid of the explicit time-dependence when we transform to a new variable (with tilde)

$$A = \tilde{A} e^{-i\omega t} \,, \tag{G.50}$$

corresponding to a change

$$\alpha \to \tilde{\alpha} = \alpha_r \left(1 + i \frac{\beta_i}{\beta_r}\right). \tag{G.51}$$

The transformed equation has the solution

$$\tilde{A} = A_0 > 0 \,. \tag{G.52}$$

Its stability can be easily calculated. Linearizing with respect to Eq. (G.52) leads to

$$\left(\partial_t - \tilde{\alpha} - \gamma_y \partial_{yy} - \gamma_z \partial_{zz} + 2\beta A_0^2\right) A_1 + \beta A_0^2 A_1^* = 0 \,. \tag{G.53}$$

Making the ansatz

$$A_1 = b e^{i(k_y y + k_z z - \omega t)} + c^* e^{i(-k_y y - k_z z + \omega^* t)} \,, \tag{G.54}$$

we find the quadratic dispersion equation

$$\begin{aligned} &\omega^2 + 2i\omega \,\mathrm{Re}\left(-\tilde{\alpha} + \gamma_y k_y^2 + \gamma_z k_z^2 + 2\beta A_0^2\right) \\ &- |-\tilde{\alpha} + \gamma_y k_y^2 + \gamma_z k_z^2 + 2\beta A_0^2|^2 + |\beta A_0^2|^2 = 0 \,. \end{aligned} \tag{G.55}$$

A necessary and sufficient condition for instability ($\mathrm{Im}\,\omega > 0$) is

$$-\tilde{\alpha}_r + \gamma_{yr} k_y^2 + \gamma_{zr} k_z^2 + 2\beta_r A_0^2 = \alpha_r + \gamma_{yr} k_y^2 + \gamma_{zr} k_z^2 < 0 \,, \tag{G.56}$$

or

$$|\beta A_0^2| > |-\tilde{\alpha} + \gamma_y k_y^2 + \gamma_z k_z^2 + 2\beta A_0^2|^2 \,. \tag{G.57}$$

The first condition cannot be satisfied when we assume $\alpha_r, \gamma_{yr}, \gamma_{zr} > 0$. On the other hand, the second one is equivalent to

$$1 > \left|1 + \frac{\gamma_y}{\tilde{\alpha}} k_y^2 + \frac{\gamma_z}{\tilde{\alpha}} k_z^2\right| \,. \tag{G.58}$$

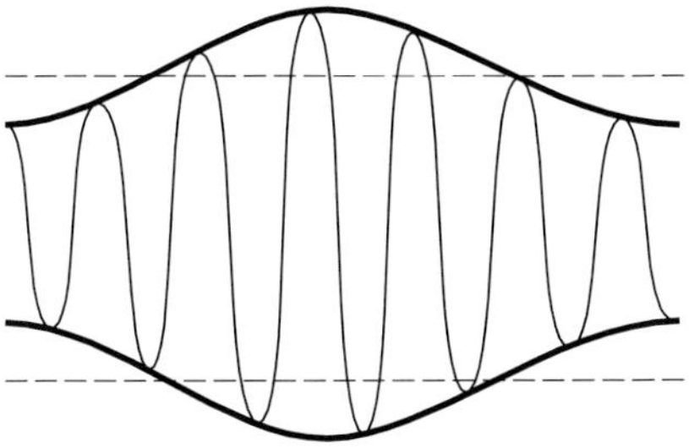

Figure G.2 Sketch of a modulationally unstable situation.

Obviously, a long-wavelength modulational instability can occur, provided

$$\mathrm{Re}\left(\frac{\gamma_y}{\tilde{\alpha}}\right) < 0 \quad \text{or} \quad \mathrm{Re}\left(\frac{\gamma_z}{\tilde{\alpha}}\right) < 0\,. \tag{G.59}$$

The amplitude modulation in such an unstable situation is sketched in Figure G.2.

Appendix H
Liapunov Stability

Stability problems play a central role in plasma theory. In the simplest, but mathematically not rigorous manner, stability is estimated from a normal mode analysis. Mathematically more correct are direct Liapunov calculations and variational principles (the latter will also be outlined in the next section). Here, we discuss the mathematically correct stability concept originally introduced by Liapunov [252]. The subsequent literature is huge, see, for example, [253–256] and references therein. In this section, we present a short summary from the mathematical literature.

In general, the stability regions are more difficult to find than sufficient criteria for instability. In the latter case, the presentation of one unstable mode is sufficient, whereas for stability, we have to prove the stability with respect to all possible perturbations.

Let us start with a few remarks regarding the stability definition. For an autonomous system

$$\dot{\boldsymbol{x}} = \boldsymbol{X}(\boldsymbol{x}), \tag{H.1}$$

we investigate the stability of a solution

$$\boldsymbol{x}_0(t) \quad \text{for} \quad t \geq 0 \quad \text{with} \quad \boldsymbol{x}_0(0) = \boldsymbol{a}_0 \,. \tag{H.2}$$

We need a measure to determine the distance between another solution $\boldsymbol{x}(t)$ and $\boldsymbol{x}_0(t)$. Taking a metric induced by a norm, that is,

$$d(\boldsymbol{x}, \boldsymbol{x}_0) = \| \boldsymbol{x} - \boldsymbol{x}_0 \|^{1/2} \,, \tag{H.3}$$

we have some freedom. We can use, for example, the so called L^q-norm or a $W^{1,2}$-norm [257]. For the latter, we introduce the notations

$$\int d\Gamma \equiv \int_{\mathcal{R}^d} d^d x \,, \tag{H.4}$$

$$\| f \|_q \equiv \left(\int d\Gamma \mid f(x) \mid^q \right)^{\frac{1}{q}} \quad (L^q\text{-norm}) \,, \tag{H.5}$$

$$\| f \|_{W^{1,2}} \equiv \left(\| f \|_2^2 + \| \nabla f \|_2^2 \right)^{\frac{1}{2}} \quad (W^{1,2}\text{-norm}) \,, \tag{H.6}$$

High Temperature Plasmas, Theory and Mathematical Tools for Laser and Fusion Plasmas, First Edition.
Karl-Heinz Spatschek.

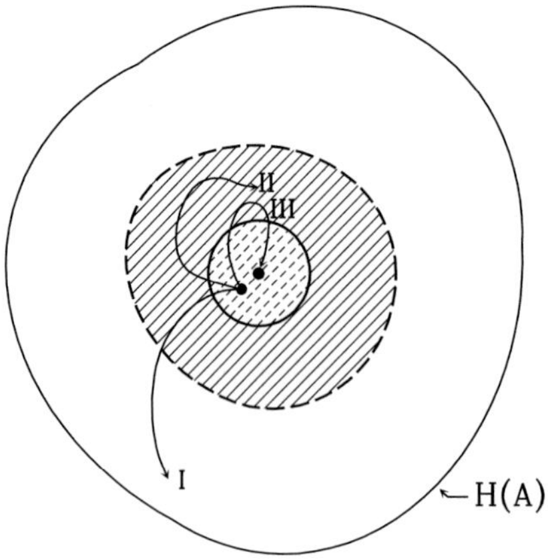

Figure H.1 Unstable (I), stable (II), and asymptotically stable (III) orbits of a dynamical system.

$$W^{1,2}\left(\mathcal{R}^d\right) \equiv \{f; \| f \|_{W^{1,2}} < \infty\} \quad \text{(Sobolev space } W^{1,2}) . \tag{H.7}$$

Let us first assume that $\boldsymbol{x}_0$ is time-independent. Then, the physically intuitive requirement for stability is reflected in the following definition.

A time-independent state $\boldsymbol{x}_0 \equiv \boldsymbol{a}_0$ is stable if and only if for any $\epsilon > 0$, there exists a $\delta(\epsilon) > 0$ such that for all initial conditions in the neighborhood

$$\| \boldsymbol{x}(t=0) - \boldsymbol{a}_0 \| < \delta , \tag{H.8}$$

$$\| \boldsymbol{x}(t) - \boldsymbol{x}_0 \| < \epsilon \tag{H.9}$$

follows for all $t \geq 0$.

We illustrate that stability concept in Figure H.1.

In principle, we could also extend that definition to time-dependent states $\boldsymbol{x}_0(t)$ by just replacing $\boldsymbol{x}_0 \to \boldsymbol{x}_0(t)$ in the above criterion. We call that the *instantaneous stability concept*. The other variant is the so called *stability of trajectory concept*:

If for any $\epsilon > 0$, there exists a $\delta(\epsilon) > 0$ such that for all initial conditions in the neighborhood

$$\| \boldsymbol{x}(t=0) - \boldsymbol{a}_0 \| < \delta , \tag{H.10}$$

$$\text{dist}\,(\boldsymbol{x}(t), \boldsymbol{x}_0) < \epsilon \tag{H.11}$$

follows for all t, we state stability of $\boldsymbol{x}_0(t)$. Here, dist $(\boldsymbol{x}(t), \boldsymbol{x}_0)$ means the distance to the trajectory $\boldsymbol{x}_0$,

$$\text{dist}\,(\boldsymbol{x}(t), \boldsymbol{x}_0) = \inf_{\tau} \| \boldsymbol{x}(t) - \boldsymbol{x}_0(\tau) \|^{1/2} . \tag{H.12}$$

One may generalize the stability of the trajectory concept for the *stability of form*, that is, for the *stability of an invariant set*. In the latter, several trajectories can be combined. Instead of an exact definition, let us only present a plausibility argument. We disturb a soliton a little bit in velocity. The disturbed, moving soliton will be compared with the original one (reference state) which has a slightly different velocity. Then, its distance from the soliton (reference state) with the original velocity will linearly grow in time. We shall not call that an instability, but include all solitons which evolve by simple translations into the invariant set (reference state). A similar behavior can occur for the phase. In simple words, the invariant set will consist of all solitons of the same form (but different positions through space translations and phase rotations).

In order to prove stability in that sense, we shall apply the following Liapunov stability theorem [253, 254]:

> *"Suppose we have a dynamical equation for Ψ. An invariant set S (in the example: all solitons of the same form) is stable if in a neighborhood U_ϵ of S a functional $L(\Psi)$ with the following properties exists:*
>
> *a) Within U_ϵ, $L(\Psi)$ can be estimated in terms of the distance $d(\Psi, S)$ between the perturbed state and the invariant set, that is,*
>
> *1. For all $c_1 > 0$, there exists a $c_2 > 0$ such that for all $\Psi \in U_\epsilon$ with $L(\Psi) < c_2$, the bound $d(\Psi, S) < c_1$ follows.*
> *2. For all $d_1 > 0$, there exists a $d_2 > 0$ such that for all $\Psi \in U_\epsilon$ with $d(\Psi, S) < d_2$, the bound $L(\Psi) < d_1$ follows.*
>
> *b) $L(\Psi)$ should be a nongrowing functional, that is, $\partial_t L \leq 0$*
> *c) $L(\Psi \in S) = 0$ should hold for $t \geq t_0$."*

Let us make the following "illustrative" comments. Condition (a) relates the Liapunov functional to the distance. It guarantees that an "instantaneous ϵ–δ-relation" can be constructed. The latter relation is known to be necessary for stability. Finally, (b) guarantees that the perturbations will stay in the neighborhood for all times (in irreversible systems $\partial_t L < 0$ might be possible (asymptotic stability), whereas in reversible systems, $L =$ const is usually the appropriate choice).

In Figure H.2, we illustrate how to construct a ϵ–δ-relation when a Liapunov functional is known. Figure H.2a and b specify what is meant by the estimate of L in terms of the distance. Thus, if we start from a contour $L = K$, where K is a constant, δ can be chosen as the radius of the inner circle and ϵ will be the radius of the outer circle.

For a (nondissipative) nonlinear Schrödinger system, we now construct a Liapunov functional out of the first constants of motion [258, 259]. For

$$i\partial_t \phi + \nabla^2 \phi + |\phi|^\nu \phi = 0 \,, \tag{H.13}$$

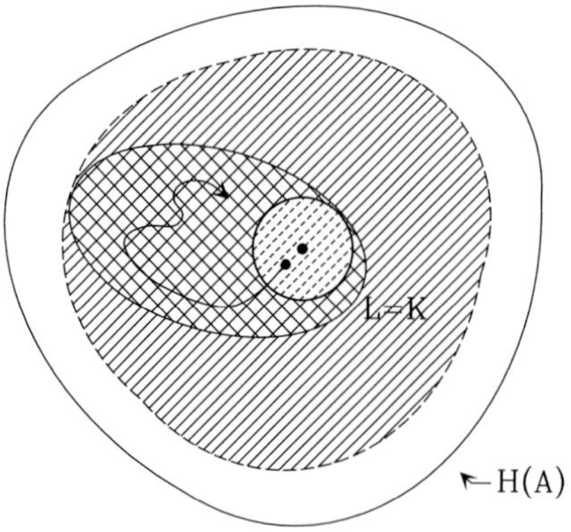

Figure H.2 Construction of the required ϵ–δ-relation when a Liapunov functional is known. $L = K$ is a contour line of L.

we have the invariants

$$I_1 = \int d\Gamma \phi \phi^* , \tag{H.14}$$

$$I_2 = \int d\Gamma \phi \nabla \phi^* , \tag{H.15}$$

$$I_3 = \int d\Gamma \left[\nabla \phi \cdot \nabla \phi^* - \left(\frac{\nu}{2} + 1 \right)^{-1} |\phi|^{\nu+2} \right] . \tag{H.16}$$

In these integrals, $d\Gamma = d^d r$ is the d-dimensional volume element, for example, $d\Gamma = dx$, or $d\Gamma = dx dy$, or $d\Gamma = dx dy dz$ in Cartesian coordinates.

We now proceed in two steps. First, we give the general outline for the Eq. (H.13), though not presenting all the details of the estimates. After that, we complete the presentation by some details for $\nu = 2$.

Because of the Galilei invariance, we shall only need I_1 and I_3 (and not the momentum) to construct

$$\begin{aligned} L = \int d\Gamma \bigg[\nabla \phi \cdot \nabla \phi^* - \left(\frac{\nu}{2} + 1 \right)^{-1} |\phi|^{\nu+2} - (\nabla \phi_s)^2 \\ + \left(\frac{\nu}{2} + 1 \right)^{-1} \phi_s^{\nu+2} \bigg] + \frac{4}{\nu^2} \eta_s^2 I_1^\theta I_{1s}^{-\theta} \int d\Gamma \left(\phi \phi^* - \phi_s^2 \right) . \end{aligned} \tag{H.17}$$

The parameter θ is a positive number; its lower limit will be determined later. Furthermore, ϕ_s is the d-dimensional (plane, cylindrical, spherical, respectively) stationary soliton solution, for example,

$$\phi_s(x) = \left[\frac{2(\nu + 2)}{\nu^2} \right]^{\frac{1}{\nu}} \eta_s^{\frac{2}{\nu}} \operatorname{sech}^{\frac{2}{\nu}} (\eta_s x) , \tag{H.18}$$

with

$$I_{1s} = \int dx \phi_s^2 . \tag{H.19}$$

Before showing that the proposed functional Eq. (H.17) is indeed a Liapunov functional for stability, let us come back to the existence of stationary solutions. In one space dimension, the existence is obvious.

Radial (cylindrical $\phi_s(\rho)$ and spherical $\phi_s(r)$) solutions in higher dimensions are also of interest. Their analytical form is not known, however existence can be guaranteed by the following theorem proved by Strauss [260].

"Assume

$$F_1(s) \geq 0\,, \quad F_2 > 0 \quad \text{for} \quad s > 0\,. \tag{H.20}$$

$$\text{As} \quad s \to 0\,, \quad \left|\frac{F_1(s)}{s}\right| \leq C < \infty \quad \text{and} \quad \frac{F_2(s)}{s} \to 0\,, \tag{H.21}$$

$$\left|\frac{F_2(s)}{s^l + F_1(s)}\right| \leq C < \infty \quad \text{and} \quad \frac{F_2(s)}{s^l + \frac{G_1(s)}{s}} \to 0 \quad \text{as} \quad s \to \infty\,, \tag{H.22}$$

where l is an arbitrary positive number for $d = 2$ and $l = 5$ for $d = 3$. G_1 and G_2 are the indefinite integrals of F_1 and F_2, respectively.
Then, there exists a $\lambda > 0$ and a solution u of

$$-\nabla^2 u + a_0 u + F_1(u) = \lambda F_2(u)\,, \quad a_0 > 0 \tag{H.23}$$

which is nonnegative, belongs to the Sobolev space $W^{1,2}$, decays exponentially as $|\mathbf{r}| \to \infty$, and $\int d\Gamma\, G_1(u) < \infty$."

Let us apply this theorem to our problem in $d = 2$ and $d = 3$. Dividing through λ, scaling the space variables and letting

$$F_1(u) = u\,, \quad F_2(u) = u^{\nu+1}\,, \tag{H.24}$$

$$\frac{a_0 + 1}{\lambda} = \frac{4\eta_s^2}{\nu^2}\,, \tag{H.25}$$

the criteria are obviously satisfied provided

$$\nu > 0 \quad \text{for} \quad d = 2\,, \tag{H.26}$$

$$4 > \nu > 0 \quad \text{for} \quad d = 3\,, \tag{H.27}$$

respectively. Here, we have used

$$G_1 = \frac{u^2}{2}\,, \quad G_2 = \frac{u^{\nu+2}}{\nu + 2}\,. \tag{H.28}$$

The starting point of the proof of the existence theorem is the variational principle

$$I = \inf \int d\Gamma \left[(\nabla u)^2 + (a_0 + 1) u^2\right] \tag{H.29}$$

under the extra condition

$$\frac{1}{\nu+2}\int d\Gamma\, u^{\nu+2} = 1\,. \tag{H.30}$$

Note that Eq. (H.23) is the corresponding Euler equation with the Lagrange parameter λ. Strauss [260] succeeded to show that the variational principle is well-defined and that the infimum will be attained.

It is interesting to note that the variational principle for existence cannot be used for stability calculations. Equation (H.30) is not a constant of motion. Therefore, the calculation of a minimum under that restriction does not imply stability under arbitrary perturbations. Nevertheless, the variational principle for existence helps to prove some of the definiteness properties of operators appearing in stability considerations.

Now, we return to the functional Eq. (H.17). Since it is built out of constants of motion,

$$\partial_t L = 0 \tag{H.31}$$

follows. Furthermore,

$$L = 0 \quad \text{for} \quad \phi = \phi_s \exp\left\{i4\eta_s^2\frac{t}{\nu^2}\right\} \tag{H.32}$$

is trivially satisfied.

Next, we have to show the estimates (a) and (b). Inserting a perturbed state

$$\phi = (\phi_s + \delta\phi)\exp\left\{i4\eta_s^2\frac{t}{\upsilon^2}\right\} \tag{H.33}$$

into the proposed Liapunov functional L, we can calculate the first order contribution which is identical to the first variation of L,

$$L_1 = \int d\Gamma\left[-\nabla^2\phi_s - \phi_s^{\nu+1} + \frac{4}{\nu^2}\eta_s^2\phi_s\right]\delta\phi^* + c.c. \tag{H.34}$$

Recalling the definition of ϕ_s, we immediately obtain

$$L_1 = 0\,. \tag{H.35}$$

For the second order term L_2, one gets after some algebra

$$L_2 = \int d\Gamma\,(aH_-a + bH_+b) + \frac{16\theta}{\nu^2}\frac{\eta_s^2}{I_{1s}}\left[\int d\Gamma\, a\phi_s\right]^2, \tag{H.36}$$

where

$$\delta\phi = a + ib\,, \tag{H.37}$$

$$H_+ = -\nabla^2 - \phi_s^\nu + \frac{4}{\nu^2}\eta_s^2\,, \tag{H.38}$$

$$H_- = H_+ - \nu\phi_s^\nu\,. \tag{H.39}$$

The higher order contributions L_3 can be handled when the Sobolev inequality is used following from the Sobolev embedding theorem. The latter theorem allows one to estimate terms of the form $\int d\Gamma a^{\mu} b^{\kappa}$ in terms of the norm N^2 of the perturbations. Here, we only mention that we can find appropriate estimates exactly in that region of ν-values where the existence of the quasistationary soliton solutions is guaranteed. Anticipating a possible estimate in the form

$$L_2 - F(N^2) \le L \le L_2 + F(N^2)\,, \tag{H.40}$$

a sufficient criterion for stability is

$$L_2 > 0 \quad \text{for nonvanishing perturbations}\,, \tag{H.41}$$

anticipating that then L_2 can be estimated in terms of the norm.

However, L_2 can still vanish. The modes $b = \phi_s$ and $a = 0$, as well as $b = 0$ and $a = (\nabla\phi_s)_i$, $i = x, y, z$ will do it. These are the so called rotation and translation modes, respectively. If we translate the original soliton in space and rotate its phase, that is,

$$\phi_s(\boldsymbol{r}) \exp\left\{i\frac{4}{\nu^2}\eta_s^2 t\right\} \to \phi_s(\boldsymbol{r} - \boldsymbol{\alpha}) \exp\left\{i\frac{4}{\nu^2}\eta_s^2 t + i\alpha_0\right\}, \tag{H.42}$$

we get a soliton of the same form. Physically, only stability with respect to form is of interest. Since a perturbed state may show in its time evolution a translation and rotation which do not affect the form, we construct from the original stationary state an invariant set by allowing for arbitrary parameters $\boldsymbol{\alpha}$ and α_0. The distance between a perturbed state

$$\phi = \left[\phi_s(\boldsymbol{r} - \boldsymbol{r}_0) + a(\boldsymbol{r} - \boldsymbol{r}_0, t) + i b(\boldsymbol{r} - \boldsymbol{r}_0, t)\right] \exp\left\{i\xi_0 + i\frac{4}{\nu^2}\eta_s^2 t\right\} \tag{H.43}$$

and the invariant set is then determined by minimizing with respect to the invariant set,

$$\partial_{\alpha_i} \parallel \phi_s + a + ib - \phi_s(\boldsymbol{r} - \boldsymbol{\alpha}) \exp\{i\alpha_0\} \parallel^2_{\boldsymbol{\alpha} = \boldsymbol{r}_0, \alpha_0 = \xi_0} = 0\,. \tag{H.44}$$

Here, we use the norm

$$N^2 \equiv \parallel A \parallel^2 = \int d\Gamma \left[|\nabla A|^2 + \frac{4}{\nu^2}\eta_s^2 |A|^2\right]. \tag{H.45}$$

The minimizing procedure leads to

$$\left\langle a | \nabla\phi_s^{\nu+1}\right\rangle = 0\,, \quad \left\langle b | \phi_s^{\nu+1}\right\rangle = 0\,, \tag{H.46}$$

where here and in the following we use the scalar product notation

$$\langle \xi | T | \eta \rangle \equiv \int dx \xi T \eta\,. \tag{H.47}$$

The two extra conditions forbid the translation and rotation modes. Thus, the concept of stability with respect to form excludes $L_2 = 0$.

By allowing for translations and rotations in phase, we have effectively constructed an invariant set. However, the following remark seems to be appropriate. Since during the time-development the shift-parameters $\boldsymbol{\alpha}$ and α_0 can also become time-dependent, the property $\partial_t L = 0$ has to be checked again. However, L is still time-independent since because of the infinite integration domain, a change of integration variables eliminates the time-dependent translation parameter $\boldsymbol{\alpha}$.

By longitudinal stability, we mean that the perturbations a and b depend in the one-dimensional case only on x, in the two-dimensional case only on ρ, and in the three-dimensional case on r (besides the time t). To evaluate longitudinal stability, we should briefly discuss the definiteness properties of the operators H_+ and H_-. Let us do it here only for the one-dimensional longitudinal stability of plane solitons.

Obviously, H_+ is positive semidefinite since it is a Schrödinger operator with $H_+ \phi_s = 0$ and ϕ_s has no node. Because of $H_- \partial_x \phi_s = 0$, H_- has one negative eigenvalue. (The same definiteness properties will be found in 2D and 3D.)

For the estimate in terms of the norm of the perturbations, the sign of $\langle a|H_-|a\rangle$ is critical. Let us first consider perturbations with

$$\langle a|\phi_s\rangle = 0\,. \tag{H.48}$$

The negative eigenvalue of H_- is denoted by λ_-; the corresponding eigenfunction is e_-. In general, a function a has a component a_- parallel to e_-, and a component $a_\perp$ perpendicular to e_-. Then, we have

$$\langle a|H_-|a\rangle = -|\lambda_-|\langle a_-|a_-\rangle + \langle a_\perp|H_-|a_\perp\rangle\,. \tag{H.49}$$

Abbreviating

$$F = H_-^{-1}\phi_s, \tag{H.50}$$

we get from $\langle a|H_-|F\rangle = 0$

$$0 = -|\lambda_-|\langle a_-|F_-\rangle + \langle a_\perp|H_-|F_\perp\rangle\,. \tag{H.51}$$

Using this and applying Schwarz inequality, one obtains

$$\langle a_\perp|H_-|a_\perp\rangle \ge |\lambda_-|^2 \frac{\langle a_-|F_-\rangle^2}{\langle F_\perp|H_-|F_\perp\rangle}\,. \tag{H.52}$$

For

$$\langle \phi_s|H_-^{-1}|\phi_s\rangle = \langle F|H_-|F\rangle < 0\,, \tag{H.53}$$

we obtain

$$\langle F_\perp|H_-|F_\perp\rangle < |\lambda_-|\langle F_-|F_-\rangle. \tag{H.54}$$

Combing Eqs. (H.49), (H.52), and (H.54), we get the desired result, that is,

$$\langle a|H_-|a\rangle \geq 0 \tag{H.55}$$

under the condition Eq. (H.53).

Next, we allow for perturbations with

$$\langle a|\phi_s\rangle \neq 0\,. \tag{H.56}$$

Then,

$$\langle a|H_-|a\rangle \geq \langle a|\phi_s\rangle^2 \left\langle \phi_s|H_-^{-1}|\phi_s\right\rangle^{-1} \tag{H.57}$$

since H_- has only one discrete eigenvalue. For the proof, we calculate

$$\min_a \frac{\langle a|H_-|a\rangle}{\langle a|\phi_s\rangle^2} \quad \text{for} \quad \langle a|\phi_s\rangle \neq 0 \tag{H.58}$$

by defining

$$a = H_-^{-1}\phi_s + \varphi \tag{H.59}$$

and choosing the normalization

$$\langle a|\phi_s\rangle^2 = \left\langle \phi_s|H_-^{-1}|\phi_s\right\rangle^2 \Rightarrow \langle \varphi|\phi_s\rangle = 0\,. \tag{H.60}$$

Then, we obtain

$$\min_a \frac{\langle a|H_-|a\rangle}{\langle a|\phi_s\rangle^2} = \min_\varphi \frac{\left\langle \phi_s|H_-^{-1}|\phi_s\right\rangle + \langle \varphi|H_-|\varphi\rangle}{\left\langle \phi_s|H_-^{-1}|\phi_s\right\rangle^2}\,. \tag{H.61}$$

Since

$$\langle \varphi|H_-|\varphi\rangle \geq 0 \quad \text{for} \quad \langle \varphi|\phi_s\rangle = 0 \tag{H.62}$$

(see before), we have proven the statement made above.

Since now

$$\langle a|H_-|a\rangle + \frac{16}{\nu^2}\theta\frac{\eta_s^2}{I_{1s}}\langle a|\phi_s\rangle^2 \geq \left[\left\langle \phi_s|H_-^{-1}|\phi_s\right\rangle^{-1} + \frac{16}{\nu^2}\theta\frac{\eta_s^2}{I_{1s}}\right]\langle a|\phi_s\rangle^2\,, \tag{H.63}$$

we can always find a critical value θ_c such that for $\theta > \theta_c$, the right-hand side is positive. From this discussion we conclude that L is nonnegative in the neighborhood of the stationary state if

$$\left\langle \phi_s|H_-^{-1}|\phi_s\right\rangle < 0\,. \tag{H.64}$$

From

$$H_+\phi_s = 0, \tag{H.65}$$

it is straightforward to show that

$$H_{-}\frac{\partial \phi_s}{\partial \eta_s^2} = -\frac{4}{\nu^2}\phi_s \,. \tag{H.66}$$

The kernel of H_{-} is given by $\nabla \phi_s$, and therefore

$$H_{-}^{-1}\phi_s = -\frac{\nu^2}{4}\frac{\partial \phi_s}{\partial \eta_s^2} \,. \tag{H.67}$$

The longitudinal stability condition is then

$$\partial_{\eta_s^2} I_{1s} > 0 \,. \tag{H.68}$$

This condition guarantees that L is indeed a Liapunov functional for stability; it formulates a sufficient criterion for stability.

Appendix I
Variational Principles

Most instability eigenvalue problems cannot be solved analytically. As has been shown in the main text, perturbative methods are possible procedures for overcoming parts of the problem, but very often they suffer from serious flaws, for example, convergence problems. Thus, in the present section, we would like to present a procedure which is quite general and helpful in many cases. Variational formulations are important in all areas of physics, see, for example, [448, 449]. We have already discussed the hydromagnetic variational principle for MHD stability. In this section, we show how variational methods can be used to determine growth rates in unstable situations, making use of complementary principles [450]. Although the method can be developed in quite general terms, we shall demonstrate it here on a specific example, namely, the transverse instability of the ZK solitons [265, 374, 375].

Let us start with the ZK equation

$$\frac{\partial \varphi}{\partial t} + \varphi \frac{\partial \varphi}{\partial x} + \frac{\partial^3 \varphi}{\partial x^3} + \frac{\partial}{\partial x} \frac{\partial^2 \varphi}{\partial y^2} = 0 , \tag{I.1}$$

which has a (plane) one-dimensional solitary wave solution

$$\varphi_s = 12\eta^2 \mathrm{sech}^2 \left[\eta \left(x - x_0 - 4\eta^2 t\right)\right] . \tag{I.2}$$

We introduce the new variables

$$\xi = \eta \left(x - x_0 - 4\eta^2 t\right) , \quad Y = \eta y , \quad \tau = 4\eta^3 t , \quad \psi = 4\eta^2 \varphi , \tag{I.3}$$

and perturb in the form

$$\psi(\xi, Y, \tau) = \psi_s(\xi) + f(\xi, \tau) e^{iKY} . \tag{I.4}$$

Splitting f into odd and even parts (with respect to ξ), that is,

$$f = a + b , \tag{I.5}$$

we find for a (or b)

$$\frac{\partial^2 a}{\partial \tau^2} = \frac{\partial}{\partial \xi} \hat{L} \frac{\partial}{\partial \xi} \hat{L} a , \tag{I.6}$$

High Temperature Plasmas, Theory and Mathematical Tools for Laser and Fusion Plasmas, First Edition.
Karl-Heinz Spatschek.

where

$$\hat{L} := -\frac{1}{4}\frac{\partial^2}{\partial\xi^2} - 3\,\mathrm{sech}^2\xi + 1 + \frac{K^2}{4}\,. \tag{I.7}$$

To be more specific, let a be odd and let us introduce

$$\varphi = \hat{L}a\,. \tag{I.8}$$

The functions φ are also odd since $\hat{L}$ does not change the symmetry. Furthermore, for $K \neq 0$, the operator $\hat{L}$ is positive definite when operating on odd functions (see the discussion on the eigenfunctions of $\hat{L}$ in the main text). Therefore, we can rewrite Eq. (I.6) in the form

$$\hat{L}^{-1}\partial_\tau^2\varphi = \partial_\xi\hat{L}\partial_\xi\varphi\,, \quad \varphi \text{ odd}\,. \tag{I.9}$$

At this stage, we leave the specific example and write Eq. (I.9) in the more general form

$$N\partial_\tau^2 a = Fa\,, \tag{I.10}$$

where N and F are self-adjoint operators. N should be positive definite (we abbreviate: $N > 0$) (and in most cases, we shall assume that F possesses only one positive (discrete) eigenvalue).

Let us define

$$\Gamma^2 = \sup_g \frac{\langle g|F|g\rangle}{\langle g|N|g\rangle} \tag{I.11}$$

and

$$\gamma^2 = \inf_g \frac{\langle g|FN^{-1}F|g\rangle}{\langle g|F|g\rangle} \quad \text{with} \quad \langle g|F|g\rangle > 0\,. \tag{I.12}$$

In the infimum principle Eq. (I.12), the restriction to a nonempty subspace of functions g with $\langle g|F|g\rangle > 0$ is important. We shall come back to this point later.

If F posseses one positive discrete eigenvalue, we shall prove that unstable perturbations exist with (exponential) growth rate Γ. If F has only one positive discrete eigenvalue, $\gamma \equiv \Gamma$ holds.

Before presenting the details of the proof, let us mention the advantages of such a variational procedure. Evaluating the infimum principle (e.g., numerically), we find upper bounds for the growth rate. Furthermore, the supremum principle leads to lower bounds for the growth rate and can be used to estimate possible instability regions.

Multiplying Eq. (I.10) from the left by $\partial_\tau a$, and integrating over ξ from $-\infty$ to $+\infty$, we obtain

$$\langle\partial_\tau a|N|\partial_\tau a\rangle - \langle a|F|a\rangle = C\,. \tag{I.13}$$

Here, C is an integration constant and

$$\langle\cdots\rangle \equiv \int_{-\infty}^{+\infty} d\xi \cdots \tag{I.14}$$

The derivation of Eq. (I.13) makes use of the Hermitean properties of the operators N and F.

Let us now consider a specific function $a(\xi,\tau)$ with

$$\partial_\tau a(\xi,\tau)|_{\tau=0} = \tilde{\Gamma} a(\xi,\tau=0) \equiv \tilde{\Gamma} a_0 \,, \tag{I.15}$$

where

$$\tilde{\Gamma}^2 = \frac{\langle a_0|F|a_0\rangle}{\langle a_0|N|a_0\rangle} > 0 \,. \tag{I.16}$$

Note that we assume that F possesses a positive eigenvalue and $N > 0$; therefore, the choice Eq. (I.16) is possible. Furthermore, the solution $a(\xi,\tau)$ will satisfy Eq. (I.13) for $C = 0$.

Next, we calculate for the chosen perturbation a

$$\begin{aligned}\partial_\tau^2\langle a|N|a\rangle &= 2\langle\partial_\tau a|N|\partial_\tau a\rangle + 2\langle a|N|\partial_\tau^2 a\rangle \\ &= 2\langle a|F|a\rangle + 2\langle a|F|a\rangle = 4\langle a|F|a\rangle \,.\end{aligned} \tag{I.17}$$

With the abbreviation

$$L := \frac{1}{2}\langle a|N|a\rangle \,, \tag{I.18}$$

we can rewrite the last equation as

$$\partial_\tau^2 L = 2\langle a|F|a\rangle = 2\langle\partial_\tau a|N|\partial_\tau a\rangle \tag{I.19}$$

since $C = 0$ holds for the perturbation under consideration.

From the triangle inequality (in symbolic form)

$$AB \geq \mathbf{A}\cdot\mathbf{B}, \tag{I.20}$$

we can estimate the right-hand side of Eq (I.19) in the form

$$\partial_\tau^2 L \equiv 2\frac{\langle\partial_\tau a|N|\partial_\tau a\rangle\langle a|N|a\rangle}{\langle a|N|a\rangle} \geq 2\frac{\langle\partial_\tau a|N|a\rangle^2}{\langle a|N|a\rangle} \,. \tag{I.21}$$

Thus,

$$\partial_\tau\left(\frac{\partial_\tau L}{L}\right) \geq 0 \tag{I.22}$$

follows. The latter inequality is valid for all $\tau > 0$. At $\tau = 0$, we have

$$\partial_\tau L = \tilde{\Gamma} L \,. \tag{I.23}$$

Therefore, we can conclude

$$\partial_\tau L \geq \tilde{\Gamma} L \quad \text{for} \quad \tau > 0 \,, \tag{I.24}$$

or

$$L \geq L(\tau = 0)e^{\tilde{\Gamma}\tau} . \tag{I.25}$$

The interpretation of the result strongly depends on the property $N > 0$. Under the latter assumption, obviously exponential instability exists. Furthermore,

$$0 < \tilde{\Gamma}^2 \leq \sup_g \frac{\langle g|F|g\rangle}{\langle g|N|g\rangle} \tag{I.26}$$

tells us that all choices $\tilde{\Gamma}^2 \leq \Gamma^2$ are possible.

However, thus far, we have only considered perturbations which satisfy Eq. (I.13) for $C = 0$. Now, we also show that for arbitrary perturbations, Γ as defined by Eq. (I.11) is the maximum exponential growth rate. Thus, let us consider a perturbation which satisfies Eq. (I.13) for $C \neq 0$. Introducing the definition

$$a = be^{\Gamma\tau} \tag{I.27}$$

into Eq. (I.13), we find

$$\begin{aligned} Ce^{-2\Gamma\tau} &= \langle\partial_\tau b|N|\partial_\tau b\rangle + 2\Gamma\langle\partial_\tau b|N|b\rangle + \Gamma^2\langle b|N|b\rangle \\ &\quad - \langle b|F|b\rangle \geq 2\Gamma\langle\partial_\tau b|N|b\rangle . \end{aligned} \tag{I.28}$$

In the last estimate, we made use of $N > 0$ and the definition Eq. (I.11). We can therefore conclude

$$\partial_\tau\langle b|N|b\rangle \leq \frac{C}{\Gamma}e^{-2\Gamma\tau} . \tag{I.29}$$

We can integrate that inequality to find

$$\langle b(\xi,\tau)|N|b(\xi,\tau)\rangle \leq \langle b(\xi,0)|N|b(\xi,0)\rangle - \frac{C}{2\Gamma^2}\left[e^{-2\Gamma\tau} - 1\right] \leq \tilde{C} . \tag{I.30}$$

With the new constant $\tilde{C} > 0$, we have therefore found

$$\langle a|N|a\rangle \leq \tilde{C}e^{2\Gamma\tau} , \tag{I.31}$$

that is, Γ is indeed the maximum exponential growth rate.

Before evaluating the supremum principle for some specific applications, we will discuss the infimum principle. Again, we start with a conservation law for the linearized Eq. (I.10). This time, we make use of the identity $F\partial_\tau a \equiv F\partial_\tau a$, multiply from the left by $\partial_\tau^2 a$, and integrate. With Eq. (I.10), we can write the final relation in the form

$$\langle\partial_\tau^2 a|F|\partial_\tau a\rangle - \langle a|FN^{-1}F|\partial_\tau a\rangle = 0 . \tag{I.32}$$

It is straightforward to derive from here

$$\partial_\tau\left[\langle\partial_\tau a|F|\partial_\tau a\rangle - \langle a|F|\partial_\tau^2 a\rangle\right] = 0 , \tag{I.33}$$

which we can rewrite in integrated form as

$$\langle \partial_\tau a|F|\partial_\tau a\rangle - \langle a|FN^{-1}F|a\rangle = C\,. \tag{I.34}$$

Thus, in the following, we start from this conservation law and derive an estimate for exponential growth in a similar way as before (when starting from Eq. (I.13)).

Let us now only consider functions which belong to the subspace

$$M := \{a|\langle a|F|a\rangle > 0\}\,. \tag{I.35}$$

This subspace is not empty if F has at least one positive eigenvalue. We now choose some initial displacement $a_0 \equiv a(\xi, \tau = 0) \in M$ and assume

$$\partial_\tau a(\xi,\tau)|_{\tau=0} = \tilde{\gamma}\, a_0\,, \tag{I.36}$$

where

$$\tilde{\gamma}^2 = \frac{\langle a_0|FN^{-1}F|a_0\rangle}{\langle a_0|F|a_0\rangle} > 0\,. \tag{I.37}$$

Because of the assumptions we have already mentioned, that choice is possible. Note that for these displacements, $C \equiv 0$ holds. It is straightforward to evaluate for perturbations satisfying the initial conditions $a_0 \in M$ and Eq. (I.36)

$$\begin{aligned}\partial_\tau^2\langle a|F|a\rangle &\equiv 2\langle\partial_\tau a|F|\partial_\tau a\rangle + 2\langle\partial_\tau^2 a|F|a\rangle\\ &= 2\langle a|FN^{-1}F|a\rangle + 2\langle a|FN^{-1}F|a\rangle = 4\langle a|FN^{-1}F|a\rangle\,.\end{aligned} \tag{I.38}$$

Let us abbreviate

$$l := \frac{1}{2}\langle a|F|a\rangle\,. \tag{I.39}$$

Then, the relation (I.38) leads to

$$\partial_\tau^2 l \equiv 2\frac{\langle a|FN^{-1}F|a\rangle\langle a|F|a\rangle}{\langle a|F|a\rangle} \geq 2\left\{\inf_{a\in M}\frac{\langle a|FN^{-1}F|a\rangle}{\langle a|F|a\rangle}\right\}\langle a|F|a\rangle = 4\gamma^2 l\,. \tag{I.40}$$

Continuity arguments allow us to assume that for a finite time T and $0 < \tau \leq T$, the perturbations $a(\xi,\tau)$ belong to M and $\langle\partial_\tau a|F|a\rangle > 0$ holds. We can therefore integrate Eq. (I.40), with the result

$$\partial_\tau l \geq 2\gamma l\,. \tag{I.41}$$

For $0 < \tau \leq T$, we find

$$l \geq \frac{1}{2}\langle a_0|F|a_0\rangle e^{2\gamma\tau}\,, \tag{I.42}$$

which means that we can even let $T \to \infty$.

Inequality Eq. (I.42) tells us that unstable perturbations exist with (exponential) growth rate γ. Since Γ is the maximum exponential growth rate, $\gamma \leq \Gamma$ follows.

Now, we prove $\gamma = \Gamma$ if F has only one positive eigenvalue. Since $\gamma \leq \Gamma$ has been already shown for the general case, we need to show $\Gamma \leq \gamma$ in the present situation. For that, we go back to the definition Eq. (I.11) and estimate

$$\Gamma^2 \leq \sup_g \frac{\langle g|F|g\rangle}{\langle g|F|f\rangle^2} \sup_g \frac{\langle g|F|f\rangle^2}{\langle g|F|g\rangle} , \tag{I.43}$$

where we have introduced some "intermediate" state f. Later, it will become important to choose f out of the subspace with $\langle f|F|f\rangle > 0$.

Each of the two suprema on the right-hand side of Eq. (I.43) can be calculated separately. We present those detailed calculations now. When calculating the second supremum on the right-hand side of Eq. (I.43), we normalize the test functions g such that

$$\langle g|F|f\rangle^2 = \langle Ff|N^{-1}|Ff\rangle^2 . \tag{I.44}$$

Introducing φ via

$$g := N^{-1}Ff + \varphi , \tag{I.45}$$

that is, φ satisfies $\langle f|F|\varphi\rangle = 0$, we find

$$\begin{aligned} \sup_g \frac{\langle g|F|f\rangle^2}{\langle g|N|g\rangle} &= \sup_g \frac{\langle f|FN^{-1}F|f\rangle^2}{\langle g|N|g\rangle} = \sup_g \frac{\langle f|FN^{-1}F|f\rangle^2}{\langle f|FN^{-1}F|f\rangle + \langle\varphi|N|\varphi\rangle} \\ &= \langle f|FN^{-1}F|f\rangle . \end{aligned} \tag{I.46}$$

This is the result for the second term on the right-hand side of Eq. (I.43).

Now, we consider the first supremum on the right-hand side of Eq. (I.43). Instead of Eq. (I.44), we now use the normalization of g in the form

$$\langle g|F|f\rangle^2 = \langle f|F|f\rangle^2 . \tag{I.47}$$

Introducing φ via

$$g := f + \varphi , \tag{I.48}$$

that is, φ satisfies $\langle f|F|\varphi\rangle = 0$, we find

$$\sup_g \frac{\langle g|F|g\rangle}{\langle g|F|f\rangle^2} = \sup_g \frac{\langle g|F|g\rangle}{\langle f|F|f\rangle^2} = \sup_\varphi \frac{\langle f|F|f\rangle + \langle\varphi|F|\varphi\rangle}{\langle f|F|f\rangle^2} . \tag{I.49}$$

The last expression can be simplified provided

$$\langle\varphi|F|\varphi\rangle \leq 0 \quad \text{for} \quad \langle f|F|\varphi\rangle = 0 . \tag{I.50}$$

Let us verify that estimate under the assumption that F possesses only one discrete positive eigenvalue $\lambda > 0$. The corresponding eigenfunction will be denoted by p. The latter should be normalized, that is, $\langle p|p\rangle = 1$. Introducing

$$f_\perp := f - \langle f|p\rangle p\,, \quad \varphi_\perp = \varphi - \langle \varphi|p\rangle p\,, \tag{I.51}$$

we can decompose f and φ in the forms

$$f = f_\parallel + f_\perp \quad \text{and} \quad \varphi = \varphi_\parallel + \varphi_\perp\,, \tag{I.52}$$

respectively. Now, we conclude

$$\langle \varphi|F|f\rangle = 0 = \langle \varphi_\perp|F|f_\perp\rangle + \lambda\langle \varphi_\parallel|f_\parallel\rangle\,, \tag{I.53}$$

and the triangle inequality leads to

$$\langle \varphi_\perp| - F|\varphi_\perp\rangle\langle f_\perp| - F|f_\perp\rangle \geq \langle \varphi_\perp| - F|f_\perp\rangle^2\,. \tag{I.54}$$

Putting the last two findings together, we obtain

$$\langle \varphi_\perp|F|\varphi_\perp\rangle \leq \lambda^2\,\frac{\langle \varphi_\parallel|f_\parallel\rangle^2}{\langle f_\perp|F|f_\perp\rangle}\,. \tag{I.55}$$

Furthermore, we have for the "intermediate" state f

$$\langle f|F|f\rangle = \lambda\langle f_\parallel|f_\parallel\rangle + \langle f_\perp|F|f_\perp\rangle > 0\,, \tag{I.56}$$

and therefore

$$\langle f_\perp|F|f_\perp\rangle^{-1} < -\frac{1}{\lambda}\langle f_\parallel|f_\parallel\rangle^{-1} \tag{I.57}$$

follows. Now, we are in a position to estimate $\langle \varphi|F|\varphi\rangle$. Using Eq. (I.55) together with Eq. (I.57), we obtain

$$\langle \varphi|F|\varphi\rangle = \lambda\langle \varphi_\parallel|\varphi_\parallel\rangle + \langle \varphi_\perp|F|\varphi_\perp\rangle \leq \lambda\langle \varphi_\parallel|\varphi_\parallel\rangle - \lambda\,\frac{\langle \varphi_\parallel|f_\parallel\rangle^2}{\langle f_\parallel|f_\parallel\rangle} \leq 0\,. \tag{I.58}$$

Thus, we can use Eq. (I.50) in Eq. (I.49), leading to

$$\sup_g \frac{\langle g|F|g\rangle}{\langle g|F|f\rangle^2} = \langle f|F|f\rangle^{-1}\,. \tag{I.59}$$

Note that we have assumed that F only has one positive eigenvalue and for f, we demand $\langle f|F|f\rangle > 0$.

Using the results (I.46) and (I.59) in the estimate Eq. (I.43), we obtain

$$\Gamma^2 \leq \frac{\langle f|FN^{-1}F|f\rangle}{\langle f|F|f\rangle} \quad \text{for} \quad \langle f|F|f\rangle > 0\,. \tag{I.60}$$

Considering all possible "intermediate" states f,

$$\Gamma^2 \le \gamma^2 \tag{I.61}$$

follows, which proves the statement that γ agrees with Γ if F has only one positive eigenvalue.

Now, we specify the variational principles for the transverse instability of the ZK equation. We have

$$N \equiv \hat{L}^{-1} ; \tag{I.62}$$

note that for $K \neq 0$, the operator $\hat{L}$ can be inverted on the function space of odd functions. Furthermore,

$$F \equiv \partial_\xi \hat{L} \partial_\xi . \tag{I.63}$$

We apply this operator to odd functions a. Since $\partial_\xi a$ is even, we conclude from our discussion of the spectrum of $\hat{L}$ that F possesses one (and only one) positive, discrete eigenvalue.

Using these identifications, we can formulate the expressions for the ZK growth rates:

$$\Gamma^2 = \sup_{\varphi \text{ odd}} \frac{\langle \varphi | \partial_\xi \hat{L} \partial_\xi | \varphi \rangle}{\langle \varphi | \hat{L}^{-1} | \varphi \rangle} , \tag{I.64}$$

$$\gamma^2 = \inf_{\varphi \in M} \frac{\langle \varphi | \partial_\xi \hat{L} \partial_\xi \hat{L} \partial_\xi \hat{L} \partial_\xi | \varphi \rangle}{\langle \varphi | \partial_\xi \hat{L} \partial_\xi | \varphi \rangle} , \tag{I.65}$$

where M is the set of odd functions φ for which $\langle \varphi | F | \varphi \rangle > 0$. By defining

$$\psi = \hat{L} \partial_\xi \varphi , \tag{I.66}$$

we can reformulate Eq. (I.65) as

$$\gamma^2 = \inf_{\psi \in \tilde{M}} \frac{\langle \psi | \partial_\xi \hat{L} \partial_\xi | \psi \rangle}{\langle \psi | \hat{L}^{-1} | \psi \rangle} , \tag{I.67}$$

where $\tilde{M}$ is the set of even localized functions ψ with $\langle \psi | \hat{L}^{-1} | \psi \rangle < 0$. The integration by parts additionally requires

$$\int_{-\infty}^{+\infty} \hat{L}^{-1} \psi \, d\xi = 0 . \tag{I.68}$$

We shall not discuss the details of the latter restriction.

Let us begin with an analytical consideration. Inspecting the sign of the numerator on the right-hand side of Eq. (I.64), we find the cut-off K_c by

$$K_c^2 = 4 \sup_{\varphi \text{ odd}} \frac{-\langle \partial_\xi \varphi | \hat{L}(K=0) | \partial_\xi \varphi \rangle}{\langle \partial_\xi \varphi | \partial_\xi \varphi \rangle} \tag{I.69}$$

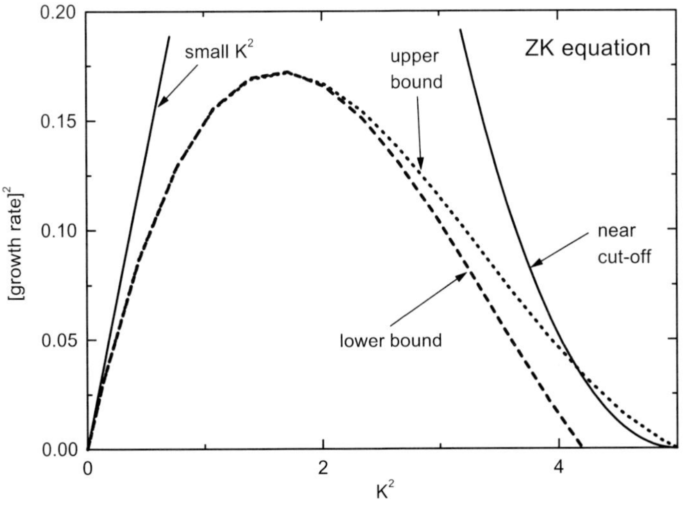

Figure I.1 Upper (dotted) and lower (dashed) bounds, respectively, for the square of the exponential growth rate. The border lines are calculated from the variational principles. In addition, the results from perturbation theory are depicted.

such that instability occurs for $K < K_c$. Since $\partial_\xi \varphi$ is an even function, we conclude from the spectral properties of $\hat{L}(K = 0)$ that $K_c^2 \leq 5$. There is a small problem related to the proof that $K_c^2 = 5$ is the exact cut-off. The reason is that $\varphi = \partial_\xi^{-1} \text{sech}^3 \xi$ belonging to the eigenvalue -5 is not vanishing for $\xi \to \pm\infty$. Nevertheless, $K_c^2 = 5$ since one can choose a series of weakly localized test functions to calculate Eq. (I.69). Details can be found in [265]. There, the variational principles Eqs. (I.64) and (I.67) were evaluated numerically. The result is reproduced in Figure I.1. Using a finite number of test functions, for example,

$$\psi_i = \hat{L} \text{sech}^{\frac{i}{2}} \xi \,, \quad i = 1, \ldots, 20 \tag{I.70}$$

in Eq. (I.67) and

$$\varphi_i = \hat{L} \partial_\xi \text{sech}^{\frac{i}{2}} \xi \,, \quad i = 1, \ldots, 20 \tag{I.71}$$

in Eq. (I.64), we obtain the shown upper and lower bounds, respectively. As can be seen from Figure I.1, the borderlines narrow the possible values for the growth rates quite successfully. In addition, we would like to remark that the mentioned difficulties with convergence problems of the perturbative treatments are not essential. The small-K^2 expansion as well as the expansion near the cut-off (also shown in Figure I.1) turn out to be meaningful.

Appendix J
Self-Adjointness of the Operator f Appearing in Hydromagnetic Variational Principles

In Section 7.3.1, we introduced the operator

$$F(\xi) = \nabla(\xi \cdot \nabla p_0 + \gamma p_0 \nabla \cdot \xi) + \frac{1}{\mu_0}\{\nabla \times [\nabla \times (\xi \times B_0)]\} \times B_0 + \frac{1}{\mu_0}(\nabla \times B_0) \times [\nabla \times (\xi \times B_0)] \,. \tag{J.1}$$

It is a linear operator being applied to displacements ξ. The operator contains the space-dependent equilibrium quantities B_0 and p_0. For realistic, for example, tokamak values of these quantities, the operator F is difficult to handle, and most of the investigations are performed numerically. The general stability problem of (ideal) magnetohydrodynamic systems was discussed under the assumption that F is a self-adjoint operator. For proving the self-adjointness property, we investigate the expression $\xi' \cdot F(\xi)$, taking into account appropriate boundary conditions. Furthermore, let us introduce the magnetic field perturbations

$$Q := \nabla \times (\xi \times B_0) \,, \quad Q' := \nabla \times (\xi' \times B_0) \,, \tag{J.2}$$

High Temperature Plasmas, Theory and Mathematical Tools for Laser and Fusion Plasmas, First Edition.
Karl-Heinz Spatschek.

which have been used already in Section 7.3.1. The following vector identity, along with Eq. (7.239), follows

$$
\begin{aligned}
&\nabla(\xi \cdot \nabla p_0) + \frac{1}{\mu_0}(\nabla \times \boldsymbol{B}_0) \times \boldsymbol{Q} \\
&= (\nabla \xi) \cdot \nabla p_0 + \xi \cdot \nabla\nabla p_0 + \frac{1}{\mu_0}(\nabla \times \boldsymbol{B}_0) \times \boldsymbol{Q} \\
&= \left[(\nabla p_0) \times \nabla\right] \times \xi + (\nabla \cdot \xi)\nabla p_0 + \xi \cdot \nabla\nabla p_0 + \left(\frac{1}{\mu_0}\right)(\nabla \times \boldsymbol{B}_0) \times \boldsymbol{Q} \\
&= \frac{1}{\mu_0}\{[(\nabla \times \boldsymbol{B}_0) \times \boldsymbol{B}_0] \times \nabla\} \times \xi + (\nabla \cdot \xi)\nabla p_0 \\
&\quad + \xi \cdot \nabla\nabla p_0 + \frac{1}{\mu_0}(\nabla \times \boldsymbol{B}_0) \times \boldsymbol{Q} \\
&= \left(\frac{1}{\mu_0}\right)\left[\boldsymbol{B}_0(\nabla \times \boldsymbol{B}_0) \cdot \nabla - (\nabla \times \boldsymbol{B}_0)\boldsymbol{B}_0 \cdot \nabla\right] \times \xi + \nabla \cdot \xi \nabla p_0 \\
&\quad + \xi \cdot \nabla\nabla \rho_0 + \frac{1}{\mu_0}(\nabla \times \boldsymbol{B}_0) \times [\boldsymbol{B}_0 \cdot \nabla\xi - \xi \cdot \nabla \boldsymbol{B}_0 - \boldsymbol{B}_0\nabla \cdot \xi] \\
&= \frac{1}{\mu_0}\{\boldsymbol{B}_0 \times \left[(\nabla \times \boldsymbol{B}_0) \cdot \nabla\xi\right] - \boldsymbol{B}_0 \times \left[\xi \cdot \nabla(\nabla \times \boldsymbol{B}_0)\right] \\
&\quad - \xi \cdot \nabla\left[(\nabla \times \boldsymbol{B}_0) \times \boldsymbol{B}_0\right] + \boldsymbol{B}_0 \times (\nabla \times \boldsymbol{B}_0)\nabla \cdot \xi\} \\
&\quad + (\nabla \cdot \xi)\nabla p_0 + \xi \cdot \nabla\nabla p_0 \\
&= \frac{1}{\mu_0}\boldsymbol{B}_0 \times \{\nabla \times \left[\xi \times (\nabla \times \boldsymbol{B}_0)\right]\} + \frac{1}{\mu_0}\boldsymbol{B}_0 \times (\nabla \times \boldsymbol{B}_0)\nabla \cdot \xi \\
&= \frac{1}{\mu_0}\boldsymbol{B}_0 \times \{\nabla \times \left[\xi \times (\nabla \times \boldsymbol{B}_0)\right]\} - (\nabla \cdot \xi)\nabla p_0 \, .
\end{aligned}
\tag{J.3}
$$

Making use of it, we may write

$$
\begin{aligned}
&\xi' \cdot \boldsymbol{F}(\xi) \\
&= \xi' \cdot \left[\nabla(\gamma p_0 \nabla \cdot \xi + \xi \cdot \nabla p_0) + \frac{1}{\mu_0}(\nabla \times Q) \times \boldsymbol{B}_0 + \frac{1}{\mu_0}(\nabla \times \boldsymbol{B}_0) \times Q \right] \\
&= \xi' \cdot \left[\nabla(\gamma p_0 \nabla \cdot \xi) + \frac{1}{\mu_0}(\nabla \times Q) \times \boldsymbol{B}_0 \right. \\
&\quad \left. + \frac{1}{\mu_0} \boldsymbol{B}_0 \times \{\nabla \times [\xi \times (\nabla \times \boldsymbol{B}_0)]\} - (\nabla \cdot \xi) \nabla p_0 \right] \\
&= \nabla \cdot \left\{ \xi' \gamma p_0 \nabla \cdot \xi - \frac{1}{\mu_0} Q \times (\xi' \times \boldsymbol{B}_0) + \frac{1}{\mu_0} [\xi \times (\nabla \times \boldsymbol{B}_0)] \times (\xi' \times \boldsymbol{B}_0)] \right\} \\
&\quad - \gamma p_0 (\nabla \cdot \xi)(\nabla \cdot \xi') - \frac{1}{\mu_0} Q \cdot Q' + \frac{1}{\mu_0} [\xi \times (\nabla \times \boldsymbol{B}_0)] \cdot Q' - (\nabla \cdot \xi) \xi' \cdot \nabla p_0 \\
&= \nabla \cdot \left(\xi' \gamma p_0 \nabla \cdot \xi - \frac{1}{\mu_0} Q \times (\xi' \times \boldsymbol{B}_0) \right. \\
&\quad \left. + \xi' \frac{1}{\mu_0} \xi \cdot (\nabla \times \boldsymbol{B}_0) \times \boldsymbol{B}_0 - \frac{1}{\mu_0} \boldsymbol{B}_0 \xi' \times \xi \cdot \nabla \times \boldsymbol{B}_0 \right\} \\
&\quad - \frac{1}{\mu_0} Q \cdot Q' + \xi \cdot \frac{1}{\mu_0} (\nabla \times \boldsymbol{B}_0) \times Q' - (\nabla \cdot \xi)(\gamma p_0 \nabla \cdot \xi' + \xi' \cdot \nabla p_0) \\
&= \nabla \cdot \left\{ \xi' \gamma p_0 \nabla \cdot \xi - \xi \gamma p_0 \nabla \cdot \xi' - \frac{1}{\mu_0} Q \times (\xi' \times \boldsymbol{B}_0) \right. \\
&\quad \left. + \frac{1}{\mu_0} Q' \times (\xi \times \boldsymbol{B}_0) + \xi' \xi \cdot \nabla p_0 - \xi \xi' \cdot \nabla p_0 - \boldsymbol{B}_0 \xi' \times \xi \cdot \nabla p_0 \right\} \\
&\quad + \nabla \cdot \left[\xi (\gamma p_0 \nabla \cdot \xi' + \xi' \cdot \nabla p_0) - \frac{1}{\mu_0} Q' \times (\xi \times \boldsymbol{B}_0) \right] \\
&\quad - \frac{1}{\mu_0} Q \cdot Q' + \xi \cdot \frac{1}{\mu_0} (\nabla \times \boldsymbol{B}_0) \times Q' - (\nabla \cdot \xi) \left(\gamma p_0 \nabla \cdot \xi' + \xi' \cdot \nabla p_0 \right) \\
&= \nabla \cdot \left\{ \xi' \left[\gamma p_0 \nabla \cdot \xi - \frac{1}{\mu_0} Q \cdot \boldsymbol{B}_0 + \xi \cdot \nabla p_0 \right] \right. \\
&\quad + \boldsymbol{B}_0 \left(\left(\frac{1}{\mu_0} \right) (\xi' \cdot Q - \xi \cdot Q') + \xi \times \xi' \cdot \nabla p_0 \right) \\
&\quad \left. - \xi \left[\gamma p_0 \nabla \cdot \xi' - \frac{1}{\mu_0} Q' \cdot \boldsymbol{B}_0 + \xi' \cdot \nabla p_0 \right] \right\} \\
&\quad + \xi \cdot \left[\nabla (\gamma p_0 \nabla \cdot \xi' + \xi' \cdot \nabla p_0) + \frac{1}{\mu_0} (\nabla \times \boldsymbol{B}_0) \times Q' + \frac{1}{\mu_0} (\nabla \times Q') \times \boldsymbol{B}_0 \right] \\
&= \nabla \cdot \left\{ \xi' \left[\gamma p_0 \nabla \cdot \xi - \frac{1}{\mu_0} (Q \cdot \boldsymbol{B}_0 + \xi \cdot (\nabla \boldsymbol{B}_0) \cdot \boldsymbol{B}_0) + \xi \cdot \nabla \left(p_0 + \frac{B_0^2}{2} \mu_0 \right) \right] \right. \\
&\quad - \xi \left[\gamma p_0 \nabla \cdot \xi' - \frac{1}{\mu_0} (Q' \cdot \boldsymbol{B}_0 + \xi' \cdot (\nabla \boldsymbol{B}_0) \cdot \boldsymbol{B}_0) + \xi' \cdot \nabla \left(p_0 + \frac{B_0^2}{2} \mu_0 \right) \right] \\
&\quad \left. + \boldsymbol{B}_0 \left(\frac{1}{\mu_0} (\xi' \cdot Q - \xi \cdot Q') + \xi \times \xi' \cdot \nabla p_0 \right) \right\} + \xi \cdot \boldsymbol{F}(\xi') \, .
\end{aligned}
\tag{J.4}
$$

The last relation shows that $\xi' \cdot F(\xi)$ differs from $\xi \cdot F(\xi')$ by a divergence term. After integration, the latter leads to surface contributions [451]. If they vanish, self-adjointness of F is shown. Boundary conditions become important. Let us assume that the plasma is confined by vacuum fields; the whole arrangement is embedded in a vessel with surface S'. Because of $\nabla \cdot B = 0$, the normal components of B are continuous in the transition from one area to the other. We shall assume $n \cdot B_0 = 0$ on S (see below) where n is the normal vector on S. Integrating Eq. (J.4) over the volume of the whole system leads to

$$\begin{aligned}
&\int d^3r \left[\xi' \cdot F\{\xi\} - \xi \cdot F\{\xi'\}\right] \\
&= \int d^2 r n \cdot \left\{ \xi' \left[\gamma p_0 \nabla \cdot \xi - \frac{1}{\mu_0} (Q \cdot B_0 + \xi \cdot (\nabla B_0) \cdot B_0) \right.\right. \\
&\quad \left. + \xi \cdot \nabla \left(p_0 + \frac{B_0^2}{2\mu_0} \right) \right] \\
&\quad \left. - \xi \left[\gamma p_0 \nabla \cdot \xi' - \frac{1}{\mu_0} (Q' \cdot B_0 + \xi' \cdot (\nabla B_0) \cdot B_0) + \xi' \cdot \nabla \left(p_0 + \frac{B_0^2}{2\mu_0} \right) \right] \right\} .
\end{aligned} \tag{J.5}$$

We have used

$$n \cdot B_0|_S = 0 , \tag{J.6}$$

which should be supported by arguments (see below).

If the plasma is bounded by a fixed surface originating from a conducting vessel, we can assume $n \cdot \xi = 0$ at the surface, and self-adjointness follows immediately. However, if we only integrate over the plasma volume (where the MHD equations apply), the boundary S to the vacuum is variable, as already discussed.

Let us first comment on Eq. (J.6). Continuity of the normal component of B is written as $n \cdot [B] = 0$. When initially $n \cdot B_{vacuum} = 0$, then also $n \cdot B_0 = 0$ holds. The time-development of the normal component of B is calculated according to

$$\frac{d}{dt}(n \cdot B) = B \cdot \frac{dn}{dt} + n \cdot \frac{dB}{dt} . \tag{J.7}$$

Note that

$$\begin{aligned}
\frac{dB}{dt} &= \frac{\partial B}{\partial t} + u \cdot \nabla B = -\nabla \times E + u \cdot \nabla B \\
&= \nabla \times (u \times B) + u \cdot \nabla B = B \cdot \nabla u - B \nabla \cdot u .
\end{aligned} \tag{J.8}$$

The time-derivative of the normal vector turns out to be

$$\frac{d}{dt} n = -n(\nabla \cdot u) + n n \cdot (\nabla u) \cdot n . \tag{J.9}$$

For the proof, we start with a surface element $d^2 r = n d^2 r = dr \times dr'$ on S. The displacement ξ changes dr by $\delta dr = dr \cdot \nabla \xi$ (and a similar formula holds for

$d\boldsymbol{r}'$). Therefore, the change of the surface element because of a displacement is to first order

$$\delta d^2\boldsymbol{r} = (d\boldsymbol{r}\cdot\nabla\boldsymbol{\xi})\times d\boldsymbol{r}' + d\boldsymbol{r}\times(d\boldsymbol{r}'\cdot\nabla\boldsymbol{\xi}) = -\left[(d\boldsymbol{r}\times d\boldsymbol{r}')\times\nabla\right]\times\boldsymbol{\xi}\ . \quad (\text{J.10})$$

Using the definition of $d^2\boldsymbol{r}$ on the left-hand side, we can write

$$d^2r\delta\boldsymbol{n} + \boldsymbol{n}\delta d^2r = -d^2r(\boldsymbol{n}\times\nabla)\times\boldsymbol{\xi}\ . \quad (\text{J.11})$$

Multiplication with $\boldsymbol{n}$ and using $\boldsymbol{n}\cdot\delta\boldsymbol{n} = 0$ (since $\boldsymbol{n}$ is a unit vector) leads to

$$\delta d^2r = -d^2r\boldsymbol{n}\cdot(\boldsymbol{n}\times\nabla)\times\boldsymbol{\xi} = -d^2r\boldsymbol{n}\times(\boldsymbol{n}\times\nabla)\cdot\boldsymbol{\xi}\ . \quad (\text{J.12})$$

Inserting that into Eq. (J.11) gives

$$\delta\boldsymbol{n} = -(\boldsymbol{n}\times\nabla)\times\boldsymbol{\xi} + \boldsymbol{n}\boldsymbol{n}\times(\boldsymbol{n}\times\nabla)\cdot\boldsymbol{\xi} = -(\nabla\boldsymbol{\xi})\cdot\boldsymbol{n} + \boldsymbol{n}\boldsymbol{n}\cdot(\nabla\boldsymbol{\xi})\cdot\boldsymbol{n}\ . \quad (\text{J.13})$$

Having in mind $\boldsymbol{\xi} = \boldsymbol{u}\delta t$ and $\dot{\boldsymbol{n}}\delta t = \delta\boldsymbol{n}$, we finally arrive at

$$\frac{d}{dt}(\boldsymbol{n}\cdot\boldsymbol{B}) = (\boldsymbol{n}\cdot\boldsymbol{B})\boldsymbol{n}\times(\boldsymbol{n}\times\nabla)\cdot\boldsymbol{u} \quad (\text{J.14})$$

for $n\cdot\boldsymbol{B}$. It is a linear differential equation for $\boldsymbol{n}\cdot\boldsymbol{B}$ at the (moving) surface S. If initially $\boldsymbol{n}\cdot\boldsymbol{B}\,|_S = \boldsymbol{n}\cdot\boldsymbol{B}_0\,|_S = 0$, it will be true for all times as has been claimed.

Next, we calculate the surface terms in Eq. (J.5). We shall show that they vanish by applying the first order relation

$$-\gamma p_0\nabla\cdot\boldsymbol{\xi} + \frac{\boldsymbol{B}_0}{\mu_0}\cdot\left[\boldsymbol{Q} + (\boldsymbol{\xi}\cdot\nabla)\boldsymbol{B}_0\right] = \frac{\boldsymbol{B}_{vacuum}}{\mu_0}\cdot\left[\delta\boldsymbol{B}_{vacuum} + (\boldsymbol{\xi}\cdot\nabla)\boldsymbol{B}_{vacuum}\right] \quad (\text{J.15})$$

at the surface S. This relation follows from the jump condition of momentum at the surface S. Let us define

$$\boldsymbol{\sigma} := \rho\boldsymbol{u}\ , \quad \mathcal{G} := \rho\boldsymbol{u}\boldsymbol{u} + \left(p + \frac{B^2}{2\mu_0}\right)I - \frac{\boldsymbol{B}\boldsymbol{B}}{\mu_0}\ . \quad (\text{J.16})$$

With these definitions, the momentum balance of ideal MHD can be written as

$$\frac{\partial\boldsymbol{\sigma}}{\partial t} + \nabla\cdot\mathcal{G} = 0\ . \quad (\text{J.17})$$

Rewriting the last equation in the form

$$\begin{aligned}\boldsymbol{n}\cdot\nabla\left[\boldsymbol{n}\cdot(\mathcal{G} - \boldsymbol{u}\boldsymbol{\sigma})\right] = &-\frac{d\boldsymbol{\sigma}}{dt} - \boldsymbol{\sigma}\boldsymbol{n}\cdot\nabla(\boldsymbol{n}\cdot\boldsymbol{u}) + \boldsymbol{n}\times(\boldsymbol{u}\times\boldsymbol{n})\cdot\nabla\boldsymbol{\sigma}\\ &+ \boldsymbol{n}\cdot(\nabla\boldsymbol{n})\cdot\mathcal{G} + \boldsymbol{n}\times(\boldsymbol{n}\times\nabla)\cdot\mathcal{G}\ ,\end{aligned} \quad (\text{J.18})$$

we can easily derive a jump condition across S since the right-hand side contains no gradients of $\boldsymbol{\sigma}$ and $\mathcal{G}$ along $\boldsymbol{n}$ (and otherwise quantities are continuous). Integrating across S over a distance ds with $\boldsymbol{n}\cdot\nabla = \partial_s$ leads to the jump condition

$$\left[\boldsymbol{n}\cdot(\mathcal{G} - \boldsymbol{u}\boldsymbol{\sigma})\right] = 0\ . \quad (\text{J.19})$$

Because of $\boldsymbol{n} \cdot \boldsymbol{B} = 0$ and the explicit forms of $\boldsymbol{\sigma}$ as well as $\mathcal{G}$, we find the pressure balance

$$\left[p + \frac{B^2}{2\mu_0}\right] = 0 . \tag{J.20}$$

The tangential component leads to

$$\boldsymbol{n} \times \nabla\left(p_0 + \frac{B_0^2}{2\mu_0} - \frac{B_{vacuum}^2}{2\mu_0}\right) = 0 . \tag{J.21}$$

When further analyzing its linearized version, we use

$$\frac{d\boldsymbol{B}}{dt} = \frac{\partial \boldsymbol{B}}{\partial t} + \boldsymbol{u} \cdot \nabla \boldsymbol{B} = \nabla \times (\boldsymbol{u} \times \boldsymbol{B}) + \boldsymbol{u} \cdot \nabla \boldsymbol{B} , \tag{J.22}$$

such that

$$\boldsymbol{B}(\boldsymbol{r}, t) - \boldsymbol{B}(\boldsymbol{r}_s, 0) \approx \boldsymbol{Q}(\boldsymbol{r}_s) + (\boldsymbol{\xi} \cdot \nabla)\boldsymbol{B}(\boldsymbol{r}_s) \tag{J.23}$$

for displacements of $\boldsymbol{r} = \boldsymbol{r}_s + \boldsymbol{\xi}$ in the time δt.

In a similar way, we obtain from

$$\frac{dp}{dt} = -\gamma p \nabla \cdot \boldsymbol{u} \tag{J.24}$$

the relation

$$p(\boldsymbol{r}, t) - p(\boldsymbol{r}_s, 0) = -\gamma p(\boldsymbol{r}_s, 0)\nabla \cdot \boldsymbol{\xi} . \tag{J.25}$$

Obviously, in vacuum $p = 0$; furthermore, we introduce the change of the vacuum magnetic field according to

$$\boldsymbol{B}_{vacuum}(\boldsymbol{r}, t) = \boldsymbol{B}_{vacuum}(\boldsymbol{r}_s, 0) + \delta \boldsymbol{B}_{vacuum} + \boldsymbol{\xi} \cdot \nabla \boldsymbol{B}_{vacuum}(\boldsymbol{r}_s) . \tag{J.26}$$

Note that for a displacement $\boldsymbol{\xi}'$ (instead of $\boldsymbol{\xi}$), we should replace $\boldsymbol{\xi}$ by $\boldsymbol{\xi}'$ and $\delta \boldsymbol{B}_{vacuum}$ by $\delta \boldsymbol{B}'_{vacuum}$ and we obtain a similar formula. Putting everything together, we have justified Eq. (J.15).

For evaluating the surface terms in Eq. (J.5), we next apply

$$p_0 + \frac{B_0^2}{2\mu_0} = \left(p_0 + \frac{B_0^2}{2\mu_0} - \frac{B_{vacuum}^2}{2\mu_0}\right) + \frac{B_{vacuum}^2}{2\mu_0} \tag{J.27}$$

and use Eq. (J.15) to obtain

$$\begin{aligned} &\mu_0 \int d^3r \left[\boldsymbol{\xi}' \cdot \boldsymbol{F}\{\boldsymbol{\xi}\} - \boldsymbol{\xi} \cdot \boldsymbol{F}\{\boldsymbol{\xi}'\}\right] \\ &= \int_S d^2r\, \boldsymbol{n} \left[-\boldsymbol{\xi}' \boldsymbol{B}_{vacuum} \cdot \nabla \times \delta \boldsymbol{A} + \boldsymbol{\xi} \boldsymbol{B}_{vacuum} \cdot \nabla \times \delta \boldsymbol{A}'\right], \end{aligned} \tag{J.28}$$

where $\delta \boldsymbol{B}_{vacuum} = \nabla \times \delta \boldsymbol{A}$ and $\delta \boldsymbol{B}'_{vacuum} = \nabla \times \delta \boldsymbol{A}'$ has been introduced. We further discuss $\delta \boldsymbol{A}$ and its relation to the electric field. The jump condition of the electric field for $\boldsymbol{u} \neq 0$ is

$$\boldsymbol{n} \times [E] = \boldsymbol{n} \cdot \boldsymbol{u}[B] . \tag{J.29}$$

In the present case, this means

$$\boldsymbol{n} \times \delta \boldsymbol{E}_{vacuum} + \boldsymbol{n} \times (\boldsymbol{u} \times \boldsymbol{B}) = (\boldsymbol{n} \cdot \boldsymbol{u}) \boldsymbol{B}_{vacuum} - (\boldsymbol{n} \cdot \boldsymbol{u}) \boldsymbol{B} , \tag{J.30}$$

or, to first order in the displacement, for $\boldsymbol{n} \cdot \boldsymbol{B} = 0$,

$$\boldsymbol{n} \times \delta \boldsymbol{E}_{vacuum} = (\boldsymbol{n} \cdot \boldsymbol{u}) \boldsymbol{B}_{vacuum} . \tag{J.31}$$

Within the Coulomb gauge $\delta \boldsymbol{E}_{vacuum} = -\delta \dot{\boldsymbol{A}}$, the relation

$$-\boldsymbol{n} \times \delta \boldsymbol{A} = (\boldsymbol{n} \cdot \boldsymbol{\xi}) \boldsymbol{B}_{vacuum} \tag{J.32}$$

follows. A similar formula is true for the primed variables resulting from $\boldsymbol{\xi}'$. This allows one to reformulate Eq. (J.28) as

$$\begin{aligned} \mu_0 \int d^3 r \left[\boldsymbol{\xi}' \cdot \boldsymbol{F}\{\boldsymbol{\xi}\} - \boldsymbol{\xi} \cdot \boldsymbol{F}\{\boldsymbol{\xi}'\}\right] \\ = \int_S d^2 r \left[(\boldsymbol{n} \times \delta \boldsymbol{A}') \cdot (\nabla \times \delta \boldsymbol{A}) - (\boldsymbol{n} \times \delta \boldsymbol{A}) \cdot (\nabla \times \delta \boldsymbol{A}')\right] . \end{aligned} \tag{J.33}$$

Adding to the right-hand side a similar term with integration over the conducting surface S' will not change the result. The reason is that the tangential component of the electric field ($\sim \boldsymbol{n} \times \delta \boldsymbol{A}$) at an ideal conductor vanishes. Thus, the last relation is identical to

$$\begin{aligned} \mu_0 \int d^3 r \left[\boldsymbol{\xi}' \cdot \boldsymbol{F}\{\boldsymbol{\xi}\} - \boldsymbol{\xi} \cdot \boldsymbol{F}\{\boldsymbol{\xi}'\}\right] \\ = \int_{S+S'} \left[(d\boldsymbol{F} \times \delta \boldsymbol{A}') \cdot (\nabla \times \delta \boldsymbol{A}) - (d\boldsymbol{F} \times \delta \boldsymbol{A}) \cdot (\nabla \times \delta \boldsymbol{A}')\right] \\ = \int_{S+S'} d\boldsymbol{F} \cdot \left[\delta \boldsymbol{A}' \times (\nabla \times \delta \boldsymbol{A}) - \delta \boldsymbol{A} \times (\nabla \times \delta \boldsymbol{A}')\right] \\ = -\int_{V_{vacuum}} d^3 r \left[\delta \boldsymbol{A}' \cdot \nabla \times \nabla \times \delta \boldsymbol{A} - \delta \boldsymbol{A} \cdot \nabla \times \nabla \times \delta \boldsymbol{A}'\right] , \end{aligned} \tag{J.34}$$

when the Gauss theorem is applied. The right-hand side vanishes since no currents ($\sim \nabla \times \delta \boldsymbol{B}_{vacuum}$) are flowing in the vacuum region. This concludes the proof that $\boldsymbol{F}$ is a self-adjoint operator.

References

1 Spitzer Jr., L. (1956) *The Physics of Fully Ionized Gases*, Interscience.

2 Stix, T. (1962) *The Theory of Plasma Waves*, McGraw-Hill.

3 Balescu, R. (1963) *Statistical Mechanics of Charged Particles*, Interscience, New York.

4 Klimontovich, Y. (1982) *Kinetic Theory of Nonideal Gases and Nonideal Plasmas*, Pergamon Press, Oxford.

5 Nicholson, D. (1983) *Introduction to Plasma Physics*, John Wiley & Sons, Inc., New York.

6 Cairns, R. (1985) *Plasma Physics*, Blackie, Glasgow, Scotland.

7 Krall, N. and Trivelpiece, A. (1986) *Principles of Plasma Physics*, San Francisco Press.

8 Nishikawa, K. and Wakatani, M. (1990) *Plasma Physics: Basic Theory with Fusion Applications*, Springer, Berlin.

9 Spatschek, K.-H. (1990) *Theoretische Plasmaphysik*, Teubner, Stuttgart.

10 Ichimaru, S. (1992) *Statistical Plasma Physics*, Addison-Wesley, New York.

11 Goldston, R.J. and Rutherford, P.H. (1995) *Introduction to Plasma Physics*, Institute of Physics, Philadelphia.

12 Itoh, K., Itoh, S.I., and Fukuyama, A. (1999) *Transport and Structural Formation in Plasmas*, Institute of Physics Publishing, Bristol.

13 Woods, L.C. (2004) *Physics of Plasmas*, Wiley-VCH Verlag GmbH, Weinheim.

14 Stacey, W.M. (2005) *Fusion Plasma Physics*, Wiley-VCH Verlag GmbH, Weinheim.

15 Bellan, P. (2006) *Fundamentals of Plasma Physics*, Cambridge U.P.

16 Chen, F. (1984) *Introduction to Plasma Physics*, Plenum, New York.

17 Hasegawa, A. (1975) *Plasma Instabilities and Nonlinear Effects*, Springer, Berlin.

18 Carroll, B. and Ostlie, D. (1996) *Modern Astrophysics*, Addison-Wesley, Reading.

19 Spatschek, K.-H. (2003) *Astrophysik*, Teubner, Stuttgart.

20 Kruer, W. (1988) *The Physics of Laser-Plasma Interaction*, Addison-Wesley.

21 Atzeni, S. and ter Vehn, J.M. (2004) *The Physics of Inertial Fusion. Beam Plasma Interaction, Hydrodynamics, Hot Dense Matter*, Oxford Univ. Press, Oxford.

22 d'Agostino, R., d'Agostino, R., Favia, P., Kawai, Y., Ikegami, H., Sato, N., and Arefi-Khonsari, F. (2007) *Advanced Plasma Technolog*, Wiley-VCH Verlag GmbH, Weinheim.

23 Kadomtsev, B. (1965) *Plasma Turbulence*, Academic Press, New York.

24 Kadomtsev, B. (1993) *Tokamak Plasma: A Complex Physical System*, IoP.

25 Hazeltine, R.D. and Meiss, J.D. (1992) *Plasma Confinement*, Addison-Wesley, Redwood City, Calif.

26 White, R. (2001) *The Theory of Toroidally Confined Plasmas*, Imperial College Press, London.

27 Reichl, L. (1980) *A Modern Course in Statistical Physics*, Edward Arnold.

28 Saha, M. (1921) On a physical theory of stellar spectra. *Proc. R. Soc. Lond.*, **99**, 135–153.

29 Book, D.L. (1990) *NRL Plasma Formulary*, Naval Research Lab, Washington.

30 McWhirter, R. (1965) Spectral intensities. *Plasma Diagnostic Techniques*, (eds

High Temperature Plasmas, Theory and Mathematical Tools for Laser and Fusion Plasmas, First Edition.
Karl-Heinz Spatschek.

R.H. Huddlestone and S.L. Leonard), Academic Press, New York.
31 Griem, H. (1966) *Plasma Spectroscopy*, Academic Press, New York.
32 Lawson, J. (1957) Some criteria for a power producing thermonuclear reactor. *Proc. Phys. Soc.*, **70**, 6–10.
33 Goldstein, H. (1963) *Klassische Mechanik*, Akademische Verlagsgesellschaft, Frankfurt.
34 Nolting, W. (1993) *Grundkurs: Theoretische Physik – 3 Elektrodynamik*, Zimmermann-Neufang, Ulmen.
35 Schätzing, F. (2009) *Limit*, Kiepenheuer and Witsch, Köln.
36 Atkinson, R.d.E. and Houtermans, F.G. (1929) Zur Frage der Aufbaumöglichkeit der Elemente in Sternen. *Z. Phys. A*, **54**, 656–665.
37 Rebhan, E. (1998) Thermonuclear burn criteria. *Transact. Fusion Technol.*, **33**, 19–27.
38 Northrop, T.G. (1963) *The Adiabatic Motion of Charged Particles*, John Wiley & Sons, Inc., New York.
39 Kruskal, M.D. (1962) *J. Math. Phys.*, **3**, 806.
40 Balescu, R. (1988) *Transport Processes in Plasmas Vol. 1: Classical Transport*, North Holland, Amsterdam.
41 Littlejohn, R.G. (1979) *J. Math. Phys.*, **20**, 2445.
42 Littlejohn, R.G. (1981) *Phys. Fluids*, **24**, 1730.
43 Littlejohn, R.G. (1983) *J. Plasma Phys.*, **29**, 111.
44 Littlejohn, R.G. (1985) Differential forms and canonical variables for drift motion in toroidal geometry. *Phys. Fluids*, **28**, 2015.
45 Ecker, G. (1972) *Theory of Fully Ionized Plasmas*, Academic Press, New York.
46 Holtsmark, J. (1919) *Ann. Phys. (Leipzig)*, **58**, 577.
47 Ecker, G. and Müller, K.G. (1958) *Z. Phys.*, **153**, 317.
48 Baranger, M. and Mozer, B. (1959) *Phys. Rev.*, **115**, 521.
49 Griem, H.R. (1964) *Plasma Spectroscopy*, McGraw-Hill.
50 Griem, H.R. (1997) *Principles of Plasma Spectroscopy*, Cambridge U.P.
51 Balescu, R. (1960) *Phys. Fluids*, **4**, 94.
52 Lenard, A. (1960) *Ann. Phys. (NY)*, **10**, 390.
53 Guernsey, R.L. (1962) *Phys. Fluids*, **5**, 322.
54 Riley, K., Hobson, M., and Bence, S. (1997) *Mathematical Methods for Physics and Engineering*, Cambridge U.P.
55 Van Kampen, N. (1996) Stochastic processes in physics and chemistry. *Phys. Fluids*, **19**, 11.
56 Scott, B. and Smirnov, J. (2010) *Phys. Plasmas*, **17**, 112302.
57 Scott, B. (2010) *Phys. Plasmas*, **17**, 102306.
58 Balescu, R. (1988) *Transport Processes in Plasmas: 2. Neoclassical Transport Theory*, North-Holland, Amsterdam.
59 Hazeltine, R.D. and Meiss, J.D. (1992) *Plasma Confinement*, Addison-Wesley, Redwood City.
60 Hasegawa, A. and Mima, K. (1978) Pseudo-three-dimensional turbulence in magnetized nonuniform plasma. *Phys. Fluids*, **21**, 87.
61 Naulin, V., Spatschek, K.-H., and Hasegawa, A. (1992) Selective decay within a one-field model of dissipative drift-wave turbulence. *Phys. Fluids B*, **4**, 2672–2674.
62 Spatschek, K.-H. and Uhlenbusch, J. (eds) (1994) *Contributions to High-Temperature Plasma Physics*, Akademie Verlag, Berlin.
63 Naulin, V., Spatschek, K.-H., Musher, S., and Piterbarg, L. (1995) Properties of a two-nonlinearity model of driftwave turbulence. *Phys. Plasmas*, **2**, 2640–2652.
64 Braginskii, S.I. (1965) *Reviews of Plasma Physics* (ed. M. Leontovich), Consultants Bureau, Vol. 1, p. 205.
65 Spatschek, K.-H., Eberhard, M., and Friedel, H. (1998) On models for magnetic field line diffusion. *Phys. Mag.*, **20**, 85–93.
66 Cowling, T. (1957) *Magnetohydrodynamics*, John Wiley & Sons, Inc., New York.
67 Biskamp, D. (1997) *Nonlinear Magnetohydrodynamics*, Cambridge U.P., Cambridge.
68 Wesson, J. (2004) *Tokamaks*, Clarendon Press, Oxford.

69 Goedbloed, J.P., Keppens, R., and Poedts, S. (2010) *Advanced Magnetohydrodynamics*, Cambridge U.P.

70 Grad, H. and Rubin, H. (1958) MHD equilibrium in an axisymmetric toroid, in *Proceedings of the 2nd UN Conf. on the Peaceful Uses of Atomic Energy, Vol. 31, Vienna.*

71 Shafranov, V.D. (1958) On magnetohydrodynamical equilibrium configurations. *Sov. Phys. JETP*, **6**, 545.

72 Solovev, L.S. (1976) Hydromagnetic stability of closed magnetic configurations. *Rev. Plasma Phys.*, **6**, 239.

73 Cross, R. (1988) *An Introduction to Alfven Waves*, Adam Hilger, Bristol.

74 Connor, J.W. (1995) Transport in tokamaks: Theoretical models and comparison with experimental results. *Plasma Phys. Control. Fusion*, **37**, 119–133.

75 Galeev, A.A. and Sagdeev, R.Z. (1984) *Handbook of Plasma Physics, Basic Plasma Physics I* (eds A.A. Galeev and R.N. Sudan), North-Holland, Amsterdam, p. 679.

76 Balescu, R. (2005) *Aspects of Anomalous Transport in Plasmas*, Institute of Physics Publishing, Bristol.

77 Diamond, P.H., Itoh, S., and Itoh, K. (2010) *Modern Plasma Physics, Vol. 1: Physical Kinetics of Turbulent Plasmas*, Cambridge U.P.

78 Jokipii, J. (1966) Cosmic ray propagation: I. charged particles in a random magnetic field. *Astrophys. J.*, **146**, 480.

79 Jokipii, J.R. (1968) Addendum and erratum to cosmic-ray propagation. I. *Astrophys. J.*, **152**, 671.

80 Jokipii, J.R. and Parker, E.N. (1969) Cosmic-ray life and the stochastic nature of the galactic magnetic field. *Astrophys. J.*, **155**, 799.

81 Jokipii, J.R. (1973) The rate of separation of magnetic lines of force in a random magnetic field. *Astrophys. J.*, **183**, 1029.

82 Casse, F., Lemoine, M., and Pelletier, G. (2001) Transport of cosmic rays in chaotic magnetic fields. *Phys. Rev. D*, **65**, 023002.

83 Qin, G., Matthaeus, W.H., and Bieber, J.W. (2002) Perpendicular transport of charged particles in composite model turbulence: Recovery of diffusion. *Astrophys. J*, **578**, L117.

84 Qin, G., Matthaeus, W.H., and Bieber, J.W. (2002) Subdiffusive transport of charged particles perpendicular to the large scale magnetic field. *Geophys. Res. Lett.*, **29**, 7.

85 Matthaeus, W.H., Qin, G., Bieber, J.W., and Zank, G.P. (2003) Nonlinear collisionless perpendicular diffusion of charged particles. *Astrophys. J.*, **590**, L000.

86 Ruffolo, D., Matthaeus, W.H., and Chuychai, P. (2004) Separation of magnetic field lines in two-component turbulence. *Astrophys. J.*, **614**, 420–434.

87 Khrapak, S.A. and Morfill, G.E. (2002) Dust diffusion across a magnetic field due to random charge fluctuations. *Phys. Plasmas*, **9**, 619.

88 Shalchi, A., Bieber, J.W., Matthaeus, W.H., and Qin, G. (2004) Nonlinear parallel and perpendicular diffusion of charged cosmic rays in weak turbulence. *Astrophys. J.*, **616**, 617.

89 Shalchi, A., Bieber, J.W., Matthaeus, W.H., and Schlickeiser, R. (2006) Parallel and perpendicular transport of heliospheric cosmic rays in an improved dynamical turbulence model. *Astrophys. J.*, **642**, 230.

90 Bieber, J.W., Matthaeus, W.H., Shalchi, A., and Qin, G. (2004) Nonlinear guiding center theory of perpendicular diffusion: General properties and comparison with observation. *Geophys. Res. Lett.*, **31**, 101029.

91 Chuychai, P., Ruffolo, D., Matthaeus, W.H., and Rowlands, G. (2005) Suppressed diffusive escape of topologically trapped magnetic field lines. *Astrophys. J.*, **633**, L49–L52.

92 Schlickeiser, R. (2002) *Cosmic Ray Astrophysics*, Springer, Berlin.

93 Kolmogorov, A.N. (1941) The local structure of turbulence in incompressible viscous fluid for very large Reynolds number. *Dokl. Akad. Nauk SSSR*, **30**, 301.

94 Dupree, T.H. (1966) Perturbation theory of strong plasma turbulence. *Phys. Fluids*, **9**, 1773.

95 Dupree, T.H. (1967) Nonlinear theory of drift-wave turbulence and enhanced diffusion. *Phys. Fluids*, **10**, 1049.

96 Weinstock, J. (1969) Formulation of a statistical theory of strong plasma turbulence. *Phys. Fluids*, **12**, 1045.

97 Weinstock, J. (1970) Turbulent plasmas in a magnetic field – a statistical theory. *Phys. Fluids*, **13**, 2308.

98 Hasegawa, A. (1985) Self-organization in continuous media. *Adv. Phys.*, **34**, 1–42.

99 Balescu, R. (1977) *Statistical Dynamics: Matter out of Equilibrium*, Imperial College Press, London.

100 McComb, W.D. (1990) *The Physics of Fluid Turbulence*, Clarendon Press, Oxford.

101 Evans, T., Moyer, R., Thomas, P., Watkins, J., Osborne, T., Boedo, J., Doyle, E., Fenstermacher, M., Finken, K., Groebner, R., Groth, M., Harris, J., LaHaye, R., Lasnier, C., Masuzaki, S., Ohyabu, N., Pretty, D.G., Rhodes, T., Reimerdes, H., Rudakov, D., Schaffer, M., Wang, G., and Zeng, L. (2004) Suppression of large edge-localized modes in high-confinement DIII-D plasmas with a stochastic magnetic boundary. *Phys. Rev. Lett.*, **92**, 235003.

102 Evans, T.E., Moyer, R.A., Burrell, K.H., Fenstermacher, M.E., Joseph, I., Leonard, A.W., Osborne, T.H., Porter, G.D., Schaffer, M.J., Snyder, P., Thomas, P., Watkins, J., and West, W. (2006) *Nat. Phys.*, **2**, 419.

103 Jakubowski, M., Schmitz, O., Abdullaev, S., Brezinsek, S., Finken, K., Kraemer-Flecken, A., Lehnen, M., Samm, U., Spatschek, K.-H., Unterberg, B., Wolf, R.C., and the TEXTOR team (2006) Effect of the change in magnetic-field topology due to an ergodic divertor on the plasma structure and transport. *Phys. Rev. Lett.*, **96**, 035004.

104 Finken, K.H., Abdullaev, S.S., de Bock, M.F.M., von Hellermann, M., Jakubowski, M., Jaspers, R., Koslowski, H.R., Kraemer-Flecken, A., Lehnen, M., Liang, Y., Nicolai, A., Wolf, R.C., Zimmermann, O., de Baar, M., Bertschinger, G., Biel, W., Brezinsek, S., Busch, C., Donnacutee, A.J.H., Esser, H.G., Farshi, E., Gerhauser, H., Giesen, B., Harting, D., Hoekzema, J.A., Hogeweij, G.M.D., Huettemann, P.W., Jachmich, S., Jakubowska, K., Kalupin, D., Kelly, F., Kikuchi, Y., Kirschner, A., Koch, R., Korten, M., Kreter, A., Krom, J., Kruezi, U., Lazaros, A., Litnovsky, A., Loozen, X., Cardozo, N.J.L., Lyssoivan, A., Marchuk, O., Matsunaga, G., Mertens, P., Messiaen, A., Neubauer, O., Noda, N., Philipps, V., Pospieszczyk, A., Reiser, D., Reiter, D., Rogister, A.L., Sakamoto, M., Savtchkov, A., Samm, U., Schmitz, O., Schorn, R.P., Schweer, B., Schueller, F.C., Sergienko, G., Spatschek, K.-H., Telesca, G., Tokar, M., Uhlemann, R., Unterberg, B., Oost, G.V., Rompuy, T.V., Wassenhove, G.V., Westerhof, E., Weynants, R., Wiesen, S., and Xu, Y.H. (2005) Toroidal plasma rotation induced by the dynamic ergodic divertor in the TEXTOR tokamak. *Phys. Rev. Lett*, **94**, 015003.

105 Finken, K.H., Abdullaev, S.S., Jakubowski, M.W., de Bock, M.F.M., Bozhenkov, S., Busch, C., von Hellermann, M., Jaspers, R., Kikuchi, Y., Kraemer-Flecken, A., Lehnen, M., Schege, D., Schmitz, O., Spatschek, K.-H., Unterberg, B., Wingen, A., Wolf, R.C., Zimmermann, O., and the TEXTOR Team (2007) Improved confinement due to open ergodic field lines imposed by the dymanic ergodic divertor in TEXTOR. *Phys. Rev. Lett.*, **98**, 065001.

106 Rechester, A.B. and Rosenbluth, M.N. (1978) Electron heat transport in a tokamak with destroyed magnetic surfaces. *Phys. Rev. Lett.*, **40**, 38–41.

107 Kadomtsev, B.B. and Pogutse, O.P. (1979) Electron heat conductivity of the plasma across a braided magnetic field. in *Plasma Physics and Controlled Nuclear Fusion Research. Proc. 7th. Int. Conf.* (Innsbruck, Austria, August 23–30, 1978), **1**, 649–662.

108 Krommes, J.A., Oberman, C., and Kleva, R.G. (1983) Plasma transport in stochastic magnetic fields. part 3. Kinetics of test particle diffusion. *J. Plasma Phys.*, **30**, 11.

109 Myra, J.R., Catto, P.J., Mynick, H.E., and Duvall, R.E. (1993) Quasilinear diffusion in stochastic magnetic fields: Reconciliation of drift-orbit modification

calculations. *Phys. Fluids B*, **5**(5), 1160–1163.
110 Laval, G. (1993) Particle diffusion in stochastic magnetic fields. *Phys. Fluids B*, **5**, 711.
111 De Rover, M., Lopes Cardozo, N.J., and Montvai, A. (1996) Hamiltonian description of the topology of drift orbits of relativistic particles in a tokamak. *Phys. Plasmas*, **3**, 4468–4477.
112 DeRover, M., Lopes Cardozo, N.J., and Montvai, A. (1996) Motion of relativistic particles in axially symmetric and perturbed magnetic fields in a tokamak. *Phys. Plasmas*, **3**, 4478.
113 DeRover, M., Schilham, A.M., Montvai, A., and LopesCardozo, N.J. (1999) *Phys. Plasmas*, **6**, 2443–2451.
114 Kubo, R. (1963) *J. Math. Phys.*, **4**, 174.
115 van Kampen, N. (1976) Stochastic differential equations. *Phys. Rep.*, **24**, 171–228.
116 Balescu, R. (2000) *Statistical Dynamics, Matter Out of Equilibrium*, Imperial College Press, London.
117 Elskens, Y. and Escande, D. (2003) *Microscopic Dynamics of Plasmas and Chaos*, Institute of Physics Publishing, Bristol.
118 Balescu, R., Wang, H.D., and Misguich, J.H. (1994) Langevin equation versus kinetic equation: Subdiffusive behavior of charged particles in a stochastic magnetic field. *Phys. Plasmas*, **1**, 3826–3842.
119 Vanden-Eijnden, E. and Balescu, R. (1996) Statistical description and transport in stochastic magnetic fields. *Phys. Plasmas*, **3**, 874.
120 Balescu, R. (1995) Anomalous transport in turbulent plasmas and continuous time random walk. *Phys. Rev. E*, **51**, 4807–4822.
121 Corrsin, S. (1959) Progress report on some turbulent diffusion research. *Adv. Geophys.*, **6**, 161–164.
122 Vanden-Eijnden, E. and Balescu, R. (1995) Liouvillian theory of magnetic fluctuations. *J. Plasma Phys.*, **54**, 185–199.
123 Wang, H.D., Vanden-Eijnden, E., Spineanu, F., Misguich, J.H., and Balescu, R. (1995) Diffusive processes in a stochastic magnetic field. *Phys. Rev. E*, **51**, 4844.
124 Vanden-Eijnden, E. and Balescu, R. (1996) Strongly anomalous diffusion in sheared magnetic configurations. *Phys. Plasmas*, **3**, 815–823.
125 Vlad, M. and Spineanu, F. (2004) Trajectory structures and transport. *Phys. Rev. E*, **70**, 056304-1.
126 Vlad, M., Spineanu, F., and Misguich, J.H. (1996) Effects of stochastic drifts and time variation on particle diffusion in magnetic turbulence. *Phys. Rev. E*, **53**, 5302–5314.
127 Vlad, M., Spineanu, F., Misguich, J.H., and Balescu, R. (1996) Effects of plasma flow on particle diffusion in stochastic magnetic fields. *Phys. Rev. E*, **54**, 791.
128 Isichenko, M.B. (1991) Effective plasma heat conductivity in braided magnetic field – I: Quasilinear limit. *Plasma Phys. Control. Fusion*, **33**, 795.
129 Isichenko, M.B. (1991) Effective plasma heat conductivity in braided magnetic field – II: Percolation limit. *Plasma Phys. Control. Fusion*, **33**, 809.
130 Isichenko, M.B. (1992) Percolation, statistical topography, and transport in random media. *Rev. Mod. Phys.*, **64**, 961.
131 Reuss, J.D. and Misguich, J.H. (1996) Low-frequency percolation scaling for particle diffusion in electrostatic turbulence. *Phys. Rev. E*, **54**, 1857–1869.
132 Vlad, M., Spineanu, F., Misguich, J.H., and Balescu, R. (1998) Diffusion with intrinsic trapping in two-dimensional incompressible stochastic velocity fields. *Phys. Rev. E*, **58**, 7359–7368.
133 Vlad, M., Spineanu, F., Misguich, J.H., and Balescu, R. (2000) Collisional effects on diffusion scaling laws in electrostatic turbulence. *Phys. Rev. E*, **61**, 3023–3032.
134 Vlad, M., Spineanu, F., Misguich, J.H., and Balescu, R. (2001) Diffusion in biased turbulence. *Phys. Rev. E*, **63**, 066304-1.
135 Vlad, M., Spineanu, F., Misguich, J.H., and Balescu, R. (2002) Electrostatic turbulence with finite parallel correlation length and radial diffusion. *Nucl. Fusion*, **42**, 157–164.
136 Vlad, M., Spineanu, F., Misguich, J.H., and Balescu, R. (2003) Magnetic line

trapping and effective transport in stochastic magnetic fields. *Phys. Rev. E*, **67**, 026406.

137 Balescu, R., Vlad, M., Spineanu, F., and Misguich, J. (2004) Anomalous transport in plasmas. *Int. J. Quantum Chem.*, **98**, 125–130.

138 Vlad, M., Spineanu, F., Misguich, J.H., Reuss, J.D., Balescu, R., Itoh, K., and Itoh, S.I. (2004) Lagrangian versus Eulerian correlations and transport scaling. *Plasma Phys. Control. Fusion*, **46**, 1051–1063.

139 Bakunin, O.G. (2004) Correlation effects and turbulent diffusion scalings. *Rep. Prog. Phys.*, **67**, 1.

140 Bakunin, O.G. (2005) Percolation models of turbulent transport and scaling estimates. *Chaos, Solitons and Fractals*, **23**, 1703.

141 Vlad, M. and Spineanu, F. (2005) Larmor radius effects on impurity transport in turbulent plasmas. *Plasma Phys. Control. Fusion*, **47**, 281–294.

142 Neuer, M. and Spatschek, K.-H. (2006) Diffusion of test particles in stochastic magnetic fields for small Kubo numbers. *Phys. Rev. E*, **73**, 026404.

143 Neuer, M. and Spatschek, K.-H. (2006) Diffusion of test particles in stochastic magnetic fields in the percolative regime. *Phys. Rev. E*, **74**, 036401.

144 Vlad, M., Spineanu, F., and Benkadda, S. (2006) Impurity pinch from the ratchet effect. *Phys. Rev. Lett.*, **96**, 085001.

145 Vlad, M., Spineanu, F., and Benkadda, S. (2008) Collision and average velocity effects on the ratchet pinch. *Phys. Plasmas*, **15**, 032306.

146 Schelin, A.B. and Spatschek, K.-H. (2010) Directed chaotic transport in the tokamap with mixed phase space. *Phys. Rev. E*, **81**, 016205.

147 Evans, T.E., Wingen, A., Watkins, J., and Spatschek, K.-H. (2010) A conceptual model for the nonlinear dynamics of edge-localized modes in tokamak plasmas. *Nonlinear Dynamics*, INTECH, Vukovar, Croatia, pp. 59–78.

148 Wingen, A., Evans, T.E., Lasnier, C., and Spatschek, K.-H. (2010) Numerical modeling of edge-localized-mode filaments on divertor plates based on thermoelectric currents. *Phys. Rev. Lett.*, **104**, 175001.

149 Lin, Z., Hahm, T., Lee, W., Tang, W., and White, R. (1998) Turbulent transport reduction by zonal flows: Massively parallel simulations. *Science*, **281**, 1835–1837.

150 Spatschek, K.-H. (2008) Basic principles of stochastic transport. *Turbulent transport in fusion plasmas: First ITER International Summer School (ed. S. Benkadda), Aix. France, 16–20 July 2007, AIP Conf. Proc.*, **1013**, 250.

151 Wingen, A. and Spatschek, K.-H. (2008) Ambipolar stochastic particle diffusion and plasma rotation. *Phys. Plasmas*, **15**, 052305.

152 Wingen, A. and Spatschek, K.-H. (2009) Sheared plasma rotation in stochastic magnetic fields. *Phys. Rev. Lett.*, **102**, 185002.

153 Wingen, A. and Spatschek, K.-H. (2010) Influence of different DED base mode configurations on the radial electric field at the plasma edge of TEXTOR. *Nucl. Fusion*, **50**, 034009.

154 Wingen, A., Abdullaev, S., Finken, K., and Spatschek, K.-H. (2006) Influence of stochastic magnetic fields on relativistic electrons. *Nucl. Fusion*, **46**, 941–952.

155 Finken, K.H., Abdullaev, S., Jakubowski, M., Jaspers, R., Lehnen, M., Schlickeiser, R., Wingen, A., Spatschek, K.-H., Wolf, R., and the TEXTOR team (2007) Runaway losses in ergodized plasmas. *Nucl. Fusion*, **47**, 91–102.

156 Wingen, A., Spatschek, K.-H., Abdullaev, S., and Jakubowski, M. (2007) Interpretation of heat losses from open chaotic systems. *Physics AUC*, **17**, 44–58.

157 Jakubowski, M., Wingen, A., Abdullaev, S., Finken, K.H., Lehnen, M., Spatschek, K.-H., Wolf, R., and the TEXTOR team (2007) Observation of heteroclinic tangles in the heat flux pattern of the ergodic divertor of TEXTOR. *J. Nucl. Mat.*, **363–365**, 371–376.

158 Wingen, A., Jakubowski, M., Spatschek, K.-H., Abdullaev, S., Finken, K., Lehnen, M., and the TEXTOR team (2007) Traces of stable and unstable manifolds in heat flux patterns. *Phys. Plasmas*, **14**, 042502.

159 Boozer, A.H. (2004) Physics of magnetically confined plasmas. *Rev. Mod. Phys.*, **76**, 1071–1141.

160 Abdullaev, S.S., Finken, K.H., Kaleck, A., and Spatschek, K.-H. (1998) Twist mapping for the dynamics of magnetic field lines in a tokamak ergodic divertor. *Phys. Plasmas*, **5**, 196–210.

161 Abdullaev, S.S. (1999) A new integration method of Hamiltonian systems by symplectic maps. *J. Phys. A: Math. Gen.*, **32**, 2745–2766.

162 Abdullaev, S.S. (2002) The Hamilton–Jacobi method and Hamiltonian maps. *J. Phys. A: Math. Gen.*, **35**, 2811–2832.

163 Abdullaev, S.S. (2004) Canonical maps near separatrix in Hamiltonian systems. *Phys. Rev. E*, **70**, 04620201–04620220.

164 Abdullaev, S.S. (2006) *Construction of Mappings for Hamiltonian Systems and Their Applications*, Springer, Berlin.

165 Boozer, A.H. (1984) Time-dependent drift Hamiltonian. *Phys. Fluids*, **27**, 2441–2445.

166 Rechester, A.B., Rosenbluth, M.N., and White, R.B. (1979) Calculation of the Kolmogorov entropy for motion along a stochastic magnetic field. *Phys. Rev. Lett.*, **42**, 1247.

167 Rechester, A.B. and White, R.B. (1980) Calculation of turbulent diffusion for the Chirikov–Taylor model. *Phys. Rev. Lett.*, **44**, 1586.

168 Rechester, A.B., Rosenbluth, M.N., and White, R.B. (1981) Fourier-space paths applied to the calculation of diffusion for the Chirikov–Taylor model. *Phys. Rev. A*, **23**, 2664–2672.

169 Rax, J.M. and White, R.B. (1992) Effective diffusion and nonlocal heat transport in a stochastic magnetic field. *Phys. Rev. Lett.*, **68**, 1523–1526.

170 Ott, E. (1994) *Chaos in Dynamical Systems*, Cambridge U.P., Cambridge.

171 Schuster, H.G. (1994) *Deterministisches Chaos*, Wiley-VCH Verlag GmbH, Weinheim.

172 Chirikov, B. (1979) A universal instability of many-dimensional oscillator systems. *Phys. Reports*, **52**, 263–379.

173 Wolf, A., Swift, J.B., Swinney, H.L., and Vastano, J.A. (1985) Determining Lyapunov exponents from a time series. *Physica D: Nonlinear Phenomena*, **16**(3), 285–317.

174 Zhang, S.C. and Elgin, J. (2004) Application of Kolmogorov entropy to the self-amplified spontaneous emission free-electron lasers. *Phys. Plasmas*, **11**(4), 1663–1668.

175 Balescu, R., Vlad, M., and Spineanu, F. (1998) Tokamap: A Hamiltonian twist map for magnetic field lines in a toroidal geometry. *Phys. Rev. E*, **58**, 951.

176 Wobig, H. and Fowler, R.H. (1988) The effect of magnetic surface destruction on test particle diffusion in the Wendelstein VII-A stellerator. *Plasma Phys. Contr. Fusion*, **30**, 721–741.

177 Misguich, J.H., Reuss, J.D., Constantinescu, D., Steinbrecher, G., Vlad, M., Spineanu, F., Weyssow, B., and Balescu, R. (2003) Noble internal transport barriers and radial subdiffusion of toroidal magnetic lines. *Ann. Phys.*, **28**(6), 87.

178 Eberhard, M. (1999) *Europhysics Conference Abstracts*, **23J**, 781.

179 Wingen, A., Spatschek, K.-H., and Abdullaev, S. (2005) Stochastic transport of magnetic field lines in the symmetric tokamap. *Contrib. Plasma Phys.*, **45**, 500–513.

180 Abdullaev, S.S. (2004) On mapping models of field lines in a stochastic magnetic field. *Nuclear Fusion*, **44**(6), 12.

181 Wittkowski, R., Schelin, A., and Spatschek, K.-H. (2009) Mean motion in stochastic plasmas with a space-dependent diffusion coefficient. *Contrib. Plasma Phys.*, **49**, 55–69.

182 Rosenbluth, M. and Kaufman, A. (1958) Plasma diffusion in a magnetic field. *Phys. Rev.*, **109**, 1–5.

183 Rosenbluth, M., Sagdeev, R., Taylor, G., and Zaslavsky, G. (1966) Destruction of magnetic surfaces by magnetic irregularities. *Nucl. Fusion*, **6**, 297.

184 Stix, T.H. (1973) *Phys. Rev. Lett.*, **30**, 833.

185 Bickerton, R. (1997) Magnetic turbulence and the transport of energy and part in tokamaks. *Plasma Phys. Control. Fusion*, **39**, 339–365.

186 Drake, J.F., Gladd, N.T., Liu, C.S., and Chang, C.L. (1980) *Phys. Rev. Lett.*, **44**, 994.

187 Rose, H.A. (1982) *Phys. Rev. Lett.*, **48**, 260.

188 Taylor, G.I. (1922) Diffusion by continuous movement. *Proc. London Math. Soc.*, **20**, 196.

189 Green, M.S. (1951) Brownian motion in a gas of noninteracting molecules. *J. Chem. Phys.*, **19**, 1036.

190 Kubo, R. (1957) Statistical-mechanical theory of irreversible processes. I. General theory and simple applications to magnetic and conduction problems. *J. Phys. Soc. Jpn.*, **12**, 570.

191 Goldeich, P. and Sridhar, S. (1995) Toward a theory of interstellar turbulence. 2: Strong Alfvenic turbulence. *Astrophys. J.*, **438**, 763.

192 Neuer, M. and Spatschek, K.-H. (2008) Pitch angle scattering and effective collision frequency caused by stochastic magnetic fields. *Phys. Plasmas*, **15**, 022304.

193 Jakubowski, M., Abdullaev, S., and Finken, K. (2004) Modelling of the magnetic field structures and first measurements of heat fluxes for the TEXTOR-DED operation. *Nucl. Fus.*, **44**, 1–11.

194 Jakubowski, M., Abdullaev, S., Finken, K., Lehnen, M., and the TEXTOR team (2005) Heat deposition patterns on the target plates dynamic ergodic divertor. *J. Nucl. Mater.*, **337–339**, 176–180.

195 da Silva, E., Caldas, I., Viana, R., and Sanjuán, M. (2002) *Phys. Plasmas*, **9**, 4917.

196 Abdullaev, S.S., Eich, T., and Finken, K.H. (2001) *Phys. Plasmas*, **8**, 2739–2749.

197 Evans, T.E., Roeder, R.K., Carter, J.A., and Rapoport, B.I. (2004) Homoclinic tangles, bifurcations and edge stochasticity in diverted tokamaks. *Contrib. Plasma Phys.*, **44**, 235–240.

198 Evans, T.E., Roeder, R.K., Carter, J.A., Rapoport, B.I., Fenstermacher, M.E., and Lasnier, C.J. (2005) Experimental signatures of homoclinic tangles in poloidally diverted tokamaks. *J. Phys.: Conf. Ser.*, **7**, 174–190.

199 Qin, G., Matthaeus, W., and Bieber, J. (2006) Parallel diffusion of charged particles in strong two-dimensional turbulence. *Astrophys. J. Lett.*, **640**(1), L103–L106.

200 Stix, T.H. (1992) *Waves in Plasmas*, AIP, New York.

201 Mikhailovskii, A.B. (1998) *Instabilities in a Confined Plasma*, IoP, Bristol.

202 Penrose, O. (1960) Electrostatic instabilities of a uniform non-Maxwellian plasma. *Phys. Fluids*, **3**, 258.

203 Buneman, O. (1958) Instability, turbulence, and conductivity in a current-carrying plasma. *Phys. Rev. Lett.*, **1**, 8–9.

204 Weibel, E.S. (1959) *Phys. Rev. Lett.*, **2**, 83.

205 Mikhailovskii, A.B. (1984) *Handbook of Plasma Physics*, North-Holland, Amsterdam, p. 587.

206 Hain, K., Lüst, R., and Schlüter, A. (1957) Zur Stabilität eines Plasmas. *Z. Naturforsch.*, **12a**, 833–841.

207 Berstein, I.B., Frieman, E.A., Kruskal, M.D., and Kusrud, R.M. (1958) An energy principle for hydromagnetic stability theory. *Proc. R. Soc. Lond.*, **244**, 17–40.

208 Laval, G., Mercier, C., and Pellat, R. (1965) Necessity of the energy principle for magnetostatic stability. *Nucl. Fusion*, **5**, 156–158.

209 Kruskal, M.D., Johnson, J.L., Gottlieb, M.B., and Goldman, L.M. (1958) Hydromagnetic stability in a stellarator. *Phys. Fluids*, **1**, 421.

210 Mercier, C. (1960) A necessary condition for hydrodynamic stability of plasma with axial symmetry. *Nucl. Fusion*, **1**, 47.

211 Mukhovatov, V.S. and Shafranov, V.D. (1971) Plasma equilibrium in a tokamak. *Nucl. Fusion*, **11**, 605.

212 Greenwald, M., Terry, J.L., Wolfe, S.M., Ejima, S., Bell, M.G., Kaye, S.M., and Nelson, G.H. (1988) A new look at density limits in tokamaks. *Nucl. Fusion*, **28**, 2199.

213 Shafranov, V.D. (1970) Hydromagnetic stability of a current-carrying pinch in a strong longitudinal magnetic field. *Sov. Phys. Techn. Phys.*, **15**, 175.

214 Wesson, J.A. (1978) Hydromagnetic stability of tokamaks. *Nucl. Fusion*, **18**, 87.

215 Richtmyer, R.D. (1960) *Commun. Pure Appl. Math.*, **13**, 297–319.

216 Scharp, D.H. (1984) An overview over Rayleigh-Taylor instability. *Physica D*, **12**, 3–10.

217 Berning, M. and Rubenchik, A.M. (1998) *Phys. Fluids*, **10**, 1564.

218 Scott Russell, J. (1844) *Report on Waves*, John Murray, London.

219 Lamb, G. (1980) *Elements of Soliton Theory*, John Wiley & Sons, Inc., New York.

220 Newell, A. (1985) *Solitons in Mathematics and Physics*, SIAM, Philadelphia.

221 Whitham, G.H. (1974) *Linear and Nonlinear Waves*, John Wiley & Sons, Inc., New York.

222 Karpman, V. (1974) *Nonlinear Waves in Dispersive Media*, Pergamon Press, New York.

223 Ablowitz, M. and Segur, H. (1981) *Solitons and the Inverse Scattering Transform*, SIAM, Philadelphia.

224 Infeld, E. and Rowlands, G. (1990) *Nonlinear Waves, Solitons and Chaos*, Cambridge U.P., Cambridge.

225 Ablowitz, M. and Clarkson, P. (1991) *Solitons, Nonlinear Evolution Equations and Inverse Scattering*, Cambridge U.P., Cambridge.

226 Hasegawa, A. and Kodama, Y. (1995) *Solitons in Optical Communications*, Oxford University Press, Oxford.

227 Akhmediev, N. and Ankiewicz, A. (1997) *Solitons, Nonlinear Pulses and Beams*, Chapman & Hall, London.

228 Kodama, Y. (2000) *On the Dispersion-Managed Soliton in Massive WDM and TDM Soliton Transmission Systems*, Kluwer Academic, Dordrecht, MA.

229 Laedke, E.W. and Spatschek, K.-H. (1982) Nonlinear ion-acoustic waves in weak magnetic fields. *Phys. Fluids*, **25**, 985–989.

230 Weberpals, H. and Spatschek, K.-H. (1987) Dynamics of solitary waves in hydrogen-bonded chains. *Phys. Rev. A*, **36**, 2946–2952.

231 Laedke, E.W. and Spatschek, K.-H. (1991) On localized solutions in nonlinear Faraday resonance. *J. Fluid Mech.*, **223**, 589–601.

232 Klauder, M., Laedke, E.W., Spatschek, K.-H., and Turitsyn, S. (1993) Pulse propagation in optical fibers near the zero dispersion point. *Phys. Rev. E*, **47**, 3844–3847.

233 Spatschek, K.-H. and Mertens, F. (eds) (1994) *Nonlinear Coherent Structures in Physics and Biology*, Plenum, New York.

234 Laedke, E.W., Spatschek, K.-H., Turitsyn, S.K., and Mezentsev, V. (1995) Analytic criterion for soliton instability in a nonlinear fiber array. *Phys. Rev. E*, **52**, 5549–5554.

235 Laedke, E.W., Spatschek, K.-H., Mezentsev, V.K., Musher, S.L., Ryzhenkova, I., and Turitsyn, S.K. (1995) Soliton stability and collapse in two-dimensional discrete systems. *JETP Lett.*, **62**, 677–684.

236 Wingen, A., Spatschek, K.-H., and Medvedev, S.B. (2003) Averaged dynamics of optical pulses described by a nonlinear Schrödinger equation with periodic coefficients. *Phys. Rev. E*, **68**, 046610.

237 Spatschek, K.-H. (2005) Dispersion management. *Encyclopedia of Nonlinear Science*, Routledge, New York, pp. 219–222.

238 Akhmediev, N. and Ankiewicz, A. (eds) (2005) *Dissipative Solitons*, Lecture Notes in Physics Vol. 661, Springer, Berlin.

239 Akhmediev, N. and Ankiewicz, A. (eds) (2008) *Dissipative Solitons: From Optics to Biology and Medicine*, Lecture Notes in Physics Vol. 751, Springer, Berlin.

240 Mihalache, D. (2008) Three-dimensional dissipative optical solitons. *Central Eur. J. Phys.*, **6**, 582–587.

241 Newell, A.C. (1974) *Nonlinear Wave Motion*, Providence, RI, Am. Math. Soc.

242 Korteweg, D.J. and de Vries, G. (1895) *Philos. Mag.*, **39**, 422.

243 Abramowitz, M. and Stegun, I. (1972) *Handbook of Mathematical Functions*, Dover, New York.

244 Ikezi, H., Taylor, R., and Baker, R. (1970) *Phys. Rev. Lett.*, **25**, 11.

245 Kadomtsev, B. and Petviashvili, V. (1970) *Sov. Phys. Dokl.*, **15**, 539.

246 Davey, A. and Stewartson, K. (1974) *Proc. Roy. Soc. Lond.*, **338**, 101–110.

247 Zakharov, V.E. and Kuznetsov, E.A. (1974) *Sov. Phys. JETP*, **39**, 285.

248 Zakharov, V. (1968) *Sov. Phys. J. Appl. Mech. Tech. Phys.*, **4**, 190–194.

249 Schamel, H., Yu, M.Y., and Shukla, P.K. (1977) *Phys. Fluids*, **20**, 1286.

250 Kim, H.C., Stenzel, R., and Wong, A.Y. (1974) *Phys. Rev. Lett.*, **33**, 886.

251 Zubov, V. (1964) *Methods of A.M. Ljapunov and Their Applications*, P. Noordhoff Ltd.

252 Lyapunov, A. (1907) *Ann. Fac. Sci. Toulouse*, **2**, 9. Reprint: Ann. Math. Studies 17, Princeton Univ. Press, 1949.

253 Malkin, J.G. (1959) *Theorie der Stabilität einer Bewegung*, R. Oldenbourg, München.

254 LaSalle, J. and Lefschetz, S. (1967) *Die Stabilitätstheorie von Ljapunov: Die direkte Methode und Anwendungen*, BI, Mannheim.

255 LaSalle, J., Lefschetz, S., and Alverson, R.C. (1962) Stability by Liapunov's direct method with applications. *Phys. Today*, **15**(10), 59–59.

256 Laedke, E.W. (1978) *Zur nichtlinearen Stabilität solitärer Wellen in Plasmen*, Dissertation, Universität Essen.

257 Adams, R.A. and Fournier, J.J.F. (1975) *Sobolev spaces*, Academic Press, New York.

258 Laedke, E. and Spatschek, K.-H. (1978) Nonlinear stability of envelope solitons. *Phys. Rev. Lett.*, **41**, 1798–1801.

259 Laedke, E.W. and Spatschek, K.-H. (1985) *Differential Geometry, Calculus of Variations, and Their Applications*, Marcel Dekker, Inc., New York.

260 Strauss, W.A. (1977) *Commun. Math. Phys.*, **55**, 149.

261 Benjamin, T.B. (1972) *Proc. R. Soc. Lond.*, **328**, 153.

262 Laedke, E.W. and Spatschek, K.-H. (1980) Liapunov stability of generalized Langmuir solitons. *Phys. Fluids*, **23**, 44–51.

263 Kuznetsov, E.A., Spector, M.D., and Falkovich, G.E. (1984) *Phys. D*, **10**, 379.

264 Alexander, J.C., Pego, R.L., and Sachs, R.L. (1997) *Phys. Lett. A*, **226**, 187.

265 Laedke, E.W. and Spatschek, K.-H. (1982) Growth rates of bending KdV solitons. *J. Plasma Phys.*, **28**, 469–484.

266 Laedke, E.W., Spatschek, K.-H., and Zocha, K. (1986) Bifurcation analysis of the transverse drift-wave-envelope instability. *Phys. Fluids*, **29**, 1127.

267 Infeld, E. and Rowlands, G. (1977) *Plasma Phys.*, **19**, 343.

268 Makhankov, V.G. (1978) Dynamics of classical solitons (in non-integrable systems). *Phys. Rep. (Sect. C of Phys. Lett.)*, **35**, 1–128.

269 Blaha, R., Laedke, E., and Spatschek, K.-H. (1987) Stability of two-dimensional finite-amplitude envelope solitons. *Phys. Rev. A*, **35**, 4273–4279.

270 Zakharov, V.E. (1972) *Sov. Phys. JETP*, **35**, 908–914.

271 Blaha, R., Kuznetsov, E.A., Laedke, E.W., and Spatschek, K.-H. (1990) *Nonlinear World* (eds V.G. Baryakhtar *et al.*), World Scientific, Singapore, **1**, 10.

272 Laedke, E.W. and Spatschek, K.-H. (1984) Stable three-dimensional envelope solitons. *Phys. Rev. Lett.*, **52**, 279–282.

273 Kaw, P.K., Nishikawa, K., Yoshida, Y., and Hasegawa, A. (1975) *Phys. Rev. Lett.*, **35**, 88.

274 Wilcox, J.Z. and Wilcox, T.J. (1975) *Phys. Rev. Lett.*, **34**, 1160.

275 Laedke, E. and Spatschek, K.-H. (1979) Exact stability criteria for finite amplitude solitons. *Phys. Rev. Lett.*, **42**, 1534–1537.

276 Laedke, E.W., Spatschek, K.-H., and Stenflo, L. (1983) Evolution theorem for a class of perturbed envelope soliton solutions. *J. Math. Phys.*, **24**, 2764–2769.

277 Lee, T.D. (1981) *Particle Physics and Introduction to Field theory*, Harwood, New York.

278 Gibbon, P. (2005) *Short Pulse Laser Interactions with Matter*, Imperial College Press.

279 Strickland, D. and Mourou, G. (1985) Compression of amplified chirped optical pulses. *Opt. Commun.*, **55**, 447–449.

280 Mourou, G.A., Tajima, T., and Bulanov, S.V. (2006) Optics in the relativistic regime. *Rev. Mod. Phys.*, **78**, 309.

281 Panofsky, W.K.H. and Phillips, M. (1977) *Classical Electricity and Magnetism*, 2nd edn, Addison-Wesley, Reading.

282 Jackson, J.D. (1999) *Classical Electrodynamics*, John Wiley & Sons, Inc., New York.

283 Landau, L.D. and Lifshitz, E.M. (1977) *Lehrbuch der theoretischen Physik II: Klassische Feldtheorie*, Akademie Verlag, Berlin.

284 Hawking, S.W. (1974) Black hole explosions? *Nature*, **248**, 30.

285 Hawking, S.W. (1975) Particle creation by black hole evaporation. *Commun. Math. Phys.*, **43**, 199.

286 Unruh, W.G. (1976) Notes on black hole evaporation. *Phys. Rev. D*, **14**, 870.

287 Unruh, W.G. (1977) Particle detectors and black hole evaporation. *Ann. NY Acad. Sci.*, **302**, 186.

288 Chen, P. and Tajima, T. (1999) Testing Unruh radiation with ultraintense lasers. *Phys. Rev. Lett.*, **83**, 256.

289 McDonalds, K.T. (1997) Positron production by laser light. http://www.hep.princeton.edu/~mcdonald/e144/ssitalk.pdf.

290 Mima, K., Jovanović, M.S., Sentoku, Y., Sheng, Z.M., Škorić, M.M., and Sato, T. (2001) Stimulated photon cascade and condensate in a relativistic laser-plasma interaction. *Phys. Plasmas*, **8**(5), 2349–2356.

291 Gorbunov, L.M. and Kirsanov, V.I. (1987) Excitation of plasma waves by an electromagnetic wave packet. *Zh. Eksp. Teor. Fiz.*, **93**, 509.

292 Zakharov, V.E. and Rubenchik, A.M. (1975) *Sov. Phys. JETP*, **38**, 494.

293 Zakharov, V.E. and Synakh, V.S. (1975) *Sov. Phys. JETP*, **41**, 465.

294 Zakharov, V.E., Kuznetsov, E.A., and Musher, S.L. (1985) *JETP Lett.*, **41**, 154.

295 Akhiezer, A.I. and Polovin, R.V. (1956) Theory of wave motion of an electron plasma. *Sov. Phys. JETP*, **3**, 696 (*Zh. Eksp. Teor. Fiz. 30, 915 (1956)*).

296 Kaw, P., Sen, A., and Valeo, E. (1983) Coupled nonlinear stationary waves in a plasma. *Physica D: Nonlinear Phenomena*, **9**, 96–102.

297 Lehmann, G. and Spatschek, K.-H. (2010) Classification and stability of plasma motion in periodic linearly polarized relativistic waves. *Phys. Plasmas*, **17**, 072102.

298 Pesch, T.C. and Kull, H.J. (2007) Linearly polarized waves with constant phase velocity in relativistic plasmas. *Phys. Plasmas*, **14**(8), 083103.

299 Grammaticos, B., Ramani, A., and Yoshida, H. (1987) The demise of a good integrability candidate. *Phys. Lett. A*, **124**, 65–67.

300 Bourdier, A. (2009) Dynamics of a charged particle in a progressive plane wave. *Physica D: Nonlinear Phenomena*, **238**, 226–232.

301 Teychenné, D., Bésuelle, E., Oloumi, A., and Saloma, R. (2000) Chaotic dynamics of coupled transverse-longitudinal plasma oscillations in magnetized plasmas. *Phys. Rev. Lett.*, **85**, 5571–5574.

302 Kaw, P. and Dawson, J. (1970) Relativistic nonlinear propagation of laser beams in cold overdense plasmas. *Phys. Fluids*, **13**, 472–481.

303 Max, C. and Perkins, F. (1971) Strong electromagnetic waves in overdense plasmas. *Phys. Rev. Lett.*, **27**(20), 1342–1345.

304 Chian, A.C.L. and Clemmow, P.C. (1975) Nonlinear, periodic waves in a cold plasma: A quantitative analysis. *J. Plasma Phys.*, **14**, 505.

305 Clemmow, P.C. and Harding, R.D. (1980) Nonlinear waves in a magnetized plasma: Periodicity analysis. *J. Plasma Phys.*, **23**, 71–89.

306 Clemmow, P.C. (1982) Further analysis of nonlinear, periodic, highly superluminous waves in a magnetized plasma. *J. Plasma Phys.*, **27**, 177–187.

307 Lehmann, G. and Spatschek, K.-H. (2011) Poincaré analysis of wave motion in ultrarelativistic electron-ion plasmas. *Phys. Rev. E*, **83**, 036401.

308 Bisai, N., Sen, A., and Jain, K. (1996) Nonlinear coupling of a superluminal electromagnetic wave to a relativistic electron beam. *J. Plasma Phys.*, **56**, 209–220.

309 Saxena, V., Sen, A., and Kaw, P. (2009) Linearly polarized superluminal electromagnetic solitons in cold relativistic plasmas. *Phys. Rev. E*, **80**(1), 016406.

310 Mahajan, S.M., Shatashvili, N.L., and Berezhiani, V.I. (2009) Asymmetry-driven structure formation in pair plasmas. *Phys. Rev. E*, **80**, 066404.

311 Sharma, A., Kourakis, I., and Shukla, P.K. (2010) Spatiotemporal evolution of high-power relativistic laser pulses in electron-positron-ion plasmas. *Phys. Rev. E*, **82**, 016402.

312 Berezhiani, V.I., El-Ashry, M., and Mofiz, U. (1994) Theory of strong-electromagnetic-wave propagation in an electron-positron-ion plasma. *Phys. Rev. E*, **50**(1), 448–452.

313 Pesch, T.C. and Kull, H.J. (2009) Large-amplitude electromagnetic waves in relativistic plasmas. *Laser Phys.*, **19**, 1753–1758.

314 Lehmann, G., Laedke, E.W., and Spatschek, K.-H. (2006) Stability and evolution of one-dimensional relativistic solitons on the ion time scale. *Phys. Plasmas*, **13**(9), 092302.

315 Spatschek, K.-H. (1976) Parametrische Instabilitäten in Plasmen. *Fortschr. Phys.*, **24**, 687–729.

316 Liu, C.S. and Tripathi, V.K. (1994) *Interaction of Electromagnetic Waves with Electron Beams and Plasmas*, World Scientific, Singapore.

317 Drake, J.F., Kaw, P.K., Lee, Y.C., Schmidt, G., Liu, C.S., and Rosenbluth, M.N. (1974) Parametric instabilities of electromagnetic waves in plasmas. *Phys. Fluids*, **17**, 778.

318 Forslund, D.W., Kindel, J.M., and Lindman, E.L. (1975) Theory of stimulated scattering processes in laser-irradiated plasmas. *Phys. Fluids*, **18**, 1002.

319 McKinstrie, C.S. and Bingham, R. (1992) Stimulated Raman scattering and the relativistic modulation instability of light waves in rarefied plasma. *Phys. Fluids B*, **4**, 2626.

320 Sakharov, A.S. and Kirsanov, V.I. (1994) Theory of Raman scattering for a short ultrastrong laser pulse in a rarefied plasma. *Phys. Rev. E*, **49**, 3274–3282.

321 Guerin, S., Laval, G., Mora, P., Adam, J.C., Heron, A., and Bendip, A. (1995) Modulational and Raman instabilities in the relativistic regime. *Phys. Plasmas*, **2**, 2807.

322 Berkhoer, A.L. and Zakharov, V.E. (1970) Self-excitation of waves with different polarizations in nonlinear media. *Sov. Phys. JETP*, **30**, 344–347.

323 Litvak, A.G. (1970) Finite-amplitude wave beams in a magnetoactive plasma. *Sov. Phys. JETP*, **3**, 696.

324 Max, C., Arons, J., and Langdon, A. (1974) Self-modulation and self-focusing of electromagnetic waves in plasmas. *Phys. Rev. Lett.*, **33**, 209–212.

325 Spatschek, K.-H. (1977) Relativistic effects on electron plasma waves. *Phys. Rev. A*, **16**, 2470.

326 Spatschek, K.-H. (1977) Self-focusing of electromagnetic waves as a result of relativistic electron-mass variation. *Plasma Phys.*, **18**, 293–303.

327 Tushentsov, M., Kim, A., Cattani, D., Anderson, D., and Lisak, M. (2001) Electromagnetic energy penetration in the self-induced transparency regime of relativistic laser-plasma interaction. *Phys. Rev. Lett.*, **87**, 275002.

328 Karle, C., Schweitzer, J., Hochbruck, M., Laedke, E.W., and Spatschek, K.-H. (2006) Numerical solution of nonlinear wave equations in stratified dispersive media. *J. Comput. Phys.*, **216**, 138–152.

329 Li, B., Ishiguro, S., Škorić, M.M., and Sato, T. (2007) Stimulated trapped electron acoustic wave scattering, electromagnetic soliton and ion intense interaction with subcritical plasmas. *Plasma Phys.*, **14**, 032101.

330 Bulanov, S.V., Califano, F., Dudnikova, G.I., Esirkepov, T.Z., Inovenkov, I.N., Kamenets, F.F., Liseikina, T.V., Lontano, M., Mima, K., Naumova, N.M., Nishihara, K., Pegoraro, F., Ruhl, H., Sakharov, A.S., Sentoku, Y., Vshivkov, V.A., and Zhakhovskii, V.V. (2001) Relativistic interaction of laser pulses with plasmas. *Reviews of Plasma Physics*, Vol. 22, Kluwer Academic/Plenum Publishers, pp. 227–335.

331 Bulanov, S., Esirkepov, T., Naumova, N., Pegoraro, F., and Vshivkov, V.A. (1999) Soliton-like electromagnetic waves behind a superintense laser pulse in a plasma. *Phys. Rev. Lett.*, **82**, 3440–3443.

332 Sentoku, Y., Esirkepov, T., Mima, K., Nishihara, K., Califano, F., Pegoraro, F., Sakagami, H., Kitagawa, Y., Naumova, N., and Bulanov, S. (1999) Bursts of super-reflected laser light from inhomogeneous plasma due to the generation of relativistic solitary waves. *Phys. Rev. Lett.*, **83**, 3434–3437.

333 Naumova, N., Bulanov, S., Esirkepov, T., Farina, F., and Nishihara, K. (2001) Formation of electromagnetic postsolitons in plasmas. *Phys. Rev. Lett.*, **87**, 185004.

334 Esirkepov, T., Nishihara, K., and Bulanov, S.; Pegoraro, F. (2002) Three-dimensional relativistic electromagnetic subcycle solitons. *Phys. Rev. Lett.*, **89**, 275002.

335 Borghesi, M., Bulanov, S., Campbell, D., Clarke, R.J., Esirkepov, T., Galimberti, M., Gizzi, L.A., MacKinnon, A.J., Naumova, N., Pegoraro, F., Ruhl, H., Schiavi, A., and Willi, O. (2002) Macroscopic evidence of soliton formation in multi-terawatt laser-plasma interaction. *Phys. Rev. Lett.*, **88**, 135002.

336 Borghesi, M., Campbell, D., Schiavi, A., Haines, M.G., Willi, O., MacKinnon, A.J., Patel, P., Gizzi, L.A., Galimberti, M., Clark, E.L., Pegoraro, F., Ruhl, H., and Bulanov, S. (2002) Electric field detection in laser-plasma interaction experiments via the proton imaging technique. *Plasma Phys.*, **9**, 214–2220.

337 Bulanov, S. and Pegoraro, F. (2002) Stability of a mass accreting shell expanding in a plasma. *Phys. Rev. E*, **65**, 066405.

338 Borghesi, M., Bulanov, S.V., Esirkepov, T.Z., Fritzler, S., Kar, S., Liseikina, T.V., Malka, V., Pegoraro, F., Romagnani, L., Rousseau, J.P., Schiavi, A., Willi, O., and Zayats, A.V. (2005) Plasma ion evolution in the wake of a high-intensity ultrashort laser pulse. *Phys. Rev. Lett.*, **94**, 195003.

339 Kaw, P.K., Sen, A., and Katsouleas, T. (1992) Nonlinear 1D laser pulse solitons in a plasma. *Phys. Rev. Lett.*, **68**, 3172–3175.

340 Esirkepov, T., Kamenets, F.F., and Bulanov, S.; Naumova, N. (1998) Low-frequency relativistic electromagnetic solitons in collisionless plasma. *JETP Lett.*, **68**, 36–41.

341 Farina, D. and Bulanov, S.V. (2001) Relativistic electromagnetic solitons in the electron-ion plasma. *Phys. Rev. Lett.*, **86**, 5289–5292.

342 Lontano, M., Bulanov, S., and Koga, J. (2001) One-dimensional electromagnetic solitons in a hot electron-positron plasma. *Phys. Plasmas*, **8**, 5113–5120.

343 Poornakala, S., Das, A., Kaw, P.K., Sen, A., Sheng, Z.M., Sentoku, Y., Mima, K., and Nishkawa, K. (2002) Weakly relativistic one-dimensional laser pulse envelope solitons in a warm plasma. *Phys. Plasmas.*, **9**, 3802–3810.

344 Poornakala, S., Das, A., and Sen, A., Kaw, P.K. (2002) Laser envelope solitons in cold overdense plasmas. *Phys. Plasmas*, **9**, 1820–1823.

345 Lontano, M., Passoni, M., and Bulanov, S. (2003) Relativistic electromagnetic solitons in a warm quasineutral electron-ion plasma. *Plasma Phys.*, **10**, 639–649.

346 Li, Z., Li, L., Tian, H., Zhou, G., and Spatschek, K.-H. (2002) Chirped femtosecond solitonlike laser pulse form with self-frequency shift. *Phys. Rev. Lett.*, **89**, 263901.

347 Neuer, M., Spatschek, K.-H., and Zhonghao, L. (2004) Chirped solitons as attractors for short light pulses. *Phys. Rev. E*, **70**, 056605.

348 Kinsler, P. and New, G. (2004) Few-cycle soliton propagation. *Phys. Rev. A*, **69**, 013805.

349 Kozlov, V.A., Litvak, A.G., and Suvorov, E.V. (1979) *Sov. Phys. JETP*, **49**, 75.

350 Farina, D. and Bulanov, S. (2005) Dynamics of relativistic solitons. *Plasma Phys, Control. Fusion*, **47**, A73–A80.

351 Hadžievski, L., Jovanović, M.S., Škorić, M.M., and Mima, K. (2002) Stability of one-dimensional electromagnetic solitons in relativistic laser plasmas. *Phys. Plasmas*, **9**, 2569.

352 Mančić, A., Hadžievski, L., and Škorić, M. (2006) Dynamics of electromagnetic solitons in a relativistic plasma. *Phys. Plasmas*, **13**, 052305.

353 Saxena, V., Das, A., Sengupta, S., Kaw, P., and Sen, A. (2007) Stability of nonlinear one-dimensional laser pulse solitons in a plasma. *Phys. Plasmas*, **14**, 072307.

354 Laedke, E. and Spatschek, K.-H. (1979) Lifetime of spikons. *Phys. Lett. A*, **74**, 205–207.

355 Laedke, E. and Spatschek, K.-H. (1979) Liapunov stability of generalized Langmuir solitons. *Phys. Fluids*, **23**, 44–51.

356 Farina, D. and Bulanov, S.V. (2001) Slow electromagnetic solitons in electron-ion plasmas. *Plasma Phys. Rep.*, **27**, 641–651.

357 Farina, D. and Bulanov, S.V. (2001) Dark solitons in electron-positron plasmas. *Phys. Rev. E*, **64**, 066401.

358 Kuehl, H.H. and Zhang, C.H. (1993) *Phys. Rev. E*, **48**, 1316.

359 Sudan, R.N., Dimant, Y.S., and Shiryaev, O.B. (1997) *Phys. Plasmas*, **4**, 1489.

360 Lehmann, G., Laedke, E.W., and Spatschek, K.-H. (2008) Two-dimensional dynamics of relativistic solitons in cold plasmas. *Phys. Plasmas*, **15**, 072307.

361 Bulanov, S., Califano, F., Esirkepov, T., Mima, K., Naumova, N., Nishihara, K., Pegoraro, F., Sentoku, Y., and Vshivkov, V.A. (2001) Generation of subcycle relativistic solitons by super intense laser pulses in plasma. *Physica D*, **152/153**, 682–693.

362 Vakhitov, M.G. and Kolokolov, A.A. (1973) *Radiophys. Quantum Electron.*, **16**, 783.

363 von Mises, R. and Pollaczek-Geiringer, H. (1929) Praktische Verfahren der Gleichungsauflösung. *Z. Angew. Math. Mech.*, **5**, 58–77, 152–164.

364 Sprangle, P., Esarey, E., and Ting, A. (1990) Nonlinear interaction of intense laser pulses in plasmas. *Rhys. Rev. A*, **41**, 4463.

365 Bulanov, S.V., Inovenkov, I.N., Kirsanov, V.I., Naumova, N.M., and Sakharov, A.S. (1992) Nonlinear depletion of ultrashort and relativistically strong laser pulses in an underdense plasma. *Phys. Fluids B*, **4**(7), 1935–1942.

366 Lontano, M., Bulanov, S., Califano, F., Esirkepov, T., Farina, D., Koga, J., Liseikina, T.V., Mima, K., Nakajima, K., Naumova, N., Nishihara, K., Passoni, M., Pegoraro, F., Ruhl, H., Sentoku, Y., Tajima, T., and Vshivkov, V.A. (2002) Relativistic electromagnetic solitons produced by ultrastrong laser pulses in plasmas. *AIP Conf. Proc.*, **634**.

367 Pegoraro, F. (2003) *Solitary Structures in Relativistic Plasmas*, Autumn college on plasma physics, Trieste, 13 October–7 November.

368 Faure, J., Glinec, Y., Pukhov, A., Kiselev, S., Gordienko, S., Lefebvre, E., Rousseau, J.P., Burgy, F., and Malka, V. (2004) A laser-plasma accelerator producing monoenergetic electron beams. *Nature*, **431**, 541.

369 Liu, C.S. and Tripathi, V.K. (2005) Ponderomotive effect on electron acceleration by plasma wave and betatron resonance in short pulse laser. *Phys. Plasmas*, **12**, 043103.

370 Weber, S., Riconda, C., and Tikhonchuk, V.T. (2005) Low-level saturation of Brillouin backscattering due to cavity formation in high-intensity laser-plasma interaction. *Phys. Rev. Lett.*, **94**, 055005.

371 Weber, S., Riconda, C., and Tikhonchuk, V.T. (2005) Strong kinetic effects in cavity-induced low-level saturation of stimulated Brillouin backscattering for high-intensity laser-plasma interaction. *Phys. Plasmas*, **12**, 043101.

372 Weber, S., Lontano, M., Passoni, M., Riconda, C., and Tikhonchuk, V.T. (2005) *Phys. Plasmas*, **12**, 112107.

373 Lontano, M., Passoni, M., Riconda, C., Tikhonchuk, V.T., and Weber, S. (2005) Electromagnetic solitary waves in the saturation regime of stimulated Brillouin backscattering. *Laser Part. Beams*, **24**, 125–129.

374 Laedke, E.W. and Spatschek, K.-H. (1982) Thresholds and growth times of some nonlinear field equations. *Physica*, **5D**, 227–242.

375 Laedke, E.W. and Spatschek, K.-H. (1982) Complementary energy principles in dissipative fluids. *J. Math. Phys.*, **23**(3), 460–463.

376 Laedke, E. and Spatschek, K.-H. (1984) The Q-theorem for charged solitons. *Advances in Nonlinear Waves*, Pittman.

377 Anderson, D., Bondeson, A., and Lisak, M. (1979) Transverse instability of soliton solutions to nonlinear Schrödinger equations. *J. Plasma Phys.*, **21**, 259–266.

378 Bridges, T.J. (2004) On the susceptibility of bright nonlinear Schrödinger solitons to long-wave transverse instability. *Proc. R. Soc. Lond.*, **460**, 2605–2615.

379 Kivshar, Y.S. and Pelinovsky, D.E. (2000) Self-focusing and transverse instabilities

of solitary waves. *Phys. Rep.*, **331**, 117–195.

380 Karle, C. and Spatschek, K.-H. (2008) Relativistic laser pulse focusing and self-compression in stratified plasma-vacuum systems. *Phys. Plasmas*, **15**, 123102.

381 Haken, H. (1996) Slaving principle revisited. *Physica D: Nonlinear Phenomena*, **97**, 95–103.

382 Goodman, J., Hou, T., and Tadmor, E. (1994) On the stability of the unsmoothed Fourier method for hyperbolic equations. *Numer. Math.*, **67**, 93–129.

383 Bingham, R., de Angelis, U., Amin, M.R., Cairns, R.A., and McNamara, B. (1992) Relativistic Langmuir waves generated by ultra-short pulse lasers. *Plasma Phys. Control. Fusion*, **34**, 557–567.

384 Faure, J., Rechatin, C., Norlin, A., Lifschitz, A., Glinec, Y., and Malka, V. (2006) Controlled injection and acceleration of electrons in plasma wakefields by colliding laser pulses. *Nature*, **444**, 737.

385 Malka, V., Faure, J., Glinec, Y., and Lifschitz, A.F. (2005) Laser-plasma accelerators: A new tool for science and for society. *Plasma Phys. Control. Fusion*, **47**, 481.

386 Geddes, C.G.R., Toth, C., van Tillborg, J., Esarey, E., Schroeder, C.B., Bruhwller, D., Nieter, C., Cary, J., and Leemans, W.P. (2004) High-quality electron beams from a laser wakefield accelerator using plasma-channel guiding. *Nature*, **431**, 538.

387 Amiranoff, F., Antonietti, A., Audebert, P., Bernard, D., Cros, B., Dorchies, F., Gauthier, J.C., Geindre, J.P., Grillon, G., Jacquet, F., Matthieussent, G., Marquès, J.R., Mine, P., Mora, P., Modena, A., Morillo, J., Moulin, F., Najmudin, Z., Specka, A.E., and Stenz, C. (1996) Laser particle acceleration: Beat-wave and wakefield experiments. *Plasma Phys. Control. Fusion*, **38**, 295.

388 Kodama, R., Sentoko, Y., Chen, Z.L., Kumar, G.R., Hatchett, S.P., Toyama, Y., Cowan, T.E., Freeman, R.R., Fuchs, J., Izawa, Y., Key, M.H., Kitagawa, Y., Kondo, K., Matsuoka, T., Nakamura, H., Nakatsutsumi, M., Norreys, P.A., Norimatsu, T., Snavely, R.A., Stephens, R.B., Tampo, M., Tanaka, K.A., and Yabuuchi, T. (2004) Plasma devices to guide and collimate a high density of MeV electrons. *Nature*, **432**, 1005.

389 Bulanov, S.V. (2006) New epoch in the charged particle acceleration by relativistically intense laser radiation. *Plasma Phys. Control. Fusion*, **48**, 29.

390 Malka, V., Fritzler, S., Lefebvre, E., Aleonard, M.M., Burgy, F., Chambaret, J.P., Chemin, J.F., Krushelnick, K., Malka, G., Mangles, S.P.D., Najmudin, Z., Pittman, M., Rousseau, J.P., Sheurer, J.N., Walton, B., and Dangor, A.E. (2002) Electron acceleration by a wakefield forced by an intense ultrashort laser pulse. *Science*, **298**, 1596.

391 Kaganovich, D., Ting, A., Gordon, D.F., Hubbard, R.F., Jones, T.G., Zigler, A., and Sprangle, P. (2005) First demonstration of a staged all-optical laser wakefield acceleration. *Phys. Plasmas*, **12**, 100702.

392 Krushelnik, K., Clark, E.L., Beg, F.N., Dangor, A.E., Najmudin, Z., Norreys, P.A., Wei, M., and Zepf, M. (2005) High intensity laser-plasma sources of ions-physics and future applications. *Plasma Phys. Control. Fusion*, **47**, 451–463.

393 Modena, A., Najmudin, Z., Dangor, A.E., Clayton, C.E., Marsh, K.A., Joshi, C., Malka, V., Darrow, C.B., Danson, C., Neely, D., and Walsh, F.N. (1995) Electron acceleration from the breaking of relativistic plasma waves. *Nature*, **377**, 606–608.

394 Pukhov, A. (2003) Strong field interaction of laser radiation. *Rep. Prog. Phys.*, **66**, 47–101.

395 Kostyukov, I., Pukhov, A., and Kiselev, S. (2004) Phenomenological theory of laser-plasma interaction in "bubble" regime. *Phys. Plasmas*, **11**(11), 5256.

396 Katsouleas, T. and Mori, W.B. (1988) Wave-breaking amplitude of relativistic oscillations in a thermal plasma. *Phys. Rev. Lett.*, **61**(1), 90–93.

397 Teychenné, D., Bonnaud, G., and Bobin, J.L. (1993) Wave-breaking limit to the wake-field effect in an underdense plasma. *Phys. Rev. E*, **48**(5), 3248–3251.

398 Esarey, E. and Pilloff, M. (1995) Trapping and acceleration in nonlinear plasma waves. *Phys. Plasmas*, **2**(5), 1432–1436.

399 Bulanov, S.V., Pegoraro, F., Pukhov, A.M., and Sakharov, A.S. (1997) Transverse-wake wave breaking. *Phys. Rev. Lett.*, **78**(22), 4205–4208.

400 Esarey, E., Hubbard, R.F., Leemans, W.P., Ting, A., and Sprangle, P. (1997) Electron injection into plasma wake fields by colliding laser pulses. *Phys. Rev. Lett.*, **79**, 2682.

401 Bulanov, S., Naumova, N., Pegoraro, F., and Sakai, J. (1998) Particle injection into the wave acceleration phase due to nonlinear wake wave breaking. *Phys. Rev. E.*, **58**, 5257.

402 Andreev, N.E., Cros, B., Gorbunov, L.M., Matthieussent, G., Mora, P., and Ramazashvili, R.R. (2002) Laser wakefield structure in a plasma column created in capillary tubes. *Phys. Plasmas*, **9**, 3999–4009.

403 Gorbunov, L.M., Mora, P., and Ramazashvili, R.R. (2003) Laser surface wakefield in a plasma column. *Phys. Plasmas*, **10**, 4563–4566.

404 Gorbunov, L.M., Mora, P., and Solodov, A.A. (2003) Dynamics of a plasma channel created by the wakefield of a short laser pulse. *Phys. Plasmas*, **10**, 1124–1134.

405 Tomassini, P., Galimberti, M., Giulietti, A., Gizzi, L.A., and Labate, L. (2003) Production of high-quality electron beams in numerical experiments of laser wakefield acceleration with longitudinal wave breaking. *Phys. Rev. Spec. Top. Accel. Beams*, **6**, 121301.

406 Fubiani, G., Esarey, E., Schroeder, C.B., and Leemans, W.P. (2004) Beat wave injection of electrons into plasma using two interfering laser pulses. *Phys. Rev. E*, **70**, 016402.

407 Sircombe, N.J., Arber, T.D., and Dendy, R.O. (2005) Accelerated electron populations formed by Langmuir wave-caviton interactions. *Phys. Plasmas*, **12**, 012303.

408 Ohkubo, T., Bulanov, S.V., Zhidkov, A.G., Esirkepov, T., Koga, J., Uesaka, M., and Tajima, T. (2006) Wave-breaking injection of electrons to a laser wake field in plasma channels at the strong focusing regime. *Phys. Plasmas*, **13**, 103101.

409 Kalmykov, S.Y., Gorbunov, L.M., Mora, P., and Shvets, G. (2006) Injection, trapping, and acceleration of electrons in a three-dimensional nonlinear laser wakefield. *Phys. Plasmas*, **13**, 113102.

410 Du, C. and Xu, Z. (2000) Positron acceleration by a laser pulse in a plasma. *Phys. Plasmas*, **7**, 1582.

411 Esirkepov, T., Bulanov, S.V., and Tajima, M.Y. (2006) Electron, positron, and photon wakefield acceleration: Trapping, wake overtaking, and ponderomotove acceleration. *Phys. Rev. Lett.*, **96**, 014803.

412 Dawson, J.M. (1959) Nonlinear electron oscillations in a cold plasma. *Phys. Rev.*, **113**, 383.

413 Davidson, R.C. and Schram, P.P. (1968) *Nucl. Fusion*, **8**, 183.

414 England, R.J., Rosenzweig, J.B., and Barov, N. (2002) Plasma electron fluid motion and wave breaking near a density transition. *Phys. Rev. E*, **66**, 016501.

415 Trines, R.M.G.M. and Norreys, P.A. (2006) Wave-breaking limits for relativistic electrostatic waves in a one-dimensional warm plasma. *Phys. Plasmas*, **13**, 123102.

416 Lehmann, G., Laedke, E.W., and Spatschek, K.-H. (2007) Localized wakefield excitation and relativistic wavebreaking. *Phys. Plasmas*, **14**(10), 103109.

417 Jackson, J. (1975) *Classical Electrodynamics*, John Wiley & Sons, Inc., New York.

418 Gardner, C., Greene, J., Kruskal, M., and Miura, R. (1967) *Phys. Rev. Lett.*, **19**, 1095.

419 Zakharov, V.E. and Shabat, A.B. (1972) *Sov. Phys. JETP*, **34**, 62–69.

420 Zakharov, V.E. and Shabat, A.B. (1974) *Funct. Anal. Appl.*, **8**, 226–235.

421 Lax, P.D. (1975) *Commun. Pure Appl. Math.*, **28**, 141.

422 Kaup, D. (1976) *J. Math. Anal. Appl.*, **54**, 849–864.

423 Zakharov, V.E. and Shabat, A.B. (1979) *Funct. Anal. Appl.*, **13**, 166–174.

424 Deift, P. and Trubowitz, E. (1979) *Commun. Pure Appl. Math.*, **32**, 121–251.

425 Miura, R. (1968) *J. Math. Phys.*, **9**, 1202.

426 Landau, L.D. and Lifshitz, E.M. (1979) *Lehrbuch der theoretischen Physik, Vol. III: Quantenmechanik*, Akademie Verlag.

427 Eickermann, T. and Spatschek, K.-H. (1990) The inverse scattering transform as a diagnostic tool for perturbed soliton systems of the Schrödinger-type. *Inverse Methods in Action*, Springer, Berlin, pp. 511–518.

428 Lie, S. (1880) Theorie der Transformationsgruppen I. *Math. Ann.*, **16**, 441–528.

429 Cary, J.R. (1981) Lie transform perturbation theory for Hamiltonian systems. *Phys. Rep.*, **79**, 129–159.

430 Spatschek, K.-H., Turitsyn, S.K., and Kivshar, Y. (1995) Average envelope soliton dynamics in systems with periodically varying dispersion. *Phys. Lett. A*, **204**, 269–273.

431 Kivshar, Y.S. and Spatschek, K.-H. (1995) Nonlinear dynamics and solitons in the presence of rapidly varying periodic perturbations. *Chaos, Solitons, and Fractals*, Pergamon Press, Oxford, pp. 2551–2569.

432 Mitchell, A.R. and Wait, R. (1977) *The Finite Element Method in Partial Differential Equations*, John Wiley & Sons, Inc., New York.

433 Karhunen, K. (1946) Zur Spektraltheorie stochastischer Prozesse. *Ann. Acad. Sci. Fennicae, Ser. A*, **1**, 34.

434 Loève, M. (1955) *Probability Theory*, Van Nostrand, Princeton, NJ.

435 Fukunaga, K. (1990) *Introduction to Statistical Pattern Recognition*, Academic Press, New York.

436 Courant, R. and Hilbert, D. (1953) *Methods of Mathematical Physics, Vol. 1*, John Wiley & Sons, Inc., New York.

437 Chorin, A.J., Hald, O.H., and Kupferman, R. (2002) Optimal prediction with memory. *Physica D: Nonlinear Phenomena*, **166**, 239–257.

438 Holmes, P.J. (1981) Center manifolds, normal forms and bifurcations of vector fields with application to coupling between periodic and steady motions. *Physica D: Nonlinear Phenomena*, **2**, 449–481.

439 Guckenheimer, J. and Holmes, P. (1983) *Nonlinear oscillations, dynamical systems, and bifurcations of vector fields*, Springer, New York.

440 Spatschek, K.-H. (2005) Center manifold reduction. *Encyclopedia of Nonlinear Science*, Routledge, New York, pp. 111–114.

441 Grosche, G., Ziegler, V., Ziegler, D., and Zeidler, E. (eds) (1995) *Teubner-Taschenbuch der Mathematik, Teil II*, Teubner, Stuttgart.

442 Marsden, J.E. and McCracken, M. (1976) *The Hopf Bifurcation and its Applications*, Springer.

443 Carr, J. (1981) *Applications of Center Manifold Theory*, Springer, Berlin.

444 Beyer, P., Grauer, R., and Spatschek, K.-H. (1993) Center manifold theory for low-frequency excitations in magnetized plasmas. *Phys Rev. E*, **48**, 4665–4673.

445 Beyer, P. and Spatschek, K.-H. (1996) Center manifold theory for the dynamics of the L-H-transition. *Phys. Plasmas*, **3**, 995–1004.

446 Newell, A.C. and Whitehead, J.A. (1969) Finite bandwidth, finite amplitude convection. *J. Fluid Mech.*, **38**(02), 279–303.

447 Schneider, G. (1995) Validity and limitation of the Newell–Whitehead equation. *Math. Nachr.*, **176**(1), 249–263.

448 Frieman, E. and Rotenberg, M. (1960) On hydromagnetic stability of stationary equilibria. *Rev. Mod. Phys.*, **32**(4), 898–902.

449 Lanczos, C. (1974) *The Variational Principles of Mechanics*, Univ. Toronto Press.

450 Srivastava, G.P. and Hamilton, R.A.H. (1978) Complementary variational principles. *Phys. Rep.*, **38**, 1–83.

451 Greene, J.M. and Johnson, J.L. (1968) Interchange instabilities in ideal hydromagnetic theory. *Plasma Phys.*, **10**, 729–745.

Index

High Temperature Plasmas, Theory and Mathematical Tools for Laser and Fusion Plasmas, First Edition.
Karl-Heinz Spatschek.